Y0-BRL-971

Applied Mathematical Sciences

Volume 161

Applied Mathematical Sciences

1. *John*: Partial Differential Equations, Fourth Edition
2. *Sirovich*: Techniques of Asymptotic Analysis
3. *Hale*: Theory of Functional Differential Equations, Second Edition
4. *Percus*: Combinatorial Methods
5. *von Mises/Friedrichs*: Fluid Dynamics
6. *Freiberger/Grenander*: A Short Course in Computational Probability and Statistics.
7. *Pipkin*: Lectures on Viscoelasticity Theory
8. *Giacaglia*: Perturbation Methods in Non-linear Systems
9. *Friedrichs*: Spectral Theory of Operators in Hilbert Space
10. *Stroud*: Numerical Quadrature and Solution of Ordinary Differential Equations
11. *Wolovich*: Linear Multivariable Systems
12. *Berkovitz*: Optimal Control Theory
13. *Bluman/Cole*: Similarity Methods for Differential Equations
14. *Yoshizawa*: Stability Theory and the Existence of Periodic Solution and Almost Periodic Solutions
15. *Braun:* Differential Equations and Their Applications, Fourth Edition
16. *Lefschetz:* Applications of Algebraic Topology
17. *Collatz/Wetterling*: Optimization Problems
18. *Grenander:* Pattern Synthesis: Lectures in Pattern Theory, Vol. I
19. *Marsden/McCracken*: Hopf Bifurcation and Its Applications
20. *Driver:* Ordinary and Delay Differential Equations
21. *Courant/Friedrichs*: Supersonic Flow and Shock Waves
22. *Rouche/Habets/Laloy*: Stability Theory by Liapunov's Direct Method
23. *Lamperti:* Stochastic Processes: A Survey of the Mathematical Theory
24. *Grenander:* Pattern Analysis: Lectures in Pattern Theory, Vol. II
25. *Davies:* Integral Transforms and Their Applications, Third Edition
26. *Kushner/Clark*: Stochastic Approximation Methods for Constrained and Unconstrained Systems
27. *de Boor*: A Practical Guide to Splines, Revised Edition
28. *Keilson:* Markov Chain Models-Rarity and Exponentiality
29. *de Veubeke*: A Course in Elasticity
30. *Sniatycki:* Geometric Quantization and Quantum Mechanics
31. *Reid:* Sturmian Theory for Ordinary Differential Equations
32. *Meis/Markowitz*: Numerical Solution of Partial Differential Equations
33. *Grenander:* Regular Structures: Lectures in Pattern Theory, Vol. III
34. *Kevorkian/Cole*: Perturbation Methods in Applied Mathematics
35. *Carr:* Applications of Centre Manifold Theory
36. *Bengtsson/Ghil/Källén*: Dynamic Meteorology: Data Assimilation Methods
37. *Saperstone:* Semidynamical Systems in Infinite Dimensional Spaces
38. *Lichtenberg/Lieberman*: Regular and Chaotic Dynamics, 2nd ed.
39. *Piccini/Stampacchia/Vidossich*: Ordinary Differential Equations in R^n
40. *Naylor/Sell*: Linear Operator Theory in Engineering and Science
41. *Sparrow:* The Lorenz Equations: Bifurcations, Chaos, and Strange Attractors
42. *Guckenheimer/Holmes*: Nonlinear Oscillations, Dynamical Systems, and Bifurcations of Vector Fields
43. *Ockendon/Taylor*: Inviscid Fluid Flows
44. *Pazy:* Semigroups of Linear Operators and Applications to Partial Differential Equations
45. *Glashoff/Gustafson*: Linear Operations and Approximation: An Introduction to the Theoretical Analysis and Numerical Treatment of Semi-Infinite Programs
46. *Wilcox:* Scattering Theory for Diffraction Gratings
47. *Hale et al.*: Dynamics in Infinite Dimensions
48. *Murray:* Asymptotic Analysis
49. *Ladyzhenskaya:* The Boundary-Value Problems of Mathematical Physics
50. *Wilcox:* Sound Propagation in Stratified Fluids
51. *Golubitsky/Schaeffer*: Bifurcation and Groups in Bifurcation Theory, Vol. I
52. *Chipot:* Variational Inequalities and Flow in Porous Media
53. *Majda:* Compressible Fluid Flow and System of Conservation Laws in Several Space Variables
54. *Wasow:* Linear Turning Point Theory
55. *Yosida:* Operational Calculus: A Theory of Hyperfunctions
56. *Chang/Howes*: Nonlinear Singular Perturbation Phenomena: Theory and Applications
57. *Reinhardt:* Analysis of Approximation Methods for Differential and Integral Equations
58. *Dwoyer/Hussaini/Voigt (eds)*: Theoretical Approaches to Turbulence
59. *Sanders/Verhulst*: Averaging Methods in Nonlinear Dynamical Systems
60. *Ghil/Childress*: Topics in Geophysical Dynamics: Atmospheric Dynamics, Dynamo Theory and Climate Dynamics

(*continued after index*)

Lu Ting Rupert Klein Omar M. Knio

Vortex Dominated Flows

Analysis and Computation for Multiple Scale Phenomena

With 139 Figures

Lu Ting
Courant Institute of Mathematical Sciences
New York University
251 Mercer Street
New York, NY 10012, USA
E-mail: ting@cims.nyu.edu

Omar M. Knio
Department of Mechanical Engineering
The Johns Hopkins University
3400 N. Charles Street
Baltimore, MD 21218, USA
E-mail: knio@jhu.edu

Rupert Klein
Mathematik und Informatik
Freie Universität Berlin
Arnimallee 2–6
14195 Berlin, Germany
and
Potsdam Institut für Klimafolgenforschung
Telegraphenberg A31
14412 Potsdam, Germany
E-mail: rupert.klein@zib.de

Library of Congress Control Number: 2007921533

ISBN-13 978-3-540-68581-4 Springer Berlin Heidelberg New York

Springer is a part of Springer Science+Business Media
springer.com

Typesetting: by the authors and aptara using a springer TEX macro package
Cover design: Joe Piliero, New York/WMXDesign, Heidelberg

Printed on acid-free paper SPIN: 10820185 55/aptara – 5 4 3 2 1 0

Preface

This monograph is a revised, updated, and enlarged edition of *Viscous Vortical Flows*, Lecture Notes in Physics 374, by Ting and Klein, now referred to as Book-I. Maintaining the spirit of Book-I, the present monograph elaborates the theoretical foundation of vortex-dominated flows, provides in-depth analyses via matched and multiscale asymptotics, and demonstrates how insight gained through these analyses can be exploited in the construction of robust, efficient, and accurate numerical techniques. The dynamics of slender vortex filaments is discussed in detail, including fundamental derivations, compressible core structure, weakly nonlinear limit regimes, and associated numerical methods. Similarly, the monograph covers asymptotic analyses and computational techniques for weakly compressible flows involving vortex-generated sound and thermoacoustics.

The present monograph is primarily intended for researchers or advanced graduate students in applied mathematics, and in theoretical and/or computational fluid dynamics. It is composed of seven chapters, 2 to 8, preceded by an introduction, Chap. 1, and ending with an epilogue, Chap. 9. Chapters 2–4 present the theoretical studies while Chapters 5–8 focus primarily on numerical simulations and applications. The theoretical studies in chapters 1 and 2, and the numerical simulations in Chapter 3 of Book-I are essentially reproduced in current Chapters 2 and 3, and in Chapter 6, respectively.

Chapter 2 treats the general theory of vortex-dominated flows, focusing on the extension of the studies in the first two sections of Chapter 1, Book-I, to account for compressible flows. It begins with the formulation of the mathematical problem of a vortex-dominated, compressible, viscous flow as an initial value problem for the compressible Navier-Stokes equations in an unbounded domain with an initial vorticity field decaying rapidly at large distance. The consistency conditions on the moments of vorticity by Truesdell (1951) are then reproduced, and the time invariants for the first and second moments of vorticity weighted by the density function are derived following Moreau (1948) for the moments of vorticity in incompressible flows. A discussion of the far-field behavior of the moments of vorticity is finally provided, reproducing the treatment of Book-I.

In Chap. 3, we study the motion of slender vortex filaments with small diffusive core structures, using matched asymptotics. The first three sections reproduce the first three sections in Chapter 2 of Book-I for incompressible flows. The discussion reflects the historical evolution of the asymptotic theory, as originally conducted in the analysis of Ting and Tung (1965) of two-dimensional vortices, which introduced the idea of avoiding a double series in ϵ and $1/|\ln\epsilon|$ by expanding in a power series in ϵ and treating $1/|\ln\epsilon|$ as order one. The evolution then included extension to the axisymmetric case by Ting and Tung (1967), Ting's (1971) generalization for three-dimensional filaments with large circumferential core velocity, and Callegari and Ting's (1978) solution of the more general case involving large circumferential and axial core velocities. The essences of Ting and Tung's analysis and of the Callegari-Ting solution were presented in detail in the first three sections of Chapter 2, Book-I, and are reproduced in Chapter 3 of this monograph. This is followed by a discussion of three recent extensions and generalizations of the analyses, including the analyses of Klein and Ting (1992) for axial core structure variation, the analysis of Liu and Ting (1987) for vortices in a background rotational flow, and the generalization of the slender filament theory, using a new version of the simpler fast track incompressible filament approach of Ting (1999), to compressible core structures with large circumferential and axial core velocity components.

Chapter 4 presents a series of theoretical studies by Klein and Majda on the weakly nonlinear dynamics, self-stretching, and interaction of nearly straight slender vortex filaments. Considered is a particular geometrical regime for vortex filament centerlines, for which simplified representations are obtained of the nonlinear and nonlocal effects of self-induction on time scales short compared with that of viscous diffusion in the vortex core. Different cases are treated using this framework, including perturbations of a straight line, the motion of a perturbed filament embedded in an outer straining field, and the interaction of nearly parallel filaments.

Chapter 5 discusses numerical simulations of slender filament dynamics, beginning with the simulations based upon the asymptotic theory in Book-I. This is followed by an elaboration of alternate numerical schemes by Knio and Klein, who develop thin-tube vortex element models that are based on, and consistent with, the asymptotic theory.

In Chap. 6, we discuss the numerical simulation of the merging or intersection of viscous vortices in two and three-dimensions. The discussion includes an updated account of the numerical solutions in Chap. 3, Book-I. Also included are recent studies of Ting and Bauer (1993) on the merging of viscous doublets.

Chapter 7 studies vortex sound generation. It begins with a summary of the asymptotic theory in Book-I, matching the acoustic far field with an incompressible vortical flow, followed by the analysis of Knio and Ting extending the theory to account for the presence of a solid body. This chapter ends with illustrations of the theoretical results based upon computations exploiting the numerical schemes outlined in Chap. 5.

In Chap. 8, recent studies by Klein and Knio are presented that focus on vortex-dominated flows generated by imposed long-wave acoustic pressure fields. The asymptotic analysis in turn guides the development of numerical schemes that overcome the limitations inherent in straightforward discretization of the equations for compressible flows. Application of these schemes is illustrated via detailed simulations of thermoacoustic devices.

The authors wish to thank Prof. Egon Krause for his foresight in unifying the literature on theoretical and numerical studies of vortical flows, and his persistence in urging L. Ting and R. Klein to prepare Book-I. This monograph materialized largely as a result of his constant encouragement to update and extend Book-I. The authors wish to renew their indebtedness to Professors Joseph B. Keller of Stanford University and Andrew J. Majda of Courant Institute of Mathematical Sciences (CIMS), then at Princeton University, for their valuable suggestions incorporated in Book-I. We gratefully thank Dr. Frances Bauer of CIMS for her help with Book-I and this monograph. Also we would like to thank Professor Denis Blackmore of New Jersey Institute of Technology (NJIT) for his useful remarks on this monograph.

Besides the indebtedness to the research grants from NASA Langley Research Center and Air Force Office of Scientific Research, noted in the preface to Book-I, L. Ting would like to acknowledge the Alexander von Humboldt Foundation Senior Scientist Award in 1995, which enabled him to visit Aachen annually for the following six years to discuss recent advances in vortical flows and to initiate this monograph.

R. Klein thanks "Deutsche Forschungsgemeinschaft" for their continuous support. A joint grant by "Deutscher Akademischer Auslandsdienst" and the US National Science Foundation (NSF) enabled cooperation with O. Knio at an early and important stage. Prof. Norbert Peters provided invaluable stimuli by assigning Hasimoto (1972) and Callegari and Ting (1978) to his graduate student R. Klein within his "Turbulenzseminar" in 1986/87. Prof. Peters' subsequent support led to many of the new contributions in the present volume. R. Klein thanks Prof. Egon Krause for his inspiring and challenging "Strömungslehre" classes at RWTH Aachen in 1981/1982, and for his strong support through various channels in subsequent years. Prof. Andrew Majda hosted R. Klein during 1988–1990 as a post-doctoral fellow. He encouraged the work on Book-I, which laid the foundation for the joint work on nearly straight vortex filaments covered in Chap. 4. Together with Prof. Ahmed Ghoniem of Massachusetts Institute of Technology (MIT), he challenged R. Klein and O. Knio to compare analytical and computational slender vortex results, thereby triggering the development of the advanced vortex element thin-tube models.

O. Knio wishes to acknowledge support of the Office of Naval Research for his work on thermoacoustics. His exchanges, interactions and visits with R. Klein at FU Berlin have been supported by the NSF under an NSF-DAAD exchange program, and by the Humboldt Foundation under a Friedrich

Wilhelm Bessel research award. He also expresses his gratitude to Prof. Ahmed Ghoniem and Dr. Habib Najm of Sandia National Laboratories, Livermore, CA for numerous contributions to his work and understanding of vortex dynamics, numerical modeling and low-Mach-number flows, to Prof. Andrew Majda for introducing him to the topic of slender filaments and for his subsequent encouragements, and to Prof. Andrea Prosperetti of The Johns Hopkins University for his insightful discussions of his work on thermoacoustics and for his unwavering support. The effort of O. Knio is dedicated to the late Prof. Robert H. Scanlan (1914-2001), with a deep sense of gratitude for interactions that have provided inspiration for a lifetime.

Finally, the authors are grateful to Ms. Grace Nasrallah for her critical and thorough editorial reading of various incarnations of this monograph.

New York, NY
Berlin, Germany
Baltimore, MD
May 31, 2006

Lu Ting
Rupert Klein
Omar Knio

Contents

1. Introduction

The central theme of Book-I by Ting and Klein (1991) was the analysis and understanding of incompressible viscous vortical flows characterized by concentrated vorticity distributions and of the associated irrotational far-field behavior. Effects of compressibility were accounted for only in the context of sound generation by vortical flows. Systematic techniques of matched asymptotic expansions provided detailed insight into both the local and far-field flow characteristics. Numerical computations played an important role in providing complementary information and independent corroboration of the analytical results.

Such numerical simulations become exceedingly demanding in the presence of extreme length and time scale separations as induced, for example, by large Reynolds or low Mach numbers. Book-I emphasized how analytical results obtained through asymptotic analyses may help identify and remedy the inefficiencies that are inherent in straightforward computational approaches that do not properly account for the scale disparties. It is this aspect of the first edition, that is, the search for systematic ways of improving computational techniques by exploiting fluid dynamical theory, which has particularly stimulated many of the new scientific developments that we discussed in this second edition, called Book-II. Since each chapter begins with a *motivation* connecting it to the preceding chapter(s) and an *extended summary*, we shall present in the Introduction abstracts of the chapters with emphasis, where appropriate, on how they extend the general schemes of Book-I into Book-II.

In Chap. 2 we first state the governing equations of viscous compressible flow, which are the basis for all subsequent developments in this volume. We then derive several "consistency conditions" and integral invariants on the moments of the curl of the product of density times the velocity, $\boldsymbol{\Xi} = \nabla \times (\rho \boldsymbol{v})$. These we present in addition to the extensions from incompressible to compressible flows of similar results for the moments of vorticity from Book-I. The consistency conditions on the vorticity field, $\boldsymbol{\Omega} = \nabla \times \boldsymbol{v}$, constitute constraints on the integrals of the Nth moments of vorticity that result directly from the fact that $\boldsymbol{\Omega}$ is divergence free and from its far-field behavior. The latter determines the upper bound of N. For an initial vorticity field of bounded support or decaying exponentially, the consistency conditions are well-defined for all N. These conditions on the integrals of the Nth moments hold for any divergence-free vector field. An example is $\boldsymbol{\Xi} = \nabla\rho \times \boldsymbol{v} + \rho\boldsymbol{\Omega}$ in a compressible flow, for which the upper bound on N is dominated by the far-field behavior of the first term when $\boldsymbol{\Omega}$ decays rapidly at large distances. In contrast, the integral invariants for the Nth moments of $\boldsymbol{\Omega}$ and $\boldsymbol{\Xi}$ are obtained as a consequence of the vorticity transport equation for incompressible flows and the corresponding equation for compressible flows, which one obtains by elimination of the pressure gradient from the momentum equation. Hence, the invariants are of inherently fluid dynamical origin. In analogy with the consistency conditions, these invariants, with $N \leq 2$ on account of the far-field behavior of $\boldsymbol{v}$, are expressed as linear combinations of moments of $\boldsymbol{\Omega}$ or $\boldsymbol{\Xi}$, for incompressible and compressible flow, respectively.

Many flows of practical relevance can be considered as flows in unbounded domains. In other words, any boundaries or sources of disturbances are far away from the vorticity field so that their effects on the flow field of interest are negligible. Thus, the far field can be considered as an assigned background potential flow plus that induced by the vorticity field. For an incompressible flow, the solution of an initial potential flow plus the flow induced by a vorticity field of bounded support or decaying exponentially in the far field can be expressed as the sum of the unsteady flow induced by the advecting and diffusing vorticity field plus the potential flow that remains unaffected by the presence of the vorticity field. Also, the velocity induced by the vorticity field, being divergence-free, can be expressed as the *curl* of a vector potential $\boldsymbol{A}$, which in turn is given by the Poisson integral of the vorticity field. Thus, we can express the far-field behavior of the vector potential in inverse powers of the distance from the vortical region, with the coefficients of the expansion defined by the integrals of moments of vorticity. Using the consistency conditions and integral invariants, the number of linearly independent nth moments of vorticity can be reduced from a total of $\frac{3}{2}(n+1)(n+2)$ to merely $2n+1$ independent combinations. With this reduction, it becomes possible and computationally efficient to exploit the moment expansions in building approximate far-field flow representations. Whereas Chap. 2 summarizes the relevant theories, we describe their systematic use in designing computational simulation schemes in Chap. 6.

Chapter 3 addresses flow fields with vorticity concentrated in tube-like regions, known as "vortex tubes" or "vortex filaments." Such flows have often been observed in nature or in man-made situations. Examples are cyclones and hurricanes, smoke rings, and trailing vortices of high-flying airplanes. See collections of examples in van Dyke (1982) and Lugt (1983). Theoretical studies of vortical flows can be found in classical textbooks and treatises, for example, Lamb (1932), Prandtl and Tietjens (1957), and Batchelor (1967), or recent symposia and special sessions, such as Krause and Gersten (1997).

In classical inviscid theory, the flow field away from a slender vortex filament is modeled by a vortex line, $\mathcal{C}$, as the filament contracts to the line, with the strength remaining constant equal to Γ. The model yields an analytical representation of the flow field induced by the filament and simplifies the studies of the interactions of the filament with the background potential flow and/or with other filaments sufficiently far apart. But the solution of the vortex line model fails to describe the flow field near and inside the filament, because the solution becomes singular on $\mathcal{C}$, and physically unacceptable. We need a solution of the flow field valid not only away from the filament but also near and inside the filament. We address this problem in Chap. 3 by the method of matched asymptotic analysis. We focus first on a flow field whose vorticity is concentrated in a slender tube-like region or vortical core, with finite total strength or circulation, Γ. We define the slenderness ϵ of the filament by the ratio of its typical core size δ^* to a charecteristic length scale ℓ of the filament geometry. In other words, $\epsilon = \delta^*/\ell \ll 1$, which in turn will be the small parameter employed to gauge the order of magnitude in our asymptotic analysis. In the length scale ℓ, the limit $\epsilon \to 0$ implies a vanishing core size, as $\delta^* = \epsilon\ell \to 0$. Thus, the filament becomes a vortex line of finite strength, Γ, and we arrive at the vortex line model for the flow in the length $O(\ell)$. Besides the two spatial scales, δ^* and ℓ, the problem could require multiple time scales. This is the case when the initial data on the velocity of the filament and/or on the core structure are inconsistent with the one-time solution in the "normal time," $T = O(\Gamma/\ell^2)$, which is associated with the filament motion and the flow outside of the filament in the length scale ℓ. Then, the inconsistencies can be removed by the admission of a shorter time scale, $T^* = O(\Gamma/\delta^{*2}) = O(\epsilon^2 T)$, to describe fast variations of the core structure on the length scale δ^*. Chapter 3 summarizes a series of studies on two- and three-dimensional slender vortices in compressible and incompressible flows. We address the two distinct length scales for inner and outer regions by the method of matched asymptotics (Friedrichs, 1942; van Dyke, 1975) and the two time scales through multiple-time asymptotics (Kevorkian and Cole, 1996). Within this analytical framework, we derive the vortex filament equations of motion as well as simplified partial differential equations for the temporal evolution of the vortical core structure.

Two results discussed in Chap. 3 are of particular value in the context of efficient numerical computation of slender vortex filament motion. The first concerns approximate solutions of the core structure equations based on asymptotic expansions in inverse powers of time. At large time, the solution is dominated by its leading term, known as the similarity solution. It becomes the optimum similarity solution, when the next term in the power series is absorbed by an optimum time shift. This, in turn, accounts for the second important characteristic of the initial profile, its polar moment, the first being its total strength. Thus, the optimum similarity solution can be employed to demonstrate the effect of nonsimilar initial profiles on the dynamics of the filament. The second result consists of an alternative derivation of the filament equation of motion that directly relates the filament centerline velocity to the leading and first order vorticity distributions in the vortical core. Both these results are exploited in Chap. 5 in the construction of efficient slender or "thin-tube" vortex element schemes.

Chapter 4 summarizes a series of investigations into the (self-)stretching of slender vortex filaments. These studies were motivated by the fact that vortex stretching is widely considered a key mechanism of three-dimensional turbulence, and it is also one of the candidate mechanisms that could lead to the spontaneous formation of singularities in incompressible flows described by the Euler or Navier–Stokes equations. The simplified asymptotic models to be discussed in this chapter shed some light onto the effects of vortex stretching in a particular, but practically important, flow regime.

In its general setting, the formula for the self-induced vortex filament motion from Chap. 3 involves a nonlinear integral expression that is too complex for direct analysis. By restricting to nearly straight filament centerlines, this nonlinear and nonlocal self-induction expression is reduced to a nonlocal but linear integral operator. At the same time, the additional selection of a particular distinguished limit for the geometrical perturbation amplitudes and wavelengths in terms of ϵ retains the essential local curvature nonlinearity that was first discovered in the seminal paper by Hasimoto (1972). The result of these steps is a simplified theory for nonlinear/local and linearized/nonlocal self-interactions of a slender vortex filament, cast in the form of a singularly perturbed cubic nonlinear Schrödinger equation. With the same approach, the behavior of a filament embedded in an external straining field as well as interactions of a collection of nearly straight, almost parallel filaments were investigated. The latter result in a nonlinear extension of the famous "Crow-Jimenez instabilities" of nearly parallel vortex pairs (Crow, 1970a; Jimenez, 1975).

In its last section, Chap. 4 reports on a different regime for vortex pair interactions, characterized by much longer wavelengths and larger amplitudes of the geometrical perturbations relative to the unperturbed filament separation distance. In this case, the local but nonlinear mutual potential vortex induction and the self-induced local curvature effects dominate, and an entirely different Hamiltonian dynamical system for the vortex filament dynamics results.

Chapter 5 explores the range of validity of the theory for slender vortex filaments established in Chap. 3 and discusses its use in developing accurate and efficient vortex element numerical schemes for slender vortex applications.

The matched asymptotic analysis of Chap. 3 provides a systematic account of the temporal evolution of the vortex core structure on the normal time scale. It includes explicitly the effects of vorticity diffusion and vortex stretching. The effect of stretching appears in the normal time scale while that of diffusion is felt at large times. The theory predicts that these changes will also affect the vortex motion at leading order so that accurate predictions should properly account for these effects. The influence of the vortex core structure evolution on the filament motion is demonstrated in the first part of Chap. 5 through a discussion of suitably chosen solutions to the vortex filament evolution equations.

The slender vortex-matched asymptotic analysis has its limits of applicability. One of these limitations is the assumption of spatial separation of all the filaments participating in a flow. Numerical solutions of the asymptotic filament equations for collections of several vortices in three dimensions demonstrate how vortex filament interactions may induce local merging and vortex reconnection, which can no longer be described by the filament theory (see also the filament interaction theory in Chaps. 4, and 6).

Vortex element schemes represent a popular and efficient approach to simulating vortex dominated flow fields. However, straightforward applications of these schemes to slender vortex filaments using a single chain of overlapping vortex blobs to represent the vortex have failed severely. In the second part of Chap. 5, we first document this failure for vortices in the nearly straight filament regime from Chap. 4. We then analyze its origins by comparing matched asymptotic expansions for the Navier–Stokes equations as described in Chap. 3 with analogous expansions applied to the straightforward vortex element representation of a filament. These comparisons allow us to identify the origin of the deviations, and they motivate improvements of the numerical scheme. A logarithmic Richardson-type extrapolation technique based on the vortex core size also allows us to overcome the severe temporal stiffness of these "thin-tube" vortex element methods.

Vortex merging and reconnection of vortex lines are considered important processes in viscous vortical and turbulent flows. In Chap. 6 we first provide a classification of vortex merging problems. We distinguish local and global merging and further subdivide the latter class into problems that are essentially laminar with the merging region comparable in size to a characteristic diffusion length, and high Reynolds number cases, for which the diffusion length scale is much smaller than the overall flow scale.

On the basis of this classification, we discuss requirements on numerical solution techniques for the Navier–Stokes equations that are to be met to efficiently address problems from each class. These include, in particular, the efficient formulation of boundary conditions in two and three space dimensions that allow one to minimize the size of the computational domain and to focus computational efforts in the region of interest.

The next two chapters address weakly compressible flows, characterized by low Mach numbers, that is, by typical flow velocities whose magnitudes are small compared to the speed of sound. Two fundamentally different phenomena are identified in this context, namely the generation and propagation of small amplitude sound waves, and the net effects of small, but nonvanishing, flow divergence in domains of strong vortical flow activity.

Chapter 7 addresses the first issue, that is, the sound generation by nearly incompressible vortical flows. We first discuss the low-Mach-number asymptotics of the compressible flow equations and identify various asymptotic regimes that differ from each other in terms of the characteristic length and time scales of pressure waves.

Next, we discuss in detail the far-field asymptotic representation of sound waves generated by a viscous and heat conducting vortical flow. Here we distinguish sound generation by coherent flow structures taking slender vortex filaments as an example, and the acoustic perturbations emanating from turbulent flow regions. Numerical simulation of sound waves induced by the interaction of slender vortex filaments with a rigid sphere are used to illustrate applications of the theory.

In Chap. 8 we consider flows belonging to a particular asymptotic regime of low-Mach-number flows involving a single time scale but multiple spatial scales for acoustics on one hand, and for vorticity, entropy, and other scalars on the other. In this regime, long-wave acoustic modes interact with small-scale, quasi-incompressible vortical flows of variable density. Differential acceleration of mass parcels of different density by long-wave pressure gradients induces small-scale fluctuations of vorticity and velocity. Two types of numerical schemes are discussed, which allow us to address the small-scale vortical flows in this regime when the long-wave acoustic modes can be considered as external driving forces.

One scheme is an extension of standard, second-order, Godunov-type compressible flow solvers to the regime of weakly compressible flows. The second builds on the vorticity-stream function formulation for incompressible flow solvers and extends the concept to include the effects of weak compressibility. The validity of both schemes is established by comparing the individual predictions of one scheme to the other, and by comparing numerical predictions to experimental data. Detailed simulations are then used to characterize the behavior of thermoacoustic refrigerators across a wide range of operating conditions.

Chapter 9, the Epilogue, provides a brief preview of ongoing and future works. Topics include an adaptation of the vortex element models of filaments to the simulation of vortex merging problems, and a unified approach to meteorological modelling that builds on the identification of scalings and the method of matched asymptotics employed in the present volume.

2. Vortex-Dominated Flows and General Theory

Motivation

The general theory for this monograph is presented in this chapter. We state the Navier–Stokes equations for unsteady compressible vicsous flows in unbounded two- or three-dimensional space and specify the initial data and far-field behavior. A number of fundamental properties or integral invariants of the flow fields are derived. These invariants shall be employed to refine the analytical description of the far-field behavior and to verify the accuracy of numerical solutions. This will provide accuracy checks in addition to, for example, the verification by the fundamental laws, that is, the conservations of mass, momentum, and energy for the computational domain. We shall show in Chap. 6 how the invariants can be employed to formulate appropriate boundary conditions on a finite computational domain consistent with the far-field behavior of the unbounded domain problem. Later, in Chap. 7, we demonstrate how the invariants provide guidelines in theoretical predictions of sound generation by vortical flows.

Chapter Summary

In Sect. 2.1, we formulate the mathematical problem of a vortical flow as an initial-value problem of a compressible viscous flow in space. We introduce the symbols used throughout this monograph: time $t \geq 0$, the inertial Cartesian coordinates and unit vectors along the axes, x_i and $\boldsymbol{e}_i$, $i = 1, 2, 3$, and the position vector $\boldsymbol{x}$. We identify the typical length scale ℓ and vorticity scale ω^* from the initial vorticity field $\boldsymbol{\Omega}(0, \boldsymbol{x}) = \boldsymbol{\Upsilon}(x_1, x_2, x_3)$. They, in turn, define the velocity scale, $U = \omega^* \ell$, and the advection time scale $t^* = \ell / U$, which we use as a reference time for nondimensionalization in the sequel. Unless specified otherwise, we set the length and velocity scales as the units, that is, $\ell = 1$ and $U = 1$, so that we can use the same symbol to denote a physical variable and the corresponding scaled variable. In the far field, $r = |\boldsymbol{x}| \gg 1$, the fluid is assumed to be at rest in the inertial coordinate system, with uniform pressure p_∞ and density ρ_∞ chosen as the scales for the corresponding state variables. The far-field behaviors of the vorticity, velocity, and perturbed states variables will then be formulated.

The speed of sound in the far field, $C_\infty = \sqrt{\gamma p_\infty/\rho_\infty}$, where γ is the ratio of specific heats, denotes a velocity scale other than U. Their ratio defines the Mach number of the flow, Ma_∞. In addition, the viscosity μ_∞, scaled by $\rho_\infty U \ell$, defines the Reynolds number Re_∞ of the flow. Unless stated otherwise, we consider the flow field in the length and velocity scales, ℓ and U, to be at low Mach number and high Reynolds number, that is,

$$\text{Reference Mach Number:} \qquad \mathrm{Ma}_\infty \equiv U/C_\infty \ll 1\,, \tag{2.1}$$

$$\text{Reference Reynolds Number:} \qquad \mathrm{Re}_\infty \equiv \rho_\infty U\ell/\mu_\infty \gg 1\,. \tag{2.2}$$

Problems with multiple length and/or time scales, associated with the singular limits $\mathrm{Re}_\infty \to \infty$ or $\mathrm{Ma}_\infty \to 0$, will be presented separately in Chap. 3 for slender vortex filaments, in Chap. 7 for far-field sound generation, and in Chap. 8 to account for the acoustic field in numerical simulations of low Mach number flows.

In Sect. 2.2 we first recount the consistency conditions on the moments of vorticity $\boldsymbol{\Omega}$ obtained by Truesdell (1951), using only the kinematic condition of $\boldsymbol{\Omega}$, being divergence free, and decaying rapidly in $\boldsymbol{x}$ (see (2.1.18)). For compressible flow, as well as incompressible, variable-density flow, these consistency conditions on vorticity remain valid but there is another divergence-free vector, the curl of the momentum vector, $\boldsymbol{\Xi} = \nabla \times (\rho \boldsymbol{v})$, giving rise to additional invariances. Because the time derivative of the momentum vector $\rho\boldsymbol{v}$ appears in the conservative form of the momentum equations, an evolution equation for $\boldsymbol{\Xi}$ is obtained from these equations via elimination of the pressure p. We define the far-field behavior of $\boldsymbol{\Xi}$, (2.2.13), which in turn limits the validity of the consistency conditions on the moments of $\boldsymbol{\Xi}$ only up to the second moments. The derivations are presented in Subsect. 2.2.1. In addition to these conditions on $\boldsymbol{\Xi}$, we derive in Subsect. 2.2.2, the time invariants for the first and second moments of $\boldsymbol{\Xi}$. The derivation follows that of the invariants for the moments of vorticity in incompressible flows by Moreau (1948a, 1948b).

The above consistency conditions for $\boldsymbol{\Omega}$ and $\boldsymbol{\Xi}$, and the time invariants of $\boldsymbol{\Xi}$ for three-dimensional flows, are reduced to those for axisymmetric flows in Subsect. 2.2.3 and for planar flows with proper far-field conditions in the plane in Subsect. 2.2.4.

In Sect. 2.3, we recover Truesdell's and Moreau's conditions on the moments of $\boldsymbol{\Omega}$ in incompressible flows directly from those derived in Sect. 2.2 by setting $\rho = \rho_\infty = 1$ and $\mu = \rho_\infty \nu_\infty = \nu_\infty$. We then study an *incompressible vortex dominated flow*, that is, a flow induced by an initial vorticity distribution $\boldsymbol{\Upsilon}$ of bounded support or decaying expotentially with no background potential flow. With the background velocity potential $\phi = 0$ at $t = 0$, we have $\phi = 0$ for $t > 0$ and thus represent the velocity $\boldsymbol{v}$ by the divergence-free vector $\nabla \times \boldsymbol{A}$. We then consider the vector potential $\boldsymbol{A}$ and the vorticity $\boldsymbol{\Omega}$ as the prime variables for the vortex-dominated flow with $\boldsymbol{A}$ related to $\boldsymbol{\Omega}$ by the Poisson integral, while $\boldsymbol{\Omega}$ is governed by the vorticity evolution equation.

In Subsect. 2.3.1, we obtain the far-field representation of $\boldsymbol{A}$ as an inverse power series in terms of r from the Poisson integral of $\boldsymbol{\Omega}$. The coefficients of the $(n+1)$-th power of r^{-1} are linear combinations of the nth moments of vorticity. In Subsect. 2.3.2 we show that the contribution to the vector potential proportional to r^{-n} yields an irrotational flow in the far field provided that the vorticity vanishes or decays faster than r^{-n-2}. Because of the $(n+3)(n+2)/2$ consistency conditions, the $3(n+2)(n+1)/2$ nth moments of vorticity are reduced to $J_n = n(n+2)$ linearly independent combinations, $\boldsymbol{A}_j^{(n)}$, $j = 1, \ldots, J_n$. Among the associated J_n contributions, $\boldsymbol{A}_j^{(n)} r^{-n-1}$, to the vector potential, $\boldsymbol{A}$, there are $n^2 - 1$ linear combinations which are curl free and do not contribute to the velocity field. The far-field velocity induced by the remaining $2n+1$ linear combinations is identified as the gradient of a scalar potential of nth order, ϕ^n, with the $2n+1$ coefficients of the spherical harmonics of ϕ^n related to those $2n+1$ linear combinations of moments of vorticity in Subsect. 2.3.3. See Klein and Ting (1990).

From the far-field solutions $\sum \phi^n$ in the length scale ℓ, we obtain in Chap. 7 the acoustic field generated by the vortical flow utilizing the method of matched asymptotic expansions. See Ting and Miksis (1990) and Knio and Ting (1997).

Using the time invariants of Moreau, we show that in the far field the leading term of $\boldsymbol{A}$ is represented by three doublets of constant strengths oriented along the three axes while the second term, $n = 2$, is represented by $n(n+2) - 8$ quadrupoles, three of which remain constant (Ting, 1983). From the asymptotic analysis, we know that only those spherical harmonics with time-dependent coefficients will generate sound. Thus we arrive at the well-known results that the acoustic field will be generated by vortical flows whose vector potential exhibits quadrupole and higher order harmonic far-field behavior proportional to $O(r^{-n})$, $n \geq 2$.

2.1 Governing Equations for Compressible Viscous Vortical Flows

We consider a flow induced by an initial vorticity distribution of length scale ℓ and velocity scale U, to be at a low Mach number,

$$\mathrm{Ma} = U/C_0 \ll 1\,, \tag{2.1.1}$$

where C_0 denotes the speed of sound at rest. Hence, the leading order solution in Ma^2 is incompressible (Janzen, 1913; Rayleigh, 1916). However, in the vorticity field, at a scale much smaller than ℓ (e.g., in the vortical core of a slender filament), the velocity can be much larger than U with Mach number finite. On the other hand, in the far field, at a length scale much larger than ℓ, the velocity is much smaller than U, the nonlinear convection terms are much smaller than the unsteady terms, and pressure fluctuations, or acoustic waves, are induced by flow-induced weak density fluctuations (Sect. 7.1). Thus the effect of compressibility has to be included in a mathematical description of acoustic wave generation. To account for the effect of compressibility in the core structures of slender filaments and in the acoustic fields, we begin from the governing equations for unsteady compressible viscous flows, the compressible Navier–Stokes equations. In Cartesian coordinates with unit vectors $\boldsymbol{e}_j$, $j = 1, 2, 3$, we denote the components of the position vector $\boldsymbol{x}$ by x_j, and of the velocity $\boldsymbol{v}$ by v_j. The state variables, density ρ, pressure p and temperature T, and the velocity $\boldsymbol{v}$ of the flow field at point $\boldsymbol{x}$ and time t, are governed by the continuity equation

$$\partial_t \rho + \sum_{j=1,2,3} \partial_{x_j}(\rho v_j) = 0 \,, \tag{2.1.2}$$

the momentum equation

$$\partial_t(\rho v_i) + \sum_{j=1}^{3} \partial_{x_j}(\rho v_i v_j) = -\partial_{x_i} p + \sum_{j=1}^{3} \partial_{x_j} \tau_{ij} \,, \quad i = 1, 2, 3 \,, \tag{2.1.3}$$

the energy equation

$$\begin{aligned} \partial_t(\rho c_p T) + \sum_{j=1}^{3} \partial_{x_j}(\rho v_j c_p T) = \; & \partial_t p + \sum_{j=1}^{3} v_j \partial_{x_j} p + \sum_{i,j=1}^{3} \tau_{ij} \partial_{x_j} v_i \\ & + \sum_{j=1}^{3} \partial_{x_j}(k \partial_{x_j} T) \,, \end{aligned} \tag{2.1.4}$$

the equation of state for an ideal gas

$$p = \rho R T \quad \text{with} \quad c_v = R/(\gamma - 1) \,, \quad c_p = \gamma c_v \,, \tag{2.1.5}$$

and the relation between the stress tensor τ_{ij} and rate of strain tensor e_{ij}

$$\tau_{ij} = -\lambda \delta_{ij} \nabla \cdot \boldsymbol{v} + \mu e_{ij} \quad \text{with} \quad e_{ij} = \partial_{x_i} v_j + \partial_{x_j} v_i \,. \tag{2.1.6}$$

Here, δ_{ij} is the Kronecker delta, which is zero when $i \neq j$ and unity when $i = j$, and R, k, and γ denote the gas constant, the thermal conductivity, and the ratio of the specific heat capacity at constant pressure to that at constant volume, c_p/c_v. In the stress–strain relationship (2.1.6), we relate the second viscosity coefficient λ to the viscosity coefficient μ by the Stokes hypothesis, $3\lambda + 2\mu = 0$. Consequently, the trace of the normal stresses vanishes, $\tau_{11} + \tau_{22} + \tau_{33} = 0$, and the rate of dilatation is attributed solely to thermodynamic pressure. Note that the hypothesis has been confirmed by an extremely large number of experiments (see Schlichting and Gersten, 2000).

For a compressible viscous and heat conducting flow, the pressure p in general is not a simple function of the density ρ. We can eliminate ∇p by applying the curl operator to the momentum equation (2.1.3), and obtain the evolution equation for the curl of the momentum vector

$$\boldsymbol{\Xi}(t,\boldsymbol{x}) = \sum_{i=1}^{3} \Xi_i \boldsymbol{e}_i = \nabla \times (\rho \boldsymbol{v}) = \rho \boldsymbol{\Omega} - \boldsymbol{v} \times \nabla \rho \,, \tag{2.1.7}$$

where

$$\boldsymbol{\Omega} \equiv \nabla \times \boldsymbol{v} \tag{2.1.8}$$

denotes the vorticity vector. The jth component of the evolution equation of $\boldsymbol{\Xi}$ is

$$\partial_t \Xi_j + \sum_{l=1}^{3} \{\partial^2_{x_k x_l}[\rho v_i v_l] - \partial^2_{x_i x_l}[\rho v_k v_l]\} = F_j \,, \tag{2.1.9}$$

$$F_j = \sum_{l=1}^{3} \{\partial^2_{x_k x_l} \tau_{il} - \partial^2_{x_i x_l} \tau_{kl}\} \,,$$

for $j = 1, 2, 3$, and i, j, k in the cyclic order. Here, F_j denotes the jth component of the viscous term $\mathcal{F}$. For example, for $j = 3$, $k = 1$, and $i = 2$, it can be rewritten as

$$F_3 = \partial^2_{x_1 x_2}[\tau_{22} - \tau_{11}] + [\partial^2_{x_1^2} - \partial^2_{x_2^2}]\tau_{12} + \partial^2_{x_3 x_1}\tau_{23} - \partial^2_{x_2 x_3}\tau_{31} \,. \tag{2.1.10}$$

In general, following Helmholtz and Hodge, the velocity field can be expressed as the sum of a potential flow, $\bar{\boldsymbol{v}}$, and a divergence free field, $\tilde{\boldsymbol{v}}$,

$$\boldsymbol{v}(t,\boldsymbol{x}) = \bar{\boldsymbol{v}} + \tilde{\boldsymbol{v}} \,, \quad \text{with} \quad \bar{\boldsymbol{v}} = \nabla \phi(t,\boldsymbol{x}) \quad \text{and} \quad \tilde{\boldsymbol{v}} = \nabla \times \boldsymbol{A}(t,\boldsymbol{x}) \,, \tag{2.1.11}$$

where ϕ and $\boldsymbol{A}$ are known as the scalar and vector potentials, respectively. On the right-hand side of the vector equation (2.1.11), there are four unknown scalar functions; therefore, an additional condition is needed. Since the addition of the gradient of any scalar function, $\nabla \chi$, to the vector potential will not change the velocity field, we may suppose χ to be chosen such that

$$\nabla \cdot \boldsymbol{A} = 0 \,. \tag{2.1.12}$$

This, in turn, serves as the additional condition.

By applying the curl operator to (2.1.11) and using (2.1.8) and (2.1.12), one obtains the vector Poisson equation

$$\triangle \boldsymbol{A} = -\boldsymbol{\Omega} \tag{2.1.13}$$

for the vector potential with the vorticity as the inhomogenous term. Now we state the initial and far-field behaviors for a flow field induced by an initial vorticity distribution, $\boldsymbol{\Upsilon}$, that is,

$$\boldsymbol{\Omega}(0,\boldsymbol{x}) = \boldsymbol{\Upsilon}(\boldsymbol{x}) \,, \tag{2.1.14a}$$

which is divergence free,

$$\nabla \cdot \boldsymbol{\Upsilon} \equiv 0 \,. \tag{2.1.14b}$$

The initial distribution, $\boldsymbol{\Upsilon}$, is usually assumed to be of bounded support or to decay exponentially in $r = |\boldsymbol{x}|$. For our purposes, it is sufficient to require that $\boldsymbol{\Upsilon}$ decays rapidly at large distance $r = |\boldsymbol{x}|$, that is,

$$\boldsymbol{\Upsilon}(\boldsymbol{x}) = O(r^{-N}) \qquad \text{for a large integer } N \,. \tag{2.1.15}$$

The initial vector velocity potential is given by the Poisson integral,

$$\boldsymbol{A}(0,\boldsymbol{x}) = \frac{1}{4\pi} \int\int\int_{-\infty}^{\infty} \frac{\boldsymbol{\Upsilon}(\boldsymbol{x}')}{|\boldsymbol{x}' - \boldsymbol{x}|} d^3\boldsymbol{x}' \,, \tag{2.1.16}$$

where $d^3\boldsymbol{x}'$ stands for $dx_1'\, dx_2'\, dx_3'$. If we consider the initial velocity field to be that induced by the vorticity $\boldsymbol{\Upsilon}$ with zero initial scalar potential, $\phi(0,\boldsymbol{x}) = 0$, or $\bar{\boldsymbol{v}}(0,\boldsymbol{x}) = 0$, then the initial data for the velocity field are

$$\boldsymbol{v}(0,\boldsymbol{x}) = \tilde{\boldsymbol{v}}(0,\boldsymbol{x}) = \nabla \times \boldsymbol{A}(0,\boldsymbol{x}) \,. \tag{2.1.17}$$

Note that $\boldsymbol{v}(0,\boldsymbol{x})$ is then divergence free and fulfills the far-field behavior of (2.1.15) by construction. In addition, we have $\boldsymbol{A} = O(r^{-1})$ and $\boldsymbol{v} = O(r^{-2})$. The latter two estimates will be refined to $O(r^{-3})$ in the following subsection, where we show that $\boldsymbol{A} = O(r^{-2})$.

The compressible flow equations describe the conservation of mass, momentum, and energy in time. Accordingly, the initial data provided as part of the specification of a concrete flow problem will have to be sufficient to define the fields of mass, momentum, and energy densities. Often, it is a natural choice to provide the initial distributions of the velocity field and of two thermodynamic state variables, such as the pressure, p, and density, ρ, or the pressure and the entropy, S, etc.

Since the vorticity $\boldsymbol{\Omega}(t,\boldsymbol{x})$ is convected at finite speed while it diffuses and decays exponentially in $r = |\boldsymbol{x}|$, $\boldsymbol{\Omega}$ will continue to decay rapidly in r, thereby satisfying the condition,

$$|\boldsymbol{\Omega}|\ell/U = O\big([\ell/|\boldsymbol{x}|]^N\big) \quad \text{as} \quad |\boldsymbol{x}|/\ell \to \infty \,, \tag{2.1.18}$$

for large integer N and $t \geq 0$. Thus, the far-field flow remains irrotational to the order $(\ell/r)^N$.

If the flow field is initially divergence free, that is, $\phi = 0$, $\bar{\boldsymbol{v}} = 0$, and $\boldsymbol{v} = \tilde{\boldsymbol{v}} = \nabla \times \boldsymbol{A}$ at $t = 0$, we can still not conclude that $\phi = 0$, $\bar{\boldsymbol{v}} = 0$, and $\nabla \cdot \boldsymbol{v} = 0$ for $t\rangle 0$ for compressible flow: In this case, the continuity equation would yield $\rho_t + \boldsymbol{v} \cdot \nabla\rho = 0$. This would constrain the density to remain constant along particle paths, and it would contradict the assumption of compressibility.

We now consider a vortical flow induced by an initial vorticity distribution $\boldsymbol{\Upsilon}$ decreasing rapidly in r, (2.1.15). In the far field, the flow remains at rest and uniform, that is,

$$\boldsymbol{v} \to 0\,, \quad p \to p_\infty\,, \quad \text{and} \quad \rho \to \rho_\infty \quad \text{with} \quad C_0 = C_\infty\,. \tag{2.1.19}$$

Thus we have completed the mathematical formulation of the viscous vortical flow problem. The differential equations (2.1.2), (2.1.3), and (2.1.4), the auxiliary conditions (2.1.5) and (2.1.6), the far-field conditions (2.1.19), and the initial data on $\boldsymbol{\Omega}$ (2.1.14a) and ϕ, or that on $\boldsymbol{v}$ (2.1.17), plus appropriate initial data on two state variables, define an initial-value problem of the vortical flow in free space.

To derive the consistency conditions and the time-invariant conditions for the moments of $\boldsymbol{\Xi}$, we need to establish behaviors of ρ and $\boldsymbol{v}$ in the far field, $r \gg \ell$, where the flow is incompressible and irrotational with an equivalent source term to account for the compressibility effect in the vortical field. If the vorticity field scaled by U/ℓ is $O(1)$, $\nabla \cdot \boldsymbol{v} = O(\mathrm{Ma}^2) = o(1)$. If there are local regions of length scale $\delta \ll 1$, having large vorticity and velocity $\boldsymbol{v}/C_0 = O(1)$, the local flow field is compressible, for example, in the core of a slender filament. For both cases, the scaled equivalent source strengths in the far field are $o(1)$. Thus we have

$$\frac{|\boldsymbol{v}|}{U} = o(r^{-2}), \quad \frac{p - p_\infty}{\rho_\infty |\boldsymbol{v}|^2} = o(r^{-4}) \tag{2.1.20a}$$

and

$$\frac{\rho - \rho_\infty}{\rho_\infty} = O\left(\frac{p - p_\infty}{p_\infty}\right) = o(r^{-4}) \tag{2.1.20b}$$

as $r \to \infty$. With these far-field behaviors, we derive the consistency conditions for the moments of $\boldsymbol{\Omega}$ and $\boldsymbol{\Xi}$ and then the time invariants in the next section.

2.2 The Consistency Conditions and Time Invariants

Using the condition that $\boldsymbol{\Omega} = \nabla \times \boldsymbol{v}$ is divergence free and that it decays rapidly, (2.1.18), Truesdell (1951, 1954) showed that there are $\frac{1}{2}(n+3)(n+2)$ linear combinations of nth moments of Ω, which must vanish for $t \geq 0$. These are known as consistency conditions, and also as kinematic conditions because they are valid for any vector function that is divergence free and decays rapidly in r, without making use of any equations of motion of the fluid. Certainly, they remain valid regardless of whether the flow is compressible or incompressible. Using the fact that the vorticity of an incompressible flow has to satisfy the vorticity evolution equation (2.3.2), Moreau (1948a, 1948b) obtained six additional time-invariant combinations of the moments, namely, three for the first moments and three for the second moments. Since these consistency conditions and time invariants are usually not available in textbooks of fluid dynamics but will be needed in Chaps. 6 and 7, we rederive the consistency conditions in Subsect. 2.2.1 for $\boldsymbol{\Omega} = \nabla \times \boldsymbol{v}$ and also for $\boldsymbol{\Xi} = \nabla \times (\rho \boldsymbol{v})$, noting their different far-field behaviors. In Subsect. 2.2.2 we derive the time invariants of the moments of $\boldsymbol{\Xi}$ for compressible flows and recover Moreau's invariants of those of $\boldsymbol{\Omega}$ for incompressible flows. The results for the general three-dimensional case are then reduced to the cases of axisymmetric flow in Subsect. 2.2.3 and to the two-dimensional case with appropriate modifications in Subsect. 2.2.4.

2.2.1 Derivation of the Consistency Conditions

To derive Truesdell's consistency conditions on the $\frac{3}{2}(n+2)(n+1)$-th moments we will first show that an nth coaxial moment of a divergence-free vector $\boldsymbol{\Omega}$ along an axis parallel to any vector $\boldsymbol{b}$ should vanish:

$$I^{(n)}(t, b_1, b_2, b_3) = \int\int\int_{-\infty}^{\infty} (\boldsymbol{x} \cdot \boldsymbol{b})^n \ \boldsymbol{\Omega} \cdot \boldsymbol{b} \, d^3\boldsymbol{x} = 0 \ , \qquad (2.2.1)$$

for $t \geq 0,\ n = 0, 1, 2, \ldots$, and for all $b_i,\ i = 1, 2, 3$, which are the components of $\boldsymbol{b}$.

A simple proof of (2.2.1) begins from the identity

$$\nabla (f\boldsymbol{\Omega}) = \boldsymbol{\Omega} \cdot \nabla f + f \nabla \cdot \boldsymbol{\Omega} = \boldsymbol{\Omega} \cdot \nabla f \ ,$$

where f is a scalar function. Letting $f = (\boldsymbol{x} \cdot \boldsymbol{b})^{n+1}/(n+1)$, the divergence theorem and the far-field behavior of Ω (2.1.18), yield

$$\langle (\boldsymbol{x} \cdot \boldsymbol{b})^n \ \boldsymbol{\Omega} \cdot \boldsymbol{b} \rangle \ = \frac{1}{n+1} \langle (\boldsymbol{x} \cdot \boldsymbol{b})^{n+1} \ \nabla \cdot \boldsymbol{\Omega} \rangle \ = 0 \ , \qquad (2.2.2)$$

for $n \leq N - 3$. Since $I^{(n)}$ is a homogeneous polynomial in b_i of degree $n + 1$ and the components b_i are arbitrary, (2.2.1) holds if and only if all the coefficients in the polynomial are equal to zero. There are $\frac{1}{2}(n+3)(n+2)$ coefficients, which are linearly independent combinations of nth moments of vorticity. These coefficients are proportional to the left-hand side of the following equation,

$$i \, \langle x_1^{i-1} x_2^j x_3^k \omega_1 \rangle + j \, \langle x_1^i x_2^{j-1} x_3^k \omega_2 \rangle + k \, \langle x_1^i x_2^j x_3^{k-1} \omega_3 \rangle \ = 0 \ , \qquad (2.2.3)$$

for $i + j + k = n + 1$ and $i, j, k \geq 0$. Consequently, the number of linearly independent combinations of the nth moments is

$$J(n) = \frac{3}{2}(n+2)(n+1) - \frac{1}{2}(n+3)(n+2) = n(n+2) \ . \qquad (2.2.4)$$

In particular, for $n = 0$, we have $J(0) = 0$ and thus obtain the three consistency conditions,

$$\langle \omega_i \rangle = 0 \ , \quad i = 1,2,3 \ , \qquad \text{that is,} \quad \langle \boldsymbol{\Omega} \rangle = 0 \ . \tag{2.2.5}$$

For $n = 1$, we have $J(1) = 3$ and consequently the six consistency conditions,

$$\langle x_i \omega_j \rangle + \langle x_j \omega_i \rangle = 0 \ , \qquad i,j = 1,2,3 : \ j \geq i \ . \tag{2.2.6}$$

Equation (2.2.6) says that the 3×3 matrix $\langle x_i \omega_j \rangle$ is skew symmetric and only three combinations of the first moments, which are the three components of $\langle \boldsymbol{x} \times \boldsymbol{\Omega} \rangle$, remain to be defined by the initial data and through the evolution of $\boldsymbol{\Omega}$ in time. For $n = 2$, the following ten linear combinations of the second moments vanish,

$$\langle x_i^2 \Omega \rangle + 2\langle x_i \omega_i \mathbf{x} \rangle = 0 \ , \ i = 1,2,3 \tag{2.2.7}$$

and

$$\langle x_1 x_2 \omega_3 \rangle + \langle x_2 x_3 \omega_1 \rangle + \langle x_3 x_1 \omega_2 \rangle = 0 \ , \tag{2.2.8}$$

and eight linear combinations remain to be defined.

In Sect. 2.3 below, we will express the vector potential $\boldsymbol{A}$ through a far-field expansion of the Poisson intergral in terms of powers of r^{-1}. For convenience, we anticipate here the results from (2.3.18) and (2.3.19), that is,

$$\boldsymbol{A}(t, \boldsymbol{x}) = \sum_{n=0}^{m} \boldsymbol{A}^{(n)}(t, \boldsymbol{x}) + O(r^{-m-2}) \ , \tag{2.2.9}$$

where

$$\boldsymbol{A}^{(n)}(t, \boldsymbol{x}) = \frac{1}{4\pi \ r^{n+1}} \int\int\int_{-\infty}^{\infty} \boldsymbol{\Omega}(t, \boldsymbol{x}')[(r')^n P_n(\hat{\boldsymbol{x}} \cdot \hat{\boldsymbol{x}}')] d^3\boldsymbol{x}' \ . \tag{2.2.10}$$

The first three terms in the expansion, $\boldsymbol{A}^{(0)}$, $\boldsymbol{A}^{(1)}$, and $\boldsymbol{A}^{(2)}$, are defined in (2.3.20)–(2.3.22), respectively. As a result of (2.2.5), we get from (2.3.20) that $\boldsymbol{A}^{(0)} \equiv 0$ so that

$$\boldsymbol{A}(t, \boldsymbol{x}) = \boldsymbol{A}^{(1)}(t, \boldsymbol{x}) + O(r^{-3}) \tag{2.2.11}$$

and

$$|\tilde{\boldsymbol{v}}| = |\nabla \times \boldsymbol{A}| = O(r^{-3}) \qquad \text{as} \quad r \to \infty \ . \tag{2.2.12}$$

The results presented above for the moments of vorticity are valid, regardless of whether the flow is compressible or incompressible.

For a compressible flow, there is another important divergence free vector, $\boldsymbol{\Xi} = \rho\boldsymbol{\Omega} - \boldsymbol{v} \times \nabla\rho$, appearing in the conservative form of the momentum equation (2.1.9). From the far-field behaviors of $\boldsymbol{v}$ and ρ, (2.1.20a) and (2.1.20b), we obtain the far-field behavior of $\boldsymbol{\Xi}$,

$$\boldsymbol{\Xi} = O(\boldsymbol{v} \times \nabla\rho) = o(r^{-7}) \ . \tag{2.2.13}$$

Therefore, the consistency conditions for the nth moments of $\boldsymbol{\Xi}$, given by (2.2.2) with $\boldsymbol{\Omega}$ replaced by $\boldsymbol{\Xi}$, hold for $n \leq 4$. Here we state the conditions for $n = 0, 1, 2$. There are three conditions for the zeroth moments

$$\langle \Xi_i \rangle = 0 \ , \quad i = 1, 2, 3 \ , \tag{2.2.14}$$

that is, the resultant vector $\langle \boldsymbol{\Xi} \rangle = 0$, and six conditions for the nine first moments,

$$\langle x_i \Xi_j \rangle + \langle x_j \Xi_i \rangle = 0 \ , \qquad i, j = 1, 2, 3 : \ j \geq i \ , \tag{2.2.15}$$

that is, three more are needed to define the first moments, $\langle x_i \Xi_j \rangle$, which form a skew symmetric 3×3 matrix. There are ten conditions for the second moments,

$$\langle {x_i}^2 \boldsymbol{\Xi} \rangle + 2 \langle x_i \Xi_i \mathbf{x} \rangle = 0 \ , \ i = 1, 2, 3 \tag{2.2.16a}$$

and

$$\langle x_1 x_2 \Xi_3 \rangle + \langle x_2 x_3 \Xi_1 \rangle + \langle x_3 x_1 \Xi_2 \rangle = 0 \ . \tag{2.2.16b}$$

Eight more conditions are needed to define the 18 second moments.

For constant density incompressible flows, the above consistency conditions for $\boldsymbol{\Xi}$ reduce respectively to (2.2.5), (2.2.6), (2.2.7), and (2.2.8) for the first three moments of Ω.

2.2.2 Derivation of the Time Invariants

The above consistency conditions imply time invariance. Additional time invariants can be obtained from the compressible Navier–Stokes equations following the steps used by Moreau for the moments of vorticity for incompressible flows. For compressible flows, it is natural to look for invariants of the moments of $\boldsymbol{\Xi}$ from the momentum equation (2.1.3). First we rewrite the jth component of the evolution equation (2.1.9) for $\boldsymbol{\Xi}$ as

$$\partial_t \Xi_j = -\sum_{l=1}^{3} \{ \partial^2_{x_k x_l} [\rho v_i v_l] - \partial^2_{x_i x_l} [\rho v_k v_l] \} + F_j \ , \tag{2.2.17a}$$

where

$$F_j = \sum_{l=1}^{3} \{ \partial^2_{x_k x_l} \tau_{il} - \partial^2_{x_i x_l} \tau_{kl} \} \ . \tag{2.2.17b}$$

We multiply (2.2.17a) by $x_k^m x_l^n$, with $2 \geq m + n \geq m \geq n \geq 0$, and integrate over the entire space, so that the left side becomes the time derivative of $\langle x_k^m x_l^n \Xi_j \rangle$. The second step is to carry out integration by parts, twice at most, for the terms on the right-hand side to remove x_k and/or x_l in the integrands or move them inside a derivative rendering the integrands integrable. For the second moments, it is necessary to carry out the third step, namely, to find linear combinations of the moments such that the right-hand sides of the combinations vanish after twofold integration by parts.

We arrive at three time invariants for the zeroth moments of $\boldsymbol{\Xi}$ after the first step. But they are implied by the three kinematic conditions (2.2.14). We obtain nine invariants for the first moments after the second step. Six of them are implied by the conditions of skew symmetry (2.2.15). There are only three linear combinations of the first moments, identified as the components of the vector $\langle \boldsymbol{x} \times \boldsymbol{\Xi} \rangle$, which are the new time invariants, that is,

$$\langle x_i \Xi_j - x_j \Xi_i \rangle = \tilde{E}_k, \quad i, j, k \text{ in cyclic order},$$

or

$$\langle \boldsymbol{x} \times \boldsymbol{\Xi} \rangle = \tilde{\mathbf{E}}, \tag{2.2.18a}$$

where $\tilde{E}_k$, $k = 1, 2, 3$, denote the components of the constant vector $\tilde{\mathbf{E}}$ equal to the initial value of $\langle \boldsymbol{x} \times \boldsymbol{\Xi} \rangle$.

For the second moments, we note that each term on the right-hand side of (2.2.17a) is differentiated twice with respect to the Cartesian coordinates. Using the far-field behavior, (2.1.20a) and (2.1.20b), we carry out integration by parts twice to remove the second derivatives and the factor $x_k x_l$ introduced in step 1. The right-hand side would then be reduced to volume integrals of the nonlinear momentum terms and the viscous terms. Therefore, we have to carry out the third step. We find linear combinations of the second moments for which the integrands on the right-hand side vanish after making use of the symmetry of $v_i v_j$ and that of τ_{ij}. To demonstrate this step, we evaluate the rates of change of four typical moments, $\partial_t \langle x_1^2 \Xi_1 \rangle$, $\partial_t \langle x_1^2 \Xi_2 \rangle$, $\partial_t \langle x_3 x_1 \Xi_1 \rangle$, and $\partial_t \langle x_1 x_2 \Xi_3 \rangle$, showing the remaining terms on the right-hand side after the second step:

$$\partial_t \langle x_1^2 \Xi_1 \rangle = 0\,, \tag{2.2.19a}$$

$$\partial_t \langle x_1^2 \Xi_2 \rangle = 2 \langle \rho v_1 v_3 - \tau_{31} \rangle\,, \tag{2.2.19b}$$

$$\partial_t \langle x_2 x_1 \Xi_1 \rangle = \langle -\rho v_1 v_3 + \tau_{31} \rangle\,, \tag{2.2.19c}$$

$$\partial_t \langle x_1 x_2 \Xi_3 \rangle = \langle \rho (v_1^2 - v_2^2) + (\tau_{11} - \tau_{22}) \rangle\,. \tag{2.2.19d}$$

Note that when x_1^2 on the right-hand side of (2.2.19b) is replaced by x_3^2, the signs on the left-hand side are changed; therefore, we get $\partial_t \langle (x_1^2 + x_3^2) \Xi_2 \rangle = 0$. Combining that with (2.2.19a), we obtain three invariants,

$$\partial_t \langle r^2 \Xi_j \rangle = 0\,, \quad j = 1, 2, 3\,, \quad \text{or} \quad \partial_t \langle r^2 \boldsymbol{\Xi} \rangle = 0\,. \tag{2.2.20}$$

Equation (2.2.19a) and linear combinations of (2.2.19b) and (2.2.19c) yield nine invariants,

$$\partial_t \{ \langle x_i x_l \Xi_k \rangle + 2 \langle x_k x_i \Xi_l \rangle \} = 0\,, \quad i \equiv l = 1, 2, 3\,, \quad k = 1, 2, 3\,. \tag{2.2.21}$$

Finally, (2.2.19d) leads to one invariant

$$\partial_t \{ \langle x_1 x_2 \Xi_3 \rangle + \langle x_2 x_3 \Xi_1 \rangle + \langle x_3 x_1 \Xi_2 \rangle \} = 0\,. \tag{2.2.22}$$

The last ten time invariants, (2.2.21) and (2.2.22), are implied by the kinematic conditions of the second moments of $\boldsymbol{\Xi}$, (2.2.16a) and (2.2.16b). Equation (2.2.20) yields the only new time invariant for the polar moment of $\boldsymbol{\Xi}$,

$$\langle r^2 \boldsymbol{\Xi} \rangle \;=\; \text{its initial value } \tilde{\mathbf{D}} \,. \tag{2.2.23}$$

Because of the zero resultant $\langle \boldsymbol{\Xi} \rangle$ of $\boldsymbol{\Xi}$ and the skew symmetry of its first moments, (2.2.14) and (2.2.15), the polar moment is independent of a displacement of the origin. In analogy with the incompressible case, no additional time invariants for the nth moments have been found for $n > 2$.

The results presented in this and the preceding subsection for three dimensional vortical flows shall be reduced to the special cases of axisymmetric flows and two-dimensional flows in the following two subsections, respectively.

2.2.3 Axisymmetric Flows

For axisymmetric flows, we replace the Cartesian coordinates, x_1, x_2, x_3, by the cylindrical coordinates σ, θ, z with $z = x_3$, and consider flows independent of θ. Let u, v, w denote the velocity components in the radial, circumferential, and axial directions with unit vectors $\boldsymbol{r}$, $\boldsymbol{\theta}$, and $\boldsymbol{k}$. For flows having zero circumferential velocity, $v = 0$, the vorticity vector has to be in the circumferential direction, that is, $\boldsymbol{\Omega} \equiv \varpi(t, \sigma, z)\boldsymbol{\theta}$ with $\boldsymbol{\Omega} \cdot \boldsymbol{r} = \boldsymbol{\Omega} \cdot \boldsymbol{k} = 0$. Conversely, if we assume $\boldsymbol{\Omega} = \varpi \boldsymbol{\theta}$, we get $v \equiv 0$. Similarly, we have

$$\boldsymbol{\Xi} = \nabla \times (\rho \boldsymbol{v}) = \Theta \boldsymbol{\theta} \quad \text{with } \Theta(t, \sigma, z) = \partial_z(\rho u) - \partial_\sigma(\rho w) \,. \tag{2.2.24}$$

Thus we consider compressible axisymmetric flows:

$$\boldsymbol{v} = \boldsymbol{r} u + \boldsymbol{k} w = u(\cos\theta \; \boldsymbol{e}_1 + \sin\theta \; \boldsymbol{e}_2) + w \; \boldsymbol{k} \,, \tag{2.2.25}$$

$$\boldsymbol{\Omega} = \varpi \; \boldsymbol{\theta} = \varpi(-\sin\theta \; \boldsymbol{e}_1 + \cos\theta \; \boldsymbol{e}_2) \,. \tag{2.2.26}$$

Note that ρ, p, u, w, and ϖ are functions of t, σ, z independent of θ. For $\boldsymbol{\Omega}$ decaying rapidly, we require ϖ to decay exponentially in $r = \sqrt{\sigma^2 + z^2}$. It is clear that the vorticity field $\varpi(t, \sigma, z)\boldsymbol{\theta}$ is divergence free and hence all the consistency conditions on $\boldsymbol{\Omega}$ and those on $\boldsymbol{\Xi}$ derived in Subsect. 2.2.1 are fulfilled automatically. For the time invariance of the moments of $\boldsymbol{\Xi}$, all three scalar invariants for the second moments of $\boldsymbol{\Xi}$ (2.2.23) become trivial because $\langle r^2 \Theta \; \boldsymbol{\theta} \rangle \equiv 0$.

For the three scalar invariants of the first moments of $\boldsymbol{\Xi}$ (2.2.18a), only the component in the axial direction $\boldsymbol{k}$ is nontrivial. It is

$$\boldsymbol{k} \cdot \langle \boldsymbol{x} \times \boldsymbol{\Xi} \rangle = \langle \sigma \Theta \rangle = \text{const.}$$

which says that the polar moment of Θ in a meridional plane is time invariant,

$$\langle \sigma \Theta \rangle = \int_{-\infty}^{\infty} \int_0^{\infty} 2\pi \sigma^2 \Theta \; d\sigma dz = \tilde{E}_3 = \text{ const.} \tag{2.2.27}$$

Note that $\Theta d\sigma dz$ represents the strength of a circular vortex ring in the plane z, of radius σ, and cross-sectional area $d\sigma dz$. The ring is equivalent to a uniform doublet distribution oriented along the z-axis over the circular disc bounded by the ring with area $\pi\sigma^2$. Thus (2.2.27) says that the total strength of the doublet weighed by the density remains equal to its initial value $\tilde{E}_3/2$.

With the vorticity vector in the circumferential direction, $\boldsymbol{\theta}$, the total strength of vorticity in a meridional plane is equal to the weighed circulation, along a large rectangular contour in the meridional plane with one edge along the z-axis and centered at the origin. By using the far-field behavior of $\boldsymbol{\Xi}$, the contour integral reduces to a line integral along the z-axis,

$$\tilde{\Gamma}(t) = \int_{\infty}^{-\infty}\int_0^{\infty} \Theta d\sigma dz = \int_{-\infty}^{\infty} \rho(t,0,z)w(t,0,z)dz \ . \tag{2.2.28}$$

From the conditions of symmetry, $u = 0$, and $\partial_\sigma w = \partial_\sigma \rho = 0$ along the z-axis, and the axial momentum equation, we obtain the rate of change of the weighed circulation,

$$\tilde{\Gamma}'(t) = \int_{-\infty}^{\infty} dz\{-\rho w \partial_\sigma u + 2\mu\partial^2_{\sigma\sigma} w\}_{\sigma=0} \ . \tag{2.2.29}$$

First, we assume that w remains positive and attains its maximum along the z-axis, that is, $w > 0$ and $\partial^2_{\sigma\sigma} w < 0$ when $\sigma = 0$. If we assume in addition $\partial_\sigma u \leq 0$ along the axis, we get $\tilde{\Gamma}'(t) < 0$ saying that $\tilde{\Gamma}$ decreases as t increases.

2.2.4 Two-Dimensional Flows

We consider a two-dimensional unsteady flow in the x_1, x_2 plane induced by an initial vorticity field $\boldsymbol{\Upsilon} = \boldsymbol{e}_3\zeta_0(x_1, x_2)$, which decays rapidly in $\sigma = \sqrt{x_1^2 + x_2^2}$. It represents a planar flow in space independent of x_3 with velocity vector $\boldsymbol{v}$ parallel to the plane, that is, $\boldsymbol{v}\cdot\boldsymbol{e}_3 = 0$. The vorticity vector $\boldsymbol{\Omega}$ and also the vector $\boldsymbol{\Xi}$ are in the direction of $\boldsymbol{e}_3$, with $\zeta = \partial_{x_1}v_2 - \partial_{x_2}v_1$ and $\Xi = \Xi_3 = \partial_{x_1}(\rho v_2) - \partial_{x_1}(\rho v_1)$, respectively. Thus we write

$$\boldsymbol{v}(t, x_1, x_2) = \boldsymbol{e}_1 v_1 + \boldsymbol{e}_2 v_2 \ , \quad \boldsymbol{\Omega} = \boldsymbol{e}_3\zeta \quad \text{and} \quad \boldsymbol{\Xi} = \boldsymbol{e}_3\Xi \ . \tag{2.2.30}$$

Since the flow field is independent of x_3, the vorticity field does not decay rapidly in $r = \sqrt{\sigma^2 + x_3^2}$, but in σ. Thus we introduce the cylindrical coordinates σ, θ, x_3 and define the far field by $\sigma \gg 1$ and replace the far-field condition for a vorticity field in space (2.1.18) by

$$\zeta(t, \sigma, \theta) = o(\sigma^{-N}) \quad \text{for a large integer N} \ . \tag{2.2.31}$$

Hence the kinematic conditions of Truesdell for the moments of vorticity and the conditions for the moments of $\boldsymbol{\Xi}$, presented in Subsect. 2.2.1 and

Subsect. 2.2.2 are not valid for two-dimensional flows. We need to reformulate the far-field conditions also for $\boldsymbol{v}$, ρ, Ξ_3, etc., and derive anew the conditions on the moments of $\boldsymbol{\Omega}$ and $\boldsymbol{\Xi}$.

For the derivation of the kinematic conditions we use Green's or Stokes' theorem, which relates area integrals of the vorticity over some domain A in the x_1x_2 plane to contour integrals around its boundary ∂A, of the velocity,

$$\int\int_A \zeta dx_1 dx_2 = \int_{\partial A} \boldsymbol{v} \cdot d\boldsymbol{x} \, . \tag{2.2.32}$$

When ∂A lies in the far field containing the vorticity field (2.2.31), the area integral of ζ denotes the total strength $\langle\zeta\rangle$ and the contour integral is the circulation Γ. Here $\langle\ \rangle$ denotes the double integral over the x_1x_2 plane. In this case, the contour ∂A lies in the region in which the flow is essentially incompressible and irrotational. We may then apply the Helmholtz theorem, which states that the contour integral is time invariant, that is,

$$\partial_t \int_{\partial A} \boldsymbol{v} \cdot d\boldsymbol{x} = 0 \, , \quad \text{hence } \langle\zeta\rangle = \text{const. } \Gamma_0 \, . \tag{2.2.33}$$

This remains valid even if the flow in the velocity field in A is compressible. Note that the total strength Γ_0 in general is nonzero in contrast to the three-dimensional case, where the total strength has to vanish on account of the far-field behavior (2.1.18) of $\boldsymbol{\Omega}$.

To obtain the time invariants corresponding to those of Moreau for incompressible flows, we eliminate pressure p from the momentum equations of a two-dimensional compressible flow and obtain an equation for the rate of change of $\boldsymbol{\Xi} = \Xi \boldsymbol{e}_3$, where $\Xi = \{\partial_{x_1}(\rho v_2) - \partial_{x_3}(\rho v_1)\}$. The equation is equivalent to the $\boldsymbol{e}_3$ component of (2.1.9) with Ξ_3 replaced by Ξ. The equation is

$$\partial_t \Xi = [\partial^2_{x_1x_1} - \partial^2_{x_2x_2}][-\rho v_1 v_2 + \tau_{12}] + \partial^2_{x_1x_2}\{\rho[v_1^2 - v_2^2] - [\tau_{11} - \tau_{22}]\} \, . \tag{2.2.34}$$

We now study the far-field behavior of Ξ as $\sigma \to \infty$. With $\Gamma_0 \neq 0$, the leading order far-field velocity is the velocity $\bar{\boldsymbol{v}}$ of an incompressible point vortex of strength Γ_0. Thus we impose the far-field condition on $\boldsymbol{v}$,

$$\boldsymbol{v} = \bar{\boldsymbol{v}} + \tilde{\boldsymbol{v}} \text{ where } \bar{\boldsymbol{v}} = \frac{\Gamma_0[-\boldsymbol{e}_1 \sin\theta + \boldsymbol{e}_2 \cos\theta]}{2\pi\sigma} \text{ and } \tilde{\boldsymbol{v}} = O(\sigma^{-2}) \, . \tag{2.2.35}$$

The higher order correction term $\tilde{\boldsymbol{v}}$ includes the effect of the volume or area flux, $O(\delta^2/\ell^2)$, when there is a local region of size $O(\delta)$ within which the local Mach number is $O(1)$ or larger, and the flow is compressible.

From (2.2.35), we obtain the far-field behaviors of other entities properly scaled,

$$p - p_\infty = p - \frac{1}{\gamma M_\infty^2} = O(\rho_\infty |\bar{\boldsymbol{v}}|^2) = O(\sigma^{-2}) \, , \tag{2.2.36a}$$

$$\rho - \rho_\infty = \rho - 1 = O(M_\infty^2 \sigma^{-2}), \quad \nabla \rho = O(M_\infty^2 \sigma^{-3}) , \tag{2.2.36b}$$
$$T - T_\infty = T - 1 = O(M_\infty^2 \sigma^{-2}) , \tag{2.2.36c}$$
$$\Xi = \mathbf{e}_3 \cdot \boldsymbol{\Xi} = O(\mathbf{e}_3 \cdot \nabla \rho \times \mathbf{v}) = O(M_\infty^2 \sigma^{-4}) \tag{2.2.36d}$$
$$\mu - \mu_\infty = O(T - T_\infty) = O(M_\infty^2 \sigma^{-2}) , \quad \nabla \mu = O(M_\infty^2 \sigma^{-3}) , \tag{2.2.36e}$$

strain:

$$\gamma_{ij} = \bar\gamma_{ij} + \tilde\gamma_{ij} = \bar\gamma_{ij} + O(\sigma^{-3}) ,$$
$$\bar\gamma_{12} = \partial_{x_2}\bar v_1 + \partial_{x_1}\bar v_2 = \frac{\Gamma_0(x_2^2 - x_1^2)}{\pi\sigma^4} = O(\sigma^{-2}) , \tag{2.2.36f}$$

stress:

$$\tau_{12} = \bar\tau_{12} + \tilde\tau_{12} = \bar\gamma_{12} + O(\sigma^{-3}) = O(\sigma^{-2}) ,$$
$$\tau_{11} - \tau_{22} = 2\mu[\partial_{x_1} v_1 - \partial_{x_2} v_2] = O(\sigma^{-2}) . \tag{2.2.36g}$$

In (2.2.36e) we made use of the fact that the viscosity coefficient μ is a function of the state variable T, or T and p. Using these far-field behaviors, we obtain from (2.2.34) the following invariants for the total strength and the first moments of Ξ,

$$\partial_t \langle \Xi \rangle = 0, \quad \text{or} \quad \langle \Xi \rangle = C_0 \tag{2.2.37}$$
$$\partial_t \langle \boldsymbol{x} \Xi \rangle = 0 \quad \text{or} \quad \langle x_1 \Xi \rangle = C_1, \quad \langle x_2 \Xi \rangle = C_2 , \tag{2.2.38}$$

where the constants C_0, C_1, and C_2 are the corresponding initial values. Now we shall prove that $C_0 = \rho_\infty \Gamma_0$. We first apply Green's theorem,

$$\int\int_A \Xi_3 dx_1 dx_2 = \int_{\partial A} \rho(v_2 dx_2 - v_1 dx_1) , \tag{2.2.39}$$

and note that as $\sigma \to \infty$ on ∂A, $\rho \to \rho_\infty + O(M^2\sigma^{-2})$, and the line integral in (2.2.39) approaches that in (2.2.32) times ρ_∞. Hence, we conclude that *the total strength of Ξ equals that of the vorticity ζ times ρ_∞*,

$$C_0 = \langle \Xi \rangle = \rho_\infty \langle \zeta \rangle = \rho_\infty \Gamma_0 , \quad \text{for } t \geq 0 . \tag{2.2.40}$$

This implies a constraint on the initial data, $\langle \partial_{x_1}[(\rho - \rho_\infty)v_2] - \partial_{x_2}[(\rho - \rho_\infty)v_1] \rangle = 0$.

Because of (2.2.35), no invariants for the third and higher moments of $\boldsymbol{\Xi}$ exist. For the second moments, we shall examine only the polar moment $\langle \sigma^2 \boldsymbol{\Xi} \rangle$ from (2.2.35), corresponding to the only invariant for the second moments, $\langle \sigma^2 \zeta \rangle$, found for incompressible flows (Poincare, 1893). See also Book-I. To apply Green's theorem, we note the identities,

$$\sigma^2[\partial^2_{x_1x_1} - \partial^2_{x_2x_2}]f(x_1,x_2) = \partial_{x_1}[\sigma^2\partial_{x_1} f - 2x_1 f]$$
$$-\partial_{x_2}[\sigma^2\partial_{x_2} f - 2x_2 f] , \tag{2.2.41}$$
$$\sigma^2[\partial^2_{x_1x_2} g(x_1,x_2)] = \partial_{x_1}[x_2^2 \partial_{x_2} g] + \partial_{x_2}[x_1^2 \partial_{x_1} g] , \tag{2.2.42}$$

and convert the polar moments of the right-hand side of (2.2.34) over a large rectangular area A: $[-L, L] \times [-H, H]$, to line integrals[1] along its boundary, ∂A. For those line integrals where $\sigma \gg 1$, we can neglect the contributions of the higher order terms involving $\tilde{\boldsymbol{v}}$; that is, we can consider the velocity field to be that of an incompressible point vortex with ρ replaced by $\rho_\infty = 1$ (see (2.2.35)).

For the line integrals of the polar moments of the nonlinear convection terms, we set $f = -\bar{v}_1\bar{v}_2 = \Gamma_0^2 x_1 x_2/[4\pi^2\sigma^4]$. Thus $\partial_{x_j} fz$ and $x_j f$ are odd functions of x_2 for $j = 1$ (x_1 for $j = 2$) so that the line integrals along the boundaries of A vanish. Likewise, we set $g = v_1^2 - v_2^2 = \Gamma_0^2[x_2^2 - x_1^2]/[4\pi^2\sigma^4]$. Therefore, $x_j^2\partial_{x_j} g$ is an odd function of x_j and with $j = 2$ ($j = 1$) the line integrals along the right and left vertical (top and bottom horizontal) boundaries of A vanish. Evaluations of the polar moments of the nonlinear inertial terms via the line integrals are equivalent to those for the incompressible flow in Book-I.

For the viscous terms, we learn from Poincaré's analysis for incompressible flows that the corresponding line integrals do not vanish. To make use of the analysis, we write $\mu = \mu_\infty + \tilde{\mu}$ and split the viscous terms into two parts, I and II, with μ replaced by $\tilde{\mu}$ and μ_∞, respectively, so that I represents the effect of the deviation of μ from μ_∞ and is smaller than II by $O(\sigma^{-2})$ in the far field. For part I, we set $f = \tilde{\mu}\gamma_{12}$ and $g = \tilde{\mu}[\partial_{x_1} v_1 - \partial_{x_2} v_2]$, and use (2.2.36a)–(2.2.36g) to show that the integrand in a line integral is $O(\sigma^{-3})$ and hence the line integral is $O(\sigma^{-2})$ and vanishes as $L^2 + H^2 \to \infty$. Thus, we have $\langle \sigma^2 I \rangle = 0$.

For part II, with μ_∞ constant, we identify its several terms with those for incompressible viscous flows,

$$\begin{aligned} II &= \mu_\infty\{[\partial^2_{x_1x_1} - \partial^2_{x_2x_2}][\partial_{x_1} v_2 + \partial_{x_2} v_1] + 2\partial^2_{x_1x_2}[\partial_{x_1} v_1 - \partial_{x_2} v_2]\} \\ &= \mu_\infty[\partial^2_{x_1x_1} + \partial^2_{x_2x_2}][\partial_1 v_2 - \partial_2 v_1] = \mu_\infty \Delta\zeta \, . \end{aligned} \tag{2.2.43}$$

We note the identity

$$\sigma^2\Delta\zeta = \sum_{j=1,2} \partial_{x_j}[\sigma^2\partial_{x_j}\zeta - 2x_j\zeta] + 4\zeta \, ,$$

and apply Green's theorem and the far-field behavior of vorticity ζ, (2.2.31), to show that $\langle \sigma^2\Delta\zeta \rangle = 4\langle \zeta \rangle$. Thus the polar moment of all the viscous terms becomes

$$\langle \sigma^2(I + II) \rangle = \langle \sigma^2 II \rangle = 4\mu_\infty\langle \zeta \rangle = 4\mu_\infty\Gamma_0 \, . \tag{2.2.44}$$

Thus we extend the result of Poincaré for incompressible flow to a compressible vortical flow at rest in the far field,

[1] Equations (2.2.41) and (2.2.42) are true because $(\partial^2_{x_1} - \partial^2_{x_2})\sigma^2 = 0$, and $\partial^2_{x_1x_2}\sigma^2 = 0$ hold when σ^2 is the quadratic form, $x_1^2 + x_2^2$.

$$\partial_t \Xi = \partial_t \langle \sigma^2 [\partial_{x_1}(\rho v_2) - \partial_{x_2}(\rho v_1)] \rangle = 4\rho_\infty \mu_\infty \Gamma_0 \, , \tag{2.2.45}$$

or

$$\langle \sigma^2 [\partial_{x_1}(\rho v_2) - \partial_{x_2}(\rho v_1)] \rangle|_{t>0} = D_3 + 4\rho_\infty \mu_\infty \Gamma_0 t \, , \tag{2.2.46}$$

where the constant D_3 is defined by the initial data,

$$D_3 = \langle \sigma^2 [\partial_{x_1}(\rho v_2) - \partial_{x_2}(\rho v_1)] \rangle|_{t=0} \, . \tag{2.2.47}$$

In the next section, we shall specialize the results presented above to those for incompressible flows by setting ρ constant and recover those derived in Book-I. The latter results were derived under the assumption that the initial vorticity field scaled by U/ℓ, and the velocity field scaled by U, are $O(1)$, and that at a low Mach number $M = U/C \ll 1$, the leading order solution is an incompressible flow.

2.3 Incompressible Vortical Flows

For an incompressible flow, we have density $\rho = \rho_\infty = 1$, $\Xi = \boldsymbol{\Omega}$ and the kinematic viscosity $\nu = \mu$. The continuity equation (2.1.2) becomes

$$\nabla \cdot \boldsymbol{v} = 0 \, . \tag{2.3.1}$$

The evolution equation of Ξ, (2.1.9), reduces to the vorticity evolution equation

$$\boldsymbol{\Omega}_t + \nabla \cdot (\boldsymbol{v} \, \boldsymbol{\Omega}) - \nabla \cdot (\boldsymbol{\Omega} \, \boldsymbol{v}) = \nu \Delta \boldsymbol{\Omega} \, . \tag{2.3.2}$$

Again, the velocity can be expressed as the sum of a potential flow, $\bar{\boldsymbol{v}} = \nabla \cdot \phi$ plus a divergence-free flow, $\tilde{\boldsymbol{v}} = \nabla \times \boldsymbol{A}$ (see (2.1.11)). The potential ϕ of an irrotational flow is governed by the continuity equation (2.3.1),

$$\Delta \phi = 0 \, , \tag{2.3.3}$$

while the vector potential $\boldsymbol{A}$ of the divergence-free flow is governed by the vorticity equation (2.3.2) and the Poisson equation, $\Delta \boldsymbol{A} = -\boldsymbol{\Omega}$, or the Poisson integral,

$$\boldsymbol{A}(t, \boldsymbol{x}) = \frac{1}{4\pi} \int\int\int_{-\infty}^{\infty} \frac{\boldsymbol{\Omega}(t, \boldsymbol{x}')}{|\boldsymbol{x} - \boldsymbol{x}'|} d\boldsymbol{x}' \quad \text{for} \quad t \geq 0 \, , \tag{2.3.4}$$

when the initial data of $\boldsymbol{\Omega}$ decays rapidly. See (2.1.15) and (2.1.16).

Note that for an incompressible flow the solution of the velocity potential ϕ in free space is *uncoupled*[2] from that of the vector potential $\boldsymbol{A}$, but solutions of $\boldsymbol{A}$ and $\boldsymbol{\Omega}$ depend on ϕ because $\bar{\boldsymbol{v}} = \nabla \phi$ appears in the velocity $\boldsymbol{v}$ in the

[2] This is not true for compressible flow for which ϕ is coupled with $\boldsymbol{A}$ via the density variation terms in the continuity equation.

convection terms of the vorticity equation (2.3.2). In particular, if $\phi = 0$ initially, we have

$$\phi(t, \boldsymbol{x}) = 0 \ , \quad \bar{\boldsymbol{v}} = 0 \ , \quad \text{and} \quad \boldsymbol{v} = \tilde{\boldsymbol{v}} \ , \quad \text{for} \quad t \geq 0 \ . \tag{2.3.5}$$

Thus we have a *vortex dominated flow*, which is a flow induced by an initial vorticity field decaying rapidly in r (2.1.15), is defined by the vorticity $\boldsymbol{\Omega}(t, \boldsymbol{x})$ and the vector potential $\boldsymbol{A}(t, \boldsymbol{x})$, and is governed by the system of integral-differential equations, (2.3.4) and (2.3.2), fulfilling the far-field conditions (2.1.18) and (2.1.19), and the initial conditions $\phi = 0$ and $\boldsymbol{\Omega} = \boldsymbol{\Upsilon}(\boldsymbol{x})$. Also note that in this case, the velocity field can be simply expressed as $\boldsymbol{v} = \nabla \times \boldsymbol{A}$, that is, it may be determined solely on the basis of the vorticity, as expressed by the well-known Biot–Savart integral:

$$\boldsymbol{v}(t, \boldsymbol{x}) = -\frac{1}{4\pi} \iiint \frac{\boldsymbol{x}' - \boldsymbol{x}}{|\boldsymbol{x}' - \boldsymbol{x}|^3} \times \boldsymbol{\Omega}(t, \boldsymbol{x}') \ d^3\boldsymbol{x}' \ . \tag{2.3.6}$$

It was pointed out in Subsect. 2.2.1 that Truesdell's consistency conditions (2.2.3) on the nth moments of vorticity for incompressible flows remain valid for compressible flows. But the new conditions on the first three moments of $\boldsymbol{\Xi}$ for compressible flows, (2.2.14), (2.2.15), (2.2.16a), and (2.2.16b), are identical to those for $\boldsymbol{\Omega}$ for incompressible flows for which $\rho = 1$ and $\boldsymbol{\Xi} = \boldsymbol{\Omega}$. Likewise, the nontrivial time invariants of the first and second moments of $\boldsymbol{\Xi}$, (2.2.18a) and (2.2.23), become Moreau's invariants for the moments of $\boldsymbol{\Omega}$ (Moreau, 1948a, 1948b). They are

$$\langle \boldsymbol{x} \times \boldsymbol{\Omega} \rangle = \langle \boldsymbol{x} \times \boldsymbol{\Upsilon} \rangle = \mathbf{E} \quad \text{and} \quad \langle r^2 \boldsymbol{\Omega} \rangle = \langle r^2 \boldsymbol{\Upsilon} \rangle = \mathbf{D} \ . \tag{2.3.7}$$

This equation says that the polar moment of vorticity with respect to the origin is time invariant. No additional invariants for $n \geq 3$ have been found from integrations of the vorticity evolution equation (Howard, 1957).

The remaining five linearly independent combinations of second moments can be chosen as

$$F_i(t) = \langle \omega_i ({x_j}^2 - {x_k}^2) \rangle \tag{2.3.8}$$

and

$$H_i(t) = \langle 2\omega_i x_j x_k - \omega_j x_k x_i - \omega_k x_i x_j \rangle \ , \tag{2.3.9}$$

where $i = 1, 2, 3$, and i, j, k are in cyclic order. Here H_i obey the constraints

$$H_1 + H_2 + H_3 \equiv 0 \tag{2.3.10}$$

due to the consistency condition (2.2.8).

The above consistency conditions and time invariants, (2.2.5)–(2.2.8) and (2.3.7)–(2.3.9), imply that $\mathbf{A}^{(0)}$, the first term in the far-field expansion (2.3.18) of $\mathbf{A}$, vanishes and that the next two terms become

$$\mathbf{A}^{(1)}(\mathbf{x}) = -\frac{1}{8\pi}\mathbf{E} \times \nabla \left[\frac{1}{r}\right] \tag{2.3.11}$$

and

$$\begin{aligned}\mathbf{A}^{(2)}(t,\mathbf{x}) &= -\frac{1}{16\pi}\nabla(\mathbf{D}\cdot\nabla)\left[\frac{1}{r}\right] \\ &+\frac{1}{16\pi}\sum_{i=1}^{3} F_i(t)[\boldsymbol{i}\,(\partial_j^2 - \partial_k^2) - \boldsymbol{j}\,\partial_j\partial_i + \boldsymbol{k}\,\partial_k\partial_i]\left[\frac{1}{r}\right] \\ &+\frac{1}{12\pi}\sum_{i=1}^{3} \boldsymbol{i}\,H_i(t)\,\partial_j\partial_k\left[\frac{1}{r}\right] .\end{aligned} \tag{2.3.12}$$

We observe that the first term in (2.3.12) is curl free and therefore the constant second moment $\mathbf{D}$ does not contribute to the far-field velocity.

In Subsect. 2.3.1, we will show that the far-field velocity $\mathbf{v}^{(n)}$ induced by $\mathbf{A}^{(n)}$ is equivalent to the gradient of a potential $\Phi^{(n)}$, that is, $\mathbf{v}^{(n)} = \nabla \times \mathbf{A}^{(n)} = \nabla\Phi^{(n)}$. Here we point out the equivalence for $n = 1$, in order to identify the physical meaning of the constant vector $\mathbf{E}$. From (2.3.11), we have

$$\mathbf{v}^{(1)} = \nabla \times \mathbf{A}^{(1)} = \frac{\mathbf{E}}{2}\cdot\nabla\left[\frac{-1}{4\pi r}\right] . \tag{2.3.13}$$

Therefore, $\mathbf{v}^{(1)}$ represents the velocity of a doublet with strength $|\mathbf{E}|/2$ located at the origin and oriented in the direction of $\mathbf{E}$. In short, we call $\mathbf{E}/2$ the doublet vector.

For an axisymmetric flow, the nontrivial time invariant (2.2.27), with $\rho = 1$, reduces to

$$\langle\sigma\varpi\rangle = \text{ its initial value } E_3 . \tag{2.3.14}$$

Equations (2.2.28) and (2.2.29) in turn define the total strength of ϖ or the circulation in a meridional plane,

$$\Gamma(t) = \int_{\infty}^{-\infty}\int_0^{\infty} \Omega d\sigma dz = \int_{-\infty}^{\infty} w(t,0,z)dz , \tag{2.3.15}$$

and the rate of change of the circulation,

$$\Gamma'(t) = -2\nu\int_{-\infty}^{\infty}[\partial^2_{\sigma\sigma}w]_{\sigma=0}\,dz . \tag{2.3.16}$$

Thus, the circulation $\Gamma(t) > 0$ will decrease if the axial velocity w remains positive along the z-axis and attends the maximum as $\sigma \to 0$.

For a two-dimensional flow, (2.2.46) becomes

$$\langle \sigma^2 \zeta \rangle = D_3 + 4\mu_\infty \Gamma_0 t \quad \text{for} \quad t \gg 1 \,. \tag{2.3.17}$$

It says that for large t the polar moment of vorticity is a linear function of t fulfilling the Poincaré formula with the constant D_3 defined by the initial value of $\varXi = \boldsymbol{e}_3 \times \boldsymbol{\varXi}$.

It was implied by the far-field behavior of $\boldsymbol{\Omega}$ (2.1.18) that the vortex-dominated flow defined by the vector potential $\boldsymbol{A}$ should become irrotational in the far field. We shall next confirm that assertion, and then express coefficients of the spherical harmonics of the corresponding potential flow in terms of the moments of vorticity in Subsect. 2.3.3.

2.3.1 Far-Field Vector Velocity Potential

For a vorticity field decaying rapidly in r (2.1.18), we obtain the far-field behavior of $\boldsymbol{A}$ by the series expansion of the Poisson integral (2.3.4) in powers of r^{-1}, which reads

$$\boldsymbol{A}(t, \boldsymbol{x}) = \sum_{n=0}^{m} \boldsymbol{A}^{(n)}(t, \boldsymbol{x}) + O(r^{-m-2}) \,, \tag{2.3.18}$$

where

$$\boldsymbol{A}^{(n)}(t, \boldsymbol{x}) = \frac{1}{4\pi \, r^{n+1}} \int\int\int_{-\infty}^{\infty} \boldsymbol{\Omega}(t, \boldsymbol{x}')[(r')^n P_n(\hat{\boldsymbol{x}} \cdot \hat{\boldsymbol{x}}')] d^3\boldsymbol{x}' \,. \tag{2.3.19}$$

Equation (2.3.18) says that if $\boldsymbol{A}$ is approximated by the $(m+1)$-th partial sum of the series, the error is of order r^{-m-2}. The number m has an upper bound $N-3$, that is, $m \le N-3$, because of the far-field condition (2.1.18) on $\boldsymbol{\Omega}$. In the integral representation for $\boldsymbol{A}^{(n)}$, $\hat{\boldsymbol{x}}$ and $\hat{\boldsymbol{x}}'$ denote unit vectors in the directions of $\boldsymbol{x}$ and $\boldsymbol{x}'$, respectively, and P_n is a Legendre polynomial. Consequently, $(r')^n P_n$ is a homogeneous polynomial in x_i' of degree n. Likewise, $r^n P_n$ and hence $r^{2n+1}\boldsymbol{A}^{(n)}$ are homogeneous polynomials in x_i of degree n while their coefficients are linear combinations of nth moments of vorticity. The first three terms of (2.3.18) are

$$\boldsymbol{A}^{(0)} = \frac{1}{4\pi} \langle \boldsymbol{\Omega} \rangle \left(\frac{1}{r} \right) \,, \tag{2.3.20}$$

$$\boldsymbol{A}^{(1)} = \frac{1}{4\pi} \sum_{j=1}^{3} \langle x_j \boldsymbol{\Omega} \rangle \partial_j \left(\frac{-1}{r} \right) \,, \tag{2.3.21}$$

and

$$\boldsymbol{A}^{(2)} = \frac{1}{8\pi} \sum_{j,k=1}^{3} \langle x_j x_k \boldsymbol{\Omega} \rangle \partial_j \partial_k \left(\frac{1}{r} \right) \,, \tag{2.3.22}$$

where $\langle\ \rangle$ denotes the volume integral over the entire space and ∂_j denotes the partial derivative with respect to x_j.

In general, an nth moment of the kth component of vorticity is defined by

$$\left\langle \omega_k \prod_{i=1}^{3} x_i^{j_i} \right\rangle \qquad \text{with} \quad k = 1,2,3\,, \quad j_i \geq 0\,, \quad \text{and} \quad \sum_{i=1}^{3} j_i = n\,. \quad (2.3.23)$$

All the $\frac{3}{2}(n+2)(n+1)$ nth moments exist for any $n \leq m \leq N-3$, on account of the far-field condition (2.1.18) on $\boldsymbol{\Omega}$.

From (2.3.18), we see that the far-field description of $\boldsymbol{A}$ to the order r^{-m-1} is defined by the nth moments of vorticity for $n \leq m$. Moments of vorticity in general are time dependent. On the other hand, it is known that these moments are not linearly independent and that some of them are time invariant.

Let us consider the nth term, $\boldsymbol{A}^{(n)}$. We note that each component of the vector $r^{2n+1}\,\boldsymbol{A}^{(n)}$, say the kth component, is a homogeneous polynomial in x_1, x_2, and x_3 of degree n and its $(n+2)(n+1)/2$ coefficients are nth moments of ω_k. We let $\mathcal{S}_n$ denote the set of all the $\frac{3}{2}(n+1)(n+2)$ nth moments of ω_k, $k = 1,2,3$. In Sect. 2.2, we employed the consistency conditions of Truesdell to identify which $\frac{1}{2}(n+3)(n+2)$ linear combinations of the nth moments have to vanish and which $J_n = n(n+2)$ nth moments can be assigned or determined from the numerical solutions of the initial value problem formulated in Sect. 2.1. We denote the set of those J_n nth moments by $\bar{\mathcal{S}}_n$.

From the definition of $\boldsymbol{A}^{(n)}$ in (2.3.18) and (2.3.19), we note that $r^{n+1}\boldsymbol{A}^{(n)}$ is a linear combination of spherical harmonics of order n while the far-field behavior (2.1.18) of $\boldsymbol{\Omega}$ ensures the existence of its nth moments appearing as the coefficients in $\boldsymbol{A}^{(n)}$. Therefore, we get $\Delta\boldsymbol{A}^{(n)} = 0$ for $r > 0$ and certainly for large r. Another way of arriving at this conclusion is by substituting the power series (2.3.18) and (2.3.19) into the vector Poisson equation (2.1.13), and then equating coefficients of like powers of r^{-1} while observing (2.1.18). Similarly, we conclude $\nabla \cdot \boldsymbol{A}^{(n)} = 0$ either by direct evaluation or by substituting the power series (2.3.18) and (2.3.19) into (2.1.12). We then have $\nabla \times (\nabla \times \boldsymbol{A}^{(n)}) = 0$ and hence the far-field velocity $\boldsymbol{v}^{(n)}$ induced by $\boldsymbol{A}^{(n)}$ is irrotational and can be expressed as the gradient of a scalar velocity potential $\Phi^{(n)}(\boldsymbol{x})$. The above statements imply

$$\boldsymbol{v} = \textstyle\sum_{n=1}^{m} \boldsymbol{v}^{(n)} + O(r^{-m-3}) = \sum_{n=1}^{m} \nabla\Phi^{(n)} + O(r^{-m-3})\,, \quad (2.3.24)$$

with

$$\boldsymbol{v}^{(n)} = \nabla \times \boldsymbol{A}^{(n)} = \nabla\Phi^{(n)}, \quad (2.3.25)$$

and

$$\Phi^{(n)} = Y_n(\theta,\phi)\; r^{-n-1}\,. \quad (2.3.26)$$

In (2.3.26), Y_n is a Laplace spherical harmonic of order n while θ and ϕ denote the spherical angles. Since there are only $2n+1$ linearly independent spherical harmonics of order n, only $2n+1$ coefficients can be assigned to define Y_n (Courant and Hilbert, 1953, pp. 511–521). Consequently, only $2n+1$ linear combinations of the nth moments in $\bar{\mathcal{S}}_n$ will contribute to the potential flow in the far field. The following questions arise: what are those $2n+1$ linear combinations and what happened to the remaining n^2-1 linear combinations of the nth moments in $\bar{\mathcal{S}}_n$? Below, we shall outline answers to these two questions, and refer to Klein and Ting (1990) for details.

In Subsect. 2.3.2 we identify n^2-1 curl-free terms in $\boldsymbol{A}^{(n)}$, associated with n^2-1 linear combinations of nth moments in $\bar{\mathcal{S}}_n$. These terms do not contribute to the velocity in the far field. In Subsect. 2.3.3 we relate the $2n+1$ coefficients in the spherical harmonics Y_n directly to linear combinations of the nth moments of vorticity.

2.3.2 Reduction of the Vector Velocity Potential in the Far Field to the Corresponding Scalar Potential

Here, we shall introduce Maxwell's representation of spherical harmonics as used by Courant and Hilbert (1953) in the far-field expansion (2.3.18) of the vector potential $\boldsymbol{A}$. We use the consistency conditions to identify the $J_n = n(n+2)$ terms in $\boldsymbol{A}^{(n)}$ associated with the nth moments in $\bar{\mathcal{S}}$ and show that each term corresponds to a potential flow. Each term, denoted by $\boldsymbol{A}_j^{(n)}$ for $j = 1, \ldots, J_n$, is a solution of Laplace's equation proportional to r^{-n-1} and will be called a divergence-free vector potential of order n. The set of these J_n vector potentials will be denoted by $\mathcal{A}^{(n)}$. Any divergence-free vector potential of order n is a linear combination of elements of $\mathcal{A}^{(n)}$. Finally, we identify the n^2-1 vector potentials in this set which are curl free.

We represent the mth component of vector potential $\boldsymbol{A}^{(n)}$ by

$$A_m^{(n)} = H_m^{(n)}(\xi, \eta, \zeta)\, \tfrac{1}{r}\,, \tag{2.3.27a}$$

with

$$H_m^{(n)} = \textstyle\sum_{i+j+k=n} C_{i,j,k,m}^{(n)}\, \xi^i \eta^j \zeta^k\,. \tag{2.3.27b}$$

Here, $H_m^{(n)}$ is a homogeneous polynomial of degree n and its variables, ξ, η, and ζ, stand for ∂_1, ∂_2, and ∂_3, respectively. From here on we suppress the superscript (n) until it is necessary. A Taylor series expansion of the Poisson integral (2.3.4) for $|\mathbf{x}'|/r \ll 1$ shows each coefficient in (2.3.27b) to be related to an nth moment by

$$C_{i,j,k,m} = \frac{(-1)^n}{4\pi\, i!\, j!\, k!} \langle x_1^i\, x_2^j\, x_3^k\, \omega_m \rangle \qquad m = 1, 2, 3 \quad \text{and} \quad i + j + k = n \tag{2.3.28}$$

so that the consistency conditions (2.2.3) yield $\frac{1}{2}(n+3)(n+2)$ constraints on the coefficients $C_{i,j,k,m}$:

$$C_{i-1,j,k,1} + C_{i,j-1,k,2} + C_{i,j,k-1,3} = 0 \qquad \text{for} \qquad i+j+k = n+1, \quad (2.3.29)$$

with the understanding that a coefficient is equal to zero when one of its subscripts is negative.

Consider first the three cases that either i, j or k is equal to $n+1$ in (2.3.29) while the remaining two are equal to zero. Then (2.3.29) reduces to

$$C_{n,0,0,1} = 0\,, \qquad C_{0,n,0,2} = 0\,, \qquad \text{and} \quad C_{0,0,n,3} = 0\,. \quad (2.3.30)$$

These three equations are equivalent to the condition that the nth coaxial moments along the three axes are equal to zero; that is, $I(t, \mathbf{b}) \equiv 0$ for $\mathbf{b} = \hat{h}_l$, where $\hat{h}_l$, $l = 1, 2, 3$, stand for the three basic unit vectors. Had we set $n = 0$, we would have recovered the conditions $C_{(0,0,0,m)} = 0$, $m = 1, 2, 3$, showing that, indeed, the series in (2.3.24) should begin with $n = 1$.

Next, we consider cases with only one of the coefficients in (2.3.29) equal to zero, say the first one, so that $i = 0$ and $n \geq j, k \geq 1$. Equation (2.3.29) then relates the second coefficient to the third and the corresponding two terms in (2.3.27b) combine to

$$\boldsymbol{U}_{01} = C_{0,j,k-1,3}\,[-\hat{h}_2\,\zeta + \hat{h}_3\,\eta]\;\eta^{j-1}\zeta^{k-1}\,\frac{1}{r}\,, \quad (2.3.31a)$$

with $j = 1, \ldots, n$, and $k = n+1-j$. Similarly, we find n vector potentials for $j = 0$ and n more for $k = 0$. They are

$$\boldsymbol{U}_{02} = C_{i,0,k-1,3}\,[-\hat{h}_1\,\zeta + \hat{h}_3\,\xi]\;\xi^{i-1}\zeta^{k-1}\,\frac{1}{r}\,, \quad (2.3.31b)$$

with $\quad i = 1, \ldots, n, \quad k = n+1-i, \quad$ and

$$\boldsymbol{U}_{03} = C_{i,j-1,0,2}\,[-\hat{h}_1\,\eta + \hat{h}_2\,\xi]\;\xi^{i-1}\eta^{j-1}\,\frac{1}{r}\,, \quad (2.3.31c)$$

with $\quad i = 1, \ldots, n, \quad j = n+1-i.$

Finally, we consider the cases with $1 \leq i, j, k \leq n-1$, when neither one of the three coefficients in (2.3.29) vanishes. We can then express the first one in terms of the second and the third. The three terms in (2.3.27b) associated with the three coefficients in (2.3.29) combine to two terms. They are

$$\boldsymbol{U}_2 = C_{i,j,k-1,3}\,[-\hat{h}_1\,\zeta + \hat{h}_3\,\xi]\;\xi^{i-1}\eta^{j}\zeta^{k-1}\,\frac{1}{r} \quad (2.3.32a)$$

and

$$\mathbf{U}_3 = C_{i,j-1,k,2}\,[-\hat{h}_1\,\eta + \hat{h}_2\,\xi]\;\xi^{i-1}\eta^{j-1}\zeta^{k}\,\frac{1}{r} \quad (2.3.32b)$$

for $1 \leq i, j, k \leq n-1$ and $i+j+k = n+1$. There are $\frac{1}{2}n(n-1)$ vector potentials of the type $\mathbf{U}_2$ and the same number of the type $\mathbf{U}_3$. All together from (2.3.31a,b,c) and (2.3.32a,b), we have a set of $3n + n(n-1) = n(n+2) = J_n$ divergence-free vector potentials of order n. We denote the set of these J_n vector potentials by $\mathcal{A}^{(n)}$.

Had we expressed the second coefficient in (2.3.29) in terms of the third and the first, we would have obtained vector potentials which are linear combinations of those in (2.3.32a,b). For later reference, we write down the combination of the vector potentials associated with the second and the third coefficients,

$$\boldsymbol{U}_1 = C_{i,j,k-1,3}\,[-\hat{h}_2\zeta + \hat{h}_3\eta]\;\xi^i\eta^{j-1}\zeta^{k-1}\,\frac{1}{r}\,, \tag{2.3.33}$$

for $1 \le i, j, k \le n-1$ and $i + j + k = n + 1$.

Now we are ready to show that the flow field defined by each vector potential in $\mathcal{A}^{(n)}$ is a potential flow. We see that the velocity corresponding to a vector potential $\boldsymbol{U}_3$ is

$$\begin{aligned}\nabla \times \boldsymbol{U}_3 &= C_{i,j-1,k,2}\;[\hat{h}_3\,(\xi^2+\eta^2) - \hat{h}_1\,\xi\zeta - \hat{h}_2\,\eta\zeta]\;\xi^{i-1}\eta^{j-1}\zeta^k\,\frac{1}{r} \\ &= \nabla\varphi_3\,,\end{aligned} \tag{2.3.34a}$$

where

$$\varphi_3 = -C_{i,j-1,k,2}\;\xi^{i-1}\eta^{j-1}\zeta^{k+1}\,\frac{1}{r}\,, \tag{2.3.34b}$$

for $n \ge i \ge 1$, $n-1 \ge k \ge 0$, and $j = n+1-i-k \ge 1$. Here, we have absorbed $\boldsymbol{U}_{03}$ into $\boldsymbol{U}_3$ by allowing $k=0$ and $i=n$. Similarly we have

$$\nabla \times \boldsymbol{U}_2 = \nabla\varphi_2 \qquad \text{with} \qquad \varphi_2 = C_{i,j,k-1,3}\;\xi^{i-1}\eta^{j+1}\zeta^{k-1}\,\frac{1}{r}\,, \tag{2.3.35}$$

for $n \ge i \ge 1$, $n-1 \ge j \ge 0$, and $k = n+1-i-j \ge 1$. For the remaining vector potentials $\boldsymbol{U}_{01}$, we have

$$\nabla \times \boldsymbol{U}_{01} = \nabla\varphi_1 \qquad \text{with} \qquad \varphi_1 = -C_{0,j,k-1,3}\;\xi\eta^{j-1}\zeta^{k-1}\,\frac{1}{r}\,, \tag{2.3.36}$$

for $n \ge j \ge 1$, and $k = n+1-j \ge 1$. Thus, each vector potential in $\mathcal{A}$ defines an irrotational velocity. Let the set of these J_n velocities be denoted by $\mathcal{V}^{(n)}$. Since there are only $2n+1$ linearly independent potential functions of the order r^{-n-1}, there can be only $2n+1$ nontrivial linear combinations of the velocities in $\mathcal{V}^{(n)}$. We should be able to find n^2-1 linear combinations of the velocities in $\mathcal{V}^{(n)}$, or combinations of the curl of the vector potentials in $\mathcal{A}^{(n)}$, which are equal to zero. In fact, we can construct them as pairwise sums of appropriate vectors in the sets of $\boldsymbol{U}_{01}\ldots\boldsymbol{U}_3$.

We find $(n-1)(n-2)/2$ such combinations of vectors $\boldsymbol{U}_2$ and $\boldsymbol{U}_3$. They are

$$\begin{aligned}\boldsymbol{L}_{23} &= \frac{C_{i,j,k,3}+C_{i,j+1,k-1,2}}{2}[-\hat{h}_1\,(\eta^2+\zeta^2) + \hat{h}_2\,\xi\eta + \hat{h}_3\,\xi\zeta]\xi^{i-1}\eta^j\zeta^{k-1}\,\frac{1}{r} \\ &= \frac{C_{i,j,k,3}+C_{i,j+1,k-1,2}}{2}\,\nabla\left\{\xi^i\eta^j\zeta^{k-1}\,\frac{1}{r}\right\},\end{aligned} \tag{2.3.37}$$

for $1 \leq i, j, k \leq n-2$ and $i+j+k=n$. The lower bounds for i, j, and k are chosen such that they are consistent with those for both $\boldsymbol{U}$'s. We include the coefficients in the combinations to identify the corresponding vector potentials in $\mathcal{A}^{(n)}$. It is obvious that $\boldsymbol{L}_{23}$ is not only curl free, but also divergence free since $[\xi^2+\eta^2+\zeta^2]\,(1/r)=0$. Similarly, the linear combinations of $\boldsymbol{U}_1$ and $\boldsymbol{U}_3$ yield

$$\mathbf{L}_{13} = -\frac{C_{i,j,k,3} - C_{i+2,j-1,k-1,2}}{2} \nabla \left\{ \xi^i \eta^j \zeta^{k-1} \frac{1}{r} \right\}, \tag{2.3.38}$$

for $1 \leq i, j, k \leq n-2$ and $i+j+k=n$.

We recall that only two of the three sets of vector potentials, $\boldsymbol{U}_1, \boldsymbol{U}_2$, and $\boldsymbol{U}_3$ are linearly independent so that combinations of $\boldsymbol{U}_1$ and $\boldsymbol{U}_2$ will not provide additional independent vectors. However, from the linear combinations of $\boldsymbol{U}_{01}$ and $\boldsymbol{U}_2$, and also $\boldsymbol{U}_{02}$, we get

$$\boldsymbol{L}_{012} = -\frac{C_{0,j+2,k,3} + C_{2,j,k,3}}{2} \nabla \left\{ \eta^j \zeta^{k+1} \frac{1}{r} \right\}, \tag{2.3.39a}$$

for $0 \leq j,\ k \leq n-2$, and $j+k=n-2$. Likewise, we obtain $2(n-1)$ additional ones,

$$\boldsymbol{L}_{023} = \frac{C_{i+1,0,k+1,3} + C_{i+1,1,k,2}}{2} \nabla \left\{ \xi^{i+1} \zeta^k \frac{1}{r} \right\}, \tag{2.3.39b}$$

for $0 \leq i, k \leq n-2,\ i+k=n-2$, and

$$\boldsymbol{L}_{031} = -\frac{C_{i+2,j,0,2} + C_{i,j+1,1,3}}{2} \nabla \left\{ \xi^i \eta^{j+1} \frac{1}{r} \right\}, \tag{2.3.39c}$$

for $0 \leq i,\ j \leq n-2$, and $i+j=n-2$. Together (2.3.37), (2.3.38) and (2.3.39a,b,c) define the desired n^2-1 curl-free linear combinations.

An alternate procedure of finding these n^2-1 linear combinations is to utilize the following facts: (1) there are $J_{n-1}=n^2-1$ independent vector potentials of the order $n-1$, $\boldsymbol{A}_j^{(n-1)}$, in $\mathcal{A}^{(n-1)}$, (2) $\nabla \times \boldsymbol{A}_j^{(n-1)} = \nabla \varphi_j$ is divergence free and curl free, and (3) $\nabla \times \boldsymbol{A}_j^{n-1}$ is a vector potential of order n. Hence it represents a linear combination of the J_n vector potentials in $\mathcal{A}^{(n)}$, and such a combination is curl free. Thus we find the n^2-1 combinations corresponding to $j=1,\ldots,n^2-1$. For details refer to Klein and Ting (1990).

In this subsection, we identified n^2-1 curl-free terms in $\boldsymbol{A}^{(n)}$ which do not contribute to the far-field velocity $\boldsymbol{v}^{(n)}$. In principle, we can find the remaining $2n+1$ terms in $\boldsymbol{A}^{(n)}$ and the corresponding $2n+1$ linear combinations of nth moments and then define $\boldsymbol{v}^{(n)}$. It is easier to define it directly from $\boldsymbol{A}^{(n)}$ by making use of (2.3.25) and (2.3.26), which say that $\boldsymbol{v}^{(n)}$ is proportional to r^{-n-2} and can be expressed as the gradient of a scalar potential of nth order. This direct method is carried out in the next subsection.

2.3.3 Relating the Coefficients in the Scalar Potentials of nth Order Directly to nth Moments of Vorticity

At the beginning Subsect. 2.3.1, we concluded from (2.3.25) and (2.3.26) that the far-field velocity $\boldsymbol{v}^{(n)}$ induced by the vector potential $\boldsymbol{A}^{(n)}$ can be expressed as the gradient of a scalar potential $\Phi^{(n)} = Y_n(\theta, \phi) r^{-n-1}$. We now relate the $2n+1$ coefficients of a basis of spherical harmonics Y_n directly to the nth moments of vorticity by making use of (2.3.26) and the radial component of (2.3.25). The latter says that the radial velocity is equal to $\hat{\boldsymbol{x}} \cdot (\nabla \times \boldsymbol{A}^{(n)}) = \partial \Phi^{(n)}/\partial r$. Using (2.3.18) and (2.3.26), we evaluate the left- and right-hand side of the equality. It becomes

$$
\begin{aligned}
\hat{\boldsymbol{x}} \cdot \nabla \times \boldsymbol{A}^{(n)} &= \frac{1}{4\pi\, r^{2n+2}} \int\int\int_{-\infty}^{\infty} P_n'(\boldsymbol{x} \cdot \boldsymbol{x}')\, \boldsymbol{x} \cdot [\boldsymbol{x}' \times \boldsymbol{\Omega}(\boldsymbol{x}')]\; d^3\boldsymbol{x}' \\
&= \frac{1}{4\pi\, r^{n+2}} \int\int\int_{-\infty}^{\infty} P_n'(\boldsymbol{x}' \cdot \hat{\boldsymbol{x}})\, \hat{\boldsymbol{x}} \cdot [\boldsymbol{x}' \times \boldsymbol{\Omega}(\boldsymbol{x}')]\; d^3\boldsymbol{x}' \\
&= \Phi_r^{(n)}(r, \theta, \phi) = -(n+1) Y_n(\theta, \phi) r^{-n-2} \,, \qquad (2.3.40)
\end{aligned}
$$

where $P_n'(u)$ stands for the derivative of $P_n(u)$. Since $\hat{\boldsymbol{x}} = \boldsymbol{x}/r$ is a function of the spherical angles θ and ϕ only, the same is true for the last volume integral in the above equation. The corresponding spherical harmonic function of nth order is

$$
Y_n(\theta, \phi) = \frac{-1}{4\pi(n+1)} \int\int\int_{-\infty}^{\infty} P_n'(\boldsymbol{x}' \cdot \hat{\boldsymbol{x}})[\boldsymbol{x}' \times \boldsymbol{\Omega}] \cdot \hat{\boldsymbol{x}} d^3\boldsymbol{x}' \,. \qquad (2.3.41)
$$

Since the integrand is a homogeneous polynomial of x_i' and x_i/r, $i = 1, 2, 3$, of degree n, the $2n+1$ coefficients in Y_n are linear combinations of nth moments of vorticity. We now show the explicit results for the leading two terms including their dependence on t, which was suppressed in this subsection. For $n = 1$, (2.3.41) becomes

$$
Y_1(\theta, \phi) = -\frac{\hat{\boldsymbol{x}} \cdot \boldsymbol{E}}{8\pi} \,, \qquad (2.3.42)
$$

where $\boldsymbol{E} = \langle \boldsymbol{x}' \times \boldsymbol{\Omega} \rangle$ is independent of time on account of condition (2.3.7). For $n = 2$, we have

$$
\begin{aligned}
Y_2(t, \theta, \phi) &= -\frac{1}{8\pi} \sum_{i=1}^{3} \hat{\boldsymbol{x}} \cdot \left[\langle \boldsymbol{x}' \times \boldsymbol{\Omega}\, x_i' \rangle \frac{x_i}{r} \right] \\
&= \frac{1}{24\pi} \sum_{i=1}^{3} \left[9F_i(t) \frac{x_j x_k}{r^2} + 2\left(H_j(t) - H_k(t)\right) \left(\frac{x_i}{r}\right)^2 \right] \,, \qquad (2.3.43)
\end{aligned}
$$

with i, j, k in cyclic order. Here, the three components, x_i/r, $i = 1, 2, 3$, of the unit radial vector are functions of the spherical angles θ and ϕ. The coefficients F_i and H_i are defined by (2.3.8) and (2.3.9) as linear combinations of the second moments of vorticity and are time dependent. There are only five linearly independent combinations since $H_1 + H_2 + H_3 \equiv 0$. In deriving the last equation in (2.3.43), we made use of the consistency condition (2.2.8). Now we shall express the first two terms of the far-field potential, Φ, in terms of doublets and quadrupoles. They are

$$\Phi^{(1)}(\boldsymbol{x}) = \frac{1}{8\pi}\boldsymbol{E}\cdot\nabla\frac{1}{r} \tag{2.3.44}$$

and

$$\Phi^{(2)}(t,\mathbf{x}) = \frac{1}{8\pi}\sum_{i=1}^{3}\partial_j\partial_k\left[\frac{F_i(t)}{r}\right] + \frac{1}{36\pi}\sum_{i=1}^{3}\partial_i^2\left[\frac{H_j(t)-H_k(t)}{r}\right] \tag{2.3.45}$$

with i, j, k in cyclic order.

For additional results on the far-field representation of the vector potential and the corresponding scalar potential, see Klein and Ting (1990). For example, it is shown that $\boldsymbol{A}^{(n)}$ can be expressed as a combination of M_n linearly independent divergence-free vector potentials of order n. The number M_n for $n = 1$ is equal to $J_1 = 3$ and for $n \geq 2$ is equal to $4n \leq J_n$. The equality holds only when $n = 2$. This means that the J_n vector potentials in $\mathcal{S}(n)$ are not linearly independent for $n \rangle 2$. In Klein and Ting (1990), derivations of the aforementioned results along two different paths are presented. The first path, which follows the one outlined in this section, goes from the motivations to the conjectures and then to their confirmations. Knowing what to prove, the authors arrive at the same results in a more straightforward fashion, using symbolic Cartesian tensor calculus. This approach, furthermore, reveals that Truesdell's consistency conditions restrict the moment of vorticity tensors of any order to have zero symmetric part.

As mentioned before, the investigations presented in this section were partially motivated by the need to determine the leading terms of the far-field scalar potential induced by a vortical flow with rapidly decaying vorticity field. Knowing the leading terms, we shall derive the acoustic field generated by such a vortical (low Mach number) flow in Chap. 7.

3. Motion and Decay of Vortex Filaments – Matched Asymptotics

Motivation

It is well known that the inviscid theory of a vortex line $\mathcal{C}$ i.e., (or a point vortex in a planar flow) with circulation Γ has been successful in modeling the flow field away from a slender vortex filament, which is a vorticity field concentrated around the line $\mathcal{C}$ (or center) with total strength Γ. The vorticity vanishes outside the filament or away from the line, where the distance r from the line is $O(\ell)$. Here, ℓ denotes a typical length scale of the background flow and the line $\mathcal{C}$, say its radius of curvature. The inviscid theory breaks down near the vortex line (or point) where $r \ll \ell$. As $r/\ell \to 0$, the velocity becomes singular. Also, the velocity of a vortex line is undefined while that of a point vortex is assumed to be the local background flow without the point vortex.

In a fluid with small kinematic viscosity ν, that is, $\nu/(U\ell) \ll 1$, the viscous terms in the momentum equation are negligible wherever the velocity gradient is $O(U/\ell)$. In the small neighborhood of $\mathcal{C}$, that is, in and near the filament where $r/\ell \ll 1$, there is another length scale $\delta^* \ll \ell$, denoting the typical size of a cross section, or a vortical core size, of the filament. In the vortical core, where $r/\delta^* = O(1)$, the vorticity has to be much larger than the background velocity gradient, by $O(\epsilon^{-2})$, where $\epsilon = \delta^*/\ell$, to have total strength $\Gamma = O(U\ell)$. For $\epsilon \ll 1$, the viscous terms in the momentum equation have to be accounted for and in effect resolve the singularity of the inviscid solution.

In contrast to the flow field of one typical length scale ℓ studied in Chap. 2, in the current chapter we study flow fields with multiple length scales, amenable by the method of matched asymptotic analysis, originated by Friedrichs (1942, 1955), and elaborated with many applications to mechanics by van Dyke (1975) and Kevorkian and Cole (1996) (see also references therein). In this chapter, we first study basic two-length scale problems in two and three dimensions, modeling a single slender vortex filament submerged in a background potential flow. The flow field is composed of two overlapping regions, an outer region and an inner region, having two distinct length scales. There is a normal length scale ℓ for the outer region, which is the flow field away from the filament, and a much smaller length scale, known as the typical core size δ^*, for the inner region, which is the flow field in and near the filament. In addition, we assume that the basic problem has only one time scale ℓ/U, and show the restrictions on the flows implied by the solutions of one time and two length scales. For flows which do not fulfill these restrictions, we identify the multiple time and/or multiple length scales of the flows and describe the analysis accounting for the multiple scales in this and subsequent chapters.

Extended Summary

In this chapter, we study flow fields composed of a finite number of *slender* vortex filaments embedded in a background flow of normal length scale ℓ and velocity scale U. A slender vortex filament, or in short a filament, models a special type of vorticity field, in which the bulk of vorticity is concentrated in a slender tubular region around a curve $\mathcal{C}$, called its centerline. We assume that the radius of curvature of $\mathcal{C}$ and the characteristic length scale of variations of the vorticity field along $\mathcal{C}$ are $O(\ell)$. But in a normal plane of $\mathcal{C}$, a cross-sectional plane of the filament, the vorticity decays rapidly as the distance r from $\mathcal{C}$ increases, in a typical decaying length $\delta^* \ll \ell$. We call δ^* the typical core size of the filament. We require the total strength or circulation Γ of the filament to be $O(U\ell)$ so that the velocity induced by the filament at distance $O(\ell)$ is $O(U)$. Since vortex lines cannot terminate inside a flow field, a filament will either (i) form a loop of finite length $S, = O(\ell)$ or (ii) originate from a solid surface and (a) extend to infinity, that is, to a length much larger than ℓ, or (b) end on a surface. For case (iia) or (iib), the variation of the core structure along the centerline $\mathcal{C}$ would, in general, require another length scale much larger than ℓ to account for the gradual modulation of the variation in the scale ℓ, or one much smaller than ℓ to describe the interaction of the filament with the boundary layer(s).

We first consider vortical flows having only two distinct length scales, an outer or normal length ℓ for the flow field outside of the filament, for the filament geometry, and for the core structure variation along $\mathcal{C}$, and an inner length, a typical core size δ^*, for the variation in a cross-sectional plane of the filament. The slenderness ratio,

$$\epsilon = \delta^*/\ell \ll 1 \,, \tag{3.1}$$

shall be employed as the small parameter in order to (i) measure the order of magnitude of flow variables inside the vortical core relative to those outside, (ii) formulate the expansion schemes for the inner and outer variables, and (iii) carry out the matched asymptotic analysis. For later reference, we notice the asymptotic scaling of the nondimensionalized centerline curvature,

$$\kappa = O(1) \qquad \text{as} \qquad (\epsilon \to 0) \,, \tag{3.2}$$

which is a consequence of our assumption that ℓ is the characteristic length scale for the filament geometry.

Since $\Gamma/(U\ell) = O(1)$, the velocity in the vortical core is $O(\epsilon^{-1}U)$ and the vorticity is $O(\epsilon^{-2}U/\ell)$. They are larger than the background velocity and vorticity by one and two orders, respectively. We then carry out the analysis for the basic problem in two length scales. We study a filament of type (i) at a distance $O(\ell)$ away from a boundary of, or a body in, the flow field. Thus, its core structure is periodic in the axial variable s along $\mathcal{C}$ with no end points, and the filament is far away from the boundary of the flow field.

To complete the scalings, we first consider one-time solutions; that is, we assume the entire flow field, both the inner and outer regions, having one-time scale, the normal time scale U/ℓ. A one-time solution imposes constraints on the initial data, say the velocity of $\mathcal{C}$ and the core structure. In case the initial data prescribed are inconsistent with the constraints, a short-time scale has to be introduced and a two-time analysis has to be carried out to resolve the inconsistencies. The identification of the one-time solution as the average of the two-time solution in the normal time scale for the first order and second-order core structures, presented in Book-I, are reproduced in Sect. 3.2 and in Appendix B, respectively.

For flow fields having a *finite* number of slender vortex filaments, but only two distinct length scales, we need to impose the aforementioned assumptions for a single filament to all of them. In addition, we assume that the distances between filaments are $O(\ell)$ pairwise, that is, far apart relative to δ^*, so that the asymptotic solution for a single filament can be employed to describe all of them individually. That is, the core structure of each filament can be described by an inner solution of scale δ^* matched with an outer solution representing the background flow plus the flow fields induced by the other filaments.

Besides the slenderness ratio ϵ characterizing the filaments, there are two parameters for the background flow. They are the typical Reynolds number Re and Mach number Ma. Here we consider the far-field flow to be uniform with velocity U, pressure P_∞, density ρ_∞, kinematic viscosity ν_∞, and speed of sound C_∞. We then define those two numbers by their far-field values, and assume that the flow is of high Reynolds number and low Mach number,

$$\mathrm{Re}_\infty = \frac{U\ell}{\nu_\infty} \ll 1 , \tag{3.3}$$

and

$$\mathrm{Ma}_\infty = U/C_\infty \ll 1 . \tag{3.4}$$

Thus the flow field in the outer region, that is, the intersection of the outer regions of all the filaments, can be considered inviscid and incompressible. If the far-field flow is uniform, then it is also irrotational and can be expressed as a background potential flow plus the flows induced by the vortex lines, defined by the Biot–Savart formula.

Since the velocity in a filament is $O(U/\epsilon)$, the Mach number is $O(\mathrm{Ma}_\infty/\epsilon)$. To assume that the flow remains incompressible in the vortical core of the filament, we require

$$\mathrm{Ma}_\infty/\epsilon \ll 1 . \tag{3.5}$$

Under the assumptions (3.3), (3.4), and (3.5), the entire flow field, including the vortical cores of the filaments, is governed by the incompressible Navier–Stokes equations. The asymptotic analyses of slender filaments with small parameter ϵ were carried out first for two-dimensional problems by Ting and Tung (1965), and then for axisymmetric problems by Tung and Ting (1967) to show the effects of curvature and stretching of the centerline. Both effects are absent in two-dimensional problems. For a slender filament in a three-dimensional flow, orthogonal coordinates s, r, and θ, intrinsic to the centerline $\mathcal{C}$ at time t, were introduced in Ting (1971), where s denotes the axial variable along $\mathcal{C}$, and r and θ denote the polar coordinates in the normal plane at point s on $\mathcal{C}$. The analysis defined the velocity of $\mathcal{C}$ coupled with the evolution of the core structure having only large circumferential velocity. Analysis to allow also for large axial flows in the core structure was presented in Callegari and Ting (1978). The above analyses were presented in a chronological order in the first three sections of Chap. 2 in Book-I, in order to illustrate the increase in the complexity of the analyses. For the same purpose, we reproduce those three sections in Sects. 3.1, 3.2, and 3.3 of this chapter. Margerit (1997) extended the analyses reproduced here to higher order in ϵ. Besides providing higher order extensions to Callegari and Ting's results, Margerit's work may be of very high interest to some readers as he developed a systematic way of doing all asymptotic calculations, using elaborate techniques of symbolic computation.

In Sect. 3.1, we review the classical inviscid theories of two- and three-dimensional vortical flows modeled, respectively, by point vortices and vortex lines and then point out their defects. In Sect. 3.2, we describe in detail the asymptotic analysis of a two-dimensional vortical spot of size δ^* removing the singularity of a vortex point; meanwhile, matching with the outer solution of length scale ℓ defines the leading-order velocity of the center of the vortical spot independent of its core structure. When we consider the flow field having only one time scale, the velocity of the vortex center is equal to the classical formula for the velocity of the vortex point. The one-time solution imposes constraints on the initial data. For initial data inconsistent with the constraints, a two-time solution with a second time scale much shorter than the normal time scale is needed to resolve the inconsistency. The original one-time solution can be identified as the average of the two-time solution in the normal time scale. In Sect. 3.3, we describe the analysis in one time scale, ℓ/U, for a single filament forming a loop (Callegari and Ting, 1978). The analysis resolves the singularities in the Biot–Savart integral for a vortex line and defines the velocity of the filament centerline $\mathcal{C}$ and the evolution of the core structure.

We present the analysis first for the relatively simple two-dimensional problem and then for the three-dimensional problem. These analyses are systematic but not as straight forward as in standard asymptotic problems for which the analysis begins with an appropriate choice of reference scales, the small parameter, and the expension scheme, and then procedes to derive from the leading-order governing equations directly a close system of equations for the leading-order solutions. Instead, for the singular perturbation problem of concentrated vortical flows, the leading-order equations degenerate to the condition that the leading-order core structure is symmetric, that is,

$$\partial_\theta f^{(0)} = 0 \,, \tag{3.6}$$

where $f^{(0)}$ stands for the leading-order core structure, that is, the pressure $p^{(0)}$ and the swirling velocity $v^{(0)}$ (and the axial velocity $w^{(0)}$ along $\mathcal{C}$ for a filament). The core structure, that is, the dependence of $f^{(0)}$ on t, r, and s, remains undefined. Because of (3.6), the homogeneous parts of the higher order equations are integrable with respect to θ, and the periodicity in θ in turn yields compatibility conditions on the corresponding inhomogeneous parts.

The conditions for the first order inhomogeneous parts asymmetric in θ yield equations for the leading-order velocity of a filament (a two-dimensional vortical spot) coupled with (independent of) its core structure. The governing equations for the core structure come from the compatibility conditions for the second-order axisymmetric inhomogeneous parts. The second-order inhomogeneous parts contain products of the first order solutions, $f^{(1)}$'s, so far undefined. These products are nonlinear and depend on θ. It is necessary, but very tedious, to show that all those unknown nonlinear terms do not appear in the axisymmetric compatibility conditions, which then yield the equations for the leading core structure, coupled with equations for the velocity of the vortex center to form a closed system of equations. See Subsects. 3.2.2 and 3.3.6.

Having outlined the reproduction of the asymptotic analyses presented in Chap. 2 of Book-I, we shall now mention the four related new studies in this chapter.

In Subsect. 3.3.4 we present an alternative derivation of the equation of motion of a three-dimensional vortex filament for the one-time asymptotic regime, in Subsubsect. 3.3.3.2. The analysis relies upon an asymptotic evaluation of the Biot–Savart integral. This alternative representation of the vortex filament equation of motion will be useful in Chap. 5 in the construction of accurate and efficient "thin-tube" numerical vortex element schemes for slender vortex simulations.

For simplicity, we have thus far assumed that the background flow is irrotational so that it remains the same, independent of the motion of the filament.[1] Thus we can concentrate our attention on the analysis of the filament. This simplification is not required for the method of matched asymptotics. It is applicable for a rotational background flow with vorticity $O(\ell/U)$ so long as it is much smaller than, say $O(\epsilon)$ of the vorticity in the filament. But then the background vorticity, and hence the background flow, will be coupled with the motion of the filament. We elaborate on this new complexity in Subsect. 3.3.8, in which we study the motion of two vortical spots with strength $O(U\ell)$ in a two-dimensional flow which is a simple shear flow upstream with vorticity U/ℓ, that is, $O(\epsilon^2)$ of the vorticity in a spot.

We recall that in the analyses of Callegari and Ting (1978), the background flow was assumed to be irrotational and the leading-order core structure was assumed to be (i) independent of s, that is, no axial variation, and (ii) incompressible, that is, at low Mach number, (3.5). Attempts to remove the first assumption on the core structure, that is, to allow for axial core variation, were made by Klein and Ting (1992) and are described in Subsect. 3.3.7. The reader is referred to more recent work by Margerit et al. (2001) and Margerit (2002) for further developments in this direction, including a two-time analysis that accounts for axial waves on the vortex core. An analysis allowing for a background rotational flow with vorticity $O(U/\ell)$ was carried out by Liu and Ting (1987) and is reported in Subsect. 3.3.8.

Attempts to remove the second assumption on the core structure, that is, to account for a compressible core structure, were reported in Ting (2002) and Knio et al. (2003) and will be described in Sect. 3.4. To this end, the governing equations for the motion of the centerline and evolution of the core structure of a compressible filament are summarized in Subsect. 3.4.1, and analyzed subsequently via matched asymptotic expansions. Although the leading-order outer solution remains incompressible, we have to begin the asymptotic analysis from the compressible Navier–Stokes equations to account for the compressible inner solution. Instead of repeating the steps and tedious derivations in Sect. 3.3, with the additional complexity of compressibility, we make use of the *fast track* method that was first reported in Ting (1999) to simplify the derivations for incompressible core structures.

3.1 Inviscid Theories of Vortical Flows and Their Deficiencies

In this section, we review the classical solutions for incompressible inviscid flows in two- or three-dimensional space where the vorticity fields are modeled

[1] This statement is true with the understanding that the flow is unbounded and has no internal boundaries. This understanding shall be implied for similar statements appearing later in this chapter. Counterexamples are presented in Chap. 7, where we consider the motion of a filament in an unbounded flow containing a solid body and show that in this situation the interaction of the filament with the body induces an additional unsteady potential flow.

by vortex points and vortex lines, respectively. We then point out the defects of those solutions, that is, where they are applicable and where they become singular and the motion of a vortex point or vortex line is undefined.

In Subsect. 3.1.1, we study two-dimensional flows in which a point vortex with circulation Γ is used to model a vorticity field concentrated in a small area or spot with total strength Γ. In the classical inviscid theory, the vortex point is assumed to move with the local background velocity without the vortex point. To demonstrate that this well-accepted assumption is not true in general, we reproduce in Subsubsect. 3.1.1.1 the quasi-steady solution of a uniform flow around a spinning disc and the equation of motion of the disc center (see Milne-Thompson, 1973). In general, the initial velocity of the disc center, $\dot{\boldsymbol{X}}(0)$, differs from that of the uniform velocity upstream, $\boldsymbol{U}_\infty$, and the center oscillates about a mean path, a straight line with constant velocity $\boldsymbol{U}_\infty$. If and only if the initial velocity $\dot{\boldsymbol{X}}(0) = \boldsymbol{U}_\infty$, the disc center moves along the straight path with no oscillation. This phenomenon happens also in the motion of a simple oscillator, in the limiting case of vanishing mass, $M \to 0$, or rather vanishing radius, $a/\ell \to 0$, with the density of the disc finite. In the normal time scale, the acceleration scaled by U^2/ℓ is $O(1)$ and the inertia force vanishes as $(a/\ell)^2$. This requires the lift to vanish as the relative velocity $\boldsymbol{Q}(t) = \boldsymbol{U}_\infty - \dot{\boldsymbol{X}}(t) \to 0$. Thus, the disc has to move with the upstream velocity initially. If a different initial velocity is prescribed, there is a lift force. We need a second-time scale that is much shorter than the normal scale ℓ/U to increase the order of magnitude of the acceleration such that the inertia term remains $O(1)$ as $a/\ell \to 0$ and balances the lift force. The two-time solution describes the low-amplitude high-frequency oscillation of the disc about the mean path. This is a classical example in which an initial data is *lost* in a one-time solution and accounted for by a two-time solution. The example demonstrates the standard technique employed later to account for the initial data *lost* in the one-time asymptotic solutions of vortical spots or slender filaments. See Sects. 3.2 and 3.3.

When the disc is replaced by the same fluid of density ρ, the flow field is extended to the entire xy plane and the total vorticity inside the disc has to be equal to the circulation Γ. With constant vorticity inside, the fluid is rotating at a constant angular velocity $\dot{\theta} = \Gamma/[2\pi a^2]$. We have a Rankine vortex. The circumferential velocity inside the circle equals that outside induced by the potential vortex of strength Γ. We now identify the potential flow outside the spinning disc of mass $M = \rho\pi a^2$ as discussed in Subsubsect. 3.1.1.1, with that outside a free Rankine vortex as considered in Subsubsect. 3.1.1.2. This identification has a defect: across the interface separating the rotational flow inside from the potential flow outside, there is a pressure jump, because the pressure variation in the potential flow due to the doublet is not balanced by the pressure inside. We shall identify the solution for a spinning disc in Subsubsect. 3.1.1.1 as the leading-order solution for a Rankine vortex submerged in a uniform stream with the correction to the pressure jump as a higher order term. This is achieved when the disc radius, or the size of the Rankine vortex, is small, $a \ll \ell$, so that the circumferential velocity at the interface is much larger than U. That is,

$$\frac{U}{a\dot{\theta}} = \frac{2\pi U a}{\Gamma} = \frac{2\pi U \ell}{\Gamma}\epsilon \ll 1 \,, \qquad \text{where} \quad \epsilon = \frac{a}{\ell} \ll 1 \,. \tag{3.1.1}$$

With ϵ as the small parameter, the motion of the rotating disc becomes the leading-order solution of the Rankine vortex with vanishing radius. We analyze small perturbations of the interface from the circle of radius a and the next-order solutions of the Rankine vortex in Subsubsect. 3.1.1.1. This will allow us to remove the higher order pressure jump across the circle. We will show in addition that the vortex, center is drifting with the background flow in the normal time scale ℓ/U_∞, as prescribed by the classical inviscid theory, plus a small-amplitude high-frequency oscillation with period $O(\epsilon^2\ell/U_\infty)$ and amplitude $O(\epsilon^2\ell)$.

In a real fluid, viscous effects become dominant wherever the second derivatives of velocity are large. In fact, the shear stress term in the momentum equation would become singular at the interface of a Rankine vortex with small core size. Thus, inviscid solutions, such as point or Rankine vortices, will not be valid in the immediate vicinity of the region of strong, concentrated vorticity. A viscous "inner" region has to be included in the analysis to resolve the singularity. The analysis will be presented in Sect. 3.2.

In Subsect. 3.1.2, we study the flow field induced by a vortex line $\mathcal{C}$ of strength Γ, which models a slender filament with vanishing core size. The Poisson integral for the vector potential $\boldsymbol{A}$ induced by the filament reduces to a line integral and the corresponding integral for the velocity $\nabla \times \boldsymbol{A}$ is known as the Biot–Savart integral. We derive the singular behavior of the vector potential near the vortex line, say at distance r from point $\boldsymbol{X}$ on $\mathcal{C}$, and its finite part at point $\boldsymbol{X}$. They, in turn, yield the singular and finite parts of the velocity. The latter reproduce those derived directly from the Biot–Savart integral by Ting (1971) and Callegari and Ting (1978). In the singular parts of the velocity, the leading term $O(1/r)$ is the circumferential velocity $\Gamma/(2\pi r)$. It is independent of the curvature of $\mathcal{C}$ as the vortex line is approximated by its local tangent line. There are two additional singular terms. One of them is a logarithmic singularity in the binormal direction, while the other one is bounded in the circumferential direction but multivalued as $r \to 0$. Both terms are proportional to the curvature of the vortex line and thus account for the essential difference between curved and straight vortex filaments. For a straight vortex line (or a vortex point in a two-dimensional flow), the local average velocity at point $\boldsymbol{X}$ on the line $(r \to 0)$ over a cross-sectional area $\gg (\delta^*)^2$ but $\ll \ell^2$, exists because the area integral of the singular circumferential component vanishes. The local average velocity is equal to the background velocity. Thus, one could define the velocity of a vortex point given by the classical two-dimensional inviscid theory by the local average velocity. This new definition is not applicable to a curved vortex line because the corresponding local average in the cross-sectional plane of C could not remove the singular $\ln r$ term in the binormal direction for nonzero curvature. It is well known that the inviscid velocity field induced by a vortex filament of finite core size is defined including the centerline of the filament. For example, the velocity of a circular vortex ring with constant vorticity inside the small but nonzero core radius was given by Lamb (1932).

Away from a slender filament, that is, in an outer region where the distance r to the filament is much larger than the typical core size δ, the flow field can be approximated by that of a vortex line with $\delta = 0$. Because of the singularities as $r \to 0$, the approximation is not valid in the neighborhood of the filament, that is, in the vortical core or the inner region where $r = O(\delta)$. This is again a two-length scale problem amenable by the method of matched asymptotics. We shall carry out the analysis in Sect. 3.3 to construct the inner solution, resolving the singularity of the outer solution, and define the velocity of the filament centerline including the contribution from the core structure.

3.1.1 Potential Flows Around a Spinning Disc and a Rankine Vortex

For a two-dimensional potential flow around a vortex point of strength Γ located at (X, Y), the stream function ψ is the sum of the stream function Ψ for the background potential flow and that for the vortex alone, that is,

$$\psi(x, y) = \Psi(x, y) - \frac{\Gamma}{2\pi} \ln r \,, \tag{3.1.2}$$

where r is the distance between (x, y) and (X, Y). Here the dependence of the stream functions and the coordinates (X, Y) on time t has been suppressed. The velocity near the point vortex behaves as $\Gamma/(2\pi r)$ and is unbounded as $r \to 0$. The second defect of the inviscid theory, that is, the fact that the vortex center velocity is undetermined, and is circumvented by the *assumption* that the vortex point moves with the local mean velocity which is also the local velocity of the flow field in absence of the vortex. Thus,

$$\dot{X} = \Psi_y(X, Y) \qquad \text{and} \qquad \dot{Y} = -\Psi_x(X, Y) \,, \tag{3.1.3}$$

where the dot denotes differentiation with respect to t. Lamb (1932) provided justification for the above assumption on the physical consideration that a vortex point, having zero mass, must move with the local fluid. If not, its velocity $\mathbf{Q}$ relative to the local fluid would produce a finite lift force $\rho|\mathbf{Q}|\Gamma$ acting at the vortex point according to Joukowski's theorem. The lift force could not be balanced by the inertia of the vortex, which is zero, and therefore the relative velocity $\mathbf{Q}$ has to vanish. The above argument is usually referred to as the limit solution of vanishing mass. To illustrate this point and also to explain the physics of a two-time solution in its simplest form, we study the motion of a spinning disc in a uniform stream. From the analogy to the motion of a small spinning disc, we get a clear physical understanding of the highly oscillatory motion of a Rankine vortex about its mean trajectory. Later on, we shall employ the same analogy to describe the motion of a vortex with a decaying vortical core.

3.1.1.1 Motion of a Spinning Disc in a Uniform Stream. Consider the motion of a disc of radius a and uniform density ρ_0 in a two-dimensional incompressible inviscid fluid of density ρ, say in the xy-plane. The disc is spinning at a constant angular velocity $\dot{\theta}$ and is submerged in a uniform stream with velocity U_∞ in the direction of the x-axis, as shown in Fig. 3.1.

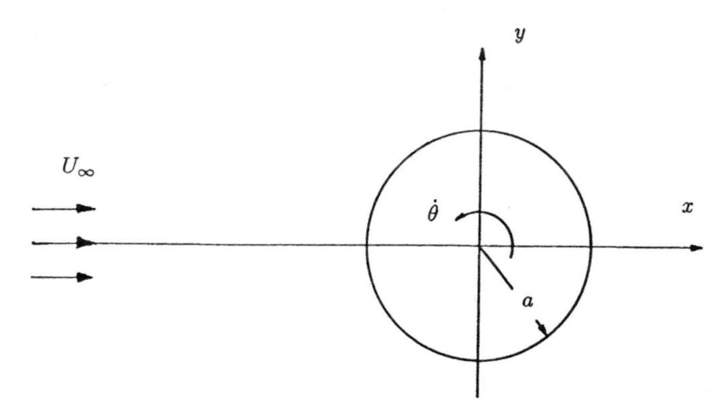

Fig. 3.1. Spinning disc in a uniform stream

For a two-dimensional potential flow, it is convenient to consider the x–y plane as the complex z plane with $z = x + iy$. Then, let the trajectory of the disc center be denoted by $Z(t) = X(t) + iY(t)$ and let the initial position $Z(0)$ and velocity $\dot{Z}(0)$ of the center be specified by

$$Z_0 = X_0 + iY_0 \qquad \text{and} \qquad \dot{Z}_0 = \dot{X}_0 + i\dot{Y}_0 \, . \tag{3.1.4}$$

The complex velocity of the uniform stream is then denoted by U_∞. The complex velocity relative to the disc is denoted by

$$Q(t) = U_\infty - \dot{Z}(t) \tag{3.1.5}$$

and the circulation around the disc by

$$\Gamma = 2\pi a^2 \dot{\theta} \, . \tag{3.1.6}$$

The complex potential of the flow field relative to the disc center is the superposition of the potentials of a uniform flow, a point vortex, and a doublet located at the vortex center,

$$w(t,z) = Q^*(t)\,(z-Z) + \frac{\Gamma}{2\pi i}\ln\frac{z-Z}{a} + \frac{Q(t)\,a^2}{z-Z}\,, \qquad (3.1.7)$$

where Q^* is the complex conjugate of Q. The flow field matches the normal velocity of the surface because of the contributions from the uniform flow and the doublet, but not the tangential velocity.

In principle, we can compute the force acting on the disc directly from the pressure distribution on the surface. However, there is a more elegant way to derive the equation of motion of the disc from a balance of momentum for a large control area containing the disc (see, e.g., Milne-Thompson, 1973). The equation of motion of the center of a disc with mass M is

$$M\,\ddot{Z} = -i\,\rho\Gamma Q - M'\ddot{Z}\,, \qquad (3.1.8)$$

where the first term on the right-hand side represents the Joukowski force, while the second term accounts for the inertia of the surrounding fluid having an equivalent mass M'. For the circular disc with density ρ_0, we have

$$M = \rho_0\pi a^2 \qquad \text{and} \qquad M' = \rho\pi a^2\,. \qquad (3.1.9)$$

The solution of (3.1.5) and (3.1.8), which is the trajectory of the center of the disc, is

$$Z(t) = \tilde{Z}(t) + A e^{i2\pi t/T}\,, \qquad (3.1.10)$$

where

$$A = iTQ(0)/(2\pi)\,, \qquad T = 2\pi/\omega = 2\pi^2 a^2(\rho+\rho_0)/(\rho\Gamma) \qquad (3.1.11)$$

and

$$\tilde{Z}(t) = Z_0 - A + U_\infty t\,. \qquad (3.1.12)$$

The last term in (3.1.10) represents the oscillatory part of the trajectory relative to the mean trajectory defined by (3.1.12). Here ω, T, and A are the frequency, period, and complex amplitude of the oscillations, respectively.

The mean trajectory (3.1.12), which is equal to the average of the real trajectory (3.1.10) over one period T, that is,

$$\tilde{Z}(t) = \frac{1}{T}\int_t^{t+T} Z(t')\,dt'\,, \qquad (3.1.13)$$

shows that the disc drifts with the background velocity U_∞. The term $-A$ in (3.1.12) represents the deviation of the initial position from the mean trajectory. If the disc is moving initially at the same velocity as the uniform stream, we have $Q(0) = U_\infty - \dot{Z}_0 = 0$ and hence $A = 0$. Then the oscillatory part of the trajectory (3.1.10) vanishes and the disc drifts along with the uniform stream with zero relative velocity, that is, $Q(t) = U_\infty - \dot{Z}(t) = 0$.

Figure 3.2 shows the oscillatory and mean trajectory of a disc that starts with zero velocity from the origin, that is, $Z_0 = 0$ and $\dot{Z}_0 = 0$. At $t = 0$,

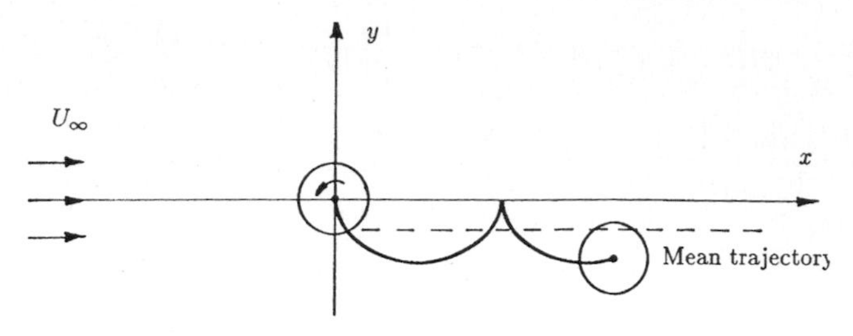

Fig. 3.2. Trajectory of a spinning disc in a uniform stream

the Joukowski force is $\rho U_\infty \Gamma$ acting in the downward direction and the disc is gaining a downward velocity, $\dot{Y} < 0$. This, in turn, induces a horizontal component of the Joukowski force, increases $\dot{X}$ from zero, and thus reduces the horizontal component of relative velocity $U_\infty - \dot{X}$. At $t = T/4$, we have $U_\infty - \dot{X} = 0$ and $-\dot{Y}$ reaches its maximum value. Afterward, $\dot{X}$ increases beyond U_∞ while $-\dot{Y}$ decreases. At the half period $\dot{Y}$ vanishes, the trajectory reaches a turning point and $\dot{X}$ reaches its maximum value of $2U_\infty$. A similar description of the trajectory holds for the second half period. The wavelength, which is the horizontal distance traveled during one period, is $U_\infty T$. The vertical shift $U_\infty T/(2\pi)$ of the mean trajectory given in (3.1.11) is equal to the average of the vertical displacement over one period. This completes the physical description of the oscillatory and mean trajectory of a spinning disc in a uniform stream.

Now we study the limiting case of "vanishing mass" or rather "vanishing radius" with finite Γ and U_∞. To set up the analogy for the motion of a Rankine vortex, and later for a vortex with decaying vortical core, we let the density of the disc equal that of the fluid,

$$\rho_0 = \rho = 1 \tag{3.1.14}$$

and introduce the *normal* length and time scales

$$\ell = \Gamma/U_\infty \,, \qquad \text{and} \qquad \ell/U_\infty = \Gamma/U_\infty^2 \,. \tag{3.1.15}$$

For the limiting case of vanishing radius, we consider the radius a to be very small relative to ℓ, that is,

$$a = \epsilon\ell = \epsilon\Gamma/U_\infty \,, \qquad \text{with} \qquad \epsilon \ll 1 \,. \tag{3.1.16}$$

To avoid the introduction of new symbols for dimensionless quantities, we can use ℓ and ℓ/U_∞ as unit length and time with $U_\infty = 1$ and $\Gamma = 1$ and thus render the quantities in (3.1.4) to (3.1.13) dimensionless. In particular, z and t become the *normal* space and time variables. Also, we note

$$T = 4\pi^2\epsilon^2 \qquad \text{and} \qquad A = 2\pi i\, \epsilon^2 Q(0) \,. \tag{3.1.17}$$

In the limit as $\epsilon = a \to 0$, the period T, the amplitude A, and the initial shift of the mean trajectory vanish as $O(\epsilon^2)$. Consequently, in the normal time scale, the inertia of the disc is $O(\epsilon^2)$ and the disc appears to be moving with the velocity of the uniform stream. However, on the small time scale of the order of ϵ^2, the velocity of the center of the disc fluctuates around U_∞ by order $O(1)$. For example, for the case of $\dot{Z}_0 = 0$, the horizontal velocity varies between 0 and $2U_\infty$ and the vertical velocity varies between $\pm U_\infty$ independent of ϵ. This is a standard example of two-time solutions when the initial velocity differs from the required initial value for the one-time solution with "vanishing area."

Now we consider the motion of a small spinning disc in a *nonuniform* potential flow with length scale ℓ. For the inner solution describing the flow field near the disc, we let $\bar{r} = r/\epsilon$ and θ denote the polar coordinates with respect to the disc center, (X, Y). The stream function of the background flow on the scale of a is given by its Taylor series at the disc center (X, Y). When the relative coordinates, $(x - X, y - Y)$, are replaced by the inner polar coordinates, the Taylor series becomes

$$\begin{aligned} \Psi = \Psi^* &+ \epsilon\left[\Psi_x^* \cos\theta + \Psi_y^* \sin\theta\right]\bar{r} \\ &+ \tfrac{\epsilon^2}{2}\left[\Psi_{xx}^* \cos 2\theta + \Psi_{xy}^* \sin 2\theta\right]\bar{r}^2 + \cdots , \end{aligned} \tag{3.1.18}$$

where the superscript * denotes the value at the disc center (X, Y). Using (3.1.18) as the far-field behavior of the stream function of the flow field outside the disc, $\bar{r} \geq 1$, we obtain

$$\begin{aligned} \psi = &-\frac{\Gamma}{2\pi}\ln\frac{\epsilon\bar{r}}{a} + \Psi^* + \epsilon\left[\Psi_x^* \bar{r}\cos\theta + \Psi_y^* \sin\theta\right](r - \bar{r}^{-1}) \\ &+ \frac{\epsilon^2}{2}\left[\Psi_{xx}^* \cos 2\theta + \Psi_{xy}^* \sin 2\theta\right](\bar{r}^2 - \bar{r}^{-2}) + \cdots . \end{aligned} \tag{3.1.19}$$

In addition to the contribution of the circulation, we have added a doublet and a quadrupole, the $\bar{r}^{-1}$ and the $\bar{r}^{-2}$ terms, to render ψ = constant on $\bar{r} = 1$. The outer solution of the flow field in the length scale ℓ, that is, away from the disc, becomes

$$\begin{aligned} \psi = \Psi &- \frac{\Gamma}{2\pi}\ln\frac{r}{\ell} - \frac{\epsilon^2}{r}\left[\Psi_x^* \cos\theta + \Psi_y^* \sin\theta\right] \\ &- \frac{\epsilon^4}{2r^2}\left[\Psi_{xx}^* \cos 2\theta + \Psi_{xy}^* \sin 2\theta\right] + \cdots . \end{aligned} \tag{3.1.20}$$

The third term on the right-hand side of (3.1.20), which is a doublet of the order of ϵ^2, represents the contribution of a disc to a uniform background flow. The last term, a quadrupole $O(\epsilon^4)$, represents the contribution to the outer flow field due to the interaction of the disc with the background velocity gradients. Similar interaction terms will be found in the outer solution away from a viscous vortex in Sect. 3.2.

3.1.1.2 Oscillations of a Rankine Vortex in a Uniform Stream. When the disc is replaced by the same fluid rotating as a rigid body with angular velocity $\dot{\theta}$ and *drifting with velocity* $U_\infty = 1$, the flow velocity in the disc area is

$$\boldsymbol{V}_- = U_\infty + \dot{\theta}(z - Z)i = U_\infty + \frac{i\Gamma(z - Z)}{2\pi a} = O(\epsilon^{-1}) \tag{3.1.21}$$

for $|z - Z| < \epsilon$. The flow is rotational with constant vorticity,

$$\zeta = 2\dot{\theta} = \frac{\Gamma}{\pi a^2} = O(\epsilon^{-2}) . \tag{3.1.22}$$

Outside the disc the flow field is that of a potential vortex

$$\boldsymbol{V}_+ = U_\infty + \frac{i\Gamma(z - Z)}{2\pi |z - Z|^2} \qquad \text{for} \qquad |z - Z| > \epsilon . \tag{3.1.23}$$

The boundary of the disc, $|z - Z| = \epsilon$, represents an interface ∂D separating the potential flow from the rotational flow. We use the subscripts $-$ and $+$ to denote the limit values inside and outside of the interface, respectively. Across the interface the velocity and pressure are continuous in this case. Only the velocity gradient is discontinuous. Equations (3.1.21) and (3.1.22) then represent the inviscid flow field of a Rankine vortex of strength Γ drifting with the uniform stream, that is, $\dot{Z} = U_\infty$.

When the vortex center has an initial velocity $\dot{Z}_0$ not equal to U_∞, the solution for a spinning disc, (3.1.7), in general is no longer applicable. Because of the doublet term, (3.1.7) yields discontinuities in the tangential velocity and pressure across the boundary of the disc, while an interface is a free boundary across which both the pressure and normal velocity have to be continuous. Since the discontinuities are of the order of $a^2 = \epsilon^2 \ll 1$, we expect the solution for the flow around a spinning disc and the same fluid rotating at constant angular velocity $\dot{\theta}$ inside to be an asymptotic solution for a small free Rankine vortex as $\epsilon \to 0$.

To set up the governing equations for the motion of a free Rankine vortex in general, we introduce the polar coordinates r and θ with respect to the moving center of mass of the vortex. Thus, with $Z = X + iY$, we have

$$z - Z = re^{i\theta} . \tag{3.1.24}$$

The inviscid flow field is composed of a rotational flow of constant vorticity $2\dot{\theta}$ inside an interface ∂D and a potential flow outside of it. Because the fluid is incompressible, the area of D remains equal to $\pi\epsilon^2$ and the center of area coincides with the center of vorticity, (X, Y), where $r = 0$. We seek solutions for which the interface deviates from a circle of radius ϵ by $O(\epsilon^2)$ and express the interface by

$$r = \epsilon\bar{r} = \epsilon + \epsilon^2 R(t, \theta, \epsilon) , \tag{3.1.25}$$

where $\bar{r}$ denotes the stretched radial coordinate. The scaled radial displacement R is decomposed in a Fourier series,

$$R = \sum_{n=2,3,\ldots} \{\, a_n(\epsilon, t) \cos n\theta + b_n(\epsilon, t) \sin n\theta \,\} \,. \tag{3.1.26}$$

The symmetric term, $(n = 0)$, is absent because the area of D is conserved. The two first harmonics are absent because the center of area coincides with the moving origin where $\bar{r} = 0$.

The boundary conditions to be imposed at the interface (3.1.25) are the kinematic condition,

$$\epsilon^2 R_t = u_\pm - \frac{\epsilon R_\theta}{1 + \epsilon R} v_\pm \,, \tag{3.1.27}$$

and the balance of pressure,

$$(p_+ - p_-)_\theta + \epsilon\, R_\theta (p_+ - p_-)_{\bar{r}} = 0 \,, \tag{3.1.28}$$

where p stands for the pressure and u and v stand for the radial and circumferential velocity components.

The flow field inside ∂D fulfills the continuity equation,

$$(ru_-)_{\bar{r}} + (v_-)_\theta = 0 \,, \tag{3.1.29}$$

and the equation of uniform vorticity

$$[(\bar{r}v_-)_{\bar{r}} - (u_-)_\theta]/\bar{r} = \epsilon^2/\pi \,. \tag{3.1.30}$$

The flow field outside ∂D fulfills the continuity equation (3.1.29) with the subscript $-$ replaced by $+$ and the equation of zero vorticity,

$$(\bar{r}v_+)_{\bar{r}} - (u_+)_\theta = 0 \,. \tag{3.1.31}$$

To relate the jump of pressure gradients in (3.1.28) to the velocity components, we use the momentum equations,

$$\epsilon u_t + u u_{\bar{r}} + v u_\theta/\bar{r} - v^2/\bar{r} = -p_{\bar{r}} - \epsilon U_t \tag{3.1.32}$$

and

$$\epsilon v_t + u v_{\bar{r}} + v v_\theta/\bar{r} + vu/\bar{r} = -p_\theta/\bar{r} - \epsilon V_t \,, \tag{3.1.33}$$

where

$$U = \dot{X} \cos\theta + \dot{Y} \sin\theta \quad \text{and} \quad V = \dot{Y} \cos\theta - \dot{X} \sin\theta \,, \tag{3.1.34}$$

denote the radial and circumferential components of the velocity of the vortex center. U should not be confused with the the reference velocity, $U_\infty = 1$, in this subsection. In (3.1.32) and (3.1.33), the subscripts $\pm$ for u, v, and p have been suppressed.

In the far field, $\bar{r} \gg 1$, the leading terms should represent the superposition of a uniform and a circulatory flow,

$$u_+ = \cos\theta - U + O(\bar{r}^{-2}) \quad \text{and} \quad v_+ = \frac{\epsilon^{-1}}{2\pi\bar{r}} - \sin\theta - V + O(\bar{r}^{-2}) \,. \quad (3.1.35)$$

These equations are far-field conditions to be imposed on solutions to (3.1.29) and (3.1.31). As we shall see later, the higher order terms will include a doublet term due to the scaled area of D. With $\bar{r}^{-2} = \epsilon^2/r^2$, the doublet term will become the term proportional to $\epsilon^2/\bar{r}$ in the stream function for the outer region. From the definition of $\dot{Z}$, we have

$$u_- = 0 \quad \text{and} \quad v_- = 0 \,, \qquad \text{at} \quad \bar{r} = 0 \,. \quad (3.1.36)$$

Thus, the general formulation of the problem is completed. We now construct the asymptotic solution for $\epsilon \ll 1$. Note that if the solution depends only on the normal time t, all the time derivative terms in the governing equations are of higher order in ϵ. Hence, the variable t will appear as a parameter in the asymptotic solution in all powers of ϵ and we are not free to impose any initial data. Should the initial data be incompatible with the one-time asymptotic solution, then it is necessary to keep the time derivative terms in the governing equations by the introduction of a short-time variable. Guided by the results for a small spinning disc, ((3.1.5), (3.1.6), (3.1.10)), and ((3.1.19), (3.1.20)), we introduce, in addition to the *normal* time variable, the *short* time,

$$\tau = t\epsilon^{-2} \,, \quad (3.1.37)$$

and replace ∂_t in (3.1.27) and (3.1.32)–(3.1.34) by

$$\partial_t + \epsilon^{-2}\partial_\tau \,. \quad (3.1.38)$$

Since the vorticity is $O(\epsilon^{-2})$, the velocity in the near field, $\bar{r} = O(1)$, is $O(\epsilon^{-1})$ and the pressure is $O(\epsilon^{-2})$. This suggests the expansion scheme,

$$f(\epsilon) = \epsilon^{-k}[f^{(0)} + \epsilon f^{(1)} + \cdots] \,, \quad (3.1.39)$$

where

$$f^{(i)} = f^{(i)}(\bar{r}, \theta, \tau, t) \quad (3.1.40)$$

and $k = 1$, when f stands for $u_\pm$ or $v_\pm$, and $k = 2$, when f stands for $p_\pm$. The assumption that the velocity of the vortex center remains $O(1)$ implies that the leading two terms in the series for Z should be independent of τ, that is,

$$Z(\tau, t, \epsilon) = Z^{(0)}(t) + \epsilon Z^{(1)}(t) + \epsilon^2 Z^{(2)}(\tau, t) + O(\epsilon^3) \quad (3.1.41)$$

and hence

$$\dot{Z}(t) = Z^{(0)}_t + Z^{(2)}_\tau + O(\epsilon) \,. \quad (3.1.42)$$

This ansatz allows us to impose any initial velocity, $\dot{Z}^{(0)}(0)$, of order one. Likewise, we expand $\dot{X}, \dot{Y}$ (or U, V) and R in a regular power series of ϵ, that is, in the form of (3.1.39) and (3.1.40) with $k = 0$.

Using the above two-time expansion scheme, it is straightforward to derive the hierarchy of perturbation equations from (3.1.29) to (3.1.31). The leading solutions are

$$u_{\pm}^{(0)} = 0 \, , \quad v_{+}^{(0)} = 1/(2\pi\bar{r}) \, , \quad v_{-}^{(0)} = \bar{r}/(2\pi) \, , \tag{3.1.43}$$

$$p_{+}^{(0)} = -\frac{1}{8\pi^2\bar{r}^2} \, , \qquad p_{-}^{(0)} = \frac{\bar{r}^2 - 2}{8\pi^2} \tag{3.1.44}$$

and the leading-order interface is the unit circle, $\bar{r} = 1$. At the next order, the kinematic condition (3.1.27) yields

$$u_{\pm}^{(1)} = R_{\tau}^{(0)} + R_{\theta}^{(0)}/(2\pi) \quad \text{on} \quad \bar{r} = 1 \, . \tag{3.1.45}$$

Corresponding to (3.1.26), we represent $R^{(0)}(t, \theta)$ by

$$R^{(0)} = \text{Re} \sum_{n=2,3,\ldots} c_n(t, \tau) e^{in\theta} \tag{3.1.46}$$

where

$$c_n = a_n^{(0)} - i b_n^{(0)} \tag{3.1.47}$$

and Re stands for "the real part of." The corresponding first order solutions are

$$u_{\pm}^{(1)} = (\phi_{\pm}^{(1)})_{\bar{r}} \, , \quad v_{\pm}^{(1)} = (\phi_{\pm}^{(1)})_{\theta}/\bar{r} \, , \tag{3.1.48}$$

where

$$\phi_{-}^{(1)} = \text{Re} \sum_{n=2,3,\ldots} \left\{ \left[\frac{(c_n)_\tau}{n} + \frac{i c_n}{2\pi} \right] \bar{r}^n e^{in\theta} \right\} \tag{3.1.49}$$

and

$$\begin{aligned} \phi_{+}^{(1)} = & \left[(1 - X_t^{(0)} - X_{\tau}^{(2)}) \cos\theta - (Y_t^{(0)} + Y_{\tau}^{(2)}) \sin\theta \right] \left[\bar{r} + \bar{r}^{-1} \right] \\ & - \text{Re} \sum_{n=2,3,\ldots} \left[\frac{(c_n)_\tau}{n} + \frac{i c_n}{2\pi} \right] \bar{r}^{-n} e^{in\theta} \, . \end{aligned} \tag{3.1.50}$$

The balance of pressure across the interface, (3.1.28), yields $[p_{+}^{(1)} - p_{-}^{(1)}]_{\theta} = 0$ and (3.1.33) in turn yields

$$[v_{+}^{(1)} - v_{-}^{(1)}]_{\tau} + [v_{+}^{(1)} - v_{-}^{(1)}]_{\theta}/(2\pi) - u_{-}^{(1)}/\pi = 0 \qquad \text{on} \qquad \bar{r} = 1 \, . \tag{3.1.51}$$

Inserting (3.1.48)–(3.1.50) into (3.1.51) and equating the coefficients of all harmonics to be zero, we obtain the equations of motion for the vortex center:

$$
\begin{aligned}
X_t^{(0)} &= 1 , & Y_t^{(0)} &= 0 , \\
Y_{\tau\tau}^{(2)} - \frac{1}{2\pi} X_\tau^{(2)} &= 0 , & X_{\tau\tau}^{(2)} + \frac{1}{2\pi} Y_\tau^{(2)} &= 0 ,
\end{aligned}
\tag{3.1.52}
$$

for $n = 1$ and equations for the variations of higher modes,

$$
2\pi(c_n)_{\tau\tau} + i(2n-1)(c_n)_\tau - \frac{n(n-1)}{2\pi} c_n = 0 , \tag{3.1.53}
$$

for $n = 2, 3, \ldots$. Equations (3.1.52) were obtained from the two equations for the first harmonics by the standard two-time method. The τ-averages of those two equations eliminate their dependence on τ and yield the first two relations in (3.1.52), which are also known as the compatibility conditions for the removal of the secular terms in $X^{(2)}$ and $Y^{(2)}$ as $\tau \to \infty$. The solution of (3.1.52) shows that the vortex center is drifting with the background flow in the normal time scale plus a small-amplitude high-frequency oscillation with period ϵ^2 and an amplitude of the order ϵ^2. Thus, the leading-order trajectory of the vortex center coincides with that of a corresponding spinning disc with the same Γ, a, U_∞, initial velocity $\dot{Z}_0$, and density $\rho_0 = \rho$. In addition, the amplitude of the nth mode of oscillation of the interface, governed by (3.1.53), is given by the general solution,

$$
c_n(\tau, t) = C_{n1}(t) e^{-in\tau/(2\pi)} + C_{n2}(t) e^{-i(n-1)\tau/(2\pi)} , \quad \text{for} \quad n \geq 2 . \tag{3.1.54}
$$

The dependence of $X^{(2)}$, $Y^{(2)}$, and the c_n's on the normal time t can be determined only by the compatibility conditions of higher order equations. The initial conditions on the amplitude and rate of change of the nth mode of oscillation of the interface define $C_{n1}(0)$ and $C_{n2}(0)$ via (3.1.46) and (3.1.47):

$$
\begin{aligned}
C_{n1}(0) + C_{n2}(0) &= a_n^{(0)}(0) - i b_n^{(0)}(0) , \\
n C_{n1}(0) + (n-1) C_{n2}(0) &= 2\pi\epsilon [i \dot{a}_n^{(0)}(0) + \dot{b}_n^{(0)}(0)] .
\end{aligned}
\tag{3.1.55}
$$

From (3.1.41)–(3.1.43), (3.1.50), and (3.1.52), we obtain the complex velocity potential in the normal length scale outside of the interface,

$$
\begin{aligned}
w(z) = {} & (1 - Z_\tau^{(2)})(z - Z) + \frac{1}{2\pi i} \ln(z - Z) - \frac{\epsilon^2 Z_\tau^{(2)}}{z - Z} \\
& - \sum_{n=2,3,\ldots} \epsilon^{n+2} \left[\frac{(c_n)_\tau}{n} + \frac{i c_n}{2\pi} \right] (z - Z)^{-n} ,
\end{aligned}
\tag{3.1.56}
$$

with the understanding that we have obtained only the leading terms in dZ/dt and in the c_n's. The first two terms on the right-hand side of (3.1.56) represent the flow field of a Rankine vortex drifting with the uniform stream. The third term is the contribution to the flow field when the initial velocity of the vortex center differs from the background velocity. This term represents a doublet, with a strength of the order ϵ^2 oscillating at a high-frequency $1/(2\pi\epsilon^2)$. The fourth term represents the flow field induced by the high-frequency oscillations of the interface when the initial shape of the interface deviates from the circle $r = \epsilon$ by $O(\epsilon^2)$. This contribution is at most of the order ϵ^4. Since the short time averages of the third and fourth terms are zero, the time average, $\bar{w}(t, z)$ of the velocity potential reduces to the first and second terms, which are equivalent to the one-time solution (3.1.23).

3.1.2 Potential Flow Induced by a Vortex Line

When the core size of a vortex filament approaches zero while its strength, the circulation $\Gamma > 0$, remains finite, the filament becomes a vortex line $\mathcal{C}$ defined by the vector function $\boldsymbol{X}(s, t)$. Here the parameter s is assumed to increase along the direction of the vorticity vector, $\Gamma\boldsymbol{\tau}$, where $\boldsymbol{\tau}$ denotes the unit tangent vector. The flow field induced by the vortex line is given by the line Biot–Savart integral

$$\boldsymbol{q}(\boldsymbol{x}; \mathcal{C}) = \frac{\Gamma}{4\pi} \int_{\mathcal{C}} \frac{\boldsymbol{X}' - \boldsymbol{x}}{|\boldsymbol{X}' - \boldsymbol{x}|^3} \times d\boldsymbol{X}' , \tag{3.1.57}$$

in which the time t can be considered as a parameter. In this section we study the singular behavior of the Biot–Savart integral with the t-dependence suppressed.

To prepare for the analysis, we recall the Serret–Frenet formulae for $\mathcal{C}$,

$$\begin{aligned} \boldsymbol{X}_s &= \sigma\boldsymbol{\tau} , & \boldsymbol{\tau}_s &= \sigma\kappa\boldsymbol{n} , \\ \boldsymbol{n}_s &= (T\boldsymbol{b} - \kappa\boldsymbol{\tau})\sigma , & \boldsymbol{b}_s &= -T\sigma\boldsymbol{n} , \end{aligned} \tag{3.1.58}$$

where

$$\sigma = |\boldsymbol{X}_s| . \tag{3.1.59}$$

Here $\boldsymbol{n}, \boldsymbol{b}, \kappa$, and T denote, respectively, the unit normal and binormal vectors and the curvature and torsion of $\mathcal{C}$. See, for example, Struik (1961).

Let us consider a closed vortex line of length S and let the parameter s be the arclength. Then $\boldsymbol{X}$ is a periodic function of s with period S, that is,

$$\boldsymbol{X}(s + S) = \boldsymbol{X}(s) \quad \text{and} \quad \sigma \equiv 1 . \tag{3.1.60}$$

For a closed vortex line, the induced vector potential given by (2.3.4) exists and becomes

$$\boldsymbol{A}(\boldsymbol{x}) = \frac{\Gamma}{4\pi} \int_{\mathcal{C}} \frac{d\boldsymbol{X}(s)}{|\boldsymbol{x} - \boldsymbol{X}|} = \frac{\Gamma}{4\pi} \int_0^S \frac{\boldsymbol{X}_s \, ds'}{|\boldsymbol{x} - \boldsymbol{X}(s')|} \tag{3.1.61}$$

and the induced velocity is given by the well-known Biot–Savart formula,

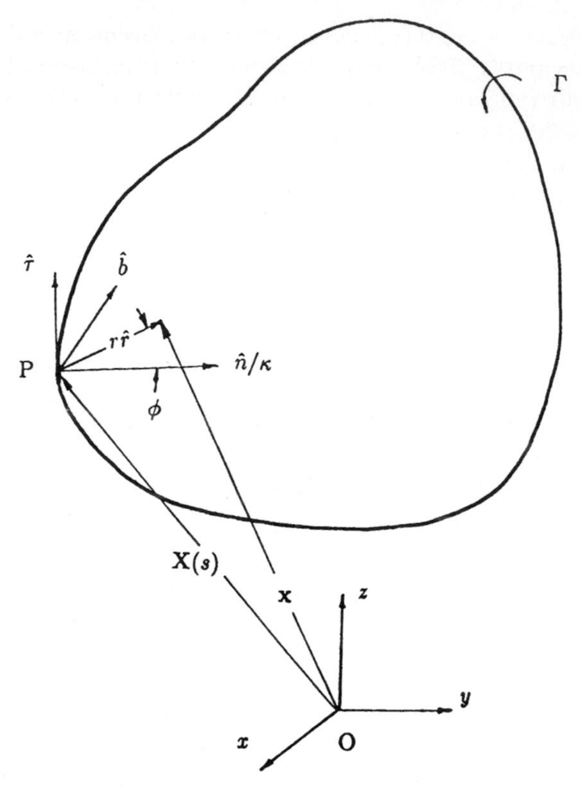

Fig. 3.3. Coordinate system for a vortex line

$$\boldsymbol{Q}(\boldsymbol{x}) = \nabla \times \boldsymbol{A} = -\frac{\Gamma}{4\pi} \int_{C} \frac{[\boldsymbol{x} - \boldsymbol{X}'] \times d\boldsymbol{X}'}{|\boldsymbol{x} - \boldsymbol{X}'|^3} . \tag{3.1.62}$$

A direct derivation of the singular and finite parts of the velocity field near the vortex line from (3.1.62) can be found in Callegari and Ting (1978). Here we shall present a similar analysis for the vector potential and then recover the results for the velocity. To study the local behavior, we introduce a local coordinate system moving with $\mathcal{C}$ as shown in Fig. 3.3. For any given point $\boldsymbol{x}$ close enough to the curve $\mathcal{C}$, we can find a point $\boldsymbol{P}$, $\boldsymbol{X}(s)$, on the curve in the neighborhood of $\boldsymbol{x}$ such that the distance $|\boldsymbol{x} - \boldsymbol{X}(s)|$ attains its minimum, denoted by r. Thus we write

$$\boldsymbol{x} = \boldsymbol{X}(s) + r\boldsymbol{r}(\phi, s) , \tag{3.1.63}$$

where $\boldsymbol{r} = \boldsymbol{n}(s)\cos\phi + \boldsymbol{b}(s)\sin\phi$ denote the unit radial vector in the normal plane and (r, ϕ) are the polar coordinates of point $\boldsymbol{x}$ relative to point $\boldsymbol{P}$. The differential of the position vector (at a fixed time t) is given by

$$d\boldsymbol{x} = \boldsymbol{r}\, dr + \boldsymbol{\phi}\, r[d\phi + T ds] + \boldsymbol{\tau}\, (1 - \kappa r \cos\phi)\, d\phi , \tag{3.1.64}$$

where $\boldsymbol{\phi} = \boldsymbol{b}\cos\phi - \boldsymbol{n}\sin\phi$ denotes the circumferential unit vector. Thus, (3.1.63) relates the Cartesian coordinates to the curvilinear coordinates r, ϕ, and s which are not orthogonal because of the term $\boldsymbol{\phi}\, rT\, ds$ in (3.1.64). This term can be removed by replacing ϕ by the new variable

$$\theta = \phi - \theta_0(s) \quad \text{with} \quad d\theta_0 = -T(s)\, ds \,. \tag{3.1.65}$$

Then r, θ, and s form a set of orthogonal coordinates associated with the orthogonal unit vectors $\boldsymbol{r}, \boldsymbol{\phi}$, and $\boldsymbol{\tau}$, and (3.1.64) becomes

$$d\boldsymbol{x} = \boldsymbol{r}\ dr + \boldsymbol{\phi}\ r d\theta + \boldsymbol{\tau}\ h_3 ds\,, \quad \text{where} \quad h_3 = 1 - \kappa r\cos\phi\,. \tag{3.1.66}$$

From the stretch ratios, in particular h_3, it follows that the coordinate transformation (3.1.63) is one to one so long as

$$0 < r < 1/\kappa\,. \tag{3.1.67}$$

This condition will be observed because the curvilinear coordinates will be employed only for points close to the curve $\mathcal{C}$, that is, $r\kappa \ll 1$.[2]

To extract the singular behavior of $\boldsymbol{A}(\boldsymbol{x})$ for points close to $\mathcal{C}$, we can work with the natural curvilinear coordinates r, ϕ, and s. Note that the singular behavior of $\boldsymbol{A}$ is generated by that of the integrand in (3.1.61) and that the integrand becomes singular as $r \to 0$ and $s' \to s$. We therefore introduce the new integration variable $\bar{s} = s' - s$ and find the first few terms of the Taylor series of the integrand in $\bar{s}$ for $r > 0$. These terms become singular as $\bar{s}^2 + r^2 \to 0$, while the remainder approaches a finite limiting value. The integral of the remainder can be evaluated numerically for any $r \geq 0$ while the integral over these first few terms is evaluated analytically for $r > 0$. The analytic result defines the singular part(s) of the Biot–Savart integral as $r \to 0$.

The Taylor series expansions of the numerator and denominator of the integrand in (3.1.61) are

$$\boldsymbol{X}_s(s') = \boldsymbol{\tau}(s') = \boldsymbol{\tau}(s) + \bar{s}\kappa\boldsymbol{n} + O(\bar{s}^2)$$

and (3.1.68)

$$\frac{1}{|\boldsymbol{x} - \boldsymbol{X}(s')|} = \frac{1}{R}\left[1 + \frac{r\bar{s}^2}{2R^2}\kappa\cos\phi + O\left(\kappa^2\frac{r\bar{s}^3}{R^2}, \kappa^2\frac{\bar{s}^4}{R^2}\right)\right],$$

respectively, where $\kappa, \boldsymbol{\tau}$, and $\boldsymbol{n}$ are evaluated at s, and $R = \sqrt{r^2 + \bar{s}^2}$. In this case we have to keep the first two terms in the expansions of the integrand to make sure that the remainder does not contribute to the singular parts of of the velocity, $\nabla \times \boldsymbol{A}$. Denoting the integrand in (3.1.61) by $\boldsymbol{F}(r, \phi, s, \bar{s})$ and using the expressions in (3.1.68), we obtain

[2] This orthogonal curvilinear coordinate system was introduced by Ting (1971) to express the Navier–Stokes equations in these coordinates for the analysis of the flow field close to the centerline $\mathcal{C}$ of a slender filament. See also Sect. 3.3.

$$\boldsymbol{F}(r,\phi,s,\bar{s}) = \frac{\boldsymbol{\tau}(s+\bar{s})}{|\boldsymbol{x}-\boldsymbol{X}(s+\bar{s})|} = \boldsymbol{G}+\boldsymbol{H} \tag{3.1.69}$$

where

$$\boldsymbol{G}(r,\phi,s,\bar{s}) = \boldsymbol{\tau}\left[\frac{1}{(r^2+\bar{s}^2)^{\frac{1}{2}}} + \frac{r\bar{s}^2\kappa\cos\phi}{2(r^2+\bar{s}^2)^{\frac{3}{2}}}\right] + \boldsymbol{n}\kappa\frac{\bar{s}}{(r^2+\bar{s}^2)^{\frac{1}{2}}} \tag{3.1.70}$$

and

$$\boldsymbol{H}(r,\phi,s,\bar{s}) = \boldsymbol{F}-\boldsymbol{G} = O\left(\frac{r\bar{s}^3,\bar{s}^4}{(r^2+\bar{s}^2)^{3/2}}\right) = O(\max[r,\bar{s}])\ . \tag{3.1.71}$$

Singular parts of the vector potential are contained in the integral of $\boldsymbol{G}$ which can be obtained explicitly as

$$\boldsymbol{A}^g(\boldsymbol{x}) = \frac{\Gamma}{4\pi}\int_{-S^-}^{S^+}\boldsymbol{G}(r,\phi,s,\bar{s})\,d\bar{s} = \boldsymbol{A}^s + \boldsymbol{A}^f\ , \tag{3.1.72}$$

where

$$\boldsymbol{A}^s = \frac{\Gamma\boldsymbol{\tau}}{2\pi}\ln\frac{S}{r} \tag{3.1.73}$$

denotes the singular part of $\boldsymbol{A}$ and

$$\begin{aligned}\boldsymbol{A}^f = \frac{\Gamma}{4\pi}\Bigg\{&\boldsymbol{\tau}\Bigg[(1+\frac{r\kappa\cos\phi}{2})\ln\frac{(S^+ + \sqrt{r^2+(S^+)^2})(S^- + \sqrt{r^2+(S^-)^2})}{S^2}\\ &-\frac{r\kappa\cos\phi}{2}\left(\ln\frac{r^2}{S^2}+\frac{S^+}{\sqrt{r^2+(S^+)^2}}+\frac{S^-}{\sqrt{r^2+(S^-)^2}}\right)\Bigg]\\ &+\boldsymbol{n}\kappa\left[\sqrt{r^2+(S^+)^2}-\sqrt{r^2+(S^-)^2}\right]\Bigg\}\end{aligned} \tag{3.1.74}$$

contributes to the finite part of the integral (3.1.61). Here $-S^-$ and S^+ with

$$S^- > 0\ ,\quad S^+ > 0 \qquad \text{and} \qquad S^- + S^+ = S \tag{3.1.75}$$

are the lower and upper limits of $\bar{s}$ in making one circuit of $\mathcal{C}$. Using (3.1.69)–(3.1.75), we can express the vector potential as

$$\boldsymbol{A}(\boldsymbol{x}) = \boldsymbol{A}^s + \boldsymbol{A}^f + \boldsymbol{A}^h \qquad \text{with} \quad \boldsymbol{A}^h(\boldsymbol{x}) = \frac{\Gamma}{4\pi}\int_{-S^-}^{S^+}\boldsymbol{H}(r,\phi,s,\bar{s})d\bar{s}\ . \tag{3.1.76}$$

Here $\boldsymbol{A}^f + \boldsymbol{A}^h$ represents the finite part of $\boldsymbol{A}$. Note that the definite integrals for $\boldsymbol{A}^f$ and $\boldsymbol{A}^h$ are independent of the choices of S^+ and S^-, so long as they fulfill (3.1.75), because of the periodicity condition (3.1.60). A natural choice is $S^+ = S^- = S/2$.

From the behavior of $\boldsymbol{A}$ near the vortex line, (3.1.72)–(3.1.74), we obtain the singular behavior of the velocity field by computing $\nabla\times\boldsymbol{A}$. The result is

$$\boldsymbol{Q}(\boldsymbol{x}) = \frac{\Gamma}{2\pi r}\hat{\theta} + \frac{\Gamma\kappa}{4\pi}\boldsymbol{b}\ln\frac{S}{r} + \frac{\Gamma\kappa}{4\pi}\hat{\theta}\cos\phi + \boldsymbol{Q}^f , \tag{3.1.77}$$

where $\boldsymbol{Q}^f$ denotes the finite single valued part of the velocity as $r \to 0$. The first singular term on the right-hand side of (3.1.77) represents the circumferential flow around a straight vortex line tangent to the curve $\mathcal{C}$ at $\boldsymbol{X}(s)$, and the second term represents a singular parallel flow in the direction of the binormal vector $\boldsymbol{b}(s)$. The third terms points in the circumferential direction and is singular because it has no unique limit as $r \to 0$. Both the second and the third terms are proportional to the local curvature κ and are, therefore, absent in the limit of a straight filament. All three terms are vectors in the normal plane of the curve $\mathcal{C}$.

Singular parts of $\boldsymbol{Q}$ are the same as those derived directly from the Biot–Savart formula (3.1.62) by Callegari and Ting (1978). The formula for the finite part $\boldsymbol{Q}^f$ will be presented in Appendix C.

It is evident that the velocity of the vortex line cannot be defined by the velocity of the flow field on the vortex line, $r = 0$, where the velocity is infinite. By intuition one might conjecture that the velocity of the vortex line is defined by the local or circumferential average of the velocity field. This conjecture yields the classical inviscid result for a straight vortex line, with $\kappa \equiv 0$, or for a vortex point in a two-dimensional flow. It does not work for a curved vortex line, $\kappa \neq 0$, because the circumferential average of the second term remains unbounded.

As we shall see in Sect. 3.3, the velocity of the center line of a slender vortex filament can be defined systematically by the method of matched asymptotics. The velocity depends on the geometry of the center line and the vortical core structure in addition to the local background velocity. The core structure defines the effective core radius that appears in the binormal velocity as $\Gamma\kappa/(4\pi)$ times $\ln(S/\delta)$ or $\ln(1/\epsilon) + \ln(\epsilon S/\delta)$ and contributes to the finite part $\boldsymbol{Q}^f$ as $r \to 0$.

To end this section, we note that the singular behavior of the vector potential $\boldsymbol{A}$ was derived for a simple closed vortex line with length $S = O(\ell)$. For an infinitely long vortex line, the integral for $\boldsymbol{A}$ is undefined in general. However, the Biot–Savart integral for the velocity is defined because its integrand decays as $|\boldsymbol{x} - \boldsymbol{X}(s')|^{-2}$ provided that the length of the segment of $\mathcal{C}$ within any finite distance from $\boldsymbol{x}$ is finite. For example, if the vortex line has the shape of a spiral, it is necessary to assume that the pitch of the spiral is not much smaller than ℓ. This condition is also necessary to ensure that the transformation (3.1.63) from the Cartesian coordinates to the curvilinear coordinates in the neighborhood of the vortex line remains one to one. For a vortex line extending to infinity or being very long relative to ℓ, we can construct the vector potential, if needed, in a finite region from the velocity field to avoid the difficulty that $\boldsymbol{A}$ is unbounded at infinity. First we derive the singular and the finite parts of the velocity field, using the formulae given by

Callegari and Ting (1978). Then, using equations $\nabla \cdot \boldsymbol{A} = 0$ and $\nabla \times \boldsymbol{A} = \boldsymbol{V}$, we solve for the vector potential $\boldsymbol{A}$ in the finite region.

3.2 Planar Motion of Vortex Spots and Their Core Structures

The motion and diffusion of viscous vortices submerged in a potential flow were studied by Ting and Tung (1965) by combining a matched asymptotic analysis in spatial variables with a two-time analysis in the short and normal time scales. The short-time average of the two-time solution was identified as a one-time solution. The physical meaning of the latter and its correspondence with the classical inviscid theory were described. In this section we present the highlights of their analyses.

Consider the Navier–Stokes equations for an unsteady incompressible flow in the x–y plane expressed in terms of the stream function $\psi(t, x, y)$ and vorticity $\zeta(t, x, y)$:

$$\zeta_t + \psi_y \zeta_x - \psi_x \zeta_y = \nu \triangle \zeta \ , \tag{3.2.1}$$

and

$$\triangle \psi = -\zeta \ . \tag{3.2.2}$$

The initial flow field, say at $t = 0$, is composed of a background potential flow defined by the stream function $\psi(0, x, y) = \Psi(x, y)$ and the flow field induced by a vorticity distribution $\zeta(0, x, y) = Z(x, y)$. The vorticity is highly concentrated in a small region, called a vortical core or spot, in which there is a large swirling flow around a point $C(X_0, Y_0)$, called the center of the core.

By a small region we imply that the length scale for the flow field in the vortical core is much smaller than that of the background flow. To characterize this separation of scales, we identify a typical length ℓ and velocity U of the background flow as the reference scales. They will also be referred to as the normal scales. For an incompressible constant density flow, we can always assign the density of the fluid to be unity. Then we choose ℓ, ℓ/U, and $\rho\ell^3$ to be the units of length, time, and mass, respectively, that is,

$$\ell = 1 \ , \qquad U = 1 \ , \qquad \text{and} \qquad \rho = 1 \ , \tag{3.2.3}$$

so that the symbol for a physical quantity in these units can also be used for the scaled quantity.

To characterize a highly concentrated vorticity distribution in a vortical core, we assume that the vorticity decays rapidly as the distance r to the center C becomes much larger than the effective core size in the sense of (2.1.18). The initial effective core size denoted by δ_0 is of the order of a small reference scale, δ^*, which in turn is much smaller than the normal length scale ℓ, that is, $\epsilon = \delta^*/\ell \ll 1$. The total strength of the vorticity is assumed to be of order one, that is, $\Gamma = O(U\ell)$. Then the scaled vorticity and swirling velocity in the core are $O(\epsilon^{-2})$ and $O(\epsilon^{-1})$, respectively. It is implied that the radial velocity relative to the center C remains $O(1)$ so that the path lines inside the core are nearly circular. This completes the description of a *small vortical core* with *highly concentrated* vorticity distribution and a *large swirling flow.*

To account for the diffusion of vorticity while confining it to a small area of the order δ_0^2 on the normal time scale, we introduce the assumption that the Reynolds number of the background flow is large in the sense that

$$Re = U\ell/\nu = 1/\nu = O(\epsilon^{-2}) \qquad \text{or} \qquad \bar{\nu} = \nu/\epsilon^2 = O(1) \ . \tag{3.2.4}$$

With this assumption the asymptotic solution remains valid for any fixed value of $\bar{\nu}$ as $\epsilon \to 0$.

Note that, although we deal with only one viscous vortex, the analysis for a single vortex can be applied to several vortices so long as the distances between them are much greater than their core sizes. Of course, the background flow of one vortex will include the potentials induced by the other vortices and will be unsteady because of the motion of the vortices.

We have assumed that the background potential flow will remain steady. This is not essential because the velocity field is governed by the Laplace equation in which the time t appears merely as a parameter. To compute the unsteady pressure field of the potential flow, that is, outside the vortical core(s), we assume that the total head, $p + \rho|\boldsymbol{v}|^2/2$, is uniform initially and hence remains constant for all t. The pressure field is then related to the velocity by the unsteady Bernoulli equation. As we shall see later, the pressure field in a vortical core is defined by the inner velocity field and the matching condition with pressure outside the core.

The evolution of vorticity is governed by (3.2.1). Its left-hand side expresses the transport of vorticity along a particle path line while its right-hand side induces a gradual diffusion of vorticity and a growth of the core size $\delta(t)$. The core size will remain $O(\delta^*)$ on the normal time scale because of the high Reynolds number assumption (3.2.4). From the discussion in Sect. 3.1, we expect that the flow field at an order one distance, $O(\ell)$, away from the vortical core, behaves as a potential flow composed of the background flow and the flow induced by a vortex point of strength Γ located at $C(X, Y)$. The question is in what sense and to what degree of accuracy is the trajectory of the vortex center C approximated by the classical inviscid formula (3.1.3). To answer this question, we need an inner solution valid in the vortical core. The solution accounts for the diffusion of vorticity and matches with the potential solution (3.1.2) away from the core. Thus we have a typical problem solvable by the method of matched asymptotics.

From our studies of inviscid flows around a spinning disc and a Rankine vortex in Subsect. 3.1.1 and from the above description of a small vortical core centered at point C, we expect that in a neighborhood of C, which is small even relative to δ^*, the circumferential velocity will be $O(1)$ instead of $O(\epsilon^{-1})$.

In other words, the fluid or any material point in this small neighborhood should move with order one velocity. We can then introduce the notion of the vortex center $C = (X(t), Y(t))$, which moves, to leading-order, as a material point with order one velocity and on the normal time scale only, that is,

$$\dot{X}(t) = \psi_y(t, X, Y) = O(1) \;, \quad \dot{Y}(t) = -\psi_x(t, X, Y) = O(1) \;. \tag{3.2.5}$$

The initial position of C is of course (X_0, Y_0), the center of the initial swirling flow.

If the flow field was considered as an initial value problem for the Navier–Stokes equations formulated in Sect. 2.1, the velocity field should be defined everywhere including the point C. Note that the initial velocity of the vortex center C is defined by (3.2.5) and can differ from the local background velocity. We now construct the solution of the initial value problem by the method of matched asymptotics, using ϵ as the small expansion parameter, and expect the velocity of the vortex center to be defined by the asymptotic solution. From the preceding inviscid studies, we see that the flow field has, in addition to the normal time scale, a much smaller time scale of the order ϵ^2 characterizing the high-frequency oscillations of the vortical core. The time scale factor ϵ^2 may be assessed from (3.1.11) with density $\rho = 1$ and $a = O(\delta) = O(\epsilon\ell)$.

In the above, we have described many properties of a vortical core based upon physical intuition. It is not clear whether all these properties are consistent with each other and how many of them should be imposed on the solution. In employing the method of matched asymptotics, we introduce ansatzes in the form of expansion schemes for the variables in the inner and outer regions to simulate only two *essential* properties of the vortical flow. The governing equations for the leading-order and higher order inner and outer solutions will then be obtained systematically. We then verify that the expansion schemes are self-consistent and show that the asymptotic solutions possess the aforementioned physical or intuitive properties of the flow field; that is, they are consistent with the two *essential* ones. Usually one states immediately the appropriate expansions for all the variables to arrive at the final results as fast as possible. We shall do the same here. But, we will single out those few variables for which the expansions are assumed and then show that the expansions for the remaining variables follow from the governing equations.

In the outer region, that is, at an order one distance away from the vortex center, the flow field is in the normal length and time scales and hence we introduce regular expansions in ϵ, for the stream function and vorticity,

$$\begin{aligned} \psi(t, x, y, \epsilon) &= \psi^{(0)}(t, x, y) + \epsilon\psi^{(1)}(t, x, y) + \cdots \;, \\ \zeta(t, x, y, \epsilon) &= \zeta^{(0)}(t, x, y) + \epsilon\zeta^{(1)}(t, x, y) + \cdots \;. \end{aligned} \tag{3.2.6}$$

In terms of the scaled variables, (3.2.2) remains the same while (3.2.1) becomes

$$\zeta_t + \psi_y \zeta_x - \psi_x \zeta_y = \epsilon^2 \bar{\nu} \triangle \zeta \ . \qquad (3.2.7)$$

After the substitution of the expansions in Eqs. (3.2.6)–(3.2.7) into Eq. (3.2.2), making use of (3.2.4) and equating the coefficients of like powers of ϵ, the governing equations for the leading-order solution $\psi^{(0)}$ and $\zeta^{(0)}$ are

$$\zeta_t^{(0)} + \psi_y^{(0)} \zeta_x^{(0)} - \psi_x^{(0)} \zeta_y^{(0)} = 0 \qquad \text{and} \qquad \triangle \psi^{(0)} = -\zeta^{(0)} \ . \qquad (3.2.8)$$

They are equivalent to the Euler equations for an inviscid flow.

The above expansion scheme permits a background rotational flow of order one vorticity. This problem, which was analyzed by Liu and Ting (1987), will be studied below in Subsect. 3.3.8. Because of the presence of the vortices, the background vorticity, if nonuniform, will be redistributed and hence the background flow has to be solved simultaneously with the motion of the vortex centers.

Here we present a somewhat simpler problem, which was treated earlier by Ting and Tung (1965), and in which the flow field away from the vortical core is assumed to be irrotational at $t = 0$. The leading outer solution, governed by (3.2.8), remains irrotational, that is,

$$\zeta^{(0)} = 0 \ , \quad \triangle \psi^{(0)} = 0 \qquad \text{for } t \geq 0 \ . \qquad (3.2.9)$$

Since the viscous term on the right-hand side of (3.2.7) is two orders higher than the inviscid terms on the left-hand side and the viscous term corresponding to a potential flow is equal to zero, we find that the nth order outer solution is irrotational if the $(n-2)$-nd is. It follows by the method of induction that

$$\zeta^{(n)} = 0 \ , \quad \triangle \psi^{(n)} = 0 \qquad \text{for } n = 0, 1, \ldots \ . \qquad (3.2.10)$$

Thus the flow in the outer region is irrotational to all orders in ϵ if it is also irrotational initially. In particular, the leading outer solution is the sum of a background potential flow and the flow induced by a vortex point in agreement with the classical result (3.1.2), that is,

$$\psi(t, x, y) = \Psi(x, y) - \frac{\Gamma}{2\pi} \ln r \ , \qquad (3.2.11)$$

$$\boldsymbol{v} = \boldsymbol{i}\psi_y - \boldsymbol{j}\psi_x \qquad (3.2.12)$$

$$\Psi_x = O(1) \ , \quad \Psi_y = O(1) \ , \qquad (3.2.13)$$

where $r = \sqrt{(x - X)^2 + (y - Y)^2}$ denotes the distance between $\boldsymbol{x}$ and the center $\big(X(t), Y(t)\big)$ of the vortical core. Here we have suppressed the superscript (0) on $\psi, \boldsymbol{v}$ and X, Y and choose the orientation of the coordinate axes to be in the same sense as the circulation so that we have $\Gamma > 0$. Equation

(3.2.13) restates the fact that the velocity of the background flow is $O(1)$ because its typical velocity has been chosen as the velocity scale in (3.2.3). Note that, although the background flow is steady, the outer stream function ψ is unsteady because of the motion of the vortex center.

The next step is to set up the expansion scheme for the inner solution in the inner spatial variables with two time scales, carry out the inner–outer matching and two-time analyses, and identify the meaning of the solution in normal time alone. This step is described next in Subsect. 3.2.1. In Subsect. 3.2.2 we present the leading-order solution of the core structure, define the velocity of the vortex center, compare the two-time solution for the trajectory of the vortex center with the one-time solution, and then describe the contribution of the inner solution to the higher order outer solution.

3.2.1 The Inner Solution

Now we study the inner region, in which there is a large swirling flow around the center $C(X,Y)$. The velocity, $\dot{X},\dot{Y}$, of the center is yet to be defined. For the inner region, we replace the coordinates (x,y) by the polar coordinates (r,θ) with respect to the center C and then stretch the radial coordinate r by ϵ. The coordinate transformations are

$$x = X + \epsilon\bar{r}\cos\theta\ , \quad y = Y + \epsilon\bar{r}\sin\theta\ , \quad \text{and}\ \bar{r} = r/\epsilon\ . \tag{3.2.14}$$

Let $\tilde{\zeta}$ denote the vorticity in the inner region and $\tilde{\psi}$ the stream function of the velocity field relative to the moving vortex center C. They are functions of two time variables, t and τ, and the stretched polar coordinates $\bar{r}$, and θ. Here τ measures on the short time scale $\epsilon^2 l/U$ and is related to the normal time t by

$$\tau = t\epsilon^{-2}\ . \tag{3.2.15}$$

To express a function of t parameterized by ϵ, say $f(t;\epsilon)$, as a function $\tilde{f}(\tau,t;\epsilon)$ of two independent variables, t and $\tau = t/\epsilon^2$, we assume that $\tilde{f}$ is bounded for all τ and that the average of $\tilde{f}$ over a large interval in τ, called the τ-average, exists. This is the standard ansatz in a two-time analysis. See, for example, Keller (1977, 1980), Schneider (1978), or Kevorkian and Cole (1996). The τ-average of $\tilde{f}$ for some $t > 0$, denoted by $\mathcal{M}\tilde{f}(t)$, is defined by

$$\mathcal{M}\tilde{f}(t) = \lim_{T\to\infty}\frac{1}{2T}\int_{t\epsilon^{-2}-}^{t\epsilon^{-2}+T}\tilde{f}(\tau,t)d\tau\ . \tag{3.2.16}$$

Note that the τ-average is a function of t only, and the averaging operator $\mathcal{M}$ is the identity for a function independent of τ.

With t and τ treated as independent variables, we must replace the partial derivative with respect to t in the original equations with

$$f_t = \frac{\partial\tilde{f}}{\partial t} + \epsilon^{-2}\frac{\partial\tilde{f}}{\partial\tau}\ . \tag{3.2.17}$$

Now the velocity $\tilde{\boldsymbol{v}}$ is related to the stream function $\tilde{\psi}$ by

$$\tilde{\boldsymbol{v}} = [\tilde{X}_t + \epsilon^{-2}\tilde{X}_\tau]\,\boldsymbol{i} + [\tilde{Y}_t + \epsilon^{-2}\tilde{Y}_\tau]\,\boldsymbol{j} + \frac{1}{\epsilon}\tilde{\boldsymbol{V}} , \tag{3.2.18}$$

where

$$\tilde{\boldsymbol{V}} = \bar{r}^{-1}\tilde{\psi}_\theta \boldsymbol{r} - \tilde{\psi}_{\bar{r}}\boldsymbol{\theta} \tag{3.2.19}$$

denotes the velocity relative to the center, and $\boldsymbol{r}$ and $\boldsymbol{\theta}$ represent the unit radial and circumferential vectors, respectively. The governing equations (3.2.1) and (3.2.2) for the stream function and vorticity as functions of $\tau, t, \bar{r}$ and θ become

$$\tilde{\zeta}_t + \epsilon^{-2}\tilde{\zeta}_\tau + \epsilon^{-2}\tilde{\boldsymbol{V}}\cdot\bar{\nabla}\tilde{\zeta} = \bar{\nu}\bar{\triangle}\tilde{\zeta} , \tag{3.2.20}$$

$$\epsilon^{-2}\bar{\triangle}\tilde{\psi} = -\tilde{\zeta} , \tag{3.2.21}$$

where $\bar{\nabla}$ and $\bar{\triangle}$ denote the gradient and Laplacian operators in the inner polar coordinates, $\bar{r}$ and θ. The relative velocity $\tilde{\boldsymbol{V}}$ vanishes at the center C because of (3.2.5), that is, C is, up to distances $O(\epsilon^2)$, a material point. This leads to two boundary conditions,

$$\tilde{\psi} = 0 \quad \text{and} \quad \tilde{\psi}_{\bar{r}} = 0 \quad \text{at } \bar{r} = 0 . \tag{3.2.22}$$

The standard matching of the inner and outer solutions from (3.2.18) and (3.2.11)–(3.2.13) in their overlap region, where $\bar{r} \gg 1$ while $r = \epsilon\bar{r} \ll 1$, yields

$$\frac{d\boldsymbol{X}}{dt} + \frac{1}{\epsilon}\left[\frac{1}{\bar{r}}\tilde{\psi}_\theta \boldsymbol{r} - \psi_{\bar{r}}\boldsymbol{\theta}\right] = \boldsymbol{i}\psi_y - \boldsymbol{j}\psi_x = \epsilon^{-1}\frac{\Gamma}{2\pi\bar{r}}\boldsymbol{\theta} + (\Psi_y\boldsymbol{i} - \Psi_x\boldsymbol{j}) + O(\epsilon) . \tag{3.2.23}$$

The stream function $\tilde{\psi}$ and vorticity $\tilde{\zeta}$ are expanded in a power series of ϵ as

$$\begin{aligned} \tilde{\psi}(\tau,t,\bar{r},\theta,\epsilon) &= \tilde{\psi}^{(0)}(\tau,t,\bar{r},\theta) + \epsilon\tilde{\psi}^{(1)}(\tau,t,\bar{r},\theta) + \cdots , \\ \tilde{\zeta}(\tau,t,\bar{r},\theta,\epsilon) &= \epsilon^{-2}[\tilde{\zeta}^{(0)}(\tau,t,\bar{r},\theta) + \epsilon\tilde{\zeta}^{(1)}(\tau,t,\bar{r},\theta) + \cdots] . \end{aligned} \tag{3.2.24}$$

Note that, once the expansion for $\tilde{\psi}$ is assumed, the expansion for $\tilde{\zeta}$ beginning with an $O(\epsilon^{-2})$ term follows from (3.2.21).

Since the flow field in a vortical core is characterized by a large, $O(\epsilon^{-1})$, swirling flow but order one radial flow, we require

$$\epsilon^{-1}\frac{1}{\bar{r}}\tilde{\psi}_\theta = O(U) \qquad \text{and hence} \qquad \tilde{\psi}_\theta^{(0)} = 0 . \tag{3.2.25}$$

This is also consistent with the leading-order matching condition (3.2.23). Since the outer solution has only a circumferential velocity component of order $O(\epsilon^{-1})$ the leading-order matching condition yields

$$\tilde{\psi}_{\bar{r}}^{(0)} \to \Gamma/(2\pi\bar{r}) \qquad \text{as } \bar{r} \to \infty \tag{3.2.26}$$

and

$$\epsilon^{-2}\tilde{X}_\tau + X_t = O(1) \;, \qquad \epsilon^{-2}\tilde{Y}_\tau + Y_t = O(1) \;. \tag{3.2.27}$$

Condition (3.2.27) in turn dictates the following expansions for the coordinates of the vortex center:

$$\begin{aligned} X(\bar{t}, t, \epsilon) &= \bar{X}^{(0)}(t) + \epsilon\bar{X}^{(1)}(t) + \epsilon^2\tilde{X}^{(2)}(\tau, t) + \cdots \;, \\ Y(\bar{t}, t, \epsilon) &= \bar{Y}^{(0)}(t) + \epsilon\bar{Y}^{(1)}(t) + \epsilon^2\tilde{Y}^{(2)}(\tau, t) + \cdots \;. \end{aligned} \tag{3.2.28}$$

Note that we are using bars and tildes over functions to distinguish a function depending only on the normal time from one that depends on both time variables. The expansion (3.2.28) is consistent with the physical expectation (3.2.5), which says that the center moves at an order one velocity. Thus the expansions for all the variables in the inner region are formulated. We want to point out once more that these expansions follow from only two basic assumptions: The regular expansion in (3.2.24) of the stream function in the inner variables and the assumption (3.2.25) of only a large swirling flow.

Now we carry out the systematic derivations of the governing equations for the leading-order and higher order solutions in two steps: In step 1 we substitute the power series (3.2.24) and (3.2.28) in the differential equations (3.2.20), (3.2.21), the auxiliary equations (3.2.18), (3.2.19), the boundary conditions (3.2.22), and the matching condition (3.2.23), and in step 2 we equate the coefficients of like powers of ϵ. The initial conditions will be introduced later when we construct the asymptotic solution.

Equation (3.2.21) yields

$$\bar{\Delta}\tilde{\psi}^{(n)} = -\tilde{\zeta}^{(n)} \;, \quad \text{for } n = 0, 1, \ldots \;. \tag{3.2.29}$$

Likewise, the boundary conditions (3.2.22) yield

$$\tilde{\psi}^{(n)} = 0 \quad \text{and} \quad \tilde{\psi}_{\bar{r}}^{(n)} = 0 \qquad \text{at} \qquad \bar{r} = 0 \;. \tag{3.2.30}$$

From (3.2.29) for $n = 0$ and (3.2.25), we obtain

$$\zeta_\theta^{(0)} = 0 \;. \tag{3.2.31}$$

Using this and (3.2.25), we equate the coefficient of ϵ^{-4} in (3.2.20) to 0 and obtain

$$\tilde{\zeta}_\tau^{(0)} = 0 \;. \tag{3.2.32}$$

Using (3.2.29) for $n = 0$ and (3.2.31), (3.2.32), and (3.2.25), we conclude that the leading-order vorticity and stream function are independent of τ and θ, that is,

$$\tilde{\zeta}^{(0)} = \bar{\zeta}^{(0)}(t, \bar{r}) \tag{3.2.33}$$

$$\tilde{\psi}^{(0)} = \bar{\psi}^{(0)}(t, \bar{r}) \ . \tag{3.2.34}$$

The stream function is related to vorticity by solving (3.2.29) and (3.2.30) for $n = 0$,

$$\bar{\psi}^{(0)}(t, \bar{r}) = -\int_0^{\bar{r}} \frac{d\xi}{\xi} \int_0^{\xi} \xi' \zeta^{(0)}(t, \xi') d\xi' = -\int_0^{\xi} \xi' \zeta^{(0)}(t, \xi') \ln \frac{\bar{r}}{\xi'} d\xi' \ . \tag{3.2.35}$$

By equating the coefficients of ϵ^{-3} and ϵ^{-2} in (3.2.20) to 0, we obtain, respectively,

$$\tilde{\zeta}^{(1)}_{\tau} + \mathcal{L}[\tilde{\psi}^{(1)}, \tilde{\zeta}^{(1)}] = 0 \ , \tag{3.2.36}$$

where $\mathcal{L}$ denotes the linear operator,

$$\mathcal{L}\left[\tilde{\psi}^{(n)}, \tilde{\zeta}^{(n)}\right] = \frac{1}{\bar{r}} \frac{\partial}{\partial \theta} \left[\bar{\zeta}^{(0)}_{\bar{r}} \tilde{\psi}^{(n)} - \bar{\psi}^{(0)}_{\bar{r}} \tilde{\zeta}^{(n)}\right] \ , \quad n = 1, 2, \ldots \tag{3.2.37}$$

and

$$\tilde{\zeta}^{(2)}_{\tau} + \mathcal{L}[\tilde{\psi}^{(2)}, \tilde{\zeta}^{(2)}] = \tilde{F}_2(\tau, t, r, \theta) + \bar{G}_2(t, \bar{r}) \ , \tag{3.2.38}$$

where

$$\tilde{F}_2 = -\frac{1}{\bar{r}} \left[\tilde{\psi}^{(1)}_{\theta} \tilde{\zeta}^{(1)}_{\bar{r}} - \tilde{\zeta}^{(1)}_{\theta} \tilde{\psi}^{(1)}_{\bar{r}}\right] \ , \tag{3.2.39}$$

$$\bar{G}_2 = -\bar{\zeta}^{(0)}_t + \bar{\nu} \bar{\Delta} \bar{\zeta}^{(0)} \ . \tag{3.2.40}$$

By using (3.2.29) for $n = 1, 2$, (3.2.36) and (3.2.38) become linear equations for $\tilde{\psi}^{(1)}$ and $\tilde{\psi}^{(2)}$. Equation (3.2.38) is inhomogeneous. The first inhomogeneous term $\tilde{F}_2$ comes from the nonlinear convective terms of the first order solution while the second term $\tilde{G}_2$ represents the normal time derivative and the linear viscous terms for the leading-order solution.

The τ-average of (3.2.36)–(3.2.38) become, respectively,

$$\mathcal{ML}\left[\tilde{\psi}^{(1)}, \tilde{\zeta}^{(1)}\right] = \frac{1}{\bar{r}} \frac{\partial}{\partial \theta} \left[\bar{\zeta}^{(0)}_{\bar{r}} (\mathcal{M}\tilde{\psi}^{(1)}) - \bar{\psi}^{(0)}_{\bar{r}} (\mathcal{M}\tilde{\zeta}^{(1)})\right] = 0 \ , \tag{3.2.41}$$

$$\begin{aligned} \mathcal{ML}\left[\tilde{\psi}^{(2)}, \tilde{\zeta}^{(2)}\right] &= \frac{1}{\bar{r}} \frac{\partial}{\partial \theta} \left[\bar{\zeta}^{(0)}_{\bar{r}} (\mathcal{M}\tilde{\psi}^{(2)}) - \bar{\psi}^{(0)}_{\bar{r}} (\mathcal{M}\tilde{\zeta}^{(2)})\right] \\ &= \mathcal{M}\tilde{F}_2 + \bar{G}_2(t, \bar{r}) \ . \end{aligned} \tag{3.2.42}$$

The next step is to show that $\tilde{F}_2$ defined by (3.2.39) can be expressed as a combination of derivatives with respect to θ and τ. By using (3.2.33) and (3.2.34), (3.2.36) is rewritten as

$$\tilde{\psi}_\theta^{(1)} = \bar{a}\tilde{\zeta}_\tau^{(1)} + \bar{b}\tilde{\zeta}_\theta^{(1)} , \qquad \text{where } \bar{a}(t,\bar{r}) = -\bar{r}/\bar{\zeta}_{\bar{r}}^{(0)} , \ \bar{b}(t,\bar{r}) = \bar{\psi}_{\bar{r}}^{(0)}/\bar{\zeta}_{\bar{r}}^{(0)} \tag{3.2.43}$$

which is then used to reduce $\tilde{F}_2$ to

$$\tilde{F}_2(\tau,t,\bar{r},\theta) = -\frac{1}{\bar{r}}\left[\frac{1}{2}[\bar{a}(\tilde{\zeta}^{(1)})^2]_{\bar{r}\tau} + \frac{1}{2}[\bar{b}(\tilde{\zeta}^{(1)})^2]_{\bar{r}\theta} - [\tilde{\zeta}^{(1)}\tilde{\psi}_{\bar{r}}^{(1)}]_\theta\right] \tag{3.2.44}$$

and the τ-average of $\tilde{F}_2$ becomes

$$\mathcal{M}\tilde{F}_2 = -\frac{1}{\bar{r}}\frac{\partial}{\partial\theta}\mathcal{M}\left[\frac{1}{2}[\bar{b}(\tilde{\zeta}^{(1)})^2]_{\bar{r}} - \tilde{\psi}_{\bar{r}}^{(1)}\tilde{\zeta}^{(1)}\right] . \tag{3.2.45}$$

Using the above equation, we integrate (3.2.42) with respect to θ from 0 to 2π to obtain the compatibility condition for the second-order equation (3.2.42). The condition is $\bar{G}_2 = 0$ or

$$\bar{\zeta}_t^{(0)} = \frac{\bar{\nu}}{\bar{r}}[\bar{r}\bar{\zeta}_{\bar{r}}^{(0)}]_{\bar{r}} . \tag{3.2.46}$$

This, in turn, serves as the equation to define the leading-order vorticity $\bar{\zeta}^{(0)}(t,\bar{r})$. For given initial data $\bar{\zeta}^{(0)}(0,\bar{r})$, the solution of the simple axisymmetric diffusion equation (3.2.46) is (Carslow and Jaeger, 1959):

$$\bar{\zeta}^{(0)}(t,\bar{r}) = \frac{1}{2\bar{\nu}t}\int_0^\infty \bar{\zeta}^{(0)}(0,\xi)e^{-(\bar{r}^2+\xi^2)/(4\bar{\nu}t)}\, I_0\left(\frac{\bar{r}\xi}{2\bar{\nu}t}\right)\xi d\xi , \tag{3.2.47}$$

where I_0 denotes the modified Bessel function. The corresponding stream function $\bar{\psi}^{(0)}$ is then defined by (3.2.35) as a weighted integral of $\bar{\zeta}^{(0)}$. These integral representations do not show clearly the behavior of the solution for large t. They are also too cumbersome to be used in the analysis of the next-order equations (3.2.36) and (3.2.37) in which they appear as coefficients in the linear operator $\mathcal{L}$ on $\tilde{\psi}^{(1)}$ and $\tilde{\zeta}^{(1)}$. In Subsect. 3.2.2, we construct an alternate solution of (3.2.46) in a power series of t^{-1} and discuss the physical meaning of the leading two terms in the series as $t \to \infty$.

Equations (3.2.33) and (3.2.34) say that the leading-order two-time inner solution is independent of the short time variable τ. The question is whether one obtains the same leading-order solution directly from a one-time analysis, in which the solution *is assumed* to depend only on the normal time t from the outset. This question was answered by Ting and Tung (1965). They carried out the one-time analysis and found that the leading-order solution is governed by the same simple axisymmetric diffusion Eq. (3.2.46) with the same initial data and boundary conditions. Thus, we conclude:

- *The leading-order solution from the two-time analysis is axisymmetric and independent of the short time variable τ and it is identical with the leading solution from the one-time analysis.*

Now we proceed to analyze the higher order solutions. Since their governing equations are linear, it is convenient to decompose the solutions into symmetric and asymmetric parts, denoted by the subscripts c and a, respectively. We write

$$f(\theta) = f_c + f_a(\theta) , \qquad \text{with} \qquad f_c = \frac{1}{2\pi}\int_0^{2\pi} f(\theta)\, d\theta \stackrel{\text{def}}{=} \{f\} , \tag{3.2.48}$$

where f stands for $\zeta^{(n)}$ or $\psi^{(n)}$ showing only their dependence on θ. The symmetric part f_c is the circumferential average or the θ-average of f. From here on we use the curly brackets, $\{\ \}$, to denote the θ-averaging operator. By using this decomposition for $n = 1$, (3.2.36) splits into separate equations for the symmetric and asymmetric parts of the first order solution. They are

$$[\tilde{\zeta}_c^{(1)}]_\tau = 0 \tag{3.2.49}$$

and

$$[\tilde{\zeta}_a^{(1)}]_\tau + \frac{1}{\bar{r}}[\bar{\zeta}_{\bar{r}}^{(0)}\tilde{\psi}_a^{(1)} - \bar{\psi}_{\bar{r}}^{(0)}\tilde{\zeta}_a^{(1)}]_\theta = 0 . \tag{3.2.50}$$

Equation (3.2.49) says that the symmetric part of the first order vorticity is independent of τ, that is,

$$\tilde{\zeta}_c^{(1)}(\tau, t, \bar{r}) = \bar{\zeta}_c^{(1)}(t, \bar{r}) , \tag{3.2.51}$$

and from (3.2.29) and (3.2.30) and the far-field conditions on $\psi^{(1)}$, we get

$$\tilde{\psi}_c^{(1)}(\tau, t, \bar{r}) = \bar{\psi}_c^{(1)}(t, \bar{r}) . \tag{3.2.52}$$

We conclude at this stage that the symmetric parts of the first order inner solutions, $\bar{\zeta}_c^{(1)}$ and $\bar{\psi}_c^{(1)}$, do not have any fast time variation. Similar to the axisymmetric parts of the leading-order solution, their dependence on $\bar{r}$ and t will be determined by compatibility conditions for the third-order solutions.

The asymmetric parts, $\tilde{\zeta}_a^{(1)}$ and $\tilde{\psi}_a^{(1)}$, are governed by the linear system (3.2.50) and (3.2.29). With the boundary conditions (3.2.30) on $\tilde{\psi}_a^{(1)}$ and initial data for $\tilde{\zeta}_a^{(1)}$, the first order asymmetric parts are defined.

Finally, consider the short time averages of the asymmetric parts,

$$\mathcal{M}\tilde{\psi}_a^{(1)}(\tau, t, \bar{r}, \theta) = \bar{\psi}_a^{(1)}(t, \bar{r}, \theta) \tag{3.2.53}$$

and

$$\mathcal{M}\tilde{\zeta}_a^{(1)}(\tau, t, \bar{r}, \theta) = \bar{\zeta}_a^{(1)}(t, \bar{r}, \theta) . \tag{3.2.54}$$

They are governed by the quasi-steady equations (3.2.29), (3.2.41), and (3.2.30), in which t appears only as a parameter through the leading-order solutions $\zeta^{(0)}(t,\bar{r})$ and $\psi^{(0)}(t,\bar{r})$.

Again we raise the question of whether the symmetric parts of the first order solution, $\bar{\zeta}_c^{(1)}$ and $\bar{\psi}_c^{(1)}$, and the τ-averages of the asymmetric parts, $\bar{\zeta}_a^{(1)}$ and $\bar{\psi}_a^{(1)}$, are equivalent to the corresponding symmetric and asymmetric parts of the first order solution obtained directly by the one-time analysis. Again, Ting and Tung (1965) provided the answer. They showed that the asymmetric parts from the one-time analysis obey the same differential equations and boundary conditions as $\bar{\zeta}_a^{(1)}$ and $\bar{\psi}_a^{(1)}$ since the leading-order solutions, which appear as coefficients in these equations, are identical. They conclude the following:

- *The asymmetric first order solution from the one-time analysis is equivalent to the τ-average of the corresponding two-time solution.*

Note that the variable t appears as a parameter in the asymmetric solution from the one-time analysis. Therefore, the solution cannot accept any initial data unless they are compatible with (3.2.29) and (3.2.41). If they are compatible, then the two-time asymmetric solution degenerates to its τ-average, that is, the one-time solution.

As regards the first order symmetric solutions, the answer to our question is negative, because they are not governed by the same set of equations. The governing equation for $\mathcal{M}\tilde{\zeta}_c^{(1)} = \mathcal{M}\{\tilde{\zeta}^{(1)}\}$, is given by the θ- and τ-average of the third-order vorticity equation. There are nonlinear inhomogeneous terms induced by the averages of the nonlinear convection terms, which involve products of the first- and second-order solutions. These terms will be absent in the corresponding equation in the one-time analysis. Those nonlinear inhomogeneous terms of the two-time solutions will be shown explicitly in Appendix A. There we present a systematic procedure to derive the governing equations for the symmetric and asymmetric parts of the nth-order two-time solution and the equations for their τ-averages.

Now we proceed to derive the equations for the second-order solutions. As before, we decompose (3.2.38) to obtain

$$[\tilde{\zeta}_c^{(2)}]_\tau = \{\tilde{F}_2\}\,, \tag{3.2.55}$$

$$[\tilde{\zeta}_a^{(2)}]_\tau + \mathcal{L}[\tilde{\psi}_a^{(2)}, \tilde{\zeta}_a^{(2)}] = \tilde{F}_2 - \{\tilde{F}_2\}\,, \tag{3.2.56}$$

where

$$\{\tilde{F}_2\} = \frac{1}{2\bar{r}}\left[\bar{a}\left[(\tilde{\zeta}_c^{(1)})^2 + \{(\tilde{\zeta}_a^{(1)})^2\}\right]\right]_{\bar{r}\tau}\,. \tag{3.2.57}$$

Here we made use of (3.2.44) and (3.2.46). The last equation (3.2.57) was obtained by taking the θ- and τ-averaging of (3.2.44).

Since (3.2.56) involves products of the first order solutions through $\tilde{F}_2$, the τ-average of the second-order asymmetric two-time solution is, in general, not equivalent to the corresponding asymmetric solution from one-time analysis. Only when the first order two-time solution is of a certain special form that renders the τ-average of the products equal to zero, are these two solutions equivalent. This exceptional case will be discussed later in Subsect. 3.2.4, where we study the first order two-time solutions.

In the following section we present a power series representation of the leading-order core structure and explain the physical meaning of the first two terms. In Subsect. 3.2.3, we obtain the τ-average of the asymmetric first order solutions and then define the velocity of the vortex center in the normal time scale. The two-time solutions, their dependence on the core structure, and their physical meaning will be presented in Subsects. 3.2.3 and 3.2.4.

3.2.2 Leading-Order Core Structure

We recall that the leading-order core structure, $\epsilon^{-2}\bar{\zeta}^{(0)}(t,\bar{r})$ and $\bar{\psi}^{(0)}(t,\bar{r})$, from the two-time analysis is identical to that from the one-time analysis. The solution is related to the initial data by the integral representations (3.2.47) and (3.2.35). These integral representations do not show clearly the behavior of the solution for small or large $\bar{r}$ or for large t, and are too cumbersome to be used as the coefficients in the governing equations for the higher order solutions. We want to approximate the solution by *an optimum similarity solution*. We shall introduce a mathematical definition of the optimum solution and explain its physical meaning.

Although we can derive the similarity solution as the leading term of an asymptotic series for large t from the integral representation (3.2.47), we find it more revealing to construct the series solution directly. By making use of the fact that the vorticity decays exponentially in $\bar{r}$, we represent the solution of (3.2.46) in a descending power series of $t + t_0$, where $t_0 > 0$ denotes a constant positive time shift so that the series solution is valid at $t = 0$. An appropriate choice of t_0 will be discussed later. The series solution is constructed in terms of two new variables,

$$\bar{t} = t + t_0 \qquad \text{and} \qquad \bar{\eta}^2 = \bar{r}^2/(4\bar{\nu}\bar{t}) \,. \tag{3.2.58}$$

Equation (3.2.46) is separable and the condition of exponential decay of $\bar{\zeta}^{(0)}$ leads to a discrete spectrum of eigenvalues.[3] The series solution is

$$\bar{\zeta}^{(0)}(t,\bar{r}) = \sum_{n=0,1\ldots} C_n\, \bar{t}^{-n-1}\, L_n(\bar{\eta}^2) e^{-\bar{\eta}^2} \,. \tag{3.2.59}$$

Here, L_n denotes the nth Laguerre polynomial (Magnus et al., 1966), and the coefficients C_n are related to the initial data at $t = 0$ or $\bar{t} = t_0 > 0$ by

[3] Note that this is the only occasion where we use explicitly the condition of exponential decay so that (2.1.18) holds for all N.

$$C_n = 2t_0^{n+1} \int_0^{\infty} \bar{\zeta}^{(0)}(0, \bar{\eta}\sqrt{4\bar{\nu}t_0}) L_n(\bar{\eta}^2)\, \bar{\eta}\, d\bar{\eta} \,. \tag{3.2.60}$$

In particular, the first coefficient is $C_0 = \Gamma/(4\pi\bar{\nu})$ and the first term in (3.2.59) is known as the *similarity solution* corresponding to a Lamb vortex created at the instant $t = -t_0$ with zero core radius. It represents the leading term of the vorticity distribution (3.2.59) for large t. We define the optimum time shift t_0^* by the condition that the second term in the series (3.2.54) vanishes, that is, that $C_1 = 0$. The condition is

$$C_1 = 2(t_0^*)^2 \int_0^{\infty} \bar{\zeta}^{(0)}(0, \bar{\eta}\sqrt{4\bar{\nu}t_0^*})\, [1 - \bar{\eta}^2]\, \bar{\eta}\, d\bar{\eta} = 0 \,, \tag{3.2.61}$$

or

$$2\pi \int_0^{\infty} \bar{\zeta}(0, \bar{r})\, \bar{r}^3\, d\bar{r} = 4\bar{\nu}t_0^*\Gamma \,, \tag{3.2.62}$$

which in turn defines the optimum time shift,

$$t_0^* = \frac{\pi}{2\Gamma\bar{\nu}} \int_0^{\infty} \bar{\zeta}^{(0)}(0, \bar{r})\, \bar{r}^3\, d\bar{r} \,. \tag{3.2.63}$$

The vorticity distribution defined by the first term of the series (3.2.59) with $t_0 = t_0^*$ is called the optimum similarity solution or the optimum Lamb vortex of age t_0^* at the instant $t = 0$ with effective core size $\delta_0 = 4\bar{\nu}t_0^*$. The reasons for this notion are twofold: From the series solution (3.2.59) and condition (3.2.61), we see the mathematical reason.

- *The optimum similarity solution is the best one term asymptotic solution in the sense that it is the only one that differs from the exact solution by $O(\bar{t}^{-3})$ instead of $O(\bar{t}^{-2})$ as $t \to \infty$, that is,*

$$\bar{\zeta}^{(0)} = \bar{\zeta}^* + O(\bar{t}^{-3}) \quad \text{where} \quad \bar{\zeta}^* = \frac{\Gamma}{4\bar{\nu}(t + t_0^*)} e^{-\bar{r}^2/[4\pi\bar{\nu}(t+t_0^*)]} \,. \tag{3.2.64}$$

From the fact that the right-hand side of (3.2.62) is equal to the polar moment of the optimum Lamb vortex, while its left-hand side is the polar moment of the initial vorticity distribution and from the linear diffusion law of the polar moment of vorticity, (2.3.17), we get the physical justification.

- *The optimum similarity solution and the exact solution (3.2.59) have not only the same total strength Γ but also the same polar moment for all $t \geq 0$.*

Of course, they always have the same first moments that are equal to zero due to symmetry. Figure 3.4 compares the optimum solution with the exact solution having an initial vorticity distribution in the shape of a top hat, that is, $\zeta(0, r) = \Gamma/(\pi\delta_0^2)H(\delta_0 - r)$, where H stands for the Heaviside function. In this case, the optimum time shift is $t_0^* = \delta_0^2/(8\nu)$. Good agreement between the optimum similarity solution and the exact one is achieved for $\bar{t}/t_0^* \sim$

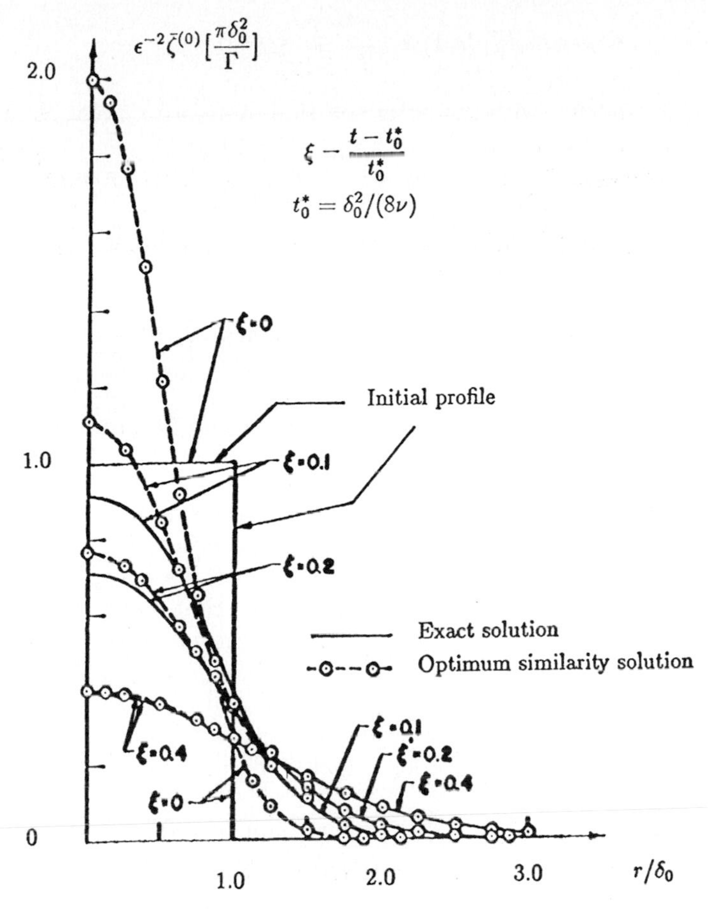

Fig. 3.4. Comparison of the optimum similarity solution and the exact solution. Adapted from Ting (1971)

3. Better agreement would be expected for a more realistic initial vorticity distribution in the shape of a bell, which is a continuous function of $r \geq 0$ decreasing monotonically from its maximum at $r = 0$ to zero. The importance of an optimum similarity solution becomes more apparent when one takes into account that in a real problem, details of the initial data beyond the total strength and an estimate of the effective size are rarely available. The optimum similarity solution could, in such cases, serve as a reasonable model for the core structure for all $t \geq 0$.

The diffusion of the leading vorticity distribution, $\epsilon^{-2}\bar{\zeta}^{(0)}$, which is axisymmetric, is defined by the exact solution (3.2.59) or the optimum similarity solution (3.2.64). The corresponding flow field is the leading circumferential flow, $-\epsilon^{-1}\bar{\psi}_{\bar{r}}^{(0)}$ with

$$-\epsilon^{-1}\bar{\psi}_{\bar{r}}^{(0)}(t,\bar{r}) = \frac{\epsilon^{-1}}{\bar{r}}\int_0^{\bar{r}} \xi\bar{\zeta}^{(0)}(t,\xi)d\xi \; . \tag{3.2.65}$$

As $\bar{r} \to \infty$, it matches with the outer solution of a vortex point as expected,

$$-\epsilon^{-1}\bar{\psi}_{\bar{r}}^{(0)}(t,\bar{r}) \to \frac{\Gamma}{2\pi r} + \cdots \quad \text{where} \quad \Gamma = 2\pi \int_0^{\infty} \bar{\zeta}^{(0)}(t,\bar{r})\,\bar{r}\,d\bar{r} \; . \tag{3.2.66}$$

In particular, the flow field corresponding to the similarity solution is that of a Lamb vortex,

$$-\epsilon^{-1}\psi_{\bar{r}}^{(0)} = \frac{\Gamma}{2\pi r}\left[1 - e^{-r^2/[4\nu(t+t_0^*)]}\right] . \tag{3.2.67}$$

3.2.3 Asymmetric First Order Solution in the Normal Time Scale and the Velocity of the Vortex

It was concluded in Subsect. 3.2.1 that an asymmetric first order one-time solution in the normal time t can be considered either as the τ-average of an asymmetric first order two-time solution or as the asymmetric solution from the one-time analysis, since these two are equivalent. The asymmetric first order vorticity and stream function are governed by the differential equations (3.2.29) and (3.2.41), the boundary conditions (3.2.30) at $\bar{r} = 0$, and the matching conditions (3.2.23), (3.2.26), and (3.2.27). In this section, we construct the solution of that system of equations, and show that in the normal time scale the leading-order velocity of the vortex center is equal to the local background velocity.

By the elimination of $\bar{\zeta}_a^{(1)}$ from (3.2.29) and (3.2.41), we obtain a single equation for the asymmetric stream function $\bar{\psi}_a^{(1)}$,

$$[\bar{\psi}_{\bar{r}}^{(0)}\bar{\Delta} + \bar{\zeta}_{\bar{r}}^{(0)}]\,(\bar{\psi}_a^{(1)})_\theta = 0 \; . \tag{3.2.68}$$

It is a linear partial differential equation in $\bar{r}$ and θ while the time t appears as a parameter. If the initial data are inconsistent with (3.2.68) and/or a matching condition, we need a two-time solution to fulfill the initial data. This will be discussed in Subsect. 3.2.4.

Since the solutions have to be periodic in θ, we can express the asymmetric solution in a Fourier series,

$$\bar{\psi}_a^{(1)}(t,\bar{r},\theta) = \sum_{j=1,2,\ldots} [\bar{\psi}_{j1}(t,\bar{r})\cos j\theta + \bar{\psi}_{j2}(t,\bar{r})\sin j\theta] \; . \tag{3.2.69}$$

Substituting (3.2.69) in (3.2.68), we obtain the equations for the Fourier coefficients,

$$\left\{\frac{\partial^2}{\partial\bar{r}^2} + \frac{1}{\bar{r}}\frac{\partial}{\partial\bar{r}} + \left[\frac{\bar{\zeta}_{\bar{r}}^{(0)}}{\bar{\psi}_{\bar{r}}^{(0)}} - \frac{j^2}{\bar{r}^2}\right]\right\}\bar{\psi}_{jk} = 0 \; , \tag{3.2.70}$$

for $k = 1, 2$, and $j = 1, 2 \ldots$.

The boundary conditions (3.2.30) at $\bar{r} = 0$ yield

$$\bar{\psi}_{jk} = 0 \qquad \text{and} \qquad (\bar{\psi}_{jk})_{\bar{r}} = 0 \, . \tag{3.2.71}$$

With the leading-order circumferential velocity matched by (3.2.66), the matching condition (3.2.23) becomes

$$\frac{1}{\bar{r}}\bar{\psi}^{(1)}_{\theta}\boldsymbol{r} - \bar{\psi}^{(1)}_{\bar{r}}\boldsymbol{\theta} \to [-\dot{X}^{(0)} + \Psi_y(t, X^{(0)}, Y^{(0)})]\,\boldsymbol{i} - [\dot{Y}^{(0)} + \Psi_x(t, X^{(0)}, Y^{(0)})]\,\boldsymbol{j} \, , \tag{3.2.72}$$

as $\bar{r} \to \infty$. (Here $\boldsymbol{i}$ and $\boldsymbol{j}$ are the unit vectors pointing in the x and y directions, respectively.) This result, in turn, yields the conditions on the Fourier coefficients,

$$\bar{\psi}_{11} \to [\dot{Y}^{(0)} + \Psi_x(t, X^{(0)}, Y^{(0)})]\,\bar{r} \, , \tag{3.2.73}$$

$$\bar{\psi}_{12} \to [-\dot{X}^{(0)} + \Psi_y(t, X^{(0)}, Y^{(0)})]\,\bar{r} \, , \tag{3.2.74}$$

and

$$\bar{\psi}_{jk} \to 0, \qquad \text{for} \quad j = 2, 3, \ldots \, , \qquad k = 1, 2 \, , \tag{3.2.75}$$

as $\bar{r} \to \infty$. With $\bar{\psi}^{(0)}_{\bar{r}\bar{r}}(t, 0) = -\bar{\zeta}^{(0)}(t, 0)/2 \neq 0$, while $\bar{\zeta}^{(0)}_{\bar{r}}(t, 0) = 0$, the ratio $\bar{\zeta}^{(0)}_{\bar{r}}/\bar{\psi}^{(0)}_{\bar{r}}$ remains finite as $\bar{r} \to 0$ and hence (3.2.70) has a regular singular point at $\bar{r} = 0$. Near this point, the solution $\bar{\psi}_{jk}$ can be expressed in terms of its two independent solutions as

$$\bar{\psi}_{jk} \sim c_1(t)\,\bar{r}^{j} + c_2(t)\,\bar{r}^{-j} \, . \tag{3.2.76}$$

For any $j \geq 2$, the first condition in (3.2.71) requires that $c_2 = 0$ while the second condition is fulfilled for all c_1. The matching condition (3.2.75), together with the fact that $\bar{\psi}_{jk}$ satisfies the *homogeneous* second-order equation (3.2.70), then requires $c_1 = 0$. Consequently, we have

$$\bar{\psi}_{jk} \equiv 0 \qquad \text{for} \quad j = 2, 3, \ldots \, , \qquad k = 1, 2 \, . \tag{3.2.77}$$

For $j = 1$, the two conditions in (3.2.71) require both $c_1 = 0$ and $c_2 = 0$ and hence

$$\bar{\psi}_{11} \equiv 0 \qquad \text{and} \qquad \bar{\psi}_{12} \equiv 0 \, . \tag{3.2.78}$$

The matching conditions (3.2.73) and (3.2.74) then become the equation for the velocity of the vortex center,

$$\boldsymbol{i}\,\dot{X}^{(0)}(t) + \boldsymbol{j}\,\dot{Y}^{(0)}(t) = \boldsymbol{i}\,\Psi_y(X^{(0)}, Y^{(0)}) - \boldsymbol{j}\,\Psi_x(X^{(0)}, Y^{(0)}) \, . \tag{3.2.79}$$

This equation says:

- *In the normal time scale, the leading-order velocity of the vortex center is given by the local background velocity.*

The velocity differs from that of the classical inviscid theory by at most $O(\epsilon)$.

From (3.2.77) and (3.2.78), and the equivalence principle restated at the beginning of this section, we see that the τ-average of the first order solution has to be symmetric, that is,

$$\mathcal{M}\tilde{\psi}_a^{(1)} = \bar{\psi}_a^{(1)} = 0 \quad \text{and} \quad \mathcal{M}\tilde{\psi}^{(1)}(\tau, t, \bar{r}, \theta) = \bar{\psi}_c^{(1)}(t, \bar{r}) \; . \tag{3.2.80}$$

It follows from (3.2.51) and (3.2.52) that $\bar{\psi}_c^{(1)}$ is equal to the symmetric part of the first order two-time solution, whose dependence on t and $\bar{r}$ is not yet defined. We recall that the governing equation (3.2.46) for the symmetric part of the leading-order solution came from the compatibility conditions, that is, the τ- and θ-averages of the Poisson equation (3.2.29) for $n = 0$ and the second-order vorticity diffusion equation (3.2.38). Similarly, we expect the compatibility conditions for the Poisson equation for $n = 1$ and the third-order vorticity diffusion equation to yield the equations for $\bar{\zeta}_c^{(1)}$ and $\bar{\psi}_c^{(1)}$.

The one-time asymmetric solution constructed in this section is defined by (3.2.77)–(3.2.79). The solution requires that the initial core structure has to be symmetric to the first order and the initial velocity of the vortex center has to be equal to the local background velocity. If the initial data are inconsistent with these requirements, a two-time asymmetric solution is needed. It is constructed in the following section.

3.2.4 Asymmetric First Order Two-Time Solution and the Oscillatory Motion of the Vortex Center

In Subsect. 3.2.1, we found that the asymmetric first order two-time vorticity and stream function are governed by two linear partial differential equations, (3.2.29) and (3.2.50), in τ, $\bar{r}$, and θ with the normal time t as a parameter. These two equations can be combined to yield one equation for the stream function, $\tilde{\psi}_a^{(1)}$. The equation is

$$[\bar{\Delta}\tilde{\psi}_a^{(1)}]_\tau - \frac{1}{\bar{r}}[\bar{\psi}_{\bar{r}}^{(0)}\bar{\Delta} + \bar{\zeta}_{\bar{r}}^{(0)}]\;[\tilde{\psi}_a^{(1)}]_\theta = 0 \; . \tag{3.2.81}$$

Since the stream function is periodic in θ, we express the function by its Fourier series in θ,

$$\tilde{\psi}_a^{(1)}(\tau, t, \bar{r}, \theta) = \sum_{j=1,2,\dots} [\tilde{\psi}_{j1}\cos j\theta + \tilde{\psi}_{j2}\sin j\theta] \; , \tag{3.2.82}$$

where the Fourier coefficients $\tilde{\psi}_{jk}$, $j = 1, 2, \dots,$ $k = 1, 2$, are functions of the variables, τ and $\bar{r}$, and the parameter t. From (3.2.81) and (3.2.82), we obtain the equations for the Fourier coefficients,

$$[\bar{\Delta}_j\tilde{\psi}_{j1}]_\tau - \frac{j}{\bar{r}}[\bar{\zeta}_{\bar{r}}^{(0)} + \bar{\psi}_{\bar{r}}^{(0)}\bar{\Delta}_j]\tilde{\psi}_{j2} = 0 \; , \tag{3.2.83}$$

$$[\bar{\Delta}_j\tilde{\psi}_{j2}]_\tau + \frac{j}{\bar{r}}[\bar{\zeta}_{\bar{r}}^{(0)} + \bar{\psi}_{\bar{r}}^{(0)}\bar{\Delta}_j]\tilde{\psi}_{j1} = 0 \; , \tag{3.2.84}$$

with

$$\bar{\Delta}_j = (1/\bar{r})\partial_{\bar{r}}(\bar{r}\partial_{\bar{r}}) - (j^2/\bar{r}^2) \ , \tag{3.2.85}$$

for $j = 1, 2, \ldots$, and $k = 1, 2$. For each Fourier coefficient, two homogeneous boundary conditions at $\bar{r} = 0$ are obtained from (3.2.30) and a matching condition with the outer solution is obtained from (3.2.23) as $\bar{r} \to \infty$. These conditions for the Fourier coefficients are given by (3.2.71) and (3.2.73)–(3.2.75) with the bar accent replaced by the tilde accent and the leading-order velocity of the vortex center, $\boldsymbol{i}\dot{X}^{(0)} + \boldsymbol{j}\dot{Y}^{(0)}$ replaced by

$$\boldsymbol{i}[X_t^{(0)}(t) + X_\tau^{(2)}(t,\tau)] + \boldsymbol{j}[Y_t^{(0)}(t) + Y_\tau^{(2)}(t,\tau)] \ . \tag{3.2.86}$$

It was noted in Subsect. 3.2.3 that a two-time solution is needed when the initial velocity of the vortex center relative to the local background flow is not equal to zero. Let the initial relative velocity be denoted by $\boldsymbol{i}\,U_0 + \boldsymbol{j}\,V_0 \neq 0$, then the initial condition on the velocity is

$$\boldsymbol{i}[X_t^{(0)} + X_\tau^{(2)}] + \boldsymbol{j}[Y_t^{(0)} + Y_\tau^{(2)}] = \boldsymbol{i}[\Psi_y(X,Y) + U_0] + \boldsymbol{j}[-\Psi_x(X,Y) + V_0] \ , \tag{3.2.87}$$

at $t = \tau = 0$. In addition, we can specify an initial asymmetric first order vorticity profile $Z(\bar{r}, \theta)$, that is,

$$\tilde{\zeta}_a^{(1)}(0, \bar{r}, \zeta) = Z(\bar{r}, \theta) \ , \tag{3.2.88}$$

provided that Z decays rapidly in $\bar{r}$. The initial asymmetric stream function is then defined by (3.2.29).

Since the first order equations are linear, we can express the solution as the sum of two solutions. The first one is a particular solution of the homogeneous equation (3.2.81) fulfilling the homogeneous boundary conditions (3.2.30) at $\bar{r} = 0$ and the inhomogeneous matching condition (3.2.23). This in turn defines the velocity of the vortex center fulfilling the initial condition (3.2.86). The initial vorticity profile of the particular solution is free to be assigned provided that it decays rapidly. The second solution of (3.2.81) fulfills the homogeneous boundary conditions (3.2.71), the homogeneous matching conditions (3.2.73)–(3.2.75), and thus the sum of the two solutions fulfills the initial condition (3.2.87).

Since the matching condition for $j \geq 2$ and the boundary conditions for all j are homogeneous, the Fourier coefficients of the stream function with the trivial initial profile vanish for all $\tau \geq 0$ and $t \geq 0$, that is,

$$\tilde{\psi}_{jk}(t, \bar{r}) \equiv 0 \qquad j = 2, 3, \ldots \ , \ k = 1, 2 \ , \qquad \text{if} \qquad Z(\bar{r}, \theta) \equiv 0 \ . \tag{3.2.89}$$

Now we proceed to construct the particular solution that is independent of the initial profile Z. We consider the particular solution having only the two first harmonics in θ. We try to account for the velocity of the local background flow by using the result (3.2.79) for the one-time solution, that is, by setting

$$\dot{X}^{(0)}(t) = \Psi_y(X^{(0)}, Y^{(0)}) \quad \text{and} \quad \dot{Y}^{(0)}(t) = -\Psi_x(X^{(0)}, Y^{(0)}) \,. \tag{3.2.90}$$

We should verify later that this choice does not lead to secular terms in $X^{(2)}$ and $Y^{(2)}$ as $\tau \to \infty$. For the construction of the short time solution, we suppress the t-dependence. The matching condition (3.2.23), or (3.2.73) and (3.2.74) with bars replaced by tildes, and (3.2.90) yield

$$\tilde{\psi}_a^{(1)} = \tilde{\psi}_{11}(\tau, \bar{r}) \cos\theta + \tilde{\psi}_{12}(\tau, \bar{r}) \sin\theta \to V(\tau)\bar{r}\cos\theta - U(\tau)\bar{r}\sin\theta \,, \tag{3.2.91}$$

as $\bar{r} \to \infty$, with $U(\tau) = X_\tau^{(2)}$ and $V(\tau) = Y_\tau^{(2)}$. The initial data for the unknowns $U(\tau)$ and $V(\tau)$ come from the prescribed initial velocity of the vortex center in (3.2.86):

$$U(0) = U_0 = X_\tau^{(2)}(0,0) \qquad \text{and} \qquad V(0) = V_0 = Y_\tau^{(2)}(0,0) \,. \tag{3.2.92}$$

Based on the matching condition (3.2.91), we propose a particular solution of the form,

$$\tilde{\psi}_a^{(1)} = W \, \sin(\theta - \bar{\omega}\tau - \alpha) \, G(\bar{r}) \,, \tag{3.2.93}$$

with

$$U(\tau) = W \cos(\bar{\omega}\tau + \alpha) \,, \qquad V(\tau) = W \sin(\bar{\omega}\tau + \alpha) \,, \tag{3.2.94}$$

where W, α, and $\bar{\omega}$ may depend on the parameter t. From the initial data (3.2.92), we require

$$W = W_0 \quad \text{and} \quad \alpha = \alpha_0 \quad \text{at} \quad t = 0 \,, \tag{3.2.95}$$

where

$$W_0 \cos\alpha_0 = U_0 \quad \text{and} \quad W_0 \sin\alpha_0 = V_0 \,. \tag{3.2.96}$$

The partial differential equations (3.2.83) and (3.2.84) then reduce to the following ordinary differential equation for $G(\bar{r})$,

$$\left[\bar{\omega} + \frac{1}{\bar{r}}\bar{\psi}_{\bar{r}}^{(0)}\right] \bar{\Delta}_1 G + \frac{1}{\bar{r}}\bar{\zeta}_{\bar{r}}^{(0)} G = 0 \,. \tag{3.2.97}$$

The boundary conditions on $\tilde{\psi}^{(1)}$ become,

$$G = 0 \,, \qquad G_{\bar{r}} = 0 \,, \qquad \text{at} \quad \bar{r} = 0 \,, \tag{3.2.98}$$

and the matching condition (3.2.91) becomes

$$G \to \bar{r} \qquad \text{as} \quad \bar{r} \to \infty \,. \tag{3.2.99}$$

Since the vorticity, $\bar{\zeta}^{(0)}$, decays rapidly in $\bar{r}$, the second term in (3.2.97) becomes negligible as $\bar{r} \to \infty$ and the solution G behaves as a first harmonic of a potential solution. Therefore, the asymptotic behavior of $G(t, \bar{r})$ is

$$G(t,\bar{r}) = \bar{r} + \frac{d(t)}{\bar{r}} + o(\bar{r}^{-N}) \tag{3.2.100}$$

for some large N as $\bar{r} \to \infty$. The coefficient of $\bar{r}$ is equal to unity on account of (3.2.99), while the coefficient $d(t)$ remains to be defined.

Since the differential operator in (3.2.97) is singular at $\bar{r} = 0$, we need to study the behavior of G near $\bar{r} = 0$. From (3.2.46), we expand the leading-order vorticity in a series of even powers of $\bar{r}$, that is, $\bar{\zeta}^{(0)}(t,\bar{r}) = z_0(t) + z_1(t)\bar{r}^2 + \cdots$ and then from (3.2.65), the circumferential velocity by a series in odd powers, $-\bar{\psi}^{(0)}_{\bar{r}} = \frac{1}{2}z_0\bar{r} + \frac{1}{4}z_1\bar{r}^3 + \cdots$. Note that $z_0(t) = \bar{\zeta}^{(0)}(t,0)$ denotes the vorticity at the vortex center. Now we represent the solution $G(\bar{r})$ by the series,

$$G(\bar{r}) = \bar{r}^{\lambda} \sum_{n=0,1,\ldots} a_n(t)\,\bar{r}^{2n}, \qquad \text{for } \bar{r} \ll 1\,. \tag{3.2.101}$$

By equating the coefficients of the leading term $\bar{r}^{\lambda-2}$ in (3.2.97), we obtain

$$[\bar{\omega} - z_0/2](\lambda^2 - 1)a_0 = 0\,. \tag{3.2.102}$$

Because of the boundary conditions (3.2.98) at $\bar{r} = 0$, we have $\lambda \neq \pm 1$. Hence for a nontrivial solution, we require

$$\bar{\omega} = z_0/2 = \bar{\zeta}^{(0)}(t,0)/2\,. \tag{3.2.103}$$

By equating the coefficients of the next term $\bar{r}^{\lambda}$, we obtain

$$\lambda = 3\,. \tag{3.2.104}$$

The coefficients a_n, with $n \geq 1$, in the series are then related to the first coefficient a_0 by recurrence formulae. It is evident that the series representation (3.2.101) fulfills the boundary conditions (3.2.98) while the unknown coefficient a_0, which can be a function of t, is to be determined by (3.2.99) or by its asymptotic behavior (3.2.100).

The standard shooting method can be used to determine a_0 for a given zeroth-order inner solution, $\bar{\zeta}^{(0)}(t,\bar{r})$, and the numerical result in turn defines the constant $d(t)$ in the asymptotic behavior (3.2.100) of $G(t,\bar{r})$.

For example, if the leading-order vorticity is given by the similarity solution (3.2.64), we have

$$\bar{\omega} = \Gamma/(2\pi\bar{\delta}^2) \qquad \text{with} \quad \bar{\delta}^2 = 4\bar{\nu}(t + t_0^*)\,. \tag{3.2.105}$$

By using the similarity variable $\eta = \bar{r}/\bar{\delta}$ and observing (3.2.103), we set $G(t,\bar{r}) = \bar{\delta}g(\eta)$. Equation (3.2.97) then becomes the following equation for the similarity solution $g(\eta)$,

$$\left[1 - \frac{1 - e^{-\eta^2}}{\eta^2}\right]\left[\frac{d^2}{d\eta^2} + \frac{1}{\eta}\frac{d}{d\eta} - \frac{1}{\eta^2}\right] g - 4e^{-\eta^2} g = 0\,. \tag{3.2.106}$$

The boundary conditions (3.2.98) and (3.2.99) become $g \to \eta$ as $\eta \to \infty$ and $g \to a_0^* \eta^3$ as $\eta \to 0$, where $a_0^* = a_0 \bar{\delta}^2$. Numerical solutions for $g(\eta)$ by means of a shooting method were explained by Ting and Tung (1965). The correct values for the coefficients $a_0(t)$ in (3.2.101) and $d(t)$ in (3.2.100) are $0.50\,\bar{\delta}^{-2}$ and $1.00\,\bar{\delta}^2$, respectively.

To determine the dependence of the amplitude W and phase α on t, we have to look at the next-order equations and find the constraints on W_t and α_t, so as to remove the secular terms as $\tau \to \infty$. Appendix B describes the analysis of the next-order solutions including the determination of the ϵ-order correction to the mean trajectory of the vortex center. We mention that Gunzburger (1973) carried the analysis to the third-order where the secular terms due to $W(t)$ and $\alpha(t)$ appear for the first time. He found that the amplitude $W(t)$ decays as $1/(t + t_0^*)$ when the leading-order vorticity is given by a similarity solution.

To summarize the results obtained in this section, we give a qualitative description of the trajectory of the vortex center defined by the two-time solution, in particular, how it differs from that defined by the one-time solution in Subsect. 3.2.3. For this purpose, it suffices to consider only the initial stage where $\tau = O(1)$ while $t \ll 1$. Thus we consider the first order asymmetric two-time solution (3.2.93) at $t = 0$.

Recall that the trajectory of the vortex center and its order one velocity are expressed in the form of (3.2.28) and (3.2.89). The trajectory defined by the first order asymmetric solution in the normal time scale is $\boldsymbol{i}\,X^{(0)}(t) + \boldsymbol{j}\,Y^{(0)}(t)$, and the velocity, (3.2.79), is equal to the local velocity of the background potential flow. As shown in Fig. 3.5, the leading-order trajectory coincides with the streamline $\mathcal{S}$ of the background flow passing through the initial position (X_0, Y_0). The equation for the streamline $\mathcal{S}$ is $\Psi(x, y) = \Psi(X_0, Y_0)$.

We note from (3.2.28) that the ϵ-order correction to the trajectory has to be in the normal time only and will be defined in Appendix B. The leading-

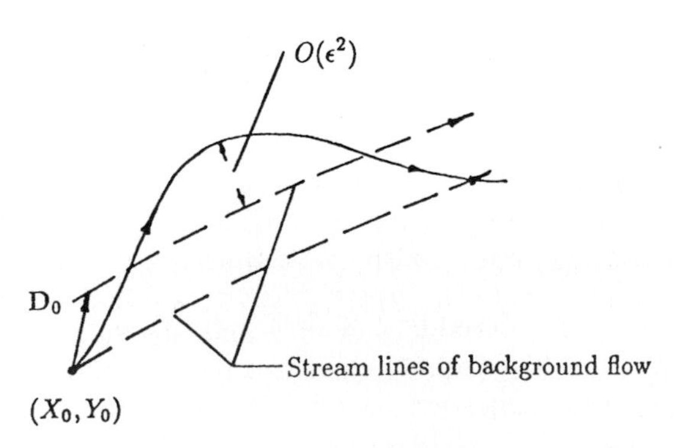

Fig. 3.5. Trajectory of a vortex center with initial velocity different from the local background velocity

order contribution of the two-time solution which accounts for the initial velocity relative to the background flow, $\boldsymbol{i}U_0 + \boldsymbol{j}V_0$, appears in the second-order displacement of the vortex center, $\epsilon^2[\boldsymbol{i}\,X^{(2)} + \boldsymbol{j}\,Y^{(2)}]$. From (3.2.86) and (3.2.91), we see that the second-order displacement has an order one velocity in two-time, $\boldsymbol{i}\,X_\tau^{(2)} + \boldsymbol{j}\,Y_\tau^{(2)} = \boldsymbol{i}\,U + \boldsymbol{j}V$. It represents the difference between the velocities defined by the two-time and one-time analyses. From the two-time solution defined by (3.2.93)–(3.2.95), (3.2.101), and (3.2.103), we see that the difference is a periodic function of τ with period $\bar{T} = 2\pi/\bar{\omega} = 4\pi/\bar{\zeta}^{(0)}(0, \bar{r} = 0)$. When the dependence of the period on t is taken into account in (3.2.94) and (3.2.103), then the velocity difference is in general an almost periodic function with a short period $\epsilon^2 2\pi/\bar{\omega}$ and an order-one amplitude, both of which modulate slowly in the normal time scale.

To be more specific, we assume that the leading-order vorticity is given by the similarity solution (3.2.64), with its initial core size in the normal length scale equal to $\delta_0 = \epsilon\bar{\delta}_0 = \sqrt{4\nu t_0^*}$. The angular frequency in the short time scale is $\bar{\omega}_0 = \Gamma/(2\pi\bar{\delta}_0^2) = \Gamma/(8\pi\bar{\nu}t_0^*)$. The two-time solution contributes to a second-order displacement of the vortex center from the streamline $\mathcal{S}$. During the initial stage the displacement is

$$\epsilon^2 \int_0^\tau d\tau' \; [\boldsymbol{i}U(\tau') + \boldsymbol{j}V(\tau')]$$

$$= \boldsymbol{D}_0 + \frac{\epsilon^2 W_0}{\bar{\omega}_0}\left[\boldsymbol{i}\cos\left(\bar{\omega}_0\tau + \alpha_0 - \frac{\pi}{2}\right) + \boldsymbol{j}\sin\left(\bar{\omega}_0\tau + \alpha_0 - \frac{\pi}{2}\right)\right] , \tag{3.2.107}$$

where

$$\boldsymbol{D}_0 = \frac{\epsilon^2}{\bar{\omega}_0}[-\boldsymbol{i}V_0 + \boldsymbol{j}U_0] = \frac{2\pi\delta_0^2 W_0}{\Gamma}\left[\boldsymbol{i}\cos\left(\alpha_0 + \frac{\pi}{2}\right) + \boldsymbol{j}\sin\left(\alpha_0 + \frac{\pi}{2}\right)\right] . \tag{3.2.108}$$

As shown in Fig. 3.5, the displacement represents an oscillation about the mean displacement vector $\boldsymbol{D}_0$ with a very short period, $T_0 = 2\pi\epsilon^2/\bar{\omega}_0$, and small amplitude, $\epsilon^2 W_0/\bar{\omega}$, in the normal time and length scales. The magnitude of the mean displacement, $|\boldsymbol{D}|$, is equal to the magnitude of initial velocity difference W_0 times $T_0/(2\pi)$ and its direction is in the direction of the "Kutta-Joukowski lifting force," obtained by rotating the velocity vector $-(\boldsymbol{i}U_0 + \boldsymbol{j}V_0)$, which is the background velocity relative to the vortex center, by 90° in the sense opposite to the circulation Γ. The amplitude of the oscillating trajectory is equal to the magnitude of the displacement vector, $|\boldsymbol{D}|$. It is interesting to note that the period of oscillation of the Lamb vortex with core size δ_0 is equal to the period of a Rankine vortex or rotating disc of radius $a = \delta_0$ and having the same circulation.

Finally, we seek the contribution of the oscillatory motion of the vortex center on the far-field flow. This comes from the far-field behavior (3.2.100) of G as $\bar{r} \to \infty$. The first term on the right-hand side of (3.2.100) matches with the leading-order outer solution to account for the difference between the initial velocity of the vortex center and the local background velocity. The second term has to be matched by or induces a doublet of second-order strength, $O(\epsilon^2)$, in the outer solution. A doublet in a two-dimensional potential flow represents the displacement effect due to the motion of a rigid body in the fluid. The doublet strength is proportional to the area of the body, which corresponds here to the effective vortical core area of order $O(\epsilon^2)$.

Thus, we conclude the following:

- *The first order asymmetric two-time solution represents a small-amplitude high-frequency oscillation of the vortex center around its mean trajectory. The latter is defined by a streamline of the background flow that passes through the initial position of the center plus a second-order displacement* $\boldsymbol{D}_0$. *The amplitude and period of the oscillation are of the order* ϵ^2.
- *The high-frequency oscillation of the vortex, which accounts for the initial difference between the velocity of the vortex center and local background flow, contributes an* ϵ^2*-order doublet to the outer flow field that accounts for the global effect of the highly oscillatory motion of the vortical core on the outer potential flow.*
- *If the initial core structure is symmetric to the first order and the initial velocity of the vortex center happens to be equal to the local background velocity, then the motion and decay of the vortex is given by the one-time solution to the first order; that is, there is no first order short-time variation, and hence its second-order contribution, as a doublet, to the outer potential does not materialize.*

We should note that for the first order inner solution, only the local velocity of the outer flow appears in the matching condition. This means that the inner flow field to the first order is in effect submerged in a uniform stream with the local background velocity. Ting and Tung (1965) showed that (i) the effect of nonuniformity of the outer flow appears in the second-order matching conditions as inhomogeneous terms proportional to the local velocity gradient and (ii) these terms are matched by a second-order one-time inner solution which in turn induced a fourth-order quadrupole to the outer solution. That is to say:

- *The interaction of the motion of a small vortex spot with the local velocity gradient of the background flow contributes an* $O(\epsilon^4)$ *quadrupole to the outer potential solution.*

Ting and Tung (1965) also showed that when the asymmetric first order two-time solution is of the special form (3.2.93)–(3.2.95), the τ-average of the nonlinear convection terms $\tilde{F}_2$ in the second-order vorticity evolution equation (3.2.38) vanishes. Furthermore, the equivalence between the first order one-time solution and the τ-average of the first order two-time solution is then extended to the second-order solutions. This is to say:

- *If $\tilde{\psi}_a^{(1)} = W(t)f(\theta - \omega\tau - \alpha(t))G(\bar{r})$, then the τ-average of the asymmetric second-order two-time solution is equivalent to the asymmetric second-order solution from the one-time analysis.*
- *The trajectory of the vortex center deviates from the streamline of the background flow passing through the initial position of the vortex center by no more than $O(\epsilon^2)$.*

Detailed studies of the second-order one-time and two-time inner solutions and their equivalence are presented in Appendix B, reproduced from Book-I.

Knowing the physical meaning of the one-time solution, its restrictions, and correspondence to the two-time solution, we are ready to study three-dimensional problems by the method of matched asymptoticis in the one-time scale. The analysis was carried out by Tung and Ting (1967) for the motion of a slender circular vortex ring submerged in an axisymmetric potential flow. Their analysis includes the evolution of the core structure due to stretching and diffusion. The equation for the core structure can be transformed to that for a two-dimensional problem and represented by an optimum similarity solution plus higher order terms for large t. When the vorticity distribution is a similarity solution, the numerical value of a constant in the velocity of the center line of the vortex ring is in error because an incorrect factor of 1/8 was used instead of 1/4 in the similarity solution. This algebraic error was corrected by Saffman (1970), (see van Dyke, 1975, p. 248). The error was also noted and corrected by Ting (1971), who presented the corresponding two-time solutions for the axisymmetric problem. He identified the one-time solutions as the averages of the two-time solutions in the normal time scale, thereby extending the results from two-dimensional to axisymmetric problems. In the same paper, the matched asymptotic analysis in one time scale was presented for a slender vortex filament with large circumferential flow. The analysis for the general case of a slender vortex filament with both large axial and circumferential velocities in the vortical core was carried out by Callegari and Ting (1978), which will be described in the following section.

3.3 Motion of Slender Vortex Filaments and Evolution of Their Core Structures

In this section, we first study the dynamics of an incompressible vortex filament in an ambient fluid, that is, the motion of the filament centerline $\mathcal{C}$ and the evolution of its core structure. The centerline, introduced at the beginning of this chapter, is now defined for a filament of finite length in the form of a torus. A spatial curve forming a loop, $\mathcal{C}(t)$, is defined by its position vector, $\boldsymbol{X}$, expressed as a function of the parameter s,

$$\mathcal{C}(t) : \boldsymbol{x} = \boldsymbol{X}(t, s) \ , \quad 0 \le s \le S_0 \ , \tag{3.3.1}$$

fulfilling the periodicity condition:

$$\boldsymbol{X}(t, s + S_0) = \boldsymbol{X}(t, s) \ . \tag{3.3.2}$$

Let $\sigma(t, s)$ and $\tilde{s}(t, s)$ denote the stretch factor and arclength of $\mathcal{C}$, as defined by the equations,

$$\sigma(t, s) = |\boldsymbol{X}_s(t, s)| \quad \text{and} \quad \tilde{s}(t, s) = \int_0^s \sigma(t, s') ds' . \tag{3.3.3}$$

We consider the length $S(t)$ of $\mathcal{C}$ finite, that is,

$$S(t) = \tilde{s}(t, S_0) = O(\ell) \ . \tag{3.3.4}$$

Usually, we define the parameter s as the initial arclength and then S_0 denotes the initial length of $\mathcal{C}$, that is,

$$S(0) = S_0 \quad \text{if} \quad \tilde{s}(0, s) = s \ . \tag{3.3.5}$$

We now identify the curve $\mathcal{C}$ as the centerline of a vortex filament, in the sense that both the components of velocity and acceleration in the plane normal to $\mathcal{C}$ at point $\boldsymbol{X}(t, s)$ remain $O(1)$, that is,

$$\begin{aligned} \frac{1}{U} \left| \boldsymbol{v}(t, \boldsymbol{X}(t, s)) \times \boldsymbol{\tau}(t, s) \right| &= O(1) \\ \frac{\ell}{U^2} \left| \frac{\partial}{\partial t} \boldsymbol{v}(t, \boldsymbol{X}(t, s)) \times \boldsymbol{\tau}(t, s) \right| &= O(1) \ , \end{aligned} \tag{3.3.6}$$

where $\boldsymbol{\tau}$ denotes the unit tangent vector to $\mathcal{C}$. Note that the first condition allows for a large axial velocity on $\mathcal{C}$. Here we follow the formulation of Callegari and Ting (1978) by setting the tangential component of $\dot{\boldsymbol{X}}$ to zero, that is,

$$\dot{\boldsymbol{X}} \cdot \boldsymbol{\tau} = 0 \quad \text{or} \quad \dot{\boldsymbol{X}} \cdot \boldsymbol{X}_s = 0 \ . \tag{3.3.7}$$

This in turn defines the parameter s for $t > 0$. Equations (3.3.6)–(3.3.7) imply that the velocity of the filament remains of order one, that is,

$$\dot{\boldsymbol{X}}/U = O(1) \qquad \text{for} \ t > 0 \ . \tag{3.3.8}$$

See Appendix E.4 for the definition of the orthogonal coordinate system attached to the filament centerline, which will be used throughout the rest of the subsequent derivations.

Here we follow the analysis of Callegari and Ting (1978), in which the order of magnitude has been expressed in powers of ϵ, and the parameter $\bar{\epsilon} = 1/\ln(1/\epsilon)$ is considered as $O(\epsilon^0) = O(1)$. For example, for $\epsilon = 0.01$, $\bar{\epsilon}$ is as large as $\bar{\epsilon} \approx 0.22$ and may be counted as $O(1)$ for all practical purposes. Noting that ϵ is exponentially small with respect to $\bar{\epsilon}$, a single power series in ϵ for the filament problem implies that we have been able to find the summation of the power series in $\bar{\epsilon}$ for each power of ϵ.[4]

[4] The summation of the power series of $1/\ln(1/\epsilon)$ for each power of ϵ was done recently for low Reynolds number flows. See Keller and Ward (1996) and references therein.

When $\bar{\epsilon}$ is small, ϵ is exponentially small. We can then approximate the leading-order solution in ϵ by the leading terms of the power series in $\bar{\epsilon}$. This is the case, for example, for the interaction of two nearly parallel straight filaments representing the trailing vortices far downstream of an aircraft. This case was studied in a sequence of papers by Klein et al. (1991a, 1991b, 1992, 1993, 1995a) and Klein (1994). Their theory provides the explicit dependence of the velocity of a filament on $\bar{\epsilon}$. It will be presented below in Chap. 4.

Equation (3.3.7) is a trivial condition for two-dimensional problems, and axisymmetric problems, without azimuthal flows. For a three-dimensional problem, we could require a point $\boldsymbol{X}(t,s)$ on $\mathcal{C}$ to be a material point, that is, replace (3.3.6) and (3.3.7) by $\dot{\boldsymbol{X}} = \boldsymbol{v}(t,\boldsymbol{X})$. This condition was used by Ting (1971) for the analysis of a vortex filament with large swirling flow but order one axial velocity so that (3.3.8) is observed, that is, $\dot{\boldsymbol{X}}/U$ remains of order one. In the analysis of a filament with both large swirling and large axial flow, Callegari and Ting (1978) imposed (3.3.7) in order to preserve condition (3.3.8), while (3.3.6) allows for large axial velocity, say $O(1/\epsilon)$.

To define the effective core size, we use results by Ting and Tung (1965), Tung and Ting (1967), and Ting (1971) in Sect. 3.2 and in the present section, which show that the leading-order vorticity distribution in a normal plane $\mathcal{P}$ of $\mathcal{C}$ at $\boldsymbol{X}$ is axisymmetric with respect to the centerline. The similarity solution of the axial vorticity $\boldsymbol{\Omega}\cdot\boldsymbol{\tau}$ in $\mathcal{P}$ is Gaussian with the maximum at the center $\boldsymbol{X}$. Thus we define the effective core size by a typical radius, δ^*, of a contour line on which $|\boldsymbol{\Omega}\cdot\boldsymbol{\tau}|$ is a fraction, say $1/e$, of its maximum value in $\mathcal{P}$, that is,

$$|\boldsymbol{\Omega}(t,\boldsymbol{x})\cdot\boldsymbol{\tau}| = |\boldsymbol{\Omega}(t,\boldsymbol{X})\cdot\boldsymbol{\tau}|/e \qquad \text{for} \qquad |\boldsymbol{x}-\boldsymbol{X}| = \delta(t,s)\,,\quad \boldsymbol{x}\in\mathcal{P}\,, \tag{3.3.9}$$

and the reference core size δ^* is chosen to be a typical initial size,

$$\delta(0,s) = O(\delta^*)\,. \tag{3.3.10}$$

For a filament submerged in a background flow, the presence of the filament will not change the background flow if it is a potential flow. The resultant flow field is given by the superposition of the flow induced by the filament and the background flow. Similar to the planar vortical flow in Sect. 3.2, we consider the background potential flow first and defer the study of the interaction of background rotational flow with vortices or filaments to Subsect. 3.3.8.

When there are several filaments submerged in a background flow, the flow field has many more distinct length scales. In addition to the length scale ℓ of the background flow, we have typical length scales for the radii of curvature of the centerlines of the filaments, the distance between a filament and the boundary of the flow field, and the distance between two filaments. We assume that all those lengths are $O(\ell)$, or rather we choose ℓ to be the smallest of all those length scales in the region away from the core structure. We shall also assume that the core sizes of the filaments are of the same order, $O(\delta^*)$.

Because of the rapidly decaying vorticity field in the sense of (2.1.18), the flow field induced by the filament at a distance $r \gg \delta^*$ approaches the potential flow induced by a vortex line along $\mathcal{C}$ with the same strength Γ but zero core radius relative to ℓ as $\epsilon = \delta^*/\ell \to 0$. The potential flow is then given by the Biot–Savart formula (3.1.57). Consider now a flow field involving N filaments ($N > 1$) that are apart from each other by distances of order $O(\ell)$. In this case, the inner vortical cores of the filaments do not overlap, and the background potential flow seen by filament j is composed of the potential flows induced by the other $N-1$ filaments plus the background potential flow common to all N filaments. Thus, to determine the motions of N filaments at distance $O(\ell)$ apart, it suffices to analyze only the motion and decay of a single filament submerged in a background potential flow. The extensions of the analysis for a single filament to several filaments were demonstrated by Liu et al. (1986b), Klein and Majda (1991a), and Klein et al. (1995a) and are reported in Subsect. 5.1.2, and in Sects. 4.3 and 4.4.

As sketched in Fig. 3.6, the bulk of vorticity of a filament is concentrated in a slender tube-like region with a centerline $\mathcal{C}$ and an effective core size δ. Also shown in the figure is an enlarged segment of the filament near a point $\boldsymbol{P}$ on $\mathcal{C}$, $\boldsymbol{x} = \boldsymbol{X}(t, s)$, and the curvilinear coordinates s, r, and ϕ with respect to $\boldsymbol{P}$ on $\mathcal{C}$, corresponding to those shown in Fig. 3.3 for a vortex line (see Appendix E.4). For a filament of circulation $\Gamma = O(U\ell)$, the assumption of slenderness, (3.1), implies that there is a large swirling flow around the centerline of the order ϵ^{-1}. That is, in the plane normal to $\mathcal{C}$ at point $\boldsymbol{P}$, the fluid has a large circumferential velocity relative to the point $\boldsymbol{P}$, and an axial (tangential to $\mathcal{C}$) velocity which can also be large of the order of ϵ^{-1}, but the radial velocity component relative to $\boldsymbol{P}$ is *assumed* to be $O(1)$, that is,

$$\boldsymbol{r} \cdot [\boldsymbol{v}(t, \boldsymbol{x}) - \boldsymbol{v}(t, \boldsymbol{X})]/U = O(1) \ . \tag{3.3.11}$$

Here $\boldsymbol{r}$ denotes the unit radial vector in the normal plane and $\boldsymbol{x} = r\boldsymbol{r} + \boldsymbol{X}$. See Fig. 3.6 and (3.1.63). Note that condition (3.3.11), which characterizes the core structure, is imposed in addition to the assumptions of (i) a rapidly decaying vorticity field, (2.1.18), (ii) a slender vortex core, (3.1), and (iii) velocities near the centerline remaining bounded as $\epsilon \to 0$, (3.3.6) and (3.3.7).

Guided by these assumptions, we apply the method of matched asymptotics to study the dynamics of a slender filament. The expansion schemes for the inner core and the outer potential flow and the matching procedure are similar to those in the two-dimensional analysis in Sect. 3.2. Again there is the need of a two-time analysis in case the initial data are incompatible with the one-time solution in the normal time scale. Here we shall assume that the initial core structure and the velocity of the centerline of the filament are compatible with the one-time solution.

The one-time analysis was carried out by Callegari and Ting (1978) for a filament with large swirling and axial flow in its core. They showed that if the leading-order swirling flow does not vary along the filament, the same is true

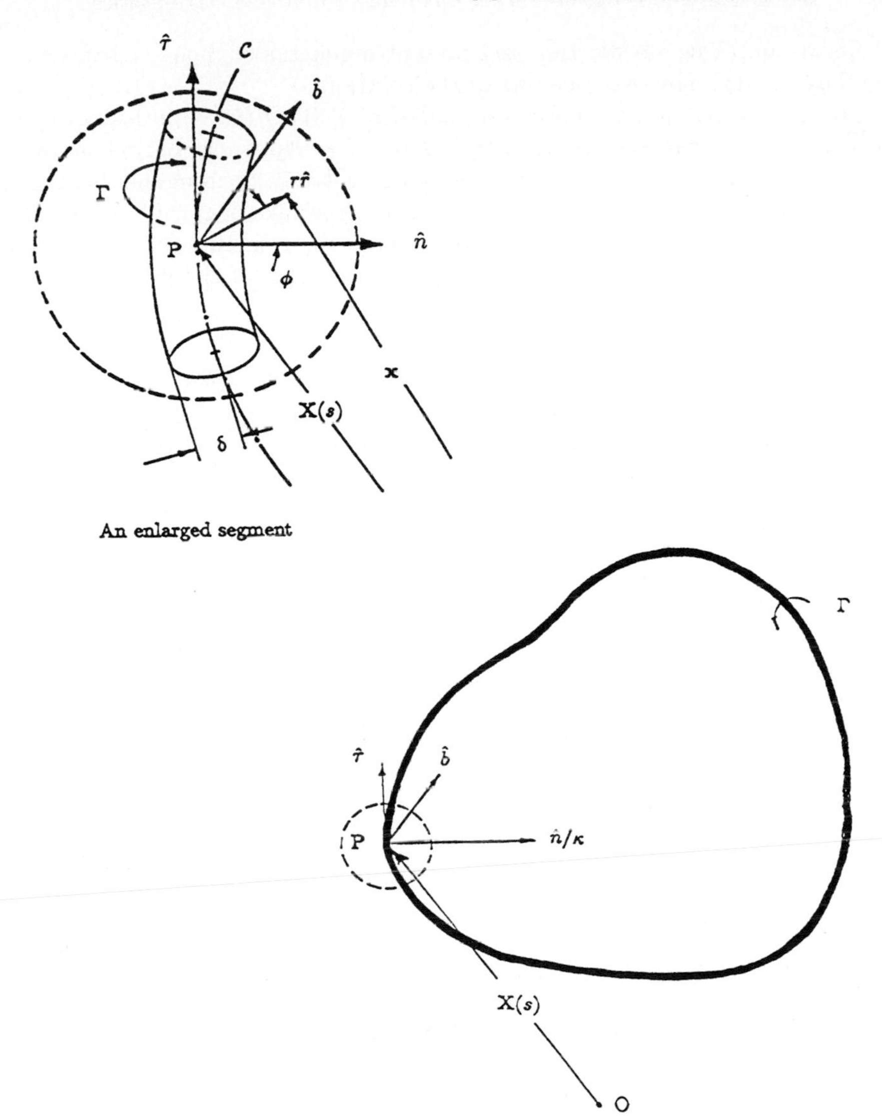

Fig. 3.6. A slender vortex filament and an enlarged segment in the neighborhood of a point $\boldsymbol{P}$ on the centerline $\mathcal{C}$

for the axial flow and vice versa. Therefore, they consider the restricted but still very complex case where the leading-order core structure has no axial variation. Here we begin our analysis with the general case that allows for axial core structure variation and derive constraints on the axial variations of the leading-order axial and circumferential velocity profiles. We then adopt

Callegari and Ting's restriction and present highlights of their analysis with emphasis on the physical meaning of the solution.

To analyze the flow field near the centerline $\mathcal{C}$, Ting (1971) introduced orthogonal curvilinear coordinates attached to $\mathcal{C}$ (see Appendix E.4). The same coordinates were employed in Subsect. 3.1.2 to study the singular behavior of the inviscid flow field near a vortex line. For the sake of clarity, we restate the definition of the coordinates. Let r denote the minimum distance from a point $\boldsymbol{x}$ to $\mathcal{C}$. Let $\boldsymbol{X}(t,s)$ denote the position vector of the point $\boldsymbol{P}$ on $\mathcal{C}$. Then we have

$$\boldsymbol{x} = \boldsymbol{X}(t,s) + r\boldsymbol{r}\ , \qquad \text{with} \qquad \boldsymbol{r} = \boldsymbol{n}\cos\phi + \boldsymbol{b}\sin\phi\ , \tag{3.3.12}$$

where $\boldsymbol{r}$ is the unit radial vector in the normal plane and ϕ is the angle between $\boldsymbol{r}$ and $\boldsymbol{n}$ (see Fig. 3.3). The unit tangent, normal and binormal vectors $\boldsymbol{\tau}$, $\boldsymbol{n}$, and $\boldsymbol{b}$, are functions of t and s and are related to $\boldsymbol{X}$ by the Serret–Frenet formulae (3.1.58). Equation (3.3.12) defines the transformation between the Cartesian coordinates, x_i $(i = 1,2,3)$, and the curvilinear coordinates, r, ϕ, and s. The latter are not orthogonal if the torsion of $\mathcal{C}$ is nonzero, that is, if $\mathcal{C}$ is not a planar curve (cf. the discussion after (3.1.63)). To facilitate the transformation of the Navier–Stokes equations, we introduce an orthogonal coordinate system by replacing the angle variable ϕ by

$$\theta = \phi - \theta_0(t,s) \tag{3.3.13}$$

with

$$\frac{\partial\theta_0}{\partial s} = -\sigma T\ , \tag{3.3.14}$$

where $\sigma = |\boldsymbol{X}_s|$ and $T = -\boldsymbol{b}_s \cdot \boldsymbol{n}\ /\sigma$ denote the linear strain and torsion of $\mathcal{C}$, respectively. Then r, θ, and s represent radial, circumferential, and tangential orthogonal coordinates while $\boldsymbol{r}$, $\boldsymbol{\theta}$, and $\boldsymbol{\tau}$ are the corresponding unit base vectors. The stretch ratios are given by

$$h_1 = 1\ , \qquad h_2 = r\ , \qquad h_3 = \sigma[1 - \kappa r\cos(\theta + \theta_0)]\ , \tag{3.3.15}$$

where $\kappa(t,s)$ is the curvature of $\mathcal{C}$. Note that conditions (3.1) and (3.2) on the core size and curvature imply that for a point near the filament in the inner region, we have $r/\ell = O(\delta/\ell) \ll 1$ or $\kappa r \ll 1$ and hence $h_3 \approx \sigma > 0$. Thus, the transformation (3.3.12) is one to one in the inner region local to $\boldsymbol{X}(t,s)$.[5]

[5] There is another condition that is usually implied in the statement that $\mathcal{C}$ is a simple curve defined by the parameter s. We shall state this condition in terms of the length scales ℓ and ϵ. We require that the distance between two points $\boldsymbol{P}$ and $\boldsymbol{P}'$ on $\mathcal{C}$ associated with different values of s has to be much larger than δ^*; that is, $|\boldsymbol{X}(t,s') - \boldsymbol{P}(t,s)| \gg \delta^*$ if $0 < s < s' \leq S_0$. This condition excludes the situation that the centerline comes close to itself at two different values of s. This situation corresponds to a local self-merging that will be considered in Chap. 6.

The velocity $\boldsymbol{v}$ in an inertial coordinate system and the associated relative velocity $\boldsymbol{V}$ in the moving frame are related by

$$\boldsymbol{v}(t, \boldsymbol{x}) = \dot{\boldsymbol{X}}(t, s) + \boldsymbol{V}(t, r, \theta, s) . \tag{3.3.16}$$

We denote the radial, circumferential, and axial components of the relative velocity $\boldsymbol{V}$ by u, v, and w, that is,

$$\boldsymbol{V} = u\boldsymbol{r} + v\boldsymbol{\theta} + w\boldsymbol{\tau} . \tag{3.3.17}$$

The continuity equation and the Navier–Stokes equations for the relative velocity $\boldsymbol{V}$ and pressure p in the curvilinear coordinates are

$$r(w_s + \dot{\boldsymbol{X}}_s \cdot \boldsymbol{\tau}) + (ruh_3)_r + (h_3 v)_\theta = 0 , \tag{3.3.18}$$

$$\ddot{\boldsymbol{X}} + \frac{1}{h_3}(w - r\boldsymbol{r}_t \cdot \boldsymbol{\tau})\dot{\boldsymbol{X}}_s + \frac{d\boldsymbol{V}}{dt} = -\nabla p + \frac{\nu}{h_3}(\frac{1}{h_3}\dot{\boldsymbol{X}}_s)_s + \nu\triangle\boldsymbol{V} . \tag{3.3.19}$$

Note that the velocity of $\mathcal{C}$, $\dot{\boldsymbol{X}}(t, s)$, which appears in these equations, is by (3.3.6) and (3.3.7) which are equivalent to

$$\dot{\boldsymbol{X}} \cdot \boldsymbol{n} = \boldsymbol{v} \cdot \boldsymbol{n} , \qquad \dot{\boldsymbol{X}} \cdot \boldsymbol{b} = \boldsymbol{v} \cdot \boldsymbol{b} \qquad \text{and} \qquad \dot{\boldsymbol{X}} \cdot \boldsymbol{\tau} = 0 , \tag{3.3.20}$$

where $\boldsymbol{v}$ stands for $\boldsymbol{v}(t, \boldsymbol{X})$. The first two equations in (3.3.20) imply

$$u = 0 \qquad \text{and} \qquad v = 0 \qquad \text{at } r = 0 . \tag{3.3.21}$$

Thus, (3.3.18)–(3.3.21) form a closed system of differential equations for $\boldsymbol{V}$, p, and $\boldsymbol{X}$. In addition to those equations, we have to impose the matching conditions with the outer potential solution and to prescribe the initial values of $\boldsymbol{V}$, $\boldsymbol{X}$, and $\dot{\boldsymbol{X}}$. Similar to the two-dimensional case treated in Sect. 3.2 and the simple examples in Sect. 3.1, we shall see next that in a one-time expansion for the three-dimensional problem we obtain a degenerate set of leading-order equations, so that we need to impose appropriate limitations on the initial data.

3.3.1 The Expansion Scheme

To study the solution of (3.3.18)–(3.3.21) in and near the vortical core of a vortex filament, we use the method of matched asymptotic expansions. We introduce the stretched variable,

$$\bar{r} = r/\epsilon . \tag{3.3.22}$$

An examination of the expansion of the Biot–Savart integral given in (3.1.77) with r replaced by $\epsilon\bar{r}$ indicates that the inner solution can be expanded in a power series in both ϵ and $\ln\epsilon$. To simplify the matching of the inner and outer solutions, we shall expand the inner solution in powers of ϵ, recognizing that the coefficients in these expansions can be power series in $\ln(\epsilon)$. After matching like powers of ϵ, we will then determine the dependence on $\ln\epsilon$. As we shall see later, logarithmic terms will appear in the expansion for $\dot{\boldsymbol{X}}(t,s)$.

Hence we assume that in the inner region, the relative velocity components have expansions in terms of powers of ϵ in the form

$$u(t,\bar{r},\theta,s,\epsilon) = u^{(1)}(t,\bar{r},\theta,s) + \epsilon u^{(2)} + \cdots , \tag{3.3.23}$$

$$v(t,\bar{r},\theta,s,\epsilon) = \epsilon^{-1}v^{(0)}(t,\bar{r},\theta,s) + v^{(1)}(t,\bar{r},\theta,s) + \epsilon v^{(2)} + \cdots , \tag{3.3.24}$$

$$w(t,\bar{r},\theta,s,\epsilon) = \epsilon^{-1}w^{(0)}(t,\bar{r},\theta,s) + w^{(1)}(t,\bar{r},\theta,s) + \epsilon w^{(2)} + \cdots . \tag{3.3.25}$$

We call $u^{(n)}, v^{(n)}$, etc., the nth-order solutions and those with $n = 0$ the leading-order solutions with the understanding that their dependence on $\ln\epsilon$ has been suppressed. In the above expansions, we allow for both large circumferential and axial components of order ϵ^{-1}. On account of the ansatz (3.3.11), the radial component remains of order one.

The radial component of the momentum equation (3.3.19) and the condition that the pressure is $O(1)$ in the outer region require that we choose

$$p(t,\bar{r},\theta,s) = \epsilon^{-2}p^{(0)}(t,\bar{r},\theta,s) + \epsilon^{-1}p^{(1)}(t,\bar{r},\theta,s) + \cdots , \tag{3.3.26}$$

in order to find the nontrivial circumferential velocity $v^{(0)}$. From the definition of vorticity, $\boldsymbol{\Omega} = \nabla \times \boldsymbol{v} = \nabla \times \boldsymbol{V}$, we see that the expansion for the vorticity should begin with an order ϵ^{-2} term:

$$\boldsymbol{\Omega}(t,\bar{r},\theta,s) = \epsilon^{-2}\boldsymbol{\Omega}^{(0)}(t,\bar{r},\theta,s) + \epsilon^{-1}\boldsymbol{\Omega}^{(1)}(t,\bar{r},\theta,s) + \boldsymbol{\Omega}^{(2)} + \cdots . \tag{3.3.27}$$

The centerline $\mathcal{C}$ of the filament also depends on the parameter ϵ. We assume that it has an expansion in the form

$$\boldsymbol{X}(t,s,\epsilon) = \boldsymbol{X}^{(0)}(t,s) + \epsilon\boldsymbol{X}^{(1)}(t,s) + \cdots , \tag{3.3.28}$$

which begins with an order one term so that the velocity of $\mathcal{C}$ remains of order one (3.3.8). The geometric parameters σ, κ and h_3, which are given in terms of the s-derivatives of $\boldsymbol{X}(t,s,\epsilon)$ by (3.1.58), become

$$h_3 = h_3^{(0)} + \epsilon h_3^{(1)} + \cdots = \sigma^{(0)} + \epsilon[\sigma^{(1)} - \sigma^{(0)}\kappa^{(0)}\bar{r}\cos\phi^{(0)}] + \cdots \tag{3.3.29}$$

with $\sigma^{(0)} = |\boldsymbol{X}_s^{(0)}|$, $\phi^{(0)} = \theta + \theta_0^{(0)}$, and $\theta_0^{(0)} = -\int \sigma^{(0)}T^{(0)}ds$.

By substituting expansions (3.3.23) and (3.3.24) into the linear homogeneous boundary conditions (3.3.21) at $\bar{r} = 0$ and equating the coefficients of like powers of ϵ, we obtain the boundary conditions for the radial and circumferential velocities,

$$v^{(n)} = 0\,, \qquad u^{(n+1)} = 0\,, \qquad n = 0,1,2,\ldots\,, \quad \text{at } \bar{r} = 0\,. \tag{3.3.30}$$

In addition, the velocity of the inner flow has to match to that of the outer inviscid flow, which is composed of the velocity $\boldsymbol{Q}$ induced by the filament itself and a background potential flow $\boldsymbol{Q}_2$,

$$\begin{aligned} \boldsymbol{v} &= \dot{\boldsymbol{X}} + \boldsymbol{V} = \boldsymbol{Q} + \boldsymbol{Q}_2 \\ &= \frac{1}{\epsilon}\frac{\Gamma}{2\pi\bar{r}}\boldsymbol{\theta} + \frac{\Gamma\kappa}{4\pi}\ln\frac{1}{\epsilon\bar{r}}\boldsymbol{b} + \frac{\Gamma\kappa}{4\pi}\cos\phi\,\boldsymbol{\theta} + \boldsymbol{Q}_f + \boldsymbol{Q}_2 + \cdots \ , \end{aligned} \tag{3.3.31}$$

as $\bar{r} \to \infty$. The first three terms on the right-hand side of (3.3.31) denote the singular parts and $\boldsymbol{Q}_f$ denotes the finite part of the Biot–Savart integral (see (3.1.62) and (3.1.77)). Note that once the leading-order outer flow is assumed to be irrotational, the outer flow is irrotational to all orders of ϵ. Thus the matching conditions on the vorticity field yield

$$\boldsymbol{\Omega}^{(n)} \to 0 \ , \quad \text{as } \bar{r} \to \infty \ , \quad \text{for all } n \ . \tag{3.3.32}$$

Since a term in the inner solution proportional to $\bar{r}^{-m}$ has to match with an $\epsilon^m r^{-m}$ term in the outer solution, the preceding matching conditions can therefore be replaced by the stronger ones,

$$\boldsymbol{\Omega}^{(n)} \to o(\bar{r}^{-m}) \ , \quad \text{as } \bar{r} \to \infty \ , \quad \text{for all } m \text{ and } n \ . \tag{3.3.33}$$

The matching conditions for the leading two orders of pressure are

$$p^{(0)} \to 0 \ , \quad p^{(1)} \to 0 \ , \qquad \text{as } \bar{r} \to \infty \ . \tag{3.3.34}$$

Thus we complete the formulation of the expansion scheme and the boundary and matching conditions. Below we derive the leading-order and higher order equations systematically by substituting these expansions into the continuity and momentum equations, (3.3.18) and (3.3.19), and equating the coefficients of like powers of ϵ.

3.3.2 The Leading-Order Equations

The leading-order terms in the continuity and momentum equations are of the order ϵ^{-1} and ϵ^{-3}, respectively. Because of the ansatz (3.3.11) or the expansion (3.3.23), which require the radial velocity u to remain of order one, the coefficient of ϵ^{-1} in the continuity equation yields

$$v_\theta^{(0)} = 0 \qquad \text{or} \qquad v^{(0)} = v^{(0)}(t, \bar{r}, s) \ . \tag{3.3.35}$$

With this result, the coefficients of ϵ^{-3} terms in the tangential, circumferential, and radial components of the momentum equation yield, respectively,

$$\begin{aligned} w_\theta^{(0)} v^{(0)} &= 0 \qquad \text{or} \qquad w^{(0)} = w^{(0)}(t, \bar{r}, s) \ , \\ p_\theta^{(0)} &= 0 \qquad \text{or} \qquad p^{(0)} = p^{(0)}(t, \bar{r}, s) \ , \\ \frac{[v^{(0)}]^2}{\bar{r}} &= p_{\bar{r}}^{(0)} \qquad \text{or} \qquad p^{(0)} = -\int_{\bar{r}}^{\infty} \frac{[v^{(0)}(t, \bar{r}', s)]^2}{\bar{r}'} d\bar{r}' \end{aligned} \tag{3.3.36}$$

and the leading vorticity components are

$$(\boldsymbol{\Omega} \cdot \boldsymbol{r})^{(0)} = 0 \,, \tag{3.3.37}$$

$$(\boldsymbol{\Omega} \cdot \boldsymbol{\theta})^{(0)} = -w^{(0)}_{\bar{r}} \,, \tag{3.3.38}$$

$$\zeta^{(0)}(t, \bar{r}, s) \equiv (\boldsymbol{\Omega} \cdot \boldsymbol{\tau})^{(0)} = \frac{1}{\bar{r}}[\bar{r}v^{(0)}]_{\bar{r}} \,. \tag{3.3.39}$$

Here ζ denotes the axial component of $\boldsymbol{\Omega}$. It will be easier to solve for $\zeta^{(0)}$ and $w^{(0)}$ instead of $v^{(0)}$ and $w^{(0)}$.

The radial component of the momentum equation $(3.3.36)_3$ expresses the balance of the pressure gradient and the centrifugal force due to the large circumferential velocity. In arriving at the equation for $p^{(0)}$, we used the matching condition (3.3.34). Besides the boundary condition $v^{(0)} = 0$ at $\bar{r} = 0$ from (3.3.30), the matching of the inner velocity with the outer velocity according to (3.3.31) yields

$$v^{(0)} = \frac{\Gamma}{2\pi\bar{r}} + o(\bar{r}^{-m}) \,, \quad w^{(0)} = o(\bar{r}^{-m}) \,, \quad \text{for all } m \text{ as } \bar{r} \to \infty \,. \tag{3.3.40}$$

Here we made use of the relations (3.3.38) and (3.3.39) between the leading-order vorticity and velocity and of the matching condition (3.3.33).

The above equations say that the leading-order flow field is axisymmetric, or symmetric, in the normal plane of $\mathcal{C}$. Equations (3.3.23), (3.3.35), and $(3.3.36)_1$ imply that the streamlines are nearly circular helices around $\mathcal{C}$. According to (3.3.37)–(3.3.39), the trajectories of the vorticity field are also nearly circular helices, but in general they differ from the streamlines. If there is no large axial flow, the streamlines become nearly circular helices with small pitch while the vortex lines are nearly parallel to $\mathcal{C}$. This qualitative description is consistent with an intuitive understanding of the flow field in the vortical core of a filament.

Up to now, we have shown that the leading-order inner solution is independent of θ. Its dependence on t, $\bar{r}$, and s as well as that of the centerline $\boldsymbol{X}^{(0)}$ on t and s are yet unknown. Similar to the two-dimensional analysis in Sect. 3.2, these dependencies will be derived from solvability or compatibility conditions of higher-order equations.

We recall the general procedure for the two-dimensional analysis presented in Sect. 3.2 and Appendices A and B. Since the higher-order equations are linear, equations for the symmetric part of the nth-order solution are separated from those for the asymmetric part by averaging the nth-order equations with respect to θ over $[0, 2\pi]$. The θ-averages of the $(n+1)$th-order equations, known also as the compatibility conditions, become the governing equations for the symmetric part of the $(n-1)$th-order solution. Given the $(n-1)$th-order symmetric solution, the asymmetric part of the nth-order solution is defined by the nth-order equations.

A similar pattern will be observed in a three-dimensional problem. However, additional compatibility conditions on the $(n+1)$th-order equations will appear leading to constraints on the dependence of the nth-order solutions on s. With these general patterns in mind, we now proceed to the next-order analysis.

3.3.3 The First Order Equations and the Asymmetric Solutions

The first order continuity and the three components of the momentum equation are

$$\frac{1}{\bar{r}}[v_\theta^{(1)} + (\bar{r}u^{(1)})_{\bar{r}}] = -\frac{1}{\sigma^{(0)}}w_s^{(0)} - (v\,\kappa\sin\phi)^{(0)}, \tag{3.3.41}$$

$$u^{(1)}w_{\bar{r}}^{(0)} + \frac{v^{(0)}}{\bar{r}}w_\theta^{(1)} = -\frac{1}{\sigma^{(0)}}(p_s^{(0)} + w^{(0)}w_s^{(0)}) - (wv\,\kappa\sin\phi)^{(0)}\ , \tag{3.3.42}$$

$$u^{(1)}v_{\bar{r}}^{(0)} + \frac{v^{(0)}}{\bar{r}}v_\theta^{(1)} + \frac{v^{(0)}u^{(1)}}{\bar{r}} + \frac{1}{\bar{r}}p_\theta^{(1)} = -\frac{w^{(0)}}{\sigma^{(0)}}v_s^{(0)} + (w^2\,\kappa\sin\phi)^{(0)} \tag{3.3.43}$$

and

$$v^{(0)}u_\theta^{(1)} - 2v^{(0)}v^{(1)} + \bar{r}p_{\bar{r}}^{(1)} = -(w^2\kappa\bar{r}\cos\phi)^{(0)}\ . \tag{3.3.44}$$

Note that the homogeneous equations of (3.3.41), (3.3.43), and (3.3.44) are identical to the equations for $u^{(1)}$, $v^{(1)}$, and $p^{(1)}$ in the two-dimensional case. Here (3.3.42) is the additional equation needed for $w^{(1)}$. Also (3.3.41) to (3.3.44) are now inhomogeneous equations because of the axial flow, the s-derivatives, and the curvature of $\mathcal{C}$. These three-dimensional effects are absent in the two-dimensional case. For the latter, the first order equations are homogeneous and solvable. Now (3.3.41) to (3.3.44) are solvable only if the compatibility conditions obtained by θ-averaging are satisfied.

3.3.3.1 The Compatibility Conditions. Using (3.2.48), we separate the asymmetric part $f_a(\theta)$ of $f(\theta)$ from its symmetric part f_c. The θ-average of (3.3.41) reads

$$\sigma^{(0)}(\bar{r}u_c^{(1)})_{\bar{r}} + \bar{r}w_s^{(0)} = 0\ . \tag{3.3.45}$$

The averages of (3.3.42) and (3.3.43) yield

$$\sigma^{(0)}w_{\bar{r}}^{(0)}u_c^{(1)} + w^{(0)}w_s^{(0)} = 2\int_{\bar{r}}^{\infty}\frac{v^{(0)}v_s^{(0)}}{\bar{r}'}\,d\bar{r}' \tag{3.3.46}$$

and

$$\sigma^{(0)}(\bar{r}v^{(0)})_{\bar{r}}u_c^{(1)} + \bar{r}w^{(0)}v_s^{(0)} = 0\ . \tag{3.3.47}$$

In arriving at (3.3.46), we used $(3.3.36)_3$, relating $p^{(0)}$ to $v^{(0)}$. The three equations above, which are the compatibility conditions of the first-order equations, define the dependence of the leading-order core structure, $v^{(0)}$ and $w^{(0)}$, and the symmetric part of the radial flow, $u_c^{(1)}$, on $\bar{r}$ and s while t appears as a parameter.

To simplify these equations and show their physical meaning, we replace the continuity equation (3.3.45) by introducing the stream function $\psi(t, \bar{r}, s)$,

$$\psi = \int_0^{\bar{r}} \bar{r}' w^{(0)}(t, \bar{r}', s)\, d\bar{r}', \tag{3.3.48}$$

and

$$\sigma^{(0)} \bar{r} u_c^{(1)} = -\psi_s\,. \tag{3.3.49}$$

Equations (3.3.46) and (3.3.47) then become two quasi-steady equations for ψ and $\bar{v}^{(0)}$. They are

$$\psi_{\bar{r}}\psi_{s\bar{r}} - \psi_s\psi_{\bar{r}\bar{r}} + \frac{1}{\bar{r}}\psi_s\psi_{\bar{r}} = 2\bar{r}^2 \int_{\bar{r}}^{\infty} \frac{v^{(0)}v_s^{(0)}}{\bar{r}'} d\bar{r}' \tag{3.3.50}$$

and

$$(\bar{r}v^{(0)})_{\bar{r}}\psi_s - (\bar{r}v^{(0)})_s\psi_{\bar{r}} = 0\,. \tag{3.3.51}$$

With ψ related to $w^{(0)}$ by (3.3.48), (3.3.50) and (3.3.51) become the governing equations for the leading-order core structure, $v^{(0)}$ and $w^{(0)}$. The boundary conditions for the quasi-steady equations come from (3.3.30) together with $v_{\bar{r}}^{(0)} \neq 0$ at $\bar{r} = 0$, and the matching condition (3.3.40). They are

$$\bar{r}v^{(0)} \sim \bar{r}^2\,, \qquad \psi_s \sim \bar{r}^2\,, \qquad \text{as } \bar{r} \to 0 \tag{3.3.52}$$

and

$$\bar{r}v^{(0)} = \frac{\Gamma}{2\pi} + o(\bar{r}^{-n})\,, \qquad \psi_{\bar{r}} = o(\bar{r}^{-n})\,, \qquad \text{for all } n \text{ as } \quad \bar{r} \to \infty\,. \tag{3.3.53}$$

Because of assumption (3.2.4) of large Reynolds number, $Re = O(\epsilon^{-2})$, which we also adopt here, the viscous terms will begin to appear only in the second-order equations. The equations we obtained in the first order analysis are valid not only for viscous flows but also for inviscid flows. Thus we conclude the following:

- *The leading-order core structure, that is, the circumferential and axial velocities, $\epsilon^{-1}v^{(0)}$ and $\epsilon^{-1}w^{(0)}$, have to fulfill the system of equations. (3.3.50)–(3.3.53). These constraints on the core structure should be observed for viscous as well as inviscid solutions.*
- *In particular, (3.3.51) implies a functional relationship between $\bar{r}v^{(0)}$ and ψ,*

$$\bar{r}v^{(0)}(t, \bar{r}, s) = \mathcal{G}(t, \psi)\,. \tag{3.3.54}$$

That is , $[\bar{r}v^{(0)}]$ depends implicitly on $\bar{r}$ and s through ψ.

Since $2\pi\psi$ represents the axial mass flux through a circular disc with radius $\bar{r}$, centered on $\mathcal{C}$ and lying in its normal plane, and $2\pi\bar{r}v^{(0)}$ is the circulation around the boundary of the disc, (3.3.54) relates the circulation to the axial mass flux.

Combining (3.3.50) and (3.3.54), we obtain an equation for $\psi(t,\bar{r},s)$,

$$\psi_{\bar{r}}\psi_{s\bar{r}} - \psi_s\psi_{\bar{r}\bar{r}} + \frac{1}{\bar{r}}\psi_s\psi_{\bar{r}} = 2\bar{r}^2\int_{\bar{r}}^{\infty}\frac{\mathcal{G}\mathcal{G}_{\psi}\ \psi_s}{(\bar{r}')^3}\,d\bar{r}' \ , \tag{3.3.55}$$

and its boundary conditions are (3.3.52) and (3.3.53) with $\bar{r}v^{(0)}$ replaced by $\mathcal{G}$.

Since t appears only as a parameter in the system (3.3.52)–(3.3.55), a two-time solution in the normal time t and a short time τ is needed in case the initial data are incompatible with the system. Following the standard two-time analysis, the dependence of the solution on t will be defined by the compatibility conditions on the second-order equations. An in-depth study of the quasi-steady system of equations for the one-time solution and the derivation of its evolution equations are not yet available. We note that the system of equations becomes trivial when the leading-order core structure does not vary along $\mathcal{C}$, that is, when it is independent of s. In this case, the dependence of the core structure on t and $\bar{r}$ is defined by the compatibility conditions for the second-order equations. From here on we study this degenerated case where the leading-order core structure is independent of s.

In the three-dimensional problem studied by Ting (1971), the filament was assumed to have a large swirl but an order one axial velocity, that is,

$$w^{(0)} \equiv 0 \ . \tag{3.3.56}$$

With this assumption built into the expansion scheme (3.3.25), it was concluded from (3.3.45), (3.3.46) and (3.3.47) that

$$u_c^{(1)} \equiv 0 \ , \qquad p_s^{(0)} \equiv 0 \tag{3.3.57}$$

and then

$$v_s^{(0)} \equiv 0 \ , \qquad \text{or} \qquad v^{(0)} = v^{(0)}(t,\bar{r}) \ . \tag{3.3.58}$$

Later Callegari and Ting (1978) allowed for both large circumferential and axial velocities in the core. It was observed from the matching condition (3.3.40) that the leading-order circumferential velocity for large $\bar{r}$ is independent of s. Motivated by this, the circumferential velocity $v^{(0)}$ in the inner region was *assumed* to be independent of s. Under this assumption, $v_s^{(0)} \equiv 0$, the compatibility condition (3.3.47) yields $u_c^{(1)} \equiv 0$ and then (3.3.45) yields $w_s^{(0)} \equiv 0$.

From (3.3.46), (3.3.48) and (3.3.49), we see that $w_s^{(0)} \equiv 0$ implies $v_s^{(0)} \equiv 0$. Thus we conclude

$$v_s^{(0)} = 0 \Longleftrightarrow w_s^{(0)} = 0 \ . \tag{3.3.59}$$

To gain a physical understanding of this result, we shall rederive it by a simple order of magnitude analysis in the spirit of Prandtl's original derivation of the boundary layer theory.

From the order of magnitude of the centripetal acceleration induced by the large circumferential velocity $v = O(\epsilon^{-1})$ in the radial component of the momentum equation (3.3.19), we see that the pressure gradient p_r in the inner region, where $r = O(\epsilon)$, has to be $O(v^2/r) = O(\epsilon^{-3})$. This is consistent with the scaled equation $(3.3.36)_3$. Thus, we have

$$\{r = O(\epsilon) \quad \text{and} \quad p_r = O(\epsilon^{-3})\} \qquad \Rightarrow \qquad p = O(\epsilon^{-2})\,. \tag{3.3.60}$$

A nonzero axial gradient of $v^{(0)}$ on the length scale ℓ then induces an axial pressure gradient, $p_s = O(\epsilon^{-2})$, which in turn will force an axial acceleration ww_s. Therefore, both w and w_s have to be $O(\epsilon^{-1})$. Thus, we have

$$\begin{aligned}
\text{If} \quad & v_s \neq 0, \;\; v_s = O(v) = O(\epsilon^{-1})\,, \\
\Rightarrow \quad & p_s = O(p) = O(\epsilon^{-2}) \\
\Rightarrow \quad & w\, w_s = O(\epsilon^{-2}) \\
\text{then} \quad & w_s = O(w) = O(\epsilon^{-1})\,.
\end{aligned} \tag{3.3.61}$$

Following the same order of magnitude analysis, we find that

$$\begin{aligned}
\text{If} \quad & v_s = O(1), \;\; \text{while } v = O(\epsilon^{-1}) \\
\Rightarrow \quad & p_s = O(\epsilon^{-1}) \\
\Rightarrow \quad & w\, w_s = O(\epsilon^{-1}) \\
\text{then} \quad & w_s = O(1) \\
\text{even if} \quad & w = O(\epsilon^{-1})\,.
\end{aligned} \tag{3.3.62}$$

It says that w_s is $O(1)$ if v_s is $O(1)$. The converse is also true. Thus we come to the same conclusion as in (3.3.59), which says

- *The leading-order circumferential velocity profile in a cross-sectional plane of the filament remains the same along the filament if the axial velocity profile remains the same, and vice versa.*

The compatibility conditions presented above came from the solvability conditions of the first order continuity and momentum equations, (3.3.41)–(3.3.44), in which the viscous diffusion terms are still absent. The special condition (3.3.59), described by the preceding statement, was rederived by a simple order of magnitude analysis. Thus, the above compatibilty conditions have to be observed regardless whether the flow is viscous or inviscid.

Prior to the matched asympotic analysis of Callegari and Ting (1978), the corresponding three-dimensional inviscid problem was treated by Moore and Saffman (1972) by the method of patching. The core structure was assumed to be composed of a large swirling flow with no axial variation, and a large axial flow with axial variation. The patched inviscid core structure that violates the consistency condition (3.3.59) is not admissible. This point was first noted by Callegari and Ting (1978).

The regime described by (3.3.62) becomes interesting for very long filaments of length, $S = O(1/\epsilon)$. Slow variations of the core structure can accumulate along the filament to considerable changes at leading order. See Book-I, Sect. 4.1, for a discussion of this issue.

In the remainder of this section we shall follow Callegari and Ting (1978) in assuming that the leading-order circumferential velocity, $\epsilon^{(-1)} v^{(0)}$, is independent of s. It is independent of θ on account of (3.3.35). From (3.3.36)–(3.3.39) and (3.3.59) we find that $w^{(0)}$ is also independent of θ and s,

$$v^{(0)} = v^{(0)}(t, \bar{r}) \qquad \text{and} \qquad w^{(0)} = w^{(0)}(t, \bar{r}) \,. \tag{3.3.63}$$

It is understood that the one-time solution, which is quasi-steady, is applicable only when the initial data happen to be compatible with the solution. We assume that the initial core structure is compatible with (3.3.63) and has a rapidly decaying vorticity field, (2.1.18). We proceed to study the asymmetric solution of the first-order equations, (3.3.41)–(3.3.44).

3.3.3.2 The Asymmetric Solution and the Velocity of the Filament. When the leading-order core structure is expressed in the form (3.3.63), we have $p_s^{(0)} = 0$ from $(3.3.36)_3$ and $u_c^{(1)} = 0$ from (3.3.47). We replace the first-order continuity equation (3.3.41) by introducing a stream function $\psi^{(1)}(t, \bar{r}, \theta, s)$ with

$$u^{(1)} = \frac{1}{\bar{r}} \psi_\theta^{(1)} \qquad \text{and} \qquad v^{(1)} = -\psi_{\bar{r}}^{(1)} + \bar{r}\,(v\,\kappa \cos\phi)^{(0)} \,. \tag{3.3.64}$$

The axial component, (3.3.42), of the momentum equation becomes an equation for $w_\theta^{(1)}$ uncoupled from the other two equations (3.3.43) and (3.3.44). We eliminate $p^{(1)}$ from the latter two by cross differentiation and obtain an equation for $\psi_\theta^{(1)}$,

$$v^{(0)} \bar{\Delta} \psi_\theta^{(1)} - \zeta_{\bar{r}}^{(0)} \psi_\theta^{(1)} = -\kappa^{(0)} \sin\phi^{(0)} \left[(2\bar{r}\zeta^{(0)} + v^{(0)}) v^{(0)} + 2\bar{r} w^{(0)} w_{\bar{r}}^{(0)} \right] . \tag{3.3.65}$$

Here

$$\zeta^{(0)}(t, \bar{r}) = \frac{1}{\bar{r}} (\bar{r} v^{(0)})_{\bar{r}} \tag{3.3.66}$$

denotes the leading axial vorticity and $\bar{\Delta}$ the Laplacian in $\bar{r}$ and θ. Equation (3.3.65) is a linear equation for the asymmetric part of $\psi^{(1)}$ because

$$\psi_\theta^{(1)} = \left(\psi_a^{(1)}\right)_\theta . \tag{3.3.67}$$

We expand $\psi_a^{(1)}$ in a Fourier series

$$\psi_a^{(1)} = \sum_{n=1}^{\infty} (\tilde{\psi}_{n1} \cos n\phi^{(0)} + \tilde{\psi}_{n2} \sin n\phi^{(0)}) \tag{3.3.68}$$

and reduce (3.3.65) to a set of ordinary differential equations in $\bar{r}$ for the Fourier coefficients $\tilde{\psi}_{nj}(t, \bar{r}, s)$, where t and s appear as parameters. These equations are

$$\left[\frac{\partial^2}{\partial \bar{r}^2} + \frac{1}{\bar{r}}\frac{\partial}{\partial \bar{r}} - \left(\frac{n^2}{\bar{r}^2} + \frac{\zeta_{\bar{r}}^{(0)}}{v^{(0)}}\right)\right] \tilde{\psi}_{nj} = \kappa^{(0)}(t,s) H(t,\bar{r}) \delta_{n1}\delta_{j1} , \tag{3.3.69}$$

for $j = 1, 2$, and $n = 1, 2, \ldots$. Here δ_{nj} is the Kronecker delta and

$$H(t,\bar{r}) = 2\bar{r}\zeta^{(0)} + v^{(0)} + 2\bar{r}w^{(0)} w_{\bar{r}}^{(0)}/v^{(0)} . \tag{3.3.70}$$

The inhomogeneous term in (3.3.69), which is present only for $n = j = 1$, represents a three-dimensional effect due to the curvature of the filament. Because of this term, the leading-order velocity of the filament will depend on the core structure. This is not the case for a two-dimensional problem for which the inhomogeneous term is absent because $\kappa \equiv 0$.

The condition of zero relative circumferential and axial velocities at $\bar{r} = 0$ requires $\psi_a^{(1)} = 0$ and $(\psi_a^{(1)})_{\bar{r}} = 0$, which in turn yield

$$\tilde{\psi}_{nj} = 0 \quad \text{and} \quad \frac{\partial \tilde{\psi}_{nj}}{\partial \bar{r}} = 0 \quad \text{at} \quad \bar{r} = 0 , \tag{3.3.71}$$

for $j = 1, 2$, and $n = 1, 2, \ldots$. The matching condition on the velocity field, (3.3.31), yields the following far-field conditions on the Fourier coefficients of the stream function as $\bar{r} \to \infty$,

$$\begin{aligned} &\tilde{\psi}_{11} \to -\frac{\Gamma \kappa^{(0)}}{4\pi} \bar{r} \ln \frac{1}{\epsilon \bar{r}} - \bar{r}\,(\boldsymbol{Q}_0 - \dot{\boldsymbol{X}}^{(0)}) \cdot \boldsymbol{b}^{(0)} , \\ &\tilde{\psi}_{12} \to \bar{r}\,(\boldsymbol{Q}_0 - \dot{\boldsymbol{X}}^{(0)}) \cdot \boldsymbol{n}^{(0)} , \\ &\tilde{\psi}_{nj} \to 0 \qquad \text{for } j = 1, 2, \text{ and } n = 2, 3, \ldots , \end{aligned} \tag{3.3.72}$$

where

$$\boldsymbol{Q}_0 = \lim_{r \to 0} (\boldsymbol{Q}_f + \boldsymbol{Q}_2) . \tag{3.3.73}$$

From the matching condition (3.3.40) on $v^{(0)}$ and the definition of $\zeta^{(0)}$ in (3.3.66), we see that $\zeta^{(0)}$ decays rapidly and

$$H(t,\bar{r}) \to \frac{\Gamma}{2\pi \bar{r}} \qquad \text{as } \bar{r} \to \infty . \tag{3.3.74}$$

Although the inhomogeneous term and the coefficient $\zeta_{\bar{r}}^{(0)}/v^{(0)}$ in (3.3.69) depend on the leading inner velocity field, which is not yet determined, we do know their behavior for small and large $\bar{r}$ from (3.3.30), (3.3.40), and (3.3.66). Using the behavior and repeating the arguments used in Subsect. 3.2.3 for the two-dimensional case, (3.2.70)–(3.2.79), we conclude that

$$\tilde{\psi}_{nj} \equiv 0 \qquad \text{for } j = 1, 2 \text{ and } n = 2, 3, \ldots \tag{3.3.75}$$

because their differential equations, boundary conditions at $\bar{r} = 0$, and far-field conditions are homogeneous. For $n = 1$ and $j = 2$, the homogeneous differential equation (3.3.69) and two boundary conditions (3.3.71) at $\bar{r} = 0$ ensure that

$$\tilde{\psi}_{12} \equiv 0 \,. \tag{3.3.76}$$

The matching condition for $\tilde{\psi}_{12}$ in (3.3.72) in turn determines the normal velocity of $\mathcal{C}$,

$$\dot{\boldsymbol{X}}^{(0)} \cdot \boldsymbol{n}^{(0)} = \boldsymbol{Q}_0 \cdot \boldsymbol{n}^{(0)} \,. \tag{3.3.77}$$

Now we solve (3.3.69) for $\tilde{\psi}_{11}$. Although $v^{(0)}$ and $w^{(0)}$ are not yet defined, we can express $\tilde{\psi}_{11}$ in terms of these functions by using the fact that $v^{(0)}$ is a solution of the homogeneous equation of (3.3.69) for $n = 1$. The solution of (3.3.69) for $n = j = 1$ fulfilling the boundary conditions (3.3.71) at $\bar{r} = 0$ is

$$\tilde{\psi}_{11}(t, \bar{r}, s) = \kappa^{(0)}(t, s) v^{(0)}(t, \bar{r}) \int_0^{\bar{r}} \frac{1}{z[v^{(0)}(t, z)]^2} \left[\int_0^z \xi v^{(0)}(t, \xi) H(t, \xi) d\xi \right] dz \,. \tag{3.3.78}$$

Note that both t and s are treated as parameters in this expression and that the dependence of $\tilde{\psi}_{11}$ on s comes only from the curvature $\kappa^{(0)}$.

From the behavior of $v^{(0)}$ and $H^{(0)}$ for large $\bar{r}$, we obtain the behavior of $\tilde{\psi}_{11}$, namely

$$\tilde{\psi}_{11} = \bar{r}\kappa^{(0)}(t, s) C^*(t) + \frac{\Gamma \kappa^{(0)}}{4\pi} \bar{r} \ln \bar{r} + O\left(\frac{1}{\bar{r}}\right) \qquad \text{as } \bar{r} \to \infty \,, \tag{3.3.79}$$

where

$$\begin{aligned} C^*(t) &= \frac{\Gamma}{4\pi} \lim_{\bar{r} \to \infty} \left[\left(\frac{2\pi}{\Gamma} \right)^2 \int_0^{\bar{r}} \xi [v^{(0)} H](t, \xi) d\xi - \ln \bar{r} \right] \\ &= \frac{\Gamma}{4\pi} \left[\lim_{\bar{r} \to \infty} \left(\frac{4\pi^2}{\Gamma^2} \int_0^{\bar{r}} \xi [v^{(0)}(t, \xi)]^2 d\xi - \ln \bar{r} \right) + \frac{1}{2} \right] \\ &\quad - \frac{2\pi}{\Gamma} \int_0^\infty \xi [w^{(0)}(t, \xi)]^2 d\xi \,. \end{aligned} \tag{3.3.80}$$

The matching condition (3.3.72) then yields the binormal velocity of the filament,

$$\dot{\boldsymbol{X}}^{(0)} \cdot \boldsymbol{b}^{(0)} = \frac{\Gamma \kappa^{(0)}}{4\pi} \ln \frac{1}{\epsilon} + \boldsymbol{Q}_0 \cdot \boldsymbol{b}^{(0)} + \kappa^{(0)} C^* . \tag{3.3.81}$$

With $\dot{\boldsymbol{X}} \cdot \boldsymbol{\tau} = 0$, the leading-order velocity of the filament, $\dot{\boldsymbol{X}}^{(0)}$, is specified by its normal and binormal components, (3.3.77) and (3.3.81). Since $\boldsymbol{Q}_0$ is the sum of the local background velocity and the finite part of the Biot–Savart integral defined by (3.1.77), the velocity of the filament depends on the local background velocity as well as on the geometry of the centerline $\mathcal{C}$. In addition, the binormal component depends on the leading-order core structure, $v^{(0)}$ and $w^{(0)}$, through the term $\kappa^{(0)} C^*$.

Recall that in our expansion of the inner solution in powers of ϵ, terms in $\ln \epsilon$ can appear in the coefficients of the power series. Here we see that a logarithmic term does appear in the leading-order binormal velocity. We do not consider this term to be the dominant one in (3.3.81), because in many practical problems $\ln(1/\epsilon)$ is not much larger than unity. For example, with $\epsilon = 0.05$ we have $\ln(1/\epsilon) \approx 3.0$. This observation will be confirmed later by numerical examples that show the effect of the core structure on the motion of the filament via the term $\kappa^{(0)} C^*$ in (3.3.81).

However, there is an asymptotic regime involving short wavelength distortions and rapid oscillations of the filament centerline in which the temporal evolution of the core structure, and hence that of the core coefficient C^*, may be neglected. Then, the logarithmic term dominates the filament motion, and a simplified theory for nonlinear–nonlocal self-interactions of a vortex filament can be developed. A detailed theoretical account will be given in Chap. 4.

In contrast, for filament motions on the normal time scale, the evaluation of the filament velocity via (3.3.80) and (3.3.81) requires the leading-order core structure to be known as it evolves in time. The core structure evolution equations needed to this end will be derived in Subsect. 3.3.6 below.

Before we do so, we consider next an alternative derivation of the filament equation of motion that is based solely on an analysis of the vorticity transport equation. This derivation will turn out to be useful later in Chap. 5 in the construction of accurate and efficient vortex element methods for slender filament simulations.

3.3.4 Alternative Derivation: Vorticity-Based Analysis

A self-suggesting problem in the days of computational fluid mechanics is a direct comparison between analytical and computational results. Two of the authors embarked on a comparison of the "nearly straight filament asymptotics" from Klein and Majda (1991a) with "thin-tube vortex element" simulations according to Knio and Ghoniem (1990, 1991, 1992) (see also Chap. 4 and Sect. 5.2 below).

A first comparison of filament motions based on several "quick-and-dirty" computations yielded disastrous results. There was no match between the two models whatsoever. As we will see in Sect. 5.2 below, vortex element methods evolve an approximate vorticity field by advecting it according to its self-induced velocity, which is evaluated, in turn, via the Biot–Savart integral. Therefore, to get at the roots of the mismatch, it was necessary to understand subtle differences in the description of the filament core vorticity distribution between the asymptotic and numerical models, and how these affect the filament motion. To this end, an analysis of the filament equation of motion was developed in Klein and Knio (1995), which is based solely on the asymptotics of the Biot–Savart integral and of the vorticity transport equation.

3.3.4.1 The Strategy for Determining the Filament Motion. The vorticity distribution for slender filaments is given by an axisymmetric leading-order structure, centered on a time-dependent curve $\mathcal{C}(t)$, and superimposed higher-order perturbations. Given such a vorticity distribution, the Biot–Savart integral (2.3.6) determines the local flow velocities. In this context, two central questions will be addressed in this section, namely:

1. How can we extract the equation of motion of the entire filament structure from the local velocity information provided by the Biot–Savart integral?
2. What information regarding the core vorticity distribution is needed to obtain the filament motion up to the desired leading order accuracy in ϵ, and how to calculate it?

To answer the first question, we recall that the circumferential velocity in the core of a slender filament is of order $O(1/\epsilon)$ if the core thickness is of order $O(\epsilon)$ and the total circulation of order $O(1)$. Yet, viscosity enforces a smooth vorticity distribution and we assume axisymmetry at leading order. As a consequence, there must be an inner region of thickness $O(\epsilon^2)$ within which velocities are small of order $O(1)$. An observer moving with the filament will even see a stagnation point of the relative motion in a plane normal to $\mathcal{C}$ at the center of that inner core. The local induced velocity at this center point will equal the velocity of the observer and, hence, that of the total filament except for an irrelevant tangential component. If we can identify this central point within each cross section of the core with sufficient accuracy, local velocity evaluations via the Biot–Savart integral will yield the desired filament centerline equation of motion. This observation is crucial in the context of vortex element numerical methods as these advect "vorticity blobs" based on the local induced velocity at their center.

In answer to the second question we find that the leading-order filament velocity depends on the first order axial vorticity perturbations, $\zeta^{(1)}$. These are determined below by solving the first order vorticity transport equation within the vortex core.

These two ingredients, namely the evaluation of the three-dimensional Biot–Savart integral at the correct location and the determination of the first order axial core vorticity, constitute the announced alternative derivation of the filament equation of motion. By comparing with an analogous asymptotic analysis of "thin-tube" vortex element methods, Klein and Knio (1995) were able to identify the origin of the mismatch between theory and numerics and to devise an improved vortex element scheme that gives asymptotically correct results (see Sect. 5.2).

3.3.4.2 Induced velocities in the inner core of thickness $O(\epsilon^2)$. Following Klein and Knio (1995), we reconsider the three-dimensional Biot–Savart integral from (2.3.6), which we recount here for convenience

$$\boldsymbol{v}(t,\boldsymbol{x}) = -\frac{1}{4\pi}\iiint \frac{(\boldsymbol{x}-\boldsymbol{x}')}{|\boldsymbol{x}-\boldsymbol{x}'|^3} \times \boldsymbol{\Omega}(t,\boldsymbol{x}')\, d^3\boldsymbol{x}' \,. \tag{3.3.82}$$

We evaluate the velocities induced by a filament-like vorticity distribution for points $\boldsymbol{X} = \boldsymbol{X}_1 = \boldsymbol{X}(\tilde{s}_1, t)$ lying on any curve $\mathcal{C}_1(t)$ within the inner $O(\epsilon^2)$ core of the vortex filament. The integration over $\mathbb{R}^3$ in (3.3.82) is decomposed into a local and a nonlocal contribution as

$$\boldsymbol{v}_{\text{ind}}(\boldsymbol{X}_1) = \int_{R^3\setminus B_{\delta_c}} \boldsymbol{F}(\boldsymbol{x}',\boldsymbol{X}_1)\, d^3\boldsymbol{x}' + \int_{B_{\delta_c}} \boldsymbol{F}(\boldsymbol{x}',\boldsymbol{X}_1)\, d^3\boldsymbol{x}' \,. \tag{3.3.83}$$

Here $\boldsymbol{F}(\boldsymbol{x}',\boldsymbol{x})$ is the integrand from (3.3.82) and B_{δ_c} is a Ball centered on $\boldsymbol{X}_1$ as shown in Fig. 3.7. Its radius satisfies the asymptotic constraints $1 \gg \delta_c \gg \epsilon^{\frac{1}{2}}$. Here and below, the subscript "1" indicates evaluation at $\tilde{s} = \tilde{s}_1$.

The Inner Integral. Within the Ball B_{δ_c} all geometrical variations of the filament reference line are weak and may be represented asymptotically by Taylor series expansions. We expand these terms in the integrand that involve the curve structure about $\tilde{s}_1$ as

$$\begin{aligned} \boldsymbol{P} &= \boldsymbol{x}_1 \\ \boldsymbol{x} &= \boldsymbol{x}_1 + \epsilon z \boldsymbol{t}_1 + \frac{1}{2}\epsilon^2 z^2 (k\boldsymbol{n})_1 + \epsilon \overline{r} \boldsymbol{e}_{r,1} + \epsilon^2 \overline{r} z \left(\boldsymbol{e}_{r,\tilde{s}}\right)_1 + \cdots \end{aligned} \tag{3.3.84}$$

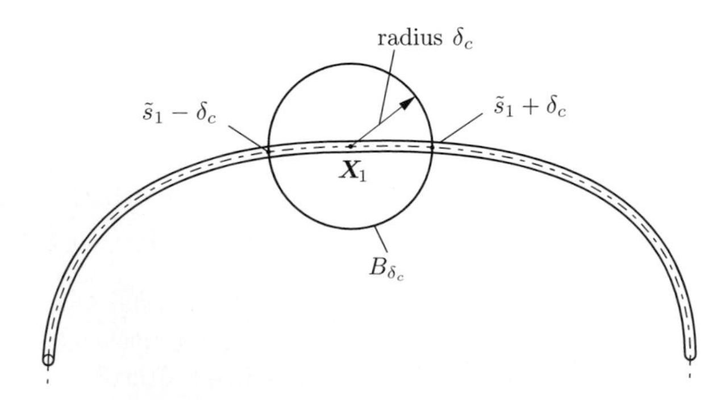

Fig. 3.7. Inner–outer decomposition of the Biot–Savart integral. Adapted from Klein (1994)

$$\boldsymbol{x} - \boldsymbol{P} = \epsilon\left(z\boldsymbol{t}_1 + \overline{r}\boldsymbol{e}_{r,1}\right) + \epsilon^2\left(\frac{z^2}{2}(k\boldsymbol{n})_1 - \overline{r}z\cos\varphi k\boldsymbol{t}_1\right) + O(\epsilon^3)$$

$$\frac{1}{|\boldsymbol{x}-\boldsymbol{P}|^3} = \frac{1}{\epsilon^3\left(\overline{r}^2+z^2\right)^{3/2}}\left(1 - \frac{3}{2}\epsilon\frac{\overline{r}z^2}{\overline{r}^2+z^2}k\cos\varphi + O\left(\epsilon^2\right)\right),$$

where

$$\overline{r} = \frac{1}{\epsilon}r \quad \text{and} \quad z = \frac{1}{\epsilon}\left(\tilde{s} - \tilde{s}_1\right), \tag{3.3.85}$$

and we have used

$$\boldsymbol{e}_{r,\tilde{s}} = \frac{\partial \boldsymbol{e}_r}{\partial \tilde{s}} = \frac{\partial}{\partial \tilde{s}}\left(\cos(\phi)\,\boldsymbol{n}(\tilde{s}) + \sin(\phi)\,\boldsymbol{b}\right) = -\cos(\phi)\,(k\boldsymbol{t})\,(\tilde{s}), \tag{3.3.86}$$

with ϕ related to the $(\overline{r}, \theta, \tilde{s})$ as in (3.1.65). Note that in this section we use the arclength variable $\tilde{s}$ to parameterize the filament centerline for convenience of notation.

Consider next the asymptotic expansion of the vorticity

$$\boldsymbol{\Omega} = \frac{1}{\epsilon^2}\left(\eta^{(0)}\boldsymbol{e}_\theta + \zeta^{(0)}\boldsymbol{t}\right) + \frac{1}{\epsilon}\left(\xi^{(1)}\boldsymbol{e}_r + \eta^{(1)}\boldsymbol{e}_\theta + \zeta^{(1)}\boldsymbol{t}\right) + O(1). \tag{3.3.87}$$

We Taylor-expand about $\tilde{s} = \tilde{s}_1$ and take into account the explicit assumption of a leading-order axisymmetric vorticity distribution without axial variation, that is, $\eta^{(0)} = \eta^{(0)}(\overline{r}, t)$ and $\zeta^{(0)} = \zeta^{(0)}(\overline{r}, t)$. Then

$$\begin{aligned}\boldsymbol{\Omega} = &\frac{1}{\epsilon^2}\left(\eta^{(0)}(\overline{r}, t)\,\boldsymbol{e}_{\theta,1} + \zeta^{(0)}(\overline{r}, t)\,\boldsymbol{t}_1\right)\\ &+\frac{1}{\epsilon}\Bigg(\xi_1^{(1)}(\overline{r}, \theta, t)\,\boldsymbol{e}_{r,1} + \eta_1^{(1)}(\overline{r}, \theta, t)\,\boldsymbol{e}_{\theta,1} + \zeta_1^{(1)}(\overline{r}, \theta, t)\,\boldsymbol{t}_1\\ &+z\frac{\partial}{\partial\tilde{s}}\left(\eta^{(0)}\boldsymbol{e}_\theta + \zeta^{(0)}\boldsymbol{t}\right)_1\Bigg) + O(1).\end{aligned} \tag{3.3.88}$$

Collecting (3.3.84)–(3.3.88), the leading contribution to the Biot–Savart integrand reads

$$\left(\frac{\boldsymbol{x}-\boldsymbol{P}}{|\boldsymbol{x}-\boldsymbol{P}|^3} \times \boldsymbol{\Omega}\right)^{(0)} = \frac{\left(-z\eta^{(0)}\boldsymbol{e}_r - \overline{r}\zeta^{(0)}\boldsymbol{e}_\theta + \overline{r}\eta^{(0)}\boldsymbol{t}\right)_1}{\left(\overline{r}^2+z^2\right)^{\frac{3}{2}}}. \tag{3.3.89}$$

We have to introduce the coordinate transformation $(\boldsymbol{x} = (x, y, z) \leftrightarrow (\overline{r}, \theta, \tilde{s}))$ in the Biot–Savart integral. From Eq. (A.12) in Callegari and Ting (1978), we extract

$$\mathrm{d}^3x \leftrightarrow \epsilon^2\overline{r}\left(1 - \epsilon\overline{r}k\cos\varphi\right)\mathrm{d}\overline{r}\mathrm{d}\theta\mathrm{d}\tilde{s}\,. \tag{3.3.90}$$

(Since we define $\tilde{s}$ to be an arclength variable, Callegari and Ting's stretching factor, σ, is identical to unity here.)

The $O(1/\epsilon)$ contribution, $\boldsymbol{q}^{(0)}$, of the induced velocity at $\boldsymbol{x}_1 = \boldsymbol{x}(\tilde{s}_1, t)$ due to the inner integration now becomes

$$\boldsymbol{q}^{(0)}_{\text{inner}} = \frac{1}{4\pi}\int_{-\delta_c/\epsilon}^{\delta_c/\epsilon}\int_0^\infty\int_0^{2\pi} \frac{\left(-z\eta^{(0)}\boldsymbol{e}_r - \overline{r}\zeta^{(0)}\boldsymbol{e}_\theta + \overline{r}\eta^{(0)}\boldsymbol{t}\right)_1}{\left(\overline{r}^2+z^2\right)^{3/2}}\,\overline{r}\mathrm{d}\theta\mathrm{d}\overline{r}\mathrm{d}z \tag{3.3.91}$$

$$= \frac{1}{2}\int_0^\infty\int_{-\delta_c/\epsilon}^{\delta_c/\epsilon} \frac{\overline{r}^2\eta^{(0)}}{\left(\overline{r}^2+z^2\right)^{3/2}}\mathrm{d}z\mathrm{d}\overline{r}\,\boldsymbol{t}_1 \tag{3.3.92}$$

$$= \left(\int_0^\infty \eta^{(0)}\left(\overline{r}\right)\mathrm{d}\overline{r}\,\boldsymbol{t}_1 + O\left(\left(\epsilon/\delta_c\right)^2\right)\right). \tag{3.3.93}$$

The step from (3.3.91) to (3.3.92) involves elimination of the vector contributions in $\boldsymbol{e}_r$ and $\boldsymbol{e}_\theta$ direction due to integration in θ. Notice that the coefficients of $\boldsymbol{e}_\theta$ and $\boldsymbol{e}_r$ in the intergrand are θ-independent. The step from (3.3.92) to (3.3.93) uses the fact that $\delta_c/\epsilon \to \infty$ as $\epsilon \to 0$, and that, according to integral no. 242 in Bronstein and Semendjajew (2000), $\int_{-\infty}^{\infty}(\overline{r}^2/\left(\overline{r}^2+z^2\right)^{3/2})\,dz = 2$. The additional error estimate in (3.3.93) is straightforward and takes into account that the integration in $\tilde{s}$ only involves the Ball B_{δ_c} of radius δ_c according to Fig. 3.7.

By similar calculations we find the first order contribution $\boldsymbol{q}^{(1)}_{\text{inner}}$, expressed in terms of its $(\boldsymbol{t},\boldsymbol{n},\boldsymbol{b})_1$-components (see Klein and Knio (1995) and Klein (1994) for details)

$$\begin{aligned}\boldsymbol{q}^{(1)}_{\text{inner}} &= \boldsymbol{t}_1\langle w^{(1)}\rangle_1(0)\\ &+ (k\boldsymbol{b})_1\,\frac{1}{4}\int_0^\infty\int_{-\delta_c/\epsilon}^{\delta_c/\epsilon}\left[\left(\frac{\overline{r}}{R}-\frac{3}{2}\frac{\overline{r}^3z^2}{R^5}\right)\zeta^{(0)} - \frac{\overline{r}^2}{kR^3}\,\zeta^{(1)}_{11}\right]\mathrm{d}z\mathrm{d}\overline{r}\\ &+ (k\boldsymbol{n})_1\,\frac{1}{4}\frac{1}{k}\int_0^\infty \zeta^{(1)}_{12}\mathrm{d}\overline{r} + o(1)\quad\text{as}\quad(\delta_c/\epsilon)\to\infty\,,\end{aligned} \tag{3.3.94}$$

where $R = \sqrt{\overline{r}^2+z^2}$ and $\zeta^{(1)}_{11}$, $\zeta^{(1)}_{12}$ are the first order Fourier modes of $\zeta^{(1)}(\overline{r},\theta,s_1)$, that is,

$$\zeta^{(1)}_{11} + i\zeta^{(1)}_{12} = \frac{1}{\pi}\int_0^{2\pi}\zeta^{(1)}(\overline{r},\theta,s_1)e^{i\theta}\mathrm{d}\theta\,. \tag{3.3.95}$$

The term multiplying $\boldsymbol{t}_1$ is the first order axial velocity on the filament centerline. This term is analogous to the corresponding leading-order term, and it does not contribute to the motion of the filament. The first term multiplying $(k\boldsymbol{b})_1$ diverges logarithmically as $\delta_c/\epsilon \to \infty$. This term will match a corresponding contribution from the outer integral, and it will provide the leading logarithmic binormal term in the filament equation of motion. This term also involves the first order vorticity via $\zeta^{(1)}_{11}$ and provides for the finite local induction in the binormal direction, which depends on the detailed structure of the vortical core. Analogously, the contribution from $\zeta^{(1)}_{12}$ yields motion in the direction of the principal normal.

Notice that the above expressions are valid for the induced velocity at *any* point in the inner core of the filament with diameter of order $O\left(\epsilon^2\right)$ so that the appearance of a term proportional to $\boldsymbol{n}_1$ from local induction does not contradict the earlier results that the local induction produces filament motion in binormal direction only. Furthermore, we observe that the velocity expressions are all *linear* in $\zeta^{(0)}$, $\zeta^{(1)}$, $\eta^{(0)}$, $\eta^{(1)}$, which at a first glance also seems to contradict, for example, Callegari and Ting (1978), who find a quadratically nonlinear influence of the leading-order core structure on the binormal velocity (via the core-constant C). The apparent contradiction will be resolved below where we analyze the vorticity transport equation to leading and first order in a filament-attached frame of reference. We will be able to show that $\zeta_{12}^{(1)} \equiv 0$, and we will find a quadratically nonlinear expression for $\zeta_{11}^{(1)}$ in terms of $\zeta^{(0)}$ and $\eta^{(0)}$.

The key conclusion from the preceding calculations, needed later in Sect. 5.2, is that one needs explicit expressions for the first-order vorticity modes $\zeta_{11}^{(1)}$, $\zeta_{12}^{(1)}$, in order to accurately compute the leading-order filament velocity.

The Outer Integral. Here we consider integration over points $\boldsymbol{x}$ with $|\boldsymbol{x}-\boldsymbol{P}| \geq \delta_c \gg \epsilon$. We can then write

$$\boldsymbol{x} - \boldsymbol{P} = \boldsymbol{x}(\tilde{s}, t) + \epsilon\overline{r}\boldsymbol{e}_r(\overline{r}, \theta, \tilde{s}, t) \tag{3.3.96}$$

and expand

$$\begin{aligned} \boldsymbol{x} - \boldsymbol{P} &= \boldsymbol{x}(\tilde{s}, t) - \boldsymbol{x}(\tilde{s}_1, t) + O(\epsilon) \\ \frac{1}{|\boldsymbol{x} - \boldsymbol{P}|^3} &= \frac{1}{|\boldsymbol{x}(\tilde{s}, t) - \boldsymbol{x}(\tilde{s}_1, t)|^3}\,(1 + O(\epsilon))\,, \end{aligned} \tag{3.3.97}$$

and therefore

$$\begin{aligned} &\frac{1}{4\pi}\int\!\!\int_{\mathbb{R}^3\setminus B_{\delta_c}}\!\!\int \frac{\boldsymbol{x} - \boldsymbol{P}}{|\boldsymbol{x} - \boldsymbol{P}|^3} \times \boldsymbol{\Omega}\,(\boldsymbol{x}, t)\,\mathrm{d}^3x \\ &= \frac{1}{4\pi}\int_{C\setminus \mathrm{i}_C} \frac{\boldsymbol{x} - \boldsymbol{P}}{|\boldsymbol{x} - \boldsymbol{P}|^3} \times \int_0^\infty\!\!\int_0^{2\pi} \boldsymbol{\Omega}^{(0)}\overline{r}\mathrm{d}\theta\mathrm{d}\overline{r}\mathrm{d}\tilde{s}\;\;(1 + O(\epsilon))\,, \end{aligned} \tag{3.3.98}$$

where $\mathrm{i}_C = [\tilde{s}_1 - \delta_c, \tilde{s}_1 + \delta_c]$. From

$$\boldsymbol{\Omega}^{(0)}(\overline{r}, \theta, \tilde{s}, t) = \eta^{(0)}(\overline{r}, t)\boldsymbol{e}_\theta(\theta, \tilde{s}, t) + \zeta^{(0)}(\overline{r}, t)\boldsymbol{t}(\tilde{s}, t) \tag{3.3.99}$$

we conclude that the contribution to (3.3.98) from the circumferential vorticity $\eta^{(0)}\boldsymbol{e}_\theta$ will be cancelled by integration in θ. Since, furthermore, the constant total circulation of the filament is given by $\mathit{\Gamma} = \int_0^\infty 2\pi\overline{r}\zeta^{(0)}(\overline{r}, t)\mathrm{d}\overline{r}$, we find

$$\boldsymbol{q}_{\mathrm{outer}} = \frac{\mathit{\Gamma}}{4\pi}\int_{C\setminus \mathrm{i}_C} \frac{\boldsymbol{x}(\tilde{s}, t) - \boldsymbol{x}_1}{|\boldsymbol{x}(\tilde{s}, t) - \boldsymbol{x}_1|^3} \times \boldsymbol{t}(\tilde{s}, t)\mathrm{d}\tilde{s}\;\;(1 + O(\epsilon))\,. \tag{3.3.100}$$

This expression depends on the matching length δ_c through the excluded interval i_C. In fact, as $|\mathrm{i}_C| \to 0$, the integral diverges logarithmically. We extract the explicit leading-order dependence of $\boldsymbol{q}_{\text{outer}}$ on δ_c in the following fashion: Close to $\boldsymbol{X}(\tilde{s}_1, t)$ the integrand can be expanded according to (3.3.84) yielding the asymptotic representation of the integrand in (3.3.100) for $|\tilde{s} - \tilde{s}_1| \ll 1$,

$$\boldsymbol{F}(\tilde{s}, t) = \frac{\boldsymbol{x}(\tilde{s}, t) - \boldsymbol{x}_1}{|\boldsymbol{x}(\tilde{s}, t) - \boldsymbol{x}_1|^3} \times \boldsymbol{t} = \boldsymbol{F}_1(\tilde{s}) + O(1) \quad \text{as} \quad (\tilde{s} - \tilde{s}_1) \to 0 \,, \tag{3.3.101}$$

where

$$\boldsymbol{F}_1(\tilde{s}) = \frac{1}{2} \frac{1}{|\tilde{s} - \tilde{s}_1|} (k\boldsymbol{b})_1 \,. \tag{3.3.102}$$

With this expansion we may rewrite the leading term in $\boldsymbol{q}_{\text{outer}}$ as

$$\boldsymbol{q}_{\text{outer}}^{(1)} = \int_{\mathcal{C}} \boldsymbol{F}(\tilde{s}, t) - H(1 - |\tilde{s} - \tilde{s}_1|)\, \boldsymbol{F}_1(\tilde{s}, t) \mathrm{d}\tilde{s} + \int_{\mathrm{i}_1 \setminus \mathrm{i}_{\mathcal{C}}} \boldsymbol{F}_1(\tilde{s}, t) \mathrm{d}\tilde{s} + O(\delta_c) \,, \tag{3.3.103}$$

where $\mathrm{i}_1 = [\tilde{s}_1 - 1, \tilde{s}_1 + 1]$. The error term $O(\delta_c)$ is the result of carrying the first integration over all of $\mathcal{C}$ instead of over $\mathcal{C} \setminus \mathrm{i}_{\mathcal{C}}$ only. This introduces an error no larger than $O(|\mathrm{i}_{\mathcal{C}}|) = O(\delta_c)$, because of the error estimate in (3.3.101). The final result is

$$\boldsymbol{q}_{\text{outer}}^{(1)} = \boldsymbol{q}^f(\tilde{s}_1, t) + \frac{\Gamma}{4\pi} (k\boldsymbol{b})_1 \ln \frac{1}{\delta_c} \,, \tag{3.3.104}$$

with the precise definition of $\boldsymbol{q}^f$ given by the first integral in (3.3.103).

Summarizing (3.3.93), (3.3.94), and (3.3.104), we have the following general representation of the self-induced velocity inside an inner $O(\epsilon^2)$-core of the vortex filament, that is,

$$\begin{aligned} \boldsymbol{q} = \frac{1}{\epsilon} \boldsymbol{q}^{(0)} + \boldsymbol{q}^{(1)} + o(1) = \boldsymbol{t} & \left[\frac{1}{\epsilon} w^{(0)} + \langle w^{(1)} \rangle \right]_{\overline{r}=0} \\ & + \boldsymbol{n} \left[\frac{1}{4} k \int_0^\infty \zeta_{12}^{(1)} \mathrm{d}\overline{r} + \boldsymbol{n} \cdot \boldsymbol{q}^f \right] \\ & + \boldsymbol{b} \left[q_{\text{inner}}^{(1)}(\delta_c) + \frac{\Gamma k}{4\pi} \ln \frac{1}{\delta_c} + \boldsymbol{b} \cdot \boldsymbol{q}^f \right] + o(1) \,, \end{aligned} \tag{3.3.105}$$

where

$$q_{\text{inner}}^{(1)} = \frac{k}{4} \int_0^\infty \int_{-\delta_c/\epsilon}^{\delta_c/\epsilon} \frac{\overline{r}}{(\overline{r}^2 + z^2)^{3/2}} \left[\left(\overline{r}^2 + z^2 - \frac{3}{2} \frac{\overline{r}^2 z^2}{\overline{r}^2 + z^2} \right) \zeta^{(0)} - \frac{\overline{r}}{k} \zeta_{11}^{(1)} \right] \mathrm{d}z \mathrm{d}\overline{r} \tag{3.3.106}$$

and where $\delta_c = o(1), \epsilon/\delta_c = o(1)$ as as $\epsilon \to 0$.

Consider the $\boldsymbol{t}$-component in (3.3.105). Here $w^{(0)}$ and $\langle w^{(1)} \rangle$ are the leading- and first-order axial velocities in the inner $O(\epsilon^2)$-core of the vortex, both averaged with respect to the circumferential angle. However, the induced motion in the axial direction does not contribute to the evolution of the filament geometry so that, in calculating the filament velocity, this component will be eliminated by projection onto planes normal to the filament axis.

As regards the principal normal $\boldsymbol{n}$-component, we will show below that $\zeta_{12}^{(1)} \equiv 0$.

For the binormal $\boldsymbol{b}$-component we will first demonstrate that the somewhat arbitrary choice of the cutoff length δ_c does not influence our results at leading order, that is, that there is a cancellation of the logarithmic term in $q_{\text{inner}}^{(1)} + \frac{\Gamma k}{4\pi} \ln \frac{1}{\delta_c}$. Second, we derive in Subsect. 3.3.5 explicit expressions for $\zeta_{11}^{(1)}, \zeta_{12}^{(1)}$, in terms of the leading-order core structure by solving the first-order vorticity transport equation in the vortex core. This will allow us to reproduce the filament equation of motion from (3.3.80) and (3.3.81) as derived via matched asymptotic expansions for the momentum equation by Callegari and Ting (1978).

Next we evaluate $\boldsymbol{q}_{\text{inner}}^{(1)} + \frac{\Gamma k}{4\pi} \ln \frac{1}{\delta_c}$ from (3.3.105) and (3.3.106). First we observe that, as $\epsilon/\delta_c \to 0$ we have

$$\int_{-\delta_c/\epsilon}^{\delta_c/\epsilon} \frac{1}{\left(\overline{r}^2 + z^2\right)^{1/2}} \mathrm{d}z = 2\left[\ln \frac{2\delta_c}{\epsilon} - \ln \overline{r}\right] + O\left(\frac{\epsilon}{\delta_c}\right), \tag{3.3.107}$$

$$\frac{3}{2} \int_{-\delta_c/\epsilon}^{\delta_c/\epsilon} \frac{\overline{r}^2 z^2}{\left(\overline{r}^2 + z^2\right)^{5/2}} \mathrm{d}z = 1 + O\left(\left(\frac{\epsilon}{\delta_c}\right)^2\right), \tag{3.3.108}$$

$$\int_{-\delta_c/\epsilon}^{\delta_c/\epsilon} \frac{\overline{r}^2}{\left(\overline{r}^2 + z^2\right)^{3/2}} \mathrm{d}z = 2 + O\left(\left(\frac{\epsilon}{\delta_c}\right)^2\right), \tag{3.3.109}$$

and that

$$\overline{r}\zeta^{(0)} = \left(\overline{r}v^{(0)}\right)_{\overline{r}} \quad \text{and} \quad \lim_{\overline{r}\to\infty} \overline{r}v^{(0)} = \frac{\Gamma}{2\pi}. \tag{3.3.110}$$

With these results inserted in (3.3.106) we find that

$$\boldsymbol{q}_{\text{inner}}^{(1)} + \frac{\Gamma k}{4\pi} \ln \frac{1}{\delta_c} = \frac{\Gamma k}{4\pi}\left[\ln \frac{2}{\epsilon} - \frac{1}{2}\right] - \frac{k}{2} \int_0^\infty \left[\frac{1}{k}\zeta_{11}^{(1)} + \ln \overline{r}\left(\overline{r}v^{(0)}\right)_{\overline{r}}\right] \mathrm{d}\overline{r}\,, \tag{3.3.111}$$

which shows that the cutoff length, δ_c, in fact does not affect the leading-order approximation of the induced velocity.

To proceed with the evaluation of the filament velocity, we anticipate from the next subsection that

$$\zeta_{12}^{(1)} \equiv 0 \tag{3.3.112}$$

$$\zeta_{11}^{(1)} = -\frac{\zeta_{\overline{r}}^{(0)}}{v^{(0)}} \tilde{\psi}_{11} - \overline{r}k\left(\frac{2w^{(0)}w_{\overline{r}}^{(0)}}{v^{(0)}} - \zeta^{(0)}\right), \tag{3.3.113}$$

with

$$\tilde{\psi}_{11} = kv^{(0)} \int_0^{\overline{r}} \frac{1}{r' v^{(0)\,2}} \int_0^{r'} r'' v^{(0)} H \mathrm{d}r'' \mathrm{d}r' \,, \tag{3.3.114}$$

where

$$v^{(0)} H = 2\overline{r} v^{(0)} \zeta^{(0)} + v^{(0)\,2} + 2\overline{r} w^{(0)} w^{(0)}_{\overline{r}} \,. \tag{3.3.115}$$

Inserting these expressions in (3.3.111), integrating by parts several times, and using the results in (3.3.105) we reproduce, after projection onto the normal plane, the filament velocity from (3.3.160)–(3.3.165), that is,

$$\boldsymbol{X}_t = \frac{\Gamma k}{4\pi} \left(\ln \frac{2}{\epsilon} + C^* \right) k \boldsymbol{b} + \boldsymbol{q}^f \cdot (\boldsymbol{1} - \boldsymbol{t} \circ \boldsymbol{t}) \,, \tag{3.3.116}$$

with the core constant

$$C^* = \lim_{\overline{r} \to \infty} \left(\frac{4\pi^2}{\Gamma^2} \int_0^{\overline{r}} \overline{r} v^{(0)\,2} \mathrm{d}\overline{r} - \ln \overline{r} \right) + \frac{1}{2} - \frac{1}{2} \left(\frac{4\pi}{\Gamma} \right)^2 \int_0^{\infty} \overline{r} w^{(0)\,2} \mathrm{d}\overline{r} \,. \tag{3.3.117}$$

Note that, because of a slightly different way of regularizing the Biot–Savart integration, we have an additional ln(2) in the leading binormal term in (3.3.116). In the earlier derivations, this contribution to the filament velocity is hidden in $\boldsymbol{Q}^f = \nabla \times \boldsymbol{A}^f$, with $\boldsymbol{A}^f$ from (3.1.74). To see this, consider the limit $r \to 0$ in that equation and note the behavior of the first logarithm.

3.3.5 Asymptotic Solution of the Vorticity Transport Equation

Here we derive explicit expressions for the first order vorticity components $\zeta_{11}^{(1)}$ and $\zeta_{12}^{(1)}$, as needed in (3.3.94) and (3.3.111) by solving the vorticity transport equation asymptotically to leading- and first order. Thus we consider (2.3.2), slightly rewritten taking into account the divergence constraint, $\nabla \cdot \boldsymbol{v} = 0$, that is,

$$\partial \boldsymbol{\Omega} / \partial \overline{t} + (\boldsymbol{v} \cdot \nabla) \, \boldsymbol{\Omega} = (\boldsymbol{\Omega} \cdot \nabla) \, \boldsymbol{v} + \frac{1}{\mathrm{Re}} \epsilon \boldsymbol{\Omega} \,. \tag{3.3.118}$$

In terms of filament-attached coordinates (r, θ, s, t), the particle time derivative becomes (Callegari and Ting, 1978):

$$\partial / \partial \overline{t} + \boldsymbol{v} \cdot \nabla = \partial / \partial t + (\boldsymbol{v} - r \partial \boldsymbol{e}_r / \partial t) \cdot \nabla \,, \tag{3.3.119}$$

where $V = \boldsymbol{v} - \boldsymbol{X}_t$ is the relative velocity component in the normal plane. Notice that we use in this section the moving coordinates as defined by (Callegari and Ting, 1978) so that s is not an arclength variable, but defined by $\boldsymbol{t} \cdot \boldsymbol{X}_t \equiv 0$ (cf. (3.3.2) and (3.3.3)).

As before, we expand the vorticity and relative velocity as

$$\boldsymbol{\Omega} = \frac{1}{\epsilon^2} \left(\eta^{(0)} \boldsymbol{e}_\theta + \zeta^{(0)} \boldsymbol{t} \right) + \frac{1}{\epsilon} \left(\xi^{(1)} \boldsymbol{e}_r + \eta^{(1)} \boldsymbol{e}_\theta + \zeta^{(1)} \boldsymbol{t} \right) + O(1) \tag{3.3.120}$$

$$V = \frac{1}{\epsilon} \left(v^{(0)} \boldsymbol{e}_\theta + w^{(0)} \boldsymbol{t} \right) + \left(u^{(1)} \boldsymbol{e}_r + v^{(1)} \boldsymbol{e}_\theta + w^{(1)} \boldsymbol{t} \right) + O(\epsilon) \,, \tag{3.3.121}$$

with axisymmetric leading-order distributions $\left(v^{(0)}, w^{(0)}, \eta^{(0)}, \zeta^{(0)}\right)(\overline{r}, t)$, but generally asymmetric first-order components. Order-of-magnitude estimates then show that the leading- and first-order solutions are determined by a balance of advection and vortex stretching. Thus we have

$$(V \cdot \nabla)\,\boldsymbol{\Omega} = (\boldsymbol{\Omega} \cdot \nabla)\,V\ (1 + O(\epsilon))\,, \tag{3.3.122}$$

where

$$\nabla = \frac{1}{\epsilon}\left(\boldsymbol{e}_r \partial_{\overline{r}} + \frac{1}{\overline{r}}\boldsymbol{e}_\theta \partial_\theta\right) + \boldsymbol{t}\frac{1}{\sigma}\partial_s\,, \tag{3.3.123}$$

with $\sigma = |\boldsymbol{X}_s|$.

Using (3.3.120)–(3.3.123), we find that the leading-order vorticity transport equation is identically satisfied for any axisymmetric leading-order core structure, because

$$(V \cdot \nabla\boldsymbol{\Omega})^{(0)} = (\boldsymbol{\Omega} \cdot \nabla V)^{(0)} = -\frac{\eta^{(0)} v^{(0)}}{\overline{r}}\boldsymbol{e}_r\,. \tag{3.3.124}$$

At first order, one has

$$\begin{aligned}(\boldsymbol{V} \cdot \nabla\boldsymbol{\Omega})^{(1)} = \boldsymbol{e}_r &\left(\frac{v^{(0)}}{\overline{r}}\left(\xi_\theta^{(1)} - \eta^{(1)}\right) - \frac{v^{(1)}\eta^{(0)}}{\overline{r}} + k\cos\varphi w^{(0)}\zeta^{(0)}\right)\\ +\boldsymbol{e}_\theta &\left(\frac{v^{(0)}}{\overline{r}}\left(\xi^{(1)} + \eta_\theta^{(1)}\right) + u^{(1)}\eta_{\overline{r}}^{(0)} - k\sin\varphi w^{(0)}\zeta^{(0)}\right)\\ +\boldsymbol{t} &\left(\frac{v^{(0)}}{\overline{r}}\zeta_\theta^{(1)} + u^{(1)}\zeta_{\overline{r}}^{(0)} - k\sin\varphi w^{(0)}\eta^{(0)}\right)\end{aligned} \tag{3.3.125}$$

and analogously

$$\begin{aligned}(\boldsymbol{\Omega} \cdot \nabla\boldsymbol{V})^{(1)} = \boldsymbol{e}_r &\left(\frac{\eta^{(0)}}{\overline{r}}\left(u_\theta^{(1)} - v^{(1)}\right) - \frac{v^{(0)}\eta^{(1)}}{\overline{r}} + k\cos\varphi w^{(0)}\zeta^{(0)}\right)\\ +\boldsymbol{e}_\theta &\left(\frac{\eta^{(0)}}{\overline{r}}\left(u^{(1)} + v_\theta^{(1)}\right) + \xi^{(1)} v_{\overline{r}}^{(0)} - k\sin\varphi w^{(0)}\zeta^{(0)}\right)\\ +\boldsymbol{t} &\left(\frac{\eta^{(0)}}{\overline{r}} w_\theta^{(1)} + \xi^{(1)} w_{\overline{r}}^{(0)} - k\sin\varphi\zeta^{(0)} v^{(0)}\right)\,.\end{aligned} \tag{3.3.126}$$

We are interested particularly in the first order axial vorticity $\zeta^{(1)}$ since it is responsible for $O(1)$-contributions to the filament motion. Thus we equate the $\boldsymbol{t}$-components of (3.3.125) and (3.3.126) and find

$$v^{(0)}\zeta_\theta^{(1)} + \overline{r}u^{(1)}\zeta_{\overline{r}}^{(0)} - \left(\eta^{(0)} w_\theta^{(1)} + \overline{r}\xi^{(1)} w_{\overline{r}}^{(0)}\right) - \overline{r}k\sin\varphi\left(w^{(0)}\eta^{(0)} - \zeta^{(0)} v^{(0)}\right) = 0\,. \tag{3.3.127}$$

A straightforward calculation yields

$$\eta^{(0)} = -w_{\overline{r}}^{(0)} \quad \text{and} \quad \xi^{(1)} = \frac{1}{\overline{r}} w_\theta^{(1)} + k\sin\varphi w^{(0)}\,, \tag{3.3.128}$$

and with these results we eliminate the first bracket in (3.3.127) to obtain

$$v^{(0)}\zeta^{(1)}_{\theta} + \bar{r}u^{(1)}\zeta^{(0)}_{\bar{r}} - \bar{r}k\sin\varphi\left(2w^{(0)}w^{(0)}_{\bar{r}} - \zeta^{(0)}v^{(0)}\right) = 0\,. \tag{3.3.129}$$

It is shown in Klein (1994, pp. 21–22) that this equation is equivalent to the second-order Poisson-type equation (4.18) of Callegari and Ting (1978), or (3.3.65) above, which determines the stream function for the first order flow in the normal plane. Multiplication of (3.3.129) by $\sin\varphi$, $\cos\varphi$ and averaging over ϕ yields, after integrations by parts,

$$\zeta^{(1)}_{11} = \frac{\bar{r}\zeta^{(0)}_{\bar{r}}}{v^{(0)}}u^{(1)}_{12} - \bar{r}k\left(\frac{2w^{(0)}w^{(0)}_{\bar{r}}}{v^{(0)}} - \zeta^{(0)}\right) \tag{3.3.130}$$

and

$$\zeta^{(1)}_{12} = -\frac{\bar{r}\zeta^{(0)}_{\bar{r}}}{v^{(0)}}u^{(1)}_{11}\,. \tag{3.3.131}$$

At this point we go back to Callegari and Ting (1978) as recounted in Subsubsect. 3.3.3.2 above and replace $u^{(1)}_{11}$, $u^{(1)}_{12}$ by the Fourier components of the perturbation stream function. From (3.3.76), (3.3.78), and the Fourier decomposition of (3.3.64), we obtain

$$u^{(1)}_{11} = \frac{1}{\bar{r}}\tilde{\psi}_{12} = 0 \tag{3.3.132}$$

and

$$\bar{r}u^{(1)}_{12} = -\tilde{\psi}_{11} = -kv^{(0)}\int_0^{\bar{r}}\frac{1}{r'{v^{(0)}}^2}\int_0^{r'} r''v^{(0)}H\mathrm{d}r''\mathrm{d}r'\,, \tag{3.3.133}$$

with $\tilde{\psi}_{11}$ and H as given in (3.3.114) and (3.3.115) above. This completes the vorticity-based analysis of the filament equation of motion.

Note that this alternative derivation began from the Biot-Savart integral, which is valid only for divergence free flows, so that it is not applicable to a filament with compressible core structure without modification. The direct derivation for the compressible case by matched asymptotics will be presented in Sect. 3.4.

3.3.6 The Second-Order Equations and the Evolution of the Core Structure

In this Subsection, we derive the compatibility conditions for the second-order equations by evaluating their θ-averages. These conditions in turn provide the evolution equations for the leading-order core structure, $\epsilon^{-1}v^{(0)}(t,\bar{r})$ and $\epsilon^{-1}w^{(0)}(t,\bar{r})$. It is these equations that we need to solve in order to be able to determine the core coefficient C^* in the filament evolution equation from (3.3.81). See also the brief discussion at the end of Subsubsect. 3.3.3.2 above.

3.3.6.1 The Compatibility Conditions. The second-order continuity equation is

$$(\bar{r}u^{(2)})_{\bar{r}} + v_{\theta}^{(2)} + \frac{v^{(0)}}{\sigma^{(0)}}(h_3^{(2)})_{\theta}$$

$$= -\frac{\bar{r}}{\sigma^{(0)}}(w_s^{(1)} + \dot{\boldsymbol{X}}_s^{(0)} \cdot \boldsymbol{\tau}^{(0)}) - \frac{\sigma^{(1)}}{\sigma^{(0)}}v_{\theta}^{(1)} + \bar{r}\kappa^{(0)}(v^{(1)} \cos\phi^{(0)})_{\theta} \qquad (3.3.134)$$

$$-\frac{\sigma^{(1)}}{\sigma^{(0)}}(\bar{r}u^{(1)})_{\bar{r}} + \kappa^{(0)} \cos\phi^{(0)} (\bar{r}^2 u^{(1)})_{\bar{r}}$$

and its θ-average yields the average radial mass flux $u_c^{(2)}$:

$$u_c^{(2)} = -\frac{1}{\bar{r}\sigma^{(0)}} \int_0^{\bar{r}} [(w_s^{(1)})_c + \dot{\boldsymbol{X}}_s^{(0)} \cdot \boldsymbol{\tau}^{(0)}]\bar{r}'d\bar{r}' \ . \qquad (3.3.135)$$

Here we used the following results on the first order radial velocity,

$$u_c^{(1)} = 0\ , \quad \bar{r}u_a^{(1)} = \psi_{\theta}^{(1)} = -\psi_{11}\sin\phi^{(0)} \quad \text{and} \quad (u^{(1)}\cos\phi^{(0)})_c = 0\ . \qquad (3.3.136)$$

Carrying out the θ-average of the axial and circumferential components of the second-order momentum equations and eliminating $u_c^{(2)}$ by (3.3.135), we arrive at the following two compatibility conditions:

$$w_t^{(0)} - \bar{\nu}\frac{1}{\bar{r}}(\bar{r}w_{\bar{r}}^{(0)})_{\bar{r}} = -\frac{1}{\sigma^{(0)}}(w^{(0)}(w_c^{(1)})_s + (p_c^{(1)})_s) - \frac{w^{(0)}}{\sigma^{(0)}}\left[\dot{\boldsymbol{X}}_s^{(0)} \cdot \boldsymbol{\tau}^{(0)}\right]$$

$$+ \frac{w_{\bar{r}}^{(0)}}{\bar{r}\sigma^{(0)}} \int_0^{\bar{r}} \left[(w_c^{(1)})_s + \dot{\boldsymbol{X}}_s^{(0)} \cdot \boldsymbol{\tau}^{(0)}\right] \bar{r}'d\bar{r}' \qquad (3.3.137)$$

and

$$v_t^{(0)} - \bar{\nu}\left[\frac{1}{\bar{r}}(\bar{r}v_{\bar{r}}^{(0)})_{\bar{r}} - \frac{1}{\bar{r}^2}v^{(0)}\right] = -\frac{w^{(0)}}{\sigma^{(0)}}(v_c^{(1)})_s$$

$$+ \frac{(\bar{r}v^{(0)})_{\bar{r}}}{\bar{r}^2\sigma^{(0)}} \int_0^{\bar{r}} \left[(w_c^{(1)})_s + \dot{\boldsymbol{X}}_s^{(0)} \cdot \boldsymbol{\tau}^{(0)}\right] \bar{r}'d\bar{r}' \ . \qquad (3.3.138)$$

Here $\bar{\nu} = \nu/(U\ell\epsilon^2) = O(1)$ is the scaled diffusion coefficient introduced in (3.2.4). The details for the derivation of these two equations were presented in Appendix D of Callegari and Ting (1978).

Since $w^{(0)}$ and $v^{(0)}$ are functions of $\bar{r}$ and t only, the above two equations imply that

$$(p_c^{(1)})_s + 2w^{(0)}(w_c^{(1)})_s - \frac{1}{\bar{r}}\left[w^{(0)} \int_0^{\bar{r}} (w_c^{(1)})_s\, \bar{r}'d\bar{r}'\right]_{\bar{r}} - \frac{1}{2}\bar{r}^3\left(\frac{w^{(0)}}{\bar{r}^2}\right)_{\bar{r}} \tilde{s}_{st}^{(0)}$$

$$= -\sigma^{(0)}F_1(t,\bar{r}), \qquad (3.3.139)$$

$$w^{(0)}(v_c^{(1)})_s - \frac{(\bar{r}v^{(0)})_{\bar{r}}}{\bar{r}^2}\int_0^{\bar{r}} (w_c^{(1)})_s \ \bar{r}' \ d\bar{r}' - \frac{(\bar{r}v^{(0)})_{\bar{r}}}{2}\tilde{s}_{st}^{(0)}$$

$$= -\sigma^{(0)}F_2(t,\bar{r}) \ , \qquad (3.3.140)$$

where

$$F_1(t,\bar{r}) = w_t^{(0)} - \bar{\nu}\,\frac{1}{\bar{r}}(\bar{r}w_{\bar{r}}^{(0)})_{\bar{r}} \qquad (3.3.141)$$

and

$$F_2(t,\bar{r}) = v_t^{(0)} - \bar{\nu}\left[\frac{1}{\bar{r}}(\bar{r}v_{\bar{r}}^{(0)})_{\bar{r}} - \frac{v^{(0)}}{\bar{r}^2}\right] . \qquad (3.3.142)$$

Here we used the identity $\dot{\boldsymbol{X}}_s^{(0)} \cdot \boldsymbol{\tau}^{(0)} = [\boldsymbol{X}_s^{(0)} \cdot \boldsymbol{\tau}^{(0)}]_t = \sigma_t^{(0)} = \tilde{s}_{st}^{(0)}$, where $\tilde{s}^{(0)}(t,s) = \int_0^s \sigma^{(0)}(t,s')ds'$ denotes the leading-order arclength at time t.

Following Callegari and Ting (1978), we consider the case in which the filament forms a slender torus so that the centerline $\mathcal{C}$ is a simple closed curve of length $S(t,\epsilon)$ with initial length $S(0) = O(\ell)$. The case of an infinite vortex filament or one with length much larger than the scaling length, ℓ, has been discussed in Book-I, Sect. 4.1.

3.3.6.2 Filament in the Form of a Slender Torus. When the centerline $\mathcal{C}$ forms a simple closed curve, all physical variables have to be periodic functions of s with period S_0,

$$\boldsymbol{X}(t,s+S_0,\epsilon) = \boldsymbol{X}(t,s,\epsilon) \quad \text{and} \quad f(t,\bar{r},\phi,s+S_0,\epsilon) = f(t,\bar{r},\phi,s,\epsilon) \ , \qquad (3.3.143)$$

where f stands for p, u, v etc. Using this, we can eliminate the s-derivatives of the first order terms, $w_c^{(1)}$, $v_c^{(1)}$, and $p_c^{(1)}$, in (3.3.139) and (3.3.140) by integrating with respect to s over the period S_0,

$$w_t^{(0)} - \bar{\nu}\,\frac{1}{\bar{r}}\left(rw_{\bar{r}}^{(0)}\right)_{\bar{r}} = \frac{1}{2}\bar{r}^3\left(\frac{w^{(0)}}{\bar{r}^2}\right)_{\bar{r}}\frac{\dot{S}^{(0)}}{S^{(0)}} \ , \qquad (3.3.144)$$

$$v_t^{(0)} - \bar{\nu}\left[\frac{1}{\bar{r}}(\bar{r}v_{\bar{r}}^{(0)})_{\bar{r}} - \frac{v^{(0)}}{\bar{r}^2}\right] = \frac{1}{2}(\bar{r}v^{(0)})_{\bar{r}}\frac{\dot{S}^{(0)}}{S^{(0)}} \ . \qquad (3.3.145)$$

These compatibility conditions in turn are the governing equations for the leading-order core structure, $w^{(0)}(t,\bar{r})$ and $v^{(0)}(t,\bar{r})$. The length of the centerline, $S^{(0)}(t)$, which appears in the coefficients on the right-hand side, couples the evolution of the core structure with the equations of motion of the centerline, (3.3.77) and (3.3.81). The terms involving $S^{(0)}(t)$ yield the overall effect of the stretching of the filament while the terms proportional to $\bar{\nu}$ account for the effects of viscous diffusion.

Since it will be easier to work with the axial vorticity $\zeta^{(0)}$ instead of $v^{(0)}$, which are related by (3.3.39), we convert (3.3.145) to an equation for the axial vorticity by differentiating $\bar{r}$ times (3.3.145) with respect to $\bar{r}$. The result is

$$\zeta_t^{(0)} - \bar{\nu}\,\frac{1}{\bar{r}}(\bar{r}\zeta_{\bar{r}}^{(0)})_{\bar{r}} = \frac{1}{2}(\bar{r}^2\zeta^{(0)})_{\bar{r}}\frac{\dot{S}^{(0)}}{S^{(0)}}\frac{1}{\bar{r}} \ . \tag{3.3.146}$$

Equations (3.3.144), (3.3.146), (3.3.77), and (3.3.81) form a closed system of equations for the leading-order solutions $\zeta^{(0)}$, $w^{(0)}$, and $\boldsymbol{X}^{(0)}$ with $v^{(0)}$ and C^* defined by (3.3.39) and (3.3.80). These equations exhibit the coupling between the inner and outer flows, and in particular indicate the effect of the core structure on the motion of the filament. The coupling is given by the integral expressions in $C^*(t)$ in (3.3.80), which depend on both $v^{(0)}(t, \bar{r})$ and $w^{(0)}(t, \bar{r})$.

Before stating the initial and boundary conditions that are compatible with the above system of equations and describing the construction of the solutions, we derive two invariants from (3.3.144) and (3.3.146) and explain their physical meaning including a discussion of the inviscid case.

Noting the rapid decay of $\zeta^{(0)}$ as $\bar{r} \to \infty$, we recover from (3.3.146) the invariance condition,

$$\frac{\partial}{\partial t}\int_0^\infty \bar{r}\zeta^{(0)}(t, \bar{r})\, d\bar{r} = 0 \ , \tag{3.3.147}$$

and hence

$$\int_0^\infty \bar{r}\zeta^{(0)}(t, \bar{r})\, d\bar{r} = \frac{\Gamma}{2\pi} = \text{const} \ . \tag{3.3.148}$$

This shows that the leading-order solutions satisfy the global constraint of conservation of circulation. Similarly from (3.3.144) we obtain the second invariance condition,

$$m(t)(S^{(0)})^2 = \text{const} \ , \tag{3.3.149}$$

where

$$m(t) = 2\pi\int_0^\infty \bar{r}w^{(0)}(t, \bar{r})\, d\bar{r} \ , \tag{3.3.150}$$

denotes a scaled axial mass flux. In dimensional notation the axial mass flux reads $U\ell^2\epsilon m$.

One can identify the above two invariants as constraints on the average axial velocity and axial vorticity in the filament. We introduce an average cross-sectional area $A(t)$ of the vortex filament such that the conservation of mass yields:

$$AS^{(0)} = \text{const} \tag{3.3.151}$$

and then define the average axial velocity $\bar{w}$ and vorticity $\bar{\zeta}$ as

$$\bar{w} = \frac{2\pi}{A}\int_0^\infty \bar{r}w^{(0)}\, d\bar{r} \quad \text{and} \quad \bar{\zeta} = \frac{2\pi}{A}\int_0^\infty \bar{r}\bar{\zeta}^{(0)}\, d\bar{r} \ . \tag{3.3.152}$$

The conservation of circulation (3.3.148) now reads as a constraint on the stretching of average vorticity

$$\bar{\zeta}/S^{(0)} = \text{const} , \tag{3.3.153}$$

while the invariant (3.3.149) becomes one on the average axial velocity,

$$\frac{\bar{w}}{A} \sim \bar{w}S^{(0)} = \text{const} . \tag{3.3.154}$$

It should be noted that these two invariants do not involve the viscosity; therefore, they are also valid for an inviscid flow. To be more specific, we can obtain the variation of $w^{(0)}$ and $\zeta^{(0)}$ for this case from (3.3.144) and (3.3.146) by simply dropping the viscous terms. We find that quantities $w^{(0)}/\bar{r}^2$ and $\zeta^{(0)}\bar{r}^2$ both obey the first order equation

$$\left(\frac{\partial}{\partial t} - \frac{\bar{r}}{2}\frac{\dot{S}^{(0)}}{S^{(0)}}\frac{\partial}{\partial \bar{r}}\right)\chi = 0 . \tag{3.3.155}$$

Its general solution is of the form $\chi(\alpha)$, where $\alpha = \bar{r}\sqrt{S^{(0)}(t)/S_0}$. Here s denotes the arclength at $t = 0$ and hence $S^{(0)}(0) = S_0$. With the initial data $w_0(\bar{r})$ and $\zeta_0(\bar{r})$ for the axial velocity and vorticity, the solutions are

$$w^{(0)}(t, \bar{r}) = w_0(\alpha)S_0/S^{(0)}(t) \tag{3.3.156}$$

and

$$\zeta^{(0)}(t, \bar{r}) = \zeta_0(\alpha)S^{(0)}(t)/S_0 . \tag{3.3.157}$$

These results are stronger than those given in (3.3.153) and (3.3.154) for the viscous case. Instead of integral averages, they relate the detailed distributions of $w^{(0)}$ and $\zeta^{(0)}$ with respect to $\bar{r}$ to the initial data through a simple scaling rule by the length of the filament. These results demonstrate once more that it is inconsistent to arbitrarily assign the instantaneous core quantities in an inviscid analysis.

From the above, it is straightforward to determine the core evolution in the inviscid case, given the time history $S^{(0)}(t)$ of the filament length. However, even with viscosity the evolution equations (3.3.144) and (3.3.146) are linear partial differential equations for $w^{(0)}(t, \bar{r})$ and $\zeta^{(0)}(t, \bar{r})$, given $S^{(0)}(t)$. Solutions are available in the form of eigenfunction expansions shown next.

3.3.6.3 Evolution of the Core Structure. We begin by rewriting the system of equations for the leading-order core structure and the motion of the filament as

$$w_t^{(0)} - \bar{\nu}\,\frac{1}{\bar{r}}(\bar{r}w_{\bar{r}}^{(0)})_{\bar{r}} = \frac{1}{2}\bar{r}^3\left(\frac{w^{(0)}}{\bar{r}^2}\right)_{\bar{r}}\frac{\dot{S}^{(0)}}{S^{(0)}} , \tag{3.3.158}$$

$$\zeta_t^{(0)} - \bar{\nu}\,\frac{1}{\bar{r}}(\bar{r}\zeta_{\bar{r}}^{(0)})_{\bar{r}} = \frac{1}{2}\frac{(\bar{r}^2\zeta^{(0)})_{\bar{r}}}{\bar{r}}\frac{\dot{S}^{(0)}}{S^{(0)}} , \tag{3.3.159}$$

$$\dot{\boldsymbol{X}}^{(0)} \cdot \boldsymbol{\tau}^{(0)} = 0, \tag{3.3.160}$$

$$\dot{\boldsymbol{X}}^{(0)} \cdot \boldsymbol{n}^{(0)} = \boldsymbol{Q}_0 \cdot \boldsymbol{n}^{(0)} , \tag{3.3.161}$$

$$\dot{\boldsymbol{X}}^{(0)} \cdot \boldsymbol{b}^{(0)} = \boldsymbol{Q}_0 \cdot \boldsymbol{b}^{(0)} + \frac{\Gamma \kappa^{(0)}}{4\pi} \ln \frac{1}{\epsilon} + \kappa^{(0)} C^*(t,s) , \tag{3.3.162}$$

where

$$C^*(t) = \frac{\Gamma}{4\pi} [C_v(t) + C_w(t)] , \tag{3.3.163}$$

$$C_v(t) = \lim_{\bar{r} \to \infty} \left(\frac{4\pi^2}{\Gamma^2} \int_0^{\bar{r}} \bar{r}' (v^{(0)})^2 \, d\bar{r}' - \ln \bar{r} \right) + \frac{1}{2} , \tag{3.3.164}$$

$$C_w(t) = \frac{-8\pi^2}{\Gamma^2} \int_0^{\infty} \bar{r}' (w^{(0)})^2 \, d\bar{r}' . \tag{3.3.165}$$

Here, $C_v(t)$ and $C_w(t)$ represent the global contributions of the circumferential and axial velocities in the core to the binormal velocity of the filament. For completeness, we recall the relationship between the circumferential velocity and the axial vorticity,

$$v^{(0)} = \frac{1}{\bar{r}} \int_0^{r} z \zeta^{(0)}(t,z) \, dz \tag{3.3.166}$$

and the definition of the length of the filament,

$$S^{(0)}(t) = \int_0^{S_0} \mid \boldsymbol{X}_s^{(0)}(t,s) \mid ds. \tag{3.3.167}$$

We now discuss the restrictions on the initial and boundary conditions for the above system. Equations (3.3.158) and (3.3.159) imply that $w^{(0)}(0,\bar{r}) = w_0(\bar{r})$ and $\zeta^{(0)}(0,\bar{r}) = \zeta_0(\bar{r})$, or equivalently $v^{(0)}(0,\bar{r}) = v_0(\bar{r})$, must be prescribed. We require that w_0 and ζ_0 decay sufficiently rapidly with $\bar{r}$ while v_0 should behave like $\Gamma/(2\pi\bar{r})$ as $\bar{r} \to \infty$.

Next, we observe that (3.3.160)–(3.3.162) contain only first-time derivatives of $\boldsymbol{X}^{(0)}$ while second-time derivatives of $\boldsymbol{X}$ appeared in the full Navier–Stokes equations (3.3.19). Thus, our expansion scheme has introduced a singular perturbation in the time variable. This implies that we cannot prescribe both the initial velocity $\dot{\boldsymbol{X}}^{(0)}(0,s)$ and the initial position or shape $\boldsymbol{X}^{(0)}(0,s)$ of $\mathcal{C}$. Instead, we prescribe only the initial position of $\mathcal{C}$ and let the initial velocity be determined by (3.3.160)–(3.3.162) with $v^{(0)}$, $w^{(0)}$, and $\boldsymbol{X}^{(0)}$ replaced by their initial data.

To treat the general case, where the initial velocity $\dot{\boldsymbol{X}}^{(0)}(0,s)$ is not compatible with (3.3.161) and (3.3.162), we have to construct a two-time solution to account for the high-frequency fluctuation of the filament velocity. This was done for a vortex in two dimensions by Ting and Tung (1965) and is rederived in Subsect. 3.2.1. Similar results were derived for a circular vortex ring in axisymmetric flows by Ting (1971).

The first pair of equations, (3.3.158) and (3.3.159), governs the evolution of the core structure $w^{(0)}(t,\bar{r})$ and $\zeta^{(0)}(t,\bar{r})$. The coupling with the motion of the filament appears only in the coefficient $\dot{S}^{(0)}(t)/S^{(0)}(t)$, which represents the overall rate of stretching of the centerline. The second set of equations, (3.3.160)–(3.3.162), defines the centerline velocity, $\dot{\boldsymbol{X}}^{(0)}(t,s)$. Only two integrals of the core structure over the cross-sectional plane contribute to the filament motion, and they appear only in the binormal component of $\dot{\boldsymbol{X}}^{(0)}$. The relevant integral expressions, $C_v(t)$ and $C_w(t)$, are defined in terms of $v^{(0)}$ and $w^{(0)}$ in (3.3.164) and (3.3.165).

To construct explicit solutions, we shall decouple these two sets of equations. We solve the system (3.3.158) and (3.3.159) for the core structure in terms of the initial data with the coefficient $\dot{S}^{(0)}/S^{(0)}$ considered a given function of t. Once this is achieved, we obtain the global contributions of the core structure to the filament motion, $C_v(t)$ and $C_w(t)$, as functionals of this coefficient and of the initial data. In this way, the second system, (3.3.160)–(3.3.162), becomes a closed system of equations for $\boldsymbol{X}^{(0)}(t,s)$, defining the motion of the filament.

An appropriate transformation of the dependent and independent variables in the first system of equations allows us to eliminate $S^{(0)}(t)$. Afterward, the function appears only in the transformation rules, but no longer explicitly in the equations themselves. Solutions to this new system are obtained by means of eigenfunction expansions.

We introduce the new independent variables η and τ_1 by the transformations,

$$\eta = \frac{\bar{r}}{\sqrt{4\bar{\nu}\tau_2(t)}}\,, \qquad \tau_2 = \tau_1/S^{(0)}(t) \tag{3.3.168}$$

and

$$\tau_1 = \int_0^t S^{(0)}(t')\,dt' + \tau_{10}\,. \tag{3.3.169}$$

Here τ_{10} is a positive constant of integration whose choice is at our disposal. A rule for optimizing τ_{10} will be discussed at the end of this section.

Next, the axial velocity, $w^{(0)}$, and the axial vorticity, $\zeta^{(0)}$, are replaced, respectively, by

$$W(\eta,\tau_1) = \tau_1 w^{(0)}\,S^{(0)} \qquad \text{and} \qquad Z(\eta,\tau_1) = \tau_1\zeta^{(0)}/S^{(0)}\,. \tag{3.3.170}$$

In terms of these two new variables, both (3.3.158) and (3.3.158) reduce to the common form

$$4\tau_1 \chi_{\tau_1} = \frac{1}{\eta}[(\eta\chi_\eta)_\eta + (2\eta^2\chi)_\eta] , \tag{3.3.171}$$

where χ stands for W and Z.

Equation (3.3.171) is separable and thus, letting $\chi(\tau_1, \eta) = T(\tau_1)\Psi(\eta)$, we find

$$\tau_1 \frac{dT}{d\tau_1} + \lambda T = 0 \quad \text{or} \quad T = \tau_1^{-\lambda} \tag{3.3.172}$$

and

$$\frac{d^2\Psi}{d\eta^2} + \left(\frac{1}{\eta} + 2\eta\right)\frac{d\Psi}{d\eta} + 4(\lambda+1)\Psi = 0 , \tag{3.3.173}$$

where λ is the separation constant. Equation (3.3.173) can be transformed to Laguerre's equation by introducing $\xi = \eta^2$ and $B(\xi) = e^{\xi}\Psi(\sqrt{\xi})$. This leads to

$$\xi \frac{d^2B}{d\xi^2} + (1-\xi)\frac{dB}{d\xi} + \lambda B = 0 . \tag{3.3.174}$$

Since both $w^{(0)}$ and $\zeta^{(0)}$ are bounded at $\bar{r} = 0$, B has to be bounded at $\xi = 0$. To satisfy the matching conditions, (3.3.40), which imply that the axial velocity and vorticity decay *exponentially* in $\bar{r}$, λ must be restricted to be a positive integer. The solutions of (3.3.174) for such values $\lambda = n$, are the Laguerre polynomials (Magnus et al., 1966):

$$B(\xi) = L_n(\xi) = \sum_{m=0}^{n}(-1)^m \binom{n}{m}\frac{\xi^m}{m!} , \tag{3.3.175}$$

and we can represent the functions $\chi = \{W, Z\}$ by a series in $L_n(\xi)\tau_1^{-n}$. Denoting the related expansion coefficients for W and Z by C_n and D_n, respectively, and inverting the transformations in (3.3.170), we obtain $w^{(0)}$ and $\zeta^{(0)}$ in terms of (τ_1, η):

$$w^{(0)} = \frac{1}{S^{(0)}}e^{-\eta^2}\sum_{n=0}^{\infty} C_n L_n(\eta^2)\tau_1^{-(n+1)} \tag{3.3.176}$$

and

$$\zeta^{(0)} = S^{(0)}e^{-\eta^2}\sum_{n=0}^{\infty} D_n L_n(\eta^2)\tau_1^{-(n+1)} . \tag{3.3.177}$$

From the relationship between $v^{(0)}$ and $\zeta^{(0)}$, (3.3.39), we get

$$v^{(0)} = \frac{2\bar{\nu}}{\bar{r}}\left[D_0(1-e^{-\eta^2}) + e^{-\eta^2}\sum_{n=1}^{\infty}\frac{D_n}{\tau_1^n}\left(L_{n-1}(\eta^2) - L_n(\eta^2)\right)\right] . \tag{3.3.178}$$

The coefficients C_n and D_n are determined in terms of the initial profiles, $\zeta_0(\bar{r})$ and $w_0(\bar{r})$, the initial length $S^{(0)}(0) = S_0$, and the arbitrary parameter τ_{10} as

$$C_n = 2S_0\tau_{10}^{n+1}\int_0^\infty w^{(0)}(0,\eta\sqrt{4\bar\nu\tau_{20}})L_n(\eta^2)\eta\,d\eta\ , \tag{3.3.179}$$

$$D_n = \frac{2\tau_{10}^{n+1}}{S_0}\int_0^\infty \zeta^{(0)}(0,\eta\sqrt{4\bar\nu\tau_{20}})L_n(\eta^2)\eta\,d\eta\ . \tag{3.3.180}$$

In particular, $C_0 = m_0S_0^2/(4\pi\bar\nu)$ and $D_0 = \Gamma/(4\pi\bar\nu)$. The appearance of the common factor, the exponential function $e^{-\eta^2}$, in (3.3.176) and (3.3.177) for the axial velocity and vorticity and the definition of η in (3.3.168) and (3.3.169) suggest the definition of an effective core size for a filament in three dimensions

$$\delta(t) = \epsilon\sqrt{4\bar\nu\tau_2} = \sqrt{4\nu\tau_2}\ . \tag{3.3.181}$$

This is a generalization of the definition introduced in Subsect. 3.2.2 for the two-dimensional cases in which $t + t_0$ appeared for τ_2.

Note that (3.3.176) and (3.3.177) express the solutions for the core structure, $w^{(0)}S^{(0)}$ and $\zeta^{(0)}/S^{(0)}$, as functions of two new variables, η and $\tau_1(t)$, with coefficients C_n and D_n depending on the initial data. The dependence of the solutions on the motion of the filament appears implicitly via the new variables, which are related to the physical variables t and $\bar r$ and to the length of the filament, $S^{(0)}(t)$, by the transformations (3.3.168) and (3.3.169). The contribution of the core structure to the velocity of the filament is contained in $C_v(t)$ and $C_w(t)$, defined by (3.3.164) and (3.3.165). These quantities can be expressed in terms of $S^{(0)}(t)$ when we replace the integration variable $\bar r$ in the integrals in (3.3.164) and (3.3.165) by $\eta/\sqrt{\delta_2(t)}$ and then carry out the integrations in η. Consequently, (3.3.160)–(3.3.165) form a closed system defining the motion of the filament through its centerline velocity $\boldsymbol{X}^{(0)}(t,s)$. A list of all relevant equations of the system will be given in Appendix C.

A computational code for this system was developed by Liu et al. (1986b) and employed to study the interaction of filaments. Numerical examples were presented to show the motion of the filaments leading to the merging or intersection of filaments and employed to identify different types of merging problems for which the asymptotics solutions were no longer applicable. Highlights of these numerical examples and their physical meanings will be reported later in the first section of Chap. 5.

Before doing so, we discuss the determination of an optimal value for the positive constant of integration, τ_{10}, introduced in the definition (3.3.169) of the new time like variable τ_1. Such an optimum choice of τ_{10} was introduced by Ting (1971) for the case in which there is no large axial velocity $w^{(0)} = 0$. The constant τ_{10} is chosen to eliminate the coefficient D_1, in the series representation (3.3.177) of $\zeta^{(0)}$. In this way, the first term alone becomes an optimum approximation to the series for large t or τ_1 with an error $O(\tau_1^{-3})$ instead of $O(\tau_1^{-2})$. The condition $D_1 = 0$ also ensures that the first term yields the correct polar moment, $\langle\bar r^2\zeta^{(0)}\rangle$, for all $t \ge 0$.

When there is also an initial large axial velocity, we cannot expect a single choice of τ_{10} to eliminate both D_1 and C_1 in (3.3.176) and (3.3.177). This dilemma was noted by Callegari and Ting (1978). It can be resolved when it is realized that the governing equations (3.3.158) and (3.3.159) for $w^{(0)}$ and $\zeta^{(0)}$ are coupled only indirectly through the stretching of the filament length, $S^{(0)}(t)$. In the transformations of these two equations to (3.3.171) there is no reason to use the same constant τ_{10} in the transformations of independent variables. We assign two different constants τ_{10}^w and τ_{10} to set $C_1 = 0$ and $D_1 = 0$ for $w^{(0)}$ and $\zeta^{(0)}$, respectively. Noting that $L_1(\xi) = 1 - \xi$, we obtain

$$C_1 = 0 \,, \qquad \tau_{10}^w = \frac{\pi S^{(0)}}{2\bar{\nu} m(0)} \int_0^\infty w^{(0)}(0, \bar{r})\, \bar{r}^3 \, d\bar{r} \tag{3.3.182}$$

and

$$D_1 = 0 \,, \qquad \tau_{10} = \frac{\pi S^{(0)}}{2\bar{\nu} \Gamma} \int_0^\infty \zeta^{(0)}(0, \bar{r})\, \bar{r}^3 \, d\bar{r} \,, \tag{3.3.183}$$

with the mass flux m defined by (3.3.150). Now the two series, (3.3.176) and (3.3.177), are approximated by their first terms accurate to $O(1/t^{-3})$ for large t, provided that we use τ_{10}^w as the constant in (3.3.176) and τ_{10} in (3.3.177).

3.3.6.4 Velocity of a Filament with Similar Core Structure. Equations (3.3.148) and (3.3.149) imply that a vortex filament has two invariant quantities, the circulation Γ and the axial mass flux times the length square, $m[S^{(0)}]^2$. Let us consider the case that the initial axial velocity and vorticity profiles are consistent with the first terms of (3.3.176) and (3.3.177), and that they have the same effective core size δ_0. The latter defines the same initial value for τ_1; that is, $\tau_1(0) = \tau_{10}^w = \tau_{10} = S(0)\delta_0^2/(4\nu)$. Then we have $C_n = D_n = 0$ for $n \geq 1$ and the axial velocity and vorticity in the vortical core are given respectively by the two first terms in (3.3.176) and (3.3.177) for $t \geq 0$. We call them the *similarity solution.* The special core structure, called the *similar* core structure, is a generalization of a rectilinear Lamb vortex with the same initial core size δ_0, circulation Γ but no axial flow to a toroidal filament with circulation Γ and axial mass flux $m(t) = m(0)[S_0/S^{(0)}(t)]^2$. For a Lamb vortex, $t_0 = \delta_0^2/(4\nu) = \tau_{10}/S(0)$, denotes the initial age of the vortex if it was created with zero core size. Assuming that the length of the filament remains of order one, then τ_{10}/S_0 measures the age of the filament at $t = 0$.

We recall that, for a filament with nonzero initial core size, the first terms in the series (3.3.176) and (3.3.177) provide the long-time behavior of the core structure. In particular, when the optimum choices of τ_{10} for $w^{(0)}$ and $\zeta^{(0)}$ are the same, we call their one-term solutions the *optimum similarity solution.* In practical applications and experiments, it is most likely that the circulation, Γ, the axial mass flux, m(0), the effective core size $\delta(0)$ and the initial configuration $\boldsymbol{X}(0, s)$ are the only initial data available. On the other hand, they are just the initial data needed for the motion of a filament with a similar core structure. For this reason, we assemble below the complete set of equations for the motion and decay of the filament.

Keeping only the first terms in (3.3.176)–(3.3.178), we obtain the similar core structure,

$$w^{(0)} = \frac{C_0}{\tau_1 S^{(0)}} e^{-\eta^2} = \frac{m(t)}{\pi \bar{\delta}^2} e^{-(\bar{r}/\bar{\delta})^2} , \tag{3.3.184}$$

$$\zeta^{(0)} = \frac{S^{(0)} D_0}{\tau_1} e^{-\eta^2} = \frac{\Gamma}{\pi \bar{\delta}^2} e^{-(\bar{r}/\bar{\delta})^2} , \tag{3.3.185}$$

$$v^{(0)} = \frac{\Gamma}{2\pi \bar{r}} \left[1 - e^{-(\bar{r}/\bar{\delta})^2} \right] , \tag{3.3.186}$$

where

$$\bar{\delta} = \delta/\epsilon = \sqrt{4\bar{\nu}\tau_2} \, , \qquad \tau_2(t) = \tau_1(t)/S^{(0)}(t) \tag{3.3.187}$$

and

$$\tau_1 = \int_0^t S^{(0)}(t')dt' + \tau_{10} \, , \quad m(t) = m(0) \left[\frac{S_0}{S^{(0)}(t)} \right]^2 . \tag{3.3.188}$$

The global contributions of the core structure to the motion of the filament are

$$C_w = -\frac{2m^2(0)}{\Gamma^2 \bar{\delta}^2} \left[\frac{S_0}{S^{(0)}(t)} \right]^4 \tag{3.3.189}$$

and

$$C_v = -\ln \bar{\delta} + \frac{\gamma + 1 - \ln 2}{2} = -\ln \bar{\delta} + 0.442 \, , \tag{3.3.190}$$

where $\gamma = 0.577$ denotes Euler's constant. The three components of the velocity of the centerline of the filament given by (3.3.160)–(3.3.162) become

$$\dot{\boldsymbol{X}}^{(0)} \cdot \boldsymbol{\tau}^{(0)} = 0 \, , \tag{3.3.191}$$

$$\dot{\boldsymbol{X}}^{(0)} \cdot \boldsymbol{n}^{(0)} = \boldsymbol{Q}_0 \cdot \boldsymbol{n}^{(0)} \, , \tag{3.3.192}$$

and

$$\dot{\boldsymbol{X}}^{(0)} \cdot \boldsymbol{b}^{(0)} = \boldsymbol{Q}_0 \cdot \boldsymbol{b}^{(0)} + \frac{\Gamma \kappa^{(0)}}{4\pi} \left[\ln \frac{1}{\delta} + 0.442 + C_w \right] . \tag{3.3.193}$$

Recall that the sign of the parameter s or the direction of the tangent vector of $\mathcal{C}$ is chosen so that the circulation, Γ, is positive. From (3.3.193), we observe that the contribution of the swirling flow to the binormal velocity is always positive and proportional to Γ. The contribution of the axial flow, C_w, is always negative, according to (3.3.189), regardless whether the axial mass flux is positive or negative. This says,

• *A large axial flow in the core retards the binormal velocity of the filament.*

As mentioned before, $\tau_{10}/S(0)$ measures the initial age of the filament. In a case where $\tau_{10} = 0$, the filament is created at $t = 0$ as a vortex line with zero core radius. The contributions, C_v and C_w, to the binormal velocity of the filament are then singular at $t = 0$. From (3.3.189) and (3.3.190) we see that $C_v \sim \ln t$ is integrable from $t = 0$ while $C_w \sim 1/t$ is not. Therefore, (3.3.193) for the binormal velocity is not valid near the creation of the filament unless the axial velocity is much smaller than the swirling flow, that is, $w^{(0)} = 0$ or $m(0) = 0$. The reason can be traced to the similarity solutions for the axial and swirling velocities, (3.3.184) and (3.3.186). Note that $w^{(0)} \sim \bar{\delta}^2 \sim 1/t$ while $v^{(0)} \sim \bar{\delta} \sim \sqrt{t}$. Therefore, $w^{(0)} \gg v^{(0)}$ as $t \to 0$. To derive a formula for the binormal velocity valid near $t = 0$, we would have to carry out an initial layer analysis with a short-time variable and expansions for the velocity components different from (3.3.23)–(3.3.25) so that the axial velocity can be much larger than the swirling velocity. We shall not pursue this analysis here.

From (3.3.161), (3.3.191)–(3.3.193), (3.3.167) and (3.3.168), (3.3.169) we see that the motion of the filament with similar core structure is completely determined by the initial configuration of its centerline, $\boldsymbol{X}(0, s)$, and three constants, the circulation Γ, the initial axial flux $m(0)$, and the initial core size $\bar{\delta}(0)$ or its initial age τ_{10}/S_0.

Thus, the matched asymptotic analysis for a filament is completed. We have the equations of motion of the centerline of the filament and the inner solution for the core structure. The outer solution for the vector potential $\boldsymbol{A}$ and the velocity $\boldsymbol{v}$ induced by the filament are given by the Biot–Savart formulae (3.1.61) and (3.1.62).

In the next chapter, we find occasions where a representation for the vector potential $\boldsymbol{A}$ uniformly valid in and away from the filament is needed. Therefore, we describe here the construction of a uniformly valid composite solution.

3.3.6.5 The Leading-Order Composite Solution for the Vector Potential. The behavior of the outer vector potential near the filament, that is, in the overlapping region, is described by (3.1.72)–(3.1.76). In particular, the singular part of $\boldsymbol{A}$ is

$$\boldsymbol{A}^s = \frac{\Gamma \boldsymbol{\tau}}{2\pi} \ln \frac{S}{r} \ . \tag{3.3.194}$$

We need the inner solution for the vector potential and its behavior in the far field as $\bar{r} \to \infty$, that is, in the overlapping region, so that we can construct the composite solution.

Since the leading-order core structure is independent of θ and s, the inner vector potential can be related to the velocity components by simple quadratures. The results are

$$\bar{\boldsymbol{A}} = \boldsymbol{\theta} \bar{A}_2 + \boldsymbol{\tau} \bar{A}_3 \ , \tag{3.3.195}$$

with

$$\bar{A}_2 = \frac{1}{\bar{r}} \int_0^r \epsilon^{-1} w^{(0)}\, r'\, dr' = \frac{1}{\bar{r}} \int_0^{\bar{r}} w^{(0)}(t, \bar{r}')\, \bar{r}'\, d\bar{r}' \qquad (3.3.196)$$

and

$$\bar{A}_3 = -\int_0^r \epsilon^{-1} v^{(0)}\, dr' = -\int_0^{\bar{r}} v^{(0)}(t, \bar{r}')\, d\bar{r}' \,. \qquad (3.3.197)$$

Here $\bar{\boldsymbol{A}}$ denotes the leading-order solution, which is of order one, and it has only circumferential and axial components, $\bar{A}_2$ and $\bar{A}_3$. Since $w^{(0)}$, given by (3.3.176), decays exponentially in $\bar{r}$, the far-field behavior of $\bar{A}_2$ is

$$\bar{A}_2 = \frac{m(t)}{2\pi\bar{r}} + o(\bar{r}^{-m}) \,, \qquad \text{for all } m \,. \qquad (3.3.198)$$

Therefore, its contribution to the overlapping region is null. With $v^{(0)}$ given by (3.3.178), the far-field behavior of $\bar{A}_3$ is

$$\bar{A}_3 = \frac{\Gamma}{2\pi} \ln\frac{\bar{\delta}}{\bar{r}} + c_3 + o(\bar{r}^{-m}) \,, \qquad \text{for all } m \,, \qquad (3.3.199)$$

where

$$c_3(t) = -\int_0^{\bar{\delta}} v^{(0)}(t, \bar{r}')\, d\bar{r}' - \int_{\bar{\delta}}^{\infty} \left[v^{(0)}(t, \bar{r}') - \frac{\Gamma}{2\pi\bar{r}'} \right] d\bar{r}' \,. \qquad (3.3.200)$$

Note that the singular part of the outer solution, (3.3.194), is removed or matched by the inner solution. The first two terms on the right-hand side of (3.3.199) represent the contribution of the inner solution in the overlapping region and should be removed in the composite solution. The composite vector potential is

$$\boldsymbol{A} = \tilde{\boldsymbol{A}} + \bar{\boldsymbol{A}} - \boldsymbol{\tau} \left[\frac{\Gamma}{2\pi} \ln\frac{\bar{\delta}}{\bar{r}} + c_3 \right] \,, \qquad (3.3.201)$$

where $\tilde{\boldsymbol{A}}$ denotes the outer solution defined by the Biot–Savart formula (3.1.61).

When the filament has a similar core structure, defined by (3.3.184)–(3.3.186), we obtain from (3.3.195) and (3.3.196) the formulae for the inner vector potential

$$\bar{A}_2 = \frac{m(t)}{2\pi\bar{r}} \left[1 - e^{-(\bar{r}/\bar{\delta})^2} \right] \qquad (3.3.202)$$

and

$$\bar{A}_3 = -\frac{\Gamma}{4\pi} \int_0^{\eta^2} \frac{1 - e^{-\xi}}{\xi} d\xi = -\frac{\Gamma}{4\pi} \left[E_1(\eta^2) - 2\ln\frac{\bar{\delta}}{\bar{r}} + \gamma \right] \,, \qquad (3.3.203)$$

where $\eta = \bar{r}/\bar{\delta} = r/\delta$ and E_1 and γ denote the exponential integral and the Euler's constant, respectively. From (3.3.200), we find that the term $c_3(t)$ in the composite solution (3.3.201) becomes a constant,

$$c_3(t) = -\frac{\gamma \Gamma}{4\pi} = -0.004434\ \Gamma\ . \tag{3.3.204}$$

We have now completed the derivation of the asymptotic solution by Callegari and Ting (1978), described in Sect. 2.3 of Book-I. We recall that the solution was obtained for a slender filament submerged in a background flow under the restrictions that the background flow is irrotational, its centerline $\mathcal{C}$ forms a simple closed curve of length $O(\ell)$, and the flow in its vortical core, the inner solution, remains incompressble and does not vary along $\mathcal{C}$. These three restrictions will be removed one by one in Subsects. 3.3.7 and 3.3.8 and in Sect. 3.4. Next we highlight the analysis of Klein and Ting (1992) accounting for axial core structure variation.

3.3.7 Filament with Axial Core Structure Variation

The assumption of no axial core structure variation, (3.3.63), was introduced in Subsect. 3.3.3 to simplify the solution of the core structure. The equations and conclusions, obtained prior to the assumption (3.3.63), remain applicable for a core structure having axial variations. We shall recount some of these equations and conclusions noting the dependence on the axial variable s. An extensive study, not recounted here, that includes fast axial waves on the vortex core was presented by Margerit (2002) (see also Margerit et al. 2001).

From the leading-order continuity equation and the radial and circumferential components of the momentum equation, we conclude that the core structure has to be axisymmetric, that is,

- $v^{(0)}, w^{(0)}, p^{(0)}$ *are independent of* θ *with* $p^{(0)}$ *related to* $v^{(0)}$ *by* $p^{(0)}_{\bar{r}} = [v^{(0)}]^2/\bar{r}$. *But* $v^{(0)}$ *and* $w^{(0)}$ *are unknown functions of* $t, \bar{r}, s$ *to be determined by the compatibility conditions on the higher-order equations.*

The higher-order equations are linear; again, we represent the solutions by Fourier series in θ or ϕ and obtain the equations for their Fourier coefficients. We denote the symmetric and asymmetric parts by the subscripts c and a, respectively. The θ-averages of these equations yield the equations for the symmetric parts and the differences of the equations from their θ-averages become the equations for the asymmetric parts. Note that in the first order equations, (3.3.41)–(3.3.44), all the ∂_s terms $v^{(0)}_s$, $w^{(0)}_s$, and $p^{(0}_s$, are independent of θ and appear only in the axisymmetric parts of the equations. In their asymmetric parts, s can be treated as a parameter. For example, the symmetric and asymmetric parts of the first-order continuity equation (3.3.41) are

$$\sigma^{(0)}(\bar{r}u^{(1)}_c)_{\bar{r}} + \bar{r}w^{(0)}_s = 0 \quad \text{and} \quad (v^{(1)}_a)_\theta + (\bar{r}u^{(1)}_a)_{\bar{r}} = \bar{r}v^{(0)}\kappa^{(0)}\sin\phi^{(0)}\ . \tag{3.3.205}$$

Thus the asymmetric parts of the first order equations, (3.3.64)–(3.3.81) after the assumption (3.3.63), remain valid with variable s treated as a parameter.

Note also that in the leading- and first order equations, ∂_t does not appear; hence, time t can also be treated as a parameter. The relevant conclusions are as follows:

- *the asymmetric solutions,* $v_a^{(1)}, w_a^{(1)}$*, and* $p_a^{(1)}$*, have only a* $\cos\phi$ *term and* $u_a^{(1)}$ *has only a* $\sin\phi$ *term, and their Fourier coefficients depend on* $v^{(0)}$ *and* $w^{(0)}$*.*

The velocity of the centerline is given by (3.3.77) and (3.3.81). It is

$$\dot{\boldsymbol{X}}^{(0)} = \hat{n}^{(0)}\boldsymbol{Q}_f \cdot \hat{n}^{(0)} + \hat{b}^{(0)}\frac{\Gamma\kappa^{(0)}}{4\pi}\ln\frac{1}{\epsilon} + \boldsymbol{Q}_f \cdot \hat{b}^{(0)} + \kappa^{(0)}C^* \ , \tag{3.3.206}$$

where C^*, which denotes the global contribution of the core structure, defined by (3.3.80), is now a function of t and s. Because of C^*, the velocity of the filament, given by (3.3.206), is coupled with the core structure.

The symmetric part of the continuity equation, (3.3.205), relates the scaled axial mass flux, $\mathcal{M}^{(0)}$, and the θ-averaged radial velocity, $u_c^{(1)}$, via

$$\mathcal{M}^{(0)}(t,\bar{r},s) = \int_0^{\bar{r}} w^{(0)}\bar{r}'d\bar{r}' = \bar{M} \quad \text{and} \quad \sigma^{(0)}\bar{r}u_c^{(1)} = -\partial_s\mathcal{M}^{(0)} \ . \tag{3.3.207}$$

We use $\bar{M}$ instead of $\mathcal{M}^{(0)}$, when we introduce the von Mises variables, that is, replace the independent variables $(t,\bar{r},s)$ by $(t,\bar{M},s)$. We define the s-derivative along a streamline, $\bar{r} = R(t,s,\bar{M})$, at instant t by $D_s = \partial_s + R_s\partial_{\bar{r}}$ with $R_s = -\mathcal{M}_s^{(0)}/\mathcal{M}_{\bar{r}}^{(0)} = \sigma^{(0)}u^{(1)}/w^{(0)}$. The first order circumferential and axial momentum equations yield one condition (3.3.54) for the leading-order circulation, $\mathcal{G}^{(0)}(t,\bar{r},s) = \bar{r}v^{(0)}$, around a stream tube and another condition, (3.3.55), which can be reduced to an equation for the leading-order total head, $\mathcal{H}^{(0)} = p^{(0)} + [(v^{(0)})^2 + (w^{(0)})^2]/2$, or the leading-order total head in the moving frame, $\mathcal{H} = p + \boldsymbol{V}^2/2$. These two compatibility conditions are

$$\begin{aligned} &D_s\mathcal{G}^{(0)} = 0, &&\text{and} \quad D_s\mathcal{H}^{(0)} = 0, &&\text{or} \\ &\mathcal{G}^{(0)} = \bar{\mathcal{G}}(t,\bar{M}) &&\text{and} \quad \mathcal{H}^{(0)} = \bar{\mathcal{H}}(t,\bar{M}) \ . \end{aligned} \tag{3.3.208}$$

These conditions hold for any quasi-steady inviscid flow. They are valid for the leading-order solutions, because the unsteady and the viscosity terms begin to appear only in the higher-order equations. Thus, we have

- *the leading-order circulation* $\mathcal{G}^{(0)}$ *around and the total head* $\mathcal{H}^{(0)}$ *on a stream tube should remain constant along the tube. The temporal variations of* $\mathcal{G}^{(0)}$ *and* $\mathcal{H}^{(0)}$ *are yet to be defined by the compatibility conditions for the symmetric second-order equations using the periodicity condition (3.3.2) on* s*.*

Now we proceed to derive these two compatibility conditions in the same manner as in Callegari and Ting (1978) and Ting and Klein (1991). The symmetric part of the second-order continuity equation is fulfilled by the introduction of the first-order stream function $\mathcal{M}^{(1)}(t,\bar{r},s)$ with

$$\sigma^{(0)}\bar{r}u_c^{(2)} + \sigma^{(1)}\bar{r}u_c^{(1)} + \frac{\bar{r}^2}{2}(\dot{\boldsymbol{X}}^{(0)})_s \cdot \hat{\tau} = -\mathcal{M}_s^{(1)} \quad \text{and} \quad \bar{r}w^{(1)} = \mathcal{M}_{\bar{r}}^{(1)} \,. \tag{3.3.209}$$

The symmetric second-order circumferential momentum equation yields

$$\sigma^{(0)}u_c^{(1)}(\mathcal{G}_c^{(1)})_{\bar{r}} + w^{(0)}(\mathcal{G}_c^{(1)})_s + \left[\sigma^{(0)}u_c^{(2)} + \sigma^{(1)}u_c^{(1)}\right]\mathcal{G}_{\bar{r}}^{(0)} + w_c^{(1)}\mathcal{G}_s^{(0)}$$
$$= \sigma^{(0)}\left\{-\mathcal{G}_t^{(0)} + \bar{\nu}\left[\mathcal{G}_{\bar{r}\bar{r}}^{(0)} - \mathcal{G}_{\bar{r}}^{(0)}/\bar{r}\right]\right\} . \tag{3.3.210}$$

Here $\mathcal{G}^{(0)}$ and $\mathcal{G}^{(1)}$ are considered functions of $t, \bar{r}$, and s. Using (3.3.207), (3.3.208), and (3.3.209), we write the left-hand side of (3.3.210) as $wD_s[\mathcal{G}_c^{(1)} - \mathcal{G}_{\mathcal{M}^{(0)}}^{(0)}\mathcal{M}^{(1)}] - (\bar{r}/2)\mathcal{G}_{\bar{r}}^{(0)}(\dot{\boldsymbol{X}}^{(0)})_s \cdot \hat{\tau}$. Now we make use of the periodicity condition (3.3.2), integrate (3.3.210) along a streamline to eliminate the first-order solutions on its left-hand side, and arrive at an equation for the evolution of leading-order circulation $\mathcal{G}^{(0)}$,

$$\int_0^{S_0} \frac{\sigma^{(0)}ds}{w^{(0)}}\left\{-\mathcal{G}_t^{(0)} + \frac{\bar{r}}{2}\mathcal{G}_{\bar{r}}^{(0)}\left(\dot{\boldsymbol{X}}^{(0)}\right)_s \cdot \hat{\tau} + \bar{\nu}\left[\mathcal{G}_{\bar{r}\bar{r}}^{(0)} - \frac{\mathcal{G}_{\bar{r}}^{(0)}}{\bar{r}}\right]\right\} = 0 \,, \tag{3.3.211}$$

with $\bar{r} = R(t,s,\bar{M})$. To identify the second compatibility condition as an evolution equation for the total head $\mathcal{H}^{(0)}$, it is clearer to derive the condition from the energy equation for an incompressible flow in the moving frame instead of the axial momentum equation. As expected, the leading- and first order energy equations yield $\mathcal{H}_0^{(0)} = 0$ and the second equation in (3.3.208). The second-order energy equation yields

$$\sigma^{(0)}u_c^{(1)}(\mathcal{H}_c^{(1)})_{\bar{r}} + w^{(0)}(\mathcal{H}_c^{(1)})_s + [\sigma^{(0)}u_c^{(1)} + \sigma^{(1)}u_c^{(1)}]\mathcal{H}_{\bar{r}}^{(0)} + w_c^{(1)}\mathcal{H}_s^{(0)}$$
$$= \sigma^{(0)}\{-\mathcal{H}_t^{(0)} + p_t^{(0)} + \bar{\nu}[\frac{\mathcal{G}^{(0)}}{\bar{r}^2}(\mathcal{G}_{\bar{r}\bar{r}}^{(0)} - \frac{\mathcal{G}_{\bar{r}}^{(0)}}{\bar{r}}) + \frac{w^{(0)}}{\bar{r}}(\bar{r}\bar{w}_{\bar{r}}^{(0)})_{\bar{r}}]\} . \tag{3.3.212}$$

The terms on the left-hand side are equal to those in (3.3.210) with $\mathcal{G}^{(0)}$ and $\mathcal{G}^{(1)}$ replaced by $\mathcal{H}^{(0)}$ and $\mathcal{H}^{(1)}$, respectively. Thus, we can likewise remove the first order solutions on the left-hand side of (3.3.212) and arrive at an evolution for $\mathcal{H}^{(0)}$. It is

$$\int_0^{S_0} ds \;\; [\sigma^{(0)}/w^{(0)}]\left\{-\mathcal{H}_t^{(0)} + p_t^{(0)} + \tfrac{1}{2}\left[\bar{r}\mathcal{H}_{\bar{r}}^{(0)} - (w^{(0)})^2\right]\left(\dot{\boldsymbol{X}}_s^{(0)} \cdot \hat{\tau}\right)\right.$$
$$\left. + \bar{\nu}\bar{r}^{-3}\left[\bar{r}\mathcal{G}^{(0)}\left(\bar{r}\mathcal{G}_{\bar{r}\bar{r}}^{(0)} - \mathcal{G}_{\bar{r}}^{(0)}\right) + w^{(0)}\bar{r}^2\left(\bar{r}\bar{w}_{\bar{r}}^{(0)}\right)_{\bar{r}}\right]\right\} = 0 \,, \tag{3.3.213}$$

along a streamline, $\bar{r} = \bar{R}(t,s,\bar{M})$. Note that (3.3.211) and (3.3.213) reduce to (3.3.144) and (3.3.145) in Ting and Klein (1991) for $v^{(0)}$ and $w^{(0)}$ independent of s. Here they do depend on s; therefore, we prefer to work with $\mathcal{G}^{(0)}$ and $\mathcal{H}^{(0)}$, which are independent of s in the von Mises variables, $(t, \bar{M}, s)$. We can replace $\mathcal{G}_t^{(0)}$ by $\bar{\mathcal{G}}_t + \bar{\mathcal{G}}_{\bar{M}} \bar{M}_t$ and the integral of the first term in (3.3.211) by $\bar{\mathcal{G}}_t \int \sigma^{(0)} ds/w^{(0)} + \bar{\mathcal{G}}_{\bar{M}} \int \mathcal{M}_t^{(0)} \sigma^{(0)} ds/w^{(0)}$ and use (3.3.211) to evaluate $\bar{\mathcal{G}}_t(t, \bar{M})$. Likewise, (3.3.213) can be considered as an explicit equation for $\bar{\mathcal{H}}_t$. Thus, we have

- *a closed system of equations for the motion of the center line of the filament and the evolution of the core structure. The equations are (3.3.206), (3.3.211), and (3.3.213) coupled with the equations relating the circulation $\bar{\mathcal{G}}$, the total head $\bar{\mathcal{H}}$, the pressure $p^{(0)}$, and the axial mass flux $\mathcal{M}^{(0)}$ to the velocity components $v^{(0)}$ and $w^{(0)}$.*

We recall that the asymptotic analyses for small vortex spots presented in Sect. 3.2 and for slender vortex filaments in Subsects. 3.3.2–3.3.7 assumed that the background flow is irrotational. Under this assumption, the motion and decay of the filaments (or vortex spots) induced only higher-order corrections to the background flows. Therefore, we could concentrate on the analysis of the leading-order inner solution, that is, the velocity of the center line and the evolution of the core structure. But this assumption is not necessary for the analysis. It needs only the assumption that the ratio of the length scales of the inner and outer regions is small, $\epsilon = \delta^*/\ell$, which serves as the small parameter for the expansion scheme. It was shown by Liu and Ting (1987) that the method of matched asymptotics remains applicable even if the outer solution is rotational with velocity and vorticity $O(1)$. The analyses for the leading-order inner solution, the core structure, and the formula for the velocity of the filament remain the same as those for a potential outer flow. But now the velocity induced by the vortices changes the convection of the outer vorticity field and hence the leading-order outer solution of a rotational flow will be unsteady and coupled with the motion of the vortices even if the initial outer solution, in absence of the vortices, would remain steady. Numerical examples demonstrating the coupling of a two-dimensional rotational flow with the motion of vortices were presented by Liu and Ting (1987). Examples of two-dimensional unsteady flows were presented to simulate the drift of the vortex lines trailing an airplane during landing or take-off in the presence of a cross wind near the ground. The results demonstrate that the coupling with a background rotational flow changes drastically the motion of the vortices. Highlights of the studies conducted by Liu and Ting (1987) are presented below.

3.3.8 Vortices in a Background Flow with Order One Vorticity

Here we consider a two-dimensional incompressible flow with length and velocity scales ℓ and U in the outer region, that is, away from the vortices of typical core size $\delta^* = \epsilon\ell$, but the scaled vorticity in the background flow can be of order one, that is,

$$|\boldsymbol{\Omega}_b(t,\mathbf{x})|/(U/\ell) = O(1) , \tag{3.3.214}$$

instead of zero. At a high Reynolds number, $U\ell/\nu = O(\epsilon^{-2})$, the outer solution obeys the Euler equations up to order ϵ^2. The vorticity evolution equation for an incompressible flow is

$$\boldsymbol{\Omega}_t + \mathbf{u}\cdot\nabla\boldsymbol{\Omega} - \boldsymbol{\Omega}\cdot\nabla\mathbf{u} = \epsilon^2\bar{\nu}\,\Delta\boldsymbol{\Omega} . \tag{3.3.215}$$

If the background flow without the filament is irrotational initially, $\boldsymbol{\Omega}_b(0.\boldsymbol{x}) = 0$, so is the outer solution. Then (3.3.215) yields $\boldsymbol{\Omega} \equiv 0$ for the leading two orders and then for all orders of ϵ.

If the outer solution is rotational initially, the leading-order vorticity field in the outer region will be unsteady due to the motion of the filament, even if the initial background flow without the filament(s) corresponds to a steady rotational flow. Since the velocity $\mathbf{u}$ in the convection terms in (3.3.215) includes the velocity induced by the filament(s), the leading outer solution is coupled with the inner solution(s) of the filament(s).

With the strength of a filament being of order one, that is, $\Gamma/(U\ell) = O(1)$, the velocity and vorticity in the core (the inner region) are of order ϵ^{-1} and ϵ^{-2}, respectively. The moving curvilinear coordinate system and the expansion scheme for the inner solution introduced in Sect. 3.2 and those in Subsect. 3.3.1 remain applicable. In particular, we mention the expansion (3.3.27) for the vorticity,

$$\boldsymbol{\Omega}(t,\bar{r},\theta,s) = \epsilon^{-2}\boldsymbol{\Omega}^{(0)}(t,\bar{r},\theta) + \epsilon^{-1}\boldsymbol{\Omega}^{(1)} + \boldsymbol{\Omega}^{(2)} + \cdots , \tag{3.3.216}$$

and its matching conditions with the outer solution,

$$\boldsymbol{\Omega}^{(0)} \to 0 , \qquad \boldsymbol{\Omega}^{(1)} \to 0 , \qquad \boldsymbol{\Omega}^{(2)} \to \boldsymbol{\Omega}_b(t,\mathbf{X}) , \tag{3.3.217}$$

and higher-order conditions as $\bar{r} \to \infty$. Noting that in the matching conditions the local outer vorticity appears only in the third condition for $\boldsymbol{\Omega}^{(2)}$, the first two conditions in (3.3.217) are the same as those for a background potential flow and can be replaced by the stronger conditions (3.3.33) for $n = 0, 1$. Similarly, the matching conditions for the leading two terms of the velocity in the inner region remain the same as given by (3.3.31). Consequently, the leading- and the first order solutions $\boldsymbol{\Omega}^{(0)}$ and $\boldsymbol{\Omega}^{(1)}$ for the core structure are independent of the outer vorticity distribution regardless whether it is of order one or identically 0. Now we infer directly from the scalings the effect of order one vorticity in an outer solution on an inner solution. In the inner region of the length scale $\epsilon\ell$, the background vorticity in a normal plane of the centerline $\mathcal{C}$ deviates from the value on $\mathcal{C}$ by an order ϵ, even though the circumferential component can be of order ϵ^{-1}. The axial flow and the motion of $\mathcal{C}$ can only induce an order one variation in the background vorticity. This is true even for a large axial flow so long as $\mathcal{C}$ is a closed curve of order one length, scaled by ℓ. We note that an order one variation in the core structure changes the strength of a filament by an order ϵ^2. Here we define the strength of a filament by the area integral of the vorticity given by the inner solution minus the local value of the outer solution over the normal plane of $\mathcal{C}$. The order one strength is given by the area integral of $\epsilon^{-2}\boldsymbol{\Omega}^{(0)}$, which is the same as that with a background potential flow and hence is time invariant.

Matched asymptotic analysis was carried out for two-dimensional problems by Liu and Ting (1987) and employed to model the drift of trailing vortices in a spanwise shear flow, that is, a cross wind near the ground.[6] The analysis was carried out for axisymmetric problems by Ishii and Liu (1987) and employed to simulate the interactions of vortex rings submerged in a jet. We note that the background vorticity field will remain undisturbed by the presence of vortex filament(s) if the background flow is a simple shear flow with constant vorticity.

Liu and Ting (1987) carried out the matched asymptotic analysis in a single time scale and confirmed the results infered above from scalings that the leading-order core structure and the velocity of the vortex spot or filament are the same as those in Sect. 3.2 or Subsect. 3.3.3–3.3.6 for a background potential flow. In general, there is an order one temporal variation of the background vorticity and hence an order one variation of the outer solution. The motion and decay of the filament are now coupled with the Euler solution of the outer region. A numerical scheme for two-dimensional vortical flows was developed by Liu and Ting (1987). The grid size and time step for the Euler solver of the background flow and the motion of the vortex spot(s) depend only on the length and velocity scales of the background flow and not on the effective core size of the vortical spots. The contributions of the velocity field of a vortex spot at nearby grid points are accounted for by a method of average because the spatial variation of the background vorticity in the length of a core size is of higher order.

Here we quote a numerical example from Liu and Ting (1987) to demonstrate the interaction of a Lamb vortex defined by (3.2.64) with a background rotational flow in the upper half xy plane (above the ground $y = 0$). Initially, the center of the vortex is located at $X = 0$, $Y = 1$, with core size $\delta_0 = 10^{-3}$ and the background shear flow is $\mathbf{Q}(0, \mathbf{x}) = \hat{\imath}U_0(y) = \hat{\imath}[1 - e^{-y}]$ with vorticity $\zeta_b(x, y) = -e^{-y}$ with $\ell = 1$, $U = 1$, and $4\nu = 10^{-6}$. The background shear flow fulfills the nonslip condition along the wall $y = 0$ but the inviscid outer solution induced by the vortex does not, and requires the addition of a boundary layer along the wall. As long as the distance from the vortex center to the wall is much greater than the sum of the core size and the boundary layer thickness, we can consider the outer solution to be an inviscid flow with the

[6] In this model, the strength of a trailing vortex is the circulation at the root section of a wing, $\Gamma = O(Vc_0\alpha)$, where V, c_0, and α denote respectively a typical landing speed, the root chord, and an effective angle of attack. V is much larger than the maximum speed U of the cross wind while c_0 is much smaller than the semispan ℓ or the vertical scale L of the ground wind shear. Thus we could consider $\Gamma = O(UL)$ instead of $O(Vc_0)$.

boundary condition of $v = 0$ on $y = 0$. The Euler solution of the outer region is coupled with the motion and decay of the inner solution, the Lamb vortex. Shown in Fig. 3.8 are the trajectories of the Lamb vortices with strengths ranging from $\Gamma = -3$ to 3.

The results show that a Lamb vortex with positive circulation, $\Gamma > 0$, drifts downstream (x-direction) and upward (y-direction) while a vortex with negative circulation drifts downstream and downward. The latter eventually ($t > t^*$) turns around and drifts upstream. These contrasting phenomena are more pronounced as $|\Gamma|$ increases. To explain this, we consider the case of a vortex with $\Gamma > 0$. The fluid moves downward behind the vortex ($x < X$) and upward ahead of it ($x > X$). For an initial background vorticity ζ_b with $\zeta_b'(y) > 0$, which is the case in Fig. 3.8, the disturbed flow increases the vorticity behind the vortex and decreases the vorticity ahead of it, that is, $\tilde{\zeta} > 0$ for $x < X$ and $\tilde{\zeta} < 0$ for $x > X$. The background vorticity variation $\tilde{\zeta}$, in turn, induces an upward motion of the vortex. Applying similar arguments for $\Gamma < 0$, we find that the background vorticity variation $\tilde{\zeta}$ induces a downward motion of the vortex. Note that a vortex of negative strength turns around and drifts upstream as it gets closer to the ground because its forward velocity induced by the background shear flow is eventually overcome by the backward velocity induced by its image with respect to the wall as $Y \to 0^+$. It should be pointed out once more that these phenomena will appear only when the background vorticity is nonuniform. When the background flow is either a uniform flow ($\zeta_b = 0$) or a constant shear flow (ζ_b = constant), the vortex drifts horizontally with no change in the background ($\tilde{\zeta} \equiv 0$).

Additional examples on the motion of a vortex pair in a nonuniform shear flow and examples on the motion and decay of the vortex ring(s) in an axisymmetric jet were presented in Liu and Ting (1987) and Ishii and Liu (1987). Those examples show that the trajectories of the (ring) vortices and the variation of the outer solution depend strongly on the signs of the circulation Γ and the (effective) vorticity gradient of the background flow.

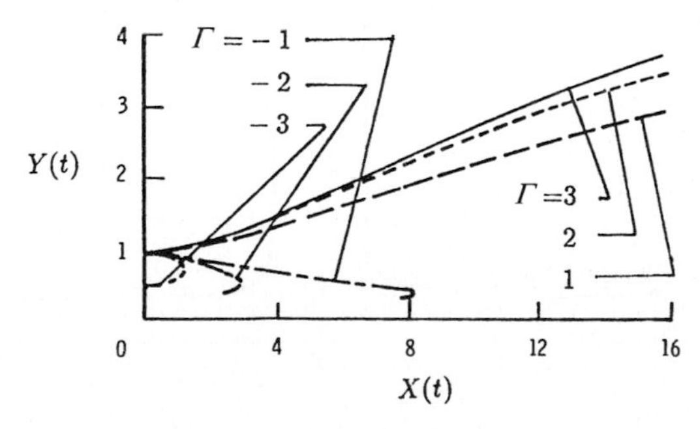

Fig. 3.8. The effect of the strength of a Lamb vortex on its trajectory in a shear flow. Adapted from Liu and Ting (1987)

Figure 3.9 shows the trajectories of a vortex pair in the upper half xy plane with a background shear flow parallel to the x-axis. This unsteady two-dimensional flow models the drifts of a pair of vortex lines trailing an airplane moving with velocity $-W_\infty \boldsymbol{k}$ above the ground, the z^*x plane. The latter is a steady three-dimensional flow in the Cartesian coordinates x, y, z^* moving with the airplane. The model changes the coordinate z^* to the stationary coordinate z by the transformation $z = z^* - W_\infty t$ with the cross section of the trailing vortex lines in plane $z = z^*$ representing the initial vortex pair in the xy plane. The subscript K of the coordinates (X_k, Y_k) stands for L and R, the left and the right vortex, respectively. This model implies that the length scale of the flow variation in z^* is much larger than that in the vertical plane. The latter is of the order of the span of the wing.

The ground shear flow used in Fig. 3.9 is the exponential profile $U_\infty(y) = 1 - e^{-y}$ with scale height equal to 1. The vortical cores of the pair are similar solutions with initial effective radii $\delta(0) \ll 1$. Figure 3.9 shows the relative trajectories of a vortex pair whose strengths are $\Gamma_K = \pm 1$. By a relative trajectory we mean that the ordinate denotes Y_K while the abscissa denotes the deviation of X_K from the mean value, $[X_R(t) - X_L(t)]/2$.

The solid curves shows the well-known trajectories of the vortex pair without a background flow, or their relative trajectories in a uniform flow. The trajectories are initially dominated by their mutual interactions and later by their images with respect to the ground $y = 0$, with the assumption that the minimum height of their trajectories remains above the boundary layer of the slip flow along $y = 0$ induced by the vortex pair and their images. Curves 3, 4 and 5, 6 in dash lines show the relative trajectories located initially at $(\pm 0.5, 3.0)$ and at $(\pm 0.5, 4)$, respectively. The deviations of those curves from the solid curves show the influence of the ground shear flow. In particular, we

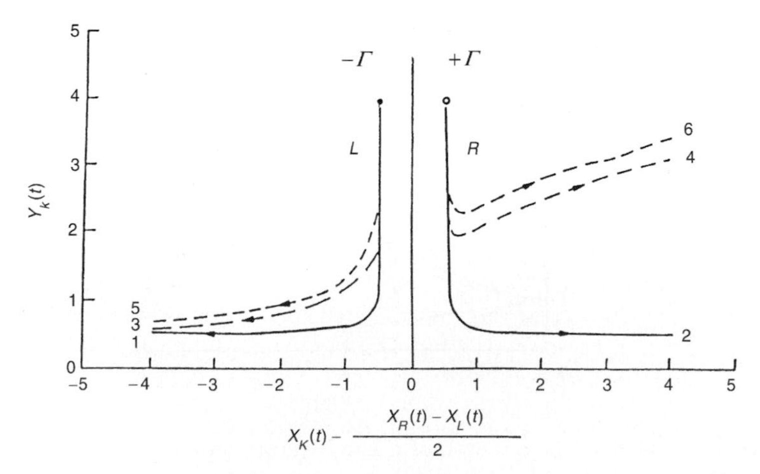

Fig. 3.9. Comparison of the trajectories of vortex pair in a shear flow with that in a uniform flow. Adapted from Liu and Ting (1987)

note that the trajectories without background vorticity are symmetric with the y-axis but asymmetric with nonuniform[7] vorticity. The relative trajectories of the vortex pair in the shear flow, the dashed curves, are asymmetric with the y-axis because they are of opposite signs and interact differently with the background vorticity redistribution (see Fig. 3.8). The left vortex of strength $-\Gamma$ descends monotonically. The right vortex of strength Γ descends initially, reaches a minimum height, and then drifts upward. The upward drift begins when upward velocity of one vortex induced by the background vorticity redistribution overcomes the downward velocity induced by the other vortex and its image. Additional numerical results were presented in Liu and Ting (1987) showing the true trajectories corresponding to Fig. 3.9 and the insensitivity of the relative trajectories to their initial height $Y(0)$, when the initial pair are well above the shear layer, for example, $Y(0) \geq 7$ times the scale height of the shear layer. Also presented were the trajectories of the pair for different strengths of the pair and for different background shear flows.

Next, we present another application of matched asymptotic analysis for a vortex embedded in a complex background flow. We consider a model for large-scale atmospheric flows that includes a dominant balance of the Coriolis force with horizontal pressure gradients, known as the "geostrophic balance" (see also Sect. 9.2). The extension of the classical inviscid point vortex in a planar flow to the geostrophic flow regime is known to have two defects. These defects were again removed by an inner solution for its viscous core structure in Subsect. 4.2.2 of Book-I, which is reproduced below.

3.3.9 Motion and Core Structure of a Geostrophic Vortex

Discrete rectilinear geostrophic (Bessel) vortices were first proposed by Stewart (1943) to model the periodic motion in semipermanent atmospheric pressure systems and by Morikawa (1962) for the prediction of hurricane tracks. Numerical studies of the motion of a geostrophic vortex submerged in a background flow and the interaction of several vortices were made by Morikawa and Swenson (1971) and Bauer and Morikawa (1976). The inviscid theory of a geostrophic vortex has two defects, namely (i) the velocity and the effective height of the atmosphere become infinite at the vortex point and (ii) the velocity of the vortex point is assumed to be moving at the local velocity of the background flow. These defects are the same as those for a potential point vortex, and can be removed by the method of matched asymptotic analysis. We consider the Bessel vortex as the outer solution of a geostrophic vortex with turbulent flow in its core or *eye*. The turbulence is accounted for by introducing an eddy viscosity coefficient ν_e. In Subsubsect. 3.3.9.1, we explain the

[7] If the background flow is a simple shear flow $U_0(y) = \omega_0 y$, with uniform vorticity ω_0, there is no redistribution of the background vorticity due to the vortex pair and their relative trajectories remain symmetric with respect to the y-axis but their real trajectories, $Y_k(t)$ vs $X_k(t)$, $k = L$ or R, are ω_0 dependent because their horizontal drifts depend on $U_0 = \omega_0 Y_k(t)$ with $Y_L(t) = Y_R(t)$.

reference scales, the small parameters, the equations of geostrophic motion in the tangent plane approximation (Pedlosky, 1982), and their axisymmetric singular solution (Morikawa, 1960), the Bessel vortex. In Subsubsect. 3.3.9.2, we present the inner solution for the core structure following Ting and Ling (1983). For more recent work on concentrated vortices in the atmosphere, see Reznik and Dewar (1994), Reznik and Grimshaw (2001), and references therein.

3.3.9.1 Geostrophic Equations of Motion. We consider the large-scale motion of the atmosphere over the rotating earth with the horizontal scale ℓ much larger than the typical effective thickness H of the atmosphere but much smaller than the radius of the earth. The ratio H/ℓ then defines the small parameter $\epsilon \ll 1$ for the expansion schemes. The motion of the thin atmospheric layer can be approximately described by that of an equivalent two-dimensional incompressible flow in the tangent plane, the local horizontal xy plane, provided the vertical shear is small. Let h and $\boldsymbol{u}$ denote the effective height and horizontal velocity scaled by H and U, respectively. They are governed by the following equations:

$$h_t + \nabla \cdot (h\mathbf{u}) = 0 \,, \tag{3.3.218}$$

$$\alpha^2[\partial_t + \mathbf{u} \cdot \nabla]\mathbf{u} + \alpha\aleph\, \hat{k} \times \mathbf{u} = -\nabla h \tag{3.3.219}$$

with

$$\alpha = \frac{U}{\sqrt{gH}}\,, \qquad \aleph = \frac{2\,\ell\,\omega\,\sin\varphi}{\sqrt{gH}}\,, \tag{3.3.220}$$

where g, ω, and φ denote the gravitational acceleration, the angular velocity of the earth, and the local latitude, respectively. $\aleph$ is known as the Coriolis parameter, and the independent variables x, y, and t are scaled by ℓ and ℓ/U. We assume $\varphi \neq 0$; that is, the local latitude is away from the equator, and $\aleph = O(1)$. By applying the curl operator to (3.3.219), we get

$$[\partial_t + \mathbf{u} \cdot \nabla][\alpha\,\zeta - \aleph\,h] = 0 \,, \tag{3.3.221}$$

where ζ denotes the vorticity. Now we consider the flow field in the length scale ℓ to be a small perturbation from equilibrium and identify the reference velocity as

$$U = \epsilon\,\sqrt{gH}\,, \quad \text{that is,} \quad \alpha = \epsilon\,. \tag{3.3.222}$$

The expansion schemes for the outer solutions are

$$\begin{aligned} h(t,x,y,\epsilon) &= 1 + \epsilon h'(t,x,y) + O(\epsilon^2)\,, \\ \mathbf{u}(t,x,y,\epsilon) &= \mathbf{u}'(t,x,y) + O(\epsilon)\,, \\ \zeta(t,x,y,\epsilon) &= \zeta'(t,x,y) + O(\epsilon)\,. \end{aligned} \tag{3.3.223}$$

From (3.3.218) to (3.3.222), we obtain the leading-order equations,

$$\nabla \cdot \mathbf{u}' = 0 , \tag{3.3.224}$$

$$\aleph \, \hat{k} \times \mathbf{u}' = -\nabla h' \tag{3.3.225}$$

and

$$\partial_t \left[\Delta h' + \aleph^2 h'\right] = 0 . \tag{3.3.226}$$

Equations (3.3.224) and (3.3.225) imply that there is a stream function ψ' related to h' by the equation,

$$\Delta[\aleph \, \psi' + h'] = 0 , \tag{3.3.227}$$

and hence ψ' can differ from $-h'/\aleph$ by a harmonic function. Note that (3.3.226) has an axisymmetric singular solution,

$$h^*(t,r) = -\aleph \, \psi^*(t,r) \quad \text{with} \quad \psi^* = \frac{\Gamma}{2\pi} K_0(\aleph \, r) , \tag{3.3.228}$$

where $r^2 = (x - X)^2 + (y - Y)^2$ and $K_0(\aleph \, r)$ denotes the modified Bessel function. The stream function ψ^* represents a geostrophic or Bessel vortex with strength Γ and centered at (X, Y). It is the solution of the equation

$$\Delta \psi^* - \aleph^2 \psi^* = -\frac{\Gamma \, \delta(r)}{2\pi r} , \tag{3.3.229}$$

where δ denotes the delta function. The dependence of ψ^* on t comes from that of $X(t)$ and $Y(t)$. The solution reduces to that of a potential point vortex when $\aleph \to 0$ and has the same logarithmic singularity as $r \to 0$. We can separate the regular parts of h' and ψ' from their singular parts and write

$$\begin{aligned} h'(t,x,y) &= h^*(t,r) + h_1(t,x,y) , \\ \psi'(t,x,y) &= \psi^*(t,r) + \psi_1(t,x,y) , \end{aligned} \tag{3.3.230}$$

where h_1 and ψ_1 are regular at the vortex center, $r = 0$. Note that ψ_1, which can differ from $-h_1 \, / \, \aleph$ by a harmonic function, represents the stream function of the background flow. To close the system, it is assumed that the vortex point moves at the local background velocity, that is,

$$\dot{X} = -\partial_y \, \psi_1(t,X,Y) \quad \text{and} \quad \dot{Y} = \partial_x \, \psi_1(t,X,Y) . \tag{3.3.231}$$

In the following, we employ the matched asymptotic analysis in the normal time scale to obtain the core structure and remove the singular part of the outer solution (3.3.230), and show that the leading-order velocity of the vortex center is given by (3.3.231).

3.3.9.2 Inner Solution of a Geostrophic Vortex. For the inner solution, we use the small length scale $\epsilon\ell$ and the normal time scale ℓ/U (see Subsects. 3.2.2 and 3.2.3). Let r, θ denote the polar coordinates relative to the vortex center $X(t), Y(t)$ and $\bar{u}, \bar{v}$ the radial and circumferential components of the velocity relative to the moving vortex center. The velocity is written as

$$\mathbf{u} = [\hat{\imath}\dot{X} + \hat{\jmath}\dot{Y}] + [\hat{r}\bar{u} + \hat{\theta}\bar{v}] \ . \tag{3.3.232}$$

Again we introduce the stretched radial variable $\bar{r} = r/\epsilon$ for the inner solution. We assume that the velocity of the vortex center $(\dot{X}, \dot{Y})$ remains order one while the circumferential velocity will be $O(\epsilon^{-1})$ to match with the singular part of the outer solution. The expansions for the inner solutions are

$$\bar{u}(t, \bar{r}, \theta, \epsilon) = \bar{u}^{(1)}(t, \bar{r}, \theta) + O(\epsilon) \ , \tag{3.3.233a}$$

$$\bar{v}(t, \bar{r}, \theta, \epsilon) = \epsilon^{-1}[\bar{v}^{(0)}(t, \bar{r}, \theta) + O(\epsilon)] \ , \tag{3.3.233b}$$

$$\bar{h}(t, \bar{r}, \theta, \epsilon) = \bar{h}^{(0)}(t, \bar{r}, \theta) + O(\epsilon) \ , \tag{3.3.233c}$$

$$\bar{\zeta}(t, \bar{r}, \theta, \epsilon) = \epsilon^{-2}[\bar{\zeta}^{(0)}(t, \bar{r}, \theta) + O(\epsilon)] \ , \tag{3.3.233d}$$

$$(X, Y)(t, \epsilon) = \left(\bar{X}^{(0)}, \bar{Y}^{(0)}\right)(t) + O(\epsilon) \ . \tag{3.3.233e}$$

To account for the turbulence in the core structure, we add the eddy diffusion term $\nu_e \triangle \mathbf{u}$ to the right-hand side of the momentum equation (3.3.219). Here ν_e stands for the small eddy viscosity coefficient scaled by $U\ell$. The order of magnitude of ν_e relative to ϵ will be assigned later. We now repeat the procedures of the asymptotic analysis carried out in Subsects. 3.2.2 and 3.2.3 and conclude that the leading-order vorticity, circumferential velocity, and effective height are axisymmetric, that is, independent of θ, and are related by the following equations:

$$\bar{h}^{(0)}_{\bar{r}}(t, \bar{r}) = [\bar{v}^{(0)}(t, \bar{r})]^2 / \ \bar{r}, \qquad \text{or}$$

$$\bar{h}^{(0)}(t, \bar{r}) = 1 - \int_{\bar{r}}^{\infty} \frac{[\bar{v}^{(0)}(t, \xi)]^2}{\xi} \, d\xi \tag{3.3.234}$$

and

$$\bar{\zeta}^{(0)}(t, \bar{r}) = [\bar{v}^{(0)} \ \bar{r}]_{\bar{r}} / \ \bar{r} \ . \tag{3.3.235}$$

In (3.3.234) we used the matching condition, $\bar{h}^{(0)} \to 1$ as $\bar{r} \to \infty$.

From the leading-order asymmetric part of the momentum equation, the boundary conditions $\bar{u} = \bar{v} = 0$ at $\bar{r} = 0$, and the matching conditions, we conclude that the velocity of the vortex center is given by (3.3.231). To obtain the leading-order (axisymmetric) solution from the circumferential averages of the higher-order equations, we have to specify the order of magnitude of the scaled eddy viscosity ν_e relative to ϵ. We consider two models:

$$\text{Model A} \, , \qquad \bar{\nu}_e = \nu_e / \epsilon^2 = O(1) \tag{3.3.236}$$

and

$$\text{Model B} \, , \qquad \tilde{\nu}_e = \nu_e / \epsilon = O(1) \ . \tag{3.3.237}$$

For model A, the leading-order inner solution is defined by the circumferential averages of the second-order equations in the same manner as in Subsect. 3.2.2. The vorticity fulfills the simple diffusion equation,

$$\bar{\zeta}_t^{(0)} = \frac{\bar{\nu}_e}{\bar{r}} [\bar{r}\bar{\zeta}_{\bar{r}}^{(0)}]_{\bar{r}} . \tag{3.3.238}$$

The vorticity and circumferential velocity in the *eye* of the geostrophic vortex are identical to those in the core of a potential vortex with $\bar{\nu}_e$ replacing $\bar{\nu}$. The deviation of the effective height, $\bar{h}^{(0)} - 1$, is proportional to the pressure variation in the core.

For model B, the eddy viscosity is one order larger than that in model A and hence we have a larger diffusion rate. To account for this fact, we introduce a short time scale $\ell/\sqrt{gH}$ and the corresponding stretched time variable,

$$\tilde{t} = t/\epsilon , \tag{3.3.239}$$

for the inner solutions. We then obtain from the circumferential averages of the first order equations two conditions relating $\tilde{v}^{(0)}$, $\tilde{h}^{(0)}$, and $\{\tilde{u}^{(1)}\}$, the axisymmetric part of $\tilde{u}^{(1)}$. These conditions are

$$[\bar{r}\tilde{h}^{(0)}]_t + [\bar{r}\tilde{h}^{(0)}\{\tilde{u}^{(1)}\}]_{\bar{r}} = 0 \tag{3.3.240}$$

and

$$\tilde{v}_t^{(0)} + \{\tilde{u}^{(1)}\}[\tilde{v}_{\bar{r}}^{(0)} + \frac{\tilde{v}^{(0)}}{\bar{r}}] = \tilde{\nu}_e[\tilde{v}_{\bar{r}\bar{r}}^{(0)} + (\frac{\tilde{v}^{(0)}}{\bar{r}})_{\bar{r}}] . \tag{3.3.241}$$

These two equations and (3.3.234) form a closed system for $\tilde{h}^{(0)}, \tilde{v}^{(0)}$, and $\{\tilde{u}^{(1)}\}$. The boundary conditions are $\tilde{v}^{(0)} = 0$ and $\{\tilde{u}^{(1)}\} = 0$ at $\bar{r} = 0$ and the matching condition is $\tilde{v}^{(0)} \to \Gamma/(2\pi\bar{r})$, as $\bar{r} \to \infty$.

Since the system of equations does not have a t-derivative of $\{\tilde{u}^{(1)}\}$, we need only the initial profile of either $\tilde{v}(0)$ or $\tilde{h}^{(0)}$ on account of (3.3.234). Numerical solutions of the system for an initial bell-shaped profile of $\tilde{h}^{(0)}(0, \bar{r})$ were constructed, and the temporal variations of the leading-order circumferential velocity and effective height in the *eye* of the geostrophic vortex were presented by Ting and Ling (1983).

Note that for both models, the r^{-1} singularity of the outer solution $\mathbf{u}'$ as $r \to 0$ is removed by the matching condition with the leading-order inner circumferential velocity. The logarithmic singularity of the outer solution $\epsilon h'$ will be removed by the matching condition with the next-order inner solution $\epsilon\tilde{h}^{(1)}$. The leading-order velocity of the vortex center is defined by (3.3.231), that is, by the local background velocity. But the structure of the *eye* for model A and that for model B are different from each other. In particular, they have different time scales.

In case the initial velocity of the vortex center disagrees with that given by (3.3.231), or when there is a sudden change of background flow, we have to introduce a two-time analysis similar to that in Subsect. 3.2.1. Details of the analysis were presented by Ling and Ting (1988). For both models, the additional short time variable τ is scaled by $\epsilon^2\ell/U$ and is equal to t/ϵ^2 or $\tilde{t}/\epsilon$. Many results of the two-time analysis in Subsect. 3.2.1 remain valid here. For example, the τ-averaging of the leading-order velocity of the vortex center given by the two-time solution remains equal to that of the one-time solution defined by (3.3.231), and the trajectory of the former differs from that of the latter by an oscillation in τ with small period and amplitude of the order of $\epsilon^2\ell/U$ and $\epsilon^2\ell$, respectively. Of course, the details of the oscillatory motion do depend on the choice of model A or B.

It should be noted that the typical scales ℓ and U are of the order of 100 miles and 10 miles/h, respectively. The normal time scale is of the order of 10 hours; therefore, the oscillatory motion should be observable. Of course, the two-dimensional geostrophic vortex is only a strongly simplified model for the dynamics of the thin atmospheric layer. Additional analyses are needed to account for the vertical stratification, the compressibility effect, and the energy transfer (see also Sect. 9.2 in this context).

3.4 Vortex Filaments with Compressible Core Structure

In Sects. 3.2 and 3.3, analyses were carried out for vortices or filaments with incompressible core structures. The extension of the analyses to account for compressible core structures was outlined by Ting (2002) and carried out later by Knio et al. (2003). This extension is of practical importance because the velocity in a vortical core, which is an order of magnitude $1/\epsilon$ larger than that in the background flow, will render the flow compressible in many practical applications. For example, even if the background flow is at a low Mach number say 0.15, that is, $\approx 50\,\mathrm{m/s}$, and the filament is not too slender, say $\epsilon = 0.2$, the velocity in the core is $\approx 250\,\mathrm{m/s}$ at Mach number $\overline{\mathrm{Ma}} \approx 0.75$. In such a case, the compressibility effect has to be accounted for.

In this section, we present the extension; that is, we study the dynamics of slender vortex filaments with compressible core structures in a low-Mach-number background flow. We allow the typical Mach number $\overline{\mathrm{Ma}}_*$ in the core, the inner region $\mathcal{R}_i$, to be $O(1)$, while the typical Mach number Ma_* in the outer region $\mathcal{R}_o$ remains low,

$$\mathrm{Ma}_* = \frac{U}{C_\infty} = O(\epsilon)\,, \qquad \text{so that} \qquad \overline{\mathrm{Ma}}_* = \frac{\mathrm{Ma}_*}{\epsilon} = O(1)\,. \tag{3.4.1}$$

Here, U and C_∞ are a reference velocity and a typical speed of sound of the outer flow. We account for the effects of viscous and thermal diffusion in the core structure by balancing the diffusion terms with the convection terms in the momentum and energy equations such that

$$\frac{1}{\mathrm{Re}} = \frac{\mu_\infty}{\rho_\infty U \ell} = O(\epsilon^2) \qquad \text{or} \qquad \bar{\nu} \equiv \frac{1}{\epsilon^2 \mathrm{Re}} = O(1)\,, \tag{3.4.2}$$

$$\mathrm{Pr} = \frac{C_p \mu}{k} = O(1)\,, \qquad \text{where} \qquad C_p = \frac{\gamma R}{\gamma - 1}\,. \tag{3.4.3}$$

Here, ρ_∞, μ_∞, and Re denote, respectively, a reference density and viscosity, and the Reynolds number in the outer region $\mathcal{R}_o$, and $\bar{\nu}$ is a normalized viscosity. For an ideal gas, the ratio of specific heats γ, the gas constant R, the specific heat at constant pressure C_p, and the Prandtl number Pr are constants, and the viscosity μ and thermal conductivity k are generally temperature-dependent, with $k(T) = [C_p/\mathrm{Pr}]\mu(T)$. The state variables p, ρ, and T and the viscosity μ and thermal conductivity k are nondimensionalized by their ambient values, p_∞, ρ_∞, T_∞ and μ_∞, k_∞, respectively. In the sequel, the specific entropy $\mathcal{E}$ is nondimensionalized by the gas constant R, and the stress components τ_{ij} by $\mu_\infty U/\ell$. The distinguished limit for large Re defined by (3.4.2) is then equivalent to that defined by (3.2.4) for incompressible vortex filaments.

3.4.1 Evolution Equations for Compressible Vortex Filaments

Here we summarize the coupled evolution equations for the centerline and core structure of a compressible vortex filament on the normal time scale ℓ/U. These equations involve, with one exception, only leading order functions, which is why we drop the superscript (0) for convenience of notation in this summary. The exception is the circumferentially and axially averaged second-order radial velocity $\{\langle u\rangle\}^{(2)}$, which we abbreviate here by u'. Details of the derivations leading to these evolution equations are provided in the rest of this section.

The filament centerline evolves according to

$$\boldsymbol{X}_t = \boldsymbol{b}\frac{\Gamma\kappa}{4\pi}\left[\ln(1/\epsilon) + C_v + C_w\right] + (\boldsymbol{1} - \boldsymbol{\tau}\circ\boldsymbol{\tau})\cdot\boldsymbol{Q}_0 \,. \tag{3.4.4}$$

Here $\boldsymbol{Q}_0(t,\boldsymbol{X}) = \nabla\Phi + \boldsymbol{Q}^f$ is the superposition of an outer potential flow, $\nabla\Phi$, and the self-induced velocity of the filament, $\boldsymbol{Q}^f$, which is given by the finite part of the Biot–Savart integral just as in the case of a filament in an incompressible flow, $\boldsymbol{1}$ is the identity tensor, so that $\boldsymbol{1} - \boldsymbol{\tau}\circ\boldsymbol{\tau}$ denotes projection onto the normal plane of the unit vector τ. The "core constants" C_v and C_w relate the filament centerline motion to the core structure distributions of the circumferential, $\bar{v}$, and axial, $\bar{w}$, velocities via

$$C_v = \frac{\Gamma}{4\pi}\left\{\lim_{\bar{r}\to\infty}\left[\left(\frac{2\pi}{\Gamma}\right)^2\int_0^{\bar{r}}\bar{\rho}(z)\bar{v}^2(z)z\,dz - \ln\bar{r}\right] + \frac{1}{2}\right\}, \tag{3.4.5}$$

$$C_w = -\frac{2\pi}{\Gamma}\int_0^\infty \bar{\rho}(z)\bar{w}^2(z)z\,dz \,. \tag{3.4.6}$$

The density $\bar{\rho}$, entropy $\bar{\mathcal{E}}$, circumferential velocity $\bar{v}$, axial velocity $\bar{w}$, and second-order radial velocity u' obey

$$
\begin{aligned}
\frac{\bar r}{S}\,(\bar\rho S)_t + (\bar r\bar\rho u')_r &= 0\,, \\
\frac{\bar r}{S}\,(\bar\rho\bar{\mathcal{E}} S)_t + (\bar r\bar\rho\bar{\mathcal{E}} u')_r &= \bar q'_{\mathcal{E}}\,, \\
\frac{\bar r}{S}\,(\bar\rho\bar v S)_t + (\bar r\bar\rho\bar v u')_{\bar r} + \bar\rho\bar v u' &= \frac{\bar\nu}{\bar r}\frac{\partial}{\partial\bar r}\left[\bar\nu\bar r^3\frac{\partial}{\partial\bar r}\left(\frac{\bar v}{\bar r}\right)\right] \\
\frac{\bar r}{S}\,(\bar\rho\bar w S)_t + (\bar r\bar\rho\bar w u')_r + (\bar\rho\bar w)\dot S &= \frac{\partial}{\partial\bar r}\left[\bar\nu\bar r\frac{\partial\bar w}{\partial\bar r}\right]\,,
\end{aligned}
\tag{3.4.7}
$$

where $S(t)$ is the total length of the (closed) filament. The pressure $\bar p$ and temperature $\bar T$ are determined through two constitutive relations,

$$
\bar p = \bar\rho\bar T\,, \qquad \bar{\mathcal{E}} = \frac{1}{\gamma-1}\ln\left(\frac{\bar p}{\bar\rho^\gamma}\right)\,, \tag{3.4.8}
$$

and the quasi-steady radial momentum balance,

$$
\bar p_{\bar r} = \bar\rho\,\frac{\bar v^2}{\bar r} \qquad \text{with} \qquad \bar p \to 1 \quad \text{as} \quad \bar r \to \infty\,. \tag{3.4.9}
$$

Finally,

$$
\bar q'_{\mathcal{E}} = \frac{1}{\bar\rho\bar T}\left\{\gamma\overline{\mathrm{Ma}}_*^2\bar\mu\left(\left[\bar r\frac{\partial}{\partial\bar r}\left(\frac{\bar v}{\bar r}\right)\right]\frac{\partial\bar v}{\partial\bar r} + \left[\frac{\partial\bar w}{\partial\bar r}\right]^2\right) + \frac{1}{(\gamma-1)\mathrm{Pr}}\frac{\partial}{\partial\bar r}\left[k\frac{\partial\bar T}{\partial\bar r}\right]\right\} \tag{3.4.10}
$$

is the entropy production term including viscous dissipation of kinetic energy and the effects of heat conduction.

The logic of these equations is as follows: Given the radial distributions of the circumferential velocity $\bar v$ and specific entropy $\bar{\mathcal{E}}$, the equations of state (3.4.8) and the radial momentum balance (3.4.9) determine the pressure, temperature, and density fields. The temporal evolution of velocity and entropy due to the effects of viscous diffusion, heat conduction, and changes of the filament length S as described in $(3.4.7)_{2,3}$ induce variations of density $\bar\rho$. These in turn fix the second-order radial velocity u' through the continuity equation $(3.4.7)_1$. The axial velocity distribution at the same time evolves according to $(3.4.7)_4$, thereby adjusting the vortex core structure coefficient C_w, just as the evolution of $\bar v$ modifies C_v. These coefficients in turn affect the filament motion through (3.4.4), which then leads to overall changes of the filament length S, which again enters all of the structure evolution equations in (3.4.7).

This completes the summary of the compressible filament evolution equations, whose derivation is described below.

3.4.2 Outer Solution for a Compressible Filament

For the low-Mach-number flow in $\mathcal{R}_o$, we scale the dynamic pressure, that is, the pressure deviation from p_∞, by $\rho_\infty U^2 = \gamma\mathrm{Ma}_*^2 p_\infty$. We write:

$$p = 1 + \gamma \mathrm{Ma}_*^2 p', \quad \text{likewise} \quad \rho = 1 + \mathrm{Ma}_*^2 \rho' \text{ and } T = 1 + \mathrm{Ma}_*^2 T'. \tag{3.4.11}$$

The scaled density and temperature deviations, ρ' and T', are introduced here to obtain the scalings $d\rho = \mathrm{Ma}_*^2 d\rho'$ and $dT = \mathrm{Ma}_*^2 dT'$, and to show the effects of compressibility being $O(\mathrm{Ma}_*^2)$. Note that in the limit of vanishing Mach number, the perturbation pressure p' approaches the "incompressible pressure p"as used in the previous sections.

From the preceding analyses for incompressible core structures for the simpler case of two-dimensional vortices in Sect. 3.2 and then for the general case of spatial filaments in Sect. 3.3, we have learned how to carry out the analysis directly for the general case of compressible filaments. We consider first a slender compressible vortex filament of strength $\Gamma = O(U\ell)$, in the form of a torus as sketched in Fig. 3.6. Again we consider the solution in one time scale $t^* = \ell/U$. We shall then denote both the physical and dimensionless variables in the outer region $\mathcal{R}_o$ by the same symbols. In particular, we have the typical core size $\delta^* = \epsilon$.

From the compressible Navier–Stokes equations, (2.1.2)–(2.1.6), we obtain the governing equations for the scaled variables in the outer region:

Continuity: $$\mathrm{Ma}_*^2 \mathcal{D}\rho' + \rho \nabla \cdot \mathbf{v} = 0, \quad \text{with } \mathcal{D} = \partial_t + \mathbf{v} \cdot \nabla, \tag{3.4.12}$$

Momentum: $$\rho \mathcal{D} v_i = -\partial_{x_i} p' + \frac{1}{\mathrm{Re}} \partial_{x_j} \tau_{ij}, \quad i = 1, 2, 3, \tag{3.4.13}$$

Energy: $$\frac{\rho}{\gamma - 1} \mathcal{D}T' - \mathcal{D}p' = \frac{1}{\mathrm{Re}} \left\{ \tau_{ij} \partial_{x_j} u_i + \frac{\partial_{x_i} [k \partial_{x_i} T']}{(\gamma - 1)\mathrm{Pr}} \right\}, \tag{3.4.14}$$

State: $$p = \rho T \text{ and } T d\mathcal{E} = \gamma \mathrm{Ma}_*^2 \left\{ \frac{dT'}{\gamma - 1} - \frac{dp'}{\rho} \right\}, \tag{3.4.15}$$

Stress tensor: $$\tau_{ij} = \mu \left[\partial_{x_i} v_j + \partial_{x_j} v_i - \frac{2}{3} \delta_{ij} \partial_{x_l} v_l \right]. \tag{3.4.16}$$

The energy equation (3.4.14) can be expressed as an equation for the specific entropy scaled by the gas constant.

Entropy: $$\rho T \mathcal{D}\mathcal{E} = \frac{\gamma \mathrm{Ma}_*^2}{\mathrm{Re}} \left\{ \tau_{ij} \partial_{x_j} u_i + \frac{1}{(\gamma - 1)\mathrm{Pr}} \partial_{x_j} (k \partial_{x_j} T') \right\}. \tag{3.4.17}$$

Thus the material derivative of $\mathcal{E}$ is small $O(\mathrm{Ma}_*^2 \epsilon^2)$.

In the limit of $\epsilon \to 0$, we have $1/\mathrm{Re} \to 0$, $\mathrm{Ma}_* \to 0$, from (3.4.1) and (3.4.2). From the above governing equations, (3.4.12)–(3.4.13), we conclude that the leading-order outer solutions, the velocity $\boldsymbol{v}$, pressure p, and perturbation pressure p' are identical to those for an inviscid incompressible flow. In particular, we may assume a potential flow, with a filament modelled by a vortex line $\mathcal{C}$ of zero core radius. Thus we shall keep the same symbols used for incompressible filaments in Sect. 3.3, whenever possible. Also we can define the geometry of a filament and the relevant orthogonal coordinates of a position vector $\boldsymbol{x}$ by the same symbols fulfilling the same set of equations, (3.3.1)–(3.3.7).

The leading-order velocity $\boldsymbol{Q}$ induced by $\mathcal{C}$ is given by the Biot–Savart formula (3.1.62). The singular behavior of $\boldsymbol{Q}$ as the point $\boldsymbol{x}$ approaches $\mathcal{C}$, that is, as its distance $r \to 0$, is defined by (3.1.25), which in turn defines the leading singularities of $\boldsymbol{v}$ in (3.3.40) and that of p from (3.3.36). The latter yields the behavior of the scaled pressure perturbation,

$$p' = -\frac{\Gamma^2}{8\pi^2} r^{-2} + O(r^{-1}) \quad \text{as } r \to 0 \,, \tag{3.4.18}$$

where Γ and r are scaled by $U\ell$ and ℓ. The leading-order outer flows are identical for the present weakly compressible and the incompressible cases, and the perturbation pressure p from the incompressible case is equivalent to the present p'. Here, however, we are interested in the effects of compressibility and, using the equation of state, we evaluate in addition the density and temperature perturbations. From the entropy equation (3.4.17) we conclude that the leading-order outer flow is adiabatic and homentropic if it is initially. Thus, using the fact that $\mathrm{Ma}_* \sim \epsilon$ according to $(3.4.1)_1$, we have

$$\mathcal{E} - \mathcal{E}_\infty = O(\epsilon^4) \,, \quad \rho' = p' + O(\epsilon^2) \,, \quad \text{and} \quad T' = (\gamma - 1) p' + O(\epsilon^2) \,, \tag{3.4.19}$$

and then the singularities of ρ' and T' derive from those of p' in (3.4.18). These singular behaviors of $\boldsymbol{v}$, p, ρ, and T in the outer region shall be matched with the inner solution, that is, to the solution in the inner region $\mathcal{R}_i$, which is an $O(\epsilon)$-neighborhood of the centerline $\mathcal{C}$.

3.4.3 Inner Solution for a Compressible Filament

Here we study the inner solution of a compressible filament, emphasizing the analyses and physical meaning of the results which differ from or do not arise in the study of an incompressible core structure in Sect. 3.3.

To construct the inner solution, we again employ the orthogonal coordinates (r, θ, s) relative to $\mathcal{C}$, see (3.3.12)–(3.3.15), and introduce the stretched radial variable,

$$\bar{r} = r/\epsilon \,. \tag{3.4.20}$$

In particular, we note the position vector relative to $\mathcal{C}$

$$\boldsymbol{x} = \boldsymbol{X}(t, s) + \epsilon \bar{r} \boldsymbol{r} \,, \tag{3.4.21}$$

and the stretch ratios for $\bar{r}, \theta, s$,

$$h_1 = \epsilon, \quad h_2 = \epsilon \bar{r}, \quad h_3 = \sigma \lambda \qquad \text{with} \tag{3.4.22}$$

$$\sigma = |\boldsymbol{X}_s| \qquad \text{and} \quad \lambda = 1 - \epsilon \kappa \bar{r} \cos(\theta + \theta_0) \,. \tag{3.4.23}$$

Below we will frequently use $\theta + \theta_0 = \phi$. Again we assume the velocity of $\mathcal{C}$,

$$\boldsymbol{X}_t = O(1) \;\; \text{normal to } \mathcal{C}, \text{ that is,} \;\; \boldsymbol{X}_t \cdot \hat{\tau} = 0 \,, \tag{3.4.24}$$

and distinguish the inner solution from the outer solution by a bar. For instance, we denote the velocity in the inner region by $\bar{\boldsymbol{v}}(t, \bar{r}, \theta, s)$. We then introduce the velocity $\bar{\boldsymbol{V}}(t, \bar{r}, \theta, s)$ relative to the velocity of $\mathcal{C}$,

$$\bar{\boldsymbol{v}} - \boldsymbol{X}_t(t, s) = \bar{\boldsymbol{V}} = \boldsymbol{r}\bar{u} + \boldsymbol{\theta}\bar{v} + \boldsymbol{\tau}\bar{w}, \tag{3.4.25}$$

and impose the conditions on $\mathcal{C}$,

$$\bar{u} = \bar{v} = 0\,, \qquad \text{at} \quad \bar{r} = 0\,. \tag{3.4.26}$$

We cannot impose $\bar{w} = 0$ on $\mathcal{C}$, because of $\boldsymbol{X}_t \cdot \hat{\boldsymbol{\tau}} = 0$ in (3.4.24).

To keep the total strength $\Gamma = O(1)$ in the core of size $O(\epsilon)$, the orders of magnitude of the vorticity $\bar{\Omega}$ and velocity $\bar{\boldsymbol{V}}$ in the inner region $\mathcal{R}_i$ have to be

$$\bar{\Omega} = O(\epsilon^{-2}) \quad \text{and} \quad (\bar{\boldsymbol{v}} - \boldsymbol{X}_t) = O(\epsilon^{-1})\,. \tag{3.4.27}$$

It follows that the dynamic pressure $\bar{p}'$ has to be $O(\mathrm{Ma}_*^2 \bar{\boldsymbol{V}}^2) = O(\mathrm{Ma}_*^2/\epsilon^2) = O(1) = O(\bar{p})$, from (3.4.1). Since the density and temperature deviations are of the order of the pressure deviation, we have, $\bar{\rho}' = O(\bar{\rho}) = O(1)$ and $\bar{T}' = O(\bar{T}) = O(1)$. Therefore, for the inner solutions, it is simpler to work directly with the scaled state variables $\bar{p}$, ρ, and $\bar{T}$. The equations of state, (3.4.15), become

$$\bar{p} = \bar{\rho}\bar{T} \quad \text{and} \quad d\bar{\mathcal{E}} = \frac{\gamma d\bar{T}}{(\gamma - 1)\bar{T}} - \frac{d\bar{p}}{\bar{p}}\,. \tag{3.4.28}$$

Using the above estimates of the various orders of magnitude, we arrive at the expansion scheme,

$$\bar{f}(t, \bar{r}, \theta, s, \epsilon) = \epsilon^{-m}\{\bar{f}^{(0)}(t, \bar{r}, \theta, s) + \epsilon\bar{f}^{(1)}(t, \bar{r}, \theta, s) + O(\epsilon^2)\}\,. \tag{3.4.29}$$

Again we consider $\ln(1/\epsilon)$ to be $O(1)$ with respect to ϵ; therefore, $\bar{f}^{(j)}, j = 0, \cdots$, could be functions of $\ln(1/\epsilon)$. If $\bar{f}$ stands for $\bar{\mathbf{v}}$, we have $m = 1$, with the radial component $\bar{u}^{(0)} \equiv 0$ to model a slender filament (Ting and Klein, 1991). If $\bar{f}$ stands for the state variables, $\bar{p}$, $\bar{\rho}$, $\bar{T}$, and $\bar{\mathcal{E}}$, we have $m = 0$.

3.4.3.1 Conservation Laws in Filament-Attached Coordinates. The conservation laws for mass and momentum, and the transport equation for entropy, all written in terms of the filament-attached orthogonal coordinates $t, \bar{r}, \theta, s$, (see Appendix E.4) are derived in Appendix E. We recount them here for convenience:

$$\begin{aligned}(h_3\,\bar{\rho})_t &= -\frac{1}{\epsilon\bar{r}}(\bar{r}h_3\,\bar{\rho}\bar{u})_{\bar{r}} - \frac{1}{\epsilon\bar{r}}(h_3\,\bar{\rho}(\bar{v} - \epsilon\bar{r}\boldsymbol{r}_t \cdot \boldsymbol{\theta}))_\theta \\ &\quad -(\bar{\rho}(\bar{w} - \epsilon\bar{r}\boldsymbol{r}_t \cdot \boldsymbol{\tau}))_s\end{aligned} \tag{3.4.30}$$

$$
\begin{aligned}
\left(h_3\,\bar\rho\,(\bar{\boldsymbol V}+\boldsymbol X_t)\right)_t = &-\frac{1}{\epsilon\bar r}\left(\bar r h_3\left[\bar\rho\,(\bar{\boldsymbol V}+\boldsymbol X_t)\,\bar u+\left(\frac{\bar p\,\mathbf 1}{\mathrm M_*^2}+\frac{\bar{\boldsymbol T}}{\mathrm{Re}}\right)\cdot\boldsymbol r\right)_{\bar r}\\
&-\frac{1}{\epsilon\bar r}\left(h_3\left[\bar\rho\,(\bar{\boldsymbol V}+\boldsymbol X_t)\,(\bar v-\epsilon\bar r\boldsymbol r_t\cdot\boldsymbol\theta)+(\frac{\bar p\,\mathbf 1}{\mathrm M_*^2}+\frac{\bar{\boldsymbol T}}{\mathrm{Re}})\cdot\boldsymbol\theta\right]\right)_\theta\\
&-\left(\bar\rho\,(\bar{\boldsymbol V}+\boldsymbol X_t)\,(\bar w-\epsilon\bar r\boldsymbol r_t\cdot\boldsymbol\tau)+\left(\frac{\bar p\,\mathbf 1}{\mathrm M_*^2}+\frac{\bar{\boldsymbol T}}{\mathrm{Re}}\right)\cdot\boldsymbol\tau\right)_s
\end{aligned}
\tag{3.4.31}
$$

$$
\begin{aligned}
\left(h_3\,\bar\rho\mathcal E\right)_t = &-\frac{1}{\epsilon\bar r}\left(\bar r h_3\,\bar\rho\mathcal E\bar u\right)_{\bar r}-\frac{1}{\epsilon\bar r}\left(h_3\,\bar\rho\mathcal E\,(\bar v-\epsilon\bar r\boldsymbol r_t\cdot\boldsymbol\theta)\right)_\theta\\
&-\left(\bar\rho\mathcal E\,(\bar w-\epsilon\bar r\boldsymbol r_t\cdot\boldsymbol\tau)\right)_s+h_3\,\rho\,q_{\mathcal E}
\end{aligned}
\tag{3.4.32}
$$

where

$$
q_{\mathcal E}=\frac{\gamma\mathrm{Ma}_*^2}{\mathrm{Re}\,\bar\rho\bar T}\left\{\boldsymbol T:\nabla\boldsymbol v-\frac{1}{(\gamma-1)\mathrm{Pr}}\nabla\cdot\boldsymbol h\right\}. \tag{3.4.33}
$$

In (3.4.30) we have used the abbreviations $\bar{\boldsymbol v}=\bar{\boldsymbol V}+\boldsymbol X_t$, $U=\boldsymbol r\cdot\boldsymbol X_t$, $V=\boldsymbol\theta\cdot\boldsymbol X_t$, $W=\boldsymbol\tau\cdot\boldsymbol X_t$, and $\mathrm M_*^2=\gamma\mathrm{Ma}_*^2$. Furthermore, $\mathbf 1$ is the unit tensor in three space dimensions, $\boldsymbol T$ denotes the stress tensor, and $\boldsymbol h=-k\nabla T'$ the heat flux density. The latter two will be specified in terms of the primary unknowns shortly. In scaling $\boldsymbol T$ and $\boldsymbol h$ we have taken into account that the heat flux term has been expressed in the outer solution in terms of the perturbation temperature $T'=(T-1)/\gamma\mathrm{Ma}_*^2$.

Equations (3.4.30)–(3.4.33) are formulated in terms of nondimensional velocities that are scaled according to the characteristic outer flow velocities and perturbation amplitudes according to (3.4.11). Inside the vortex core, however, the circumferential and axial flow velocities are by $O(1/\epsilon)$ larger than outside, and they are comparable in magnitude to the speed of sound. Also, the temperature variations are of leading order in the core, and thus by $O(1/\mathrm M_*^2)=O(1/\epsilon^2)$ larger than outside. In assessing the orders of magnitude of the various terms, we must also account for the fact that the stress tensor and heat flux are enhanced by another factor of $O(1/\epsilon)$, because of the smaller characteristic length scale of the core flow. Systematically rescaling all variables so as to obtain unknowns that are $O(1)$ as $\epsilon\to 0$, we let

$$
\left.\begin{aligned}
\bar v &= \hat v/\epsilon,\\
\bar w &= \hat w/\epsilon,\\
\bar{\boldsymbol V} &= \hat{\boldsymbol V}/\epsilon\ = \hat{\boldsymbol v}/\epsilon+\bar u\,\boldsymbol r,\\
&\qquad\ \hat{\boldsymbol v}\ = \hat v\,\boldsymbol\theta+\hat w\,\boldsymbol\tau,\\
\bar{\boldsymbol T} &= \hat{\boldsymbol T}/\epsilon^2,\\
\bar{\boldsymbol h} &= \hat{\boldsymbol h}/\epsilon^3,
\end{aligned}\right\}
\begin{pmatrix}\bar u\\ \hat v\\ \hat w\\ \hat{\boldsymbol v}\\ \hat{\boldsymbol T}\\ \hat{\boldsymbol h}\end{pmatrix}=O(1)\quad(\epsilon\to 0)\,. \tag{3.4.34}
$$

Density $\bar{\rho}$ and pressure $\bar{p}$, and all other thermodynamic variables derived from these, are $O(1)$ as $\epsilon \to 0$.

With these scalings we rewrite the conservation laws for mass and momentum in such a way that they directly reveal the proper orders of magnitude of all terms in the present slender vortex regime. To make the further derivations somewhat more transparent, we keep only those terms that will contribute to leading, first, and second order. Specifically, all terms involving the expression $\epsilon \bar{r} \boldsymbol{r}_t$ may be neglected. This is verified through tedious but straightforward calculations, taking into account the scalings of the velocity components in (3.4.34) and observing that $-\boldsymbol{\theta}_\theta \cdot \boldsymbol{r}_t = \boldsymbol{r} \cdot \boldsymbol{r}_t \equiv 0$ and $\theta \cdot \boldsymbol{r}_{\theta t} = \theta \cdot \boldsymbol{\theta}_t \equiv 0$.

Using the abbreviation $\overline{\mathrm{M}}_*^2 \equiv \gamma \mathrm{Ma}_*^2/\epsilon^2$, which is $O(1)$ as $\epsilon \to 0$, we note the rescaled and truncated conservation laws

$$
\left(h_3\,\bar{\rho}\right)_t = -\frac{1}{\epsilon\bar{r}}\left(\bar{r}h_3\,\bar{\rho}\bar{u}\right)_{\bar{r}} - \frac{1}{\epsilon^2\bar{r}}\left(h_3\,\bar{\rho}\hat{v}\right)_\theta - \frac{1}{\epsilon}\left(\bar{\rho}\hat{w}\right)_s \tag{3.4.35}
$$

$$
\begin{aligned}
\left(h_3\,\bar{\rho}\,\hat{\boldsymbol{V}}\right)_t = &-\frac{1}{\epsilon^2\bar{r}}\left(\bar{r}h_3\left[\epsilon\,\bar{\rho}\,(\hat{\boldsymbol{V}}+\epsilon\boldsymbol{X}_t)\,\bar{u} + \left(\frac{\bar{p}\,\mathbf{1}}{\overline{\mathrm{M}}_*^2} + \epsilon^2\bar{\nu}\hat{\boldsymbol{T}}\right)\cdot\boldsymbol{r}\right]\right)_{\bar{r}} \\
&-\frac{1}{\epsilon^2\bar{r}}\left(h_3\left[\bar{\rho}\,(\hat{\boldsymbol{V}}+\epsilon\boldsymbol{X}_t)\,\hat{v} + \left(\frac{\bar{p}\,\mathbf{1}}{\overline{\mathrm{M}}_*^2} + \epsilon^2\bar{\nu}\hat{\boldsymbol{T}}\right)\cdot\boldsymbol{\theta}\right]\right)_\theta \\
&-\frac{1}{\epsilon}\left(\left[\bar{\rho}\,(\hat{\boldsymbol{V}}+\epsilon\boldsymbol{X}_t)\,\hat{w} + \left(\frac{\bar{p}\,\mathbf{1}}{\overline{\mathrm{M}}_*^2} + \epsilon^2\bar{\nu}\hat{\boldsymbol{T}}\right)\cdot\boldsymbol{t}\right]\right)_s
\end{aligned} \tag{3.4.36}
$$

$$
\left(h_3\,\bar{\rho}\bar{\mathcal{E}}\right)_t = -\frac{1}{\epsilon\bar{r}}\left(\bar{r}h_3\,\bar{\rho}\bar{\mathcal{E}}\bar{u}\right)_{r} - \frac{1}{\epsilon^2\bar{r}}\left(h_3\,\bar{\rho}\bar{\mathcal{E}}\hat{v}\right)_\theta - \frac{1}{\epsilon}\left(\bar{\rho}\bar{\mathcal{E}}\hat{w}\right)_s + \bar{q}_{\mathcal{E}} \tag{3.4.37}
$$

where

$$
\bar{q}_{\mathcal{E}} = \frac{\bar{\nu}}{\bar{\rho}\bar{T}}\left\{\overline{\mathrm{M}}_*^2\,\boldsymbol{T} : \nabla\boldsymbol{v} + \frac{1}{(\gamma-1)\mathrm{Pr}}\nabla\cdot(k\nabla\bar{T})\right\}, \tag{3.4.38}
$$

with $\bar{\nu}$ from (3.4.2). The equation of state needed to close the system of equations is

$$
\bar{\mathcal{E}} = \frac{1}{\gamma-1}\ln\left(\frac{\bar{p}}{\bar{\rho}^\gamma}\right). \tag{3.4.39}
$$

Equations (3.4.35)–(3.4.39) will be the basis for the subsequent asymptotic expansions based on the scheme described in (3.4.29). Substituting these expansions and equating the coefficents of like powers of ϵ, we obtain the leading- and higher-order equations. The results deduced from the first three orders of equations will be described shortly.

Before we do so, we complete the formulation of the inner flow problem by providing the appropriate matching conditions, and we recount several properties of the asymptotics for incompressible vortices that will be helpful to have in mind.

3.4.3.2 Matching Conditions. From (3.4.11), (3.4.19), (3.1.77), (3.4.18), and (3.4.29), we have the matching conditions for the leading-order velocity components and state variables in $\mathcal{R}_i$ as $\bar{r} \to \infty$,

$$\text{Velocity} \qquad \bar{v}^{(0)} \to \frac{\Gamma}{2\pi\bar{r}}, \quad \bar{w}^{(0)} \to o(\bar{r}^{-n}) \ \text{ for all } n \tag{3.4.40}$$

$$\text{State variables} \quad \bar{p}^{(0)} \to 1 - \frac{\overline{\mathrm{M}}_*^2 \Gamma^2}{8\pi^2\bar{r}^2}, \quad \bar{\rho}^{(0)} \to 1 - \frac{1}{\gamma}\frac{\overline{\mathrm{M}}_*^2 \Gamma^2}{8\pi^2\bar{r}^2}. \tag{3.4.41}$$

In the analysis of the incompressible core structure in Sect. 3.3, we used the matching conditions (3.4.40) to determine $\bar{v}^{(0)}$ and $\bar{w}^{(0)}$, while $\bar{p}^{(0)}$ or rather $\bar{p}^{(0)} - 1$ was *defined*, for the case of an incompressible flow, by $\bar{v}^{(0)}$ to fulfill the matching condition in (3.4.41). The situation is somewhat different for compressible core structures. Again, we will need only the matching conditions on the velocity components to determine the core structure, but now the matching conditions on $\bar{p}^{(0)}, \bar{\rho}^{(0)}$ constitute compatibility conditions on the initial data for the flow field. If the matching conditions are fulfilled initially, then we can construct consistent asymptotic normal time solutions. If they are not, then more complex asymptotic constructions would be required, possibly involving intermediate regions to connect $\mathcal{R}_i$ to $\mathcal{R}_o$ or multiple time scales.

3.4.3.3 Properties of the Expansions for Incompressible Flow. Here we recount several main features in the analysis of incompressible core structures in Sect. 3.3 that are applicable to compressible cores as well:

[1] In the curvilinear coordinates, $\bar{r}$, θ, and s, the core structure is 2π periodic in θ and periodic in s with period S_0.
[2] The leading-order equations require that the leading-order core structure should be axisymmetric, that is, independent of θ.
[3] The jth-order equations for $j \geq 1$ are linear equations for the jth-order solutions appearing as the homogeneous terms, while the inhomogeneous (and nonlinear) terms involve only the lower order terms, $f^{(k)}$, $0 \leq k < j$. Thus, each jth-order equation can be decomposed into a system of equations for the coefficients of the Fourier series in θ. In particular, we obtain an equation for the symmetric part of the solution by averaging in θ, whereas the remainder determines the asymmetric part.
[4] The symmetric part of the jth-order pressure, $\langle p^{(j)} \rangle$, appears only in the the jth-order radial momentum equation and hence can be decoupled from the system, as a direct equation for $\langle p^{(j)} \rangle$.
[5] Because of the assumption of one time scale, $t^* = \ell/U$, the leading- and first order equations are quasi-steady and the ∂_t terms appear as inhomogeneous terms in the second-order and higher order equations.
[6] Because of the distinguished limits on Ma_* and Re in (3.4.1) and (3.4.2), the viscous and thermal diffusion terms do not appear in the leading- and first order equations, but will appear in the second-order and higher

order equations only. This is also seen directly in the scaled equations (3.4.35)–(3.4.35).

[7] The evolution equations for the leading-order core structure come from the compatibility conditions on the inhomogeneous terms of the second-order equations because of [5] and [6] and the periodicity conditions on θ and s stated in [1].

[8] Because of [2], [5], and [6], the θ-averages of the inhomogeneous terms in the second-order equations, the ∂_t terms and the viscous and thermal diffusion terms are identical transforms.

[9] If the leading-order core structure is independent of s, which was stated in (3.3.63) and assumed in Callegari and Ting (1978) as well as in Subsect. 3.3.6, then [8] holds also for the s-averages. That is, the s-averages of the inhomogeneous terms in the second-order equations are identical transforms.

[10] The momentum conservation law in (3.4.36) states conservation of momentum *in an inertial frame of reference.* Our velocity coordinates, u, v, and w are, however, velocity components in a non-Cartesian and time-dependent coordinate system. To derive from (3.4.36) separate equations for these velocity components, we have to project the equations onto the unit vectors $\boldsymbol{r}$, $\boldsymbol{\theta}$, and $\boldsymbol{\tau}$, thereby taking into account that $\boldsymbol{r} = \boldsymbol{r}(t, \theta, s)$ $\boldsymbol{\theta} = \boldsymbol{\theta}(t, \theta, s)$, and $\boldsymbol{\tau} = \boldsymbol{\tau}(t, s)$. This is to be remembered especially in expressions such as $\boldsymbol{r} \cdot (\boldsymbol{f})_\theta = (\boldsymbol{r} \cdot \boldsymbol{f})_\theta - \boldsymbol{f} \cdot \boldsymbol{r}_\theta$, where $\boldsymbol{f}$ is a generic flux vector.

3.4.4 The Leading-Order Equations

The leading-order equations are quasi-steady and can be cast into the form:

$$\partial_\theta \left[\bar{\rho}^{(0)} \bar{v}^{(0)}\right] = 0, \tag{3.4.42}$$

$$\partial_{\bar{r}} \bar{p}^{(0)} - \bar{\rho}^{(0)} (\bar{v}^{(0)})^2 / \bar{r} = 0\,, \tag{3.4.43}$$

$$\partial_\theta \left[\bar{\rho}^{(0)} [\bar{v}^{(0)}]^2 - \bar{p}^{(0)}\right] = 0\,, \tag{3.4.44}$$

$$\bar{\rho}^{(0)} \bar{v}^{(0)} \partial_\theta \bar{w}^{(0)} = 0\,, \tag{3.4.45}$$

$$[\partial_\theta \bar{p}^{(0)}]/\bar{p}^{(0)} - \gamma [\partial_\theta \bar{\rho}^{(0)}] = \partial_\theta \bar{\mathcal{E}}^{(0)} = 0\,, \tag{3.4.46}$$

$$\bar{p}^{(0)} - \bar{\rho}^{(0)} \bar{T}^{(0)} = 0\,. \tag{3.4.47}$$

From (3.4.43) and (3.4.45) we conclude immediately that

$$\bar{p}^{(0)} = 1 - \int_{\bar{r}}^{\infty} \bar{\rho}^{(0)} [\bar{v}^{(0)}]^2 \frac{d\bar{r}}{\bar{r}} \qquad \text{and} \qquad \partial_\theta \bar{w}^{(0)} = 0\,. \tag{3.4.48}$$

To confirm {2}, we combine (3.4.43) and (3.4.44) to get $[\bar{r}\partial_{\bar{r}} + 1]\partial_\theta \bar{p}^{(0)} = 0$, or $\partial_\theta \bar{p}^{(0)} = c_0(\theta) + c_1(\theta)\bar{r}^{-1}$. Since $\bar{p}^{(0)}$ has to be continuous and single valued in $\mathcal{R}_i$, particular at $\bar{r} = 0$, we obtain $c_1 = c_2 = 0$ and hence $\partial_\theta \bar{p}^{(0)} = 0$. It follows that $\partial_\theta \bar{\rho}^{(0)} = 0$ from (3.4.46) and then $\partial_\theta \bar{v}^{(0)} = 0$ and $\partial_\theta \bar{T}^{(0)} = 0$ from (3.4.42) and (3.4.47). We conclude that the leading-order inner solutions are *axisymmetric* with respect to $\mathcal{C}$, that is,

$$\bar{v}^{(0)}, \bar{w}^{(0)}, \bar{p}^{(0)}, \bar{\rho}^{(0)}, \bar{T}^{(0)}, \text{ and } \bar{\mathcal{E}}^{(0)} \text{ are independent of } \theta \,. \tag{3.4.49}$$

Again their dependencies on t, $\bar{r}$, and s, and the velocity of the filament centerline have to be determined by the compatibility conditions of the first- and second-order Navier–Stokes equations.

3.4.5 The First Order Equations and the Velocity of the Filament

The first order equations are again quasi-steady,

$$[\bar{v}^{(0)}\bar{\rho}^{(1)} + \bar{\rho}^{(0)}\bar{v}^{(1)}]_\theta + [\bar{\rho}^{(0)}\bar{r}\bar{u}^{(1)}]_{\bar{r}} \\ = -\bar{r}[\bar{\rho}^{(0)}\bar{w}^{(0)}]_s/\sigma - \bar{\rho}^{(0)}\bar{r}\bar{v}^{(0)}\kappa\sin\phi \,, \tag{3.4.50}$$

$$-\bar{\rho}^{(1)}(\bar{v}^{(0)})^2/\bar{r} + \bar{\rho}^{(0)}\{\bar{v}^{(0)}\bar{u}^{(1)}_\theta - 2\bar{v}^{(0)}\bar{v}^{(1)}\} + \bar{p}^{(1)}_{\bar{r}} \\ = -\bar{\rho}^{(0)}(\bar{w}^{(0)})^2\bar{r}\kappa\cos\phi \,, \tag{3.4.51}$$

$$\bar{\rho}^{(0)}\{\bar{v}^{(0)}\bar{v}^{(1)}_\theta + \bar{u}^{(1)}(\bar{r}\bar{v}^{(0)})_{\bar{r}}\} + \bar{p}^{(1)}_\theta \\ = \bar{\rho}^{(0)}(\bar{w}^{(0)})^2\bar{r}\kappa\sin\phi - \bar{r}\bar{\rho}^{(0)}\bar{w}^{(0)}\bar{v}^{(0)}_s/\sigma \,, \tag{3.4.52}$$

$$\bar{\rho}^{(0)}\{\bar{u}^{(1)}\bar{w}^{(0)}_{\bar{r}} + \bar{w}^{(1)}_\theta\bar{v}^{(0)}/\bar{r}\} \\ = -\bar{\rho}^{(0)}\bar{w}^{(0)}\bar{v}^{(0)}\kappa\sin\phi - (\bar{p}^{(0)}_s + \bar{\rho}^{(0)}\bar{w}^{(0)}\bar{w}^{(0)}_s)/\sigma \,, \tag{3.4.53}$$

$$\bar{v}^{(0)}\bar{\mathcal{E}}^{(1)}_\theta/\bar{r} + \bar{u}^{(1)}\bar{\mathcal{E}}^{(0)}_{\bar{r}} = -\bar{w}^{(0)}\bar{\mathcal{E}}^{(0)}_s/\sigma \,, \tag{3.4.54}$$

$$\bar{p}^{(1)}/\bar{p}^{(0)} - \bar{\rho}^{(1)}/\bar{\rho}^{(0)} - \bar{T}^{(1)}/\bar{T}^{(0)} = 0 \,. \tag{3.4.55}$$

These are linear equations for the first order solutions, $\bar{f}^{(1)}$, appearing as the homogeneous terms, with coefficients depending only on the leading-order solutions $\bar{f}^{(0)}$, which are independent of θ. The inhomogeneous terms are nonlinear in $\bar{f}^{(0)}$ (see item [3] listed in Subsubsect. 3.4.3.3). The first order solutions $\bar{f}^{(1)}$ can be written as the sum of the symmetric part $\langle \bar{f}^{(1)} \rangle$, the θ-average of $\bar{f}^{(1)}$, and the asymmetric part $\tilde{f}^{(1)} = \bar{f}^{(1)} - \langle \bar{f}^{(1)} \rangle$.

The θ-averages of (3.4.50), (3.4.53), (3.4.52), and (3.4.54) eliminate $\bar{v}^{(1)}$, $\bar{w}^{(1)}$, $\bar{p}^{(1)}$, $\bar{\rho}^{(1)}$, and $\bar{\mathcal{E}}^{(1)}$, and yield a system of partial differential equations in $\bar{r}$ and s with t as a parameter for $\bar{v}^{(0)}$, $\bar{w}^{(0)}$, $\bar{p}^{(0)}$, $\bar{\rho}^{(0)}$, and $\langle \bar{u}^{(1)} \rangle$. They represent a quasi-steady axisymmetric compressible inviscid flow in a meridional plane, the s-$\bar{r}$ plane (see items [3], [5], and [6]).

Instead of repeating almost one to one the studies for the incompressible case in Subsubsect. 3.3.3.1 with appropriate addition of density $\bar{\rho}$, we present a direct approach via the von Mises variables. For the quasi-steady axisymmetric flow we introduce the stream function,

$$\mathcal{M} = \int_0^{\bar{r}} \bar{\rho}^{(0)} \bar{w}^{(0)} \bar{r}' d\bar{r}' \quad \text{with} \quad \bar{\rho}^{(0)} \left\langle \bar{u}^{(1)} \right\rangle \bar{r} = -\mathcal{M}_s / \sigma \tag{3.4.56}$$

to replace the continuity equation. We replace the variables $\bar{r}, s$ by the von Mises variables $\mathcal{M}, s_*$ with $s_* = s$.

From the symmetric part of (3.4.52), we get $\partial_{s_*} \bar{v}^{(0)} = 0$, from (3.4.54), we get $\partial_{s_*} \bar{\mathcal{E}}^{(0)} = 0$. The latter says that $\mathcal{E}^{(0)}$, and thus also $p^{(0)}/(\rho^{(0)})^\gamma$, remains constant along a streamline in the s_*-$\bar{r}$-plane at an instant t. Therefore, entropy can be expressed as a function only of an appropriate stream function in the s_*-$\bar{r}$-plane. Combining (3.4.52), (3.4.43), (3.4.44), and (3.4.54), we get $\partial_{s_*} \bar{\mathcal{H}}^{(0)} = 0$, where $\bar{\mathcal{H}}^{(0)} = [(\bar{v}^{(0)})^2 + (\bar{w}^{(0)})^2]/2 + \bar{\mathcal{P}}$ with $d\bar{\mathcal{P}} = d\bar{p}^{(0)}/\bar{\rho}^{(0)}$. Notice that the differential is taken *along a stream surface* so that $d\bar{p}^{(0)}/\bar{\rho}^{(0)}$ is a total differential, parameterized by the value of entropy on that stream surface. Consequently, the core structure for $t \geq 0$ has to be consistent with the following three classical relationships, regardless of whether the viscous and thermal diffusion effects are included or not:

C. 1) The circulation $\mathcal{G} = 2\pi \bar{r} v^{(0)}$ around an axisymmetric stream tube is independent of s_*; that is, it remains constant along the streamline in the s_*-$\bar{r}$ plane (Helmholtz' Theorem).

C. 2) The specific entropy $\mathcal{E}^{(0)}$ is constant along the streamline.

C. 3) The total head $\mathcal{H} = [(v^{(0)})^2 + (w^{(0)})^2]/2 + \mathcal{P}$ is constant along a streamline (Bernoulli's equation generalized to flows with a simple $p - \rho$ relationship).

From (3.4.56) and *C.1)*, we see that

C. 4) $\mathcal{M}$, $\bar{v}^{(0)}$, and $\bar{\rho}^{(0)}\bar{w}^{(0)}$ are independent of s, that is, have no axial variation, and $\langle \bar{u}^{(1)} \rangle = 0$, if any one of them does.

From (3.4.43) and *C. 2)*, it then follows that $\bar{p}^{(0)}/\bar{\rho}^{(0)}$ and $\bar{p}^{(0)}/(\bar{\rho}^{(0)})^\gamma$ are independent of s. Consequently, the state variables $\bar{p}^{(0)}, \bar{\rho}^{(0)}$, and $\bar{T}^{(0)}$ are then also independent of s. As an extension of the case studied by Callegari and Ting Callegari and Ting (1978), we cite a special case of *C. 4)*.

C. 5) If $\bar{v}^{(0)}$ is independent of s, the core structure has no axial variation, that is, $\bar{w}^{(0)}, \langle \bar{u}^{(1)} \rangle, \bar{p}^{(0)}, \bar{\rho}^{(0)}, \bar{T}^{(0)}$ are independent of s.

Combined with (3.4.49), we then have the leading-order core structure independent of θ and s. The dependence on t and $\bar{r}$ will be governed by the compatibility conditions of the second-order Navier–Stokes equations, presented in Subsect. 3.4.6.

Now we study the first order asymmetric inner solutions, $\tilde{f}^{(1)}; f = \{u, v, p, \rho\}$. They obey, dropping the superscript $\{\cdot\}^{(0)}$ on the leading-order functions,

$$[\bar{\rho}\tilde{u}^{(1)}\bar{r}]_{\bar{r}} + [\tilde{\rho}_\theta^{(1)}\bar{v} + \bar{\rho}\tilde{v}_\theta^{(1)}] = -\bar{r}\bar{\rho}\bar{v}\kappa\sin\phi\,, \quad (3.4.57)$$

$$\bar{\rho}\tilde{u}^{(1)}\bar{r}\bar{v}_{\bar{r}} + \bar{\rho}\bar{v}\tilde{v}_\theta^{(1)} + \bar{\rho}\bar{v}\tilde{u}^{(1)} + \tilde{p}_\theta^{(1)} = \bar{\rho}\,\bar{w}^2\,\kappa\sin\phi\,, \quad (3.4.58)$$

$$\bar{\rho}\bar{v}\tilde{u}_\theta^{(1)} - 2\bar{v}(\bar{\rho}\tilde{v}^{(1)} + \tilde{\rho}^{(1)}\bar{v}) + \bar{r}\tilde{p}_{\bar{r}}^{(1)} = -\bar{\rho}\bar{w}^2\kappa\bar{r}\cos\phi\,, \quad (3.4.59)$$

$$(\bar{r}\tilde{u}^{(1)}\bar{p}_{\bar{r}} + \bar{v}\tilde{p}_\theta^{(1)})/\bar{p} = \gamma[\bar{r}\tilde{u}^{(1)}\bar{\rho}_{\bar{r}} + \bar{v}\tilde{\rho}_\theta^{(1)}]/\bar{\rho}\,, \quad (3.4.60)$$

while $\tilde{w}^{(1)}$, uncoupled from the above four equations, is governed by the axial momentum equation,

$$\bar{\rho}[\tilde{u}^{(1)}\bar{w}_{\bar{r}} + \bar{v}\tilde{w}_\theta^{(1)}] = -\bar{\rho}\bar{w}\bar{v}\,\kappa\sin\phi\,. \quad (3.4.61)$$

The boundary conditions at $\bar{r} = 0$ are the homogeneous conditions,

$$\tilde{u}^{(1)} = \tilde{v}^{(1)} = 0\,. \quad (3.4.62)$$

As $\bar{r} \to \infty$, the inner solution matches with the behavior of the outer solution near the vortex line $\mathcal{C}$ given by (3.3.31). For the leading-order solutions, $\bar{v}^{(0)}$ and $\bar{w}^{(0)}$, the conditions were recounted in (3.4.40). For the next-order asymmetric solutions, they are as follows:

$$\tilde{u}^{(1)} \to \frac{\Gamma\kappa}{4\pi}\ln\frac{1}{\epsilon\bar{r}}\sin\phi + [\boldsymbol{Q}_0 - \dot{\boldsymbol{X}}]\cdot\boldsymbol{r} \quad (3.4.63)$$

$$\tilde{v}^{(1)} \to \frac{\Gamma\kappa}{4\pi}\ln\frac{1}{\epsilon\bar{r}}\cos\phi + \frac{\Gamma\kappa}{4\pi}\cos\phi + [\boldsymbol{Q}_0 - \dot{\boldsymbol{X}}]\cdot\boldsymbol{\theta}\,. \quad (3.4.64)$$

Equations (3.4.57)–(3.4.61) are a system of linear equations with variable coefficients for the first-order asymmetric solutions as functions of $(\bar{r}, \theta)$, with s and t as parameters. The coefficients depend on the leading-order inner solutions, which are not yet defined. The inhomogeneous terms and matching conditions have only the first harmonics in ϕ or θ, and the conditions at $\bar{r} = 0$ are homogeneous. Thus, the asymmetric parts of the first-order solutions have only the first harmonics and should fulfill the identity

$$\partial_\theta^2\tilde{f}^{(1)} = -\tilde{f}^{(1)}\,. \quad (3.4.65)$$

For an incompressible filament, the system of equations was reduced to a second-order equation in Subsect. 3.3.3. A general solution was found and employed to reduce the second-order equation to first order for which the solution is available. This procedure does not work here because we have not been able to find a general solution for the compressible system. We shall reduce the system directly to a first order equation making use of the condition (3.4.65).[8]

[8] This direct method can, of course, be employed to replace the derivation made in Subsect. 3.3.3 for an incompressible filament.

In fact, introducing the first order stream function $\tilde{\psi}^{(1)}$, defined via $\bar{\rho}\bar{r}\tilde{u}^{(1)} = \tilde{\psi}^{(1)}_{\theta}$, we find $\bar{\rho}\bar{r}\tilde{u}^{(1)}_{\theta} = \tilde{\psi}^{(1)}_{\theta\theta} = -\tilde{\psi}^{(1)}$. With this result we can eliminate $\tilde{u}^{(1)}$ and $\tilde{u}^{(1)}_{\theta}$ from the above four equations and obtain

$$\tilde{\psi}^{(1)}_{\bar{r}} + [\bar{\rho}\tilde{v}^{(1)} + \tilde{\rho}^{(1)}\bar{v}] = \bar{\rho}\bar{v}\kappa\cos\phi. \tag{3.4.66}$$

$$\tilde{\psi}^{(1)}\bar{v}_{\bar{r}} + \bar{\rho}\bar{v}\tilde{v}^{(1)} + \tilde{\psi}^{(1)}\bar{v}/\bar{r} + \tilde{p}^{(1)} = -\bar{\rho}\bar{w}^2\bar{r}\kappa\cos\phi, \tag{3.4.67}$$

$$(\bar{v}/\bar{r})\tilde{\psi}^{(1)} + \bar{v}[2\bar{\rho}\tilde{v}^{(1)} + \tilde{\rho}^{(1)}\bar{v}] - \bar{r}\tilde{p}^{(1)}_{\bar{r}} = \bar{\rho}\bar{w}^2\bar{r}\kappa\cos\phi, \tag{3.4.68}$$

$$\bar{v}[\bar{C}^{-2}\tilde{p}^{(1)} - \tilde{\rho}^{(1)}] = -(\gamma - 1)\bar{r}\tilde{\psi}^{(1)}\bar{\mathcal{E}}_{\bar{r}}/\gamma\,. \tag{3.4.69}$$

Here $\bar{C}^2 = \gamma\bar{p}/\bar{\rho}$ denotes the leading-order speed of sound. Note that with $\bar{\rho} = 1$, $\tilde{\psi}^{(1)}$ reduces to the first order asymmetric stream function for an incompressible filament, $\psi_{11}(\bar{r})\cos\phi + \psi_{12}(\bar{r})\sin\phi$. Equations (3.4.66)–(3.4.69) can be considered as ordinary differential equations in $\bar{r}$, for $\tilde{v}^{(1)}, \tilde{\rho}^{(1)}, \tilde{p}^{(1)}$, and $\tilde{u}^{(1)}_{\theta}$, or $\tilde{\psi}^{(1)}$.

To reduce the system to a first order equation for $\tilde{\psi}^{(1)}$, we first eliminate $\tilde{\rho}^{(1)}$ and $\tilde{v}^{(1)}$. We get from [$\bar{v}$ (3.4.66) + (3.4.67) – (3.4.68)],

$$(\bar{v}\tilde{\psi}^{(1)})_{\bar{r}} + (\bar{r}\tilde{p}^{(1)})_{\bar{r}} = -[2\bar{w}^2 - \bar{v}^2]\bar{\rho}\bar{r}\kappa\cos\phi \tag{3.4.70}$$

and from $\bar{r}[\bar{v}(3.4.66) - (3.4.67) + \bar{v}(3.4.69)]$,

$$\bar{v}\bar{r}\tilde{\psi}^{(1)}_{r} - \tilde{\psi}^{(1)}(\bar{r}\bar{v})_{\bar{r}} + \tfrac{\gamma-1}{\gamma}\tilde{\psi}^{(1)}\bar{r}\bar{v}\bar{\mathcal{E}}_{\bar{r}} - \bar{r}\tilde{p}^{(1)}(1 - \bar{M}_1^2) = [\bar{v}^2 + \bar{w}^2]\bar{\rho}\bar{r}^2\kappa\cos\phi, \tag{3.4.71}$$

where $\bar{M}_1 = \bar{v}/\bar{C}$. From the matching conditions, we have the behavior of $\tilde{\psi}^{(1)}$ and $\tilde{p}^{(1)}$, as $\bar{r} \to \infty$. We then integrate (3.4.70) from $\bar{r}$ to η and let $\eta \to \infty$. The result is

$$\bar{v}\tilde{\psi}^{(1)} + \bar{r}\tilde{p}^{(1)} = G(\bar{r})\kappa\cos\phi, \tag{3.4.72}$$

where

$$G(\bar{r}) = -\int_0^{\bar{r}} \bar{\rho}[2\bar{w}^2(\xi) - \bar{v}^2(\xi)]\xi d\xi. \tag{3.4.73}$$

Next we eliminate $\tilde{p}^{(1)}$ from (3.4.71) and (3.4.72) and obtain a linear equation of the first order for $\tilde{\psi}^{(1)}$,

$$\tilde{\psi}^{(1)}_{\bar{r}} + I(\bar{r})\tilde{\psi}^{(1)} = J(\bar{r})\kappa\cos\phi, \tag{3.4.74}$$

where

$$I(\bar{r}) = -\frac{\bar{v}_{\bar{r}}}{\bar{v}} - \frac{\bar{\rho}_{\bar{r}}}{\bar{\rho}}, \tag{3.4.75}$$

$$J(\bar{r}) = [(1 - \bar{M}_1^2)G + \bar{\rho}\bar{r}^2(\bar{v}^2 + \bar{w}^2)]/(\bar{r}\bar{v})\,. \tag{3.4.76}$$

The integration factor is

$$U = \exp\{\int I(\bar{r})d\bar{r}\} = 1/[\bar{\rho}\bar{v}]\,, \tag{3.4.77}$$

and the solution fulfilling the condition $\psi^{(1)} = 0$ at $\bar{r} = 0$ is

$$\tilde{\psi}^{(1)}(\bar{r},\phi) = \kappa\cos\phi\bar{\rho}(\bar{r})\bar{v}(\bar{r})\int_0^{\bar{r}} \frac{J(\bar{r}')}{\bar{\rho}(\bar{r}')\bar{v}(\bar{r}')}d\bar{r}'\,. \tag{3.4.78}$$

In determining the behavior of $\tilde{\psi}$ as $\bar{r}\to\infty$, we use the far-field behavior of $\bar{v},\bar{w}$ from (3.4.40) and obtain

$$\tilde{\psi}^{(1)} \to \kappa\cos\phi\bar{r}\left\{C^*(t) + \frac{\Gamma}{4\pi}\ln\bar{r}\right\}, \tag{3.4.79}$$

where

$$\begin{aligned}
C^*(t) &= \lim_{\bar{r}\to\infty}\left\{\frac{\bar{\rho}\bar{v}}{\bar{r}}\int_0^{\bar{r}}\frac{J(t,\bar{r}')}{\bar{\rho}(\bar{r}')\bar{v}(\bar{r}')}d\bar{r}' - \frac{\Gamma}{4\pi}\ln\bar{r}\right\}\\
&= \frac{\Gamma}{4\pi}\lim_{\bar{r}\to\infty}\left\{\left(\frac{2\pi}{\Gamma}\right)^2\int_0^{\bar{r}}\bar{\rho}(z)[\bar{v}^2(z) - 2\bar{w}^2(z)]z dz - \ln\bar{r}\right\} + \frac{\Gamma}{8\pi}\\
&= C_v + C_w\,,
\end{aligned} \tag{3.4.80}$$

and where we have defined

$$C_v(t,s) = \frac{\Gamma}{4\pi}\left\{\lim_{\bar{r}\to\infty}\left[\left(\frac{2\pi}{\Gamma}\right)^2\int_0^{\bar{r}}\bar{\rho}(z)\bar{v}^2(z)z dz - \ln\bar{r}\right] + \frac{1}{2}\right\}, \tag{3.4.81}$$

$$C_w(t,s) = -\frac{2\pi}{\Gamma}\int_0^{\infty}\bar{\rho}(z)\bar{w}^2(z)z\,dz\,. \tag{3.4.82}$$

The matching conditions for the asymmetric solutions, (3.4.63) and (3.4.64), then yield the far-field behavior of $\tilde{\psi}^{(1)}$

$$\tilde{\psi} \to \frac{\Gamma\kappa}{4\pi}\bar{r}\cos\phi\ln(\epsilon\bar{r}) - [\boldsymbol{Q}_0 - \dot{\boldsymbol{X}}]\cdot\hat{\theta}\,. \tag{3.4.83}$$

By matching the asymmetric inner solution of the core structure with the outer solution for $r\to 0$, the second and the third singular terms in the Biot–Savart integral (3.3.31) are removed and the leading-order velocity of the centerline, $\boldsymbol{X}_t$, is defined by comparing (3.4.80) and (3.4.83)

$$\boldsymbol{X}_t^{(0)}(t,s) = \boldsymbol{b}\frac{\Gamma\kappa}{4\pi}(\ln(1/\epsilon) + C_v + C_w) + (\boldsymbol{1} - \boldsymbol{\tau}\circ\boldsymbol{\tau})\cdot\boldsymbol{Q}_0\,. \tag{3.4.84}$$

Here, $\boldsymbol{Q}_0(t,\boldsymbol{X}) = \nabla\Phi + \boldsymbol{Q}^f$ denotes the background velocity without the filament plus the finite part of the Biot–Savart integral. The global contribution of the core structure C^* is partitioned into two parts, $C_v(t,s)$ and

$C_w(t,s)$, representing, respectively, the contributions of the circumferential and axial velocity weighed by the scaled density. They reduce to those for the incompressible case, (3.3.163)–(3.3.165), when $\bar{\rho}^{(0)} \equiv 1$.

The compressible core structure, $\bar{v}^{(0)}$, $\bar{w}^{(0)}$, and $\bar{\rho}^{(0)}$ as functions of t, s, and $\bar{r}$, is needed to define the velocity of $\mathcal{C}$, (3.4.84), (3.4.81), and (3.4.82). The governing equations for an incompressible core structure were derived in Subsect. 3.3.6 from the compatibility conditions for the symmetric parts of the second-order Navier–Stokes equations. From the systematic but tedious derivation, an alternative but simpler procedure was presented by Ting (1999). We shall apply the alternative procedure below to derive the governing equations for the evolution of a compressible core structure.

Again we note that there are restrictions on the initial data for the pressure distribution in the vortex core. Given the velocity and density distributions, the radial distribution of pressure is fixed immediately by (3.4.48). Similarly, the initial velocity field must agree with the matching condition, and thus be set appropriately depending on the initial filament geometry. These compatibility conditions for the initial data result from the assumption of only one time scale t^*. In case the initial core structure is inconsistent with those restrictions, we should introduce a short time scale ϵt^* so that unsteady terms will appear in the first order equations. If the assigned initial velocity of the filament differs from (3.4.84), we need an even shorter time scale $\epsilon^2 t^*$ (see Subsect. 3.2.4).

Remarks: In preparation for the derivation of the core structure evolution equations, we collect here several orthogonality relations on the first order solutions. We note that the first order solutions do not have second or higher harmonics and can be written as

$$\bar{u}^{(1)} = \left\langle \bar{u}^{(1)} \right\rangle + \bar{u}_{12}^{(1)} \sin\phi \quad \text{and} \tag{3.4.85}$$

$$\bar{g}^{(1)} = \left\langle \bar{g}^{(1)} \right\rangle + \bar{g}_{11}^{(1)} \cos\phi\,, \tag{3.4.86}$$

where $\bar{g}$ stands for $\bar{v}, \bar{w}, \bar{p}, \bar{\rho}, \bar{T}$, or $\bar{\mathcal{E}}$, and $\bar{u}_{12}$ and $\bar{g}_{11}$ stand for the Fourier coefficients, that is, independent of ϕ or θ. It follows that

$$\left\langle \bar{u}^{(1)} \cos\phi \right\rangle = 0 \quad \text{and} \quad \left\langle \bar{g}^{(1)} \sin\phi \right\rangle = 0\,. \tag{3.4.87}$$

In the case analyzed by Callegari and Ting (1978), the incompressible core structure had no axial variation, which followed when either the leading circumferencial velocity $\bar{v}^{(0)}$ or the axial velocity $\bar{w}^{(0)}$ was assumed to be independent of s. We shall extend their analysis below to a compressible core structure with no axial variation (see item *C.5)* above). With $\langle u^{(1)} \rangle = 0$, (3.4.85) becomes

$$\bar{u}^{(1)} = \bar{u}_{12} \sin\phi\,, \tag{3.4.88}$$

and with (3.4.86), we have the orthogonality conditions,

$$\left\langle \bar{u}^{(1)}\bar{g}^{(1)} \right\rangle = 0, \quad \text{for} \quad g = v, w, p, \rho, T, \text{ and } \mathcal{E}. \tag{3.4.89}$$

These conditions will be employed to remove the products of first order solutions, $\bar{u}^{(1)}\bar{g}^{(1)}$, appearing in the second-order equations.

3.4.6 The Second-Order Equations and the Evolution of the Core Structure

For the analysis of a compressible core structure, we intend to follow the procedure of Ting (1999), which is somewhat more straightforward than that of Callegari and Ting (1978). To illustrate the essence of this procedure, we present first a trivial example, and then a well-known example for shock conditions. In these two examples, the original system of equations are rearranged to a new system so that the leading-order solutions are defined by the leading-order equations of the new system.

3.4.6.1 A Trivial Example. Let us consider the solution of the system of equations,

$$x - y = \epsilon \qquad \text{and} \qquad x^2 - y^2 = 2\epsilon\ . \tag{3.4.90}$$

If we use the perturbation method and expand $x(\epsilon)$ and $y(\epsilon)$ in powers of ϵ, we obtain from (3.4.90) only one leading-order equation $x^{(0)} - y^{(0)} = 0$. The next-order equations are $x^{(1)} - y^{(1)} = 1$ and $x^{(0)}[x^{(1)} - y^{(1)}] = 1$. They yield $x^{(1)} - y^{(1)} = 1$ and the compatibility condition $x^{(0)} = 1$. Thus we complete the leading-order solution with $y^{(0)} = 1$. To complete the first order solution, we have to introduce the second-order equations of (3.4.90). They are $x^{(2)} - y^{(0)} = 0$ and $x^{(0)}[x^{(2)} - y^{(2)}] = [y^{(1)}]^2 - [x^{(1)}]^2$. The latter becomes the compatibility condition $x^{(1)} + y^{(1)} = 0$, and we complete the first order solution $x^{(1)} = 1/2$ and $y^{(1)} = -1/2$. We could continue to higher-order equations and show that $x^{(j)} = y^{(j)} = 0$ for $j \geq 2$. Thus, the solution is $x = 1 + \epsilon/2$ and $y = 1 - \epsilon/2$.

It is obvious that we can get the result directly by replacing the factor $x - y$ in the second equation in (3.4.90) by ϵ and then cancelling the common factor ϵ. The equivalent system of equations is $x - y = \epsilon$ and $x + y = 2$. We immediately recover the solution $x = 1 + \epsilon/2$ and $y = 1 - \epsilon/2$.

3.4.6.2 Shock Conditions. Consider a one-dimensional flow in which a shock is advancing with speed V toward the gas at rest with pressure p_1, density ρ_1, and speed of sound $C_1 = \sqrt{\gamma p_1/\rho_1}$ with $V > C_1$. Behind the shock, the pressure p, density ρ, and the velocity u are governed by the equations of mass, momentum, and energy conservation across a normal shock:

$$\begin{aligned} \rho(V - u) &= m_1\,, \\ p - p_1 &= m_1 u\,, \\ (V - u)^2/2 + \gamma p/[(\gamma - 1)\rho] &= H_1\,, \end{aligned} \tag{3.4.91}$$

where $m_1 = \rho_1 V$ and $H_1 = [V^2/2] + \{\gamma p_1/[(\gamma-1)\rho_1]\}$. As $V - C_1 \to 0^+$, the shock becomes weaker and weaker and numerical solution of the shock equations would require higher and higher precision and would break down eventually. Of course, an aeronautical engineer would avoid this predicament by replacing the energy equation in (3.4.91) with the Prandtl relationship,

$$V(V-u) = 2(\gamma-1)H_1/(\gamma+1) \ . \tag{3.4.92}$$

We can explain the predicament and its resolution by the perturbation method. If we introduce the perturbation state variables $p' = p - p_1$ and $\rho' = \rho - \rho_1$, the shock equations (3.4.91) yield three linear homogeneous equations for p', ρ', and u, with only two of them linearly independent. The missing third equation would come from the compatibility condition on the inhomogeneous terms of the next-order equations. Now we trace the derivation of the Prandtl relationship (3.4.92). First we use the mass conservation equation in (3.4.91) to rewrite the energy equation as $(V-u)/2 + \gamma p/[(\gamma-1)m_1] = H_1/(V-u)$ and $V/2 + \gamma p_1/[(\gamma-1)m_1] = H_1/V$. Their difference, after using the momentum equation, becomes $u(\gamma+1)/[2(\gamma-1)] = uH_1/[V(V-u)]$. We cancel the common factor u, which is small but nonzero, and arrive at the Prandtl relationship, (3.4.92). Now we see why an appropriate combination of the conservation equations (3.4.91) resolves the numerical predicament and also avoids the need of the compatibility condition for the second-order perturbation equations.

3.4.7 Evolution Equations for the Compressible Core Structure

Based upon what we learned about the main features of the analyses of the core structure, items [1]–[10] as listed in Subsubsect. 3.4.3.3, we speed up the derivation of the evolution equations by adopting the conservative form of the governing equations and deriving the system of axisymmetric equations by taking the circumferential average of the Navier–Stokes equations, (3.4.35)–(3.4.39), to remove the ∂_θ terms. Next, we take the s-average of these equations, that is, the average along the length of the centerline $\mathcal{C}$ of the closed loop filament over its total length $S(t)$, to remove the ∂_s terms. Finally, we introduce the expansion schemes from (3.4.29) into the new system of the averaged Navier–Stokes equations yielding the evolution equations of the leading-order core structure. That is, we obtain the evolution equations for the large velocity components, $\bar{v}^{(0)}$ and $\bar{w}^{(0)}$, and the state variables $\bar{\rho}^{(0)}$ and $\bar{p}^{(0)}$.[9]

We introduce $\langle f \rangle$ and $\{f\}$ to denote, respectively, the θ- and s-averages of some function f, that is,

[9] In Callegari and Ting (1978) and in Book-I, the expansion schemes were introduced into the Navier–Stokes equations first, the leading-order and higher order equations were derived systematically and finally the compatibility conditions of the second-order equations yielded the evolution equations for the leading-order core structure.

$$
\begin{aligned}
\langle f \rangle &= \frac{1}{2\pi} \int_0^{2\pi} f \, d\theta \,, \\
\{f\} &= \frac{1}{S(t)} \int_0^{S(t)} f \, d\tilde{s} = \frac{1}{S} \int_0^{S_0} f \, \sigma ds \,,
\end{aligned} \tag{3.4.93}
$$

where

$$
\tilde{s}(t,s) = \int_0^s \sigma \, ds' \qquad \text{and} \qquad S(t) = \int_0^{S_0} \sigma \, ds \tag{3.4.94}
$$

denote the arclength and the length of the filament at instant t, respectively. We choose s to be the initial arclength such that at $t = 0$, $\sigma = 1$, and $S = S_0$.

The θ- and s-averages of the continuity equation truncated to terms of order $O(1)$ from (3.4.35) are

$$
\langle h_3 \bar{\rho} \rangle_t = -\frac{1}{\epsilon \bar{r}} \big(\bar{r} \, \langle h_3 \bar{\rho} \bar{u} \rangle \big)_{\bar{r}} - \big(\langle \bar{\rho} \bar{w} \rangle \big)_s \,, \tag{3.4.95}
$$

$$
\{ \langle h_3 \bar{\rho} \rangle \}_t = -\frac{1}{\epsilon \bar{r}} \big(\bar{r} \, \{ \langle h_3 \bar{\rho} \bar{u} \rangle \} \big)_{\bar{r}} \,. \tag{3.4.96}
$$

Next, we recall that $\bar{\rho}^{(0)}$ depends only on $(t, \bar{r})$, and that $h_3 = \sigma(t,s)(1 + \epsilon \bar{r} \kappa \cos\phi)$, that σ satisfies (3.4.94), and that $\langle u^{(1)} \rangle = \{ \langle u^{(1)} \rangle \} = 0$ according to (3.4.56) for s-independent leading-order core structure, and we recall the orthogonality relationships for the first order solutions from (3.4.87). The leading-order version of (3.4.96) then relates the second-order s- and θ-averaged radial velocity $\{ \langle \bar{u}^{(2)} \rangle \}$ to temporal changes of the leading-order density and the overall stretching of the filament,

$$
\frac{\bar{r}}{S^{(0)}} \left(\bar{\rho}^{(0)} S^{(0)} \right)_t + \left(\bar{r} \bar{\rho}^{(0)} \{ \langle \bar{u} \rangle \}^{(2)} \right)_r = 0 \,. \tag{3.4.97}
$$

Analogous considerations for the entropy equation (3.4.37) yield

$$
\frac{\bar{r}}{S^{(0)}} \left(\bar{\rho}^{(0)} \bar{\mathcal{E}}^{(0)} S^{(0)} \right)_t + \left(\bar{r} \bar{\rho}^{(0)} \bar{\mathcal{E}}^{(0)} \{ \langle \bar{u} \rangle \}^{(2)} \right)_r = \{ \langle \bar{q}_{\mathcal{E}} \rangle \}^{(0)} \,, \tag{3.4.98}
$$

where, omitting the superscript (0) for the moment,

$$
\{ \langle \bar{q}_{\mathcal{E}} \rangle \} = \frac{1}{\bar{\rho} \bar{T}} \left\{ \gamma \overline{\mathrm{Ma}}_*^2 \bar{\mu} \left(\left[\bar{r} \frac{\partial}{\partial \bar{r}} \left(\frac{\bar{v}}{\bar{r}} \right) \right] \frac{\partial \bar{v}}{\partial \bar{r}} + \left[\frac{\partial \bar{w}}{\partial \bar{r}} \right]^2 \right) + \frac{1}{(\gamma - 1)\mathrm{Pr}} \frac{\partial}{\partial \bar{r}} \left[k \frac{\partial \bar{T}}{\partial \bar{r}} \right] \right\} . \tag{3.4.99}
$$

In these equations, the leading-order density, temperature, entropy, and circumferential velocity are related through (i) the equations of state,

$$
\bar{p}^{(0)} = \bar{\rho}^{(0)} \bar{T}^{(0)} \,, \qquad \bar{\mathcal{E}}^{(0)} = \frac{1}{\gamma - 1} \ln \frac{p^{(0)}}{\rho^{(0)\gamma}} \,, \tag{3.4.100}
$$

and (ii) the leading-order radial momentum balance,

$$\bar{p}_{\bar{r}}^{(0)} = \frac{\rho^{(0)} v^{(0)2}}{\bar{r}} . \tag{3.4.101}$$

The extraction of the second-order compatibility conditions from the mass and entropy balances was straightforward, because the s and θ averages immediately cancelled the $(\cdot)_\theta$ and $(\cdot)_s$ terms in (3.4.35) and (3.4.37). Extracting the evolution equations for the leading-order circumferential and axial velocities from the momentum balances (3.4.36) is somewhat more involved. We cannot simply average in θ and s and rely on the same cancellation effect, because the momentum balance is written in vector notation. To extract equations for the *vector components*, $\bar{v}$ and $\bar{w}$, we need to project the equation onto the circumferential and axial unit vectors $\boldsymbol{\theta}$ and $\boldsymbol{\tau}$, respectively. In doing so, we must account for the fact that for any vector $\boldsymbol{f}$, we have $\boldsymbol{n}_i \cdot \partial_{\xi_j} \boldsymbol{f} = \partial_{\xi_j}(\boldsymbol{n}_i \cdot \boldsymbol{f}) - (\partial_{\xi_j}\boldsymbol{n}_i) \cdot \boldsymbol{f}$, and generally $\partial_{\xi_j}\boldsymbol{n}_i \neq 0$.

For example, in the axial momentum equation, obtained by projecting (3.4.36) onto $\boldsymbol{\tau}$, we find a term,

$$\boldsymbol{\tau} \cdot \left(\bar{\rho}\hat{w}\hat{\boldsymbol{V}}\right)_s = \left(\bar{\rho}\hat{w}^2\right)_s - \bar{\rho}\hat{w}\hat{\boldsymbol{V}} \cdot \boldsymbol{\tau}_s . \tag{3.4.102}$$

Using $\boldsymbol{\tau}_s = \sigma\kappa\boldsymbol{n} = \sigma\kappa(\cos\phi\, \boldsymbol{r} - \sin\phi\, \boldsymbol{\theta})$, this reads

$$\boldsymbol{\tau} \cdot \left(\bar{\rho}\hat{w}\hat{\boldsymbol{V}}\right)_s = \left(\bar{\rho}\hat{w}^2\right)_s - \sigma\kappa\bar{\rho}\hat{w}\left(\epsilon\bar{u}\cos\phi - \hat{v}\sin\phi\right) . \tag{3.4.103}$$

Employing the orthogonality relations (3.4.87) and (3.4.89), we can show that the θ average of the last term in this equation is no larger than $O(1)$ as $\epsilon \to 0$. Therefore we have

$$\left\langle \boldsymbol{\tau} \cdot \left(\bar{\rho}\bar{w}\boldsymbol{V}\right)_s \right\rangle = \left\langle \bar{\rho}\bar{w}^2 \right\rangle_s + O(1) , \tag{3.4.104}$$

and the first term on the right-hand side will be eliminated by averaging in s.

Another helpful observation is that all the terms in (3.4.36) involving $\epsilon\boldsymbol{X}_t$ can be combined with the continuity equation to yield

$$-\frac{1}{\epsilon\bar{r}}\left[(\bar{r}h_3\bar{\rho}\boldsymbol{X}_t\bar{u})_{\bar{r}} + (h_3\bar{\rho}\boldsymbol{X}_t\hat{v})_\theta\right] - \left(\bar{\rho}\boldsymbol{X}_t\hat{w}\right)_s = -\bar{\rho}\hat{w}\boldsymbol{X}_{ts} + o(1) . \tag{3.4.105}$$

Then, making use of (3.4.87) and (3.4.89), we obtain the desired evolution equation for the leading-order axial velocity,

$$\frac{\bar{r}}{S^{(0)}}\left(\bar{\rho}\bar{w}S\right)_t^{(0)} + \left(\bar{r}(\bar{\rho}\bar{w})^{(0)}\{\langle\bar{u}\rangle\}^{(2)}\right)_r + (\bar{\rho}\bar{w})^{(0)}\dot{S}^{(0)} = \frac{\partial}{\partial\bar{r}}\left[\bar{\nu}\bar{r}\frac{\partial}{\partial\bar{r}}\bar{w}\right]^{(0)} . \tag{3.4.106}$$

Similarly, the average of the circumferential momentum equation yields

$$\frac{\bar{r}}{S^{(0)}}\left(\bar{\rho}\bar{v}S\right)_t^{(0)} + \left(\bar{r}(\bar{\rho}\bar{v})^{(0)}\{\langle\bar{u}\rangle\}^{(2)}\right)_{\bar{r}} + (\bar{\rho}\bar{v})^{(0)}\{\langle\bar{u}\rangle\}^{(2)} = \frac{\bar{\nu}}{\bar{r}}\frac{\partial}{\partial\bar{r}}\left[\bar{r}^3\bar{\nu}\frac{\partial}{\partial\bar{r}}\left(\frac{\bar{v}}{\bar{r}}\right)\right]^{(0)} . \tag{3.4.107}$$

This completes the derivation of the compressible core structure evolution equations summarized in (3.4.4)–(3.4.10) above.

For incompressible flows we were able to provide efficient low-order approximations to the solution of the core structure equations. We found similarity solutions that could be determined uniquely by two parameters, the initial polar moments of the axial components of velocity and vorticity. These similarity solutions represent the leading-order terms in an expansion of the core structure evolution equations in terms of Laguerre polynomials, with the nth term proportional to $1/t$.

For a compressible core structure, the solutions of the governing equations would depend on the initial axial velocity, axial vorticity, and temperature profiles with varying viscosity and thermal conductivity. Again, we would like to be able to approximate the numerical solutions of the initial value problem by similarity solutions for the axial vorticity and velocity and the temperature so that we can reduce the effects of initial profiles on the velocity of the filament in terms of a few parameters. It is well known that similarity solutions for compressible boundary layers and slender jets (Kleinstein, 1962; Schlichting and Gersten, 2000) are available only with appropriate approximations on the variations of the viscosity and thermal conductivity. We would need analogous approximations to render similarity solutions admissible here. The accuracies of the approximations would have to be verified by comparing the approximate solutions with numerical solutions for various initial data. The construction of the approximate similarity solutions and the validation of their accuracies shall be carried out in future work.

4. Nonlinear Dynamics of Nearly Straight Vortex Filaments

Motivation

The phenomenon of *vortex stretching* is expressed by the last term on the left-hand side of the vorticity transport equation (2.3.2). Here, we rewrite that equation as

$$\boldsymbol{\Omega}_t + \boldsymbol{v} \cdot \nabla \boldsymbol{\Omega} - \boldsymbol{\Omega} \cdot \nabla \boldsymbol{v} = \nu \triangle \boldsymbol{\Omega} \,, \tag{4.1}$$

using $\nabla \cdot \boldsymbol{\Omega} = \nabla \cdot \boldsymbol{v} \equiv 0$. Vortex stretching does not exist in two space dimensions, because the vorticity vector $\boldsymbol{\Omega} = \frac{1}{2} \nabla \times \boldsymbol{v}$ is orthogonal to the flow plane, whereas the gradient operator generates vectors that lie *within* that plane. Thus $\boldsymbol{\Omega} \cdot \nabla \equiv 0$. We conclude that, from the perspective of vorticity evolution, vortex stretching is the most prominent phenomenon distinguishing three-dimensional flow from two-dimensional flow. The vortex stretching term is quadratically nonlinear and, because of the integral relationship between velocity and vorticity, also nonlocal. Its effect on the dynamics of vorticity is extremely difficult to analyze in all generality.

In the present chapter we discuss asymptotic theories, developed in a series of papers by one of the authors in collaboration with A.J. Majda and coworkers (1991a, 1991b, 1992, 1993, 1995a) (see also Klein, 1994; Majda, 1998), which elucidate the influences of vortex stretching nonlocality in a somewhat simplified setting. The idea is to analyze slender vortex filaments in analogy with the considerations in Chap. 3, but to restrict to filament geometries given by small deviations from straight-line background configurations (Fig. 4.1). Thus we consider generalized binormal propagation laws

$$\dot{\boldsymbol{X}}(\tilde{s}, \bar{t}) = (\kappa \boldsymbol{b})(\tilde{s}, \bar{t}) + \tilde{\delta} \boldsymbol{v}(\tilde{s}, \bar{t}) \tag{4.2}$$

in analogy with the filament equation of motion from (3.3.77) and (3.3.81). Various new effects associated, for example, with the influence of an outer straining field (Sect. 4.2), or with the mutual interaction of two almost parallel, almost straight neighboring filaments (Sect. 4.3) will be considered through suitable definitions of the perturbation velocity $\tilde{\delta} \boldsymbol{v}$.

In this chapter, the distinction between various possible parameterizations of the filament centerline will be important. In (4.2) we have introduced, in particular, an arclength coordinate, $\tilde{s}$, which will play a central role when we introduce the Hasimoto transformation (Hasimoto, 1972) below.

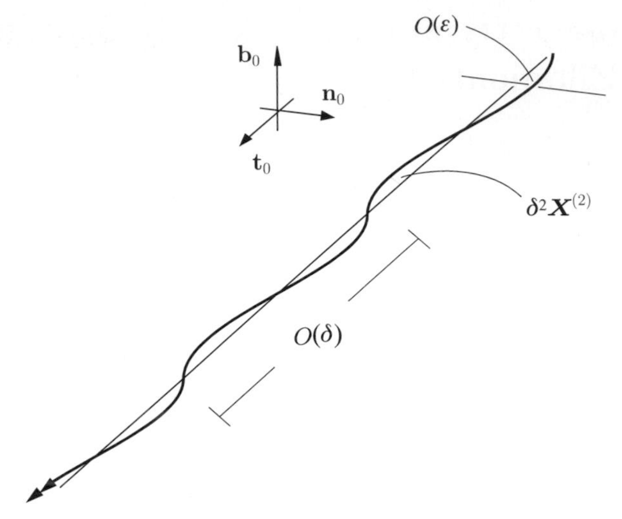

Fig. 4.1. A slender vortex with small deviations from a straight line. Adapted from Klein (1994)

One of the most striking results of this theory is a simplified asymptotic nonlinear Schrödinger-type equation for a complex "filament function," $\psi(\tau, \tilde{\sigma})$. The simplest version of this equation, called the *filament equation with self-stretch*, reads

$$\frac{1}{i}\,\psi_\tau = \psi_{\tilde{\sigma}\tilde{\sigma}} + \delta^2 \left(\frac{1}{2} |\psi|^2 \psi - \mathcal{I}[\psi] \right) . \tag{4.3}$$

Details of its derivation, which utilizes Hasimoto's ingenious transformation (Hasimoto, 1972), will be given shortly. Here we merely emphasize that the term in brackets on the right-hand side describes the direct interaction of the well-known curvature–binormal nonlinearity of slender vortex motion, first revealed in this form in Hasimoto's original work, with nonlocal self-induction. The latter is represented by the (nonlocal) pseudo-differential operator $\mathcal{I}[\cdot]$ defined in (4.1.8) below.

Chapter Summary

In Sect. 4.1 we consider an isolated, nearly straight filament of infinite length in a quiescent environment following Klein and Majda, (1991a, 1991b). This section provides the key ideas that form the foundation of this chapter, namely the Klein–Majda asymptotic scaling regime, the linearization of the Biot–Savart integral, and the Hasimoto transformation. The Klein–Majda regime for geometric disturbances of a nearly straight filament is chosen so as to ensure a direct interaction between the effects of the curvature nonlinearity and the nonlocal effects of the self-induced filament motion. Another consequence of this asymptotic scaling regime is a linearization of the Biot–Savart integral which, after Hasimoto's transformation, takes the form of a pseudo-differential operator representing a singular, nonlocal perturbation of the filament equation of motion. Further subsections provide a number of theoretical results for this simplest case, such as analytical exact solutions and their linear and nonlinear stability, and discuss various results of numerical solutions of the reduced filament equation with self-stretch.

Section 4.2 considers vortex filaments within the same perturbation regime, but embedded in an outer straining field (Klein et al. 1992). Again we describe the linear and rigorous nonlinear stability of solutions of the resulting simplified asymptotic equation, and report on the results of numerical solutions demonstrating mode coupling in wavenumber space.

An important practical as well as theoretical issue is the interaction of separated vortices. In Sect. 4.3 we extend the Crow–Jimenez stability theory for parallel and antiparallel pairs of nearly straight vortices to the Klein–Majda regime (Klein and Majda, 1993). (Antiparallel vortices are parallel and of equal strength but opposite in direction.) While the main innovation in Sect. 4.1 was the inclusion of nonlocal effects in Hasimoto's nonlinear theory, the present section emphasizes the inclusion of the curvature nonlinearity in the earlier linear theories for filament pair interactions. After the asymptotic derivations we discuss again a number of numerical simulation results.

In the Crow–Jimenez linear theories, as well as in the weakly nonlinear analyses described in the previous paragraph, the displacements of the participating vortex filaments from their unperturbed straight-line configurations are assumed to be asymptotically small compared with their separation distances. Thus, these theories are not able to account for the ultimate touching (and subsequent reconnection) of unstable vortex pairs. Section 4.4 addresses this point following Klein et al. (1995a). We now allow for order one displacements on the filament separation scale, but maintain the geometrical properties of the Klein–Majda regime. The dominant nonlinearity in this regime is no longer the curvature effect of a single filament. Rather, it is the consequence of the leading order mutually induced velocities. These are due to the fact that each of the vortex filaments is embedded in the potential vortex velocity field of the others. The resulting leading-order equations describe the direct interaction of curvature effects with this, essentially two-dimensional, potential vortex interaction. Their structure differs from the equations of the extended Crow–Jimenez theory, revealing an interesting Hamiltonian structure and the possibility of strong singularity formation. Numerical simulations show that these singularities correspond to local contact between pairs of filaments. Clearly, as the local filament distance decreases to become comparable to the vortex filament core size, the slenderness assumptions that form the basis of the present theories break down. As a consequence, the theory of this section still falls short of describing vortex reconnection.

The reader is also referred to work by Lions and Majda (2000), who investigate the statistical mechanics of ensembles of nearly straight vortex filaments in the above-mentioned large displacement regime.

4.1 Self-Induced Stretching and the Curvature Nonlinearity for a Single Filament

4.1.1 The Klein–Majda Regime

The challenge in constructing a simplified asymptotic theory for the self-stretching of slender vortex filaments may be formulated as follows: Find a particular regime of solutions in which the dominant curvature nonlinearity is retained, and in which the nonlocal effects of self-induction appear in a simplified, yet prominent, fashion.

To this end, Klein and Majda (1991a) introduce a new small parameter $\delta \ll 1$ and define the perturbed filament geometry through

$$\boldsymbol{X}(s,\bar{t};\delta) = s\,\boldsymbol{t}_0 + \delta^2 \boldsymbol{X}^{(2)}\left(\tfrac{s}{\delta}, \tfrac{\bar{t}}{\delta^2}\right) + o(\delta^2)\,. \tag{4.1.1}$$

Here, s denotes an arclength coordinate on the unperturbed straight line, while

$$\bar{t} = \frac{\Gamma}{4\pi}\left(\ln\left(\frac{2\delta}{\epsilon}\right) + C\right) t\,, \tag{4.1.2}$$

with the core structure parameter C in (3.3.162), is an appropriately scaled time variable. Its particular dependence on δ and ϵ will be derived in the context of (4.1.21) below. Meanwhile, $\boldsymbol{t}_0$ is a constant unit vector defining the orientation of the unperturbed filament as displayed in Fig. 4.1. Without loss of generality we require

$$\boldsymbol{t}_0 \cdot \boldsymbol{X}^{(2)} \equiv 0\,. \tag{4.1.3}$$

In the sequel we abbreviate the arguments of the expansion function $\boldsymbol{X}^{(2)}$ as

$$s/\delta \equiv \sigma\,, \qquad \bar{t}/\delta^2 \equiv \tau\,. \tag{4.1.4}$$

4.1.1.1 Scaling Considerations. The order of magnitude, $O(\delta^2)$, of the "small deviations from a straight line" is linked to the vortex filament core size through the judiciously chosen distinguished limit

$$\delta^2 = \left(\ln\frac{2\delta}{\epsilon} + C\right)^{-1} \ll 1 \qquad \text{as} \qquad \epsilon \to 0\,. \tag{4.1.5}$$

We will see below that, under this distinguished limit, the effects of nonlinearity and nonlocality attain comparable orders of magnitude in the simplified asymptotic filament evolution equation.

Equation (4.1.1) is a dimensionless representation of the filament centerline, and we have used a characteristic radius of curvature for the nondimensionalization of both the deviation from the straight line and the arclength, s, on the unperturbed curve. In fact, the scalings induce curvatures $\kappa \sim \boldsymbol{X}_{ss} = \boldsymbol{X}^{(2)}_{\sigma\sigma} = O(1)$ as $\delta \to 0$. The time scaling in (4.1.1) follows from the earlier general results for slender filament dynamics from Sect. 3.3. From the binormal component of the asymptotic expression for the filament motion in (3.3.193) and the definition of $\bar{t}$ in (4.1.2), we find

$$\partial_{\bar{t}} \boldsymbol{X} = \partial_\tau \boldsymbol{X}^{(2)} = \kappa \boldsymbol{b} + o(1) = \boldsymbol{X}^{(2)}_{\sigma\sigma} + o(1) \quad \text{as} \quad \delta \to 0\,. \tag{4.1.6}$$

Thus the spatiotemporal scalings in the perturbation ansatz from (4.1.1) are appropriate.

4.1.2 Linearization of Nonlocal Self-Induction in the Klein–Majda Regime

Here we will derive a simplified expression for the nonlocal velocity contribution, $\boldsymbol{Q}_0$, in the filament equation of motion, (3.3.81), valid in the regime of geometrical perturbations of a straight filament as given in (4.1.1). Since we consider a filament embedded in a quiescent background, $\boldsymbol{Q}_0 = \boldsymbol{Q}^f$, with the finite, single-valued part of the Biot–Savart integral from (3.1.77). We will show here that in the present regime the self-induced motion of a vortex filament is given by

$$\boldsymbol{X}_{\bar{t}} = \kappa \boldsymbol{b} + \delta^2\, \mathcal{I}[\boldsymbol{X}^{(2)}] \times \boldsymbol{t}_0\,, \tag{4.1.7}$$

where $\mathcal{I}[\cdot]$ is the linear pseudo-differential operator (see Hörmander, 1984)

$$\mathcal{I}[w](\sigma) = \int_{-\infty}^{\infty} \frac{1}{|h^3|} \left[w(\sigma+h) - w(\sigma) - h w_\sigma(\sigma+h) + \frac{h^2}{2} \mathcal{H}(1-|h|) w_{\sigma\sigma}(\sigma) \right] \, dh\,. \tag{4.1.8}$$

We consider the finite part of the Biot–Savart line integral along the centerline, taking into account the particular asymptotic scalings from (4.1.1). As these scalings include geometrical variations on length scales of order $O(\delta)$, we have to modify the regularization of the integral in comparison with the derivations in Subsect. 3.3.4, see (3.3.103). Let

$$\begin{aligned} \boldsymbol{q}_{\text{outer}} = & \int\limits_{\mathcal{L}} \left[\boldsymbol{F}\left(\frac{\tilde{s}}{\delta}, \frac{\bar{t}}{\delta^2}\right) - H\left(1 - \frac{|\tilde{s} - \tilde{s}_1|}{\delta}\right) \boldsymbol{F}_1\left(\frac{\tilde{s}}{\delta}, \frac{\bar{t}}{\delta^2}\right) \right] \mathrm{d}\tilde{s} \\ & + \int\limits_{\mathrm{i}_\delta - \mathrm{i}_C} \boldsymbol{F}_1\left(\frac{\tilde{s}}{\delta}, \frac{\bar{t}}{\delta^2}\right) \mathrm{d}\tilde{s} + \mathrm{O}\left(\delta_C\right)\,. \end{aligned} \tag{4.1.9}$$

Here, $\mathrm{i}_\delta = [\tilde{s}_1 - \delta, \tilde{s}_1 + \delta]$ and i_C and δ_C have the same meaning as in the context of the earlier vorticity-based analysis. They indicate the truncation of the line Biot–Savart by some inner interval near $\tilde{s}_1$. The matching procedure to fit the δ_C-dependence of the outer integration to that of the detailed inner three-dimensional integration is equivalent to that employed in Subsect. 3.3.4.

In (4.1.9) we have used, in addition, the abbreviations

$$\boldsymbol{F}\left(\frac{\tilde{s}}{\delta},\frac{\bar{t}}{\delta^2}\right) = \frac{\boldsymbol{X}(\tilde{s},t;\delta)-\boldsymbol{X}_1}{|\boldsymbol{X}(\tilde{s},t;\delta)-\boldsymbol{X}_1|^3}\times\boldsymbol{t}$$
$$\boldsymbol{F}_1\left(\frac{\tilde{s}}{\delta},\frac{\bar{t}}{\delta^2}\right) = \frac{1}{2}\frac{1}{|\tilde{s}-\tilde{s}_1|}\left(k\boldsymbol{b}\right)_1 . \tag{4.1.10}$$

The subscript $(\cdot)_1$ indicates evaluation at $\tilde{s}=\tilde{s}_1$ as before. Next we expand these expressions in the limit $\delta\to 0$.

For convenience, we introduce the stretched coordinates

$$\sigma = \frac{s}{\delta} \quad \text{and} \quad \hat{\sigma} = \frac{s-s_1}{\delta} \tag{4.1.11}$$

and take into account the following relation between the arclength, $\tilde{s}$, and the linear coordinate, s, on the unperturbed straight line,

$$s = \tilde{s}\,(1+O(\delta^2)) . \tag{4.1.12}$$

This allows us to deduce

$$\boldsymbol{X}-\boldsymbol{X}_1 = \delta\,\hat{\sigma}\boldsymbol{t}_0 + \delta^2\Delta\boldsymbol{X}^{(2)}(\hat{\sigma},\sigma_1;\delta) , \tag{4.1.13}$$

where

$$\Delta\boldsymbol{X}^{(2)}(\hat{\sigma},\sigma_1;\delta) = \boldsymbol{X}^{(2)}(\hat{\sigma}+\sigma_1;\delta) - \boldsymbol{X}^{(2)}(\sigma_1;\delta) . \tag{4.1.14}$$

The unit tangent, $\boldsymbol{t}=\boldsymbol{X}_s/|\boldsymbol{X}_s|$, obeys

$$\boldsymbol{X}_s = \boldsymbol{t}_0 + \delta\boldsymbol{X}^{(2)}_\sigma . \tag{4.1.15}$$

Because of (4.1.13) and the orthogonality between $\boldsymbol{t}_0$ and $\boldsymbol{X}^{(2)}$, we then have

$$\frac{1}{|\boldsymbol{X}-\boldsymbol{X}_1|^3} = \frac{1}{(\delta\,|\hat{\sigma}|)^3}\left(1+O(\delta^2)\right) . \tag{4.1.16}$$

This yields the asymptotic representation

$$\boldsymbol{F} = \frac{1}{\delta|\hat{\sigma}|^3}\left[\Delta\boldsymbol{X}^{(2)}(\hat{\sigma}+\sigma_1;\delta) - \hat{\sigma}\boldsymbol{X}^{(2)}_\sigma(\hat{\sigma}+\sigma_1;\delta)\right]\times\boldsymbol{t}_0\;\left(1+O\left(\delta,\frac{\delta^2}{\hat{\sigma}^2}\right)\right) \tag{4.1.17}$$

and

$$\boldsymbol{F}_1 = -\frac{1}{2\,\delta|\hat{\sigma}|}\boldsymbol{X}^{(2)}_{\sigma\sigma}\times\boldsymbol{t}_0\,(1+O(\delta)) \tag{4.1.18}$$

so that (4.1.9) becomes

$$\boldsymbol{q}_{\text{outer}} = \mathcal{I}[\boldsymbol{X}^{(2)}] \times \boldsymbol{t}_0 + \frac{\Gamma}{4\pi} (k\boldsymbol{b})_1 \ln \frac{\delta}{\delta_C} . \tag{4.1.19}$$

Here, $\mathcal{I}[\cdot]$ is the linear integral operator from (4.1.8).

Matching (4.1.19) to the result of the three-dimensional inner integration as in Subsect. 3.3.4, we obtain the filament equation of motion for the Klein–Majda regime,

$$\boldsymbol{X}_{\bar{t}} = \kappa \boldsymbol{b} + \tilde{\delta}\, \mathcal{I}[\boldsymbol{X}^{(2)}] \times \boldsymbol{t}_0 , \tag{4.1.20}$$

where we have introduced the slow time variable announced in (4.1.2)

$$\bar{t} = \left(\ln \frac{2\delta}{\epsilon} + C \right) t \tag{4.1.21}$$

and an effective perturbation parameter

$$\tilde{\delta} \equiv \left(\ln \frac{2\delta}{\epsilon} + C \right)^{-1} . \tag{4.1.22}$$

That $\tilde{\delta}$ should equal δ^2, as anticipated in (4.1.7), will be demonstrated below. There, we impose the additional requirement that nonlocal and nonlinear effects in the filament evolution equation should compete at the same order in δ.

In addition, we have anticipated that on the time scale of $\bar{t}$, and certainly on the time scale of $\tau = \bar{t}/\delta^2$, the vortex core structure will be quasi-stationary, and that it is homogeneous along the vortex centerline. The latter two assumptions are not essential for the present derivations. Dropping them, however, will generate a considerably more complex system that, to the best of our knowledge, has not been studied thus far. This completes the linearization of the nonlocal self-induction term for vortex filaments in the Klein–Majda regime.

4.1.3 Hasimoto's Transformation

An asymptotic expansion of the fluid dynamics equations based on the solution ansatz (4.1.1) for the vortex filament geometry leads to quite unwieldy nonlinear expressions. Their mathematical structure is very hard to analyze, and the resulting simplified asymptotic equations yield little insight beyond what is already known from the theory for general filament geometries discussed in the previous chapter.

Hasimoto (1972) analyzed vortex motion under the "local induction approximation" which neglects the nonlocal self-induction and external flow field effects represented by $\boldsymbol{Q}_0$ in (3.3.161) and (3.3.162). (Recall that $\boldsymbol{Q}_0$ is the sum of any imposed external flow velocity and the finite, single-valued part, $\boldsymbol{Q}^f$, of the Biot–Savart integral in (3.1.77).) Thus, he considered the motion of space curves $\mathcal{L}(t)$: $t \to \boldsymbol{X}(t,t)$ governed by the evolution equation

$$\boldsymbol{X}_t = \kappa \boldsymbol{b}\,, \tag{4.1.23}$$

where, as before, κ and $\boldsymbol{b}$ are the local curvature and binormal unit vector, respectively, t is an appropriately scaled time variable, and s denotes an arclength coordinate on the vortex centerline $\mathcal{L}$.

Hasimoto invented an ingenious transformation of variables that turns the curvature–binormal nonlinearity from (4.1.23) into the cubic nonlinear Schrödinger equation (abbreviated from here on by “NLS” for nonlinear Schrödinger where convenient)

$$\frac{1}{i}\psi_t = \psi_{ss} + \frac{1}{2}|\psi|^2\psi\,. \tag{4.1.24}$$

Here,

$$\psi(s,t) = \kappa(s,t)\exp\left(i\int_0^s T(s',t)\,ds'\right)\,, \tag{4.1.25}$$

with T the torsion of the space curve $\mathcal{L}$, is the new unknown dubbed the “Hasimoto's filament function” in Klein and Majda (1991a). Through the distributions of curvature and torsion, the filament function contains, for fixed time, $\bar{t}$, the entire geometrical information needed to reconstruct the filament centerline, except for translations and rigid rotations.

The cubic nonlinear Schrödinger equation is well-investigated in the context of inverse scattering theory (Newell, 1985). An interesting property is that it allows for soliton solutions, that is travelling wave solutions of the form

$$\psi(\tilde{s},\bar{t}) = \tilde{\psi}(\tilde{s} - c\bar{t})\,, \tag{4.1.26}$$

where $\tilde{\psi}(\xi)$ is rapidly decaying for large $|\xi|$. The soliton velocity c depends nonlinearly on the structure of the shape function $\tilde{\psi}$.

Hasimoto restricted his analysis to the local induction approximation, (4.1.23). However, any motion that is directed only in the binormal direction will preclude vortex stretching – a process generally considered of importance in three-dimensional fluid dynamics. In fact, it is easily verified that the local rate of stretching of the filament centerline in the general case is given by (Klein and Majda, 1991a),

$$S(\tilde{s},\bar{t}) = -\kappa\boldsymbol{n}(\tilde{s},\bar{t})\cdot\boldsymbol{X}_{\bar{t}}\,, \tag{4.1.27}$$

and this expression vanishes identically under the local induction approximation. Here, $\boldsymbol{n}(\tilde{s},\bar{t})$ is the principal normal to $\mathcal{L}(\bar{t})$ in $\boldsymbol{X}(\tilde{s},\bar{t})$.

The effects of an externally imposed flow field and of nonlocal self-induction may be accounted for by including a perturbation term in the filament equation of motion such that

$$\dot{\boldsymbol{X}}(\tilde{s},\bar{t}) = (\kappa \boldsymbol{b})(\tilde{s},\bar{t}) + \tilde{\delta}\boldsymbol{v}(\tilde{s},\bar{t}) \,. \tag{4.1.28}$$

Here, $\boldsymbol{v}$ combines the additional influences and $\tilde{\delta}$ is a perturbation parameter of order $O\left((\ln(1/\epsilon))^{-1}\right)$ as the dimensionless core size, ϵ, vanishes (see (4.1.22)).

Starting from such an equation of motion, Klein and Majda (1991a) used Hasimoto's transformation to obtain the following generalization of the cubic Schrödinger equation:

$$\frac{1}{i}\psi_{\bar{t}} = \psi_{\tilde{s}\tilde{s}} + \frac{1}{2}|\psi|^2\psi - \tilde{\delta}\left(i(\boldsymbol{N}\cdot\boldsymbol{v}_{\tilde{s}})_{\tilde{s}} + \psi\int_{\tilde{s}_0}^{\tilde{s}} \mathrm{Im}(\psi\bar{\boldsymbol{N}})\cdot\boldsymbol{v}_{\tilde{s}}\, d\tilde{s}\right). \tag{4.1.29}$$

Here, $\boldsymbol{N}(\tilde{s},\bar{t})$ is a complex combination of the principal and binormal unit vectors,

$$\boldsymbol{N} = (\boldsymbol{n} + i\boldsymbol{b})\exp\left(i\int_{\tilde{s}_0}^{\tilde{s}} T\, ds\right). \tag{4.1.30}$$

Equation (4.1.29) is quite general. It allows for arbitrary values of the parameter $\tilde{\delta}$; that is, no assumption is made yet regarding smallness in the sense of an asymptotic expansion parameter. Via the general velocity contribution $\boldsymbol{v}(\tilde{s},\bar{t})$, any modification of Hasimoto's original binormal propagation law is conceivable. In particular, one may include the effects of an outer imposed straining field (see Sect. 4.2) or of the interaction among two or more neighboring vortex filaments (see Sect. 4.3). While this kind of generality is valuable, (4.1.29) is still rather complex and unclosed. The vector function $\boldsymbol{N}(\tilde{s},\bar{t})$ as well as the contributions to $\boldsymbol{v}(\tilde{s},\bar{t})$ from nonlocal self-induction will have to be expressed in terms of the filament function itself to obtain a mathematically closed equation for $\psi(\tilde{s},\bar{t})$.

We will be able to derive such a closed equation for the simplified situation of perturbed, nearly straight filaments considered in this chapter (see (4.1.1)). Specifically, we will obtain perturbed nonlinear Schrödinger equations, in which the cubic curvature nonlinearity interacts direcly with nonlocal, yet linearized terms representing vortex self-induction. Following Klein and Majda (1991a, 1991b), we will show in particular how this interaction produces vortex self-stretching and a nonlinear energy transfer in wavenumber space.

Before we specialize to nearly straight filaments, we recount the Hasimoto transformation for the general filament propagation law in (4.1.28), leaving the perturbation velocity, $\boldsymbol{v}$, unspecified for the moment.

4.1.3.1 Derivation of the Generalized Filament Equation. In this section we follow Klein and Majda (1991a) in deriving the general filament equation from (4.1.29). The starting point will be the perturbed binormal propagation law (4.1.28). The main analytical tool is Hasimoto's transformation (4.1.25). All calculation follows closely Hasimoto's original exposition in Hasimoto (1972) for the case $\tilde{\delta}\boldsymbol{v} \equiv 0$.

First we recall the Serret–Frenet formulae,

$$\begin{aligned} \boldsymbol{X}_{\tilde{s}} &= \boldsymbol{t}\,, & \boldsymbol{t}_{\tilde{s}} &= \kappa \boldsymbol{n}\,, \\ \boldsymbol{n}_{\tilde{s}} &= T\boldsymbol{b} - \kappa \boldsymbol{t}\,, & \boldsymbol{b}_{\tilde{s}} &= -T\boldsymbol{n}\,, \end{aligned} \tag{4.1.31}$$

which describe the variation of the attached basis $\boldsymbol{t}, \boldsymbol{n}, \boldsymbol{b}$, along the curve $\mathcal{L}(\bar{t})$. In the subsequent calculations we will heavily use the complex vector $\boldsymbol{N}$ from (4.1.30) and its conjugate $\overline{\boldsymbol{N}}$ in place of the original basis $\boldsymbol{n}, \boldsymbol{b}$ in the normal planes of the filament centerline. These vectors satisfy the orthogonality relations

$$\boldsymbol{t}\cdot\boldsymbol{t} = 1\,, \qquad \boldsymbol{t}\cdot\boldsymbol{N} = 0\,, \qquad \boldsymbol{N}\cdot\boldsymbol{N} = 0\,, \qquad \boldsymbol{N}\cdot\overline{\boldsymbol{N}} = 2\,. \tag{4.1.32}$$

Using the Serret–Frenet formulae, one derives

$$\begin{aligned} \boldsymbol{N}_{\tilde{s}} &= -\psi \boldsymbol{t} \\ \boldsymbol{t}_{\tilde{s}} &= \frac{1}{2}(\overline{\psi}\boldsymbol{N} + \psi\overline{\boldsymbol{N}}) \end{aligned} \tag{4.1.33}$$

for the variations of $\boldsymbol{N}$ and $\boldsymbol{t}$ along $\mathcal{L}$. Differentiation of the perturbed binormal propagation law (4.1.28) w.r.t. $\tilde{s}$ yields the temporal variation of the tangent vector,

$$\boldsymbol{t}_{\bar{t}} = \kappa_{\tilde{s}}\boldsymbol{b} - \kappa T\boldsymbol{n} + \tilde{\delta}\boldsymbol{v}_{\tilde{s}}\,. \tag{4.1.34}$$

In terms of the new basis, $\boldsymbol{N}, \overline{\boldsymbol{N}}$, this reads

$$\boldsymbol{t}_{\bar{t}} = \frac{1}{2}i(\psi_{\tilde{s}}\overline{\boldsymbol{N}} - \overline{\psi}_{\tilde{s}}\boldsymbol{N}) + \tilde{\delta}\boldsymbol{v}_{\tilde{s}}\,. \tag{4.1.35}$$

Next, we obtain an evolution equation for the filament function, ψ, by comparing the $\boldsymbol{t}$ and $\boldsymbol{N}$ components of two independent representations of the cross-derivative $\boldsymbol{N}_{\tilde{s}\bar{t}}$ of $\boldsymbol{N}$. Using the decomposition

$$\boldsymbol{N}_{\bar{t}} = \alpha\boldsymbol{N} + \beta\overline{\boldsymbol{N}} + \gamma\boldsymbol{t}\,, \tag{4.1.36}$$

and the orthogonality relations from (4.1.32), we find

$$\begin{aligned} \alpha + \bar{\alpha} &= \frac{1}{2}(\boldsymbol{N}\cdot\overline{\boldsymbol{N}}_{\bar{t}} + \boldsymbol{N}_{\bar{t}}\cdot\overline{\boldsymbol{N}}) = \frac{1}{2}(\boldsymbol{N}\cdot\overline{\boldsymbol{N}})_{\bar{t}} \equiv 0 \\ \beta &= \frac{1}{4}(\boldsymbol{N}\cdot\boldsymbol{N})_{\bar{t}} \equiv 0 \\ \gamma &= \boldsymbol{t}\cdot\boldsymbol{N}_{\bar{t}} = -\boldsymbol{t}_{\bar{t}}\cdot\boldsymbol{N} = -i\psi_{\tilde{s}} - \tilde{\delta}\,\boldsymbol{N}\cdot\boldsymbol{v}_{\tilde{s}}\,. \end{aligned} \tag{4.1.37}$$

As a consequence,

$$\boldsymbol{N}_{\bar{t}} = i\big(R\boldsymbol{N} - (\psi_{\tilde{s}} - i\tilde{\delta}\,\boldsymbol{N}\cdot\boldsymbol{v}_{\tilde{s}})\,\boldsymbol{t}\big)\,, \tag{4.1.38}$$

where $R = R(\tilde{s}, \bar{t})$ is a yet unknown function. The two desired representations of $\boldsymbol{N}_{\tilde{s}\bar{t}}$ follow after differentiation of $(4.1.33)_1$ and (4.1.38) w.r.t. $\bar{t}$ and $\tilde{s}$:

$$\boldsymbol{N}_{\tilde{s}\bar{t}} = -\psi_{\bar{t}}\boldsymbol{t} - \frac{1}{2}\, i\left(\psi\psi_{\tilde{s}}\overline{\boldsymbol{N}} - \psi\overline{\psi}_{\tilde{s}}\boldsymbol{N}\right) - \tilde{\delta}\psi\boldsymbol{v}_{\tilde{s}}$$

$$\boldsymbol{N}_{\bar{t}\tilde{s}} = i\Big[R_{\tilde{s}}\boldsymbol{N} - R\psi\boldsymbol{t} - \left[\psi_{\tilde{s}\tilde{s}} - i\tilde{\delta}(\boldsymbol{N}\cdot\boldsymbol{v}_{\tilde{s}})_{\tilde{s}}\right]\boldsymbol{t} \tag{4.1.39}$$

$$-\frac{1}{2}\left(\psi_{\tilde{s}} - i\tilde{\delta}\,\boldsymbol{N}\cdot\boldsymbol{v}_{\tilde{s}}\right)\left(\overline{\psi}\boldsymbol{N} + \psi\overline{\boldsymbol{N}}\right)\Big]\,.$$

Equating the coefficients of $\boldsymbol{t}$ and $\boldsymbol{N}$ in these equations, we find

$$\psi_{\bar{t}} + \tilde{\delta}\,\psi\,\boldsymbol{v}_{\tilde{s}}\cdot\boldsymbol{t} = i\Big[R\psi + \psi_{\tilde{s}\tilde{s}} - i\,\tilde{\delta}\,(\boldsymbol{N}\cdot\boldsymbol{v}_{\tilde{s}})_{\tilde{s}}\Big] \tag{4.1.40}$$

and

$$R_{\tilde{s}} = \frac{1}{2}(\psi_{\tilde{s}}\overline{\psi} + \psi\overline{\psi}_{\tilde{s}}) + \frac{1}{2}\,i\,\tilde{\delta}\,(\psi\overline{\boldsymbol{N}} - \overline{\psi}\boldsymbol{N})\cdot\boldsymbol{v}_{\tilde{s}}\,. \tag{4.1.41}$$

Here, we reproduce Hasimoto's result for $\tilde{\delta} = 0$. We have

$$R = \frac{1}{2}|\psi|^2 \quad \text{for} \quad \tilde{\delta} = 0\,. \tag{4.1.42}$$

Insertion into (4.1.40) yields the cubic Schrödinger equation (4.1.24). Hasimoto showed that an arbitrary, time-dependent integration constant, $A(\bar{t})$, that may be introduced in the context of (4.1.41) will not affect the filament geometry, and may be eliminated without restriction of generality. One may verify this by applying a phase shift

$$\tilde{\Phi}(\tilde{s}, \bar{t}) = \int_{\tilde{s}_0}^{\tilde{s}} T(s,t)\,ds - \frac{1}{2}\int_0^{\bar{t}} A(t)\,dt \tag{4.1.43}$$

to the filament function. The entire geometrical information regarding curvature and torsion of $\mathcal{L}$, that is,

$$|\psi| = \kappa \qquad \text{and} \qquad T = \left[\arg(\psi)\right]_{\tilde{s}} = \Phi_{\tilde{s}} = \tilde{\Phi}_{\tilde{s}}\,, \tag{4.1.44}$$

will not be affected.

In the more general situation where $\tilde{\delta} \neq 0$, integration of (4.1.41) will lead to the generalized filament equation stated earlier in (4.1.29). We notice for later reference that $\boldsymbol{v}_{\tilde{s}}\cdot\boldsymbol{t} \equiv 0$ if $\tilde{s}$ is an arclength variable. See the Appendix in Klein and Majda (1993) for details.

4.1.3.2 Specialization for the Klein–Majda Regime. Here we demonstrate how the velocity perturbation in (4.1.28) enters the filament equation (4.1.29) via the Hasimoto transformation.

In a first step, we introduce the scaled space and time coordinates

$$\tilde{\sigma} = \frac{\tilde{s}}{\delta} \quad \text{and} \quad \tau = \frac{\bar{t}}{\delta^2} \tag{4.1.45}$$

into (4.1.29) and neglect all contributions that are small compared to δ^2 and $\tilde{\delta}$. The intermediate result reads

$$\frac{1}{i}\psi_\tau = \psi_{\tilde{\sigma}\tilde{\sigma}} + \left[\delta^2 \frac{1}{2}|\psi|^2\psi - i\,\tilde{\delta}\,(\boldsymbol{N}\cdot\boldsymbol{v}_{\tilde{\sigma}})_{\tilde{\sigma}}\right]. \tag{4.1.46}$$

This is not yet a closed equation for the filament dynamics, because the perturbation velocity, $\boldsymbol{v}$, still appears explicitly. Only after expressing this velocity in terms of $\boldsymbol{X}^{(2)}$, and relating the resulting terms to ψ, will a closed equation emerge.

An important side effect of the above relation is that it suggests the particular distinguished limit for the two perturbation parameters δ^2 and $\tilde{\delta}$ announced in (4.1.5). If we equate these parameters letting $\epsilon^2 = \tilde{\delta}$, that is,

$$\delta^2 = \left(\ln\frac{2\delta}{\epsilon} + C\right)^{-1} \tag{4.1.47}$$

according to (4.1.22), then we obtain a direct competition of nonlinear and nonlocal terms at the same perturbation order, represented by $\frac{1}{2}|\psi|^2\psi$ and $(\boldsymbol{N}\cdot\boldsymbol{v}_{\tilde{\sigma}})_{\tilde{\sigma}}$, respectively. For a single isolated vortex, we conclude that in this regime the local induction approximation is inappropriate, as it favors the nonlinear effects of the local binormal motion over the effects of nonlocal self-induction.

Notice that the relation in (4.1.47) is a particular distinguished limit connecting the core size parameter, ϵ, and the geometric perturbation parameter δ from (4.1.1). It is this distinguished limit that defines the Klein–Majda regime as announced earlier in this chapter.

In a second step, we verify the relation

$$\mathcal{N} \equiv -i\left(\boldsymbol{N}\cdot\boldsymbol{v}_{\tilde{\sigma}}\right)_{\tilde{\sigma}} = -\mathcal{I}[\psi] \tag{4.1.48}$$

for the leftover $\boldsymbol{v}$-dependent expression in (4.1.46). To this end we first differentiate (4.1.1) w.r.t. σ and notice that the tangent vector, $\boldsymbol{t}$, satisfies $\boldsymbol{t} = \boldsymbol{t}_0(1 + O(\delta))$. This allows us to deduce the following leading-order results:

$$-i\boldsymbol{N}\cdot(\boldsymbol{I}\times\boldsymbol{t}_0)_\sigma = i(\boldsymbol{N}\times\boldsymbol{t})\cdot\boldsymbol{I}_\sigma = -\boldsymbol{N}\cdot\boldsymbol{I}_\sigma\,. \tag{4.1.49}$$

To verify the second equality, one inserts the definition (4.1.30) for $\boldsymbol{N}$ and uses the fact that $(\boldsymbol{t},\boldsymbol{n},\boldsymbol{b})$ form a right-handed, orthonormal basis.

Next, we show that $\boldsymbol{N}$ is constant up to order $O(\delta)$ on the length and time scales considered here, which are represented by the (σ,τ)-variables. Noticing further that σ equals the arclength $\tilde{\sigma}$ except for errors of order

$O(\delta^2)$, one concludes from (4.1.33) and (4.1.38), that is, from $\boldsymbol{N}_{\tilde{s}} = -\psi \boldsymbol{t}$ and $\boldsymbol{N}_{\tilde{t}} = i(R\boldsymbol{N} - (\psi_{\tilde{s}} - i\tilde{\delta}\,\boldsymbol{N}\cdot\boldsymbol{v}_{\tilde{s}})\boldsymbol{t})$, that

$$\begin{aligned} \boldsymbol{N}_\sigma &= -\delta\,\psi\,\boldsymbol{t}\,(1+O(\delta)) \\ \boldsymbol{N}_\tau &= -\delta\,i\psi_\sigma\,\boldsymbol{t}\,(1+O(\delta))\,. \end{aligned} \tag{4.1.50}$$

Thus, one has

$$\boldsymbol{N} = \boldsymbol{N}_0 + O(\delta) \quad \text{for} \quad (\sigma, \tau = O(1))\,. \tag{4.1.51}$$

The results in (4.1.49) and (4.1.51) help transform the expression in (4.1.48) into

$$\mathcal{N} = -(\boldsymbol{N}\cdot\mathcal{I}[\boldsymbol{X}^{(2)}]_{\sigma\upsilon}) + O(\delta)\,. \tag{4.1.52}$$

Using again (4.1.51) and the definition of the $\mathcal{I}$-Operator, $\mathcal{I}[\cdot]$, in (4.1.8), one finds

$$\mathcal{N}(\sigma) = -\int_{-\infty}^{\infty} \frac{1}{|h|^3}\mathcal{K}(\sigma,h)\,dh + O(\delta)\,, \tag{4.1.53}$$

where

$$\begin{aligned} \mathcal{K}(\sigma,h) = {} & \boldsymbol{N}_0\cdot\boldsymbol{X}^{(2)}_{\sigma\sigma}(\sigma+h) - \boldsymbol{N}_0\cdot\boldsymbol{X}^{(2)}_{\sigma\sigma}(\sigma) \\ & -h\,\boldsymbol{N}_0\cdot\boldsymbol{X}^{(2)}_{\sigma\sigma\sigma}(\sigma+h) + \mathcal{H}(1-|h|)\,\frac{h^2}{2}\,\boldsymbol{N}_0\cdot\boldsymbol{X}^{(2)}_{\sigma\sigma\sigma\sigma}(\sigma)\,. \end{aligned} \tag{4.1.54}$$

Finally, we show that

$$\boldsymbol{N}_0\cdot\partial^j\boldsymbol{X}^{(2)}/\partial\sigma^j = \partial^{j-2}\psi/\partial\tilde{\sigma}^{j-2} + O(\delta) \quad \text{for} \quad j = 2,3,4 \tag{4.1.55}$$

which, due to (4.1.53), implies that

$$\mathcal{N} = -\mathcal{I}[\psi]\,. \tag{4.1.56}$$

For the sequel we notice that (4.1.55) involves the *arclength* coordinate, $\tilde{\sigma}$.

To verify (4.1.55), we use the Serret–Frenet formulae (4.1.31) and recall that the torsion, T, may be expressed in terms of derivatives of the filament function via $T = (\arg\psi)_{\tilde{s}}$. We conclude that

$$\boldsymbol{N}\cdot\boldsymbol{X}_{\tilde{s}\tilde{s}} = (\boldsymbol{n}+i\boldsymbol{b})e^{i\Phi}\cdot\kappa\boldsymbol{n} = \kappa e^{i\Phi}\,, \tag{4.1.57}$$

$$\boldsymbol{N}\cdot\boldsymbol{X}_{\tilde{s}\tilde{s}\tilde{s}} = (\boldsymbol{n}+i\boldsymbol{b})e^{i\Phi}\cdot\big(\kappa_{\tilde{s}}\boldsymbol{n} + \kappa(T\boldsymbol{b}-\kappa\boldsymbol{t})\big) = (\kappa_{\tilde{s}} + i\kappa\Phi_{\tilde{s}})e^{i\Phi}\,, \tag{4.1.58}$$

and

$$\begin{aligned} \boldsymbol{N}\cdot\boldsymbol{X}_{\tilde{s}\tilde{s}\tilde{s}\tilde{s}} &= (\boldsymbol{n}+i\boldsymbol{b})\,e^{i\Phi}\cdot \\ &\qquad \Big(\kappa_{\tilde{s}\tilde{s}}\boldsymbol{n} + 2\kappa_{\tilde{s}}\Big[T\boldsymbol{b}-\kappa\boldsymbol{t}\Big] + \kappa\Big[T_{\tilde{s}}\boldsymbol{b} - \kappa_{\tilde{s}}\boldsymbol{t}\Big] - \kappa\Big[T^2+\kappa^2\Big]\,\boldsymbol{n}\Big) \\ &= \Big(\kappa_{\tilde{s}\tilde{s}} + 2i\kappa_{\tilde{s}}T + i\kappa T_{\tilde{s}} - \kappa(\kappa^2+T^2)\Big)e^{i\Phi}\,. \end{aligned} \tag{4.1.59}$$

This yields

$$\begin{aligned} \boldsymbol{N} \cdot \boldsymbol{X}_{\tilde{s}\tilde{s}} &= \psi \\ \boldsymbol{N} \cdot \boldsymbol{X}_{\tilde{s}\tilde{s}\tilde{s}} &= \psi_{\tilde{s}} \\ \boldsymbol{N} \cdot \boldsymbol{X}_{\tilde{s}\tilde{s}\tilde{s}\tilde{s}} &= \psi_{\tilde{s}\tilde{s}} - |\psi|^2 \psi \,, \end{aligned} \tag{4.1.60}$$

and with the asymptotic scalings of the filament geometry in (4.1.1) we find

$$\boldsymbol{N} = \boldsymbol{N}_0 + O(\delta) \,, \tag{4.1.61}$$

$$\boldsymbol{X}_{\tilde{s}\tilde{s}} = \boldsymbol{X}^{(2)}_{\sigma\sigma} + O(\delta) \,, \tag{4.1.62}$$

$$\boldsymbol{X}_{\tilde{s}\tilde{s}\tilde{s}} = \frac{1}{\delta} \boldsymbol{X}^{(2)}_{\sigma\sigma\sigma} + O(1) \,, \tag{4.1.63}$$

$$\boldsymbol{X}_{\tilde{s}\tilde{s}\tilde{s}\tilde{s}} = \frac{1}{\delta^2} \boldsymbol{X}^{(2)}_{\sigma\sigma\sigma\sigma} + O\big(\frac{1}{\delta}\big) \,, \tag{4.1.64}$$

$$\partial \tilde{s} / \partial \sigma = \delta \left(1 + O(\delta^2)\right) \,. \tag{4.1.65}$$

With these results, (4.1.55) then follows from (4.1.60). Furthermore, using the results in (4.1.53) yields (4.1.48), and from (4.1.46) follows the filament equation with self-stretch in (4.3).

4.1.4 Interaction of Curvature and Self-Induction Effects

The above analysis showed that the local induction approximation from (4.1.23) approximates the motion of slender fluid vortices only up to logarithmic corrections in terms of the slenderness parameter ϵ. The missing term expresses the effects of nonlocal self-induction. We have seen that in the present perturbation regime it takes the form of the linear pseudo-differential operator, $\mathcal{I}[\cdot]$, defined in (4.1.8) so that

$$\boldsymbol{X}_{\bar{t}} = \kappa \boldsymbol{b} + \delta^2 \, \mathcal{I}[\boldsymbol{X}^{(2)}] \times \boldsymbol{t}_0 \,. \tag{4.1.66}$$

Using the Hasimoto transformation and exploiting the asymptotic structure of the ansatz for $\mathcal{L}(t)$ in (4.1.1), Klein and Majda (1991a) derived the filament equation with self-stretching (4.3). The key point of this equation is that it describes an interaction of nonlocal self-induction and curvature nonlinearity at the same perturbation order.

4.1.4.1 Summary of Main Results. Detailed analyses of the mathematical structure of this equation, of some general solution properties, and of a wide range of numerical solutions can be found in Klein and Majda (1991a), Klein et al. (1992). Here we emphasize three particularly interesting aspects:

Vortex self-stretching. A detailed analysis, summarized below, shows that $\mathcal{I}[\psi]$ is the only term in the filament equation that induces vortex self-stretching. In fact, the total rate of elongation of the filament centerline can be expressed analytically as a quadratic functional of the filament function, that is,

$$\frac{1}{\ell}\dot{\ell} = \delta^4 \int_{-\infty}^{\infty} \frac{1}{|h|} \Big[\overline{\psi}(\sigma + h, \tau) \psi(\sigma, \tau) - \overline{\psi}(\sigma, \tau) \psi(\sigma + h, \tau) \Big] \, dh \,. \tag{4.1.67}$$

Although the order of magnitude of self-stretching is small, this expression serves as a valuable diagnostic tool when analyzing numerical solutions. The stretching formula allows one, for example, to check easily whether "dramatic events" in the evolution of the filament geometry are associated with intense vortex stretching or not.

The Linear Spectrum of the Filament Equation. The cubic nonlinear Schrödinger equation from (4.1.24) has been known to feature soliton solutions since Hasimoto's seminal paper (Hasimoto, 1972). The stabilization of these nonlinear travelling wave solutions depends crucially on the interplay between the linear Schrödinger operator

$$\frac{1}{i}\partial_t - \partial_{ss}$$

and the cubic nonlinearity,

$$\frac{1}{2}|\psi|^2\psi\,.$$

Important questions in the present context thus concern properties of the linear part of the filament equation, that is, of the pseudo-differential operator

$$\frac{1}{i}\partial_t - \partial_{ss} + \delta^2\,\mathcal{I}[\,\cdot\,]$$

and of its interplay with the cubic nonlinearity in large amplitude solutions. An efficient characterization of the properties of a pseudo-differential operator is its "symbol" (Hörmander, 1984): Consider the action of a spatial pseudo-differential operator $\mathcal{L}[\cdot]$ on the harmonic exponential function $\exp(i\,\xi s)$. The general result reads

$$\mathcal{L}[\exp(i\,\xi s)] = \hat{L}(\xi)\exp(i\,\xi s)\,,$$

where $L(\xi)$ is the so-called operator symbol. For $\mathcal{I}[\cdot]$ the result reads

$$\hat{I}(\xi) = -\xi^2(\ln|\xi| - c_0)\,. \tag{4.1.68}$$

Here $c_0 = \frac{1}{2} - \mathcal{E} \approx -0.07721\ldots$, with $\mathcal{E}$ Euler's constant. For a detailed derivation see Klein and Majda (1991a) and Klein (1994).

As a consequence, the linear spectrum of the filament equation reads

$$\omega(\xi) = -\xi^2((1-\delta^2 c_0) - \delta^2\ln|\xi|)\,. \tag{4.1.69}$$

Clearly, as long as $\delta \ll 1$ and $\ln(\xi) = O(1)$, this dispersion relation resembles very much that of the Schrödinger equation. Yet, for sufficiently high frequencies the logarithm $\delta^2\ln|\xi|$ may dominate over $(1-\delta^2 c_0)$, and then the sign of the right-hand side of (4.1.69) changes. This corresponds to the ill-posed nonlinear Schrödinger equation with a reversed sign of the second derivative.

Below we discuss, in the context of numerical solutions, how this property of the linear dispersion relation of the filament equation can induce an energy transfer in wavenumber space.

Stability of travelling helix solutions. Suppose the complex function ψ satisfies a cubic nonlinear Schrödinger-type equation

$$\frac{1}{i}\psi_t = \mathcal{L}[\psi] + \frac{1}{2}|\psi|^2\psi \tag{4.1.70}$$

where $\mathcal{L}[\cdot]$ has a real-valued symbol $\hat{L}(\cdot)$ (of course, in the case of the filament equation, $\mathcal{L}[\cdot] = \partial_{ss} - \delta^2\mathcal{I}[\cdot]$). Any functions

$$\psi_H(\sigma, \tau; A, \xi) = A\ \exp\big(i\,(\xi\,\sigma - \Omega(\xi, A; \delta)\,\tau)\big) \tag{4.1.71}$$

are exact travelling wave solutions of the nonlinear generalized filament equation from (4.1.70), provided that Ω satisfies the *nonlinear* dispersion relation

$$\Omega(\xi, A) = \hat{L}(\xi) + \frac{\delta^2}{2}A^2\,. \tag{4.1.72}$$

That these solutions describe travelling constant torsion helices is easily verified, considering the definition of the filament function from (4.1.25).

These nonlinear exact solutions potentially constitute valuable tests for computational codes aiming at numerically solving the filament equation. To this end, however, the (nonlinear) stability properties of these solutions should be analyzed and taken into account when evaluating numerical results. An efficient ansatz for analyzing such stability properties considers modulations of the background traveling wave solution. It was proposed by Benjamin and Feir (1967) and reads

$$\psi(\sigma, \tau) = \psi_H(\sigma, \tau; A, \xi)\,\left(1 + b^+\,e^{i\,(\beta\sigma + \alpha\tau)} + b^-\,e^{-i\,(\beta\sigma + \overline{\alpha}\tau)}\right) \tag{4.1.73}$$

with $\beta, b \in I\!R$, $b \ll 1$ and $\alpha = \alpha(\beta, \xi, A; \delta) \in \boldsymbol{C}$ remaining to be determined. As usual, the overbar $\overline{(\cdot)}$ denotes complex conjugation. The coefficients α, β will be called "relative frequency" and "relative wavenumber," respectively.

A tedious, but straightforward, analysis yields

$$\alpha_{1,2} = \frac{1}{2}\left[\left(\Delta^+ + \Delta^-\right) \pm \sqrt{(\Delta^+ - \Delta^-)\left((\Delta^+ - \Delta^-) + 4a\right)}\right], \tag{4.1.74}$$

where

$$\begin{aligned} \Delta^+(\beta, \xi) &= \hat{L}(\xi + \beta) - \hat{L}(\xi) \\ \Delta^-(\beta, \xi) &= \hat{L}(\xi) - \hat{L}(\xi - \beta) \\ a(A; \delta) &= \frac{\delta^2}{2}A^2\,. \end{aligned} \tag{4.1.75}$$

Associated with the eigenvalues $\alpha_{1,2}$ are eigenvectors $b^{\pm}_{1,2}$, which are related by

$$b^+ = \frac{1}{a}\left(\Omega - \alpha - \hat{L}(\xi - \beta) - a\right)\overline{b^-}\,. \tag{4.1.76}$$

We compare the stability properties of travelling helices under the nonlinear Schrödinger equation (4.1.24) ($\mathcal{L}[\cdot] = \partial_{ss}$) with those obtained for the filament equation from (4.3) ($\mathcal{L}[\cdot] = \partial_{ss} - \delta^2 \mathcal{I}[\cdot]$). A convenient variable for this comparison is

$$G = \operatorname{sgn}(R)\sqrt{|R|}\,, \tag{4.1.77}$$

where R is the radicand in the expression (4.1.74) for the complex modulation frequency α. For $R > 0$ the helix solutions are neutrally stable under the given perturbation, and G is their relative oscillation frequency. For $R < 0$ the perturbations are unstable and $-G$ is the growth rate of the unstable mode. Figure 4.2a shows that long wavelength perturbations are unstable under both equations. However, in contrast to the nonlinear Schrödinger equation, the filament equation features a second short wavelength band of instability as seen in Figs. 4.2b and c.

Let us analyze the dispersion relations in more detail, particularly in the vicinity of these unstable wavenumber bands. First we realize that a straight vortex is neutrally stable, because for R to be negative, we must have $a \neq 0$. The largest occuring growth rates are

$$-\alpha_{i,\max} = -\alpha_{i,\max}^{\text{NLS}} = \frac{\delta^2}{2} A^2\,, \tag{4.1.78}$$

where we have used the complex variable representation $\alpha = \alpha_r + i\alpha_i$.

Instability is associated with negative radicands R. For the cubic nonlinear Schrödinger equation we find

$$R^{\text{NLS}} < 0 \quad \text{for} \quad -\delta A < \beta < \delta A\,. \tag{4.1.79}$$

Thus for this equation there is only one band of instability at long perturbation wavelengths.

The situation is quite different for the filament equation. Asymptotic representations of the dispersion relation for long-wave ($\beta \ll 1$) and short-wave ($\beta \gg 1$) perturbations yield approximate descriptions of *two* instability bands as seen in Fig. 4.2b. Using primes to denote the derivative $d/d\xi$, we find that

long-wave instabilities arise for

$$\frac{-\delta A}{1 + \delta^2 \hat{I}''(\xi)} < \beta < \frac{\delta A}{1 + \delta^2 \hat{I}''(\xi)}\,, \tag{4.1.80}$$

whereas the short-wave instability band is characterized by

$$\exp\left(\frac{1}{\delta^2}\right) - A^2 \exp\left(\frac{-1}{\delta^2}\right) < \beta < \exp\left(\frac{1}{\delta^2}\right)\,. \tag{4.1.81}$$

Both the filament equation and the nonlinear Schrödinger equation thus yield long wavelength instabilities with perturbation wavenumbers of order $O(\delta)$. This corresponds to wavelengths in terms of the stretched coordinate σ

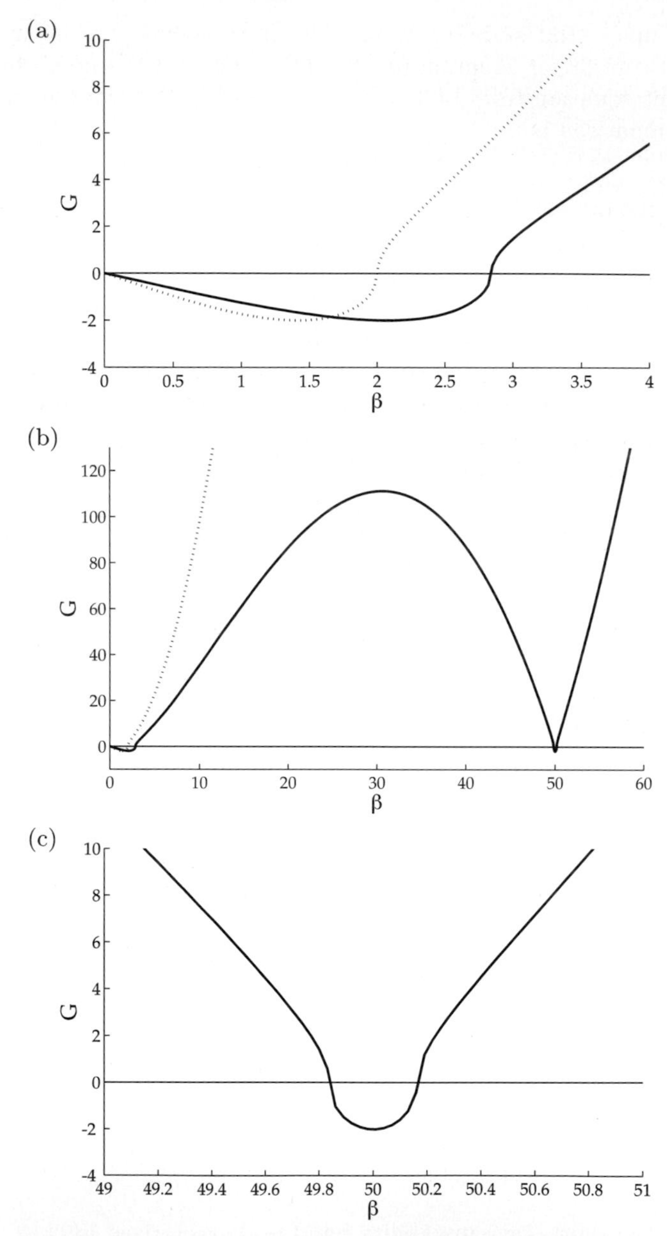

Fig. 4.2. Stability diagrams for the helix configuration (4.1.71) with $\xi = 2.0, A = 4.0, \delta^2 = 0.25$, for the cubic nonlinear Schrödinger equation (4.1.24) $(\cdots\cdots)$, and for the filament equation with self-stretching (4.3) (———). Instability is associated with values $G < 0$, and growth rates equal $|G|$, (a) long-wave instabilities, (b) spectrum up to the vortex core size scale, (c) short-wave instability for the filament equation. Adapted from Klein and Majda (1991b)

of order $O(1/\delta)$, which is the radius of curvature scale of the vortex filament centerline. The associated growth rates are of order $O(\delta^2)$, corresponding to $O(1/\delta^2)$ times in terms of the stretched time variable τ. This is the characteristic time scale of the overall evolution of a three-dimensional curved vortex filament according to Hasimoto (1972) or Callegari and Ting (1978). These length and time scales are formally outside the range of validity of the present perturbation analysis. Nevertheless, these results hint at the possibility of strong interactions between such perturbations and the overall motion of a general three-dimensional vortex filament.

For helices under the filament equation, (4.1.81) reveals another band of instability at systematically short wavelengths. In fact, the associated wavelengths are of the order $O(\exp(1/\delta^2))$. Considering the definition of δ in (4.1.5), we conclude that the associated perturbation wavelengths are of order $O(\epsilon)$, that is, comparable to the vortex core size. Obviously, these axial scales are again outside the range of validity of the asymptotics. Thus, any (computational) solution of the filament equation loses its physical significance once perturbations within this wavenumber band are excited. In computational approximate solutions of the filament equation, to be discussed shortly (Klein and Majda, 1991b), the temporal evolution of the solution spectrum in wavenumber space has been used to assess the temporal range of validity of the simulations.

The existence of the short wavelength instability band is of interest in another context. It arises because of the presence of the nonlocal operator $\mathcal{I}[\cdot]$ and its singular contribution to the linear dispersion relation for the filament equation. Similar short wave unstable modes have been observed by Crow (1970b) for antiparallel perturbed vortex pairs and by Widnall and Sullivan (1973) for circular vortex rings. Their results had also been derived on the basis of an asymptotic slenderness assumption for the vortex core size, which is why they dismissed these unstable modes correctly as spurious. However, Widnall et al. (1974) developed an extended theory for short wave instabilities, taking into account the detailed vortex core structure and dynamics. They also discovered comparable short wave modes, but with modified dispersion relations and, of course, with much richer spatial structure. Thus, the high wavenumber band of instabilities, revealed above for the travelling wave helical vortices, is indicative of a physical instability, but the present slender vortex theory is insufficient to provide an accurate description of the nature of these modes.

Despite this criticism, the nonlinear Klein–Majda theory presented in this section provides a new aspect in that it elucidates a mechanism for the self-excitation of these short wave modes. In the linearized setting, the unstable modes correspond to isolated, decoupled pairs of Fourier modes. In contrast, the cubic nonlinearity of the filament equation (4.3) couples the Fourier solution components, and thus induces dispersion in Fourier space.

Ultimately this leads to excitation of the unstable modes, and thus to a nontrivial energy transfer in wavenumber space.

4.1.4.2 Numerical Solution of the Filament Equation. Klein and Majda (1991b) used an efficient split step numerical technique for solving the cubic nonlinear Schrödinger and the filament equation. The key observation in the construction of a numerical method is that the linear and nonlinear parts of these equations, that is,

$$\frac{1}{i}\psi^{\mathrm{l}}_{\tau} = \mathcal{L}[\psi^{\mathrm{l}}] \qquad \text{and} \qquad \frac{1}{i}\psi^{\mathrm{nl}}_{\tau} = \frac{\delta^2}{2}|\psi^{\mathrm{nl}}|^2\psi^{\mathrm{nl}} \tag{4.1.82}$$

are easily solved explicitly. In fact, given any distribution of a complex periodic function $\psi^{\mathrm{l}}(\sigma,\tau)$ at time τ, the exact solution to the linear part of the filament equation at a later time $\tau+\Delta\tau$ is obtained as the inverse Fourier transform of the exact solution in Fourier space:

$$\psi^{\mathrm{l}}(\sigma,\tau+\Delta\tau) = \frac{1}{2\pi}\int_{-\infty}^{\infty}\hat{\psi}^{\mathrm{l}}(\xi,\tau+\Delta\tau)\exp(-i\,\xi\sigma)\,d\xi\,, \tag{4.1.83}$$

where

$$\hat{\psi}^{\mathrm{l}}(\xi,\tau+\Delta\tau) = \hat{\psi}^{\mathrm{l}}(\xi,\tau)\exp(i\,\hat{L}(\xi)\Delta\tau)\,. \tag{4.1.84}$$

This exact solution is easily approximated numerically by using a finite number of Fourier modes and utilizing fast Fourier transform (FFT) techniques.

The exact solution of the nonlinear part of the equations (second equation in (4.1.82)) is equally straightforward because, given $\psi^{\mathrm{nl}}(\sigma,\tau)$, we have

$$\psi^{\mathrm{nl}}(\sigma,\tau+\Delta\tau) = \psi^{\mathrm{nl}}(\sigma,\tau)\exp(i\,a(\sigma)\Delta\tau)\,, \tag{4.1.85}$$

where

$$a(\sigma) = \frac{\delta^2}{2}|\psi^{\mathrm{nl}}(\sigma,\tau)|^2 = \frac{\delta^2}{2}|\psi^{\mathrm{nl}}(\sigma,\tau+\Delta\tau)|^2\,. \tag{4.1.86}$$

Notice the time invariance, under the nonlinear part of the equation, of the local value of the modulus $|\psi^{\mathrm{nl}}|$!

Let $\Phi^{l}_{\Delta\tau}[\cdot], \Phi^{nl}_{\Delta\tau}[\cdot]$ denote the exact time evolution operators from (4.1.83), (4.1.84), and (4.1.85), respectively, so that

$$\begin{aligned}\psi^{l}(\sigma,\tau+\Delta\tau) &= \Phi^{l}_{\Delta\tau}[\psi^{l}(\sigma,\tau)]\\ \psi^{nl}(\sigma,\tau+\Delta\tau) &= \Phi^{nl}_{\Delta\tau}[\psi^{nl}(\sigma,\tau)]\,.\end{aligned} \tag{4.1.87}$$

Then, according to Strang (1968), one finds a second-order accurate numerical method for the full nonlinear Schrödinger equation and for the filament equation by a judicious interleaving of the operators, that is,

$$\Phi_{\Delta\tau} = \Phi^{l}_{\Delta\tau/2}\Phi^{nl}_{\Delta\tau}\Phi^{l}_{\Delta\tau/2} + O\left((\Delta\tau)^3\right) \quad \text{as} \quad \Delta\tau\to 0\,. \tag{4.1.88}$$

A very efficient implementation of this scheme is obtained by systematically using the explicit partial exact solutions as explained above, and employing standard FFT algorithms to alternate between physical and Fourier space. This scheme is also stable in the L_2-norm because (i) in each of the two exact solution operators, the local amplitudes in Fourier and physical space, respectively, remain constant and (ii) the FFT preserves the L_2-norm.

4.1.5 Structure of Solutions

Here we first demonstrate the performance of the numerical technique sketched above for the travelling wave helix solutions from (4.1.71). Following the arguments in Klein and Majda (1991a), we show that the numerical scheme correctly reproduces the predicted linear instabilities. Nonlinear effects will lead to saturation of the exponential growth of perturbations, and subsequently nonlinear effects take over. By comparing the evolution from comparable unstable initial data for the filament and the nonlinear Schrödinger equations, we find next that the nonlocal self-induction term in the latter is responsible for triggering what may be called a "nonlinear energy transfer in wavenumber space."

Helix Instabilities. Application of the scheme from Subsubsect. 4.1.4.2 is illustrated here on the basis of solutions of the filament and/or the nonlinear Schrödinger equation with perturbed helix initial data

$$\psi(\sigma, 0) = Ae^{i\xi\sigma}(1.0 + b_+e^{i\beta\sigma} + b_-e^{i\beta\sigma}), \tag{4.1.89}$$

with

$$A = 4.0, \quad \beta = 1.0, \quad \text{and} \quad b_+ = 0.001 \tag{4.1.90}$$

The coefficient b_- is determined from (4.1.76) such that the combination of the two perturbation terms in (4.1.89) corresponds to an unstable eigenmode of the filament or the cubic nonlinear Schrödinger equations. The perturbation parameter for the subsequent calculations is $\delta^2 = 0.25$.

To test our numerical scheme, we follow the solution during a short-time interval during which we expect exponential growth of the perturbation with the growth rate predicted by linear stability theory. Figure 4.3 shows the amplitude of the wavenumber $\xi + \beta = 3$ Fourier mode as a function of time in comparison with the analytical result.

The split-step numerical method described above properly reproduces the predicted linear instability. As announced above, nonlinear effects affect the solution once the perturbation amplitude has grown sufficiently.

How does nonlocal self-induction affect the fully developed nonlinear dynamics of a perturbed vortex filament? To answer this question, we compare next numerical long-time solutions of the cubic Schrödinger and the filament equations on the basis of simplified initial data

$$\psi(\sigma, 0) = 4.0e^{i2\sigma} + 0.004e^{i3\sigma}. \tag{4.1.91}$$

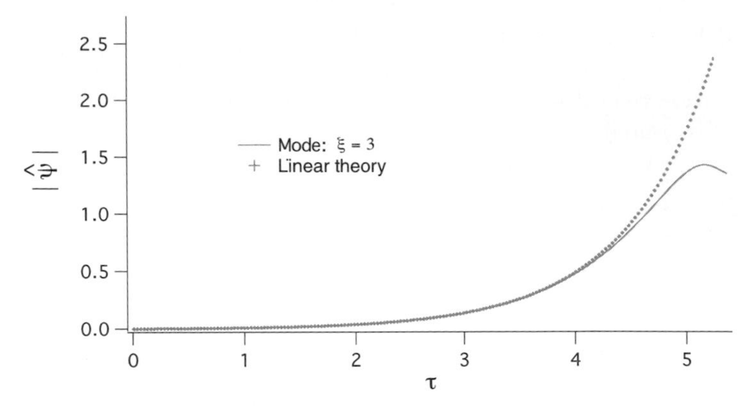

Fig. 4.3. Temporal evolution of the wavenumber 3 Fourier mode for the perturbed helix inititial data from (4.1.89) and (4.1.90) under linear theory (++++) and under the filament equation (———). Numerical solution with 128 grid points and Fourier modes. Adapted from Klein (1994)

The initial perturbation again contains a nonzero component of an unstable eigenmode for the helix with amplitude (curvature) 4.0 and wavenumber (torsion) 2.0. Thus we expect, after an initial nearly exponential growth, nonlinear saturation and fully nonlinear long-time dynamics. It is well known that the cubic nonlinear Schrödinger equation is completely integrable in the sense of inverse scattering theory (Newell, 1985) so that the regular nearly time periodic behavior of the curvature maximum within the domain (Fig. 4.4a) does not come as a surprise.

In contrast, under the filament equation, quite irregular bursts of the maximum curvature are observed immediately after the first deviation from the exponential perturbation growth has occurred (Fig. 4.4b). Examination of the spatial solution structures at the times of occurance of the local extrema of the maximum curvature in Fig. 4.5b reveals that under the filament equation there is a systematic generation and amplification of small scale fluctuations. This is in marked contrast with the cubic Schrödinger dynamics (Fig. 4.5a).

We emphasize that the small scale fluctuations seen in Fig. 4.5b are not due to underresolved numerical representations. The present results have been obtained using 256 grid points and Fourier modes. Analogous computations based on the same initial data but using up to 1024 points and modes showed that the presented computations are, in fact, converged.

An independent indication that these computations had sufficient spatial resolution is seen in Fig. 4.6, which displays a snapshot at time $\tau = 5.5$ of the Fourier spectrum for the solution of the filament equation from Fig. 4.4. The highest wavenumbers with noticeable Fourier amplitudes are smaller than $\lambda = 50$, whereas the 256 mode computation has an effective highest represented wavenumber of $\lambda_{\max} = 128$. Comparable spectra for the nonlinear Schrödinger solution are even narrower (not shown).

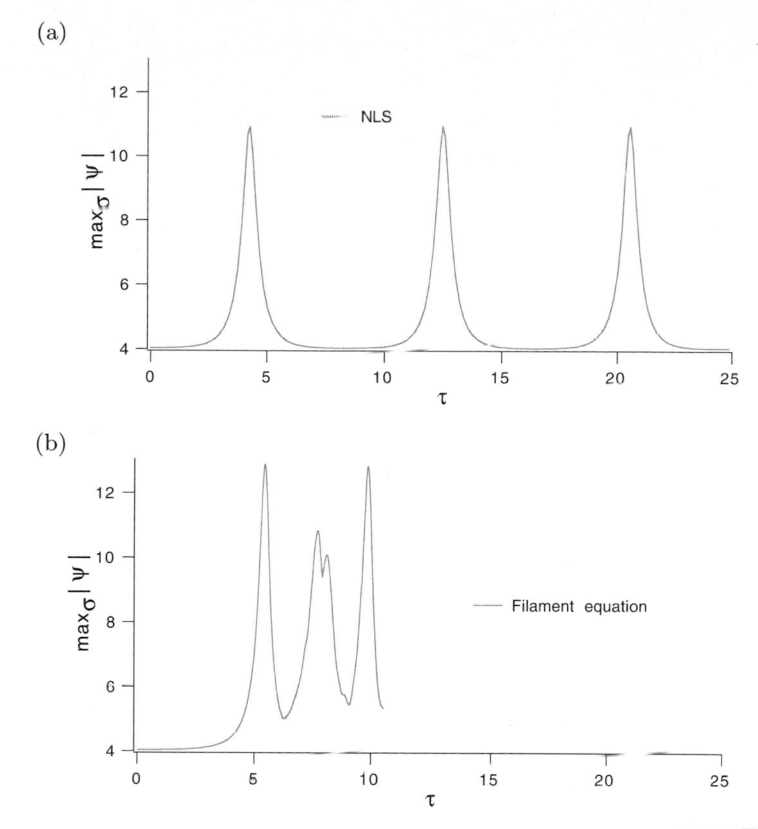

Fig. 4.4. Comparison of the maximum curvature within a space period versus time for (a) the cubic Schrödinger equation (4.1.24), and (b) the filament equation (4.3). Numerical solutions with 256 grid points and Fourier modes. Adapted from Klein (1994)

The spectrum generated by the filament dynamics also exhibits a pronounced local maximum near $\lambda \approx 50$. A straightforward calculation shows that the associated axial wavelength corresponds with the vortex core size scale for the present value of $\delta^2 = 0.25$. We conclude that the nonlinear generation of small scales, because of the nonlinear–nonlocal interaction described by the filament equation, rapidly leads to excitation of perturbation modes on the filament core size scale. Of course, the asymptotic filament equation cannot be expected to yield a faithful representation of the vortex filament dynamics beyond the time of generation of these small scales.

Another indication that the asymptotic theory reaches its limits is obtained by comparing the minimum radius of curvature, that is, the inverse of the maximum modulus of the filament function, with the filament core size. According to (4.1.5), a value of $\delta^2 = 0.25$ corresponds with a core size parameter $\epsilon \approx 0.05$. The minimum nondimensional radius of curvature

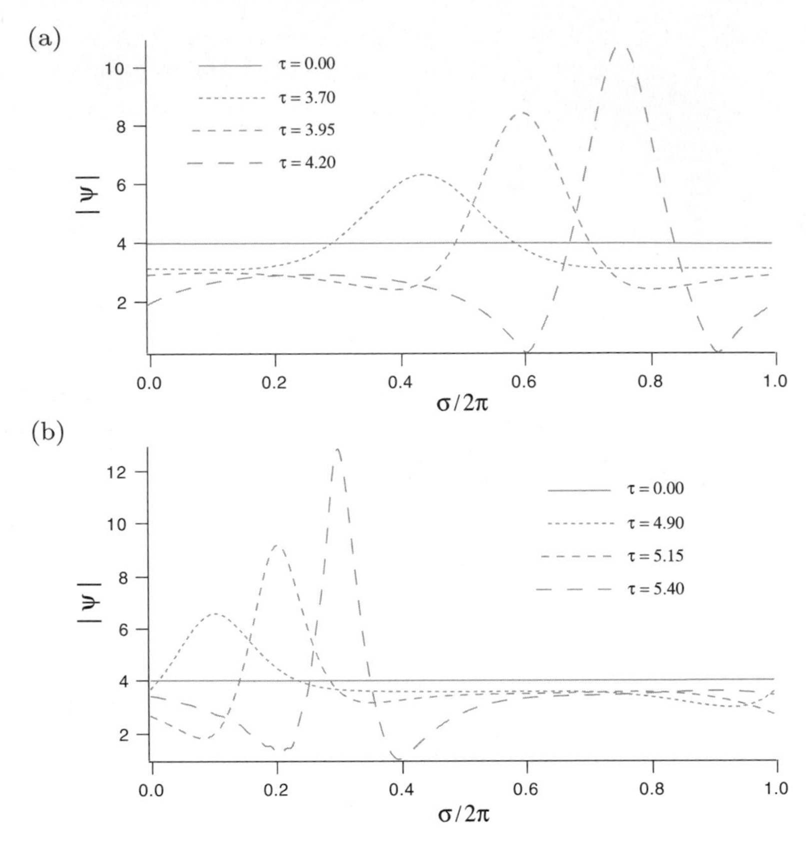

Fig. 4.5. Comparison of spatial structures of curvature shortly before and at the times of occurance of the first local curvature extrema in Fig. 4.4. Adapted from Klein and Majda (1991b) and Klein (1994)

of the filament centerline at the time of peak values for $|\psi|$ in Fig. 4.4 is $R_{\min} = 1/\max_{\tau,\sigma}(|\psi(\sigma,\tau)|) \approx 0.075 \sim \frac{3}{2}\epsilon$. This indicates that the slenderness assumption, which is the basis of the present asymptotic theories, gets violated locally. Carrying a consistent asymptotic description beyond that point in time would require matched asymptotic expansions, with a local resolution of the strongly distorted vortex core near the point of minimum radius of curvature, and a far-field expansion that coincides with the filament asymptotics described above.

Figure 4.7 displays the time evolution of the Fourier mode $\hat{\psi}(40)$ for the same calculation. We find that there is a sudden generation of small scales around time $\tau = 5.4$, the same time at which the snapshot of the Fourier spectrum in Fig. 4.6 was taken. The Fourier spectrum still exhibits rapid decay for wavenumbers beyond the band of instabilities (near wavenumber 50) so that only wavenumbers in the intermediate range between the bands of instability

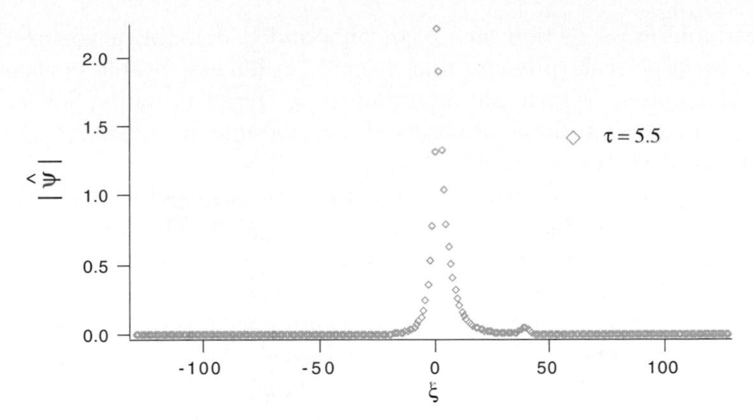

Fig. 4.6. Fourier spectrum of filament solution from Fig. 4.4b at time $\tau = 5.5$. Adapted from Klein (1994)

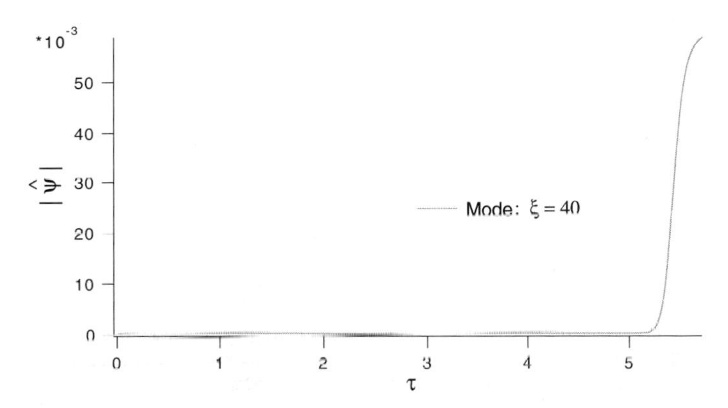

Fig. 4.7. Time evolution of Fourier mode $\hat{\psi}(40)$ for filament solution from Fig. 4.4b. Adapted from Klein (1994)

get excited through nonlinear–nonlocal interactions. In fact, in all our computations we never observed sizeable Fourier amplitudes for wavenumbers larger than those corresponding to the high wavenumber unstable band.

4.2 Linear and Nonlinear Stability of a Filament in an Imposed Straining Field

In the previous section we discussed the effects of self-induction for a perturbed, nearly straight vortex filament in a quiescent environment. Many applications, however, are characterized by the presence of an overall external straining field, which is the result of a large scale nontrivial flow in which the vortex is embedded. Examples are longitudinal vortices in three-dimensional shear and boundary layers, a longitudinal vortex in a channel with variable cross section, or, in an unbounded domain, a vortex that is subject to large-scale pressure and velocity gradients. Stable concentrated vortex filament structures that are strained by the surrounding flow can also be found in turbulent flows, as observed, for example, in the direct numerical simulations of She et al. (1990).

In the present section we report on studies of a model problem that was first introduced and analyzed by Klein et al. (1992). This problem was designed to provide some analytical and quantitative numerical insight into the effects of such outer straining fields on short wavelength perturbations of a slender vortex. Here we follow closely the summary given in Klein (1994).

Again we consider a slender vortex filament whose centerline has a parameterization of the form (4.1.1). The filament is now considered to be embedded in a large-scale flow field which essentially imposes a linearized straining motion in the vicinity of the vortex. The outer field is assumed not to affect an exactly straight vortex so that its net effect on the filament can be described by linearization of the outer field for small deviations from the unperturbed straight centerline. Thus, the outer flow gradient evaluated on the unperturbed straight line, that is,

$$\nabla \boldsymbol{Q}_2 = \begin{pmatrix} s_{11} & s_{12}+\omega & 0 \\ s_{12}-\omega & s_{22} & 0 \\ 0 & 0 & s_{33} \end{pmatrix} \tag{4.2.1}$$

determines the deformation of the filament by the outer flow. Here $\boldsymbol{Q}_2$ is the velocity field of the external flow, s_{33} is the strain rate in the axial direction, s_{11}, s_{12}, and s_{22} are the components of the symmetric part of the strain tensor in a plane normal to the unperturbed straight line, and ω is the rotation rate of the imposed field. Notice that we assume $\omega = O(1)$ as $\epsilon \to 0$ so that the outer field's vorticity is by two orders of magnitude in ϵ smaller than the vorticity found within the vortex core (cf. (3.3.27)). Again, ϵ is the core size parameter, related to the geometric perturbation parameter, δ, through (4.1.5) as in the previous section.

In this section we will derive and analyze the modified filament equation with strain

$$\frac{1}{i}\psi_\tau = \psi_{\sigma\sigma} + \delta^2\left[\frac{1}{2}|\psi|^2\psi - I[\psi]\right] + \delta^2\left[(\omega + \frac{5}{2}is_{33})\psi + is_{33}\sigma\psi_\sigma + \gamma\bar{\psi}\right], \tag{4.2.2}$$

where

$$\gamma = s_{12} - i\frac{s_{11}-s_{22}}{2}. \tag{4.2.3}$$

The first three terms on the right-hand side in (4.2.2) represent the self-induction discussed extensively in the previous sections, whereas the second bracket includes the new effects of the outer straining field. Before embarking on detailed derivations, we discuss here qualitatively the physical interpretation of these terms.

The general influence of these terms is understood best by briefly analyzing the split set of equations

$$
\begin{aligned}
\frac{1}{i}\psi_\tau &= \omega\psi \\
\frac{1}{i}\psi_\tau &= \frac{5}{2} i\, s_{33}\, \psi \\
\frac{1}{i}\psi_\tau &= \gamma\bar{\psi} \\
\frac{1}{i}\psi_\tau &= i\, s_{33}\, \sigma\psi_\sigma\,.
\end{aligned}
\tag{4.2.4}
$$

Equation $(4.2.4)_1$ describes the rotation of ψ in the complex plain with rotation frequency ω. Intuitively, one associates the corresponding vorticity with an effective solid body rotation. The second equation $(4.2.4)_2$ represents for positive s_{33} (positive axial straining), an exponential decay of ψ, while for negative s_{33} (compressive axial strains), it predicts exponential growth. This, too, corresponds with intuition provided we take into account the zero divergence constraint of incompressible flows. Positive axial strain must be associated with a net compressive strain in the normal plane, and this compression would tend to decrease the amplitudes of ψ. This latter consideration, however, does not account for asymmetries of the straining field in the normal plane, which is reflected in $(4.2.4)_3$.

This third term is probably the most interesting one to be considered here. It describes, as stated, the distribution of strain in the plane normal to the unperturbed straight line. Splitting the equation into its real and imaginary parts, we obtain

$$
\boldsymbol{\psi}_\tau = \mathbf{A}\boldsymbol{\psi}\,; \quad \boldsymbol{\psi}(\tau) = \exp(\mathbf{A}\tau)\boldsymbol{\psi}(0)\,, \tag{4.2.5}
$$

where $\boldsymbol{\psi} \equiv (\psi_1, \psi_2)$, $\psi_1 + i\psi_2 \equiv \psi$ and the $2{\times}2$-Matrix $\mathbf{A}$ is given by

$$
\mathbf{A} = \begin{pmatrix} -\gamma_2 & \gamma_1 \\ \gamma_1 & -\gamma_2 \end{pmatrix} \quad \text{with} \quad \begin{aligned} \gamma_2 &= \tfrac{1}{2}(s_{22} - s_{11}) \\ \gamma_1 &= s_{12}\,. \end{aligned} \tag{4.2.6}
$$

The matrix has eigenvalues $\lambda_{1/2} = \pm\sqrt{\gamma_1^2 + \gamma_2^2}$ and, depending on the initial direction of the two-dimensional vector $\boldsymbol{\psi}(0)$ in relation to the right eigenvectors of $\mathbf{A}$, (4.2.5) describes an exponential growth or decay of the initial data. Generally, a vortex filament will rotate under the influence of all the other terms already mentioned, as well as due to self-induction, such that the normal strain effects will alternatingly amplify and diminish the filament function.

Finally $(4.2.4)_4$ results from the advection of vorticity in the axial direction that occurs when there is a nonzero axial gradient of the axial velocity. In fact, this term is in the form of a classical advection term with advection velocity $s_{33}\,\sigma$.

The subsequent sections will summarize first the derivation of the modified filament equation. Next we will discuss one particularly interesting result from a rigorous nonlinear stability theory for this equation when $s_{33} \neq 0$. We then also provide a linear stability analysis and the associated construction of our split-step numerical method for the modified equation. Finally we present and interpret a number of results from numerical solutions for the case of pure two-dimensional strain in the normal plane, that is, for $s_{33} \equiv 0$.

4.2.1 Derivation of the Filament Equation with External Strain

In previous sections we outlined the general procedure for deriving perturbed nonlinear Schrödinger equations for slender vortex filaments. When a filament is embedded in an outer strain field, the self-induction velocity from (4.1.28), $\widetilde{\delta \boldsymbol{v}}$, is modified. The nonlocal self-induced velocity is complemented additively by the local velocity of the strain field.

A class of outer imposed velocity fields that implies an order $O(1)$ contribution to this perturbation velocity, and which is also compatible with the length and time scalings assumed in our derivations for the perturbed nonlinear Schrödinger equations is given by Klein et al. (1992),

$$\boldsymbol{Q}_2(\boldsymbol{x}, t; \delta) = \frac{1}{\delta^2}\boldsymbol{q}\left(s, \boldsymbol{x}_\perp, \frac{t}{\delta^2}\right) . \tag{4.2.7}$$

Here, s is the axial coordinate on the unperturbed filament centerline and

$$\boldsymbol{x}_\perp = \boldsymbol{x} - s\boldsymbol{t}_0 \,, \quad \boldsymbol{x}_\perp \cdot \boldsymbol{t}_0 \equiv 0 \tag{4.2.8}$$

is the component of the position vector $\boldsymbol{x}$ normal to that axis. We assume further that the outer field will not displace an unperturbed straight filament so that

$$\boldsymbol{q}(s, \boldsymbol{0}, \tau) = q_0(s, \tau)\boldsymbol{t}_0 \,, \tag{4.2.9}$$

and, without loss of generality, we let

$$q_0(0, \tau) \equiv 0 \,. \tag{4.2.10}$$

The external velocity field from (4.2.7) is then evaluated asymptotically, as $\delta \to 0$, at locations on the perturbed filament centerline, yielding

$$\begin{aligned} \boldsymbol{v}_2^*(\tilde{s}, \bar{t}; \delta) &= \boldsymbol{Q}_2\left(s(\tilde{s}, \bar{t}; \delta),\ \delta^2 \boldsymbol{X}^{(2)}\left(\frac{\tilde{s}}{\delta}, \frac{\bar{t}}{\delta^2}\right), \frac{\bar{t}}{\delta^2}\right) \\ &= \frac{1}{\delta}\sigma q_0'(\tau)\boldsymbol{t}_0 + \boldsymbol{X}^{(2)} \cdot \boldsymbol{A}_0(\tau) + \frac{1}{2}\sigma^2 q_0''\boldsymbol{t}_0 + O(\delta) \,. \end{aligned} \tag{4.2.11}$$

Here

$$\boldsymbol{A}_0(\tau) = \frac{\partial \boldsymbol{q}}{\partial \boldsymbol{x}_\perp}(0, \boldsymbol{0}, \tau) \qquad \text{and} \qquad q_0'(\tau) = q'(0, \tau) \,, \tag{4.2.12}$$

and $' \hat{=} \partial/\partial s$. Except for one last correction, $\boldsymbol{v}_2^*$ from (4.2.11) represents the perturbation velocity $\boldsymbol{v}_2$ needed in the perturbed binormal law (4.1.28). There, the time derivative *at constant* $\tilde{s}$ is applied to $\boldsymbol{X}(\tilde{s}, \bar{t})$, that is, the time derivative at constant arclength. This implies the following constraint for the perturbation velocity $\boldsymbol{v}_2(\tilde{s}, \bar{t}; \delta)$

$$\boldsymbol{v}_{2,\tilde{s}} \cdot \boldsymbol{t} \equiv 0 \,. \tag{4.2.13}$$

A detailed derivation of this relation is given by Klein et al. (1992). Since a tangential motion along the filament centerline will not affect its geometry, the correct perturbation velocity is obtained by adding to $\boldsymbol{v}_2^*$ from (4.2.11) a suitable tangential component so that (4.2.13) is satisfied. The result is

$$\boldsymbol{v}_2 = \boldsymbol{v}_2^* - \left(\int_{\tilde{s}_0}^{\tilde{s}} \boldsymbol{t} \cdot \boldsymbol{v}_{2,\tilde{s}}^* \, d\tilde{s} \right) \boldsymbol{t} \,, \tag{4.2.14}$$

and, with $\boldsymbol{X}_0^{(2)} = \boldsymbol{X}^{(2)}(0, \tau)$, the asymptotic representation of $\boldsymbol{v}_2$ is

$$\boldsymbol{v}_2 = \boldsymbol{X}^{(2)} \cdot \boldsymbol{A}_0 \cdot (\mathbf{1} - \boldsymbol{t}_0 \circ \boldsymbol{t}_0) - q_0' \sigma \boldsymbol{X}_\sigma^{(2)} + (\boldsymbol{X}_0^{(2)} \cdot \boldsymbol{A}_0 \cdot \boldsymbol{t}_0)\boldsymbol{t}_0 + O(\delta) \,. \tag{4.2.15}$$

The correct equation of motion for a perturbed vortex filament embedded in an outer straining field is thus

$$\begin{aligned} \frac{\partial \boldsymbol{X}}{\partial t}(\tilde{s}, \bar{t}) = \kappa \boldsymbol{b} + \delta^2 \Big\{ & \mathcal{I}[\boldsymbol{X}^{(2)}] \times \boldsymbol{t}_0 \\ & + \boldsymbol{X}^{(2)} \cdot \boldsymbol{A}_0(\tau)(\mathbf{1} - \boldsymbol{t}_0 \circ \boldsymbol{t}_0) - q_0'(\tau) \sigma \boldsymbol{X}_\sigma^{(2)} \\ & + (\boldsymbol{X}_0^{(2)}(\tau) \cdot \boldsymbol{A}_0(\tau) \cdot \boldsymbol{t}_0)\boldsymbol{t}_0 \Big\} + O(\delta^3) \,. \end{aligned} \tag{4.2.16}$$

The perturbation velocity in (4.2.15), together with the self-induced velocity $\mathcal{I}[\boldsymbol{X}^{(2)}] \times \boldsymbol{t}_0$ from (4.1.20), now yields the perturbation velocity to be inserted into the perturbed binormal law. Application of the Hasimoto transformation and calculations analogous to those in Sect. 4.1 yield the filament equation with strain effects (4.2.2).

4.2.2 Rigorous Nonlinear Stability Theory

For a perturbed, nearly straight vortex filament in a quiescent environment, it was shown in Klein and Majda (1991a) that the energy norm of the filament function, that is,

$$\|\psi_0\|^2 = \int |\psi|^2 \, d\sigma \,, \tag{4.2.17}$$

is conserved. This quantity is an integral measure of the filament curvature, as $|\psi| = \kappa$ is the local filament curvature. Under general outer strain fields, we do not expect the energy norm to still be conserved. Instead, depending on the orientation of the strain tensor $\boldsymbol{A}$ in (4.2.1), one would expect amplification or damping of geometric perturbations of a filament.

In fact, Klein et al. (1992) derive the following exact evolution equation:

$$\frac{d}{d\tau}\int |\psi|^2\, d\sigma = 2\delta^2 \left[-2s_{33}\int |\psi|^2\, d\sigma + \boldsymbol{\psi}\cdot\boldsymbol{M}\cdot\boldsymbol{\psi}\right]. \tag{4.2.18}$$

Here $\boldsymbol{\psi}$ is the two-dimensional vector (ψ_1, ψ_2) if the real and imaginary parts of the complex filament function are defined through $\psi = \psi_1 + i\psi_2$, and

$$\boldsymbol{M} = \begin{pmatrix} \frac{1}{2}(s_{11} - s_{22}) & s_{12} \\ s_{12} & -\frac{1}{2}(s_{11} - s_{22}) \end{pmatrix}. \tag{4.2.19}$$

By decomposing $\boldsymbol{\psi}$ into its components with respect to the eigenvectors of $\boldsymbol{M}$, one may show that

$$\boldsymbol{\psi}\cdot\boldsymbol{M}\cdot\boldsymbol{\psi} \le \lambda_+ |\psi|^2\,, \tag{4.2.20}$$

where

$$\lambda_\pm = \pm\sqrt{\frac{(s_{11} - s_{22})^2}{4} + s_{12}^2} \tag{4.2.21}$$

are the eigenvalues of $\boldsymbol{M}$. Thus

$$\frac{d}{d\tau}\|\psi_0\|^2 \le 2\delta^2\left[-2s_{33} + \lambda_+\right](\tau)\|\psi_0\|^2\,. \tag{4.2.22}$$

Using Gronwall's inequality (Gronwall, 1919), which says that $\|\psi_0\|^2$ remains bounded if there is a constant $\overline{\alpha} > 0$ such that

$$\frac{d}{d\tau}\|\psi_0\|^2 \le -\overline{\alpha}\|\psi_0\|^2\,, \tag{4.2.23}$$

we will be able to prove stability of the solution in the energy norm provided the following sufficient (but not necessary) stability criterion for the components s_{33}, s_{11}, s_{22}, s_{12} of the strain tensor is satisfied,

$$(2s_{33} - \lambda_+)(\tau) \ge 0\,, \quad s_{33} > 0\,. \tag{4.2.24}$$

This criterion has a geometrical interpretation. By rotation of the reference coordinate system about the unperturbed filament centerline, the external strain tensor $\boldsymbol{A}$ from (4.2.1) is diagonalized with the new representation

$$\boldsymbol{A}^* = \begin{pmatrix} \gamma_1 & 0 & 0 \\ 0 & \gamma_2 & 0 \\ 0 & 0 & s_{33} \end{pmatrix}. \tag{4.2.25}$$

Here,

$$\gamma_1 = -\frac{1}{2}s_{33} + \lambda_+ > 0 \qquad \text{and} \qquad \gamma_2 = -\frac{1}{2}s_{33} - \lambda_+ < 0 \tag{4.2.26}$$

are the compressive and expansive contributions to the external velocity gradient in the plane normal to the unperturbed filament centerline and along the principal axes of strain. Taking into account that the outer flow is assumed to be incompressible, that is, divergence free, we obtain $\gamma_1 + \gamma_2 + s_{33} = 0$ and the above stability criterion turns into a constraint on the ratio of the two strain rates in that normal plane:

$$\gamma_1 < \frac{3}{5}|\gamma_2| \, . \tag{4.2.27}$$

Interestingly, there is a regime for the outer straining field characterized by

$$\frac{1}{2}|\gamma_2| \; < \gamma_1 < \frac{3}{5}|\gamma_2| \tag{4.2.28}$$

for which

1. a perturbed slender vortex is stable,
2. the strain along the unperturbed vortex axis is positive, and
3. its value, s_{33}, corresponds with the middle eigenvalue of the strain tensor:

$$\gamma_2 < 0 < s_{33} < \gamma_1 \, . \tag{4.2.29}$$

Thus, in this regime the vortex is oriented along the principal axis of positive, intermediate size strain. This regime might be of considerable relevance for the interpretation of direct simulations of turbulence, as in She et al. (1990), in which long-lived, relatively slender vortex filament configurations were observed with the same orientation relative to the surrounding flow fields.

4.2.3 Linear Stability Theory

Here, and in the next section on numerical solutions, we consider the special case of plane strain normal to the unperturbed filament centerline. Figure 4.8 displays this situation, which typically arises when a concentrated vortex forms within an otherwise nearly two-dimensional flow. The filament equation (4.2.2) for nearly straight vortices in this case reads

$$\frac{1}{i}\psi_\tau = \psi_{\sigma\sigma} + \delta^2 \left(\frac{1}{2}|\psi|^2\psi - I[\psi] + i\gamma\overline{\psi} \right) , \tag{4.2.30}$$

where γ is the strain rate of the outer field. Through a suitable rotation of the coordinate system about the unperturbed filament centerline, we can always achieve $\gamma \in I\!R$, $\gamma > 0$. As for the filament in a quiescent environment, we solve (4.2.30) in the next section by operator splitting, that is, the linear part of the equation,

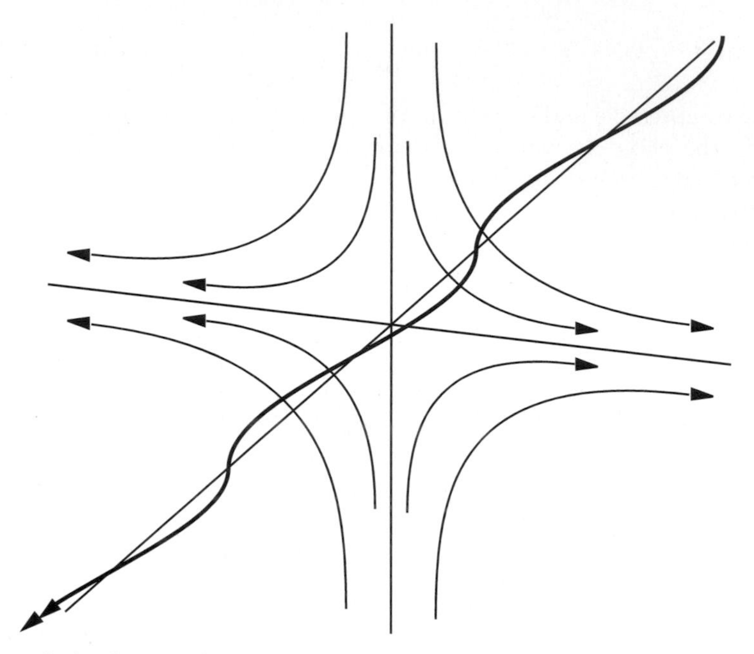

Fig. 4.8. A slender, nearly straight vortex embedded in a plane normal strain field. Adapted from Klein (1994)

$$\frac{1}{i}\psi_\tau = \psi_{\sigma\sigma} + \delta^2\left[i\gamma\overline{\psi} - I[\psi]\right] , \tag{4.2.31}$$

is solved alternatingly with the cubic equation from $(4.1.82)_2$. Again, the linear part is solved exactly using FFT techniques (see Klein et al. 1992).

In the present section we discuss this solution with particular emphasis on the influence of the strain term, keeping in mind that (4.2.31) is also the limit equation obtained from (4.2.30) for infinitesimal amplitudes of the filament function. Thus, we develop here the linear stability theory for that equation.

Immediately, we observe that the nonlinear, sufficient, but not necessary stability criterion (4.2.24) from the last section is not satisfied here at all times, because $s_{33} = 0$. In fact, we will find that a straight filament is unconditionally unstable under normal plane strains. Thus, the importance of the present linear theory will not be to determine whether a vortex filament would be stable, but to determine exactly what wavenumber instabilities would or would not occur.

The Fourier transformation will turn (4.2.31) into a system of coupled ordinary differential equations for the Fourier modes $\widehat{\psi}(\xi,\tau)$ and $\widehat{\psi}(-\xi,\tau)$ via

$$\begin{pmatrix}\widehat{\psi}(\xi,\tau)\\ \overline{\widehat{\psi}}(-\xi,\tau)\end{pmatrix}_\tau = i\left\{\widehat{\mathcal{L}}(\xi)\begin{bmatrix}1 & 0\\ 0 & -1\end{bmatrix} - i\delta^2\gamma\begin{bmatrix}0 & 1\\ 1 & 0\end{bmatrix}\right\}\begin{pmatrix}\widehat{\psi}(\xi,\tau)\\ \overline{\widehat{\psi}}(-\xi,\tau)\end{pmatrix} , \tag{4.2.32}$$

where $\widehat{\mathcal{L}}(\xi)$ is given by

$$\widehat{\mathcal{L}}(\xi) = -\xi^2 - \delta^2 \widehat{I}(\xi)\,. \tag{4.2.33}$$

The Fourier symbol $\widehat{I}(\xi)$ of the I-Operator from (4.1.8) has been given earlier in (4.1.68). The exact solution of (4.2.31) may be formulated in Fourier space variables as

$$\begin{pmatrix} \widehat{\psi}(\xi,\tau) \\ \overline{\widehat{\psi}}(-\xi,\tau) \end{pmatrix} = \left(\cos(\tau p(\xi))\, \boldsymbol{I} + i \sin(\tau p(\xi))\, \boldsymbol{P}\right) \begin{pmatrix} \widehat{\psi}(\xi,0) \\ \overline{\widehat{\psi}}(-\xi,0) \end{pmatrix}, \tag{4.2.34}$$

where

$$p(\xi) = (\widehat{\mathcal{L}}^2(\xi) - \delta^4\gamma^2)^{\frac{1}{2}}\,, \tag{4.2.35}$$

$\boldsymbol{I}$ is the 2×2 unit tensor, and

$$\boldsymbol{P} = (p(\xi))^{-1} \left[i\delta^2\gamma \begin{bmatrix} 0 & 1 \\ 1 & 0 \end{bmatrix} + \widehat{\mathcal{L}}(\xi) \begin{bmatrix} 1 & 0 \\ 0 & -1 \end{bmatrix} \right]. \tag{4.2.36}$$

The functions $\cos(z)$ and $\sin(z)$ are to be evaluated for complex z in this equation. As a consequence, solutions to the linearized equation (4.2.31) exhibit oscillatory behavior for large wavenumbers ξ with

$$\widehat{\mathcal{L}}^2(\xi) > \delta^4\gamma^2\,, \tag{4.2.37}$$

and instability for long-wave perturbations with

$$\widehat{\mathcal{L}}^2(\xi) < \delta^4\gamma^2\,. \tag{4.2.38}$$

We observe in addition that the outer strain influences the filament dynamics in two distinct ways. On the one hand it induces long-wave instability, and on the other it is responsible for a mode-coupling in Fourier space which was absent in the linear part of the filament equation for vortices in a quiescent environment. In fact, Fourier modes with opposite wavenumbers $\widehat{\psi}(-\xi)$ and $\widehat{\psi}(\xi)$ interact. Both these effects will now be studied further on the basis of numerical solutions of the full nonlinear filament equation with outer strain (4.2.30).

4.2.4 Numerical Solutions for the Filament Equation with External Strain

The earlier investigations of the dynamics of an isolated filament revealed two interesting mechanisms. Nonlocal–nonlinear interactions induced an energy transfer in wavenumber space so that the evolution of long-wave initial data would ultimately lead to the excitation of short wavelength perturbations on the vortex core size scale. Of course, this would then signal the limit of applicability of the theory. The second mechanism was self-induced vortex stretching, which we were able to explicitly express as a quadratic functional of the filament function (Klein and Majda, 1991b; Klein et al., 1992).

Here, we study how an outer strain field will modify these mechanisms inherent to the dynamics of an isolated filament. The linear and nonlinear stability theories of the previous sections showed that sufficiently large strain rates of the surrounding flow, $|\gamma| = O(\delta^{-2})$, will always generate unstable behavior; however, this order of magnitude of the strain rates is formally outside the regime of validity of our theory, which was derived under the assumption that $|\gamma| = O(1)$ as $(\delta \to 0)$. In this regime, Fourier modes with wavenumbers of order $O(1)$ are stable, and the observed effects will be more subtle. The imposed strain will not immediately force a dramatic response, but only its cumulative influence will be seen. (For studies that also address the regime of larger strain rates, see Klein et al. 1992).

4.2.4.1 A Helix Under Plane Strain. In a first example we consider helix initial data

$$\psi(\sigma, 0) = 4.0 \exp{(i\, 2\sigma)} \; . \tag{4.2.39}$$

These data would imply a travelling wave helix solution in the absence of the outer straining field. Since such a helix is, however, not an exact solution of the filament equation with strain (4.2.30), we may expect nontrivial modifications.

As an example, Fig. 4.9 displays the maximum self-stretching rate within the computational domain versus time. While, for symmetry reasons, the helix has zero self-stretch, we observe that substantial self-stretch is generated immediately in the graph. High-frequency oscillations and self-stretching rates $\dot{\ell}/\ell \approx 10 \ldots 15$ imply that the helix configuration is rapidly destroyed. Because of the accumulation of nonlinear effects, the solution ultimately leaves the regime of validity of the asymptotics as in earlier simulations for an

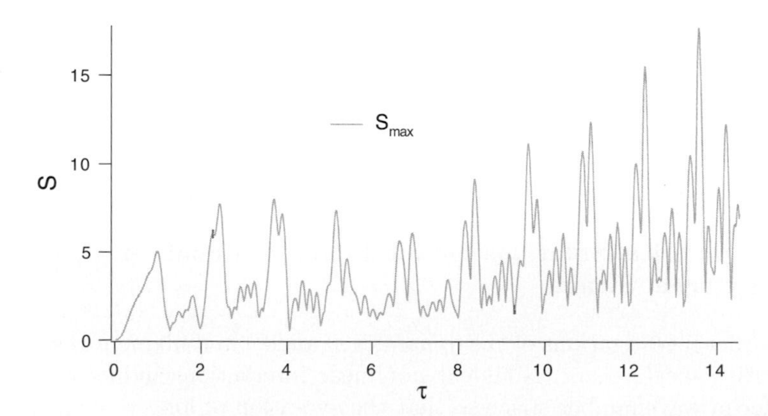

Fig. 4.9. History of the self-induced stretching of an initially helical filament with filament function, $\psi(\sigma, 0) = 4.0 \exp(i\, 2\sigma)$ in a (weak) outer straining field with $\gamma = 4.0 + i0.0$. Adapted from Klein et al. (1992) and Klein (1994)

isolated filament through the already-known "energy transfer in wavenumber space."

This example shows that a relatively weak outer strain will not, in accordance with the results of linear theory, have dramatic consequences immediately. Nevertheless, through mode coupling it can excite the mechanisms of self-induction and their interaction with the local nonlinear curvature effects. The result is considerable overall self-stretch, and ultimately a breakdown of the asymptotic description through excitation of perturbations on the vortex core size scale.

4.2.4.2 Triggering Instabilities Through Mode Coupling. The next example involves perturbed helix initial data

$$\psi(\sigma, 0) = 4.0 \exp(2i\sigma) + 0.4 \exp(-3i\sigma) \ . \tag{4.2.40}$$

For a vortex in a quiescent environment, such a perturbation with wavenumber $\xi = -3$ is stable as we have previously seen. However, a perturbation of wavenumber $\xi = +3$ would be unstable. Under the influence of an outer straining field we have a coupling of the modes $\xi = \pm 3$ so that the unstable mode will quickly be generated from the above initial data, and the solution will become unstable. Subsequently, we will observe the inherent helix instability in action, with rapid energy transfer to short wavelength modes.

Figure 4.10 verifies these claims, as the displayed distributions of curvature arc, except for minor details and the precise timing, equivalent with the dashed line in Fig. 4.5, which was the result of a simulation in the absence of strain. We emphasize again that in the present case, the breakdown of the asymptotics is the result of self-induction as well. The outer strain field merely triggers these inherent instability mechanisms. It has no essential influence

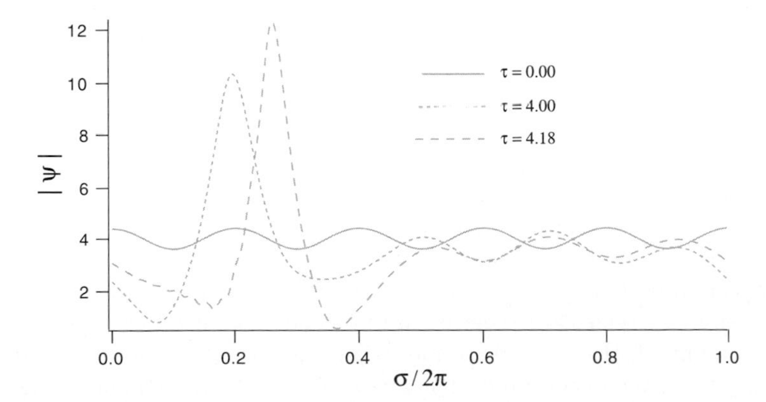

Fig. 4.10. Solution structure for a perturbed helix $\Psi(\sigma, 0) = 4.0 \exp(i\,2\sigma) + 0.4 \exp(-i\,3\sigma)$ under weak outer strain $\gamma = 1.0 + i0.0$ at a sequence of times close to the critical time of "energy transfer to the core size scale." Adapted from Klein et al. (1992) and Klein (1994)

on the rapid subsequent nonlinear evolution, and on how the energy transfer is taking place in detail. The next example will show, however, that these conclusions do not hold in general, and that there are other configurations in which the influence of the outer straining field is felt throughout the entire filament evolution.

4.2.4.3 Plane Geometrical Perturbations and the Influence of Outer Strain on the Energy Transfer in Wavenumber Space. We consider a vortex filament which is initially perturbed by plane geometric displacements. The associated filament function initial data are

$$\psi(\sigma, 0) = 8.0 \cos(2\sigma) + 0.8 \cos(3\sigma) \ . \tag{4.2.41}$$

We study the influence of the orientation of an outer straining field by considering complex strain parameters

$$\gamma_0 = 4. + i0.0 \qquad \text{and} \qquad \gamma_1 = 4.0 \exp\left(i\frac{2}{3}\pi\right) . \tag{4.2.42}$$

The second corresponds to a rotation of the outer strain field with respect to the first by $2\pi/3$.

Figures 4.11a and b show the spatial distributions of curvature for these two cases at selected instances in time. Early on in the first example, one observes the generation of small-scale structures, and this solution leaves the regime of validity of the asymptotics at about $\tau \approx 0.65$. In contrast, the solution for $\gamma = 4.0 \exp(i2\pi/3)$ remains smooth and stable for long times, and the excitation of short wave modes occurs abruptly much later at $\tau \approx 1.6$.

Thus, not only does the presence of the outer strain considerably influence the solution structures, but we also conclude that the detailed arrangement of the strain field relative to the filament geometry can affect the vortex evolution substantially. As before, the orientation of the outer strain influences, in particular, the generation of small scales. With $\gamma_0 = 4.0 + i0.0$ a gradual, nearly monotonous excitation of the high wavenumber mode ($\xi = 50$) occurs as shown in Fig. 4.12a. In contrast, for $\gamma = 4.0 \exp\left(i\frac{2}{3}\pi\right)$ we find a completely different behavior as shown in Fig. 4.12b. Over a considerable period, the short wave components are essentially unaffected, until a sudden burst at time $\tau \approx 1.6$ essentially destroys the asymptotic approximation.

The present example illustrates that the inherent mechanisms of self-induction of a vortex filament, which ultimately generate strong disturbances on the core size scale, are sensitively influenced by an outer straining field. On the one hand, the outer field may trigger instabilities through mode coupling, while on the other hand one can stabilize such a filament by judicious application of an outer strain.

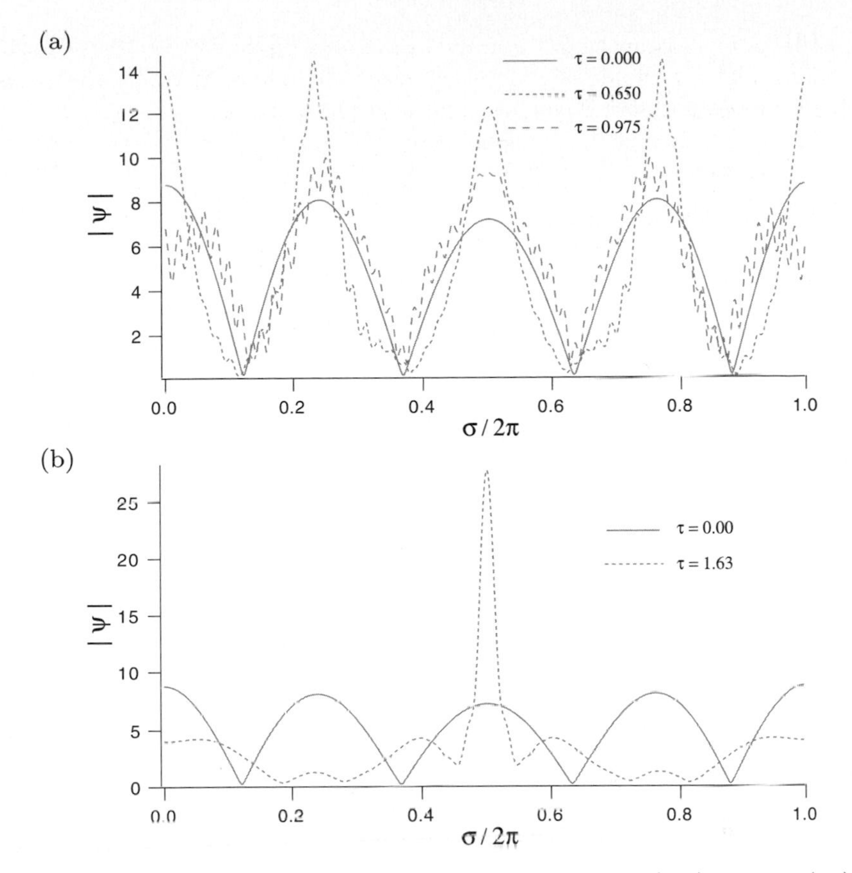

Fig. 4.11. Solution structures developing from initial data: $\psi(\sigma, 0) = 8.0\cos(2\sigma) + 0.8\cos(3\sigma)$, under the influence of outer straining fields with, (**a**) $\gamma = 4.0 + i0.0$, and (**b**) $\gamma = 4.0\exp(i\frac{2}{3}\pi)$. Adapted from Klein (1994)

4.3 Nonlinear Extension of the Crow/Jimenez Instability Theory for Parallel Vortex Filaments

The dynamical behavior of parallel, counterrotating slender vortices is of major interest as a standard model for aircraft trailing vortices. In addition, it represents the simplest abstract framework within which one may study the interaction of concentrated, well-separated vortices in three space dimensions. The present section follows closely the earlier derivations in this chapter by identifying, for each of the vortices in a pair, the induced velocity from the respective "other" vortex as a perturbation velocity $\tilde{\delta}\boldsymbol{v}$ as in the earlier subsections. Thus we consider the coupled evolutions equations

$$\partial \boldsymbol{X}_\alpha / \partial \bar{t} = \Gamma_\alpha \left(\kappa \boldsymbol{b}\right)_\alpha + \tilde{\delta}\boldsymbol{v}_\alpha\,, \quad \alpha = I, II. \tag{4.3.1}$$

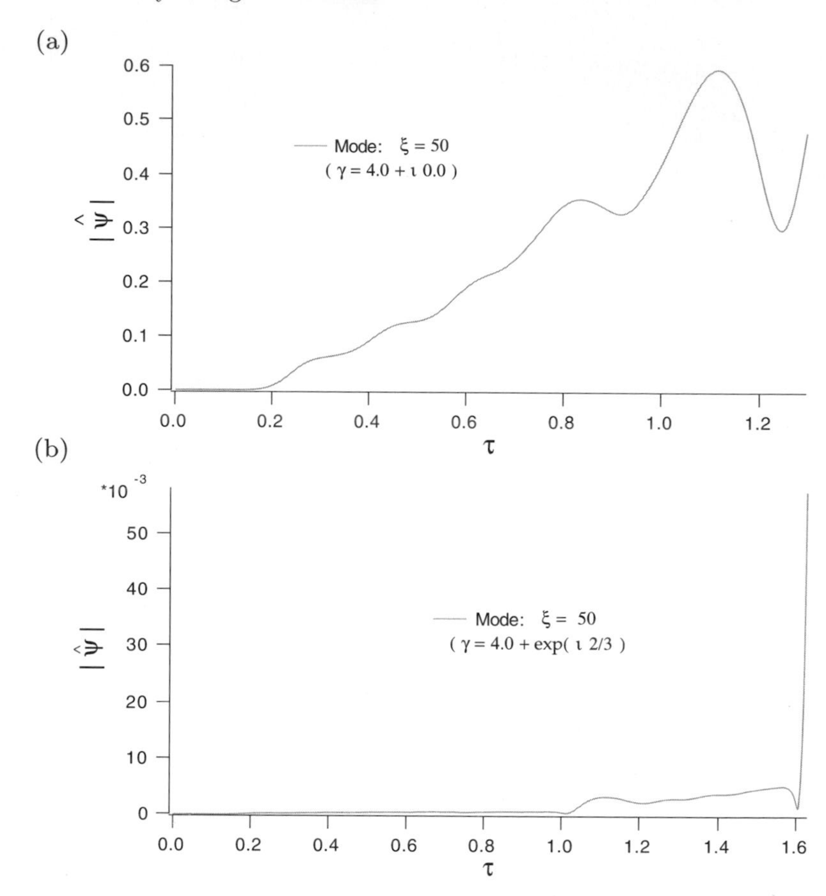

Fig. 4.12. Temporal evolution of the Fourier mode $|\hat{\psi}(\xi = 50)|$, whose wavelength is comparable to the vortex core size for $\delta^2 = 0.25$, for the examples from Figs. 4.11a and b. Adapted from Klein et al. (1992) and Klein (1994)

Here, $\Gamma_{I/II}$ are the, possibly different, circulations of the two vortices. Coupled equations of motion for the vortices arise due to the fact that the perturbation velocity $\tilde{\delta \boldsymbol{v}}_I$, for example, of the first vortex, is nontrivially influenced by the geometrical perturbations of the second vortex and vice versa. Following the general pattern for the derivation of filament equations in the context of (4.1.29), we next have to express the mutually induced perturbation velocities $\boldsymbol{v}_{I/II}$ in terms of the filament geometries, and then obtain the desired coupled filament equations by applying the Hasimoto transformation to the position vector fields $\boldsymbol{X}^{(2)}_{I/II}$.

The result of these transformations is the following coupled equation system for the filament functions $\psi(\sigma, \tau)$ (Filament I) and $\phi(\sigma, \tau)$ (Filament II). Details of the derivations will be given below.

$$
\begin{aligned}
\frac{1}{i}\psi_\tau &= \left(1-\epsilon^2\lambda^\psi\right)\psi_{\sigma\sigma} + \epsilon^2\left[\frac{1}{2}|\psi|^2\psi - I[\psi]\right] \\
&\quad + \epsilon^2\left[-\frac{2}{b^2}\left((1+\Gamma^*)\psi + \Gamma^*\overline{\psi}\right) - \Gamma^*\left(I^b[\phi] - M^b[Re\phi]\right)\right] \\
\frac{1}{i}\phi_\tau &= \Gamma^*\left[\left(1-\epsilon^2\lambda^\phi\right)\phi_{\sigma\sigma} + \epsilon^2\left[\frac{1}{2}|\phi|^2\phi - I[\phi]\right]\right] \\
&\quad + \epsilon^2\left[-\frac{2}{b^2}\left((\Gamma^*+1)\phi + \overline{\phi}\right) - \left(I^b[\psi] - M^b[Re\psi]\right)\right] .
\end{aligned}
\tag{4.3.2}
$$

Here $I^b[\cdot]$ and $M^b[\cdot]$ are linear integral operators, whose structure and interpretation will be explained in detail below, and

$$
\Gamma^* = \frac{\Gamma_{II}}{\Gamma_I} \tag{4.3.3}
$$

is the circulation ratio.

Notice that these equations for two vortices of *different* strengths were first correctly stated in Klein (1994), whereas in the earlier discussion in Klein and Majda (1993) the rotation terms $\epsilon^2\, 2/b^2\,(1+\Gamma^*)\psi$, etc., had been erroneously omitted. These terms are, however, important in describing the stability features of, for example, a corotating vortex pair as analyzed by Jimenez (1975). The specialized results in Klein and Majda (1993) for counterrotating vortices of equal absolute strength, that is, for ($\Gamma^* = -1$), are nevertheless correct, because in this case the overall rotation effects mentioned above cancel exactly.

Before embarking on the discussion of technical details of the derivations leading to (4.3.2), we consider first the physical meaning of the various expressions in these equations:

The terms

$$
\psi_{\sigma\sigma} + \epsilon^2\left(\frac{1}{2}|\psi|^2\psi - \mathcal{I}[\psi]\right) \tag{4.3.4}
$$

represent the vortex self-induction discussed in the previous sections.

The expression

$$
\frac{2}{b}\overline{\psi} \tag{4.3.5}
$$

accounts for the effect of a plane external strain normal to the unperturbed vortex axes. This effect arises, because the considered vortex filaments are embedded *in their mutual induced potential vortex velocity fields.* Their leading-order contributions correspond with the induced velocities of a pair of straight point vortices of strengths $\Gamma_{I/II}$. As the vortices are considered to be parallel in their unperturbed configurations, their induced velocities in fact have components only in the normal planes of the unperturbed centerlines.

An additional nonlocal coupling of the filament functions occurs through the terms involving

$$I^b[\phi] + M^b[\mathrm{Re}\phi]\,, \tag{4.3.6}$$

where I^b and M^b are integral operators given by

$$\begin{aligned} I^b[w](\sigma) &= \int_{-\infty}^{\infty} \frac{w(\sigma+h) - hw'(\sigma+h)}{(h^2+b^2)^{3/2}}\,dh \\ M^b[w](\sigma) &= \int_{-\infty}^{\infty} \frac{3b^2 w(\sigma+h)}{(h^2+b^2)^{5/2}}\,dh\,. \end{aligned} \tag{4.3.7}$$

They capture the modifications of the mutually induced velocities due to the geometric perturbations of the filament centerlines. These displacements, of course, also influence the induced velocities, and it turns out that these effects are, in fact, felt at an order of magnitude that makes them relevant for the filament dynamics equations. In particular, we will find that $I^b[\cdot]$ is the result of perturbations of the tangent vector of the vortex axes, that is, of the axial direction of the centerline, whereas $M^b[\cdot]$ is caused by local deviations of the filament distance from their unperturbed value.

As before, much of the effect of these linear nonlocal (coupling) terms can be understood by considering their Fourier symbols $\widehat{I^b}[\xi]$ and $\widehat{M^b}[\xi]$. A straightforward calculation shows that

$$\begin{aligned} \widehat{I^b}(\xi) &= 3b^2 \widehat{I^b_5}(\xi) - \widehat{I^b_3}(\xi) \\ \widehat{M^b}(\xi) &= 3b^2 I^b_5(\xi)\,, \end{aligned} \tag{4.3.8}$$

where the operators I^b_j are given by

$$I^b_j[w](\sigma) = \int_{-\infty}^{\infty} \frac{w(\sigma+h)}{(h^2+b^2)^{j/2}}\,dh\,. \tag{4.3.9}$$

The Fourier symbols, $\widehat{I^b_j}(\xi)$, are the Fourier transforms of the functions $1/(\sigma^2+b^2)^{j/2}$, and in Abramowitz and Stegun (1972) one finds the exact expressions

$$\begin{aligned} \widehat{I^b_3}(\xi) &= \frac{2}{b^2}(b|\xi|)K_1(b|\xi|) \\ \widehat{I^b_5}(\xi) &= \frac{2}{3b^4}\left(2(b|\xi|)K_1(b|\xi|) + (b|\xi|)^2 K_0(b|\xi|)\right)\,. \end{aligned} \tag{4.3.10}$$

Here, $K_0(z)$ and $K_1(z)$ are the modified Bessel functions of the second kind. The symbols $\widehat{I^b}$ and $\widehat{M^b}$ are displayed in Fig. 4.13, where we observe that the strongest interactions occur for long wavelengths with wave numbers $\xi \lessapprox 8.0$. A closer analysis of the modified Bessel functions for large $|b\xi|$ further shows that the interaction strengths vanish as $(\xi \to \infty)$.

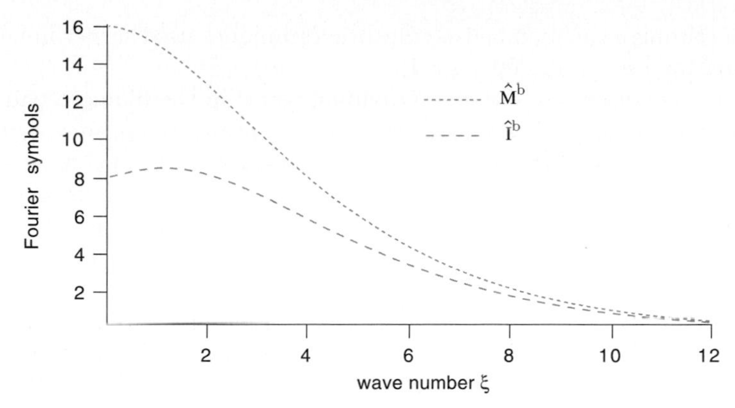

Fig. 4.13. Graph of the Fourier symbols of the filament interaction operators, $\hat{I}^b$ and $\hat{M}^b$ from (4.3.8). Adapted from Klein (1994)

The expressions

$$\frac{2}{b^2}(1+\Gamma^*)\psi \qquad \text{and} \qquad -\frac{2}{b^2}(\Gamma^*+1)\phi \tag{4.3.11}$$

account for the fact that two vortices of unequal strength will rotate around their common center of vorticity with a dimensionless rotation frequency $\delta^2\, 2/b^2(1+\Gamma^*)$. This becomes particularly apparent by comparison of these terms with the rotation terms in the filament equation for a single vortex in an outer straining field from the previous section (see (4.2.2)).

Finally, the terms

$$\lambda_\psi \psi_{\sigma\sigma} \tag{4.3.12}$$

are symmetry-breaking arclength corrections, which arise when the geometrical perturbations of the filaments do not have bounded support but extend over large distances in the axial direction. These corrections become necessary here: In the derivation of the general filament equation from (4.1.29) in Sect. 4.1 all the nonlinearities collapse into the single cubic expression in terms of the filament function only, if the spacial coordinate is, in fact, an arclength *on the actual filament centerline.* If the two vortices considered here have independent initial data, then their arclengths over one spatial period will be different, and it is then necessary to rescale at least one of these coordinates in order to obtain periodic data with the same period for both filaments.

We observe that (4.3.2) is equivalent to Crow's linearized equations from Crow (1970a) when we neglect the cubic terms. Thus, the present theory is a natural nonlinear extension of Crow's stability theory. A technical, but important, detail concerns the regularization of the singular Biot–Savart integral for a singular vortex line in three dimensions. While Crow employs an ad hoc regularization, we will later on show explicitly how the free parameter of his regularization term can be systematically related to the vortex core structure through the matched asymptotic techniques used earlier in Sect. 3.3 (see also Margerit et al. 2001, 2004).

From these discussions of the individual terms in the filament pair equations, we derive a hypothesis regarding the nonlinear mechanisms of instability for a pair of counterrotating vortices: Long-wave instabilities, as already predicted by Crow (1970a), lead to exponential growth of the individual filament amplitudes until the cubic nonlinearities begin to play a role. Nonlinear–nonlocal interactions, as induced by the terms in (4.3.4), then again lead to an energy transfer in wave number space, and ultimately to nontrivial modifications of the vortex cores. We will verify these predictions on the basis of numerical solutions of the filament pair equations in the next subsection for the case relevant to aircraft trailing vortices, that is, for counterrotating vortices of equal strength, $\Gamma^* = -1$.

4.3.1 Derivation of the Filament Pair Equations

Here we consider perturbed, nearly straight, nearly parallel vortex filaments with parameterizations of their filament centerlines according to

$$\begin{aligned} \boldsymbol{X}_I &= s_I \boldsymbol{t}_0 + \frac{\delta}{2} b \boldsymbol{n}_0 + \delta^2 \boldsymbol{X}_I^{(2)} \left(\frac{s_I}{\delta}, \frac{t}{\delta^2} \right) + O(\delta^2) \\ \boldsymbol{X}_{II} &= s_{II} \boldsymbol{t}_0 - \frac{\delta}{2} b \boldsymbol{n}_0 + \delta^2 \boldsymbol{X}_{II}^{(2)} \left(\frac{s_{II}}{\delta}, \frac{t}{\delta^2} \right) + O(\delta^2) \end{aligned} \tag{4.3.13}$$

as $\delta \to 0$. This representation uses a coordinate system, $(\boldsymbol{t}_0, \boldsymbol{n}_0, \boldsymbol{b}_0)$, which is oriented relative to the unperturbed filaments as sketched in Fig. 4.14.

The distance between the unperturbed vortices is chosen to be δb with $b = O(1)$. With this choice, the perturbation wavelengths for our theory and

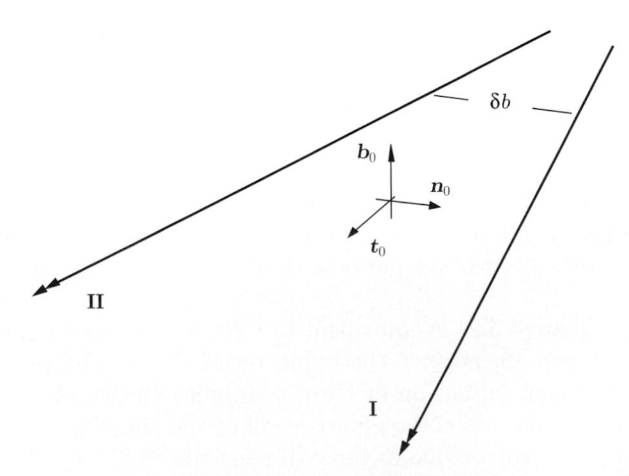

Fig. 4.14. Orientation of the reference system $(\boldsymbol{t}_0, \boldsymbol{n}_0, \boldsymbol{b}_0)$ for the analysis of nearly parallel vortex filaments. Adapted from Klein (1994)

the filament separation are comparable in order of magnitude. This is the same regime that was also considered in Crow (1970a), and we must require that we reproduce his result in the limit of infinitesimal geometric deviations from the leading-order straight lines.

To derive the coupled filament equations summarized in the previous section, we first have to obtain an expression for the perturbation velocities $\boldsymbol{v}_{I/II}$ in (4.3.1) in terms of the geometrical perturbations of the two vortices from (4.3.13). These velocities result from a combination of self-induction and mutual induction between the filaments as follows:

$$\boldsymbol{v}_\alpha = \mathcal{I}[\boldsymbol{X}_\alpha^{(2)}] \times \boldsymbol{t}_0 + \boldsymbol{v}_2^{(\alpha)} \tag{4.3.14}$$

where

$$\begin{aligned} \boldsymbol{v}_2^I &= \Gamma^* \left\{ -\frac{2}{\delta b}\boldsymbol{b}_0 - \frac{2}{b^2}\boldsymbol{A}\cdot\boldsymbol{X}_I^{(2)} + I^b\left[\boldsymbol{X}_{II}^{(2)}\right]\times\boldsymbol{t}_0 + \left(\boldsymbol{n}_0\cdot M^b\left[\boldsymbol{X}_{II}^{(2)}\right]\right)\boldsymbol{b}_0 \right\} \\ \boldsymbol{v}_2^{II} &= \frac{2}{\delta b}\boldsymbol{b}_0 - \frac{2}{b^2}\boldsymbol{A}\cdot\boldsymbol{X}_{II}^{(2)} - I^b\left[\boldsymbol{X}_I^{(2)}\right]\times\boldsymbol{t}_0 - \left(\boldsymbol{n}_0\cdot M^b\left[\boldsymbol{X}_I^{(2)}\right]\right)\boldsymbol{b}_0\,, \end{aligned} \tag{4.3.15}$$

and

$$\boldsymbol{A} = (\boldsymbol{n}_0 \circ \boldsymbol{b}_0 + \boldsymbol{b}_0 \circ \boldsymbol{n}_0)\,. \tag{4.3.16}$$

4.3.1.1 Mutually Induced Perturbation Velocities. The formulae from (4.3.15) are derived here through asymptotics analyses for the Biot–Savart integral applied to a filament pair. The induced velocity from a vortex filament of circulation Γ, the centerline of which is a space curve $\mathcal{L} : s \to \boldsymbol{X}(s,t)$, is given by the Biot–Savart formula

$$\boldsymbol{v}_{\text{ind}}(\boldsymbol{P};\mathcal{L}) = \Gamma \int_{\mathcal{L}} \frac{\boldsymbol{P}-\boldsymbol{X}}{|\boldsymbol{P}-\boldsymbol{X}|^3} \times \frac{\partial \boldsymbol{X}}{\partial s}\, ds\,. \tag{4.3.17}$$

To determine the contribution $\boldsymbol{v}_2^I$ to the perturbation velocity of the first filament as induced by the second, we replace $\mathcal{L}$ with $\mathcal{L}_{II}$: $s_{II} \to \boldsymbol{X}_{II}(s_{II},\bar{t};\delta)$ according to (4.3.13) and consider a reference point $\boldsymbol{P} = \boldsymbol{X}_I(s_I,\bar{t};\delta)$ on $\mathcal{L}_I$. Taylor expansion of the integrand then yields

$$\begin{aligned} \boldsymbol{P} &= \delta\sigma\boldsymbol{t}_0 + \frac{\delta}{2}b\boldsymbol{n}_0 + \delta^2\boldsymbol{X}_I^{(2)}(\sigma) + O(\delta^2) \\ \boldsymbol{X} &= \delta(\sigma+h)\boldsymbol{t}_0 - \frac{\delta}{2}b\boldsymbol{n}_0 + \delta^2\boldsymbol{X}_{II}^{(2)}(\sigma+h) + O(\delta^2) \\ \frac{\partial\boldsymbol{X}}{\partial s} &= \boldsymbol{t}_0 + \delta\boldsymbol{X}_{II,\sigma}^{(2)}(\sigma+h)\,, \end{aligned} \tag{4.3.18}$$

where $\sigma = s_I/\delta$ is the arclength coordinate of the reference point on the first filament, and $h = (s_{II} - s_I)/\delta$ is the integration variable. The leading terms in the integrand are

$$
\begin{aligned}
\delta \frac{\boldsymbol{P}-\boldsymbol{X}}{|\boldsymbol{P}-\boldsymbol{X}|^3} \times \frac{\partial \boldsymbol{X}}{\partial s} = & \frac{1}{(h^2+b^2)^{3/2}} \\
& \left\{ \frac{1}{\delta} b\, \boldsymbol{b}_0 + \boldsymbol{X}_I^{(2)}(\sigma) \times \boldsymbol{t}_0 \right. \\
& + \frac{3b^2}{h^2+b^2} \left(\boldsymbol{n}_0 \cdot \boldsymbol{X}_I^{(2)}(\sigma) \right) \boldsymbol{b}_0 \\
& - \left(\boldsymbol{X}_{II}^{(2)} - h \boldsymbol{X}_{II,\sigma}^{(2)} \right) (\sigma + h) \times \boldsymbol{t}_0 \\
& \left. - \frac{3b^2}{h^2+b^2} \left(\boldsymbol{n}_0 \cdot \boldsymbol{X}_{II}^{(2)}(\sigma+h) \right) \boldsymbol{b}_0 \right\} .
\end{aligned}
\tag{4.3.19}
$$

The first term in curly braces is the induced velocity from an unperturbed vortex. In fact, for a pair of exactly straight and parallel vortices, these terms yield the well-known rules for two-dimensional vortices. Counterrotating vortices of equal strength will move under their mutual induction at constant velocity in the direction normal to the plane they are embedded in. If the vortices have unequal strength, they will rotate around each other. This background motion defines the moving frame of reference that we will use to pursue the subsequent analysis as shown in Fig. 4.14. The second and fourth terms in (4.3.19) result from the cross-product $(\boldsymbol{P} - \boldsymbol{X}) \times \partial \boldsymbol{X}/\partial s$ in the Biot–Savart integral. Thus, they characterize the influence of geometrical perturbations of the vortex filaments. The remaining contributions in the direction of $\boldsymbol{b}_0$ result from the asymptotic expansion of the denominator, $1/|\boldsymbol{P} - \boldsymbol{X}|^3$ within the Biot–Savart integral and represent the effect of changes of the net intervortex distance.

Integrating in h and taking into account that

$$\int_{-\infty}^{\infty} 1/(h^2+b^2)^{3/2} dh = 2/b^2 \tag{4.3.20}$$

and

$$\int_{-\infty}^{\infty} 3b^2/(h^2+b^2)^{5/2} dh = 4/b^2 \ , \tag{4.3.21}$$

we obtain from the first three terms in (4.3.19) the first two terms in the expressions for the induced velocities in (4.3.15). Comparing the result with the equation for a single vortex filament in a straining field from (4.2.15), we find that the expression $\frac{2}{b^2} \boldsymbol{A} \cdot \boldsymbol{X}_I^{(2)}$ expresses the geometrical displacement, $\delta^2 \boldsymbol{X}_I^{(2)}$, of Filament I within the potential vortex field of Filament II.

The terms involving $\boldsymbol{X}_{II}^{(2)}$ in (4.3.19) reduce to the linearized expressions $I^b[\boldsymbol{X}_{II}^{(2)}] \times \boldsymbol{t}_0$ and $(\boldsymbol{n}_0 \cdot M^b[\boldsymbol{X}_{II}^{(2)}])\boldsymbol{b}_0$ in (4.3.15). Their physical interpretation is clear from the previous discussions: The term $-I^b[\boldsymbol{X}_{II}^{(2)}] \times \boldsymbol{t}_0$ results from geometrical perturbations of the central line of Filament II, whereas $-\boldsymbol{n}_0 \cdot M^b[\boldsymbol{X}_{II}^{(2)}]\, \boldsymbol{b}_0$ goes back to changes of the intervortex distances.

4.3.1.2 Rotating Frame of Reference. The base vectors $(\boldsymbol{t}_0, \boldsymbol{n}_0, \boldsymbol{b}_0)$ from Fig. 4.14 are defined by the *instantaneous* configuration of the filament pair. For counterrotating vortices of equal strength, that is, for $\Gamma_{\mathrm{I}} = -\Gamma_{\mathrm{II}}$ this system will be time independent, but for any $\Gamma_{\mathrm{I}} \neq -\Gamma_{\mathrm{II}}$ it will rotate about the joint center of vorticity of the two filaments.

When both filaments are undisturbed so that $\boldsymbol{X}_{\mathrm{I}}^{(2)} = \boldsymbol{X}_{\mathrm{II}}^{(2)} \equiv \boldsymbol{0}$, then the distance $\boldsymbol{X}_{\mathrm{II}} - \boldsymbol{X}_{\mathrm{I}} = \delta b \boldsymbol{n}_0$ between them is given by

$$\begin{aligned} \frac{\partial}{\partial t}\left(\boldsymbol{X}_{\mathrm{II}} - \boldsymbol{X}_{\mathrm{I}}\right) = \delta b \frac{\mathrm{d}\boldsymbol{n}_0}{\mathrm{d}t} &= \left(\frac{\partial \boldsymbol{X}_{\mathrm{II}}}{\partial t} - \frac{\partial \boldsymbol{X}_{\mathrm{I}}}{\partial t}\right)_{\boldsymbol{X}_{\mathrm{I}}^{(2)}=\boldsymbol{X}_{\mathrm{II}}^{(2)}\equiv \boldsymbol{0}} \\ &= \frac{2}{\delta b}\boldsymbol{b}_0 - \frac{2\Gamma^*}{\delta b}\left(-\boldsymbol{b}_0\right) = \frac{2}{\delta b}\left(1+\Gamma^*\right)\boldsymbol{b}_0 \,, \end{aligned} \tag{4.3.22}$$

where t is our usual "normal time." As a consequence,

$$\frac{\mathrm{d}\boldsymbol{n}_0}{\mathrm{d}t} = \frac{2}{(\delta b)^2}\left(1+\Gamma^*\right)\boldsymbol{b}_0 \,. \tag{4.3.23}$$

Thus $\boldsymbol{n}_0$ and $\boldsymbol{b}_0$ rotate, on that time scale, with a frequency

$$\omega_3 = \frac{2\left(1+\Gamma^*\right)}{(\delta)^2} \,. \tag{4.3.24}$$

Next, we translate this rotation frequency into a rate of rotation measured on the evolution time scale of short wave perturbations of the vortex filaments: Taking into account the scalings between the variables $t, \bar{t}, \tau$ as introduced earlier in Callegari and Ting (1978), Hasimoto (1972), and Klein and Majda (1991a), that is,

$$\tau = \frac{1}{\delta^2}\,\bar{t} = \frac{1}{\delta^2}\left(\ln\frac{2\delta}{\epsilon} + C\right) t = \frac{1}{\delta^4}\,t \,, \tag{4.3.25}$$

we obtain the rotation rate, measured on the τ time scale, as

$$\omega = \delta^2\,\frac{2\left(1+\Gamma^*\right)}{b^2} \,. \tag{4.3.26}$$

Relative to the fixed frame of reference $\boldsymbol{t}^{(0)}, \boldsymbol{n}^{(0)}, \boldsymbol{b}^{(0)}$ with $\boldsymbol{t}^{(0)} \equiv \boldsymbol{t}_0$, we thus find

$$\begin{aligned} \boldsymbol{n}_0(\tau) &= \cos\left(\omega\tau\right)\boldsymbol{n}^{(0)} + \sin\left(\omega\tau\right)\boldsymbol{b}^{(0)} \,, \\ \boldsymbol{b}_0(\tau) &= -\sin\left(\omega\tau\right)\boldsymbol{n}^{(0)} + \cos\left(\omega\tau\right)\boldsymbol{b}^{(0)} \,, \end{aligned} \tag{4.3.27}$$

where $\omega = O(1)$ as $\delta \to 0$. We will see that this coordinate rotation will have to be accounted for carefully in the Hasimoto transformation, which we will again introduce to obtain a mathematically transparent form of the evolution equation.

4.3.1.3 The Hasimoto Transformation for a Vortex Pair. First we take into account that (4.1.33) and (4.1.38) still hold so that

$$\boldsymbol{N}_{\tilde{s}} = -\psi \boldsymbol{t} \tag{4.3.28}$$

and

$$\boldsymbol{N}_{\tilde{t}} = i \left[R\boldsymbol{N} - \left(\psi_{\tilde{s}} - i\tilde{\delta} \left(\boldsymbol{N} \cdot \boldsymbol{v}_{\tilde{s}} \right) \right) \boldsymbol{t} \right] . \tag{4.3.29}$$

As a consequence, we have

$$\boldsymbol{N}_{\tilde{\sigma}} = -\delta \psi \boldsymbol{t}^{(0)} = O\left(\delta\right) \tag{4.3.30}$$

and

$$\boldsymbol{N}_{\tau} = \delta \left(-i\psi_{\tilde{\sigma}} - \left(\boldsymbol{N} \cdot \boldsymbol{v}_{\tilde{\sigma}} \right) \right) + o(\delta) = O(\delta) . \tag{4.3.31}$$

Notice that $\boldsymbol{v}_{\tilde{\sigma}} = O(1)$, even though (4.3.15) implies $\boldsymbol{v} = O(1/\delta)$. This is a result of $\frac{2}{\delta b} \boldsymbol{b}_0(\tau)$ being independent of $\tilde{\sigma}$, and we find that the complex vector $\boldsymbol{N}$ is constant to leading order despite the rotation of the entire system, that is,

$$\boldsymbol{N} = \boldsymbol{N}^{(0)} + O(\delta) = \underbrace{\boldsymbol{n}^{(0)} + i\boldsymbol{b}^{(0)}}_{\text{const}} + O(\delta) . \tag{4.3.32}$$

For the Hasimoto transformation of the mutually induced velocities $\boldsymbol{v}_2^{\mathrm{I}}$ and $\boldsymbol{v}_2^{\mathrm{II}}$ from (4.3.15), following (4.1.46) we have to evaluate the expressions

$$-i \left(\boldsymbol{N}^{(0)} \cdot \boldsymbol{v}_{2,\tilde{\sigma}\tilde{\sigma}}^{\mathrm{I/II}} \right) . \tag{4.3.33}$$

Consider first $\boldsymbol{v}_2^{\mathrm{I}}$, that is, the velocity of Filament I with filament function ψ. The first term in (4.3.15) yields

$$-i\boldsymbol{N}^{(0)} \cdot \left(-\frac{2}{\delta b} \boldsymbol{b}_0(\tau) \right)_{\sigma\sigma} \equiv 0 . \tag{4.3.34}$$

If we denote the components of $\boldsymbol{X}_{\mathrm{I}}^{(2)}$ with respect to the fixed frame of reference by

$$\boldsymbol{X}_{\mathrm{I}}^{(2)} = \left(\alpha^{\mathrm{I}} \boldsymbol{n}^{(0)} + \beta^{\mathrm{I}} \boldsymbol{b}^{(0)} \right) \tag{4.3.35}$$

we have

$$\begin{aligned} &-i\boldsymbol{N}^{(0)} \cdot \left(-\frac{2}{b^2} \boldsymbol{A} \cdot \boldsymbol{X}_{\mathrm{I},\sigma\sigma}^{(2)} \right) \\ &= i\frac{2}{b^2} \left(\boldsymbol{n}^{(0)} + i\boldsymbol{b}^{(0)} \right) \left(\boldsymbol{n}_0 \circ \boldsymbol{b}_0 + \boldsymbol{b}_0 \circ \boldsymbol{n}_0 \right)(\tau) \cdot \left(\alpha_{\sigma\sigma}^{\mathrm{I}} \boldsymbol{n}^{(0)} + \beta_{\sigma\sigma}^{\mathrm{I}} \boldsymbol{b}^{(0)} \right) . \end{aligned} \tag{4.3.36}$$

With $\boldsymbol{n}_0(\tau)$ and $\boldsymbol{b}_0(\tau)$ from (4.3.27) we obtain after a few intermediate steps:

$$-i\boldsymbol{N}^{(0)} \cdot \left(-\frac{2}{b^2} \boldsymbol{A} \cdot \boldsymbol{X}_{\mathrm{I},\sigma\sigma}^{(2)} \right) = \frac{2i}{b^2} e^{i\omega\tau} \left(\alpha_{\sigma\sigma}^{\mathrm{I}} i e^{i\omega\tau} + \beta_{\sigma\sigma}^{\mathrm{I}} e^{i\omega\tau} \right) = -\frac{2i}{b^2} e^{2i\omega\tau} \overline{\psi} . \tag{4.3.37}$$

The next term analogously becomes

$$-i\boldsymbol{N}^{(0)} \cdot \mathrm{I}^b\left[\boldsymbol{X}^{(2)}_{\mathrm{II},\sigma\sigma}\right] \times \boldsymbol{t}_0 = -\boldsymbol{N}^{(0)} \cdot \mathrm{I}^b\left[\boldsymbol{X}^{(2)}_{\mathrm{II},\sigma\sigma}\right] = -\mathrm{I}^b\left[\phi\right] \,, \tag{4.3.38}$$

and finally,

$$-i\boldsymbol{N}^{(0)} \cdot \left(\boldsymbol{n}_0(\tau) \cdot M^b\left[\boldsymbol{X}^{(2)}_{\mathrm{II},\sigma\sigma}\right] \boldsymbol{b}_0(\tau)\right) = e^{i\omega\tau} M^b\left[Re\left(e^{+i\omega\tau}\overline{\phi}\right)\right] . \tag{4.3.39}$$

Collecting the results from (4.3.34)–(4.3.39), and taking into account that the self-induction terms remain unchanged, we obtain the following preliminary result for the motion of Filament I:

$$\begin{aligned}\frac{1}{i}\psi_\tau &= \psi_{\sigma\sigma} + \delta^2\left[\frac{1}{2}|\psi|^2\psi - \mathrm{I}[\psi]\right] \\ &\quad + \delta^2\Gamma^*\left[-\frac{2}{b^2}e^{2i\omega\tau}\overline{\psi} - \mathrm{I}^b[\phi] + e^{i\omega\tau}M^b\left[e^{+i\omega\tau}\overline{\phi}\right]\right] .\end{aligned} \tag{4.3.40}$$

Similarly we have for Filament II,

$$\begin{aligned}\frac{1}{i}\phi_\tau &= \Gamma^*\left[\phi_{\sigma\sigma} + \delta^2\left[\frac{1}{2}|\phi|^2\phi - \mathrm{I}[\phi]\right]\right] \\ &\quad + \delta^2\left[-\frac{2}{b^2}e^{2i\omega\tau}\overline{\phi} - \mathrm{I}^b[\psi] + e^{i\omega\tau}M^b\left[e^{i\omega\tau}\overline{\psi}\right]\right] .\end{aligned} \tag{4.3.41}$$

To arrive at (4.3.2), we recall that the filament functions ϕ, ψ, according to (4.1.60) and (4.1.62) obey

$$\phi,\, \psi = \boldsymbol{N}^{(0)} \cdot \left(\boldsymbol{X}^{(2)}_{\mathrm{II},\sigma\sigma}, \boldsymbol{X}^{(2)}_{\mathrm{I},\sigma\sigma}\right) + o(1) \quad \text{as} \quad \delta \to 0 \,. \tag{4.3.42}$$

In these definitions, we replace $\boldsymbol{N}^{(0)}$ with

$$\begin{aligned}\boldsymbol{N}_0(\tau) &= \left(\cos(\omega\tau)\,\boldsymbol{n}^{(0)} + \sin(\omega\tau)\,\boldsymbol{b}^{(0)}\right) \\ &\qquad + i\left(-\sin(\omega\tau)\,\boldsymbol{n}^{(0)} + \cos(\omega\tau)\,\boldsymbol{b}^{(0)}\right) \\ &= e^{-i\omega\tau}\left(\boldsymbol{n}^{(0)} + i\boldsymbol{b}^{(0)}\right) = e^{-i\omega\tau}\boldsymbol{N}^{(0)}\end{aligned} \tag{4.3.43}$$

and consider the transformed functions

$$\tilde{\psi} = \boldsymbol{N}_0(\tau) \cdot \boldsymbol{X}^{(2)}_{\mathrm{I},\sigma\sigma} = (\boldsymbol{n}_0(\tau) + i\boldsymbol{b}_0(\tau)) \cdot \boldsymbol{X}^{(2)}_{\mathrm{I},\sigma\sigma} = e^{-i\omega\tau}\boldsymbol{N}^{(0)} \cdot \boldsymbol{X}^{(2)}_{\mathrm{I},\sigma\sigma} = e^{-i\omega\tau}\psi \tag{4.3.44}$$

which are formulated for the rotating frame of reference $(\boldsymbol{t}_0, \boldsymbol{n}_0, \boldsymbol{b}_0)(\tau)$. Next we notice that

$$\frac{1}{i}\tilde{\psi}_\tau = \frac{1}{i}\left(e^{-i\omega\tau}\psi_\tau - i\omega\tilde{\psi}\right) \tag{4.3.45}$$

and multiply (4.3.40) with $e^{-i\omega\tau}$ to obtain

$$\frac{1}{i}\tilde{\psi}_\tau = \tilde{\psi}_{\sigma\sigma} + \delta^2\left[\frac{1}{2}|\tilde{\psi}|^2\tilde{\psi} - \mathrm{I}[\tilde{\psi}]\right]$$
$$+\delta^2\left[-\frac{\omega}{\delta^2}\tilde{\psi} - \frac{2}{b^2}\Gamma^*\overline{\tilde{\psi}} - \Gamma^*\left(\mathrm{I}^b\left[\tilde{\phi}\right] - M^b\left[Re\tilde{\phi}\right]\right)\right]. \tag{4.3.46}$$

Using (4.3.26) for the rotation frequency of the moving frame of reference, we perform an analogous calculation for the second filament and obtain (after dropping all "~"s) the equations in (4.3.2) except for the arclength corrections $-\delta^2\lambda^\psi\,\psi_{\sigma\sigma}, -\delta^2\lambda^\phi\,\phi_{\sigma\sigma}$ the derivation of which follows next.

4.3.1.4 Arclength Corrections. The Hasimoto transformation allowed us to elegantly represent the nonlinear effects of local self-induction, associated with the curvature-binormal term ($\kappa\boldsymbol{b}$), in the form of the classical cubic Schrödinger nonlinearity in the general filament equation from (4.3). Yet, this transformation relies on the axial centerline coordinate being an arclength variable. Considering a single vortex, it is always possible to scale this arclength variable to a period of 2π in the cases of a closed vortex or, as in this chapter, of a spacially periodic vortex. A period of 2π is useful within the pseudo-spectral numerical solution framework described earlier, which makes intense use of FFTs.

Here, we consider vortex pairs and allow for quite general initial data within the scaling regime from (4.1.1). The two vortices may be initiated with different arclengths over one spacial period. The following transformation, introduced first by Klein and Majda (1993), maintains period 2π for the centerline coordinates of both filaments, but leads to the arclength correction terms announced earlier,

$$\tilde{\sigma}_x = 2\pi\frac{\tilde{\sigma}_I}{\ell_I} = 2\pi\frac{\tilde{\sigma}_{II}}{\ell_{II}}. \tag{4.3.47}$$

The arclengths $\ell_{I/II}$ of both filaments along one spacial period may be represented asymptotically (for $\delta \to 0$) as

$$\ell_{I/II} = 2\pi\left(1 + \frac{1}{2}\delta^2\lambda_{I/II}\right), \tag{4.3.48}$$

where we have assumed that the original reference length had been chosen such that the period on the unperturbed filaments was scaled to period 2π. The deviations $\lambda_{I/II}$ from this reference period are

$$\lambda_{I/II} = \int_0^{2\pi}\left|\boldsymbol{X}^{(2)}_{I/II,\sigma}\right|^2 d\sigma\,. \tag{4.3.49}$$

Taking into account the relation between the displacements $\boldsymbol{X}^{(2)}_{I/II,\sigma\sigma}$ and the filament functions according to (4.1.60) and (4.1.62), we obtain a representation of $\lambda_{I/II}$ in terms of Fourier modes $\hat{\psi}^{I/II}(\xi,\tau)$ of the filament functions,

$$\lambda_{I/II} = \sum_{\xi} \left| \frac{1}{\xi} \hat{\psi}^{I/II}(\xi, \tau) \right|^2 . \tag{4.3.50}$$

Introducing the joint coordinate σ^* and neglecting terms of higher order than $O(\delta^2)$, we identify the leading second derivatives with the arclength corrections from (4.3.2) and (4.3.12), respectively.

For the (numerical) implementation, it is important to notice that the local stretching rates of the filament centerlines obey

$$\lim_{\delta\ell \to 0} \frac{1}{\delta\ell} \partial(\delta\ell)/\partial\tau(\tilde{\delta}_*, \tau) = -\delta^2 \kappa \boldsymbol{n} \cdot (\tilde{\delta}\boldsymbol{v}) , \tag{4.3.51}$$

where $\delta\ell$ is an infinitesimal line element on one of the filament centerlines, and $\boldsymbol{v}$ is the perturbation velocity in the binormal propagation law (4.1.28). Since $\tilde{\delta} = \delta^2$, the local stretching rate is of order $O(\delta^4)$ and one concludes that the arclength corrections $\lambda_{I/II}$ will be asymptotically constant up to (nondimensional) times of order $\tau = O(\frac{1}{\delta^2})$. As a consequence, we may replace the Fourier modes in (4.3.50) with their initial values $\hat{\psi}^{I/II}(\xi_j, 0)$, and one has

$$\lambda_{I/II} = \sum_{\xi_j} \left| \frac{1}{\xi_j} \hat{\psi}^{I/II}(\xi_j, 0) \right|^2 . \tag{4.3.52}$$

4.3.2 Linear Stability of Parallel Vortices of Opposite Strength and Crow's Theory

In this section, we recount the exact solutions for the linearized version of the vortex pair equations from (4.3.2) in the case of corotating vortices of equal strength ($\Gamma^* = -1$). The solutions are needed, on one hand, as a building block for an operator-split numerical scheme. On the other hand, we show in this section that the theory from Klein and Majda (1993) reproduces Crow's classical stability theory for a vortex pair (Crow, 1970a). An interesting aspect of the present derivations is the justification of Crow's ad hoc regularization parameter: We will derive an explicit relation between his "cutoff length," δ_{CROW}, and the present expansion parameter δ. As the latter can be related to the vortex core structure through the derivations earlier in this chapter (see (4.1.5)), we have thus a relation between physical properties of the vortex filaments and Crow's cutoff parameter.

Here, we consider the linearized equations from (4.3.2) for antiparallel vortices of equal absolute strength, to be obtained formally by considering the limit equations for small $|\psi|$ and $|\phi|$, that is,

$$\begin{aligned} \frac{1}{i}\psi_\tau &= \psi_{\sigma\sigma} + \delta^2 \left\{ -I[\psi] + \frac{2}{b^2}\overline{\psi} + I^b[\phi] - M^b[Re\phi] \right\} \\ -\frac{1}{i}\phi_\tau &= \phi_{\sigma\sigma} + \delta^2 \left\{ -I[\phi] + \frac{2}{b^2}\overline{\phi} + \left(I^b[\psi] - M^b[Re\psi] \right) \right\} . \end{aligned} \tag{4.3.53}$$

It turns out to be convenient to combine ψ and ϕ into a symmetric and antisymmetric part, that is,

$$\psi_{\mathrm{S}} = \frac{1}{2}(\psi + \phi) , \qquad \text{and} \qquad \psi_{\mathrm{A}} = \frac{1}{2}(\psi - \phi) , \tag{4.3.54}$$

as this leads to two decoupled equations

$$\frac{1}{i}\psi_{\mathrm{S/A}_\tau} = \psi_{\mathrm{S/A}_{\sigma\sigma}} + \delta^2 \left[\frac{2}{b^2}\overline{\psi_{\mathrm{S/A}}} - I[\psi_{\mathrm{S/A}}]\right] \pm \delta^2 \left[I^b\left[\psi_{\mathrm{S/A}}\right] - M^b\left[Re\psi_{\mathrm{S/A}}\right]\right] . \tag{4.3.55}$$

After Fourier transformation, these equations can again be solved analytically. As the complex conjugates $\overline{\psi_{\mathrm{S/A}}}$ appear, we obtain a coupled system of ordinary differential equations for the Fourier modes $\hat{\psi}_{\mathrm{S/A}}(\xi, \tau)$ and $\hat{\psi}_{\mathrm{S/A}}(-\xi, \tau)$, namely

$$\frac{\partial}{\partial \tau}\begin{pmatrix} \hat{\psi}(\xi,\tau) \\ \bar{\hat{\psi}}(-\xi,\tau) \end{pmatrix}_{S/A} = i\boldsymbol{A}_{S/A}(\xi)\begin{pmatrix} \hat{\psi}(\xi,\tau) \\ \bar{\hat{\psi}}(-\xi,\tau) \end{pmatrix}_{S/A} , \tag{4.3.56}$$

where

$$\boldsymbol{A}_{S/A}(\xi) = \begin{pmatrix} \alpha & \beta \\ -\alpha & -\beta \end{pmatrix}_{S/A} (\xi) \tag{4.3.57}$$

and

$$\begin{aligned} \alpha_{S/A}(\xi;\delta) &= -\xi^2 + \delta^2\left[-\hat{I}(\xi) \pm \frac{3}{2}b^2 \hat{I}_5^b(\xi)\right] \\ \beta_{S/A}(\xi;\delta) &= \ \delta^2\left[\frac{2}{b^2} \pm \left(\hat{I}_3^b(\xi) - \frac{3}{2}b^2\hat{I}_5^b(\xi)\right)\right] , \end{aligned} \tag{4.3.58}$$

and $\hat{I}, \hat{I}_3^b$ and $\hat{I}_5^b$ are the Fourier symbols of the nonlocal operators in (4.1.8) and (4.3.9). The exact solution to (4.3.56) and (4.3.57) reads

$$\begin{pmatrix} \hat{\psi}(\xi,\tau) \\ \bar{\hat{\psi}}(-\xi,\tau) \end{pmatrix}_{S/A} = (\cos(\tau p(\xi))I + i\sin(\tau p(\xi))P)\begin{pmatrix} \hat{\psi}(\xi,0) \\ \bar{\hat{\psi}}(-\xi,0) \end{pmatrix}_{S/A} , \tag{4.3.59}$$

where I is the 2×2 identity matrix,

$$P(\xi) = \frac{1}{p(\xi)}\begin{pmatrix} \alpha & \beta \\ -\alpha & -\beta \end{pmatrix} , \tag{4.3.60}$$

and

$$p(\xi) = \sqrt{\alpha^2 - \beta^2}(\xi) , \tag{4.3.61}$$

and where we have dropped the S/A subscripts on all α, β, p, etc. The expressions $\cos(z)$ and $\sin(z)$ have complex-valued arguments as before. These explicit solutions of the linear part of the filament equations for a vortex pair are used in an operator-split numerical solution scheme for the *nonlinear* pair equations as a building block.

From (4.3.61) we infer that a *linear instability* will occur, when

$$\beta^2_{S/A}(\xi;\delta) > \alpha^2_{S/A}(\xi;\delta)\,. \tag{4.3.62}$$

This stability result matches exactly with Crow's (1970b) result. To verify this claim, we introduce the definitions

$$\begin{aligned}
\beta &= b\xi\,,\\
\omega &= \frac{1}{2\delta^2}\left[1-\delta^2\left(\frac{1}{2}\ln(\xi^2)-c_0\right)\right],\\
\chi &= \beta K_1(\beta)\,,\\
\psi &= \beta^2 K_0(\beta)+\beta K_1(\beta)\,.
\end{aligned} \tag{4.3.63}$$

Now the quantities

$$G_{S/A}(\xi) = \frac{\delta^2}{b^2}\left[\mathrm{sgn}(-p^2)\sqrt{|p^2|}\right]_{S/A}(\xi) \tag{4.3.64}$$

with $p_{S/A}$ from (4.3.61) determine the growth of perturbations of a given wave number ξ. Inserting the above definitions, we obtain

$$G_{S/A}(\xi) = \frac{\delta^2}{b^2}\left[\left(1\mp\psi+\beta^2\omega\right)\left(1\pm\chi-\beta^2\omega\right)\right]. \tag{4.3.65}$$

This is in agreement with Crow's formula, Eq. (11) in Crow (1970b), except for a prefactor of $2\delta^2/b^2$, which accounts for a different nondimensionalization of time in his analysis. We emphasize a technical, but important, difference between the present analysis and Crow's approach. The result in (4.3.65) has been derived here on the basis of a systematic analysis of the Navier–Stokes equations. In particular, we have accounted for the influence of the vortex core structure on the filament dynamics. In contrast, Crow introduces, in line with other work in the field, an ad hoc regularization of the Biot–Savart integral. When evaluated for points close to the filament centerline, the Biot–Savart integral is singular with terms proportional to $1/r$ and $\ln(1/r)$, when r is the distance of the point of reference to the filament centerline. On the line itself, the Biot–Savart integral is undefined. Crow regularizes it by terminating the integration at some cutoff distance d to both sides from the reference point when it lies on the filament centerline. His results, in the form of (4.3.65), include the effect of regularization through his version of the parameter ω. A direct comparison with the present definition yields

$$\left(\frac{d}{b}\right) = \frac{1}{b}\exp\left(-\frac{1}{\delta^2}\right), \tag{4.3.66}$$

where b is the (nondimensional) unperturbed filament distance. Using the relation between δ, the vortex core size parameter, ϵ, and the core structure parameter C from (4.1.5), then we find

$$\delta_{\mathrm{CROW}} = \frac{d}{R} = \frac{1}{2}\epsilon \exp\left(-C\right) , \tag{4.3.67}$$

where R is a reference radius of curvature of the filament, which was used here originally for nondimensionalization. This result is an aposteriori justification of the ad hoc regularization of the line Biot–Savart integral. Figure 4.15 shows the growth parameters G_S (Fig. 4.15a) and G_A (Fig. 4.15b) as functions of wave number for values of $\delta^2 = 0.2, 0.25$, and for $b = 0.5$. This corresponds with $\delta_{\mathrm{CROW}} \approx 0.063$, which Crow used on the basis of a realistic estimate based on experimental data. As in previous sections, $G > 0$ signals instability with growth rate G, whereas negative G signals oscillatory behavior with $|G|$ the oscillation frequency. We find that antisymmetric perturbations are stable for all wavelengths whereas long-wave symmetric perturbations are unstable. We emphasize this here, prominently, to contrast it with the *nonlinear stability* results to be discussed shortly! These will reveal that the antisymmetric modes play a crucial role in triggering instabilities through mode interactions in Fourier space.

4.3.3 Numerical Solutions for an Antisymmetric Vortex Pair

Here we demonstrate, through numerical solutions of the full nonlinear filament pair equations from (4.3.2) for an antiparallel vortex pair ($\Gamma^* = -1$), how the linear stability discussed in the previous section is modified by the cubic curvature-binormal nonlinearities. First we observe that, as in the linear case, symmetric and antisymmetric perturbations of the filament pair are also possible solutions of the nonlinear equations. In a first example we demonstrate that long-wave symmetric disturbances will grow exponentially in an early stage as expected from linear stability analysis, but that nonlinear saturation will ultimately occur. The pair will then oscillate with finite amplitude for a while before nonlinear energy transfer in wave number space sets in and triggers the generation of high-frequency modes.

The antisymmetric modes are neutrally stable according to linear theory. A second example will show, however, that – in contrast – they have a strongly destabilizing influence in the nonlinear case. Superposition of a 10% antisymmetric perturbation of a given symmetric mode leads to acceleration of the nonlinear energy transfer in wave number space. The solution with antisymmetrically perturbed symmetric initial data experiences an energy transfer to the very high wave number band corresponding to the core size scale much earlier than a purely symmetric solution.

In a third example we impose helix initial data for one of the vortices, whereas the centerline of the second is set up with a plane sinusoidal displacement. The helix would be neutrally stable if isolated. Yet when interacting with the second vortex it undergoes a rapid amplification and the formation of small-scale structures. The latter signal that the solution is about to leave the regime of validity of the asymptotics.

4.3.3.1 Nonlinear Evolution of Symmetric Perturbations. For this first test case we initiate the filament functions for the vortex pair as

$$\psi^I(\sigma) = A\cos(\sigma)\exp\left(i\tfrac{\pi}{4}\right), \qquad A = 10^{-2}$$
$$\psi^{II}(\sigma) = \overline{\psi^I(\sigma)}. \tag{4.3.68}$$

The unperturbed filament distance b and the perturbation parameter δ are

$$b = \frac{1}{2} \qquad \delta^2 = 0.2. \tag{4.3.69}$$

According to linear stability analysis, as displayed in Fig. 4.15, this symmetrical perturbation is linearly unstable. Therefore, we may expect an exponential growth at first, and nonlinear saturation after sufficient growth of the perturbations. Figure 4.16 shows the time evolution of the curvature maximum, $\max_\sigma |\psi^I(\sigma,\tau)|$, during the initial stages of evolution. For comparison, we have added to the graph the prediction of linear stability analysis. Very good agreement is seen up until $\tau \sim 2.5$, which may serve as one contribution to validating the numerical method. As the filament function amplitudes grow beyond $\max|\psi^I| \geq 3.5$, a clear departure from linear theory is observed. The exponential growth is capped, and the amplitudes begin to oscillate.

The result of a long-term simulation for this same case is shown in Fig. 4.17. The linear instability of the symmetric mode obviously does not play a role anymore after the initial growth phase. Instead, the cubic nonlinearity enforces long-time oscillations. These ultimately produce a burst of curvature locally, associated with wavelengths on the core size scale, and hence with a departure of the solution from the regime of validity of the asymptotics.

Figure 4.18a shows the spatial structure close to the critical time of energy transfer to the high wave number range. The long-wave initial filament displacements have developed two intense peaks, and in the neighboring local curvature extrema we observe superimposed short wave perturbations that signal the excitation of high wave number Fourier modes. The associated Fourier spectrum is shown in Fig. 4.18b. The local peaks near wave numbers $\sim \pm 40 \ldots 50$ corroborate that the solution has left the regime of validity of the asymptotic equations. This is the case, because for $\delta^2 = 0.2$, wave numbers $\xi \sim 40 \ldots 60$ correspond to wavelengths comparable with the vortex core size.

4.3.3.2 The Role of Antisymmetric Modes in the Nonlinear Dynamics. Linear stability theory indicates that antisymmetric filament pair

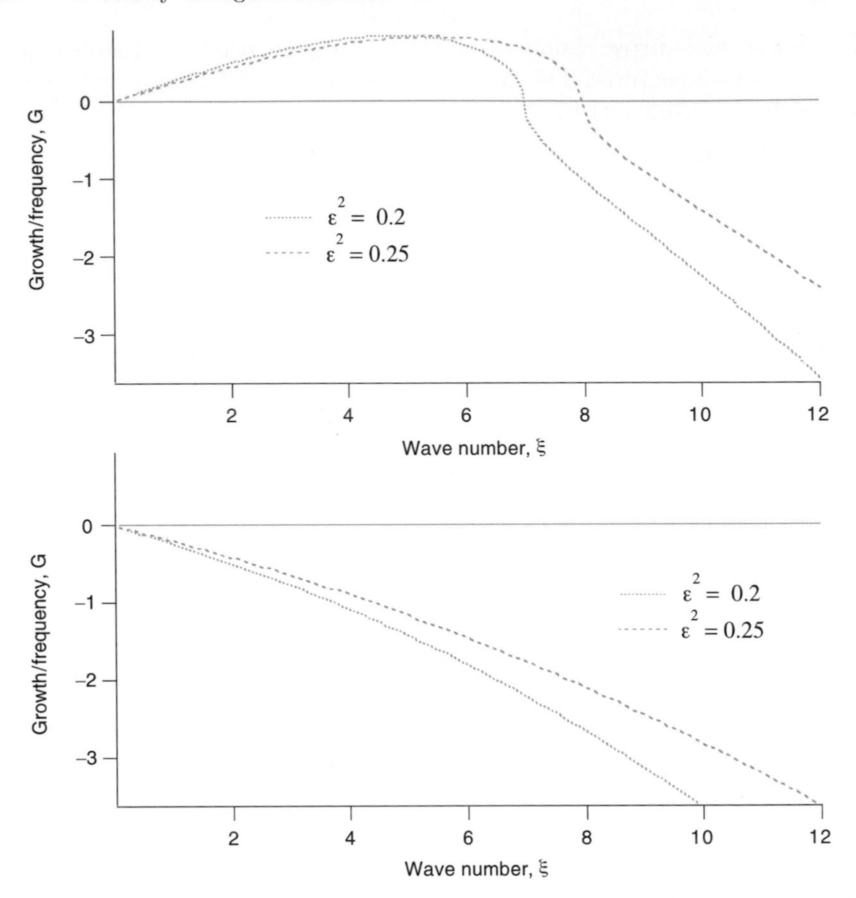

Fig. 4.15. Stability diagram for infinitesimal perturbations of a vortex pair: (**a**) symmetric and (**b**) antisymmetric perturbations. Adapted from Klein (1994)

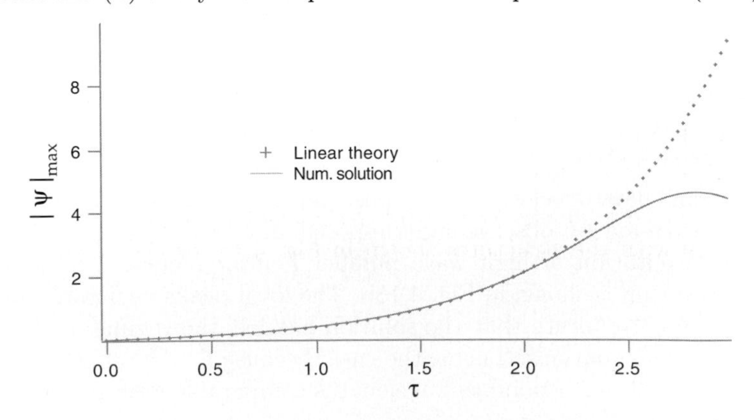

Fig. 4.16. Temporal evolution of the maximum curvature during the initial phase of a symmetric filament pair interaction with initial data from (4.3.68), and comparison with the prediction of linear stability analysis. Adapted from Klein and Majda (1993) and Klein (1994)

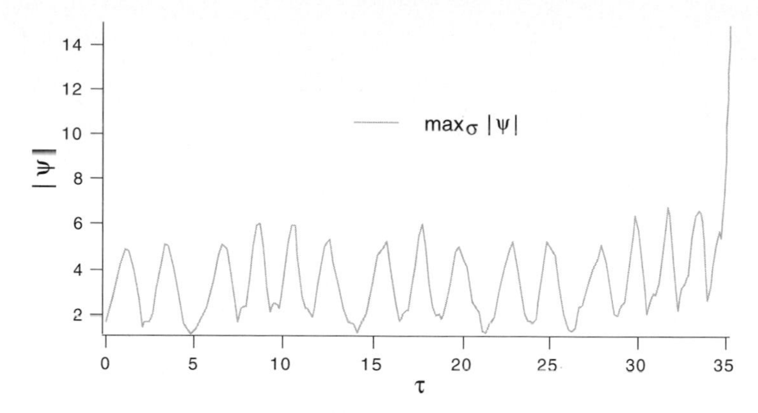

Fig. 4.17. Long-time oscillation of the maximum curvature for the symmetric filament pair from Fig. 4.16, up to the critical point of energy transfer to the core size scale. Adapted from Klein and Majda (1993) and Klein (1994)

displacements are neutrally stable. The following example demonstrates, however, that they do play a highly nontrivial *destabilizing* role when nonlinear effects are nonnegligible.

The example involves a symmetric initial displacement as in the last example, albeit with a larger initial amplitude, superimposed by an antisymmetric perturbation, the amplitude of which amounts to about 10% of that of the symmetric part. The filament functions are initiated with

$$\begin{aligned} \psi^I(\sigma) &= 1.6\cos(\sigma)\exp\left(i\tfrac{\pi}{4}\right), \qquad A = 10^{-2} \\ \psi^{II}(\sigma) &= 1.44\cos(\sigma)\exp\left(-i\tfrac{\pi}{4}\right). \end{aligned} \tag{4.3.70}$$

The temporal evolution for the curvature maximum is plotted in Fig. 4.19. The critical time of energy transfer is now reached at $\tau = 9.19$. This is less than one third of the time needed for this amplification in the case of a symmetric perturbation. The fact that we had started the symmetric mode with much smaller initial curvature is not crucial here, as the linear instability drives the symmetric mode to order $O(1)$ amplitudes very quickly as seen in Fig. 4.17

Comparison of the solution structures of Filaments I and II at time $\tau = 9.19$ in Fig. 4.20 with the structures shown in Fig. 4.18 for the symmetric case, reveals that the initial asymmetry is strongly amplified by nonlinear effects. The "critical" filament in this case is Filament II, which again exhibits two extreme and localized curvature peaks, but otherwise has an entirely different structure than that observed in the symmetric setting. The prominent neighboring local extrema seen in the symmetric case are still visible on Filament I, but they have not developed on the second "critical" filament. This example emphasizes that the linear stability of a vortex pair is of limited relevance for the actual evolution. Through nonlinear effects, antisymmetric

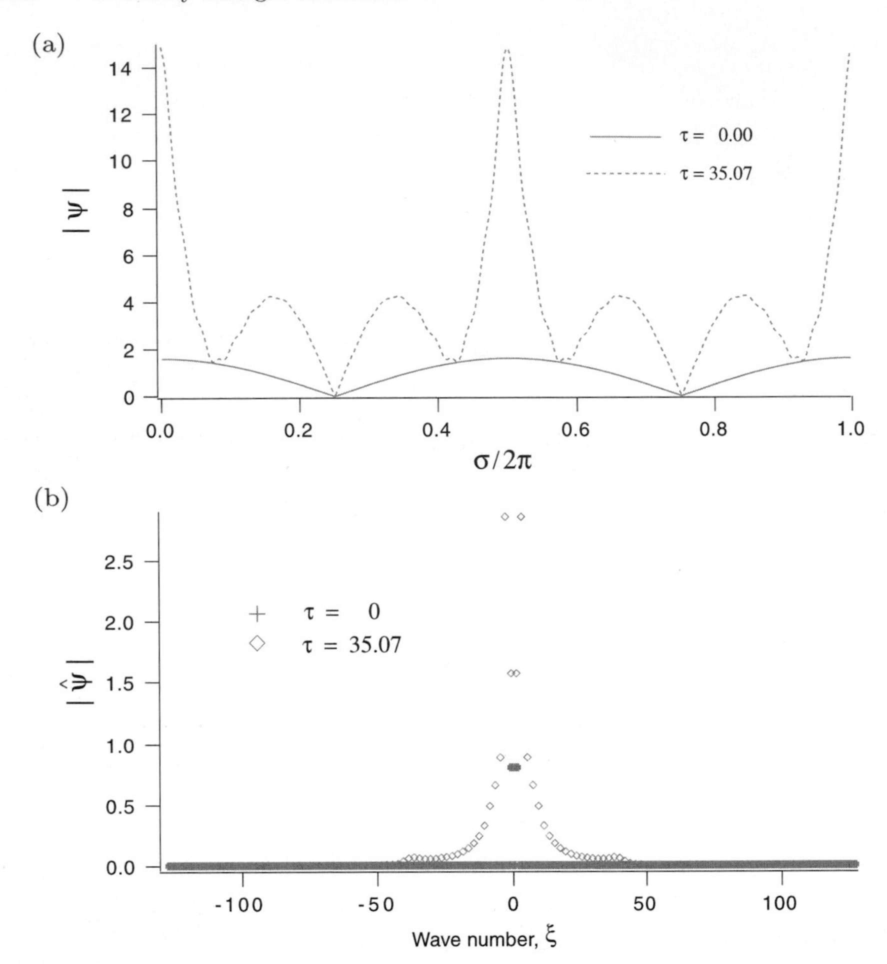

Fig. 4.18. (**a**) Spatial structure and (**b**) Fourier spectrum of the symmetric solution from Fig. 4.17 close to the critical time $\tau = 35.07$. The short wave oscillations seen superimposed on the subextrema in (a) correspond with the excitation of a band of higher wave numbers up to $|\xi| \sim 50$ in (b). Adapted from Klein and Majda (1993) and Klein (1994)

perturbations can trigger strong excursions to the limit of validity of the asymptotics, whereas they play the role of "tame" neutrally stable modes in the linear theory.

4.3.3.3 Triggering the Nonlinear Helix Instability by Vortex Pair Interactions. In the last example we consider entirely unrelated initial data for both filaments. Filament I is only slightly displaced from its unperturbed straight line position by two Fourier modes,

$$\psi^I(\sigma, 0) = 0.1\,(\exp(i\sigma) + \exp(-i4\sigma)) \;. \tag{4.3.71}$$

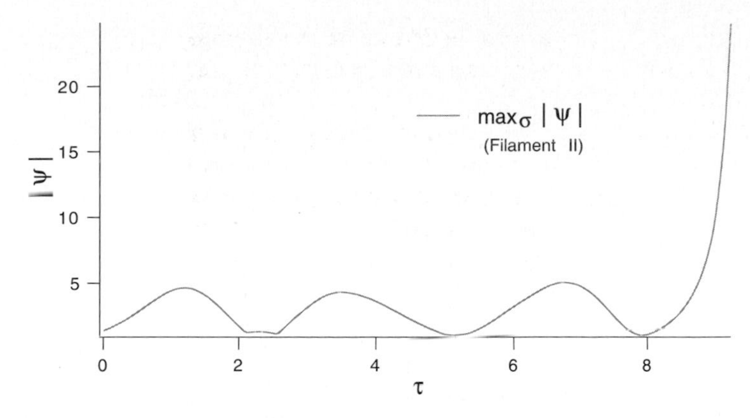

Fig. 4.19. Temporal evolution of the curvature maximum of Filament II under the antisymmetrically perturbed initial data from (4.3.70). The critical time of energy transfer to the core size scale is $\tau \sim 9.2$, and thus only a third of the time that is needed for a purely symmetric mode. Adapted from Klein (1994)

Filament II starts as a helix,

$$\psi^{II}(\sigma, 0) = 1.0 \exp(i\sigma) . \tag{4.3.72}$$

Filament II would be neutrally stable in a quiescent environment with respect to perturbations with wave numbers $\xi = 1, 2, 3, \ldots$, that is, with respect to those wave numbers that we can realize with our numerical method. Interactions with the other filament will therefore *not* induce a direct initiation of the first filament's linear instability, and large excursions of the solution must be due to nonlinear–nonlocal interactions. The spacial structures of ψ^I and ψ^{II} are shown for times $\tau = 0$ and $\tau = 5.4$ in Figs. 4.21a and 4.21b, respectively.

Filament I, which initially is only slightly displaced from the leading order straight line, exhibits the largest curvatures with $\max_\sigma |\psi^{II}| \approx 20.0$ at the critical time of departure from the regime of validity. The characteristic length of perturbations on Filament I is at the same time relatively large, and an analysis of the Fourier spectrum (not shown) reveals that there is no excitation of Fourier modes corresponding to the core size scale. In contrast, Filament II does develop oscillations on the core size scale, albeit with smaller absolute value than the extrema seen on Filament I. We conclude that the development of extreme local amplitudes of the filament function(s) does not necessarily coincide with very short wave perturbations.

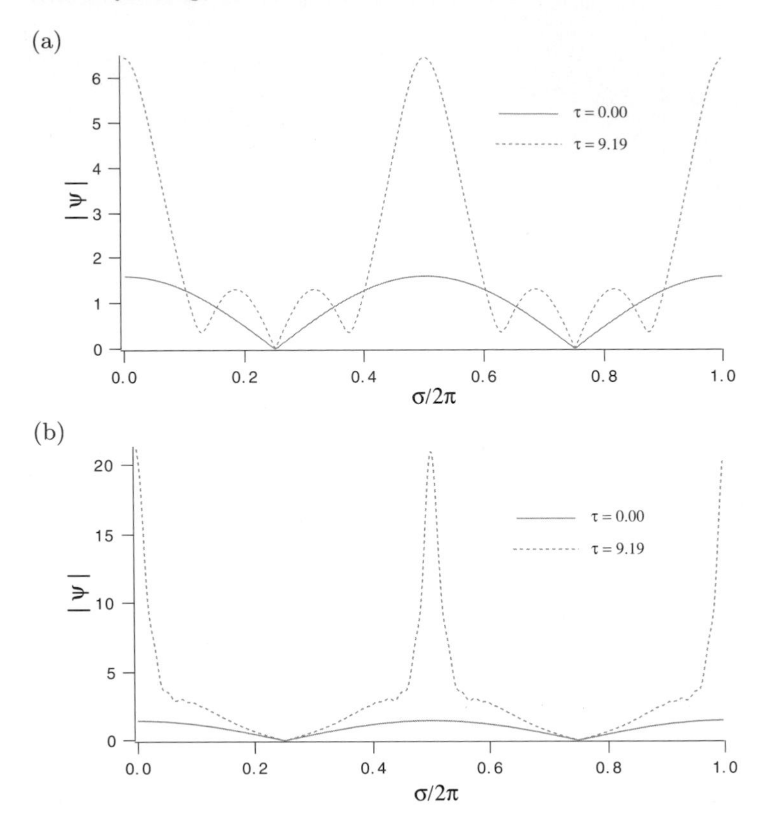

Fig. 4.20. Spacial structure of the solutions close to the critical time $\tau \sim 9.2$ for the antisymmetrically perturbed initial data from (4.3.70): (**a**) Filament I, (**b**) Filament II. Adapted from Klein and Majda (1993) and Klein (1994)

4.4 Large Amplitude and Long Wavelength Displacements of a Filament Pair

In the previous section we discussed the interaction of pairs of vortex filaments. In the course of these developments, we recovered the linear stability theories of Crow (1970a) and Jimenez (1975). These theories predict a long wavelength linear instability for perturbations on counterrotating vortex pairs. As we will see shortly, the characteristic wavelength of these unstable perturbations is systematically large of order $O(\delta^{-1})$ when measured in terms of the characteristic vortex separation distance. As a consequence, the associated eigenmodes violate the formal regime of validity of the weakly nonlinear asymptotic theory developed by Klein and Majda (1993) and summarized in the previous section.

To analyze the nonlinear, large amplitude evolution of these long-wave modes, Klein et al. (1995a) derived an asymptotic theory based on suitably

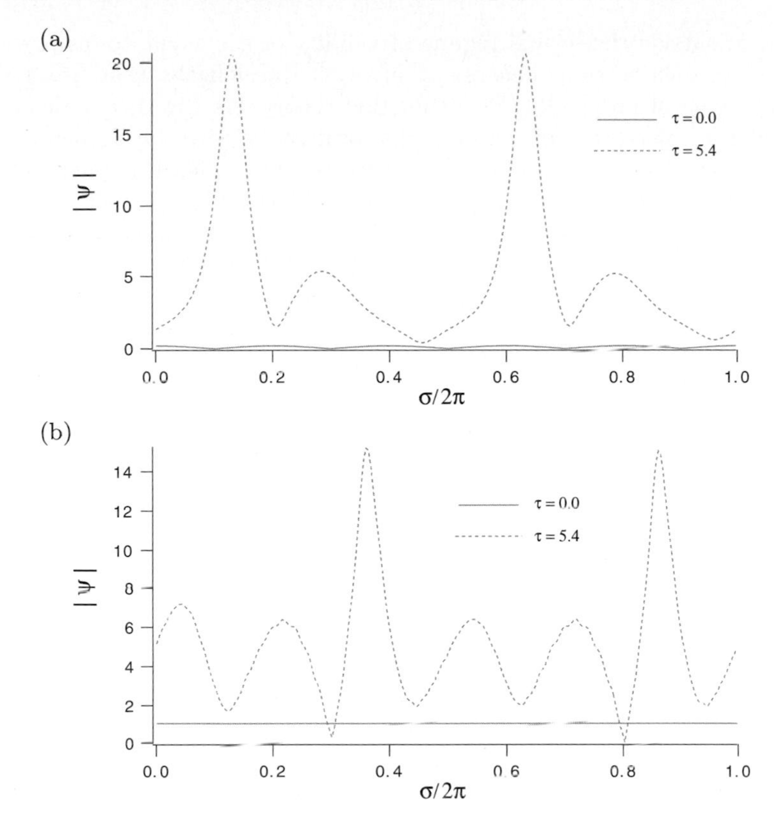

Fig. 4.21. Spacial structure of solutions for the uncorrelated initial data from (4.3.71) and (4.3.72): (**a**) Filament *I* develops large amplitudes of curvature, while (**b**) Filament *II* has already left the regime of validity of the asymptotic expansions through energy transfer to the core size scale. Adapted from Klein and Majda (1993) and Klein (1994)

adjusted asymptotic scalings of the geometrical filament perturbations. We summarize their developments briefly in this section (see also Majda, 1998).

4.4.1 Amplitude–Wavelength Scalings

A pair of antisymmetric vortex filaments (of equal strength but opposite rotation direction) is unstable with respect to perturbations for which $\beta^2 > \alpha^2$ according to (4.3.62). That is, with the definitions in (4.3.58),

$$\left|\delta^2\left[\frac{2}{b^2} \pm \left(\hat{I}_3^b(\xi) - \frac{3}{2}b^2\hat{I}_5^b(\xi)\right)\right]\right| > \left|-\xi^2 + \delta^2\left[-\hat{I}(\xi) \pm \frac{3}{2}b^2\hat{I}_5^b(\xi)\right]\right| . \quad (4.4.1)$$

There is one regime with $\xi = O(\delta)$ for which this condition can be met, and which corresponds with the long-wave instabilities discussed above. This regime is outside the formal regime of validity of the asymptotics presented in the previous section, because it involves wavelengths that are systematically large of order $O(1/\delta)$ within this theory. As the theory dealt with characteristic perturbation wavelengths comparable with the filament separation distance, b, we should now consider perturbations featuring perturbation wavelenths $\lambda \gg b$. An appropriate theory was derived by Klein et al. (1995a), who considered perturbation wavelengths of order b/δ, while allowing at the same time for filament displacements $|\boldsymbol{X} - \boldsymbol{X}^{(0)}| = O(b)$. Also, they generalized the theory from filament pairs to entire arrays of N nearly parallel vortex filaments with individual circulations of order $O(1)$.

The flow regime just mentioned is captured by the following asymptotic scalings of the filament centerlines:

$$\boldsymbol{X}_i^\delta(s,t) = s\boldsymbol{t}_0 + \delta^2 \boldsymbol{X}_i^{(2)}\left(\frac{s}{\delta}, \frac{t}{\delta^4}\right) . \tag{4.4.2}$$

Notice that all filaments are described relative to a common unperturbed straight line, $s\boldsymbol{t}_0$, so that their characteristic separation distance is comparable with their displacement amplitudes, and both are of order $O(\delta^2)$ as $\delta \to 0$.

The spatial scaling for the axial coordinate is chosen to match the preceding discussion for the scaling of the perturbation wavelengths relative to the filament separation distances. The rescaling of time by δ^{-4} results from $O(\delta^2)$ amplitudes of the filaments and mutually induced displacement velocities of order $O(\delta^{-2})$, which are due to both the filaments' potential vortex interactions and the leading curvature term from their self-induced velocities (see (4.4.3)).

4.4.2 Nonlinear Curvature – Potential Flow Interactions

The relevant equation of motion for the individual filaments has already been derived in Chap. 3. We found that the ith filament should move according to

$$\frac{\partial \boldsymbol{X}_i}{\partial t} = \left(\ln \frac{2\delta}{\epsilon} + C_i\right) \frac{\Gamma_i}{4\pi} (\kappa \boldsymbol{b})_i + \boldsymbol{Q}_i^f + \boldsymbol{Q}_i^{\text{outer}} , \tag{4.4.3}$$

where $\boldsymbol{Q}_i^f$ denotes the filament's nonlocal self-induction through the (linearized) finite part of the Biot–Savart integral as before, and $\boldsymbol{Q}_i^{\text{outer}}$ is the velocity induced at the location of the ith filament by the other vortices of the filament array.

According to (4.1.5), the factor multiplying the binormal term in (4.4.3) equals $1/\delta^2$, whereas the nonlocal self-induction term $\boldsymbol{Q}_i^f$ is merely of order $O(1)$. In the previous developments for a filament embedded in an outer straining field (Sect. 4.2) and for the weakly nonlinear Crow–Jimenez theories (Sect. 4.3), the externally imposed velocity component $\boldsymbol{Q}_i^{\text{outer}}$ was also of order $O(1)$ only as $\delta \to 0$.

Here, however, we have the following estimate for the velocity induced by all the neighboring filaments at the location of the ith one (Klein et al. 1995a)

$$\boldsymbol{Q}_i^{\text{outer}} = \sum_{j \neq i} \boldsymbol{Q}_{j/i}^{\text{outer}}, \tag{4.4.4}$$

with

$$\boldsymbol{Q}_{j/i}^{\text{outer}} = \frac{\Gamma_j}{2\pi r_{j/i}} \boldsymbol{\theta}_{j/i} + \frac{\Gamma_j}{4\pi} \left\{ \ln\left(\frac{1}{r_{j/i}}\right) (\kappa \boldsymbol{n})_{j/i} + (\kappa \cos\phi\, \boldsymbol{\theta})_{j/i} + \boldsymbol{Q}_{j/i}^f \right\}. \tag{4.4.5}$$

Here the subscript $(\cdot)_{j/i}$ denotes the location on filament j that has the shortest distance from the considered point $\boldsymbol{X}_i = s_i \boldsymbol{t}_0 + \delta^2 \boldsymbol{X}_i^{(2)}$ on Filament i, with

$$r_{j/i} = \delta^2 \left| \boldsymbol{X}_{j/i}^{(2)} - \boldsymbol{X}_i^{(2)} \right| \tag{4.4.6}$$

denoting this shortest distance. $\boldsymbol{n}_{j/i}$ is the principal normal at that location on Filament j, $\boldsymbol{\theta}_{j/i}$ is the circumferential unit vector for Filament j, erected at $\boldsymbol{X}_i$, and $\phi_{j/i}$ is the circumferential angle in the normal plane at $\boldsymbol{X}_{j/i}$ measured with respect to the principal normal $\boldsymbol{n}_{j/i}$. $\boldsymbol{Q}_{j/i}^f$ is the finite part of the Biot–Savart integral for Filament j evaluated at $\boldsymbol{X}_{j/i}$. Finally, $(\kappa \boldsymbol{b})_{j/i}$ is the curvature times the binormal unit vector of Filament j in $\boldsymbol{X}_{j/i}$. Figure 4.22 provides an illustration of these various quantities.

We are interested here in the leading-order contribution to the filaments' motion only. From (4.4.6) we conclude that the first (and largest) term in the foreign-induced velocity, (4.4.5), is of order $O(\delta^{-2})$. At the same time, the leading term of the self-induced velocity, the first two terms in (4.4.3), is of order $O\big(\ln(2\delta/\epsilon) + C_i\big)$, which according to (4.1.5) is also of order $O(\delta^{-2})$. Collecting only these leading terms, we have

$$\frac{\partial \boldsymbol{X}_i}{\partial t} = \frac{1}{4\pi\delta^2} \left\{ \Gamma_i \,(\kappa \boldsymbol{b})_i + \sum_{j \neq i} \frac{2\Gamma_j}{\left| \boldsymbol{X}_{j/i}^{(2)} - \boldsymbol{X}_i^{(2)} \right|} \boldsymbol{\theta}_{j/i} \right\}. \tag{4.4.7}$$

Next we observe that

$$\frac{\partial \boldsymbol{X}_i}{\partial t} = \frac{1}{\delta^2} \boldsymbol{X}_{i,\tau}^{(2)} \left(1 + O(\delta)\right), \tag{4.4.8}$$

$$(\kappa \boldsymbol{b})_i = (\boldsymbol{t} \times \boldsymbol{t}_{\tilde{s}})_i = \boldsymbol{t}_0 \times \boldsymbol{X}_{i,\sigma\sigma}^{(2)} \left(1 + O(\delta)\right), \tag{4.4.9}$$

and

$$\boldsymbol{\theta}_{j/i} = \boldsymbol{t}_0 \times \frac{\boldsymbol{X}_i^{(2)} - \boldsymbol{X}_{j/i}^{(2)}}{\left| \boldsymbol{X}_i^{(2)} - \boldsymbol{X}_{j/i}^{(2)} \right|} \left(1 + O(\delta)\right). \tag{4.4.10}$$

Collecting all these expressions, we obtain a closed and coupled system of equations for the displacements $\left\{\boldsymbol{X}_i^{(2)}\right\}_{i=1}^N$,

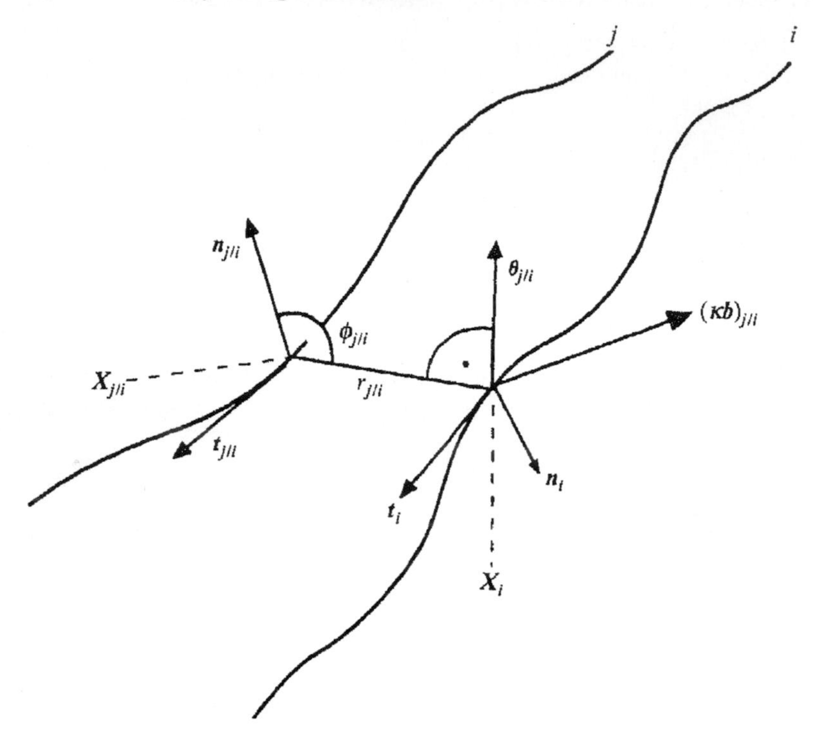

Fig. 4.22. Associated points on neighboring filaments and auxiliary quantities for the description of vortex–vortex interactions. Adapted from Klein et al. (1995a) and Klein (1994)

$$\boldsymbol{X}_{i,\tau}^{(2)} = \frac{1}{4\pi}\, \boldsymbol{t}_0 \times \left\{ \Gamma_i \boldsymbol{X}_{i,\sigma\sigma}^{(2)} + \sum_{j\neq i} 2\Gamma_j \frac{\boldsymbol{X}_i^{(2)} - \boldsymbol{X}_{j/i}^{(2)}}{\left|\boldsymbol{X}_i^{(2)} - \boldsymbol{X}_{j/i}^{(2)}\right|^2} \right\} . \tag{4.4.11}$$

This completes the derivation of the equations of motion for long-wave perturbations on an array of nearly parallel vortex filaments.

4.4.3 Mathematical Properties

4.4.3.1 Hamiltonian Structure. The simplified dynamics of a filament array, as derived in the previous section, has Hamiltonian structure (Klein et al., 1995a). This is revealed by first rewriting the cross product $\boldsymbol{t}_0 \times ...$ in (4.4.11) as

$$\boldsymbol{t}_0 \times \boldsymbol{X} = \boldsymbol{J}\boldsymbol{X} \tag{4.4.12}$$

where

$$\boldsymbol{J} = \begin{pmatrix} 0 & -1 \\ 1 & 0 \end{pmatrix} . \tag{4.4.13}$$

Then the evolution equation for $\boldsymbol{X}_j^{(2)}$ may be rewritten as

$$\Gamma_j \boldsymbol{X}_{j,\tau} = \boldsymbol{J} \left[\Gamma_j^2 \, \boldsymbol{X}_{j,\sigma\sigma} \right] + \boldsymbol{J} \left[\sum_{k \neq j} 2\Gamma_j \Gamma_k \, \frac{\boldsymbol{X}_j - \boldsymbol{X}_k}{|\boldsymbol{X}_j - \boldsymbol{X}_k|^2} \right] , \tag{4.4.14}$$

and this is equivalent with the Hamiltonian law

$$\Gamma_j \boldsymbol{X}_{j,\tau} = \boldsymbol{J} \frac{\delta \mathcal{H}}{\delta \boldsymbol{X}_j} . \tag{4.4.15}$$

Here, $\delta\mathcal{H}/\delta\boldsymbol{X}_j$ denotes the functional derivative of the Hamiltonian $\mathcal{H}$ with respect to the curve $\boldsymbol{X}_j$ compared through the L^2 inner product for curves (Logan, 1987), and

$$\mathcal{H} = -\sum_{j=1}^{N} \frac{1}{2} \Gamma_j^2 \int |\boldsymbol{X}_{j,\sigma}|^2 \, d\sigma + 2 \sum_{j<k}^{N} \Gamma_j \Gamma_k \, \ln\left(|\boldsymbol{X}_j - \boldsymbol{X}_k|\right) \, d\sigma . \tag{4.4.16}$$

The reader is referred to Klein et al. (1995a) for details of the derivation.

The Hamiltonian consists of two contributions, each of which represents one of the terms in the equation of motion from (4.4.11). The first is the Hamiltonian for the linearized binormal propagation law, whereas the second is the well-known Hamiltonian for an array of exactly *two-dimensional* vortices (Lamb, 1932).

Because $\boldsymbol{v}^t \boldsymbol{J} \boldsymbol{v} = 0$ for any two-dimensional vector $\boldsymbol{v}$, the Hamiltonian is conserved under the dynamics of a filament array.

4.4.3.2 Additional Conserved Quantities. One can also verify that the curvature–potential vortex interaction described by (4.4.11) induces the following additional exact conserved quantities. They are

Center of Vorticity (conserved)

$$\boldsymbol{M} = \int \sum_{j=1}^{N} \Gamma_j \, \boldsymbol{X}_j \, d\sigma , \tag{4.4.17}$$

Mean Angular Momentum (conserved)

$$A = \int \sum_{j=1}^{N} \Gamma_j \, |\boldsymbol{X}_j|^2 \, d\sigma , \tag{4.4.18}$$

Translational Invariant (conserved)

$$W = \int \sum_{j=1}^{N} \Gamma_j \, \boldsymbol{X}_{j,\sigma}^t \, (\boldsymbol{J} \boldsymbol{X}_j) \, d\sigma . \tag{4.4.19}$$

These quantities are conserved independent of any particular choice of the individual filaments' circulations. The (nonobvious) existence of the last quantity, W, was concluded in Klein et al. (1995a) from Noether's theorem and the fact that the Hamiltonian $\mathcal{H}$ is invariant under a translation $\sigma \to \sigma + h$ with h chosen arbitrarily.

There is, however, at least one more conservation law in the special case of an array of vortex filaments with equal strength and rotation direction, so that $\Gamma_j \equiv \Gamma$ for $j = 1, \ldots, N$. In this case, the

Circulation-weighted distance

$$I = \frac{1}{2} \int \sum_{j,k=1}^{N} \Gamma_j \Gamma_k \, |\boldsymbol{X}_j - \boldsymbol{X}_k|^2 \, d\sigma \tag{4.4.20}$$

reduces to the standard

Distance (conserved for $\Gamma_j \equiv \Gamma$, $j = 1, \ldots, N$)

$$I/\Gamma^2 = \frac{1}{2} \int \sum_{j,k=1}^{N} |\boldsymbol{X}_j - \boldsymbol{X}_k|^2 \, d\sigma \,, \tag{4.4.21}$$

and is an additional conserved quantity.

Lions and Majda (2000) explore the rich mathematical structure of these equations to derive a rigorous theory of equilibrium statistical mechanics for an array of nearly parallel vortices of equal strength (see also Majda, 1998).

4.4.4 Long-Wave Interactions of Filament Pairs

As mentioned earlier, we are interested particularly in the behavior of vortex pairs in the regime of long-wave instabilities as predicted both by the linear theories by Crow (1970a) and Jimenez (1975) and by the weakly nonlinear theory in Klein and Majda (1993).

Here, we first recount the linearized stability theory for long-wave perturbations that results from the curvature–potential vortex interactions described by (4.4.3), and we compare it with the long-wave limits of the earlier theories.

Second, we report on numerical simulations based on the full nonlinear system in (4.4.3), which allows for order $O(1)$ displacements of the filaments relative to their unperturbed separation distance. As a consequence, the theory incorporates the possibility of describing the onset of vortex reconnection, which in the present setup arises when two of the filaments touch somewhere along their axes.

Before doing so, however, it is useful to present a "streamlined version" of the dynamics in (4.4.3) valid for filament pairs. Let $\boldsymbol{X}_j = (x_j, y_j)$, and consider the complex representation

$$\psi_j(\sigma, \tau) = (x_j + i\, y_j)(\sigma, \tau)\,. \tag{4.4.22}$$

Let us assume further that, by nondimensionalization, $\Gamma_1 = 1$ and $\Gamma_2 = \Gamma \in [-1, 1]$. Then, the long-wave dynamics of the filament pair may by represented by

$$\begin{aligned} \frac{1}{i}\, \psi_{1,\tau} &= \psi_{1,\sigma\sigma} + 2\Gamma\, \frac{\psi_1 - \psi_2}{|\psi_1 - \psi_2|^2}\,, \\ \frac{1}{i}\, \psi_{2,\tau} &= \Gamma\, \psi_{1,\sigma\sigma} - 2\frac{\psi_1 - \psi_2}{|\psi_1 - \psi_2|^2}\,. \end{aligned} \tag{4.4.23}$$

It is also revealing to consider the dynamics in the symmetric/antisymmetric coordinates

$$\phi = \psi_1 + \psi_2\,, \qquad \psi = \psi_1 - \psi_2\,. \tag{4.4.24}$$

In terms of these variables, we find

$$\begin{aligned} \frac{1}{i}\, \phi_\tau &= \frac{1}{2}(1 + \Gamma)\, \phi_{\sigma\sigma} + \frac{1}{2}(1 - \Gamma) \left[\psi_{\sigma\sigma} - 4\frac{\psi}{|\psi|^2}\right], \\ \frac{1}{i}\, \psi_\tau &= \frac{1}{2}(1 - \Gamma)\, \psi_{\sigma\sigma} + \frac{1}{2}(1 + \Gamma) \left[\psi_{\sigma\sigma} + 4\frac{\psi}{|\psi|^2}\right]. \end{aligned} \tag{4.4.25}$$

This last set of equations makes it particularly obvious to distinguish the two limiting cases $\Gamma = \pm 1$, that is, the cases of filaments of equal strength and the same (opposite) rotation direction. These cases will be of particular interest below, where we compare the present nonlinear theory with Crow's and Jimenez' linear stability theories.

4.4.4.1 Linearization of the Long-Wave Dynamics Versus Long-Wave Limits of the Earlier Theories.

Corotating Filaments of Equal Strength. For this setting, Jimenez (1975) predicted stability in the long wavelength limit, and this is consistent with the Klein Majda weakly nonlinear theory described in Sect. 4.3 above. In the present setting, using the symmetric/antisymmetric mode amplitude representation from (4.4.25), we immediately corroborate this conclusion: For corotating filaments of equal strength we have $\Gamma = 1$ so that

$$\begin{aligned} \frac{1}{i}\, \phi_\tau &= \phi_{\sigma\sigma}\,, \\ \frac{1}{i}\, \psi_\tau &= \psi_{\sigma\sigma} + 4\frac{\psi}{|\psi|^2}\,. \end{aligned} \tag{4.4.26}$$

Obviously, the two modes decouple. The symmetric mode (ϕ) simply oscillates according to the linear Schrödinger equation, while the second mode undergoes nonlinear oscillations with controlled energy. This is seen by observing that

$$\frac{1}{i}\,\overline{\psi}_\tau = -\overline{\psi}_{\sigma\sigma} - 4\frac{\overline{\psi}}{|\psi|^2}\,, \tag{4.4.27}$$

and by deriving an evolution equation for $|\psi|^2$, namely,

$$\frac{1}{i}\,|\psi|^2_\tau = \overline{\psi}\psi_{\sigma\sigma} - \psi\overline{\psi}_{\sigma\sigma}\,. \tag{4.4.28}$$

Integrating by parts in σ both terms on the right-hand side, we find

$$\frac{d}{d\tau}\int |\psi|^2\, d\sigma \equiv 0\,. \tag{4.4.29}$$

Thus, in agreement with the earlier theories, we find long-wave stability for corotating filaments of equal strength.

Counterrotating Filaments of Equal Strength. As discussed earlier, counterrotating filaments of equal strength are long-wave unstable. The present theory, when specialized to $\Gamma = -1$, yields the coupled nonlinear evolution equations for the symmetric and antisymmetric modes,

$$\begin{aligned} \frac{1}{i}\,\phi_\tau &= \psi_{\sigma\sigma} - 4\frac{\psi}{|\psi|^2}\,, \\ \frac{1}{i}\,\psi_\tau &= \psi_{\sigma\sigma}\,. \end{aligned} \tag{4.4.30}$$

This is a system of equations very close to that proposed by Zakharov (1988) as a model for singularity formation on a vortex pair (see also Klein et al. 1995a). Zakharov included an additional (weak) logarithmic nonlinearity as a prefactor in front of the second derivatives, $\phi_{\sigma\sigma}, \psi_{\sigma\sigma}$, yet this is not found from the systematic asymptotic derivations summarized here.

Klein et al. (1995a) derive the detailed linear stability theory for the present regime with arbitrary values of the circulation ratio, $-1 \le \Gamma \le 1$. Here we are interested in negative circulation ratios, in particular, for $\Gamma = -1$. The following formula for the growth rate/oscillation frequency in analogy with the earlier representations in this chapter is obtained for general values of Γ,

$$G = \operatorname{sgn}(2\sqrt{P} - R)|2\sqrt{P} - R|^{1/2}\,, \tag{4.4.31}$$

where

$$\begin{aligned} R &= 8a^2(1+\xi^2) + (a^2+b^2)\,\xi^4 \\ P &= 16a^4 + 32a^2b^2\,\xi^2 + \left(20a^2b^2 + 4b^4\right)\,\xi^4 + 8a^2b^2\,\xi^6 + a^2b^2\,\xi^8\,, \end{aligned} \tag{4.4.32}$$

with $a = 1+\Gamma$ and $b = 1-\Gamma$. (Note, when comparing with Klein et al. (1995a), that we have set the unperturbed filament separation distance to $d = 1$ by our choice of a reference length in the nondimensionalization.)

Obviously, for the special case of counterrotating filaments with equal strength, that is, for $\Gamma = -1$, these formulae reduce to $R = 4\,\xi^4$ and $P = 64\,\xi^4$ so that

$$G = \operatorname{sgn}\left(4 - \xi^2\right) 2\xi \sqrt{|4 - \xi^2|} \qquad \text{for} \qquad \Gamma = -1\,. \tag{4.4.33}$$

Klein et al. (1995a) verify that that is also the long-wave asymptotic behavior for wavelengths of order $O(1/\delta)$ that can be extracted from Crow's linear stability analysis (Crow, 1970a).

4.4.4.2 Numerical Solutions and Nonlinear Collapse for a Counterrotating Vortex Pair. Here we summarize the numerical solutions and associated scaling analysis from Klein et al. (1995a), which demonstrate that the long-wave linearly unstable modes described in the previous section tend to evolve toward a characteristic pattern of collapse of a counterrotating filament pair. We focus here on the case of filaments of equal strength, that is, on $\Gamma = -1$.

Numerical Solution Technique. The numerical solution technique used by Klein et al. (1995a) takes advantage of the particular mathematical structure of the long-wave interaction equations from (4.4.23) in analogy with the techniques used in Sects. 4.1, 4.2, and 4.3. An operator splitting scheme is devised on the basis of a linear/nonlinear decomposition of the equations into

$$\begin{aligned} \frac{1}{i}\,\psi_{1,\tau} &= \psi_{1,\sigma\sigma}\,, \\ \frac{1}{i}\,\psi_{2,\tau} &= \Gamma\,\psi_{2,\sigma\sigma}\,, \end{aligned} \tag{4.4.34}$$

and

$$\begin{aligned} \frac{1}{i}\,\psi_{1,\tau} &= 2\Gamma\,\frac{\psi_1 - \psi_2}{|\psi_1 - \psi_2|^2}\,, \\ \frac{1}{i}\,\psi_{2,\tau} &= -2\frac{\psi_1 - \psi_2}{|\psi_1 - \psi_2|^2}\,. \end{aligned} \tag{4.4.35}$$

It turns out that these equations again have explicitly computable exact solutions, which are coupled as usual via Strang splitting (Strang, 1968). This method automatically conserves, up to machine accuracy, the mean center of vorticity, $\boldsymbol{M}$, from (4.4.17), the total angular momentum, A, from (4.4.18), and the distance functional I from (4.4.20) for $\Gamma = -1$, because they are exactly conserved in each substep. Conservation of the Hamiltonian in (4.4.16) up to a specified accuracy is used to control the time step.

Collapse of a Counterrotating Filament Pair. Figure 4.23 displays the time evolution of a counterrotating vortex pair of equal strength ($\Gamma = -1$) from initial data

$$\begin{aligned}\psi_1(\sigma,0) &= -\frac{1}{2} + A\left(-\frac{1}{2} + \frac{1}{2\sqrt{3}}i\right)\left(e^{i\sigma} - e^{-i\sigma}\right),\\ \psi_2(\sigma,0) &= \frac{1}{2} + A\left(-\frac{1}{2} - \frac{1}{2\sqrt{3}}i\right)\left(e^{i\sigma} - e^{-i\sigma}\right),\end{aligned} \tag{4.4.36}$$

with the amplitude A adjusted such that the initial minimal filament separation distance is $d_{\text{min}} = 0.4$. Clearly, there is a rapid approach of the two filaments in the left half of the period, and a rapid stretching in the vertical direction as the local filament distance decreases. The physics of the process is quite transparent: the mutual potential vortex induction in this case drives the two vortices in the vertical, and this effect is stronger the closer the filaments get. As this stretching is most intense near the tip of the bends on the vortex centerlines, their curvatures are systematically enhanced in the process. At the same time, the local binormal self-induction, which is proportional to the local curvature, drives the two filaments toward each other, thereby enhancing in turn the stretching due to the potential vortex interaction.

The compounding of these two mechanisms ultimately leads to a collapse of the vortices, and thus to the breakdown of the present asymptotic approximation. The asymptotics must assume well-separated filament centerlines whose separation distance is large compared with the vortex core size, because it presupposes nearly axisymmetric vortex cores.

There is a rich series of tests and simulations in Damodaran (1994) and Klein et al. (1995a) demonstrating that this mechanism for singularity formation is active whenever $\Gamma < 0$, that is, whenever the filaments are counterrotating. Also, there is a strong indication that the collapse may locally be characterized by a self-similar singular structure, and this will be discussed briefly in the next subsection.

The positive feedback between the mutual potential vortex induction and the local binormal self-induction becomes inactive for $\Gamma > 0$. It is consistent with this fact that no collapse was ever observed for corotating filaments in any of the simulations recorded in the cited publications. There might be self-similar singular solutions as discussed below even in that case, but it is argued in Klein et al. (1995a) that these solutions will likely be dynamically unstable, in which case they would not materialize in practice.

4.4.4.3 Self-Similar Structure of the Nonlinear Collapse for a Counterrotating Vortex Pair. Zakharov (1988) identified the possibility of self-similar collapsing solutions for $\tau \nearrow \tau_0$ with a time-dependent rescaling of the spacial structure and solution amplitudes according to $\sigma = \sqrt{\tau_0 - \tau}\,\eta$, and $\psi_i(\sigma,\tau) \sim \sqrt{\tau_0 - \tau}\,\tilde{\psi}_i(\tau_0 - \tau, \eta)$, respectively. Here, we verify that this kind of scaling actually leads to a consistent description of locally collapsing solutions for the system in (4.4.23).

Following Klein et al. (1995a), we insert the general ansatz for similarity transformations,

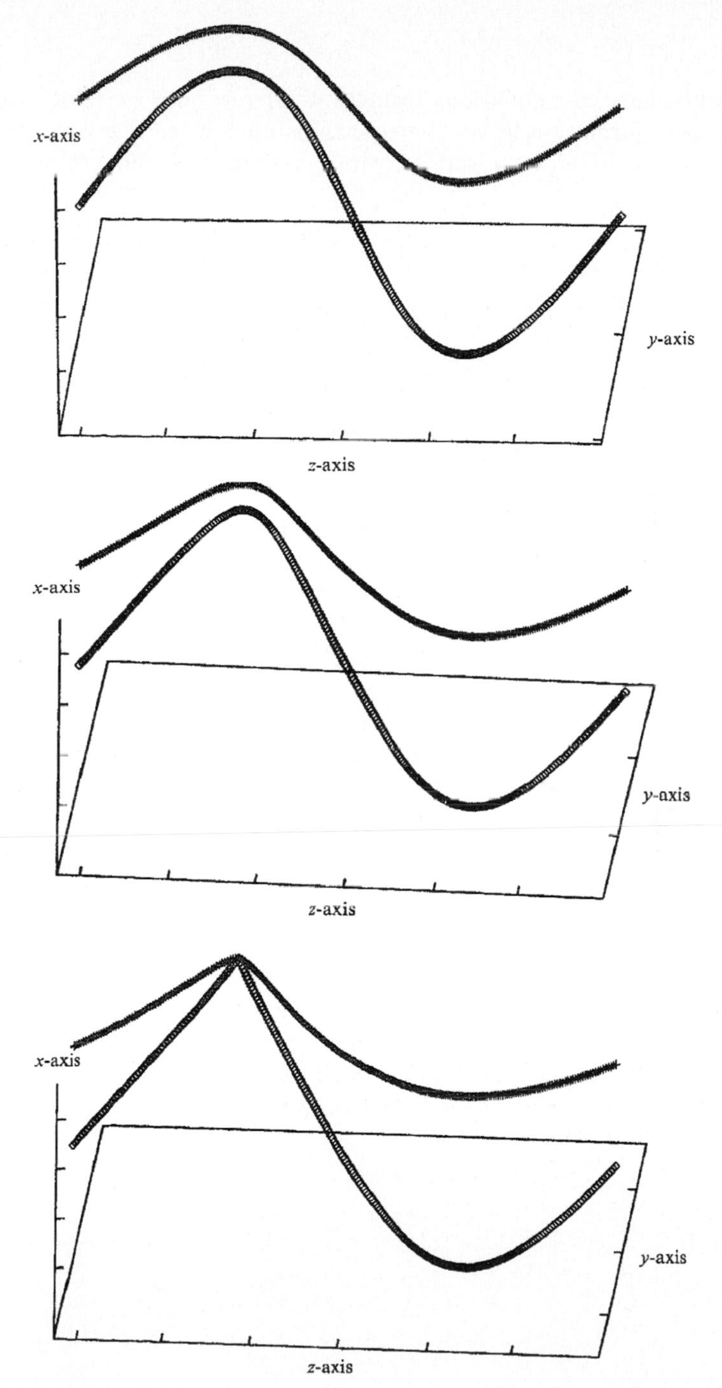

Fig. 4.23. Snapshots of the filament geometries during the long-wave collapse of a vortex pair with initial data from (4.4.36) at times $\tau = 0, 0.08, 0.1146$. Adapted from Klein et al. (1995)

$$\tau' = \lambda\tau\,, \quad \sigma' = \lambda^a\sigma\,, \qquad \psi' = \lambda^c\psi\,, \quad \phi' = \lambda^d\phi \tag{4.4.37}$$

into the filament pair equations from (4.4.25) (see also (4.4.30)), and find that this transformation leaves the equations unchanged if $a = c = d = 1/2$. This induces the following ansatz self-similar solutions in time (Logan, 1987),

$$\begin{pmatrix}\psi\\ \phi\end{pmatrix}(\sigma,\tau) = \sqrt{\tau_0 - \tau}\begin{pmatrix}f\\ g\end{pmatrix}(\eta) \qquad \text{where} \qquad \eta = \frac{\sigma}{\sqrt{\tau_0 - \tau}}\,, \tag{4.4.38}$$

and the associated equations for their spatial structure,

$$\begin{aligned}
\frac{1}{i}\,(\eta f' - f) &= (1-\varGamma)g" + (1+\varGamma)\left(f'' + \frac{4f}{|f|^2}\right)\,,\\
\frac{1}{i}\,(\eta g' - g) &= (1+\varGamma)g" + (1-\varGamma)\left(f'' - \frac{4f}{|f|^2}\right)\,.
\end{aligned} \tag{4.4.39}$$

These equations have not been studied thus far. However, the numerical results to be presented shortly indicate that, together with suitable far-field conditions as $\eta \to \pm\infty$, they will describe the solution structure near the location of collapse of counterrotating filaments quite accurately.

For the solutions displayed in Fig. 4.23 (first case) and for a counterrotating pair with $\varGamma = -1/2$ and symmetric helix initial data for both filaments (second case), Klein et al. (1995a) monitored the minimal distance between the filaments as a function of time. In a double-logarithmic plot the result is reproduced here in Fig. 4.24. The graph shows quite convincing agreement with the square-root scaling law for the first case (see Fig. 4.23), and good agreement over a considerable intermediate time horizon in the second case.

It is not quite understood as yet whether the deviation from the self-similar behavior seen in the late stages of the second example (Fig. 4.24b) is due to a truly non-self-similar solution structure or due to underresolved numerical calculations. Notice in this context that the filament separation distance has decreased by several orders of magnitude by the time the deviation occurs. The solution structure near the point of collapse will by then also have developed extremely small scales that would require grid-adaptive numerical computations to be resolved properly.

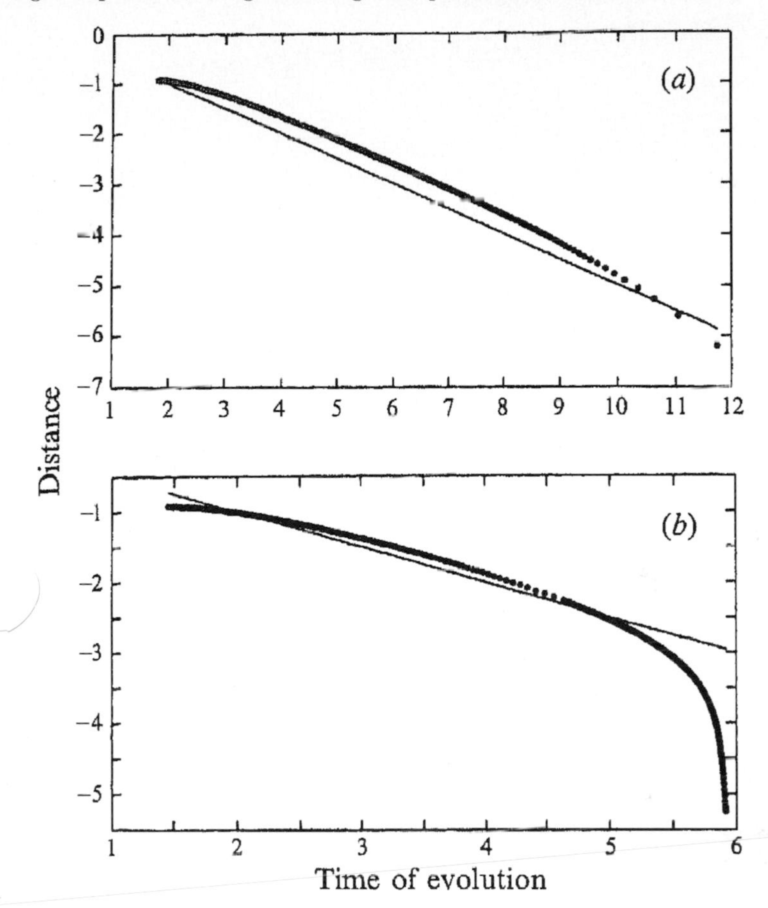

Fig. 4.24. Double-logarithmic plot of the time history of the minimal filament distance for (**a**) the same solution as displayed in Fig. 4.23 and for (**b**) another case with $\Gamma = -1/2$ and symmetric helix initial data. Adapted from Klein et al. (1995)

5. Numerical Simulation of Slender Vortex Filaments

Motivation

In Chap. 3, an asymptotic theory was developed for the motion of slender vortex filaments at high Reynolds number. The theory is based on the assumptions that the characteristic radius of curvature of the filament centerline is much larger than the core size, and that the vortex core remains well-separated from solid boundaries. Under these assumptions, the asymptotic theory provides (1) a leading-order prediction of the filament centerline velocity and (2) evolution equations for the leading-order vorticity and axial velocity structures. The centerline velocity is given as the sum of a nonlocal velocity expressed as a convolution over the centerline with nonsingular kernel, and a local contribution that is a function of the local curvature, and of the instantaneous core structure. Meanwhile, the local core structure evolves according to stretching of the filament centerline and vortex diffusion phenomena. Thus, the filament motion is generally closely coupled with the evolution of its vortex core.

The present chapter is motivated by two main topics. The first concerns the impact of the core structure variation on the large-scale filament motion, and the practical region of validity of the asymptotic theory. This topic is addressed in the first section of this chapter. The second topic concerns the adaptation of thin-tube vortex element models to the simulation of slender filaments. Thin-tube models are based on discretization of the filament using a chain of vortex elements having finite, overlapping cores, and on tracking these elements in a Lagrangian reference frame. Application of these models has in many cases been based on ad hoc approaches, such as (1) identification of the leading-order vorticity structure inherited from the numerical regularization with the physical leading-order vorticity structure and (2) simplified descriptions for the evolution of the vortex structure. These approaches have been problematic in many respects. Fundamentally, ad hoc approaches do not enjoy the well-established theoretical foundation of the asymptotic theory, nor of the underlying smooth vortex methods that are based on detailed representation of the vorticity field. Practically, these approaches have also been known to suffer from essential deficiencies, such as their inability to predict the correct propagation velocity of a circular ring in the inviscid limit.

The origin of such deficiencies, and potential means of overcoming them, are the focus of the remainder of the chapter.

Chapter Summary

We address the first topic in Sect. 5.1 based on numerical simulations of slender vortices under different configurations. In Subsect. 5.1.1, we rely on axisymmetric computations to demonstrate the impacts of curvature (or slenderness ratio) and Reynolds number. The simulations by Wang (1970) of the axisymmetric passage of a vortex ring over a sphere, and by Gunzburger (1972) of the interaction of two coaxial vortex rings are used for this purpose. The three-dimensional computations of Liu et al. (1986b) of interacting slender filaments are then used in Subsect. 5.1.2 to highlight the effect of axial flow along the filament centerline, and to illustrate the practical region of validity of the asymptotic theory, particularly as filaments tend to merge or intersect at large simulation times.

We then turn our attention in Sect. 5.2 to the application of thin tube models to slender vortex flows. We start by analyzing the origin of the deficiency of thin-tube models. Specifically, we quote the analysis of Klein and Knio (1995), who identify the origin of the deficiency of ad hoc thin-tube models by applying the asymptotic vorticity-based analysis (Subsect. 3.3.4) to the *numerical* vorticity distribution. A spurious first order vorticity is identified in the numerical vorticity distribution, which results in an inherent O(1) error in the leading-order filament velocity. Having quantified the error term, the analysis in Klein and Knio (1995) provides different means for removing this error. Thus, corrected thin-tube models are constructed that yield predictions that are consistent with the asymptotic theory. Favorable comparison between the predictions of corrected thin-tube models and those of the Klein–Majda theory is then summarized, which demonstrates the validity of both approaches.

Attention is then focused in Subsect. 5.2.3 on generalization of the corrected thin-tube model to account for the evolution of the core structure. The generalization is based on incorporation of viscous solutions of the core evolution equations (Chap. 3) into the computations. This results in a generalized thin-tube formulation that properly accounts for stretching and diffusion effects on the axial vorticity and axial flow structures, which also enables us to identify the limitations of ad hoc core structure models. In Subsect. 5.2.4, the computational performance of thin-tube models is examined. We focus specifically on limitations arising from the requirement of overlapping vortex cores, which results in a strongly overresolved representation of the filament geometry, and consequently to stiff computations. We discuss how the asymptotic theory can be used to guide the development of various means to overcome these limitations. Of particular interest among the methods outlined in Subsect. 5.2.4 are those that do not rely on explicit evaluation of the filament curvature, as these exhibit better ability in defeating both the spatial and temporal stiffness of the equations.

5.1 Validity of the Slender Vortex Asymptotics and Applications

In Subsect. 5.1.1 we present numerical examples simulating the interactions of coaxial vortex rings in axisymmetric flows obtained by Wang (1970) and Gunzburger (1972). We use these examples to demonstrate the importance of the core structure on the motion of the rings. In Subsect. 5.1.2 we quote several numerical examples simulating the interaction of slender toroidal vortex filaments obtained by Liu et al. (1986b). These examples are chosen to show the complexity of fully three-dimensional problems. They show the motion of the filaments leading to merging or intersection and are employed to identify different types of merging problems. The examples also serve the purpose of establishing the practical region of validity of the asymptotic solutions.

5.1.1 Coaxial Vortex Rings in an Axisymmetric Flow

We use the axisymmetric examples to show the essential difference between the motion of a filament with curvature and a rectilinear filament or a vortex in two dimensions. We point out once more that in two-dimensional problems the velocity of a vortex center with a core size much smaller than the distances to other vortices is defined by the local background velocity independent of its core structure. In contrast, the leading-order velocity of a slender vortex filament does depend on its core structure. Furthermore, it is dominated by the logarithm of its effective core size. The core structure changes in time due to viscous diffusion and the stretching of the filament, the latter being an inviscid effect. To demonstrate the relative effect of viscous diffusion and inviscid stretching, we quote the numerical investigations of the motion of a slender vortex ring in an axisymmetric flow around a rigid sphere by Wang (1970).

5.1.1.1 Passage of a Vortex Ring Over a Sphere. We consider axisymmetric flows with zero circumferential velocity. As shown in Fig. 5.1 (left), a sphere with radius a is centered at the origin. The upstream velocity $U\hat{k}$ is uniform and parallel to the z-axis, which is also the axis of the ring. Here $\sigma = \sqrt{x^2 + y^2}$ denotes the radial coordinate. We consider the case that the initial radius R_0 of the ring at an upstream position is not too small relative to the radius of the sphere. Thus, the ring can pass over the sphere while its distance to the sphere remains much larger than the core size and the boundary layer thickness along the sphere. The potential flow away from the ring is composed of three parts, a uniform flow around the sphere, the flow induced by the isolated ring with zero core size, and the flow induced by the image of the ring inside the sphere. The last part cancels the normal velocity on the sphere induced by the ring (Lamb, 1932).

The image of a vortex ring with centerline radius R lying in the plane $z = Z$ is a coaxial ring inside the sphere. The strength Γ', radius R', and the plane $z = Z'$ of the image are defined by

$$R' = \frac{a^2 R}{R^2 + Z^2}\,, \quad Z' = \frac{a^2 Z}{R^2 + Z^2}\,, \quad \text{and} \quad \Gamma' = -\Gamma\sqrt{\frac{R}{R'}} = -\Gamma\frac{\sqrt{R^2+Z^2}}{a}\,. \tag{5.1.1}$$

The stream function of the background flow, that is, the uniform flow around the sphere plus the flow induced by the mirror image is

$$\Psi(t,\sigma,z) = \frac{1}{2}U\sigma^2\left[1 - \frac{a^3}{r^3}\right] - \frac{\Gamma'}{2\pi}(r'_+ + r'_-)[F(\lambda') - E(\lambda')]\,, \tag{5.1.2}$$

where $r'_\pm = \sqrt{(\sigma \mp R')^2 + (z \mp Z')^2}$, $\lambda' = (r_- - r'_+)/(r'_- + r'_+)$, and F and E are the complete elliptic integrals of the first and second kind, respectively.

The vortex ring is assumed to have a similar core structure with initial core size δ_0. Because of symmetry, the self-induced velocity of the ring given by the asymptotic theory has only a binormal component, pointing in the z-direction. Evaluating the finite part of the Biot–Savart integral for the circular centerline, we obtain

$$\dot{Z}_1(t) = \frac{\Gamma}{4\pi R}\left[\ln\frac{8R}{\delta} - 0.558\right] \tag{5.1.3}$$

where

$$\delta^2(t) = \frac{4\nu\tau_1(t)}{S(t)} = \frac{4\nu\int_0^t R(t')\,dt'}{R(t)} + \delta_0^2\,. \tag{5.1.4}$$

The velocity of the ring submerged in the uniform stream over a sphere is equal to the local velocity of the background flow (5.1.2) plus the self-induced velocity (5.1.3). The result is

$$\hat{\sigma}\dot{R} + \hat{k}\dot{Z} = \frac{1}{\sigma}\left(\hat{\sigma}\Psi_z(t,R,Z) - \hat{k}\Psi_\sigma(t,R,Z)\right) + \hat{k}\dot{Z}_1(t)\,. \tag{5.1.5}$$

In the numerical examples quoted here, we have $\Gamma = Ua/2$ and $\mathrm{Re} \equiv \Gamma/\nu = 3.125\times10^4$. At $t = 0$, the centerline of the ring lies in the upstream station $z = Z(0) = -10a$ and its radius is $R(0) = a/2$.

The variation of the core size, δ versus t, is shown in Fig. 5.1(right) for two different initial core sizes, $\delta_0/a = 0.01$ and $\delta_0/a = 0.001$. In the inviscid theory, the core size decreases to compensate for the stretching of the ring as it passes over the sphere and returns to its original size afterward. In the viscous theory, the core size tends to grow all the time due to viscous decay and hence retards the motion. Only when the ring is passing over the sphere, does the inviscid stretching effect overcome the viscous effects.

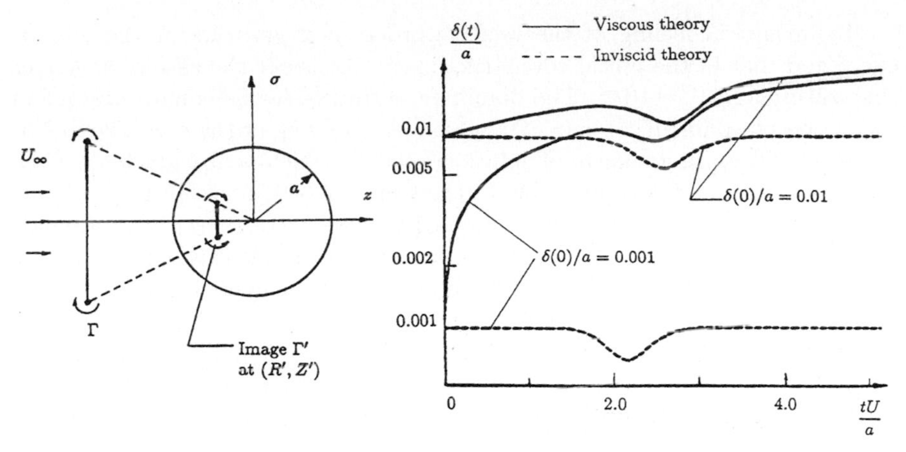

Fig. 5.1. *Left*: A vortex ring moving toward a sphere and the "image" of the ring; *right*: variation of the core size. Adapted from Wang (1970)

Figure 5.1(right) shows that the viscous effects are pronounced for the case with a much smaller initial core size especially near the initial stage. The core size grows more than tenfold from $\delta_0 = 0.001a$ before the ring begins its passage over the sphere.

The trajectories of the vortex ring, Z versus t, and the variation of the ring radius, R versus t, are shown in Fig. 5.2. From this figure we sec that the enlargement of the ring radius begins when the ring is passing over the sphere and that the forward motion of the ring is faster for the inviscid theory. When the ring is passing over the sphere, the viscous effect is enhanced since the core size becomes smaller because of the stretching of the ring. Again we see from Fig. 5.2 that the viscous effect of retarding the forward motion is larger for the ring with a smaller initial core. Similar results (Klein et al., 1996) were obtained using the improved vortex element numerical schemes to be described below in Sect. 5.2.

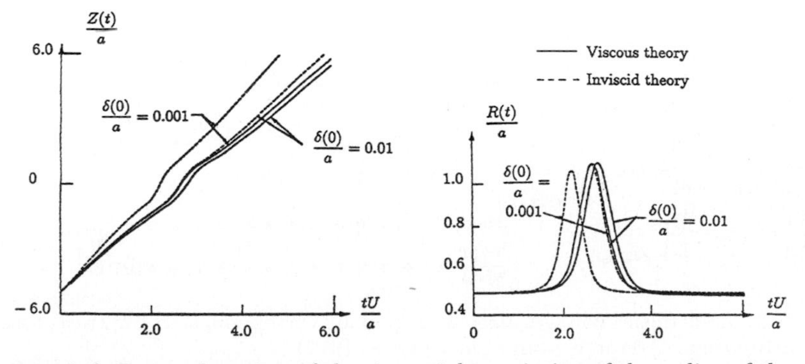

Fig. 5.2. *Left*: Forward motion of the ring; *right*: variation of the radius of the ring. Adapted from Wang (1970)

To acquire a feeling of the relative order of magnitude of the viscous effect and that of the initial core structure, we consider the case of a slender ring with $\delta_0/R(0) = 10^{-2}$. The dominant term inside the square bracket in (5.1.3) becomes $\ln(8R/\delta) = \ln 800 \sim 6.68$. A doubling of the core size due to diffusion will make a change of $-\ln 2 \sim -0.693$, which is already 10% of the dominant term and is comparable to the constant -0.558 in (5.1.3).

Note that (5.1.3) defines the motion of a single vortex ring with a similarity vorticity distribution submerged in an ambient fluid. If the initial profile is nonsimilar, we can still use (5.1.3) to define $\dot{Z}_1$ provided that the constant -0.558 is replaced by $K(t)$, which can be identified as $C_v(t) - \ln(8R/\ell)$ in (3.3.162). It was pointed out in Subsect. 3.3.6 how to define an optimum similarity solution by matching the polar moment of the actual initial vorticity distribution with that of a similarity distribution. This is done in practice by appropriately choosing the similarity solution's initial core size δ_0^*. Then we have

$$K^*(t) = -0.558 + O(t^{-2}) \tag{5.1.6}$$

and the differences between a nonsimilar and an optimum similar core structure on the forward motion of vortex rings will be small. To confirm this expectation, we quote numerical examples shown in Gunzburger (1972).

Interaction of Two Coaxial Vortex Rings with Nonsimilar Core Structure. The initial geometry of the two rings is shown in Fig. 5.3 (left). The initial vorticity distribution is assumed to be that of a Rankine vortex, namely

$$\zeta_i(t_0, r) = \varpi_i \; H(\delta_{0i} - r) \; , \quad i = 1, 2 \; , \tag{5.1.7}$$

where $\varpi_i = \Gamma_i/(2\pi\delta_{0i}^2)$ and H represents the Heaviside function. The equations of motion of the vortex rings for nonsimilar core structure were integrated numerically for the case, $\Gamma_1 = \Gamma_2 = 1$ and $R_e = 10^6$. The initial configuration at $t = t_0$ is

$$R_1 = R_2 = Z_2 \; , \quad \delta_1 = \delta_2 = R_1/100 \; . \tag{5.1.8}$$

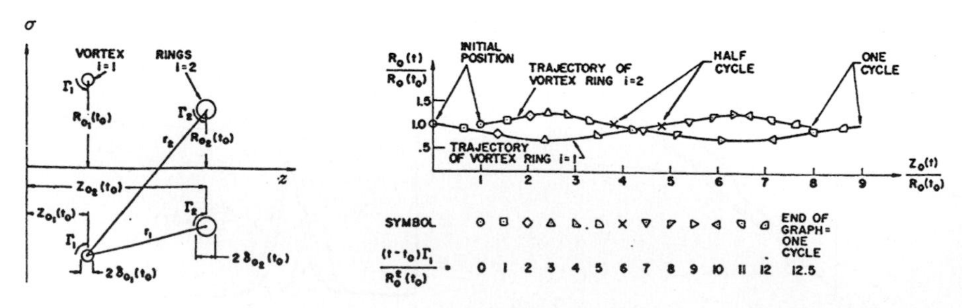

Fig. 5.3. *Left*: Initial configuration of a pair of vortex rings; *right*: trajectories of the vortex rings. Adapted from Gunzburger (1972)

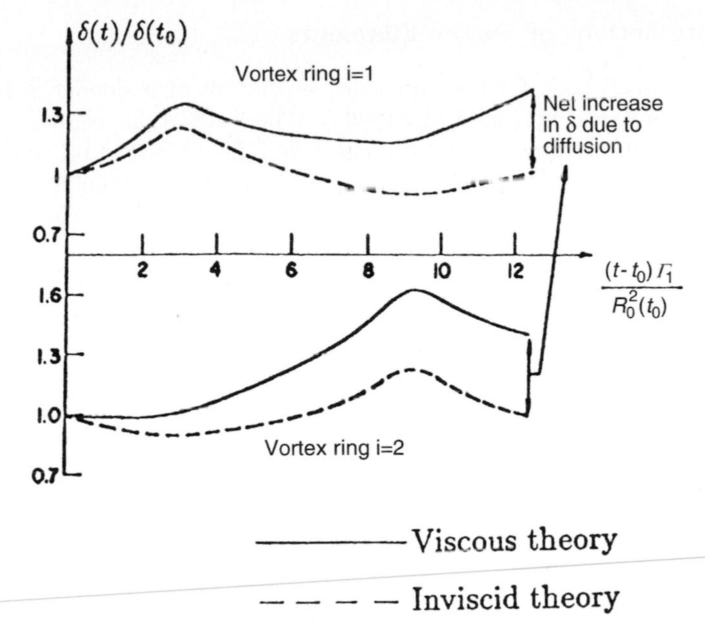

Fig. 5.4. Core sizes vs. time for one cycle. Adapted from Gunzburger (1972)

The resulting motions of the rings are presented in Fig. 5.3 (right). As t increases, the ring in the rear decreases in size, $\dot{R}_1(0) < 0$, the ring in the front increases in size, $\dot{R}_2(0) > 0$ and the first ring is gaining up on the second ring. When the ring that was initially behind the other moves ahead, the roles are reversed. The rings therefore take turns overtaking each other and going through each other. This behavior is qualitatively similar to that of the inviscid theory (Lamb, 1932). For the inviscid theory, the core size changes because of stretching only. At the end of each periodic relative motion, the cores return to their initial sizes. With viscous diffusion, the motion of the rings is aperiodic. The changes of core size during the first period are shown in Fig. 5.4.

The results displayed in Figs. 5.3 (right) and 5.4 were obtained for the nonsimilar solution by using enough terms in the series for $\bar{\zeta}^{(0)}$ to ensure an accuracy of 0.01% in $R_i(t)$, $Z_i(t)$, and $\delta_i(t)$ for $i = 1, 2$. Results were also obtained using the optimum similarity solution, and these results were within 0.5% of the "exact" (nonsimilarity) solution shown in Figs. 5.3 (right) and 5.4. These results confirm our expectations that

- *viscous effects cannot be ignored in the motion of vortex rings whereas the difference between the nonsimilar solution and the corresponding optimum similarity solution is small.*

We now study the interaction of slender filaments without axisymmetry.

5.1.2 Interactions of Vortex Filaments

A computational code for the equations of motion of a slender vortex filament was developed by Liu et al. (1986a). These equations were obtained in matched asymptotic analysis by Callegari and Ting (1978) and rederived in Sect. 3.3. They are summarized in Appendix C. For a filament with similar core structure, the required initial data are the shape of the centerline, $\mathbf{X}(0, s)$, the circulation Γ, the axial mass flux $m(0)$, and the core size δ_0. If the core structure is nonsimilar, then the initial profiles of the axial vorticity and velocity have to be specified. Since the effect of core structure, similar or nonsimilar, was demonstrated by examples of axisymmetric problems above, we shall quote here numerical examples of filaments with similar core structure to demonstrate the following effects not accounted for in the axisymmetric examples:

1. The effects of the axial mass flux, $m(0)$, and viscosity on the motion of a filament, given its initial shape, $\mathbf{X}(0, s)$, circulation, Γ, and core size, δ_0.
2. The effect of the initial orientations of two filaments relative to each other on their motion and interaction leading to merging or intersection and the onset of different types of merging problems.

It was pointed out earlier that the asymptotic analysis was formulated under the conditions that the effective core size $\delta(t)$ for a filament is much smaller than the other length scales in the flow field, and that each filament is in the form of a slender torus of finite length. For flows induced by free vortex filaments in an unbounded domain, those length scales are (1) the minimum radius of curvature $R_{\min}$ of a centerline $\mathcal{C}$, (2) the minimum distance $d_{\min}$ between two "distinct" points on $\mathcal{C}$, and (3) the minimum distance d_{ij} between two centerlines $\mathcal{C}_i$ and $\mathcal{C}_j$. The definition of a small core size implies the following conditions on the three length ratios:

$$(R_{\min}/\delta)_i \geq k_1 \gg 1 \ , \quad \text{for } i = 1, \ldots, N \ , \tag{5.1.9}$$

$$(d_{\min}/2\delta)_i \geq k_2 \gg 1 \ , \quad \text{for } i = 1, \ldots, N \ , \tag{5.1.10}$$

and

$$d_{ij}/(\delta_i + \delta_j) \geq k_3 \gg 1 \ , \text{ for } i, j = 1, \ldots, N \ , \quad i \neq j \ . \tag{5.1.11}$$

On account of (5.1.9), we define two "distinct" points of $\mathcal{C}$ by the condition that the arclength between them along $\mathcal{C}$ is finite, say, larger than $\pi R_{\min}$. These three conditions (5.1.9), (5.1.10), and (5.1.11) exclude self-merging, self-interaction, and intersection of two filaments, respectively. When there is a background potential flow, such as the flow around a body, additional conditions should be imposed so that the vortical core will not overlap with the boundary layer along the body surface.

The onset of merging is defined as the instant when the equality sign holds in any one of (5.1.9), (5.1.10), and (5.1.11). These criteria will be used in Chap. 6 to classify the types of global and local mergings. Now we have to assign the values of k_i, $i = 1, 2, 3$, and would like to assign only moderate values noting that the vorticity in the merged region is of the order of $\exp(-k_i^2)$. This formal extension of the asymptotic theory to moderate values of the k_i has been verified by comparing the asymptotic solutions with numerical solutions of the Navier–Stokes equations. Let t_0 and t_1 denote the times where the length ratios in (5.1.9)–(5.1.11) predicted by the asymptotic solutions are greater or equal to $2\,k_i$ and k_i, respectively. Then the computations were initialized with the asymptotic solution at $t = t_0$ and continued to $t = t_1$. The accuracy of the extended asymptotic solutions at $t = t_1$ was established by comparison with the computational results. These verifications were carried out by Liu and Ting (1982) for the axisymmetric problems and by Chamberlain and Weston (1984) and Chamberlain and Liu (1985) for the fully three-dimensional problems. All these computations showed that the difference between the asymptotic solution and the numerical solution remains sufficiently small, say less than 2%, if we choose $k_1 = 2$ and $k_2 = k_3 = 1.5$. These values are used in the following examples to define the practical region of validity of our asymptotic solutions.

Now we quote examples to show the effect of "large" initial mass flux in the vortical core and viscous diffusion on the motion of a filament.

5.1.2.1 Effect of Axial Mass Flux and Reynolds Number on the Motion of a Filament. From the scaling introduced in Sect. 3.3, a large axial velocity in the vortical structure is $O(\epsilon^{-1})$ while the cross-sectional area is $O(\epsilon^2)$; therefore, the "large" axial mass flux in the scales of ℓ and U is $\epsilon\, m(t)$.

The circulation of the filament used in the examples is $\Gamma = 5$. The initial shape $\mathbf{X}(0, s)$ of the centerline is an ellipse lying in the xy plane with the major axis equal to 2 along the x-axis and the minor axis equal to 1.5 along the y-axis. The initial core structure is similar with core size $\delta_0 = \sqrt{4\nu}$.

For the reference case, we choose $m(0) = 0$, that is, no initial axial flux, and $\nu = 0.0045$ and hence $\delta_0 = 0.134$ and $Re_0^{-1/2} = \sqrt{\nu/\Gamma} = 0.03$. Shown in Fig. 5.5 are the side and top views of the centerline $\mathbf{X}(t, s)$ of the filament at various instants from $t = 0$ to the final instant $t = 21.02$. The latter marks the commencement of the local self-merging as the core size becomes comparable to the minimum radius of curvature of the filament according to criterion (5.1.9).

From Fig. 5.5 we see that the centerline ceases to be planar for $t > 0$ due to the variation of the radius of curvature along its arclength. The centerline becomes almost planar again at $t = 4.75$, which we identify as "the first half-period," with its shape nearly the same as the initial one but with an interchange of the major and minor axes. At $t = 12.30$, "the first period," the centerline becomes almost planar again with its shape nearly the same as the initial one. Because of the decay of the core structure, the motion becomes more aperiodic as the core size increases with time.

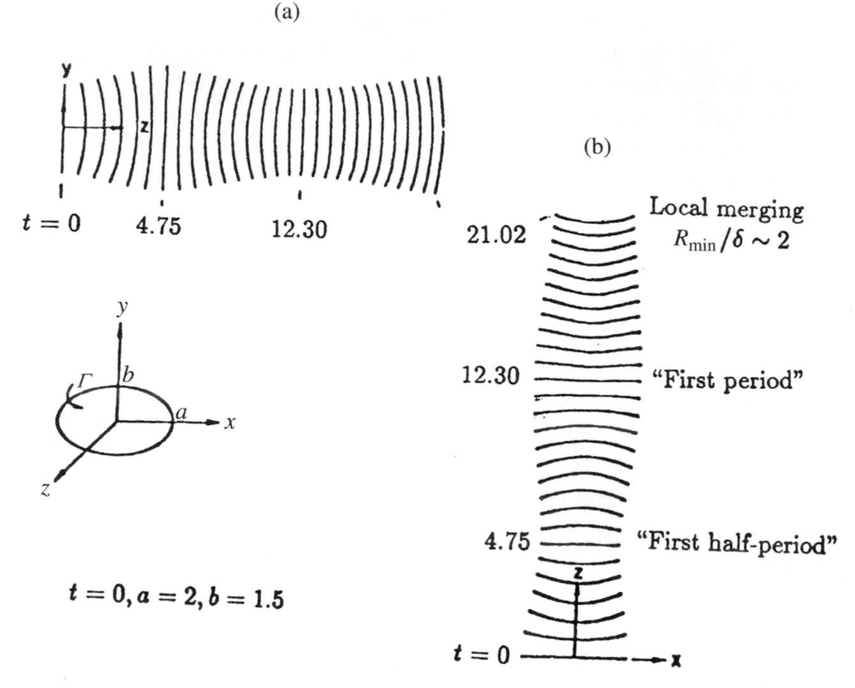

Fig. 5.5. Motion of the centerline of an initially elliptical vortex filament. (**a**) side view, (**b**) top view. Adapted from Liu et al. (1986b)

Let $x^*(t)$ and $y^*(t)$ denote the maxima of the x- and y-coordinates of the centerline, $\mathbf{X}(t,s)$. Then the envelopes in the top view and the side view of $\mathbf{X}(t,s)$ in Fig. 5.5 show the variations of $x^*(t)$ and $y^*(t)$, respectively.

The motion of the centerline for a filament with axial mass flux and a different Reynolds number look qualitatively similar to the reference case. Therefore, we use the first half period of the motion of the centerline as a quantitative measure of its difference from the reference case. Figure 5.6 illustrates the dependence of the first period of x_1^* on the Reynolds number $Re_0 = \Gamma/\nu$ and the initial axial mass flux in the core. We see from Fig. 5.6 that the movement of the filament slows down for larger viscosity, ν, or larger $1/Re$ because of the larger core size. The axial flow in the core tends to slow down the movement as expected from (3.3.189) and (3.3.193). The effect of the axial mass flux is more pronounced for larger ν/Γ.

Now we show examples of the interaction of two filaments leading to different types of merging problems.

5.1.2.2 Interaction of Two Filaments. Before our systematic study of the interaction of two filaments with different initial orientations, we show an example in which two filaments tend to touch each other with the vorticity vectors in the adjacent segments of the filaments in opposite directions. Thus

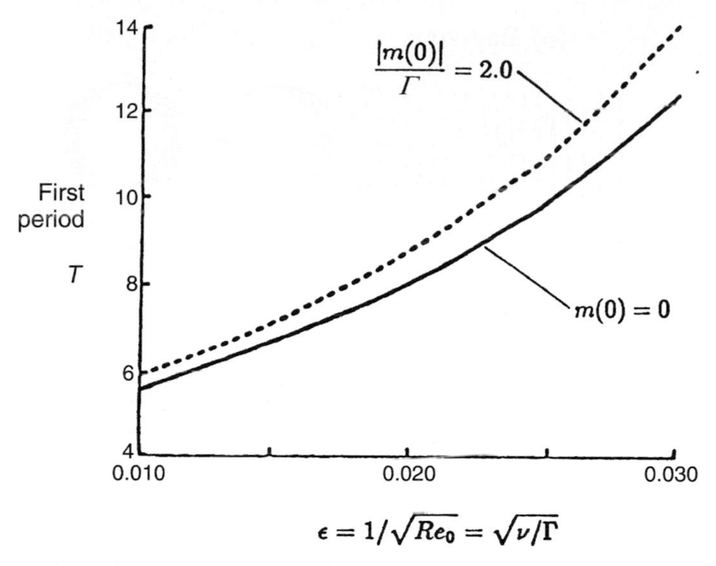

Fig. 5.6. Effect of the axial flow on the first period of motion. Adapted from Liu et al. (1986b)

in the scale of core size, these two segments appear to be merging as two rectilinear filaments of opposite strength. We expect that the subsequent local merging of the filament segments will result in the mutual cancellation of the vorticity. This example is of special interest because the subsequent merging and cancellation of vorticity in the overlapping segments of the filaments and their reconnection to a single filament were investigated experimentally by Oshima and Izutsu (1988) and by numerical solution of Navier–Stokes equations by Ishii et al. (1989). Both investigations will be reported in Sect. 6.3.

The initial configuration of two vortex rings is shown in Fig. 5.7. They simulate two vortex rings created by two coplanar orifices facing the same direction at $t = 0$. They have the same strength, $\Gamma_1 = \Gamma_2 = 5$, and their centerlines are circles of radius $R_0 = 1.5$ lying in the xy plane centered at $(\pm 2.5, 0, 0)$. Both core structures are similar profiles with the same core size $\delta_0 = \sqrt{4\nu}$ and have no large axial flow, $m(0) = 0$. With $\nu = 0.0005$, we have $\delta_0 = 0.0447$ and $Re_0 = \Gamma_1/\nu = 10{,}000$. The initial minimum distance between the rings is 2, $d_{12}(0) = 2 = 44.7\delta_0$. As time progresses, the centerlines of the rings become nonplanar and tilted toward each other. The shape of a ring deviates from a circle, and the deviation is more pronounced for the segments close to each other, where the effect of interaction is maximal. Shown in Fig. 5.7 are the three views of the centerlines at different instants from $t = 0$ to $t_f = 7.04$. At t_f, we have $\delta = 1.25$ and $d_{12} = 0.36 < 3\delta$. The two filaments touch each other according to the criterion (5.1.11). Thereafter, local merging takes place and the asymptotic solution is no longer applicable. The subsequent merging and cancellation of vorticity in the overlapping segments

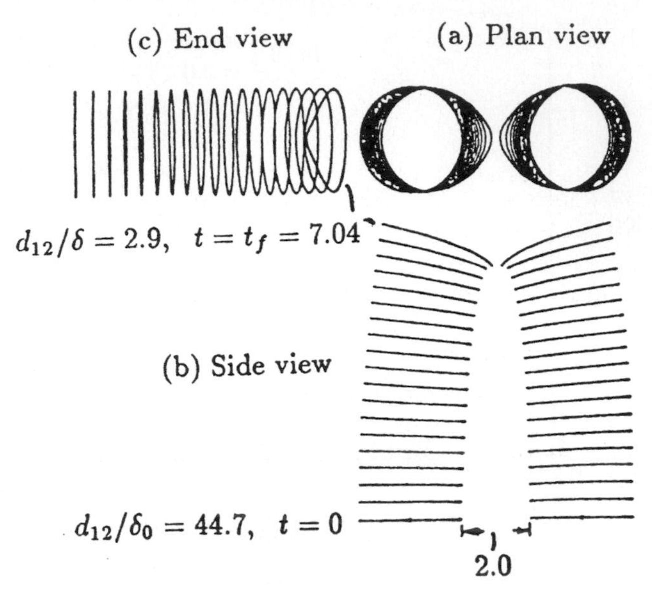

Fig. 5.7. Interaction of two filaments of the same strength, $\Gamma_1 = \Gamma_2 = 5$ with $\nu = 0.0005$, $\delta_0 = \sqrt{0.0002}$, and ring radius $R_0 = 1.5$ centered at $(\mp 2.50, 0, 0)$. Adapted from Liu et al. (1986b)

of the filaments and their reconnection to a single filament will be discussed in Sect. 6.3.

Now we present three numerical examples simulating different types of interactions of two vortex filaments. These examples differ from each other in one parameter that is the ratio of the initial distance between the two filaments and their size. By varying this parameter, we demonstrate how we can use the matched asymptotic solutions to study the interaction of filaments leading to the onset of different types of merging.

In these three examples, the two filaments have opposite strengths, $\Gamma_1 = -\Gamma_2 = 5$. We choose $\nu = 0.0005$ and hence $Re_0 = 10,000$. The initial core structures of both filaments are similar with core size $\delta_0 = \sqrt{0.002}$ and there is no axial mass flux ($m_0 = 0$). Their centerlines are initially circles of the same radius, $R_0 = 1.5$. These two circles lie in the two parallel planes, $z = \mp 0.25$, and are centered at $(\mp x_0, 0, \mp 0.25)$ respectively. The initial data in Figs. 5.8–5.10 differ only in one parameter, the ratio between the radius R_0 and the distance $2x_0$ between the centers of the two circles. The projections of these two circles onto the xy plane overlap each other for $|x| < R_0 - x_0$. We define

$$\lambda^* = (R_0 - x_0)/R_0 \tag{5.1.12}$$

as the initial overlap parameter and call the filament on the left side of the yz plane the first filament. We use the polar angle θ for each circle as the parameter s for the centerline. Because of the symmetry with respect to the

yz plane, it suffices to consider $x_0 \geq 0$, that is, $\lambda^* \leq 1$. There is no initial overlap if $\lambda^* < 0$. The choice of the signs of Γ_1 and Γ_2 are such that the two rings will begin to move toward each other.

When $\lambda^* = 1$, we have the head-on collision of two circular vortex rings. This is an axisymmetric problem. The asymptotic solution provides the solution of the problem until the criterion (5.1.11) is met; that is, the distance between these two centerlines is equal to 1.5 times the sum of their core sizes. Thereafter, the cores of the rings merge or overlap and finite difference solutions of Navier–Stokes equations will be employed in Subsect. 6.3.1 to simulate the subsequent merging process resulting in the cancellation of vorticity.

When $\lambda^* < 1$, the problem is no longer axisymmetric. For $x_0/R_0 \ll 1$, or $\lambda^* \to 1^-$, we have a nearly head-on collision problem and the merging will take place along the entire length of the rings. That means we have a global merging problem.

When $x_0/R_0 \gg 1$ or $-\lambda^* \gg 1$, these two rings are far apart laterally and the effect of interaction is weak. They will move in opposite directions, nearly parallel to the z-axis as isolated rings and pass each other when they cross over the xy plane. Eventually, the core of each ring will grow because of diffusion and become comparable to the ring radius. Then we have a global self-merging and the structure of a slender ring disappears.

In between these two limiting cases, merging or intersection of the rings can take place locally, so that only small segments of the filaments will merge or overlap each other. Because of symmetry, the merging takes place when the filaments are crossing over the xy plane. In this intermediate range, we choose three values of λ^*, namely, 0.75, 0.133, and 0.0833. The motion or interaction of the filaments corresponding to these three values of λ^* are shown in Figs. 5.8, 5.9, and 5.10, respectively. These three figures illustrate three different types of filament interaction leading to (1) intersection or merging in two local regions, (2) touching or merging in one local region, and (3) passing through the xy plane without intersection or touching with local self-merging appearing afterward, respectively.

Shown in Fig. 5.8 is the motion of the centerlines of the two filaments with $\lambda^* = 0.75$. They move toward each other, that is, toward the xy plane, with the "nonoverlapping" part of the centerline remaining nearly planar while the "overlapping" part bends backward. As $t_f = 0.2217$, the core size δ is 0.04581, and the two filaments intersect in two local regions near the yz plane at $\theta^* = \pm 75^o$, where their distance is down to $d_{12} = 0.1372 \leq 3\delta$ and the criterion (5.1.11) is met. The projection of a centerline onto the xy plane deviates from a circle, and the deviation is more pronounced near θ^*, where the interaction is the strongest and local merging is about to take place.

Shown in Fig. 5.9 is the motion of the two centerlines with $x_0 = \ell_c = 1.30$ or $\lambda^* = 0.133$. Here ℓ_c is a critical distance found through numerical experiments. For $x_0 < \ell_c$, the filaments merge, while for $x_0 > \ell_c$ they manage

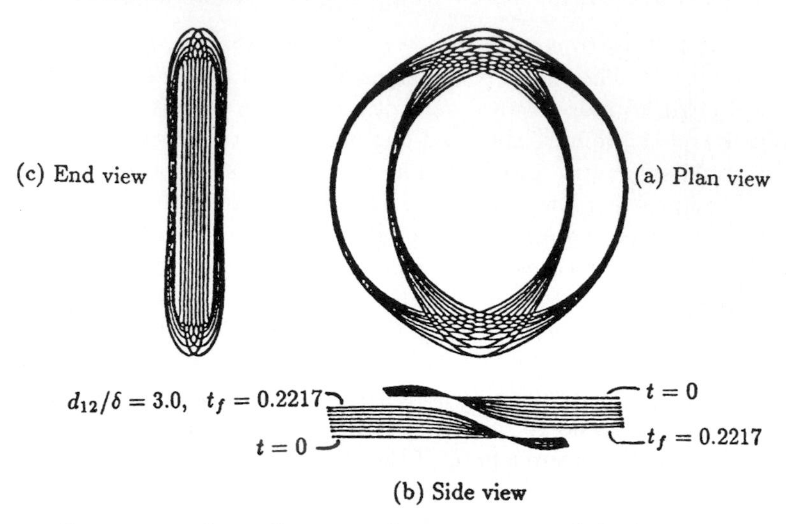

Fig. 5.8. Interaction of two filaments of opposite strength leading to intersection or merging in two local regions. Adapted from Liu et al. (1986b)

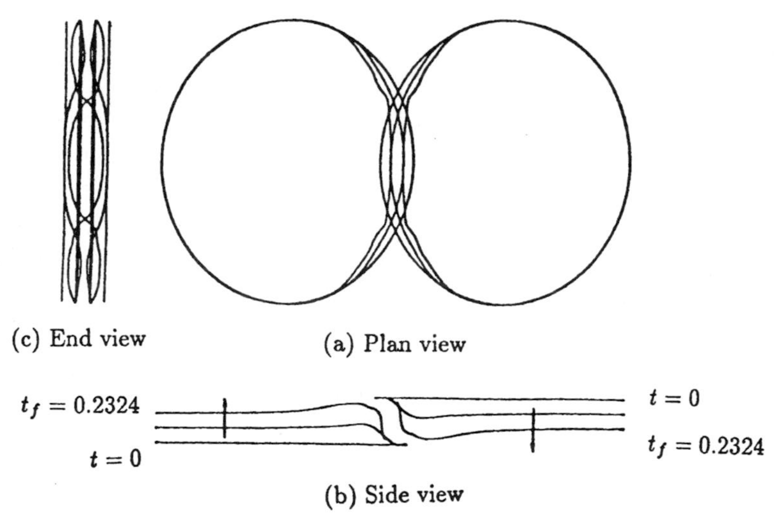

Fig. 5.9. Interaction of two filaments of opposite strength leading to merging or touching in one local region. Adapted from Liu et al. (1986b)

to pass each other without any noticeable merging or overlapping of the cores. Qualitatively, the deformation of the centerlines in Fig. 5.9 prior to merging resembles that in Fig. 5.8 except that the overlapping portion is smaller but bends backward more sharply. At $t_f = 0.2324$ the two filaments intersect at one local region near $\theta^* = 0$, where the criterion (5.1.11) is met. They are in effect touching each other. The incipient stage of merging differs from that in Fig. 5.8. Now the vorticity vectors in the adjacent segments

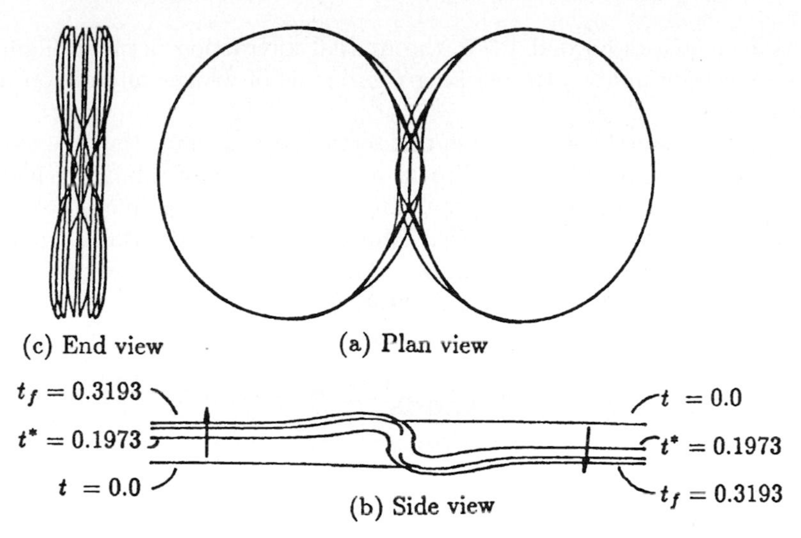

Fig. 5.10. Interaction of two filaments of opposite strength leading to local self-merging. Adapted from Liu et al. (1986b)

of the two filaments are nearly in the same direction and hence we expect the vorticity field in the merging region to intensify during the subsequent merging stage. This is completely different from the local merging region appearing in Fig. 5.7, where the vorticity vectors in the two segments are in the opposite direction and cancellation of vorticity is expected.

We note that the time t_f of intersection decreases as x_0 increases from 0 to ℓ_c or the initial overlap parameter decreases from 1 to 0.133. In addition, the critical value x_0/R_0 depends on the scaled initial core size, δ_0/R_0 and the viscous effect, Re_0. If the initial core size is not too small or Re is not too large, it is possible that the critical value x_c/R_0 exceeds 1, that is, $\lambda^* < 0$.

For $x_0 > \ell_c$, we expect the two filaments to pass each other without intersection although their projections on the xy plane in this case do overlap initially. Shown in Fig. 5.10 is the motion of the two centerlines with $x_0 = 1.375$ or $\lambda^* = 0.0833$. We see from Fig. 5.10 that the two centerlines move toward each other with the nonoverlapping portion retaining nearly the same circular shape while the small overlapping region bends backward sharply at almost a right angle to the xy plane. When $t^* \sim 0.1973$, the portions of the centerlines which were initially in the nonoverlapping region are now nearly coplanar in the xy plane, while the small portions initially in the overlapping region bend backward sufficiently to allow the mutual passage of the two filaments. For $t > t^*$, the two filaments are moving away from each other, that is, from the xy plane, and the effect of interaction is reversed. The portion that previously bent backward sharply begins to stretch forward laterally until the criterion (5.1.9) is met and local *self-merging* takes place at $t_f = 0.3193$.

As x_0 increases beyond 1.375, the mutual interaction between filaments becomes weaker and we are reaching the regime of weaker interaction mentioned above.

From the examples discussed in this section, we conclude that the asymptotic solution can (i) be extended to its practical region of validity to identify the type of merging that is going to take place, and (ii) be employed to generate the initial data for the numerical solution of the Navier–Stokes equations to study the merging process. Numerical simulations of the merging process will be presented in the following chapter.

5.2 Vortex Element Methods for Slender Vortex Filaments

In this section we discuss application of the asymptotic theories developed in Chaps. 3 and 4 to the simulation of slender vortices. The computational simulation of slender vortices in three space dimensions is extremely expensive for mainly two reasons: First, it is necessary to properly represent the concentrated vorticity distribution and its induced velocity field. This induces an *accuracy issue*. Second, within the vortex core there are extremely high circumferential velocities, which are by one order of magnitude larger in terms of the core size parameter than the characteristic speed of filament motion. The associated very short time scale, which is in fact even two orders of magnitude shorter than the characteristic vortex filament evolution time scale, induces a severe numerical stiffness.

Resolving the vortex core structure in detail is prohibitively expensive for sufficiently thin vortex cores. Thus, a numerical scheme specialized for slender vortex dynamics must address both the accuracy and the stiffness issue without explicitly resolving the vortex core structure with many computational degrees of freedom. In this section, we describe a vortex element numerical method that overcomes the accuracy issues by employing the results of asymptotic analysis (Klein and Knio, 1995), and the stiffness issue by using the asymptotic scaling relations in Richardson-type extrapolation procedure for narrow vortex cores.

In Knio and Ghoniem (1990), the authors present a thin-tube vortex element scheme specialized for slender vortex applications. The method builds on a more general vortex element method (Beale and Majda, 1982, 1985). However, in contrast to the convergence conditions described by these authors, the thin-tube scheme underresolves the vorticity distribution, and must therefore be considered as an ad hoc approximation to begin with. In fact, we will demonstrate through an *asymptotic analysis of the numerical method*, following Klein and Knio (1995) that the standard thin-tube representation of a slender vortex as a single chain of overlapping spherical vortex elements involves leading-order errors in the core vorticity distribution. The consequence will be large errors in the predicted filament velocity. On the basis of this asymptotic analysis, we are, however, in the position to propose a correction of the vortex element scheme that will guarantee its proper asymptotic behavior in comparison with a slender vortex flow. The three different approaches to improving the scheme from Klein and Knio (1995) will be recounted here. The reader may also consult Margerit et al. (2004) for an extensive comparative study of various alternative numerical schemes for slender vortex filament dynamics.

To demonstrate the improvements quantitatively, we compare numerical simulations of nearly straight vortices with short-wave perturbations with the asymptotics developed in Chap. 4. This comparison will amount to a mutual test of the proposed numerical technique and of the asymptotics in that each adopts a different set of approximations to simplify the original problem (described in Chap. 3): The thin-tube vortex element method relies on an approximate description of the net effects of vortex core structure on the filament motion, but treats the effects of nonlocal induction essentially without error. The asymptotics for perturbed, nearly straight vortices by Klein and Majda from Chap. 4 employs the asymptotically correct core structure representation, while using a linearization of the finite part of the Biot–Savart integral.

These comparisons will be particularly interesting for relatively thick vortex filaments with core size parameter $\epsilon \sim 0.015$, corresponding to expansion parameters $\delta \sim 0.2 \ldots 0.25$ in the Klein–Majda theory. The comparison will be favorable, demonstrating that the latter theory is well applicable in this parameter range.

For later reference within this section, we summarize some results from Subsect. 3.3.4 here. The equation of motion for a slender vortex filament reads

$$\boldsymbol{X}_t = \frac{\Gamma}{4\pi}\left(\ln\frac{2}{\epsilon} + C\right)\kappa\boldsymbol{b} + \boldsymbol{Q}^{f,\perp}\,. \tag{5.2.1}$$

Here the "core constant"

$$C = \lim_{\overline{r}\to\infty}\left(\frac{4\pi^2}{\Gamma^2}\int_0^{\overline{r}} r {v^{(0)}}^2 dr\right) - \frac{1}{2} - \frac{1}{2}\left(\frac{4\pi}{\Gamma}\right)\int_0^{\infty} r {w^{(0)}}^2 dr \tag{5.2.2}$$

captures the influence of the detailed vortex core structure, that is, the influence of the detailed radial distributions of the leading-order circumferential and axial velocities, $v^{(0)}$ and $w^{(0)}$, respectively. In the construction of improved thin-tube vortex element methods, to be described below, the following intermediate result from Subsect. 3.3.4 will be important,

$$C = -\left\{\frac{2\pi}{\Gamma}\int_0^{\infty}\left[\frac{1}{\kappa}\zeta_{11}^{(1)} + \overline{r}\ln(\overline{r}\zeta^{(0)})\right] d\overline{r} + \frac{1}{2}\right\}. \tag{5.2.3}$$

This equation reveals that the *nonlinear* expressions from (5.2.2), involving terms of the leading-order velocities in the core, originate from *linear* expressions in terms of the leading- and first-order vorticity components. The former equation was derived by analyzing the first-order vorticity transport equation in the core and by deriving explicit representations of $\zeta^{(0)}, \zeta_{11}^{(1)}$ in terms of $v^{(0)}, w^{(0)}$. Equation (5.2.3) will be at the core of our strategy for constructing accurate thin-tube vortex element schemes.

5.2.1 Thin-Tube Models for Weakly Stretched Vortices

5.2.1.1 The Standard Thin-Tube Model. The basis of the scheme is a discrete representation of the filament geometry by a chain of nodes $\{\boldsymbol{\chi}_i\}_{i=1}^N$. Each pair $(\boldsymbol{\chi}_i, \boldsymbol{\chi}_{i+1})$ represents a segment of the (approximate) filament centerline. Associated with these segments are spherical vorticity distributions, centered at $\boldsymbol{\chi}_{i+\frac{1}{2}} = \frac{1}{2}(\boldsymbol{\chi}_i + \boldsymbol{\chi}_{i+1})$. The computational representation of the filament vorticity distribution then results from superposition, that is,

$$\boldsymbol{\omega}^{\text{ttm}} = \sum_{i=1}^{N} \Gamma \left(\boldsymbol{\chi}_{i+1} - \boldsymbol{\chi}_i\right) f_\varepsilon \left(\left|\boldsymbol{x} - \boldsymbol{\chi}_{i+\frac{1}{2}}\right|\right), \tag{5.2.4}$$

where

$$f_\varepsilon(\boldsymbol{\xi}) = \frac{1}{\varepsilon^3} f\left(\frac{\boldsymbol{\xi}}{\varepsilon}\right) \tag{5.2.5}$$

is a spherical smoothing function, rapidly decaying for $|\boldsymbol{\xi}| > \varepsilon$. See Beale and Majda (1982, 1985) for criteria for the choice of this smoothing function. Here we use

$$f_g(r) = \frac{3}{4\pi} \exp(-r^3) \qquad \text{and} \qquad f_s(r) = \frac{3}{4\pi} \operatorname{sech}^2(r^3). \tag{5.2.6}$$

An additional constraint for the vorticity discretization is sufficient overlap of the vortex elements,

$$\max_i \left\{\left|\boldsymbol{\chi}_{i+1} - \boldsymbol{\chi}_i\right|\right\} < \alpha\varepsilon \qquad (\alpha < 1). \tag{5.2.7}$$

The discretized vorticity distribution from (5.2.4) to (5.2.6) implies an induced velocity field via the Biot–Savart integral (3.1.62). The discretized velocity field is thus given by the desingularized Biot–Savart law

$$\boldsymbol{v}^{\text{ttm}}(\boldsymbol{x}, t) = -\frac{\Gamma}{4\pi} \sum_{i=1}^{N} k_\varepsilon \left(\left|\boldsymbol{x} - \boldsymbol{\chi}_{i+\frac{1}{2}}(t)\right|\right) \frac{\left|\boldsymbol{x} - \boldsymbol{\chi}_{i+\frac{1}{2}}(t)\right|}{\left|\boldsymbol{x} - \boldsymbol{\chi}_{i+\frac{1}{2}}(t)\right|^3} \times \left(\boldsymbol{\chi}_{i+1} - \boldsymbol{\chi}_i\right)(t). \tag{5.2.8}$$

Here

$$k_\varepsilon(\boldsymbol{\xi}) = k\left(\frac{\boldsymbol{\xi}}{\varepsilon}\right) \qquad \text{with} \qquad k(r) = 4\pi \int_0^r z^2 f(z)\, dz. \tag{5.2.9}$$

The temporal evolution of the discrete vorticity distribution is given now by transport of the vortex element nodes by the induced velocity at their respective location, that is,

$$d\boldsymbol{\chi}_i/dt = \boldsymbol{u}_i \equiv \boldsymbol{v}^{\mathrm{ttm}}(\boldsymbol{\chi}(t), t)\,. \tag{5.2.10}$$

This is a system of coupled ordinary differential equations, for which accurate numerical solution techniques are readily available (Knio and Ghoniem, 1990; Klein and Knio, 1995).

Structure and Principal Approximation Error of the Standard Scheme. If the overlap condition (5.2.7) is satisfied for sufficiently small α, then the summation formula in (5.2.4) for the discretized vorticity field may be considered as a good approximation to the integral expression

$$\boldsymbol{\omega}^{\mathrm{ttm}}(\boldsymbol{x}, t) = \Gamma \int_{L^{\mathrm{ttm}}(t)} f_\varepsilon(\boldsymbol{x} - \boldsymbol{X})\, d\boldsymbol{X}\,, \tag{5.2.11}$$

where $L^{\mathrm{ttm}}(t)$ is a smooth, time-dependent curve interpolating the $\boldsymbol{\chi}_i(t)$. Equation (5.2.11) is a smooth, slender vorticity distribution of the type considered in the asymptotic analysis of Navier–Stokes vortices in Chap. 3. For small ε one may therefore evaluate the induced velocities asymptotically as shown in the "alternative derivation" of Subsect. 3.3.4. As in that section, we split the integral in a nonlocal and a local part according to

$$\boldsymbol{\omega}^{\mathrm{ttm}}(\boldsymbol{x}, t) = \Gamma \left\{ \int_{L^{\mathrm{ttm}}(t)\setminus I_b} f_\varepsilon \boldsymbol{t}\, d\tilde{s} + \int_{I_b} f_\varepsilon \boldsymbol{t}\, d\tilde{s} \right\}, \tag{5.2.12}$$

where $\boldsymbol{t} = \partial \boldsymbol{X}^{\mathrm{ttm}}/\partial \tilde{s}$ is the unit tangent to $L^{\mathrm{ttm}}(t)$, $I_b = (\tilde{s}_1 - \varepsilon_b, \tilde{s}_1 + \varepsilon_b)$, and $\tilde{s}_1$ is defined through

$$\boldsymbol{x} = \boldsymbol{X}^{\mathrm{ttm}}(\tilde{s}_1, t) + r\boldsymbol{e}_{r,1}\,. \tag{5.2.13}$$

In other words, $\tilde{s}_1$ is the value of the arclength coordinate on $L^{\mathrm{ttm}}(t)$ for which the distance $\left|\boldsymbol{x} - \boldsymbol{X}^{\mathrm{ttm}}(\tilde{s}_1, t)\right|$ is minimal. (We will be interested here in sufficiently small distances so that this specification yields a unique $\tilde{s}_1$.)

The parameter ε_b satisfies the constraint $1 \gg \varepsilon_b \gg \varepsilon$ as before, and since the smoothing function f_ε decays rapidly for large arguments, the contribution of the outer integral $\int_{L^{\mathrm{ttm}}\setminus I_b}$ will be negligible. To evaluate the inner integral asymptotically, we may now Taylor-expand the geometry of L^{ttm}, because $1 \gg \varepsilon_b$. Thus,

$$\begin{aligned} \boldsymbol{x} - \boldsymbol{X}^{\mathrm{ttm}}(\tilde{s}, t) &= \varepsilon(\overline{r}\boldsymbol{e}_{r,1} + z\boldsymbol{t}) + \varepsilon^2 \frac{z^2}{2} \left(\kappa\boldsymbol{n}\right)_1 + o(\varepsilon^2) \\ \boldsymbol{t}(\tilde{s}, t) &= \boldsymbol{t}_1 + \varepsilon z \left(\kappa\boldsymbol{n}\right)_1 + o(\varepsilon^2)\,, \end{aligned} \tag{5.2.14}$$

with the stretched coordinates $\overline{r} = r/\varepsilon$ and $z = (\tilde{s} - \tilde{s}_1)/\varepsilon$, and κ the curvature of L^{ttm}. The subscript 1 denotes evaluation at $\tilde{s} = \tilde{s}_1$. Using this expansion in the definition of f_ε in (5.2.5), we find

$$f_\varepsilon(\boldsymbol{x} - \boldsymbol{X}^{\mathrm{ttm}})\,\boldsymbol{t} = f(R)\,\boldsymbol{t} + \varepsilon\kappa\left[z f(R)\boldsymbol{n}_1 + \frac{1}{2}\cos(\phi) f'(R)\,\boldsymbol{t}_1\right] + O(\varepsilon^2)\,, \tag{5.2.15}$$

where

$$R = \sqrt{\overline{r}^2 + z^2} \qquad f'(R) = \frac{df}{dr}(R)\,. \tag{5.2.16}$$

With this asymptotic expansion we now evaluate the second integral in (5.2.12) to obtain

$$\omega^{\mathrm{ttm}} = \frac{1}{\varepsilon^2}\left(\zeta^{(0)}(\overline{r}) + \varepsilon\cos(\phi)\zeta_{11}^{(1)}(\overline{r})\right)\boldsymbol{t}_1\left(1 + O(\varepsilon\varepsilon_b, f(\varepsilon_b/\varepsilon))\right)\,. \tag{5.2.17}$$

For the present discretized vorticity distribution, the axial vorticity components $\zeta^{(0)}$ and $\zeta_{11}^{(1)}$ are given by

$$\begin{aligned} \left(\zeta^{(0)}\right)^{\mathrm{ttm}}(\overline{r}) &= \Gamma\int_{-\infty}^{\infty} f\left(\sqrt{\overline{r}^2 + z^2}\right)dz\,, \\ \left(\zeta_{11}^{(1)}\right)^{\mathrm{ttm}}(\overline{r}) &= \kappa\Gamma\int_{-\infty}^{\infty}\frac{\overline{r}z^2}{2\sqrt{\overline{r}^2 + z^2}} f'\left(\sqrt{\overline{r}^2 + z^2}\right)dz\,. \end{aligned} \tag{5.2.18}$$

To evaluate next the relevant transport velocities of the nodes $\boldsymbol{\chi}_i$, we recall that (i) they are supposed to move with their local induced velocity, and that (ii) they are located within the core of the present (numerical) vorticity distribution. Thus we can immediately adopt the formulae, derived in Subsect. 3.3.4, where we analyzed the induced velocity on the centerline of a slender vortex. Thus we identify the centerline $\mathcal{C}$ from that section with L^{ttm} introduced here and directly use the results from (5.2.1) and (5.2.3).

The nonlocal part of the Biot–Savart integral, $\boldsymbol{Q}^{f,\perp}$, depends solely on the total circulation of the filament and on the instantaneous filament geometry, and thus is the same for the Navier–Stokes vortex and its numerical approximate counterpart. If we decide to equate the physical and numerical core size parameters ϵ and ε, then the only difference left is in the core structure parameter C. From (5.2.3) we conclude that this parameter is entirely determined by the distributions $\zeta^{(0)}$ and $\zeta_{11}^{(1)}$, and herein lies the main difficulty of the thin-tube numerical scheme: Even if we succeed in choosing a numerical core smoothing function f such that the two leading-order distributions would coincide, we cannot guarantee that the core structure coefficients match. The perturbation distributions $\zeta_{11}^{(1)}$ are determined by very different mechanisms in the two cases. This is quite obvious from the observation that, according to (3.3.130), $\zeta_{11}^{(1)}$ is a quadratic functional of the leading-order velocity/vorticity distributions, whereas in the numerical approximate scheme it depends on these only linearly. In fact, in the numerical setting $\zeta_{11}^{(1)}$ is the result of a skewed linear superposition induced by curvature of L^{ttm}.

The standard thin-tube model from Knio and Ghoniem (1990) will accordingly produce errors of order $O(\ln^{-1}(1/\varepsilon))$ of the filament velocity. For realistic values of $\varepsilon \sim 0.015$ this amounts to unacceptable 30% errors.

5.2.1.2 Improved Thin-Tube Schemes Through Asymptotic Error Compensation. On the basis of the observations from the previous section, Klein and Knio (1995) propose three different strategies for overcoming the errors of the thin-tube approximation:

Direct Velocity Correction. This first method compares the formulae for local induced velocities between the vortex element nodes, $\boldsymbol{\chi}_i(t)$, and points $\boldsymbol{X}(\tilde{s}, t) \in L^{\mathrm{ttm}}(t)$ on the centerline of the physical vortex filament. For given leading-order structures, $\zeta^{(0)}$ and $w^{(0)}$, the core structure coefficients are readily evaluated, and an obvious correction scheme reads,

$$d\boldsymbol{\chi}_i/dt = \boldsymbol{v}^{\mathrm{ttm}}(\boldsymbol{\chi}_i(t), t) + \frac{\Gamma\kappa}{4\pi}\left(C - C^{\mathrm{ttm}}\right)\boldsymbol{b}\,. \tag{5.2.19}$$

Here C^{ttm} may be calculated on the basis of the chosen smoothing function in advance of a simulation. The correct determination of $C = C(\tilde{s}, t)$ will generally require the simultaneous solution of a coupled partial differential equation system for the core structure evolution following Callegari and Ting (1978). We will discuss numerical strategies for efficient incorporation of time-dependent core structures later in Subsect. 5.2.3. Here we concentrate on filaments with weak overall stretching, and on time scales short enough that viscous diffusion does not play a significant role so that the core structure coefficient may be considered a constant. This is valid, in particular, for vortices in the Klein–Majda regime of short wavelength perturbations away from a straight line as discussed in Chap. 4. We may then also evaluate C once before a simulation, and the equation system for the vortex element nodes is closed.

Evaluation of Induced Velocities at Local Stagnation Points. The second approach to thin-tube error correction again relies on direct evaluations of the Biot–Savart integral. Yet, this time the evaluation is not performed directly at the vortex element nodes, but rather at judiciously chosen locations in their vicinity. The idea is to choose these locations such that the difference of induced velocities from the numerical vorticity distribution between the nodes and the neighboring points just makes up for the error between the nodal velocities and the desired centerline speed of a physical vortex. In this second scheme the evolution equations for the vortex nodes read

$$\frac{d\boldsymbol{\chi}_i}{dt} = \boldsymbol{u}_i = \boldsymbol{v}^{\mathrm{ttm}}\left(\boldsymbol{\chi}_{i(t)+\varepsilon^2(z_i\boldsymbol{n}_i)(t)}\right) \tag{5.2.20}$$

where

$$z_i = \left[\frac{\Gamma\kappa}{2\pi}\frac{C - C^{\mathrm{ttm}}}{\zeta^{(0)}\big|_{\overline{r}=0}}\right]_i . \tag{5.2.21}$$

This formula is obtained from a Taylor expansion of (5.2.20) according to

$$\boldsymbol{v}^{\mathrm{ttm}}\left(\chi_i + \varepsilon^2 z\boldsymbol{n}, t\right) = \boldsymbol{v}^{\mathrm{ttm}}\left(\chi_i, t\right) + \varepsilon^2(z\boldsymbol{n})_i \times (\nabla\boldsymbol{v}^{\mathrm{ttm}})_i + \cdots \tag{5.2.22}$$

and considering that

$$\varepsilon^2\left(\boldsymbol{n}\times\nabla\boldsymbol{v}^{\mathrm{ttm}}\right)_{\overline{r}=0} = \frac{1}{2}\zeta^{(0)}\big|_{\overline{r}=0}\boldsymbol{b}. \tag{5.2.23}$$

This last result follows from $\overline{r}\zeta^{(0)} = \left(\overline{r}\boldsymbol{v}^{(0)}\right)_{\overline{r}}$.

The schemes described so far both have the disadvantage of requiring explicit evaluations of local differential properties of the filament centerline, such as the curvature and the local basis $(\boldsymbol{t}, \boldsymbol{n}, \boldsymbol{b})$. Curvature evaluation requires taking approximate second tangential derivatives of the nodal curve $L^{\mathrm{ttm}}(t)$. In contrast, the original scheme does not require any explicit evaluation of curve derivatives so that it is numerically much more robust. The third correction scheme described below was conceived to overcome this issue.

Scaled Standard Scheme. The last, and numerically most robust, correction method consists of a suitable *local* space and time-dependent choice of the numerical core size parameter $\varepsilon^{\mathrm{ttm}}$. Following the previous derivations, we know that the induced velocity at the computational nodes obeys

$$\boldsymbol{X}_t^{\mathrm{ttm}} = \frac{\Gamma}{4\pi}\left(\ln\frac{2}{\varepsilon^{\mathrm{ttm}}} + C^{\mathrm{ttm}}\right)\boldsymbol{b} + \boldsymbol{Q}^{f,\perp} . \tag{5.2.24}$$

Here C^{ttm} is determined through (5.2.3) from the numerical vorticity components $\zeta^{(0)}$ and $\zeta_{11}^{(1)}$ from (5.2.18). If we next relax the requirement that the numerical and physical core size parameters, ϵ and $\varepsilon^{\mathrm{ttm}}$, should match exactly, we are free to adjust the latter in such a way that its logarithmic contribution in the propagation law just makes up for the velocity discrepancy. Specifically, we set

$$\ln\frac{2}{\varepsilon^{\mathrm{ttm}}} + C^{\mathrm{ttm}} = \ln\frac{2}{\epsilon} + C , \tag{5.2.25}$$

that is,

$$\varepsilon^{\mathrm{ttm}} = \epsilon\exp\left(C^{\mathrm{ttm}} - C\right) . \tag{5.2.26}$$

This last correction is easily implemented. In particular, it does not require the additional evaluation of curve derivatives, and therefore has the desired robustness. This scheme is used in the sequel.

5.2.2 Performance in the Klein–Majda Regime

Here we compare results obtained with the thin-tube vortex element schemes of the last section with the asymptotics for nearly straight vortices from Sect. 4.1. This is a relevant comparison, because we have as yet assumed constant core structures in the numerical corrections, and we have shown earlier that this is a valid approximation for filaments in the Klein–Majda regime.

The comparison is furthermore of interest as it amounts to a mutual check of two very different modelling approaches:

1. The asymptotic theory
   - approximates the local contribution to the filament velocity up to errors of order $O(\epsilon)$, but
   - linearizes the Biot–Savart integral, thereby introducing Taylor series truncation errors of order $O(\delta^2) = O(\ln \epsilon)$.
2. The standard thin-tube model
   - is able to represent the local contribution of the induced velocity merely qualitatively, that is, with errors of order $O(1)$, while it
   - determines the nonlocal part of the Biot–Savart integral essentially with machine accuracy.
3. The corrected thin-tube models combine the advantages of both; that is,
   - they achieve the accuracy of the slender vortex asymptotics as regards the local contribution and
   - they also handle the nonlocal induction velocities with high accurary.

It follows that we can quantify the dominant errors of the standard vortex element scheme by comparison with the asymptotics, and to quantify the linearization errors in the asymptotic scheme by comparison with the improved thin-tube models.

The comparisons shown below are all based on a value of the Klein–Majda expansion parameter of $\delta^2 = 0.25$, which corresponds with a core size parameter $\epsilon \sim 0.015$, as pointed out earlier. The local errors of the asymptotics will thus be in the percent range, whereas regarding the linearization errors from the Biot–Savart integral, we operate at the limit of applicability of the asymptotics.

Choice of the Core Size Parameter. From the asymptotic analysis we have seen that the expression

$$\ln\left(\frac{1}{\epsilon}\right) + C \tag{5.2.27}$$

entirely reflects the leading-order influence of vortex core size and vortex core structure on the filament motion. In the computations of this section we assume a constant reference value for this quantity, which is appropriate in the Klein–Majda regime of geometric perturbations as pointed out earlier. For the geometry expansion parameter δ we have

$$\ln\left(\frac{1}{\epsilon}\right) + C = \frac{1}{\delta^2} - \ln(2\delta) \tag{5.2.28}$$

so that δ is constant as well. This is in line with evaluations of the asymptotics in Klein and Majda (1991b), with which we will compare the results of simulations based on the thin-tube models of this section.

In the following we intend to compare asymptotic predictions for some physical vortex with thin-tube model calculations for the same vortex, assuming that the physical vortex does have the same core structure as the numerical one. To obtain the relevant core size parameter $\epsilon_{g,s}$ for the Gaussian and "cosech" core smoothing functions in (5.2.6), respectively, when comparing with the Klein–Majda theory for given $\delta^2 = 0.25$, we must require

$$\ln\left(\frac{1}{\epsilon_{g,s}}\right) + C_{g,s} = \frac{1}{\delta^2} - \ln(2\delta) \tag{5.2.29}$$

where the values for $C_{g,s}$ follow from (3.3.117) as

$$C_g = -0.3212 \qquad C_s = -0.2201\,. \tag{5.2.30}$$

Space-Time Rescaling. There is an interesting transformation for the asymptotic filament equation (4.3), which has been crucial in obtaining the present results by helping overcome a somewhat subtle scaling issue: The asymptotic filament equation has been set up with an arclength coordinate as the independent variable parameterizing the centerline. This is in fact an important aspect in Hasimoto's transformation.

The semispectral numerical scheme used to solve the filament equation assumes periodicity of the solution with period 2π in the arclength variable. Setting up a computation with the thin-tube model, it is natural to assume periodicity in the cartesian axial direction.

The following transformation allowed us to move back and forth between a 2π-periodic arclength variable and the corresponding arclength in the thin-tube setup.

Let L be the unscaled arclength period of some given centerline initial data, and $\nu = L/2\pi$. Then, if $\psi(\sigma, \tau)$ solves the filament equation for some given parameter δ, then

$$\psi^*(\sigma^*, \tau^*) = \nu\psi(\sigma, \tau) \tag{5.2.31}$$

with

$$\sigma^* = \sigma/\nu \qquad \tau^* = (1 + \delta^2 \ln(\nu))\tau/\nu^2 \tag{5.2.32}$$

solves the exact same filament equation in the new variables (σ^*, τ^*), albeit with the rescaled expansion parameter

$$\delta^{*2} = \frac{\delta^2}{1 + \delta^2 \ln(\nu)}\,. \tag{5.2.33}$$

Arclength Versus Cartesian Coordinates. We remark in addition that, given initial data for the filament function $\psi(\sigma, 0)$, it is not a trivial task to determine the appropriate geometrical configuration in cartesian coordinates, which we need for the vortex element scheme. The nonlinear mapping between arclength and the main periodic cartesian coordinate must be accounted for to obtain accurate results.

Static Comparisons. Here we compare the asymptotically and numerically predicted filament centerline velocities for the following plane curve initial data

$$\boldsymbol{x}(s) = s\boldsymbol{t}_0 + \delta^2 \tilde{a} \sin(2s/\delta) , \tag{5.2.34}$$

with $\delta^2 = 0.25$ and varying amplitudes $\tilde{a}$. We denote by $\boldsymbol{t}_0, \boldsymbol{n}_0, \boldsymbol{b}_0$ the cartesian orthogonal, normalized basis from Fig. 4.1. The centerline according to (5.2.34) is entirely embedded in the $\boldsymbol{t}_0$–$\boldsymbol{n}_0$ plane and a symmetry consideration taking into account the leading order axisymmetry of the vorticity distribution shows that the induced velocity must be everywhere orthogonal to that plane, that is, in the direction of $\boldsymbol{b}_0$. It will thus suffice to plot only the scalar velocity component in that direction in the following comparisons.

The asymptotic theory according to Chap. 4 yields

$$v^{\mathrm{K\&M}}(\sigma) = -\frac{4\tilde{a}\ \sin(2\sigma)}{\left[1 + \delta^2 \tilde{a}^2\ \cos^2(2\sigma)\right]^{3/2}} - \delta^2 \tilde{a}\ \hat{I}(2)\ \sin(2\sigma) , \tag{5.2.35}$$

where $\sigma = s/\delta$ and $\hat{I}$ is the symbol of the integral operator, $I[\cdot]$, in (4.3). This velocity distribution will now be compared with the nodal velocities of the standard and improved thin-tube methods.

Figures 5.11 and 5.12 exhibit the results obtained for $\tilde{a} = 0.1$. Specifically, Fig. 5.11 shows the predicted filament propagation speed for (i) the local induction approximation (LIA) by Arms (see Batchelor, 1967 or Hasimoto, 1972), (ii) the standard, uncorrected thin-tube model (ttm) from Knio and Ghoniem (1990), and (iii) the asymptotic analysis from Klein and Majda (1991a) (K&M). First we observe the considerable difference between the LIA on the one hand, and the asymptotic and standard thin-tube results on the other. Evidently, nonlocal induction contributes about 7% of the full induced velocities. Second, we find a reasonable agreement between the standard thin-tube scheme and the asymptotic results. As we will see shortly, the observed discrepancy between these two distributions is due mainly to the inaccurate *local* predictions of the numerical scheme, namely, due to its misrepresentation of the first-order-vorticity component $\zeta_{11}^{(1)}$ as discussed above.

Next we consider in Fig. 5.12 the differences $(v^{\mathrm{K\&M}} - v^{\mathrm{ttm}})$ and $(v^{\mathrm{K\&M}} - v^{\mathrm{ttm}}_{\mathrm{corr}})$ between the asymptotic theory and the standard and improved thin-tube models, respectively. Figure 5.12a shows that the deviations between the theoretical and the numerical model predictions may be reduced drastically. In fact, close examination shows that the curvature-scaled velocity difference $(v^{\mathrm{K\&M}} - v^{\mathrm{ttm}})/\kappa$ is constant up to fluctuations of about 2%. This indicates

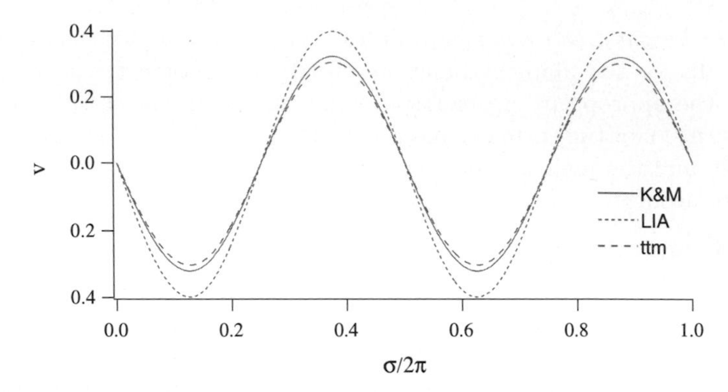

Fig. 5.11. Velocity predictions for the plane distortions from (5.2.35) for $\tilde{a} = 0.1$, $\delta^2 = 0.25$. Adapted from Klein and Knio (1995)

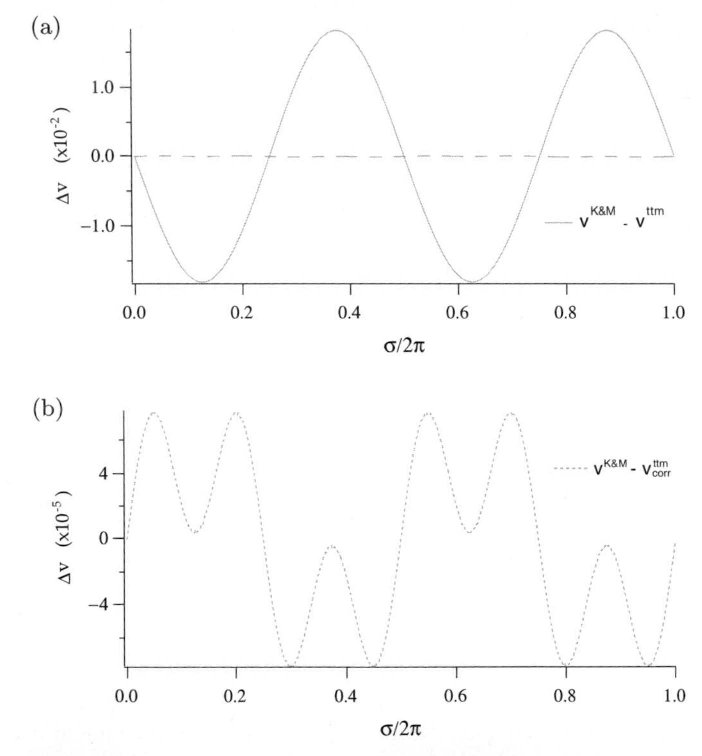

Fig. 5.12. Deviations between asymptotic and numerical predictions for the plane curve initial data from (5.2.34). (**a**) Comparison for the standard and improved thin-tube models; (**b**) comparison for the improved model alone with rescaled graph axis. Adapted from Klein and Knio (1995)

that the suggested local, curvature-based correction will likely improve the situation drastically. The remaining differences between the asymptotic prediction and the improved thin-tube model, shown in Fig. 5.12b, are of the order $O(10^{-5})$. This corresponds with remaining relative differences of less than 0.3‰ of the maximum induced velocity.

Regime of Validity of the Klein–Majda Asymptotics. With the small value for the geometry amplitudes $\tilde{a} = 0.1$ in (5.2.34), the ratio of displacement from the straight line and perturbation wavelength is just $0.1/4\pi \approx 7 \cdot 10^{-3}$. The linearization of the nonlocal part of the Biot–Savart integral in the Klein–Majda theory may well be expected to yield very good results in this case. Here we study the scaling of the leftover deviations between asymptotics and numerics as the perturbation amplitude increases. Thus we consider a sequence $\tilde{a} = 0.01$, 0.1, 1.0, and 4.0. Figure 5.13 shows the behavior of the two largest Fourier components, modes $\xi = 2$ and $\xi = 6$, of the deviations as a function of $\tilde{a}$ in a double-logarithmic diagram. We observe, for mode $\xi = 6$, a quite accurate scaling with $\tilde{a}^3$. This is true for mode $\xi = 2$ only for the larger values of the perturbation amplitude. The $\tilde{a}^3$ scaling matches with the expectation that the main effects here result from the linearization of the nonlocal contributions. The deviation from this scaling at small amplitudes for the mode $\xi = 2$ indicates that there are additional effects that are strictly coupled to the leading harmonic of the filament geometry. We suspect that these error contributions stem from the fact that the overlap between neighboring vortex elements is finite so that the assumption of an axially invariant numerical vortex core structure is not quite accurate.

For the present plane curve initial data we conclude that there should be a limit amplitude $\tilde{a}^*$ beyond which the linearization errors in the asymptotics become comparable to the local induction errors of the standard thin-tube model. A close examination of the data shows that this is the case for $\tilde{a}^* \approx 1.0$ (see Fig. 5.14 for a qualitative impression). This provides the desired

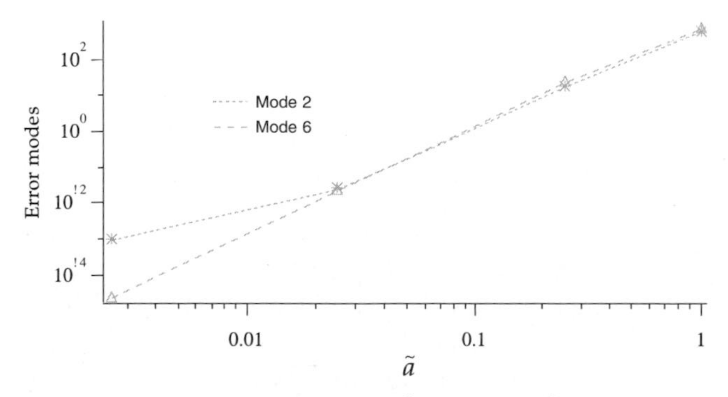

Fig. 5.13. Scaling of the Fourier modes $\widehat{\Delta v}(\xi = 2), \widehat{\Delta v}(\xi = 6)$ of the velocity deviations between the asymptotic and improved thin-tube predictions with the displacement amplitude $\tilde{a}$ from (5.2.34). Adapted from Klein and Knio (1995)

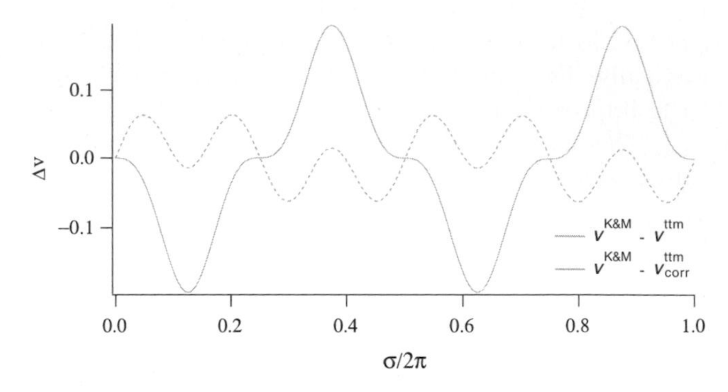

Fig. 5.14. Deviations in predicted velocities for a large displacement amplitude $\tilde{a} = 1.0$ in (5.2.34). Adapted from Klein and Knio (1995)

rough estimate of the validity of the Klein–Maja asymptotics for the core size parameter $\epsilon \approx 0.015$. As we will see in the next section, the same positive conclusion may be drawn for dynamical problems, not just for the present static comparison for a given, fixed filament geometry.

Static Comparisons. In this section we consider the temporal evolution of vortices in the Klein–Majda regime and compare again evaluations of the asymptotics and vortex element-based numerical computations. After a comparison for plane curve initial data, as described in the previous static comparison, we discuss results obtained for more general multimode initial data. This will include one case that is definitely outside the formal regime of validity of the Klein–Majda asymptotics; the results for this case are used, in particular, to check whether this theory is able to at least produce qualitatively correct results when applied to extreme data.

The results of this section have all been generated using 1024 vortex elements in the thin-tube scheme and 512 Fourier modes in the solution of the asymptotic filament equation from Chap. 4. Issues of efficiency of the schemes – as opposed to accuracy – will be addressed below in Subsect. 5.2.4.

Sinusoidal Plane Curve Initial Data. We analyze here the vortex filament dynamics given the plane curve initial data from (5.2.34) with $\tilde{a} = 1.0$. Figure 5.15a shows the temporal evolution of the maximum curvature within the computational domain. The results obtained with the vortex element scheme (ttm), with the corrected vortex element scheme (ttm_corr), and on the basis of the asymptotics (K&M) are in qualitative agreement. There is nontrivial quantitative disagreement, however. First, we observe a systematic time lag for the standard thin-tube model. Second, there remains a marked difference between the curves, even if we rescale time to eliminate the mentioned time lag. As seen in Fig. 5.15b, the standard vortex element scheme misses some of the dynamics within the intermediate period of large curvatures between rescaled times $\tau^* = 0.43$ and $\tau^* = 0.85$, where $\tau^* = \tau/\tau_I$ and τ_I is time

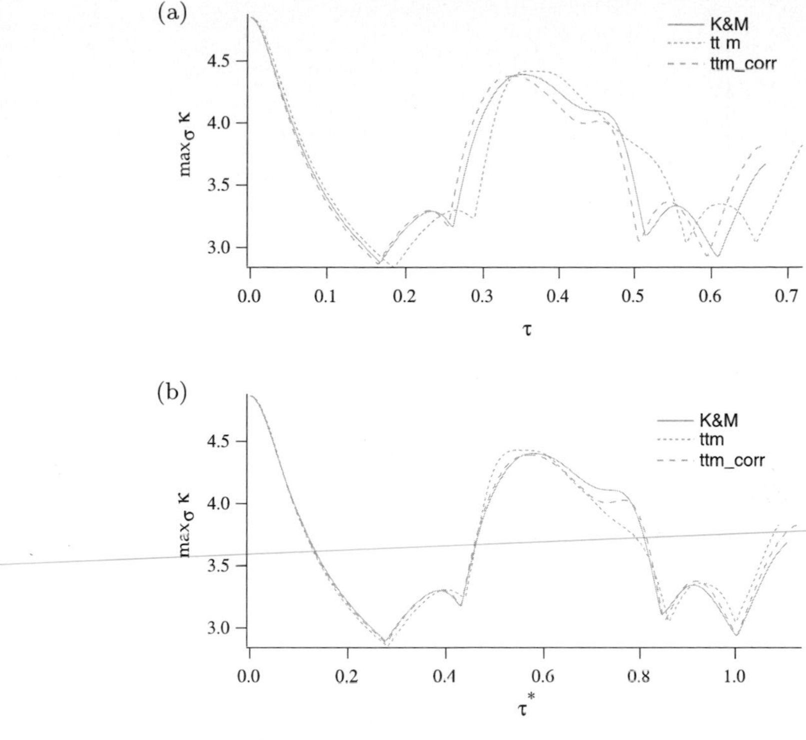

Fig. 5.15. Temporal evolution of maximum curvature of the filament centerline, computed using the standard and corrected thin-tube vortex element models, and the asymptotic theory from Chap. 4 (Klein and Majda, 1991b), for the initial data from (5.2.34). (**a**) Direct comparison of computed results, (**b**) comparison after rescaling time to match the last local minimum seen in (a). Adapted from Klein and Knio (1995)

of the last local minimum for each of the curves in Fig. 5.15a, with $I \in \{\mathsf{ttm}, \mathsf{ttm\_corr}, \mathsf{K\&M}\}$.

In Figs. 5.16a and b, we demonstrate that not only the evolution of the curvature maximum but also the spatial distributions of the curvature at some selected times match convincingly for the asymptotic and the corrected thin-tube models. Figure 5.16a shows the spatial distribution of curvature at time $\tau = 0.3968$ in the vortex element simulation and at time 0.4062 for the solution of the asymptotic filament equation. These times correspond to equal values of τ^* in Fig. 5.15b. The maximum deviation between the results is less than 1.6%.

We have seen in Chap. 4 that the nonlinear perturbation theory involves an explicit representation of vortex stretching. The quadratically nonlinear functional

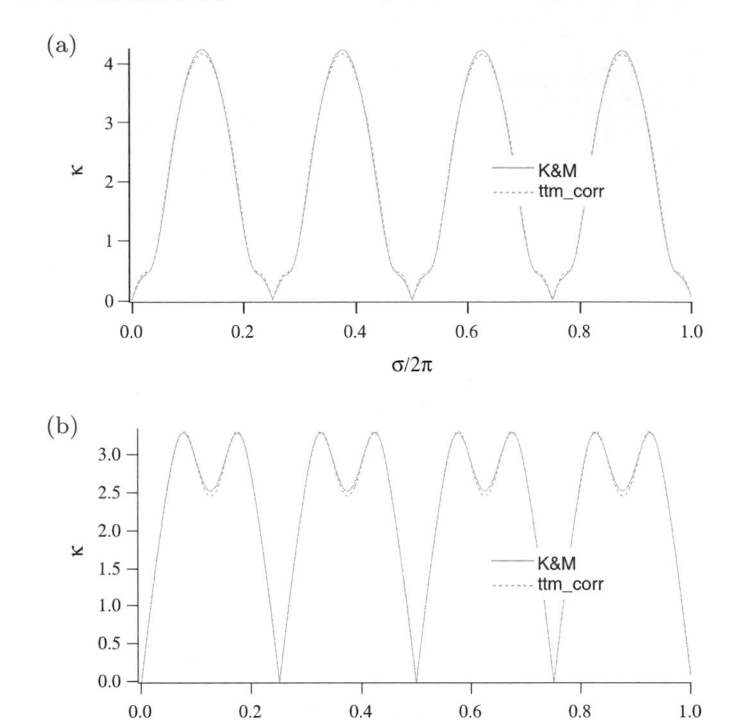

Fig. 5.16. Spatial distributions of curvature at selected times for the computations from Fig. 5.15. Times are chosen such that the rescaled time variables τ^* used in Fig. 5.15b match, (**a**) $\tau = 0.3968$ (ttm_corr), $\tau = 0.4062$ (K&M), (**b**) $\tau = 0.5284$ (ttm_corr), $\tau = 0.5409$ (K&M). Adapted from Klein and Knio (1995)

$$\left\{ \lim_{\ell \to 0} \left(\frac{\dot{\ell}}{\ell} \right) \right\} (\sigma, \tau) = \frac{\delta^2}{4} \int_{-\infty}^{\infty} \frac{1}{|h|} \left(\overline{\psi}(\sigma + h, \tau) \psi(\sigma, \tau) - \overline{\psi}(\sigma, \tau) \psi(\sigma + h, \tau) \, dh \right) \tag{5.2.36}$$

of the filament function describes the local rate of self-stretching with *relative* errors of order $O(\delta)$, provided that the exact filament function is inserted. While there is an elegant approximate expression for (5.2.36) in the framework of the asymptotics and the filament function, the stretching rate must be calculated explicitly from the motion of the individual vortex blobs when we use the thin-tube vortex element scheme. The approximation used here is simply

$$\frac{1}{\left| \boldsymbol{\chi}_{i+1} - \boldsymbol{\chi}_i \right|} \frac{d}{dt} \left| \boldsymbol{\chi}_{i+1} - \boldsymbol{\chi}_i \right| . \tag{5.2.37}$$

In Fig. 5.17 we compare the temporal evolution of the maximum rate of self-stretching within the computational domain as computed by the standard thin-tube model (ttm), the corrected scheme (ttm_corr), and the Klein–Majda asymptotics (K&M). Again there is qualitative agreement. This is true, in

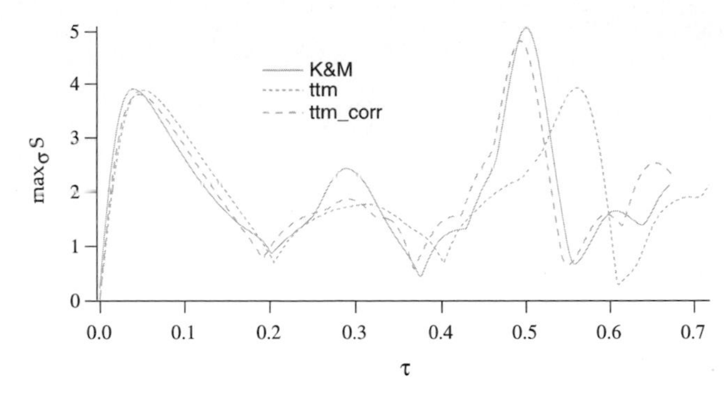

Fig. 5.17. Temporal evolution of maximum local stretching rates, calculated on the basis of the standard and improved thin-tube vortex element methods and based on the Klein–Majda asymptotics for the initial data from (5.2.34). Adapted from Klein and Knio (1995)

particular, for the corrected thin-tube model and the asymptotics are in very good agreement! The relative deviations between the different model results are larger here than those observed for the maximum curvature. The main reason should be that the expression for the stretching rate shown in (5.2.36) requires two successive linearizations, whereas the derivation of the filament equation with self-stretch relied on only a single one: In deriving the filament equation with self-stretch in (4.1.8), we have linearized the finite part of the line Biot–Savart integral. This equation determines the temporal evolution of the centerline curvature via $\kappa = |\psi|$. To obtain the present results for the local stretching rate, we have to first solve the filament equation for $|\psi|$, which involves this linearization, and then evaluate the approximate expression for the stretching rate from (5.2.36), which relies on an additional asymptotic approximation. This involvement of two successive approximation steps may explain the less convincing agreement of the stretching rate data between the asymptotics and the vortex element numerics as compared to the agreement seen in the curvature distributions.

More General Initial Data. Here we consider initial displacements of the vortex filament centerline according to

$$\boldsymbol{X}(s) = s\boldsymbol{t}_0 + \delta^2 \left[\sin(s/\delta) + \sin(2s/\delta)\right] \boldsymbol{n}_0 \tag{5.2.38}$$

and

$$\boldsymbol{X}(s) = s\boldsymbol{t}_0 + \delta^2 \left[\sin(2s/\delta) + \sin(3s/\delta)\right] \boldsymbol{n}_0 \,, \tag{5.2.39}$$

and again we let $\delta^2 = 0.25$.

Figures 5.18a and b exhibit the temporal evolution of maximum curvature for the asymptotic and vortex element models in analogy with Figs. 5.15a and b. The time interval for rescaling the graphs in Fig. 5.18b is the time of occurance of the second local minima on the individual curves. In this

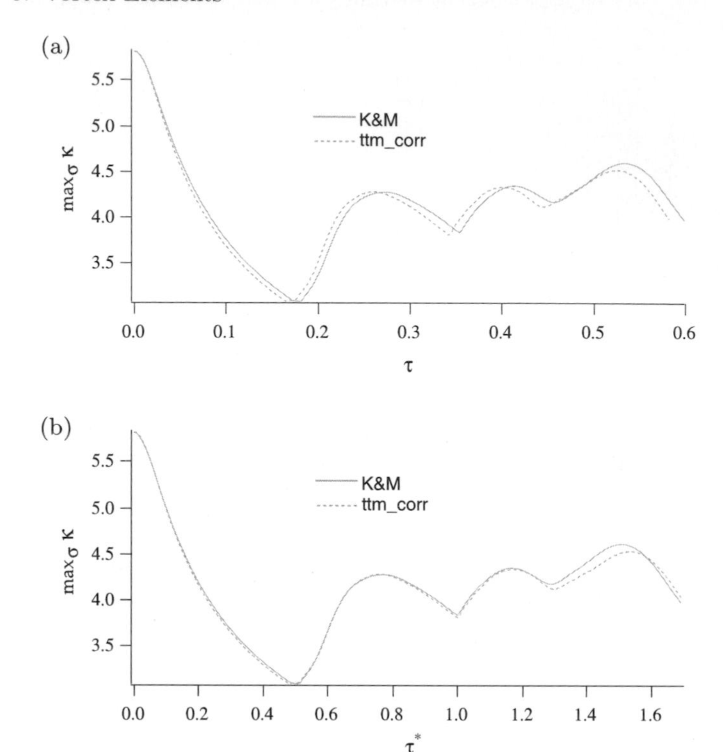

Fig. 5.18. Temporal evolution of maximum curvature of the filament centerline, computed using the standard and corrected thin-tube vortex element models, and the asymptotic theory from Chap. 4 (Klein and Majda, 1991b), for the initial data from (5.2.38). (**a**) Direct comparison of computed results, (**b**) comparison after rescaling time to match the second local minimum seen in (a). Adapted from Klein and Knio (1995)

example we again observe quite convincing agreement. Figures 5.19a and b compare the spacial distributions of curvature at times $\tau = 0.3598$ (a) and $\tau = 0.4716$ (b) in the vortex element computation, and times $\tau = 0.3719$ (a) and $\tau = 0.4874$ (b) in the solutions of the filament equation. The agreement is again quite acceptable. This is remarkable since the maximum curvature reaches values of $\kappa_{\max} \approx 4.5$, which is larger than the inverse $1/\delta^2 = 4.0$ of the expansion parameter, so that the present cases are at the limits of validity of the asymptotics.

The initial data from (5.2.39) represent an extreme test in this sense. Even at time $\tau = 0$ the maximum curvature within the domain is of the order of $\kappa_{\max} \approx 20$, thereby exceeding $1/\delta^2$ by a factor of 5. This places the *initial* data outside the formal regime of validity of the asymptotics. Nevertheless, we compare again the vortex element numerics with evaluations of the asymptotics in Figs. 5.20 and 5.21. We find good qualitative agreement, the

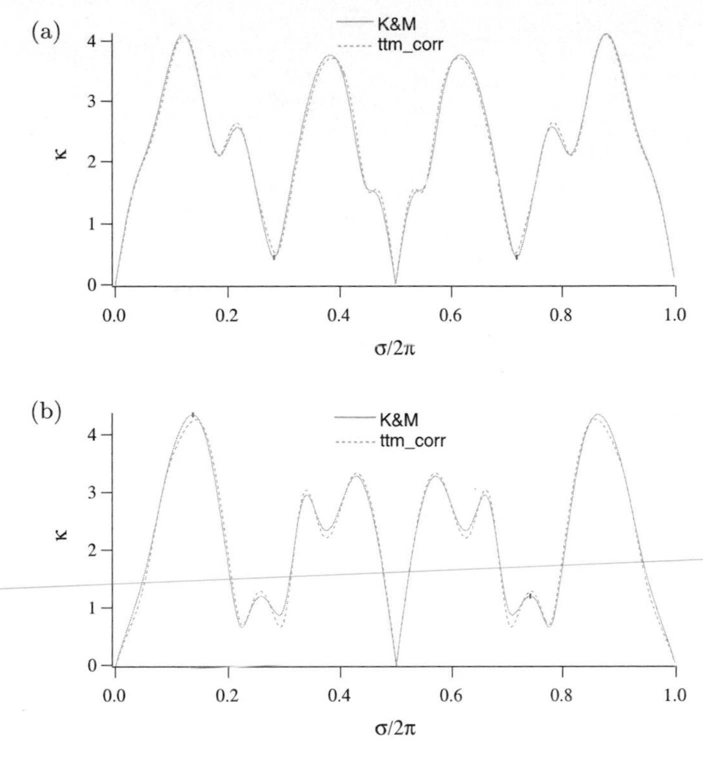

Fig. 5.19. Spatial distributions of curvature at selected times for the computations from Fig. 5.18. Times are chosen such that the rescaled time variables τ^* used in Fig. 5.18b match, (**a**) $\tau = 0.3598$ (ttm_corr), $\tau = 0.3719$ (K&M), (**b**) $\tau = 0.4716$ (ttm_corr), $\tau = 0.4874$ (K&M). Adapted from Klein and Knio (1995)

major difference being a somewhat "rougher" time evolution of the asymptotic solutions. We suspect that some dispersive effects of the Biot–Savart integral have been eliminated in the linearizations that entered the filament equation, whereas these are fully present in the vortex element numerics.

Figure 5.21 shows that the spatial curvature distributions are again in qualitative agreement. Both simulations exhibit, at times as early as $\tau \approx 0.1$, high-frequency components that correspond to perturbations on the core size scale. These oscillations are more pronounced in the solutions of the filament equation, consistent with the earlier observation that this model produces somewhat "rougher" results.

This is explained by the Fourier spectra of the filament function in Fig. 5.22. We observe nonnegligible amplitudes at both output times $\tau = 0$ and $\tau = 0.0155$ at wave numbers $\xi \approx 50$, which correspond with the core size scale. Notice that the spectra feature smooth distributions within the range of $-100 < \xi < 100$ even at the initial time. This may by surprising, considering that the initial data in (5.2.39) consist of only two Fourier modes. The

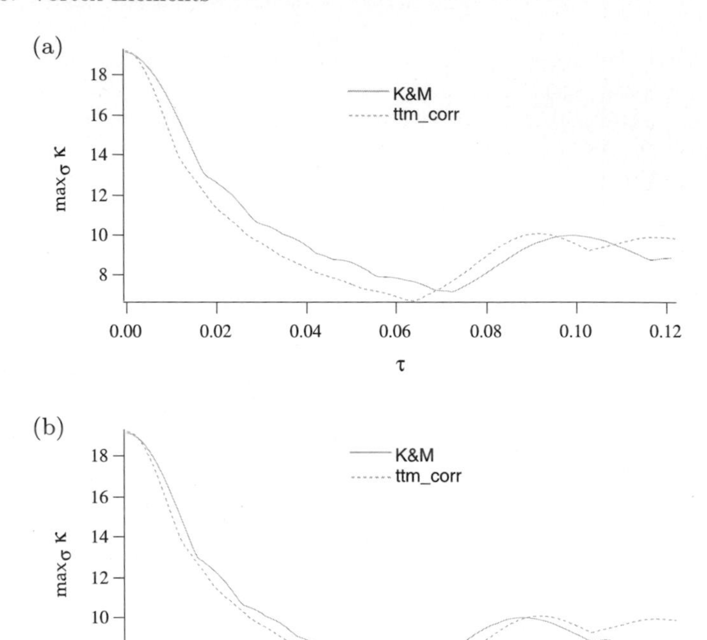

Fig. 5.20. Temporal evolution of maximum curvature of the filament centerline, computed using the standard and corrected thin-tube vortex element models, and the asymptotic theory from Chap. 4 (Klein and Majda, 1991b), for the initial data from (5.2.39). (**a**) Direct comparison of computed results, (**b**) comparison after rescaling time to match the first local minimum seen in (a). Adapted from Klein and Knio (1995)

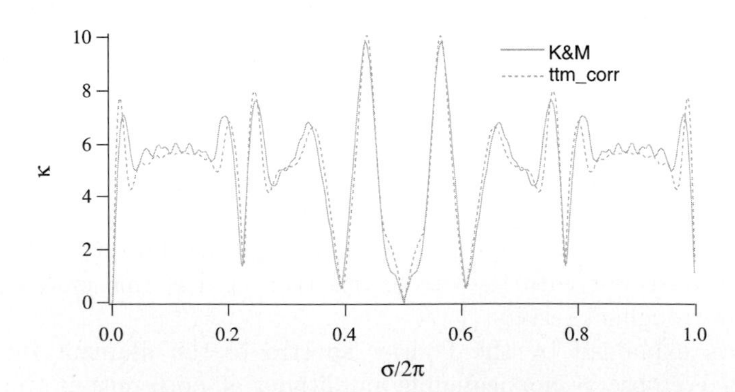

Fig. 5.21. Spatial distributions of curvature at selected times for the computations from Fig. 5.20. Times are chosen such that the rescaled time variables τ^* used in Fig. 5.20b match, $\tau = 0.09341$ (ttm_corr), $\tau = 0.1055$ (K&M). Adapted from Klein and Knio (1995)

reason is that the filament equation uses an *arclength* variable on the actual centerline whereas the initial data are prescribed in terms of the arclength on the unperturbed straight line. This coordinate transformation induces the observed spread of the Fourier spectrum.

Considering that the asymptotic theory relies on the assumption that all scales of the filament geometry are much larger than the vortex core size, we reiterate that the present initial data place the filament outside of the formal regime of validity of the asymptotics. Nevertheless the theory yields qualitative agreement with the corrected thin-tube vortex element scheme, which incorporates a more accurate representation of nonlinear self-induction.

Another interesting feature of Fig. 5.22 is that the spectrum of the data at the later time is *narrower* than that at the initial time. In this sense the asymptotic theory is robust in that it has a trend of bringing out-of-range data back into the range of validity. In addition, this narrowing of the spectrum in time corresponds with an inverse energy cascade from small to large scales. Considering the earlier observations of energy transfer from the lowest wave numbers to the core size scale in other examples, we conclude that the energy appears to be trapped within the wave number band between the spacial period (wave number $\xi = 1$) and the core size scale (wave number $\xi \approx 50$ in the present cases).

In summary:

i) The local asymptotic corrections of the thin-tube vortex element scheme from Klein and Knio (1995) lead to convincing agreement between this model and the asymptotics for slender filaments perturbed from a straight line in Klein and Majda (1991a, 1991b). This is also true, in particular, for realistic values of the core size parameter $\epsilon \approx 0.015$, which corresponds with a value for the filament asymptotics of $\delta^2 = 0.2 \ldots 0.25$.

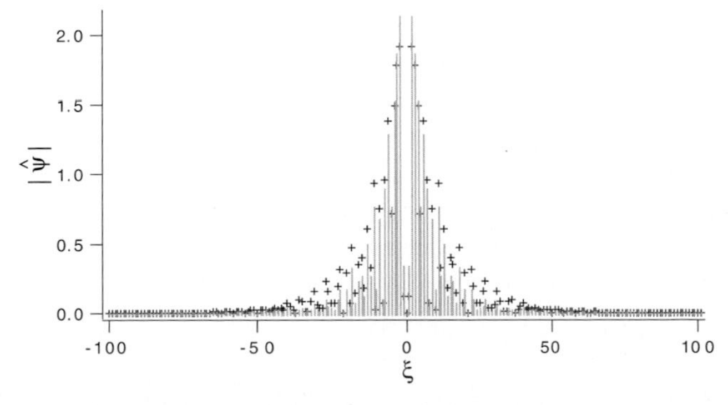

Fig. 5.22. Fourier spectra of the filament function extracted from a simulation for the initial data in (5.2.39) at $\tau = 0$ (crosses) and $\tau = 0.1055$ (vertical bars). Adapted from Klein and Knio (1995)

ii) Even outside the formal regime of validity we find qualitative agreement between the asymptotics and the corrected thin-tube model. The asymptotic predictions on the other hand are obtained at a small fraction of the computational cost of the thin-tube computations. Means to improve the efficiency of the numerical model are later addressed in Subsect. 5.2.4.

5.2.3 Representation of Core Dynamics in Thin-Tube Simulations

Several different approaches have been applied to relate the core structure evolution to the large-scale motion and deformation of slender vortex filaments. In many cases, these have been based on ad hoc relationships that govern the core structure response to stretching and diffusion. Examples include the conservation of volume approximation and the viscous core spreading approach (Leonard, 1980).

The conservation-of-volume model is obtained from the fact that for inviscid flow, the integral volume of vorticity is conserved. Approximating this constant of the motion by the product of total filament arclength and the square of the (numerical) core radius yields a simple relationship between local core structure and global filament motion, which may be easily enforced in filament computations (Leonard, 1980). Unfortunately, extension of this modeling approach to more general flow conditions is not obvious; in particular, difficulties in accommodating viscous diffusion and/or axial flow along the filament centerline may be anticipated.

The core spreading approach is based on the assumption that the (leading-order) axial vorticity distribution within the core of the filament is Gaussian (Leonard, 1985), and that it is governed by the two-dimensional heat equation. By identifying the numerical core size with the standard deviation of a Gaussian profile, its evolution is simply expressed as $\tilde{\delta}^2(\tilde{t}) = \tilde{\delta}^2(\tilde{t}_0)+4\tilde{\nu}(\tilde{t}-\tilde{t}_0)$. Application of this approach may be justified in some simple situations, for example, for an isolated slender ring with Gaussian leading-order core structure. However, the use of core spreading in general is problematic, since the vortex core does not always feature a Gaussian structure and, unlike the isolated vortex ring, may be affected by stretching, diffusion, and axial flow mechanisms evolving simultaneously.

To overcome such limitations, in this section we generalize the analysis introduced above to account for the effects of diffusion and stretching in the vortex core. To this end, it is necessary to account for variations in the core structure coefficient, $C(t)$, which as discussed in Chap. 3 is defined by

$$
\begin{aligned}
C(t) &= C_v(t) + C_w(t) \\
C_v(t) &= \lim_{\overline{r}\to\infty} \left(\frac{4\pi^2}{\Gamma^2} \int_0^{\overline{r}} \overline{r} v^{(0)\,2} d\overline{r} - \ln \overline{r} \right) - \frac{1}{2} \\
C_w(t) &= -\frac{1}{2} \left(\frac{4\pi}{\Gamma} \right)^2 \int_0^{\infty} \overline{r} w^{(0)\,2} d\overline{r} ,
\end{aligned}
\tag{5.2.40}
$$

where $\overline{r} \equiv r/\delta$.

To determine the evolution of $C(t)$, we rely on the analyses of Callegari and Ting (1978) and Klein and Ting (1995), who provide analytical solutions for the evolution of the leading-order axial vorticity and axial velocity distributions. Below, we summarize these viscous core solutions and then manipulate them to obtain suitable expressions for the core structure coefficient, $C(t)$. These expressions are later incorporated into the thin-tube model of Subsect. 5.2.1, which results in a slender filament scheme that correctly accounts for stretching and diffusion effects.

5.2.3.1 Viscous Core Solutions. Following Klein and Ting (1995), we describe the core structure in terms of the evolution equations for the leading-order axial velocity and axial vorticity distributions, respectively,

$$w_t = \frac{K^2}{\overline{r}} (\overline{r} w_{\overline{r}})_{\overline{r}} + \frac{\overline{r}^3}{2} \frac{\dot{S}}{S} \left(\frac{w}{\overline{r}^2} \right)_{\overline{r}} , \tag{5.2.41}$$

$$\zeta_t = \frac{K^2}{\overline{r}} (\overline{r} \zeta_{\overline{r}})_{\overline{r}} + \frac{\dot{S}}{S} \frac{\overline{r}^2 \zeta_{\overline{r}}}{2\overline{r}} . \tag{5.2.42}$$

The constant K is a nondimensional viscosity defined by

$$K = \frac{\nu}{\Gamma \delta^2} , \tag{5.2.43}$$

while S denotes the total arclength of the slender filament. Thus, terms involving K reflect viscous diffusion effects while terms involving S account for stretching phenomena. Inserting (5.2.41) and (5.2.42) into (5.2.40b) and (5.2.40c), respectively, leads, after some algebraic manipulation, to the following evolution equations for the core structure coefficients:

$$\dot{C}_w(t) = \frac{16\pi^2}{\Gamma^2} \left[K^2 \int_0^\infty z(w_z)^2 \mathrm{d}z + \frac{3\dot{S}}{2S} \int_0^\infty z w^2 \mathrm{d}z \right] , \tag{5.2.44}$$

$$\dot{C}_v(t) = \frac{8\pi^2}{\Gamma^2} \int_0^\infty v(z v_t) \mathrm{d}z = -\frac{8\pi^2 K^2}{\Gamma^2} \int_0^\infty z \zeta^2 \mathrm{d}z + \frac{\dot{S}(t)}{2S(t)} . \tag{5.2.45}$$

As shown by Klein and Ting (1995), solutions of (5.2.44) and (5.2.45) are more conveniently expressed in terms of the stretched time and radial coordinates

$$\varsigma = \int_0^t \frac{S(t')\mathrm{d}t'}{S_0} , \tag{5.2.46}$$

$$\xi = \overline{r} \sqrt{\frac{S(t)}{S_0}} . \tag{5.2.47}$$

Using this change of variables, and setting $\chi = w\, S/S_0$, transforms the governing equation for w into the standard axisymmetric heat equation

$$\left[\partial_\varsigma - \left(\frac{K^2}{\xi}\right)\partial_\xi(\xi\partial_\xi)\right]\chi(\varsigma,\xi) = 0\ , \tag{5.2.48}$$

with initial condition $\chi(0,\xi) = w_0(\xi)$. Similarly, (5.2.45) can be transformed into (5.2.48) if the change of variables is used in conjunction with $\chi = \zeta\, S/S_0$. In the latter case, the appropriate initial condition for (5.2.48) is $\chi(0,\xi) = \zeta_0(\xi)$.

Since χ decreases rapidly in ξ, the solution of (5.2.48) can be efficiently represented in terms of a series of Laguerre polynomials. Following Kleinstein and Ting (1971), the solution is expressed for arbitrary ς_0 as

$$\chi(\varsigma,\xi) = \frac{e^{-\xi^2}}{\varsigma+\varsigma_0}\left\{c_0 + \sum_{n=1}^{N} c_n L_n(\xi^2)\left[\frac{\varsigma_0}{\varsigma+\varsigma_0}\right]^n + \mathrm{O}\left(\left[\frac{\varsigma_0}{\varsigma+\varsigma_0}\right]^{N+1}\right)\right\}. \tag{5.2.49}$$

The coefficients c_n are obtained from the initial data exploiting the orthogonality of the Laguerre polynomials:

$$c_n = \varsigma_0\int_0^\infty \chi\left(0,\, 2K\sqrt{\lambda\varsigma_0}\right) L_n(\lambda)\mathrm{d}\lambda\ . \tag{5.2.50}$$

Note that $c_0 = \langle\chi\rangle/4\pi K^2$ is independent of ς_0, and that $\langle\chi\rangle = 2\pi\int\chi\, r\, dr$, the total strength is time invariant. This solution decays toward the leading Gaussian similarity solution as ς^{-2} for $\varsigma\to\infty$. This becomes apparent in the above representation when the degree of freedom of choosing the time shift ς_0 is used to eliminate the coefficient c_1. The full series does not change when the time shift is varied, but the approximation obtained by any truncated version is drastically improved (see Kleinstein and Ting, 1971).

In the inviscid limit, $K\to 0$, the stretched variable χ remains frozen, and the corresponding evolution of the core structure coefficients admits a simple form. The solutions, obtained by Klein and Ting (1995), are expressed as

$$C_w(t) = \left[\frac{S_0}{S(t)}\right]^3 C_w(0)\ , \tag{5.2.51}$$

$$C_v(t) = C_v(0) + \ln\sqrt{\frac{S(t)}{S_0}}\ . \tag{5.2.52}$$

If $C_w(0) = 0$, then $C(t)\equiv C_v(t)$ and the formula in (5.2.52) coincides with the simple conservation of volume ansatz. However, the presence of axial flow $(C_w(0)\neq 0)$ forbids such a simplified treatment.

For more general situations including stretching *and* diffusion, the dependence of the core structure coefficients on the stretched time variable ς cannot be simply absorbed through simple formulae. We first focus on deriving expressions for determining $C_v(t)$ and then sketch the construction of $C_w(t)$.

The starting point for constructing $C_v(t)$ is the corresponding evolution equation (5.2.45). Using the change of variables $\varsigma = \int_0^t S(t')\mathrm{d}t'$, we have

$$\frac{\mathrm{d}C_v}{\mathrm{d}\varsigma} = -\frac{8\pi^2K^2}{\Gamma^2}\frac{S_0}{S}\int_0^\infty z\zeta^2\mathrm{d}z + \frac{1}{2S}\frac{\mathrm{d}S}{\mathrm{d}\varsigma} . \tag{5.2.53}$$

Equation (5.2.53) immediately yields

$$C_v(\varsigma) - C_{v,0} = -\alpha I(\varsigma) + \ln\sqrt{\frac{S(\varsigma)}{S_0}} , \tag{5.2.54}$$

where

$$I(\varsigma) \equiv \int_0^\varsigma \frac{S_0}{S}\int_0^\infty z\zeta^2\mathrm{d}z\mathrm{d}\varsigma' \tag{5.2.55}$$

and

$$\alpha \equiv \frac{8\pi^2K^2}{\Gamma^2} . \tag{5.2.56}$$

Note that in the inviscid limit, $\alpha = 0$, and the evolution of C_v can be found without resorting to the stretched coordinate system. In this case, (5.2.54) may be simply recast as

$$C_v(t) - C_{v,0} = \ln\sqrt{\frac{S(t)}{S_0}} . \tag{5.2.57}$$

When $\alpha \neq 0$, evaluation of the time-dependent moment $I(\varsigma)$ is necessary. To this end, we first implement the change of variables

$$z = \frac{\xi}{\sqrt{S/S_0}} ; \quad \mathrm{d}z = \frac{\mathrm{d}\xi}{\sqrt{S/S_0}} \tag{5.2.58}$$

$$\chi = \frac{S_0}{S}\zeta ; \quad \zeta = \frac{S}{S_0}\chi \tag{5.2.59}$$

to recast (5.2.55) as

$$I(\varsigma) = \int_0^\varsigma \frac{S_0}{S}\int_0^\infty \frac{\xi}{\sqrt{S/S_0}}\left(\frac{S}{S_0}\right)^2 \chi^2\frac{\mathrm{d}\xi}{\sqrt{S/S_0}}\mathrm{d}\varsigma' = \int_0^\varsigma\int_0^\infty \xi\chi^2\mathrm{d}\xi\mathrm{d}\varsigma' . \tag{5.2.60}$$

Next, we note that the change of variables

$$\eta \equiv \frac{\xi}{2K\sqrt{\varsigma+\varsigma_0}} ; \quad \mathrm{d}\eta \equiv \frac{\mathrm{d}\xi}{2K\sqrt{\varsigma+\varsigma_0}} \tag{5.2.61}$$

transforms (5.2.60) into

$$I(\varsigma) = 4K^2\int_0^\varsigma(\varsigma+\varsigma_0)\left(\int_0^\infty \eta'\chi^2\mathrm{d}\eta'\right)\mathrm{d}\eta' \tag{5.2.62}$$

a form that enables us to immediately incorporate the analytical solutions summarized above. Thus, inserting (5.2.49) into (5.2.62) and rearranging, we have

$$I(\varsigma) = 2K^2 \sum_{j=0}^{N} \sum_{i=0}^{N} A_{ij} F_{ij}(\varsigma) , \tag{5.2.63}$$

where

$$\begin{aligned} F_{ij}(\varsigma) &\equiv \int_0^{\varsigma} \frac{1}{\varsigma' + \varsigma_0} \left(\frac{\varsigma_0}{\varsigma' + \varsigma_0} \right)^{i+j} \mathrm{d}\varsigma' \\ &= \begin{cases} \ln \left(\dfrac{\varsigma + \varsigma_0}{\varsigma_0} \right) , & \text{if } i = j = 0 ; \\ \dfrac{1}{i+j} \left[1 - \left(\dfrac{\varsigma_0}{\varsigma + \varsigma_0} \right)^{i+j} \right] , & \text{otherwise} \end{cases} \end{aligned} \tag{5.2.64}$$

and

$$A_{ij} \equiv c_i c_j \int_0^{\infty} \exp(-2x) L_i(x) L_j(x) \mathrm{d}x = \frac{(i+j+1)!}{i! j! 2^{i+j+1}} c_i c_j . \tag{5.2.65}$$

Equations (5.2.54), (5.2.56), and (5.2.63–5.2.65) constitute the desired construction, since these yield the evolution of C_v in terms of the stretched time variable ς, given initial data represented by the Laguerre polynomial coefficients c_i. Note that ς closely couples the effects of stretch and diffusion, and is implicitly related to the global evolution of the filament arclength.

To construct expressions for the axial flow contribution to the core structure coefficient, we start with the definition of $C_w(t)$ as stated in (5.2.40c). The latter is recast in terms of the stretched variables ς and η, respectively, defined in (5.2.46) and (5.2.61), as

$$C_w(\varsigma) = \frac{-16\pi^2 K^2}{\Gamma^2} (\varsigma + \varsigma_0) \int_0^{\infty} \chi^2 \mathrm{d}\eta^2 . \tag{5.2.66}$$

The right-hand side of (5.2.66) resembles that of (5.2.53); an analogous treatment to that explained above yields

$$C_w(\varsigma) = -2\alpha(\varsigma + \varsigma_0) \sum_{i=0}^{N} \sum_{j=0}^{N} A_{ij} \left(\frac{\varsigma_0}{\varsigma + \varsigma_0} \right)^{i+j} , \tag{5.2.67}$$

where α and A_{ij} are given in (5.2.56) and (5.2.65), respectively. This completes our construction of explicit evolution equations for the core structure coefficient.

5.2.3.2 Generalized Thin-Tube Model. Generalization of the thin-tube model of Subsect. 5.2.1 is based on the key observation that the scaling corrections in Klein and Knio (1995), conducted for filaments with frozen axial vorticity core structure, naturally extend to the more general situations, in which stretching, diffusion, and axial flow effects are simultaneously present. This is the case because, to leading order, these phenomena can be represented in terms of the evolution of a single global parameter, namely the Klein–Majda parameter $\epsilon(t)$ or, alternatively, the core structure coefficient $C(t)$. Thus, (5.2.26) may be generalized to

$$\delta^{\mathrm{ttm}}(t) = \delta \exp(C^{\mathrm{ttm}} - C(t)) . \tag{5.2.68}$$

The only remaining difficulty concerns computational tracking of the core structure coefficient $C(t)$. To this end, we take advantage of the analytical solutions for the stretched viscous core evolution, as summarized above. The objective here is to construct suitable expressions for the time-dependent behavior of $C_v(t)$ and $C_w(t)$. Approaches to the construction of each of these dynamic constants differ slightly and are thus discussed separately.

5.2.3.3 Numerical Scheme. We are now in a position to summarize the application of our generalized slender filament computational scheme.

1. In a preprocessing step, the coefficient matrix A_{ij} is computed using (5.2.65), based on the coefficients of the Laguerre function expansion of the axial vorticity distribution. The matrix is stored and fed into the main calculations.
2. A similar preprocessing procedure is performed for the coefficient matrix corresponding to the initial axial flow distribution.
3. The corrected slender filament scheme, based on the corrected desingularized Biot–Savart law:

$$\boldsymbol{v}(\boldsymbol{x},t) = -\frac{\Gamma}{4\pi}\sum_{i=1}^{N}\frac{(\boldsymbol{x}-\chi_i(t))\times\delta\chi_i(t)}{|\boldsymbol{x}-\chi_i(t)|^3}\kappa_{\delta^{\mathrm{ttm}}}(\boldsymbol{x}-\chi_i(t)) , \tag{5.2.69}$$

   is used to advance the filament geometry. In the present computations, the node velocity predictions are used in conjuction with the second-order Adams–Bashforth scheme to update the Lagrangian vector $\{\chi_i\}_{i=1}^{N}$.
4. Based on the updated locations, the new value of the filament arclength, S, is computed.
5. The second-order Crank–Nicolson scheme is then used to update the stretched time variable ς. We have

$$\varsigma^{n+1} = \varsigma^n + \frac{\Delta t}{2S_0}(S^{n+1} + S^n) , \tag{5.2.70}$$

   where Δt is the computational time step.
6. Next, (5.2.54) and (5.2.63) are used to deduce the new values of C_v, while (5.2.67) is used to update C_w.
7. Using the updated values of C_v and C_w, (5.2.68) is used to modify δ^{ttm}.

Steps (3–7) are repeated to advance the solution over successive time steps.

5.2.3.4 Numerical Examples. Here, we present computational results for a test problem which illustrate

(i) nontrivial changes of the vortex core structure due to stretching and diffusion in a setting where these changes, even though considerable, induce only minor modifications of the vortex motion, and
(ii) the fact that, in more general situations, core structure variations do have quite a dramatic effect on the motion.

As further discussed below, the test consists of the interaction between three axisymmetric vortex rings. Before entering the discussion of the computational results, we recall from the theory presented above the parameters characterizing an individual vortex and their physical meaning.

In an inviscid slender vortex flow, each vortex filament is defined by its initial geometry and by initial values for the core coefficients $C_v(t)$ and $C_w(t)$, which represent the influence of circumferential and axial flow in the vortex core, respectively. The explicit formulae in (5.2.51), (5.2.52), and (5.2.46), together with the filament geometry evolution equation (5.2.24), then completely define the flow. It is, incidentally, possible without any restriction to consider collections of interacting filaments. In these situations, the finite part of the Biot–Savart integral in (5.2.24) includes the contribution of all filaments.

For viscous flow, the redistribution of vorticity by viscous diffusion is to be taken into account. The analysis above reveals that an individual vortex must be assigned a scaled viscosity K and two, theoretically infinite, sequences of Laguerre polynomial coefficients $\{c_{v,i}\}_{i=0}^{\infty}$ and $\{c_{w,i}\}_{i=0}^{\infty}$ characterizing the initial core structure. However, we note that the higher the index i, the more rapid the decay of the associated Laguerre polynomial contribution to the vortex core structure. Hence, it is physically meaningful to carry only the first few contributions in actual computations.

As regards the physical meaning of the $c_{v,i}, c_{w,i}$, we observe from a close examination of (5.2.49) that the leading terms $i = 0$ correspond to the long-time similarity solution that remains after the more rapid decay of the contributions with $i \geq 1$. (Notice that the coefficients with index 1, $c_{v,1}, c_{w,1}$ are absorbed in the definition of the individual $\varsigma_{v,0}, \varsigma_{w,0}$ (Klein and Ting, 1995). The similarity solution corresponds directly to the Lamb-Oseen vortex with a Gaussian axial vorticity distribution. If all the $c_{v,i}, c_{w,i}$ for $i > 0$ are zero, then the structure with a minimized influence of viscous diffusion is achieved. In other words, choosing nonzero values for these coefficients enhances the net effect of diffusion for a transient period, during which the associated core structure contributions decay.

The scaled viscosity K is the inverse of the flow Reynolds number $Re = \Gamma/\nu$, scaled by a factor of δ^2 (see Subsect. 5.2.43). This scaling takes into account that the relevant diffusion time scale is two orders of magnitude (in δ) larger than the vortex core turnover time. In the examples of this section, we pick equal and constant values of K.

5.2.3.5 Interaction of Coaxial Vortex Rings. We follow the evolution of three interacting coaxial vortex rings of equal strength and sense of rotation that are placed initially at $Z_1(0) = 0.0, Z_2(0) = 0.0$, and $Z_3(0) = 0.2$, and have radii $R_1 = 1.0, R_2 = 0.5$, and $R_3(0) = 0.75$. We consider three cases:

1. inviscid vortex rings,
2. viscous rings initialized with a similarity core structure,
3. viscous rings with nonsimilar initial core structure,

and compare the results with predictions obtained with a standard thin-tube model with frozen core size. As time evolves, the three rings undergo a quite violent interaction, where pairing and divorce are taking place intermittently. During the periods of close approach of two rings, their mutual potential vortex interaction makes them spin around each other at a high rotation rate. Before discussing the aspects of these motions and their dependence on the effects of stretching and viscosity, we first consider in Fig. 5.23 the evolution of the effective vortex core sizes $\ln(1/\delta) + C(t)$ for the three rings and for all four computations. The three figures show, separately for rings 1, 2, and 3, this comparison, thereby allowing a direct assessment of the influence of vortex stretching and diffusion on the individual vortex cores. We find that the effective cores sizes undergo order O(1) changes in the course of the interaction due to stretching alone. The maximum value of the core size parameter in the inviscid run is about 0.024, which is 1.6 times the initial core size for ring 1.

For the present choice of the scaled viscosity $K = 0.1$, we find an equally strong influence of vorticity diffusion in the core. In fact, the changes of the effective core structure due to viscosity drive the maximum of the core size parameter to 0.03 so that the net effect of diffusion is very much comparable to the net effect of stretching. It remains to show that these changes of the core structure do go along with nontrivial modifications of the paths of the rings during the interaction.

Figure 5.24 shows the traces of the vortex rings in the Z–R-plane for all four computations. We observe indeed major differences in the evolution of the system depending on whether we run the standard thin-tube model (standard-ttm), take into acount vortex stretching only, consider a vortex with similarity core structure in a viscous fluid, or allow for nonsimilar contributions to the core structures.

Surprisingly, the standard-ttm computation produces a pattern that is nearly periodic in Z. This pattern is characterized by an intermittent sequence of events, where periods of close and intense interaction of rings 2 and 3 alternate with more silent phases where all ring-to-ring distances are comparable. During the phases of strong interaction, rings 2 and 3 rotate around each other rapidly, mainly driven by their mutual potential vortex interaction.

This quasi-periodic behavior is no longer observed when the influence of vortex stretching is taken into account. As in the standard-ttm predictions,

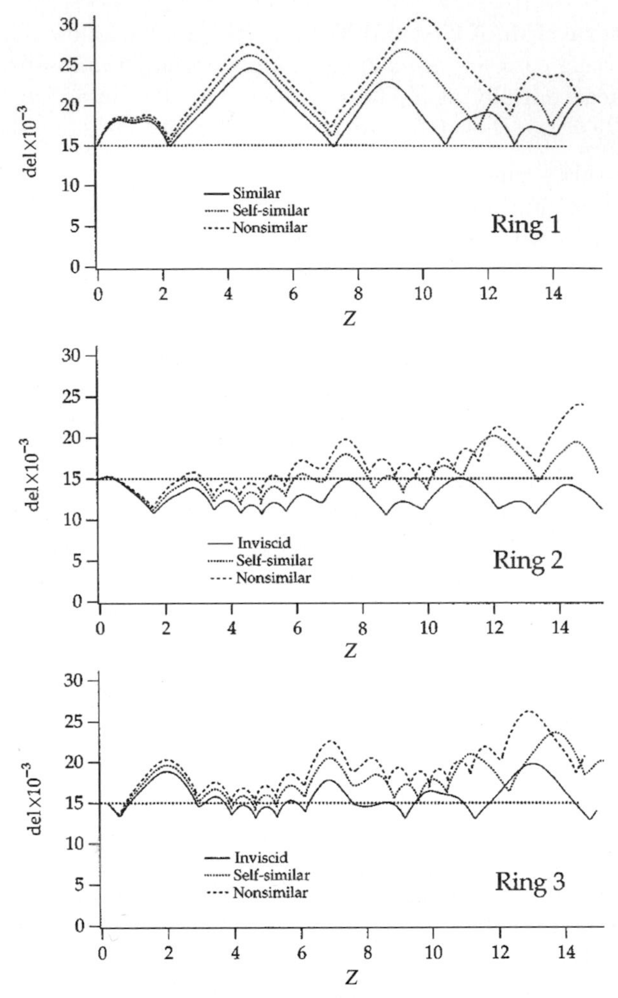

Fig. 5.23. Histories of effective core size parameters in the three-ring interaction problem. Adapted from Klein et al. (1996)

a pairing takes place shortly after the start. The coupling is more intense than observed in the standard-ttm computation as indicated by the larger number of five revolutions instead of three. Yet, after this first interaction, rings 2 and 3 depart. In contrast to the previous case they do not pair again within the time span of the computation. When viscous effects are present, we observe that the tendency of rings 2 and 3 to pair is enforced. After the first pairing sequence of revolutions, they manage to connect again near $Z = 8.0$, yet the coupling strength is reduced. While the first close interaction led to about five revolutions (Fig. 5.24), the second has four turnarounds only. The strongest coupling, again between rings 2 and 3, occurs when the effects of

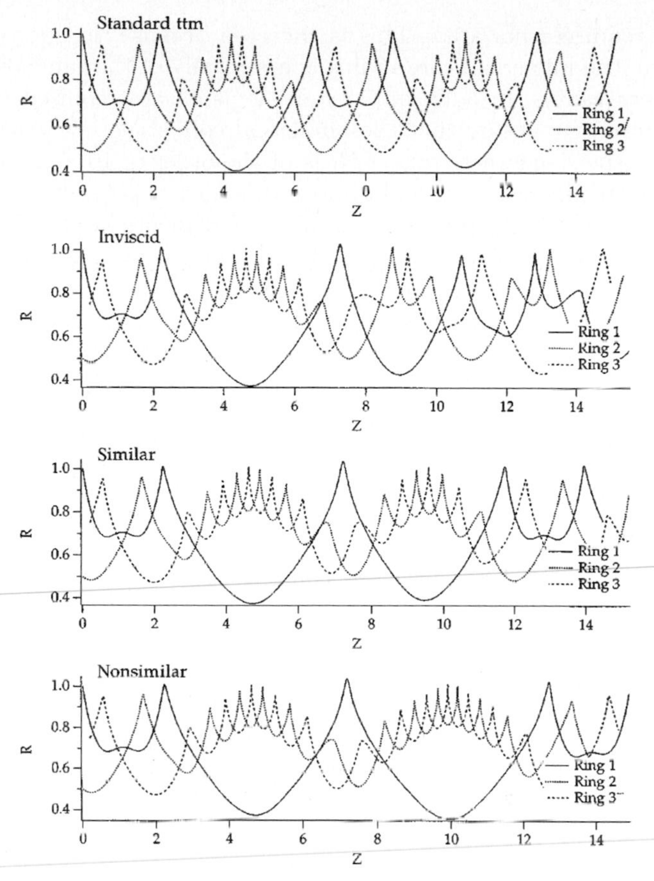

Fig. 5.24. Ring radii in the three-ring interaction problem. Adapted from Klein et al. (1996)

viscosity are further enhanced by imposing nonsimilar initial data for the core structure. In this case, the second coupling is even stronger than the first: we now observe seven revolutions in the second interaction event.

Thus, for the present flow configuration, viscosity not only triggers more rapid oscillations than are observed in the corresponding inviscid flow but also tends to concentrate these oscillations in space and time. We find these observations to be strikingly reminiscent of intermittency in turbulence, as in this case too, strong fluctuations are triggered by viscosity.

5.2.4 Enhancement of Performance

Application of thin-tube vortex element models to very thin vortices is faced with two difficulties. First, when the ratio δ/R of the core size to the characteristic radius of curvature of the filament centerline is small, high spatial resolution requirements arise. This is the case because, in approaches outlined above, the numerical core radius is essentially of the same order as the physical core radius. In addition, vortex elements are required to strongly overlap in order to ensure that the numerical core structure is well-defined. Thus, when the slenderness ratio δ/R is of the order of 10^{-2} or smaller, the number of elements needed to satisfy overlap with cores of order δ may exceed by several orders of magnitude the number of elements needed for adequate resolution of the filament centerline. This would, consequently, necessitate prohibitively large CPU resources.

A second (related) difficulty is that the small spacing between elements may impose a severe restriction on the integration step. These stiffness problems are well-known in various Lagrangian calculations in which computational elements may tend to cluster. The temporal stiffness compounds the spatial resolution problem and substantially increases the demand for computational resources.

In this section, we outline three methods for optimizing the corrected thin-tube model discussed above. In all cases, our approach is motivated by the observation that (1) in the previous constructions, the number of elements required to achieve overlap is much larger than that needed to adequately describe the centerline geometry, and (2) consequently, both the temporal stiffness and high spatial resolution problems can be effectively addressed if a coarser (or generally more efficient) discretization can be adopted. This observation suggests, in particular, the following optimization strategies. Confer Margerit et al. (2004) for a comparison of the present scheme with various alternatives.

5.2.4.1 Method 1: Extrapolation Technique. This approach is based on selecting a discretization level that is fine enough for the representation of the filament centerline *irrespective* of the filament core size δ. As previously discussed, selection of numerical core size levels using either of the approaches of the previous section becomes problematic since overlap among neighboring elements may not be satisfied. In fact, for small slenderness ratios, overlap is likely to be violated everywhere.

To avoid this issue, a large value of the numerical core size is used. Let $\tilde{\sigma}$ denote the core size parameter predicted by (5.2.26),

$$\sigma_o(t) = \max_{i=1,\ldots,N} |\delta\chi_i(t)| \tag{5.2.71}$$

be the maximum interelement separation distance, $\sigma_1 = K\sigma_o$, and $\sigma_2 = \alpha\sigma_1$. K and α are real constants. Note that for any choice $K > 1$ and $\alpha > 1$, implementation of a thin-tube model using either σ_1 or σ_2 will automatically satisfy the overlap condition.

Next, we recall that for any σ the velocity field predicted by the uncorrected thin-tube model corresponds to (Klein and Knio, 1995):

$$\boldsymbol{v}^{\mathrm{ttm}} = \frac{\Gamma}{4\pi}\left[\log\left(\frac{2}{\sigma}\right) + C^{\mathrm{ttm}}\right]\kappa\boldsymbol{b} + \boldsymbol{Q}^f \tag{5.2.72}$$

as long as strong overlap is satisfied. Consequently, if we let $\boldsymbol{v}_1$ and $\boldsymbol{v}_2$ denote the thin-tube velocity prediction based on σ_1 and σ_2, respectively, then the *corrected* velocity field prediction can be estimated on the basis of the following logarithmic extrapolation procedure:

$$\boldsymbol{v}^{\mathrm{cor}} = \boldsymbol{v}_1 + (\boldsymbol{v}_1 - \boldsymbol{v}_2)\frac{\log\left(\frac{\sigma_1}{\tilde{\sigma}}\right)}{\log \alpha} . \tag{5.2.73}$$

In other words, the corrected centerline velocity can be obtained at the cost of two "coarse mesh–fat core" evaluations. Note that the present procedure does not require estimates of the filament curvature and binormal.

5.2.4.2 Method 2: Explicit Velocity Correction. The velocity correction technique is based on a single velocity evaluation using large σ and implementing an explicit correction to the corresponding velocity prediction. This approach has been first implemented in Zhou's study of Kelvin waves on a slender vortex (Zhou, 1997). Here, we implement a variant of the velocity correction approach that is summarized by

$$\boldsymbol{v}^{\mathrm{cor}} = \boldsymbol{v}_1 + \frac{\Gamma}{4\pi}\log\left(\frac{\sigma_1}{\tilde{\sigma}}\right)\kappa\boldsymbol{b} , \tag{5.2.74}$$

where σ_1 is dynamically determined during the simulation using the same definitions discussed for Method 1. In the computations, the curvature and binormal are obtained following the procedure given in Klein and Knio (1995).

5.2.4.3 Method 3: Local Mesh Refinement. This approach is motivated by the observation that the numerical core structure along a particular location on the axis of the thin tube is actually induced by neighboring elements only. Since the core smoothing function decays rapidly, the numerical *vorticity* structure at a given point is determined by the fields of elements lying within a few numerical core radii of that point. This suggests the following approach for estimating the corrected thin-tube velocity.

As in the previous two methods, we rely on a discretization level that is fine enough to adequately represent the filament geometry. However, unlike the previous two approaches, the rescaled numerical core size

$$\sigma = \delta \exp\left(C^{\mathrm{ttm}} - C\right) \tag{5.2.75}$$

is used so that overlap is not globally satisfied. Based on this choice of parameters, a modified velocity evaluation procedure is implemented. To evaluate the velocity field at a given point, the elements are divided into two disjoint groups: a group of neighboring elements and a group of well-separated elements. A separation distance of three core radii is used as a basis for this segregation procedure. The contribution of elements belonging to the second group is computed directly on the coarse grid. Meanwhile, the contribution of neighboring elements is accounted for by locally remeshing the corresponding segments using a fine enough grid for core overlap to be locally satisfied.

In the present computations, a simplified version of the local remeshing procedure is implemented. It consists of determining, at every time step and in a global fashion, a Lagrangian grid that is fine enough for overlap to be satisfied everywhere. The fine grid is determined by Fourier transformation of the filament geometry, extending the resulting spectrum by array padding, and then inverse transforming the extended spectrum onto physical space.

Remark. The improved schemes outlined above address the stiffness of the slender filament equations using different strategies. From a practical standpoint, method 1 (M1) is very attractive as it does not require any modification of a corrected thin-tube code (Klein and Knio, 1995; Klein et al. 1996). Specifically, all the code elements that determine the element velocity remain unchanged, and the optimization operates on their output only. The attractive feature of method 2 (M2) is that it isolates the singular part of the element evolution equations and, as outlined in Zhou (1997), enables the optimization of the numerical integration using operator-splitting approaches. However, a disadvantage of M2 is that it requires explicit evaluation of higher order derivatives of the filament centerline. This has adverse effects on the numerical stability of the scheme, as observed in Klein and Knio (1995) and discussed further below. Method 3 (M3) would be quite attractive in the context of fast vortex element methods (e.g., Almgren et al. 1994; Salmon et al. 1994), which use similar clustering ideas to defeat the $O(N^2)$ cost of direct vortex interactions. The considerable book-keeping and storage management associated with these methods as compared to the direct methods would then lose its importance. A key disadvantage of M3, however, concerns the stiffness issue. As the scheme operates with the physical core size and evaluates velocities by directly accounting for neighboring (overlapping) computational elements, there will be strong cancellation of large kernels as the core size diminishes. The phenomenon is due to the fact that the local induced velocities are by a factor $O(1/\delta)$ larger than the net filament velocity (Callegari and Ting, 1978; Klein and Knio, 1995) it leads to amplification of roundoff errors and compounds the stiffness problem.

It follows from the present discussion that, from a practical standpoint, M1 appears as the most attractive alternative. This claim is further supported in the computational tests below.

5.2.4.4 Parameter Selection. The optimization methods introduced above involve various numerical parameters that are related to the discretization of the slender filament and to choice of numerical core size(s). Specifically, for M1 and M2 one needs to specify the number of elements, N, and the overlap parameter, K. For M1, one must, in addition, specify a second parameter, α. For M3, core coarsening parameters are not needed but fine and coarse discretization levels are used instead.

We now explore general guidelines for the selection of these parameters and start by emphasizing that the present constructions are based on the key assumption that the predictions of the thin-tube model agrees with the asymptotic expression. For this to hold, two conditions must be satisfied (Klein and Knio, 1995): (a) the selected (coarsened) numerical core size must still be small enough for the corresponding prediction to fall within the range of validity of the asymptotic theory and (b) the numerical Biot–Savart integral must be accurately evaluated. As discussed below, one can use these two conditions to develop guidelines for the selection of optimization parameters.

To illustrate these guidelines, we consider the case of a circular vortex ring. Variables are normalized so that the dimensionless ring radius $R = 1$ and the dimensionless circulation $\Gamma = 1$. The core smoothing function $f(r) = \text{sech}^2(r^3)$ is used. The corresponding velocity kernel is $\kappa(r) = \tanh(r^3)$ (Beale and Majda, 1985), and the numerical core structure coefficient $C^{\text{ttm}} = -0.4202$ (Klein and Knio, 1995). For brevity, we focus exclusively on static velocity predictions.

We start by addressing condition (b) above, and so we examine the role of the parameter K that enters in M1 and M2. As discussed earlier, K measures the ratio between the maximum element length and the numerical core size, and consequently of the degree of overlap among neighboring cores. Thus, K is referred to as "overlap" parameter. This aspect has been extensively discussed in theoretical convergence studies of particle methods (e.g., Beale and Majda, 1985) as well as computational studies (e.g., Knio and Ghoniem, 1990). Based on previous experiences, one would expect that the numerical evaluation of the Biot–Savart integral becomes valid when neighboring cores overlap, that is, in the range $K > 1$. In Fig. 5.25, we examine the effect of K by applying M2 to compute the self-induced velocity of a slender vortex ring with $\tilde{\sigma} = 0.001$. The results indicate that for $K > 1.5$, the prediction becomes essentially independent of the selected value of K. In all the tests below, we conservatively choose overlap parameters in the range $K \geq 2$.

We now turn our attention to the first condition, which concerns the slenderness ratio of the filament. Note that both M1 and M2 rely on numerical Biot–Savart integrals with enlarged cores, and one must consequently ensure that the corresponding slenderness ratio $\epsilon \equiv \kappa_{\max}\sigma$ is small enough to fall within the regime of validity of the theory. To explore the regime of validity of the constructions, we rely on the computed self-induced velocity of the circular ring to extract the numerical core constant; we use

$$C^{\text{ttm}} \simeq \frac{4\pi V}{R} - \log\left(\frac{8R}{\tilde{\sigma}}\right) , \tag{5.2.76}$$

where V is the computed self-induced velocity and $\tilde{\sigma}$ is the numerical core radius. In Fig. 5.26, we plot the values of C^{ttm} computed using equation (5.2.76) against the ring slenderness ratio $\epsilon = \tilde{\sigma}/R$. As can be observed in Fig. 5.26, for $\epsilon < 0.1$ the computed core constant C^{ttm} is weakly dependent on the slenderness ratio. It is also interesting to note that as ϵ decreases the predicted values tend toward the theoretical value (Klein and Knio, 1995).

The above exercise can be used, in the application of the optimization techniques, to guide the selection of appropriate grid coarsening levels, or

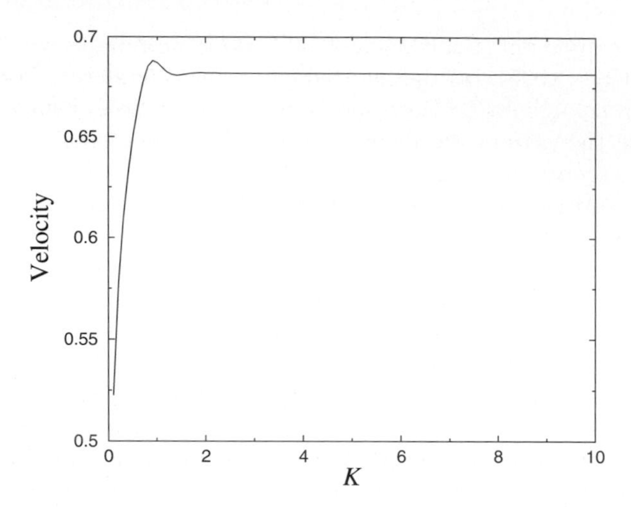

Fig. 5.25. Effect of overlap parameter K on the velocity prediction. Results are obtained using M2 with $N = 512$. The core radius $\tilde{\sigma} = 0.001$. Adapted from Knio and Klein (2000)

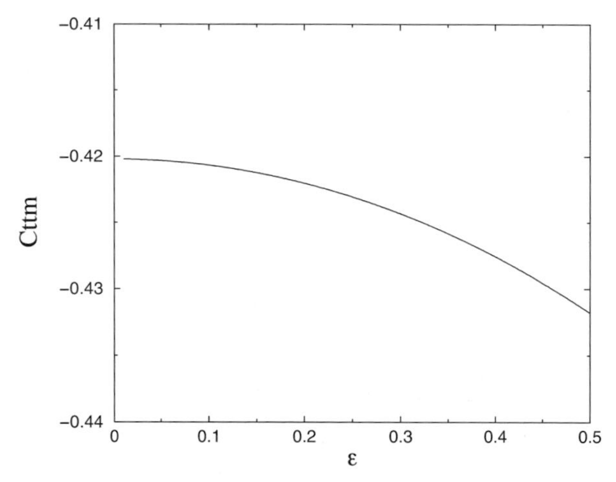

Fig. 5.26. Dependence of the numerical core constant, C^{ttm}, on the ring slenderness ratio $\epsilon \equiv \tilde{\sigma}/R$. The ring is discretized using $N = 2{,}048$ elements. Adapted from Knio and Klein (2000)

alternatively of the number of elements. For instance, for a filament with characteristic peak curvature κ_{max} the condition $\epsilon < 0.1$ can be used to estimate a maximum allowable element length, $\delta\ell_{\text{max}}$. A reasonable estimate for M2 is $\delta\ell_{\text{max}} \simeq 0.1/(K\kappa_{\text{max}})$; for M1, we set $\delta\ell_{\text{max}} \simeq 0.1/(K\alpha\kappa_{\text{max}})$. Assuming that the overall filament arclength is S, and that the lengths of vortex elements are fairly uniform, then the number of elements may be estimated using $N \simeq S/\delta\ell_{\text{max}}$. In addition to relying on this initial estimate, the computations of the following section continuously monitor ϵ to verify that it remains within acceptable levels. Our experiences indicate that the results of the optimized methods are valid when ϵ remains small; in contrast, an example is provided below which shows that the predictions deteriorate for $\epsilon > 0.1$.

We conclude this discussion with a brief comment regarding the parameter α used in M1. Following the discussion above, when K and N have been optimized, α must naturally be restricted to the range $\alpha > 1$. In order not to penalize the discretization level, it would obviously be advantageous to choose α as close to unity as possible. At the same time, one should guard against the possible amplification of extrapolation errors. To examine this issue, we use M1 to compute the self-induced velocity of a slender circular vortex ring; the results are plotted in Fig. 5.27 against the corresponding values of α. As shown in the figure, in the range $\alpha \geq 1.25$ the predictions become essentially insensitive to the selected value of α. This restriction is imposed in all the calculations below.

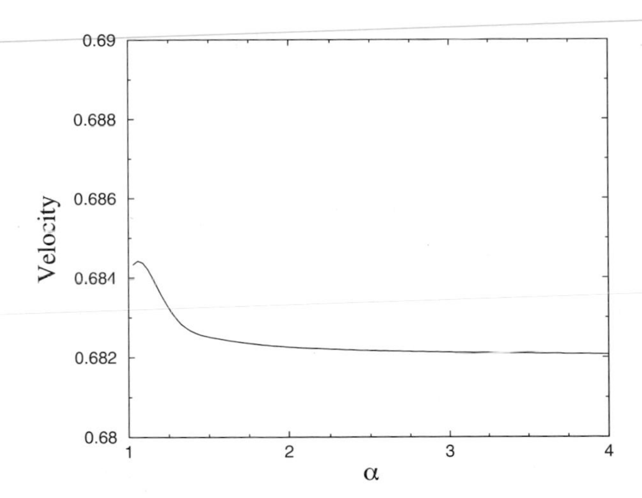

Fig. 5.27. Effect of α on the computed self-induced velocity of a circular vortex ring with $\tilde{\sigma} = 0.001$. Results are obtained using M1 with $K = 2$ and $N = 2{,}048$. Adapted from Knio and Klein (2000)

5.2.4.5 Performance Measures. To illustrate performance enhancement gained through implementation of M1, M2, and M3, we consider inviscid three-dimensional simulations of a slender vortex ring. We use the same normalization convention and the same core smoothing function used immediately above.

To observe a nontrivial slender vortex evolution, the ring centerline is perturbed using plane, azimuthal bending waves. Initially, the perturbed ring radius is given by

$$\rho(\theta) = R\left[1 + a_1 \sin(k_1\theta) + a_2 \sin(k_2\theta)\right] , \tag{5.2.77}$$

where θ is the azimuthal angle, k_1 and k_2 are integer wavenumbers, while a_1 and a_2 are the corresponding amplitudes.

The evolution of the slender ring described above is computed using the rescaled numerical core radius approach (Klein and Knio, 1995), which is referred to as original scheme, as well as the optimized schemes based on methods 1, 2, and 3 above, which are labeled M1, M2, and M3, respectively. Results are obtained for a slender vortex ring with $\tilde{\sigma} = 0.02$, $k_1 = 2$, $k_2 = 3$, and $a_1 = a_2 = 0.02$. The parameters used in the runs are summarized in Table 5.1. Note that core overlap using $\tilde{\sigma}$ would not be satisfied in M1, M2, and M3 at the selected coarse-grid resolutions.

Table 5.1. Numerical parameters and performance measures. CPU is measured on an SGI R10000 195 MHz processor, on computations using $\Delta t = 0.002$ and extending over 1500 integration steps

	Original	M1	M2	M3
σ	$\tilde{\sigma}$	(σ_1, σ_2)	σ_1	$\tilde{\sigma}$
N	1024	256	256	256/1024
K	–	2	3	–
α	–	1.5	–	–
CPU time	708.9	68.19	45.98	71.37
Performance Gain	1	10.39	15.42	9.93

The predictions of the original and optimized schemes are first illustrated in Fig. 5.28, which shows the evolution of the peak centerline curvature and velocity. The results show that the peak curvature of the filament undergoes large-amplitude periodic oscillations, and that the behavior of the peak velocity follows closely that of the peak curvature. The figures show that the predictions of the optimized schemes are in close agreement with each other, and with the predictions of the original scheme.

The agreement between the curves in Fig. 5.28 is quantified by computing the relative deviation between the predictions of the improved schemes and the results of the original method. For instance, the relative deviation in peak curvature for method $\mathrm{M_i}$ is defined as

$$D_i = \frac{\max_{[0,t]} \left| \kappa_{\max}^{\mathrm{orig}}(t) - \kappa_{\max}^{\mathrm{M_i}(t)} \right|}{\max_{[0,t]} \kappa_{\max}^{\mathrm{orig}}(t)} . \tag{5.2.78}$$

A similar definition is used on the basis of the peak velocity. Based on these definitions, the computed relative deviation in peak curvature is 0.159% for M1, 0.118% for M2, and 0.194% for M3. Meanwhile, the relative deviations in peak velocity is 0.109% for M1, 0.064% for M2, and 0.234% for M3.

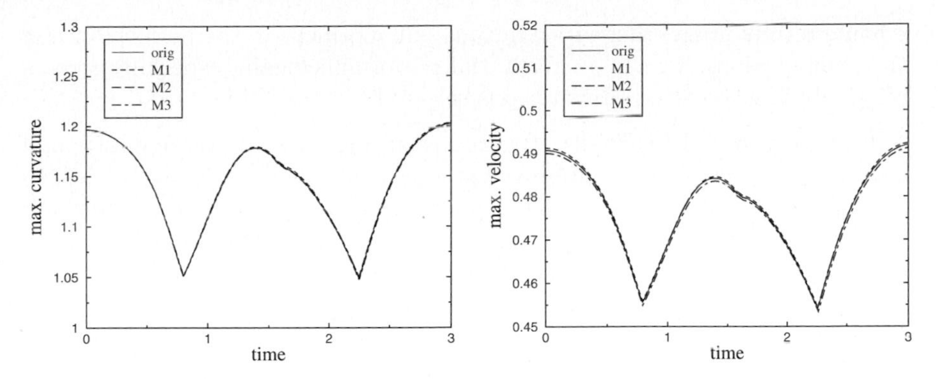

Fig. 5.28. Evolution of the maximum curvature (left) and maximum velocity (right) for a perturbed vortex ring with $\tilde{\sigma} = 0.02$. Plotted are results obtained using the original method, and schemes M1, M2, and M3. Parameters are indicated in Table 5.1. Adapted from Knio and Klein (2000)

Very close agreement between the predictions of the original and optimized schemes is also evident in Fig. 5.29, which depicts the spatial distribution of centerline curvature along the circumference of the ring. As observed earlier, the results of the optimized schemes are nearly identical to those obtained using the original model.

The performance of the original and optimized schemes is quantified in Table 5.1. The table provides the total CPU time spent and the perfor mance gain that is defined as the ratio of the current CPU time to the CPU time spent in the original scheme. The results show that the optimized

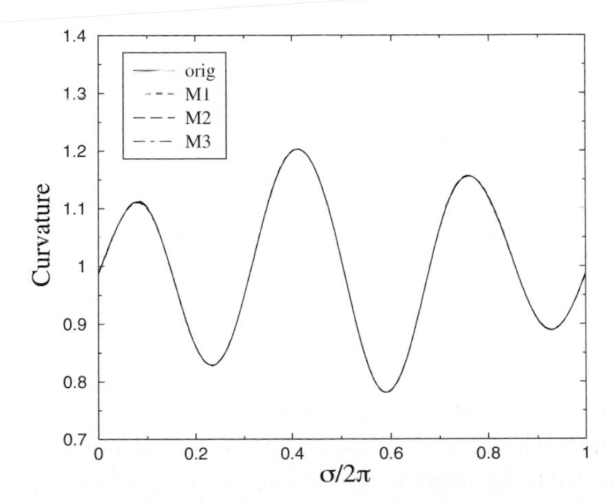

Fig. 5.29. Spatial distribution of the curvature along the filament axis at $t = 3$. Curves are generated using the results of the original method, and schemes M1, M2, and M3. Parameters are indicated in Table 5.1. Adapted from Knio and Klein (2000)

scheme results in an order-of-magnitude enhancement in the performance of the computations. This is a substantial performance gain, especially since a relatively large core-to-radius ratio is selected.

5.2.4.6 Temporal Stiffness. By only requiring adequate representation of the filament centerline, the improved schemes may also be effective in defeating the temporal stiffness of the slender filament equations of motion. We briefly explore this possibility in light of a simplified heuristic analysis of the critical time step, Δt_{crit}. Specifically, the analysis is based on performing unsteady calculations with different values of the time step, Δt, and monitoring the value at which the simulations become numerically unstable. To interpret the observations, simulations are performed with different values of the core size; results are obtained with $\tilde{\sigma} = 0.02$, 0.002, and 0.0002. Again, we restrict our attention to the initial centerline geometry of the previous section, and to improved schemes M1 and M2 with $N = 256$.

In the examples of Subsubsect. 5.2.4.5, the optimization parameters were selected following the guidelines developed in Subsubsect. 5.2.4.4. In particular, the criterion $\epsilon < 0.1$ was satisfied throughout the computations. In the present implementations, the parameters for M1 are deliberately chosen so that this criterion is violated; we use $K = 3$ and $\alpha = 2$. This enables us to observe the effects of a poor choice in optimization parameters. For M2, we still use $K = 3$.

The results of the analysis are summarized in Table 5.2, which provides the observed value of Δt_{crit} for different values of $\tilde{\sigma}$, and in Fig. 5.30, which depicts the evolution of the peak centerline curvature for $\tilde{\sigma} = 0.002$ and 0.0002. Table 5.2 also shows the dominant period of the oscillations in the filament centerline, which is deduced from the curvature evolution curves in Figs. 5.28 and 5.30.

Table 5.2. Critical time step and period

$\tilde{\sigma}$	Δt_{crit}: M1	Δt_{crit}: M2	Period
0.02	4.43×10^{-2}	1.35×10^{-2}	1.420
0.002	1.80×10^{-2}	4.17×10^{-3}	0.836
0.0002	1.10×10^{-2}	2.56×10^{-3}	0.586

The results show that for both M1 and M2, Δt_{crit} decreases slowly as $\tilde{\sigma}$ is reduced by two orders of magnitude. The behavior in Δt_{crit} appears to be consistent with the reduction in the observed period of the centerline oscillations and with the logarithmic nature of the filament self-induction law. This indicates that at a given *filament* resolution level, the critical time step is restricted by the *centerline* evolution rate. Since the optimized schemes require adequate resolution of the filament centerline only, the present tests

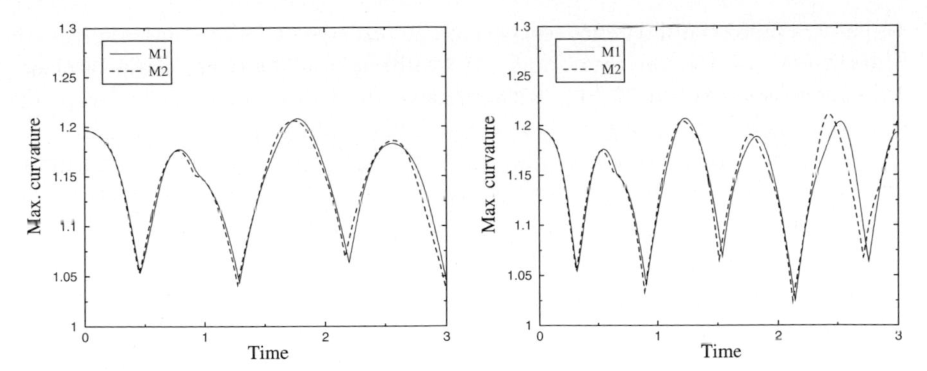

Fig. 5.30. Evolution of the maximum curvature for a perturbed vortex ring with $\tilde{\sigma} = 0.002$ (*left*) and $\tilde{\sigma} = 0.0002$ (*right*). Plotted are results obtained using schemes M1 and M2. Adapted from Knio and Klein (2000)

show that they are also effective in overcoming the stiffness of the equations of motion.

It is also interesting to note that Δt_{crit} for scheme M1 is consistently larger than that of M2. For the conditions of the present simulations, the results in Table 5.2 show that at the critical time step there are about 30–50 iterations per period for scheme M1, and about 100–200 iterations per period for scheme M2. This behavior can be attributed to the fact that scheme M1 avoids the evaluation of higher-order derivatives (Klein and Knio, 1996; Hou et al., 1998) which enter in the expressions of curvature and binormal. Combined with the performance measures of the previous section, the results in Table 5.2 indicate that if the differences in Δt_{crit} were accounted for then scheme M1 would outperform M2.

One should also note that for $\tilde{\sigma} = 0.0002$, the agreement between the predictions for M1 and M2 is evidently degraded. The deviation between the two curves in Fig. 5.30 is 3.24%, and the corresponding deviation based on peak velocity is 2.37%. To examine the origin of the deviation, we note that for M2 with $N = 256$ and $K = 3$, the peak slenderness ratio based on σ_1 is $\epsilon_1 \equiv \max_{[0,t]} \kappa_{\max}(t)\sigma_1(t) = 0.0926$. For M1 with $N = 256$, $K = 3$, and $\alpha = 2$, the peak slenderness ratio based on σ_1 is $\epsilon_1 = 0.0926$; based on σ_2, we have $\epsilon_2 = 0.1853$. As discussed in Subsubsect. 5.2.4.4, slenderness ratios above 0.1 are beyond the range of validity of the slender vortex theory and the predictions of the optimized schemes may consequently suffer from significant modeling errors. Thus, the deviations observed in Fig. 5.30 at $\sigma = \tilde{0}.0002$ appear to be due to the poor choice of K and α in M1. Note that at $\tilde{\sigma} = 0.002$ the deviations between M1 and M2 are harder to detect, which also suggests that large ϵ effects become more pronounced as one extrapolates across a larger core size range.

To verify the above claim, computations were repeated using M1 with $K = 2$ and $\alpha = 1.5$. For these values of K and α, the computed values of ϵ_1

and ϵ_2 are 0.0617 and 0.0926, respectively. Consistent with the experiences of Subsubsect. 5.2.4.4, the corresponding results (not shown) are again in close agreement with those of M2, with relative deviations of 0.978% for peak curvature and 0.912% for peak velocity. This enables us to emphasize that the predictions of the improved schemes and original method are consistent with each other and independent of the choice of numerical parameters, *as long as* these parameters obey the restrictions of the numerical construction and of the underlying slender vortex theory.

5.2.4.7 Tests with Large Deformation. In the numerical examples of Subsubsects. 5.2.4.5 and 5.2.4.6, the perturbation initially imposed had a small amplitude and the deformations of the filament centerline remained moderate. Specifically, spatial variations of the filament curvature were within $\pm 20\%$ of the mean, approximately.

This section provides a numerical example that illustrates the performance of the optimized schemes in a regime with O(1) changes in the filament curvature. To this end, we consider a slender ring with $\tilde{\sigma} = 0.002$, initially perturbed using $k_1 = 5$, $k_2 = 6$, and $a_1 = a_2 = 0.03$. Unsteady computations are performed using M1 and M2. To observe the guidelines of Subsubsect. 5.2.4.4, for M1 with use $K = 2$, $\alpha = 1.25$, and $N = 1024$; for M2, we use $K = 2.5$ and $N = 1024$.

Results of the simulations are shown in Fig. 5.31, which shows the evolution of peak velocity, and the spatial distribution of curvature along the filament arclength at the end of the computations. The results clearly show that the peak curvature changes substantially with time, and that significant differences in the curvature occur along the filament centerline. In addition, Fig. 5.31 shows that the predictions of M1 and M2 remain in close agreement

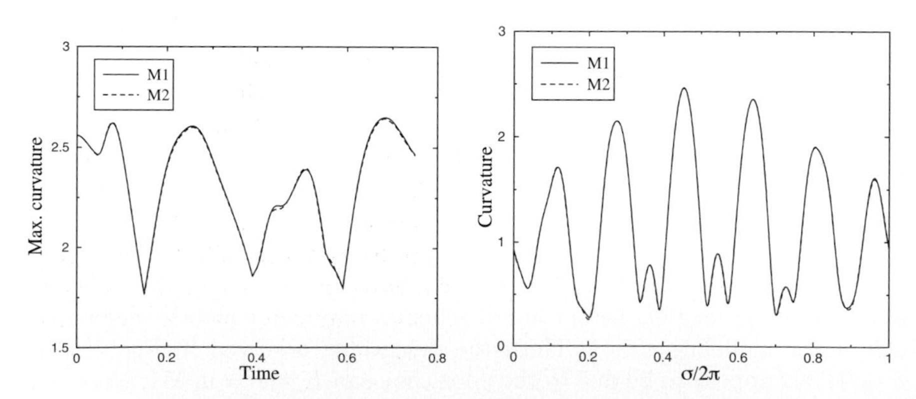

Fig. 5.31. *Left*: Evolution of the maximum velocity of a perturbed vortex ring with $\tilde{\sigma} = 0.002$, $a_1 = a_2 = 0.03$, $k_1 = 5$, and $k_2 = 6$. *Right*: Spatial distribution of the curvature along the filament axis at $t = 0.75$. Plotted are results obtained using M1 with $K = 2.5$ and $N = 1,024$, and M2 with $K = 2$, $\alpha = 1.25$, and $N = 1,024$. Adapted from Knio and Klein (2000)

with each other throughout the computations. This agreement provides a strong support to the validity of the optimized constructions.

5.2.4.8 Closing Remarks. We conclude by noting that the present constructions differ from those of Zhou (1997), who used an asynchronous splitting technique to treat stiff local terms and nonstiff nonlocal terms, and Hou et al. (1998), who relied on a recasting of the filament equations of motion using generalized curvature coordinates. It appears to be possible, in principle, to seek further enhancement of performance by combining some of the present techniques with those outlined in Zhou (1997) and Hou et al. (1998). Possible extensions also include the generalization of the present methods to accommodate core size variations along the filament centerline.

6. Numerical Simulations of the Merging of Vortices or Filaments

Motivation

In Chap. 2, we formulated the initial value problem of a flow field induced by an initial vorticity distribution of bounded support or decaying exponentially with diffusion length scale of the order of the typical scale ℓ of the flow field. Thus, the flow field, which remains at rest in the far field, is governed by the unsteady Navier–Stokes equations. In Chap. 3, we presented the asymptotic theory of slender vortex filament(s) having vorticity concentration in the filament core(s) of typical size δ with the slenderness ratio $\delta/\ell \ll 1$. The viscous effect could be significant in the evolution of the core structure. The theory yields the equation for the velocity of each filament, or rather its centerline and the equations for the evolution of the core structure(s), accounting for the viscous effect. These equations are coupled because of the dependence of the velocity on the core structure. In Chap. 4, the asymptotic theory was specialized to nearly straight and extremely slender filaments with $\tilde{\epsilon} = 1/\ln(\ell/\delta) \ll 1$ (that is, the slenderness ratio, δ/ℓ, is exponentially small with respect to $\tilde{\epsilon}$). The asymptotic theory is no longer applicable when there is a local segment of the filament, where the radius of curvature is $O(\delta)$, and we say that a local *self-merging* is taking place. The theory for a single slender filament is applicable to several filaments with or without rigid bodies, provided that the distance from one filament to another and the distance from a filament to a body are $\gg \delta$. When the distance between two filaments becomes $O(\delta)$, the theory is no longer applicable and we say that these two filaments are *merging* or intersecting with final reconnection to two filaments or a single one. When the distance of a filament from a body is $O(\delta)$, we have a different type of merging, the merging of a segment of the filament with the boundary layer along the body. Eventually one or several of these merging phenomena would occur, because the core size δ, initially small, increases with time, leading to $\delta = O(\ell)$, where ℓ measures and the break down of the asymptotic theory. Thus, we present numerical simulations of different types of merging phenomena and relevant experimental data to show that the viscous effect would play a dominant role in the merging or cancellation of vorticity, and result in global changes of the vorticity and flow field.

Extended Summary

In this chapter, numerical simulations of the merging phenomena are carried out by finite difference solution of Navier–Stokes equations in a finite computational domain $\mathcal{D}$, with the initial data provided by the asymptotic solutions valid in the premerging stage. The boundary conditions imposed on $\partial\mathcal{D}$ have to be consistent with (that is, to simulate) the far-field behavior of the original problem in unbounded space. Thus, we need to identify different types of merging problems, the corresponding computation domains, and the appropriate boundary conditions. By equating the accuracy of the numerical scheme in $\mathcal{D}$ with that of the approximate boundary conditions on $\partial\mathcal{D}$, we determine an optimal size of $\mathcal{D}$, with minimum number of grid points for a given degree of accuracy. Thus, we arrive at an efficient numerical scheme. We then present the long-time behavior of the flow field so that the numerical solution for a finite time can be extrapolated to long time.

In Sect. 6.1, we present the classification of the types of merging according to the relative order of magnitude of three typical length scales inherent in the initial data prior to merging. The scales are the size of the vortical field $L = O(\ell)$, the decay length $\ell_d = O(\delta^*)$, and the size of the merging region ℓ_m (Liu et al. 1985). We have a global merging when $\ell_m = \ell$ and a local merging when $\ell_m \ll \ell$.

For two-dimensional flows, we have either global merging of two vortices or simultaneous merging of several. For three-dimensional flows, local mergings are often found when the core size of a filament becomes comparable to its local radius of curvature or the distance between two filaments. Global merging of two initially distinct filaments can take place only by the end of local merging stages. Instantaneous global merging can take place only for special configurations, for example, the merging of two coaxial circular vortex rings of nearly equal radii.

In Sect. 6.2, we study two-dimensional vortical flows, for which the zeroth moment and first moment of vorticity are conserved, and define the total strength Γ and doublet vector $\mathbf{D}$. If $\Gamma \neq 0$, the center of the vorticity field is stationary, and the flow in the far field behaves as a potential vortex of strength Γ. We study the merging of an initial vorticity field that finally evolves to a single Lamb vortex of strength Γ, with a stationary center and a core size increasing as $\sqrt{t}$, that is, conserving the first moment and fulfilling the Poincaré condition on the polar moment.

If $\Gamma = 0$ while $\mathbf{D} \neq \mathbf{0}$, the center of the vorticity field is moving, while the first moment of vorticity, which is conserved, defines a constant vector $\mathbf{D}$. If $\mathbf{D} \neq \mathbf{0}$, the flow in the far field behaves as a doublet of constant strength $\mathbf{D}$. We then define the velocity of the vorticity field by the velocity of the doublet. We study the merging of the vorticity field simulating a viscous doublet and its long-time behavior.

For numerical simulation of the merging problem, we first explain how to find the computational domain and the appropiate boundary conditions.

We then present numerical solutions of several merging problems with initial data provided by the asymptotic solution for distinct vortices of small core sizes, for example, a superposition of Lamb vortices. In addition, we introduce *extended asymptotic solution* which is the continuation of the asymptotic solution, beyond the initial stage, disregarding the restriction that the typical core size $\delta \ll d_{ij}$, where d_{ij} denotes the distance between two adjacent vortices i and j. Note that the extended solution fulfills the conservations of the resultant and first moments of vorticity and the Poincaré invariant of the polar moment. Therefore, the extended solution would finally approach a single Lamb vortex, which is the same as in the numerical solution, if the resultant of vorticity is nonzero. We compare the extended solution with the numerical solution to find a quantitative criterion on the validity of the asymptotic solution instead of a qualitative one, $\delta(t)/d_{ij}(t) \ll 1$. Note that the extended asymptotic solution fulfills the linear Navier–Stokes equations, without the nonlinear convection terms. To improve the extended solution, we present an *approximative solution* of the Navier–Stokes equations using superposition of Lamb vortices with the velocities of the vortices as parameters minimizing the resultant of the nonlinear convection terms. By comparison with the numerical solutions of several merging problems, we show that the corresponding approximate solutions reduce significantly the errors of the extended asymptotic solutions.

In Sect. 6.3, we study three-dimensional vortical flows, for which the resultant, or zeroth moment of, vorticity is zero. The first moment of vorticity, which is conserved, defines a constant vector $\mathbf{D}$. We study the evolution of vorticity with $\mathbf{D}$ nonzero. The flow in the far field behaves as a doublet of strength $\mathbf{D}$. We then define the velocity of the vorticity field by the velocity of the doublet and the long-time behavior of the vortical flow. For axisymmetric flows, we study the self-merging of a vortex ring when its core size becomes comparable to the ring radius and the merging of two vortex rings with nearly equal radii prior to their self-merging. In a meridian plane, the cores of two rings merge to a domain with size $\ell_m \ll \ell$.

The intersection and reconnection of filaments are extremely complex three-dimensional unsteady phenomena. To initiate their study, we report a few numerical and experimental investigations of a basic problem, namely the local merging of two filaments leading to their reconnection to a single filament. In view of these numerical and experimental data, a simple model simulating the local merging process and a criterion for the reconnection are proposed.

6.1 Classification of Merging Problems and Efficient Numerical Schemes

In Subsect. 6.1.1, we describe the classification of merging problems, introduced by Liu et al. (1985), into type I, global merging, for $\ell_m = O(L)$, and type II, local merging, for $\ell_m \ll L$, where ℓ_m and L denote the size of the merged region and that of the vorticity field, respectively. For global merging problems, we have two types: I(a) for $\ell_d = O(L)$ and I(b) for $\ell_d \ll L$, where ℓ_d denotes the viscous diffusion length. We then explain in Subsect. 6.1.2 the meaning of an efficient numerical scheme and the choice of the computational domain. We elaborate on these general concepts with the formulation of the appropriate boundary data and numerical schemes for different types of merging problems in two- and three-dimensional flows in Sects. 6.2 and 6.3, respectively.

6.1.1 Classification of Merging Problems

From the structure of initial data that exhibit the merging of vortices or the commencement of their merging, we can identify three characteristic length scales. They are the size of the vortical field L, the size of the merging region(s) ℓ_m, and a vorticity decay length ℓ_d. The size L is usually taken to be the normal length scale ℓ of the flow field introduced in Chaps. 2 and 3. The decay length ℓ_d depends on the Reynolds number, Re, of the flow field and is of the order $L/\sqrt{\mathrm{Re}}$. Depending on the order of magnitude of ℓ_d relative to L, the types of merging are classified as type I, denoting a global merging with $\ell_m = O(L)$, and type II, which is a local merging with $\ell_m \ll L$. The global merging can be further subdivided into two types: type I(a) when $\ell_d = O(L)$ and type I(b) when $\ell_d \ll L$.

In summary, we identify the characteristic lengths

$$\begin{aligned} L\ &: \quad \text{overall extend of the vorticity distribution,} \\ \ell_m &: \quad \text{size of the merging region, and} \\ \ell_d\ &: \quad \text{vorticity decay length,} \end{aligned} \tag{6.1.1}$$

and then classify the different types of merging by

$$\begin{aligned} &\text{type I(a)} \ : \ \ell_m/L = O(1), \quad \ell_d/L = O(1), \\ &\text{type I(b)} \ : \ \ell_m/L = O(1), \quad \ell_d/L \ll 1, \\ &\text{type II} \quad\ \ : \ \ell_m/L \ll 1. \end{aligned} \tag{6.1.2}$$

Examples of global merging of type I(a) and of type I(b) are shown in Fig. 6.1. Figure 6.1a shows a not so slender vortex ring. The size of its vortical core is comparable to its radius that in turn is $O(L)$. Thus we have $\ell_d \approx R \approx L$. Eventually, the vorticity field covers the center of the ring, as shown in Fig. 6.1b. In the numerical simulation, the vortical field can be effectively confined inside a domain $\mathcal{D}$ with size $D \gg L$ for a finite time interval provided that the domain is allowed to move with the center of vorticity.

Figure 6.1c shows two slender filaments merging along their entire length almost simultaneously, while their core sizes remain much smaller than the minimum radius of curvature of the filaments. During merging, the resultant vorticity field of both filaments can be effectively confined in a slender toroidal

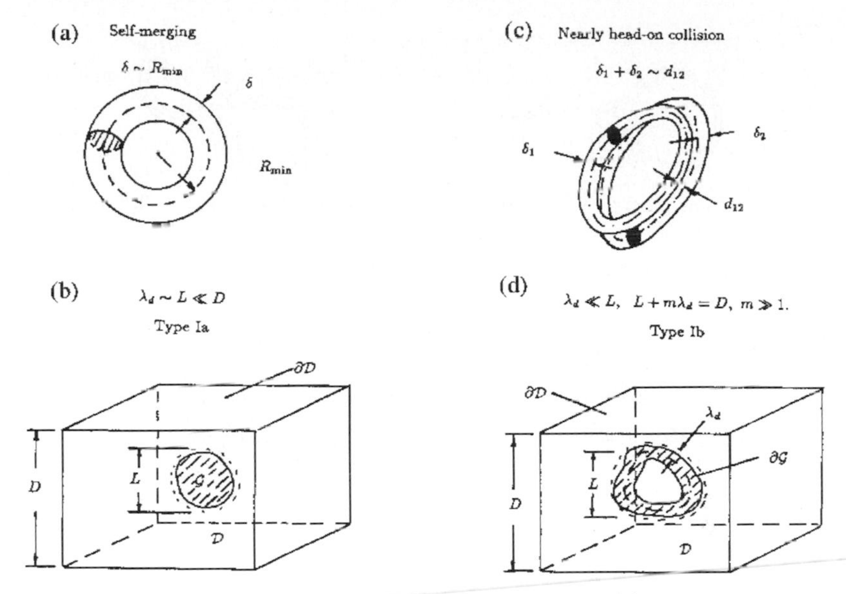

Fig. 6.1. Global merging of filament(s)

region with its cross-sectional size much smaller than the size of the torus which is $O(L)$. From these two examples, we can describe these two types of global mergings in general as follows:

For *type I(a)*, the initial vorticity distribution is assumed to decay exponentially on a length scale on the order of the size, L, of the vorticity field. Therefore, the far-field behavior of the vorticity distribution obeys

$$|\Omega(t, \mathbf{x})| = O\big(e^{-r^2/L^2}\big) \quad \text{as } r \to \infty \quad \text{for } t > 0 \;, \tag{6.1.3}$$

where $r = |\mathbf{x}|$. The dominant part of Ω is contained in a sphere of radius $L = O(\ell)$ and the size, D, of the computational domain has to be much larger than L, that is, $D \gg L$ (see Fig. 6.1b).

For *type I(b)*, the initial vorticity distribution is effectively contained in a domain $\mathcal{D}$ of size L, but it is assumed to decay exponentially outside of $\mathcal{D}$ with a much smaller decay length $\ell_d \ll L$. Therefore, the vorticity behaves outside of $\mathcal{D}$ as

$$|\Omega| = O\big(e^{-h^2/\ell_d^2}\big), \quad \text{with } \ell_d \ll L \;, \tag{6.1.4}$$

where h denotes the distance normal to the boundary $\partial\mathcal{D}$ of $\mathcal{D}$ (see Fig. 6.1d). The computational domain may then be of size $D = L + m\ell_d$ with $m \gg 1$ but $m\ell_d = O(L)$.

Examples of local mergings of type II are shown in Fig. 6.2. They demonstrate two cases of local self-merging of a filament in violation of criterion (5.1.9) and (5.1.10) and two cases of the intersection of two filaments in local region(s) in violation of (5.1.11). The size of a local merging region is of the

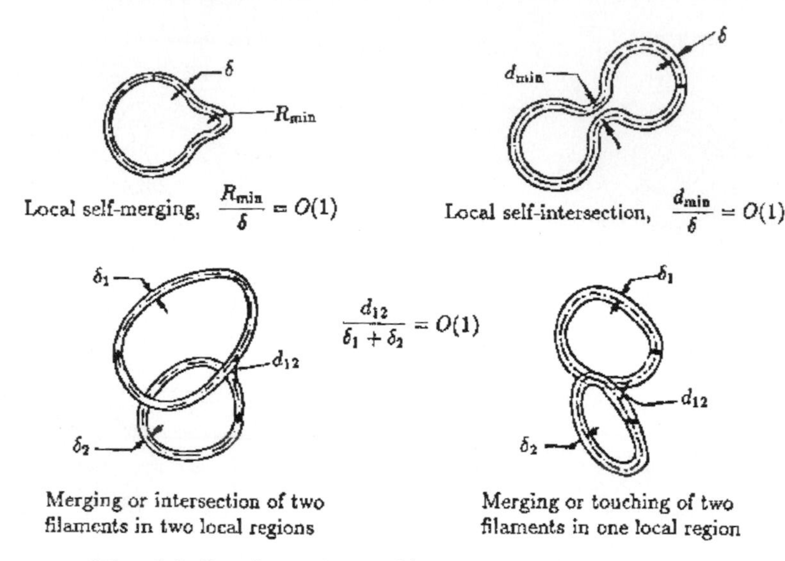

Fig. 6.2. Local merging and/or interaction of filament(s)

order of the core size and is much smaller than the size of the filament(s). Thus $\ell_d, \ell_m = O(\delta) \ll L$. The slender filament structures persist away from the local merging regions, and it is clear that it would be very inefficient to study the merging process by numerical solution of the Navier–Stokes equations on a domain containing the entire vortical field.

6.1.2 Efficient Numerical Scheme and Computational Domain

We shall first identify different sources of errors in the finite difference solutions and then describe the essence of an efficient scheme and the choice of the computational domain.

Source of Numerical Errors. For a finite-difference scheme, the errors e_t and e_s in the temporal and spatial derivatives are usually assessed in terms of the nondimensional time step Δ_t/t_{ref} and of the scaled grid size $\Delta_s/\ell_{\text{ref}}$, where t_{ref} and ℓ_{ref} are characteristic time and length scales for significant local changes of the vorticity (see, e.g., Richtmyer and Morton, 1967). Estimates for the global discretization errors can be expressed in the form

$$e_t = O\big([\Delta_t/t_{\text{ref}}]^{p_t}\big) \ll 1 \qquad \text{and} \qquad e_s = O\big([\Delta_s/\ell_{\text{ref}}]^{p_s}\big) \ll 1\,, \qquad (6.1.5)$$

where $p_s{+}1$ and $p_t{+}1$ are integers specifying the order of local approximations of the difference scheme. In our discussion, the relevant reference length is the decay length ℓ_d and, for initial data of the type I(a), it coincides with the overall extension, L, of the vorticity distribution. The relevant reference time for type I(a) is L/U, where U denotes an average velocity in the initial flow field.

For a given degree of accuracy, say e_a, we require

$$e_s \sim e_t \sim e_a \ . \tag{6.1.6}$$

This, in turn, defines the grid size Δ_s and the time step Δ_t through (6.1.5) for a given set of p_t and p_s, that is, for a particular discretization.

Since the numerical solution can be carried out only for a finite computational domain, $\mathcal{D}$, other sources of error are introduced because of the replacement of the far-field conditions by approximate conditions applied on the boundary, $\partial\mathcal{D}$. Consider first the condition of exponential decay of $|\Omega|$ in (6.1.3). If this condition is replaced by

$$\Omega = 0 \qquad \text{on} \quad \partial\mathcal{D} \ , \tag{6.1.7}$$

the error is exponentially small, namely

$$e_\Omega = O\big(e^{-D^2/L^2}\big) \ . \tag{6.1.8}$$

On the other hand, we have shown in Sect. 2.2 that the velocity and vector potential $\mathbf{A}$ decrease only in descending powers of L/D. Thus, if we impose

$$\mathbf{A} = 0 \qquad \text{on} \quad \partial\mathcal{D} \ , \tag{6.1.9}$$

the error is

$$e_A = O(L^2/D^2) \ , \tag{6.1.10}$$

because the zeroth moments of vorticity vanish for all t. Therefore, the error in this set of boundary data is dominated by e_A.

By using the far-field expansion procedure described in Sect. 2.2, (2.3.10)–(2.3.12), we can reduce the error to

$$e_A = O((L/D)^{m+2}) \ , \tag{6.1.11}$$

where m denotes the highest moments of vorticity retained in the far-field expansion. Such an improvement of the boundary data for $\mathbf{A}$ provides a considerable reduction of the computational efforts needed to achieve a given degree of accuracy. To clarify this aspect, we determine the number of grid points, N, needed to formally keep the errors in the boundary data and the errors of the discrete approximation errors in $\mathcal{D}$ at the order e_a. With

$$N = O\big(D^3/\Delta_s^3\big) \ , \tag{6.1.12}$$

(6.1.5) and (6.1.11) imply

$$N = O(e_a^{-\mu}) \qquad \text{where} \qquad \mu = \frac{3}{p_s} + \frac{3}{m+2} \ . \tag{6.1.13}$$

Thus, an increase in the accuracy of the boundary data has the same potential for reducing the numerical effort as an increase in the order of the discretization, provided that the constraints (6.1.6) on the errors are observed.

To give a concrete illustration of the effect of a more accurate boundary condition on **A**, we consider p_s fixed and show the savings in the number of grid points, N, when m is increased from 0 to 2 by retaining up to second moments of vorticity in the far-field expansion. Let μ_j denote the exponent μ in (6.1.13) with $m = j$, $j = 0, 2$, and let N_j denote the corresponding number of grid points needed. With the exponent μ_0 reduced to $\mu_2 = \mu_0 - 3/4$, the number of grid points is reduced by a factor $e_a^{-3/4}$. For example, even for a moderate value of $e_a = 1/256$, we have $N_2 = N_0/64$ leading to a considerable savings in the computational effort.

In arriving at (6.1.13), we required that the error in the boundary data is of the order of the error in the discretization, that is, $e_A \sim e_s \sim e_a$. In principle, we could allow a larger error, e_A, relative to e_s since the errors induced in the boundary conditions have a weaker influence on the global accuracy. (The reason is that the boundaries are a manifold of dimension one less than that of the computational domain.) However, we prefer the stronger condition on e_A because the decisions on the number N and the computational domain $\mathcal{D}$ are based on the estimates for the characteristic lengths in the structure of the initial data. As time proceeds, the size of the vorticity field L increases accompanied by an increase in the error of the boundary data while that of the discrete approximation diminishes. Since we do not want to change the computational domain frequently, we choose a value for e_A smaller than necessary, that is, a computational domain larger than necessary at the beginning. This practical consideration was reported by Chamberlain and Weston (1984). In a series of numerical simulations using different order boundary conditions, they observed spurious generation of vorticity at the boundaries after a finite time T for lower order approximations. This spurious generation can be delayed to a much later time by increasing the size of $\mathcal{D}$ at a few time steps before T or by using higher order boundary conditions to begin with.

Meaning of an Efficient Numerical Scheme. For a scheme to be efficient, the number of operations for the evaluation of the approximate boundary data for **A** should be at most of the same order as that of the finite-difference solution at each time step. Since we are interested in time-accurate approximations for transient phenomena (we are not seeking numerical schemes with large Δ_t), the vorticity field can be advanced in time by solving an explicit finite difference approximation of the vorticity diffusion equation (2.3.2). The number of operations per time step is then of order N for most such methods. Therefore, the overall number of operations for each time step is dominated by that for the solution of the Poisson equation for the vector potential **A**. Using a fast Poisson solver, this number is of order $N \ln N$ (Swartztrauber and Sweet, 1979). Since $\ln N$ may be considered as a finite number relative to N, our objective is to increase the accuracy of the boundary data so as to reduce the size of $\mathcal{D}$ and hence the number of grid points, while keeping the number of operations in the computation of the boundary data at $O(N)$.

It was noted by Ting (1983) that, although boundary data obtained by a direct evaluation of the Poisson integral will reduce the error e_A to that of e_Ω in (6.1.3), the number of computational steps for a straightforward numerical integration is $O(N^{5/3})$. Hence, this procedure is not acceptable and a more efficient scheme had to be devised. Such a method and related accuracy estimates were presented in Subsect. 3.1.3 of Book-I and will be illustrated here in Subsect. 6.3.1 for the simulation of the merging of two coaxial vortex rings.

In principle, we can reduce the error e_A in (6.1.10) to $O([L/R]^{m+2})$ by including additional terms in the series (2.2.9)–(2.2.10) up to the mth moment of vorticity and reduce accordingly the size of $\mathcal{D}$. In general, the number of moments that must be evaluated at each time step is $4\Sigma j - 7 = 2m(m+1) - 7$ for $m \geq 2$ (see the integral invariants discussed in Sect. 2.2). The number of computational steps for the evaluation of the boundary values of $\mathbf{A}$ then becomes $O(2m^2 N)$. Furthermore, the error induced by omitting the vorticity outside of $\mathcal{D}$ in an approximate evaluation of an mth moment of vorticity is $O([D/L]^m\, e^{-D^2/L^2})$, in which the significance of the factor $[D/L]^m$ increases as m increases. These two observations imply a practical upper bound on m for a given D/L. The upper bound increases as D/L increases.

If we increase the accuracy requirement by choosing a smaller e_a while retaining the same number of grid points, and the same discrete approximation, we have to reduce the grid size Δ_s and the time step Δ_t accordingly. Hence, we have to shrink the computational domain. If $\mathcal{D}$ and R are the domain and its size for the required accuracy, and $\mathcal{D}_1$ and R_1 are those for $e_{a1} < e_a$, we have

$$\mathcal{D}_1 \subset \mathcal{D} \qquad \text{and} \qquad R_1 < R\,. \tag{6.1.14}$$

To reduce also the error e_A in the approximate boundary conditions to the order e_{a1}, we have to increase the number of terms in the far-field expansion (2.2.9). However, in this case, the error of the numerical evaluation of an mth moment due to the neglect of the vorticity outside of $\mathcal{D}_1$ can be larger than e_{a1}.

This difficulty can be overcome by using a larger domain, say $\mathcal{D}_2$, only in updating the vorticity distribution and for the numerical evaluation of the mth moments. The boundary condition for the vorticity, $\Omega = 0$, is then imposed on $\partial\mathcal{D}_2$, while approximate boundary conditions for the vector potential are still imposed on $\partial\mathcal{D}_1$. Without lowering our requirements of efficiency, we are allowed to use a larger domain, and hence a larger number of grid points, N_2, for the vorticity calculation, because the number of steps in updating the vorticity is of order N_2. So long as $N_2 = O(N_1 \ln(N_1)$, the number of computational steps per time step remains of order $N_1 \ln(N_1)$. This *two-domain method* will be described in detail in Subsect. 6.3.1.

For initial data of the type Ib, we use the fact that the vorticity decays on a short length scale $\ell_d \ll L$. This allows us to reduce the size of the computational domain $\mathcal{D}$ from a size much larger than L to a size $D =$

$L + O(k\lambda)$ with $k \gg 1$, and $k\lambda$ being at most of order L. In this case, the vorticity on the boundary is small, of order $\exp(-k^2)$, but the velocity is of order one because $D/L = O(1)$. Hence, it is *incorrect*, not just inaccurate, to impose $\mathbf{A} = 0$ on $\partial\mathcal{D}$. It is imperative that approximate boundary data be derived for $\mathbf{A}$ such that the error e_A is reduced to say $O(k^{-4})$.

For a local merging problem of type II, as shown in Fig. 6.2, the computational domain may be much smaller than the size of the vortical field. Now it is incorrect to impose $\mathbf{A} = 0$ on the boundary $\partial\mathcal{D}$, and it is also incorrect to impose $\Omega = 0$. For this problem we have to develop a scheme that combines the asymptotic solution outside $\mathcal{D}$ with the numerical solution in $\mathcal{D}$. For the latter, we need to construct approximate boundary data for both $\mathbf{A}$ and Ω and assess their errors.

The general procedures for carrying out the efficient schemes for the types of merging problems classified above were described in Subsect. 3.1.2–3.1.4 of Book-I. Here we shall illustrate these procedures by examples for two- and three-dimensional problems in the following two sections respectively.

Before doing so, we present next the long-time behaviors of merging problems so that the numerical solutions carried out to a finite time can be extended to long time.

6.1.3 Long-Time Behavior

To develop the long-time behavior of a merging problem, we first change the independent variables from $(t, \boldsymbol{x})$ to new ones, $\bar{t}, \Xi$, with

$$\Xi = \boldsymbol{x}/\lambda(t)\ , \quad \lambda(t) = \sqrt{4\nu\bar{t}} \quad \text{and} \quad \bar{t} = t + t_0\ . \tag{6.1.15}$$

Here Ξ represents the spatial variable $\boldsymbol{x}$ nondimensionalized by the diffusion length, $\lambda(t) = \sqrt{4\nu\bar{t}} = \sqrt{4\nu(t+t_0)}$, where $t_0 > 0$ is free to be assigned, because $\bar{t} = t(1 + t_0/t) \approx t$ for long time. For example, t_0 could be identified as the age of the smallest initial core size, with $\lambda(0) = \delta(0) = \sqrt{4\nu t_0}$. In the new independent variables, we denote the gradient and Laplacian operators by $\tilde{\nabla}$ and $\tilde{\Delta}$, respectively. The vorticity evolution equation (2.3.2) becomes

$$\boldsymbol{\Omega}_{\bar{t}} - \frac{1}{2\bar{t}}[\Xi \cdot \tilde{\nabla}]\boldsymbol{\Omega} + \frac{1}{\sqrt{4\nu\bar{t}}}[\tilde{\nabla} \cdot (\boldsymbol{v}\boldsymbol{\Omega}) + \tilde{\nabla} \cdot (\boldsymbol{\Omega}\boldsymbol{v})] = \frac{1}{4\bar{t}}\tilde{\Delta}\boldsymbol{\Omega} \tag{6.1.16}$$

for $\bar{t} \geq t_0$, with

$$\tilde{\nabla} \cdot \boldsymbol{v} = 0 \quad \text{and} \quad \tilde{\nabla} \times \boldsymbol{v} = \sqrt{4\nu\bar{t}}\ \boldsymbol{\Omega}\ . \tag{6.1.17}$$

Here, we use the same symbols $\boldsymbol{v}$ and $\boldsymbol{\Omega}$ for the functions in the new independent variables. For long time $\bar{t} \gg t_0$, we write the leading terms of $\boldsymbol{\Omega}$ and that of $\boldsymbol{v}$,

$$\boldsymbol{\Omega} = \bar{t}^{-k}[\mathbf{f}_0 + \bar{t}^{-h}\mathbf{f}_1 + \cdots] \quad \text{and} \quad \boldsymbol{v} = \bar{t}^{-k+0.5}[\mathbf{g}_0 + \bar{t}^{-h}\bar{g}_1 + \cdots]\ , \tag{6.1.18}$$

where $\mathbf{f}_i,\ \ \mathbf{g}_i,\ \ i = 0, 1, \ldots$ are functions of Ξ.

The leading-order terms in (6.1.16) are the unsteady and viscous terms. They are linear and have a common factor of $\bar{t}^{-k-1}$. The nonlinear convection terms begin with a factor $\bar{t}^{-2k}$ requiring $h = k - 1$. The power $-k$ of the leading term of $\boldsymbol{\Omega}$ shall be defined by the initial data for two- and three-dimensional problems, respectively.[1]

Two-Dimensional Problems. In the x_1, x_2 plane, we have $\boldsymbol{\Omega} = \omega \boldsymbol{e}_3$. With (6.1.15), we see that an area integral in the x_1, x_2 plane, denoted by $\langle\ \rangle$, is equal to that in ξ_1, ξ_2 times $4\nu\bar{t}$. We recall that the total strength of vorticity and its first moment in the x_1, x_2 plane are time invariants,

$$\langle \omega \rangle = \Gamma_0 \quad \text{and} \quad \langle \boldsymbol{x}\omega \rangle = \mathbf{D}_0. \tag{6.1.19}$$

If $\Gamma_0 \neq 0$, the center of vorticity $\boldsymbol{X}_c = \mathbf{D}_0/\Gamma_0$ remains *stationary*. See Subsect. 2.2.4 with density $\rho = 1$. The first moment of vorticity with respect to the center $\boldsymbol{X}_c$ vanishes. In the $\xi_1\xi_2$ plane the area integral of ω is proportional to $\bar{t}^{-k}$ and should equal to $\Gamma_0 t_0/4\nu\bar{t}$. Thus, we have

$$k = 1 \quad \text{and} \quad h = k - 1 = 0. \tag{6.1.20}$$

Therefore, the leading term $\bar{t}^{-1}g(\xi_1, \xi_2)$ could be a solution of the nonlinear vorticity equation, which yields a partial differential equation for $g(\xi_1, \xi_2)$. This is consistent with the result in Subsect. 3.2.2 that at large time the vorticity field would approach the optimum similarity solution of a vortex of strength Γ centered at $\boldsymbol{X}_c$ with t_0 identified as the optimum age t_0^* to fulfill the Poincaré formula for the polar moment of vorticity. With $k = 1$, we could express the long-time solution of (6.1.16) in a power series of $\bar{t}^{-1}$, beginning with $\bar{t}^{-1}$, setting $t_0 = 0$ for simplicity or replacing t_0 by the optimum time shift t_0^* defined in Subsect. 3.2.2.

If $\Gamma_0 = 0$, we have $\Gamma(t) \equiv 0$, and the far field behaves as a doublet with constant strength $\mathbf{E}$ defined by the constant first moments of vorticity. The power k in the leading term in (6.1.18) is now defined by conservation of the first moments of vorticity

$$k = 3/2 \quad \text{and hence} \quad h = -1/2\,. \tag{6.1.21}$$

Thus, we can rewrite the series (6.1.18) for ω and $\boldsymbol{v}$ as

$$\omega = \bar{t}^{-3/2}[f_0 + \bar{t}^{-1/2} f_1 + \bar{t}^{-1} f_2 + \cdots] \tag{6.1.22}$$

and

[1] We introduced the new time variable $\bar{t}$ with $t_0 > 0$ so that we could solve the initial value problem (6.1.16) in a computational domain $\tilde{D}$ in the ξ_1, ξ_2 plane. Besides moving with the center of the vorticity distribution, if not stationary, the domain can remain the same for $t > 0$, because its image in the x_1, x_2 plane is expanding. There is no need to add t_0 for the study of long-time behaviors. We could replace $\bar{t}$ by t in the repesentations of $\boldsymbol{\Omega}$ and $\boldsymbol{v}$ in (6.1.18).

$$\boldsymbol{v} = \bar{t}^{-1}(\mathbf{g}_0 + \bar{t}^{-1/2}\bar{g}_1 + \bar{t}^{-1}\mathbf{G}_2 + \cdots] \, . \tag{6.1.23}$$

Now we can express the long time solutions of ω and $\boldsymbol{v}$ in power series of $\bar{t}^{-1/2}$, beginning with $\bar{t}^{-3/2}$ and $\bar{t}^{-1}$ respectively.

Three-Dimensional Problems. We note that a volume integral in x_1, x_2, x_3, denoted by $\langle \ \rangle$, is equal to that in ξ_1, ξ_2, ξ_3 times $(t_0/\bar{t})^{3/2}$. We recall (2.2.5) for the total vorticity and (2.3.7) for the nontrival time invariants of the first moments of vorticity. They are

$$\langle \omega \rangle = 0 \quad \text{and} \quad \langle \boldsymbol{x} \times \boldsymbol{e}_3 \omega \rangle = \text{const. } \mathbf{E}_0 \, . \tag{6.1.24}$$

In the far field, the leading-order solution represents a doublet of constant strength $\mathbf{E}_0$. To keep the first moment time invariant, we require the power $-k$ of the leading term in (6.1.18) to be

$$k = 2 \quad \text{hence} \quad h = k - 1 = 1. \tag{6.1.25}$$

Thus, we can rewrite the series (6.1.18) for $\boldsymbol{\Omega}$ and $\boldsymbol{v}$ as,

$$\boldsymbol{\Omega} = \bar{t}^{-2}[\mathbf{f}_0 + \bar{t}^{-1}\mathbf{f}_1 + \bar{t}^{-2}\mathbf{f}_2 + \cdots] \tag{6.1.26}$$

and

$$\boldsymbol{v} = \bar{t}^{-3/2}(\mathbf{g}_0 + \bar{t}^{-1}\bar{g}_1 + \bar{t}^{-2}\mathbf{G}_2 + \cdots] \, . \tag{6.1.27}$$

Now we can express the long-time solution of $\boldsymbol{\Omega}$ and $\boldsymbol{v}$ in power series of $\bar{t}^{-1}$ with leading term $\bar{t}^{-2}$ and $t^{-3/2}$, respectively.

Instead of restating the general schemes for different types of merging problems, presented in Subsects. 3.1.2–3.1.4 of Book-I, we demonstrate these schemes by numerical examples for two- and three-dimensional merging problems in Sects. 6.2 and 6.3, respectively. The numerical solutions carried for a finite time which in turn will define the unknowns in their long-time solutions. Numerical examples illustrating the procedure presented in Subsect. 3.1.4 in Book-I for the intersection of two slender filaments, their local merging and final reconnection problem have not been carried out, because it is hard to implement. Instead, we recommend analyzing the problem by the extension of vortex element method for slender filaments as outlined in Sect. 9.1.

6.2 Merging of Two-Dimensional Vortices

According to the classifications in Subsect. 6.1.1, the merging of two-dimensional vortices is always a global merging problem. Applications are presented in Subsect. 6.2.1, focusing in particular on the merging of viscous vortices and the roll-up of thin vortical layers.

In Subsect. 6.2.2 we employ the integral invariants for two-dimensional viscous flow to formulate *the rules of merging* for closely interacting vortices and to explain the physical meaning of an optimum Lamb vortex. Then we discuss the superposition of Lamb vortices moving according to the asymptotic theory for nonoverlapping vortices (vortices far apart from each other) as an approximation for a viscous vortical field. When this representation is employed, even if the cores of adjacent vortices overlap considerably, we call it the *extended asymptotic solution.*

In Subsect. 6.2.3 we present a different approach. The sum of Lamb vortices with their velocities treated as unknowns are considered as as an *approximate solution* in the sense that the Navier–Stokes equations are satisfied approximately under a minimum principle. The minimum principle in turn defines the unknown velocities. This scheme was employed to study the merging of vortices, and the results compare favorably with corresponding finite difference solutions reported in Subsect. 6.2.1.

We now restate some relevant results for two-dimensional flows from Sect. 2.3, namely, the governing equations, the integral invariants and the far-field behaviors. To study the interaction and merging of vortices in two-dimensional space, the xy plane, it is natural to use the vorticity ζ and the stream function ψ as dependent variables. For an incompressible fluid, they are governed by the equations

$$\zeta_t + \boldsymbol{V} \cdot \nabla\zeta = \nu\,\Delta\zeta \tag{6.2.1}$$

and

$$\Delta\psi = -\zeta \tag{6.2.2}$$

with the velocity

$$\boldsymbol{V} = \hat{\imath}\psi_y - \hat{\jmath}\psi_x \; . \tag{6.2.3}$$

The prescribed initial vorticity distribution

$$\zeta(0, x, y) = \zeta_0(x, y) \tag{6.2.4}$$

is assumed to be of bounded support or to decay exponentially in the radial coordinate $\sigma = \sqrt{x^2 + y^2}$. This condition is certainly fulfilled when ζ_0 is concentrated around a finite number of points (X_i, Y_i), $i = 1, \ldots, N$, so as to model the vorticity distribution of N vortices. Because of the exponential decay of the vorticity, the induced velocity vanishes at infinity for $t \geq 0$ and thus

$$\zeta(t, x, y) = o(\sigma^{-m}) \tag{6.2.5}$$

for all m and

$$|\boldsymbol{V}| \to 0 \; . \tag{6.2.6}$$

The initial data for $\boldsymbol{V}$ are uniquely defined by (6.2.2), (6.2.3), and (6.2.6).

Equations (6.2.1)–(6.2.6) define an initial value problem in two-dimensional space. The two-dimensional solution yields relations expressing the time invariance of the total strength and first moments of vorticity, and the linear growth of the polar moment. These expressions, which were first derived by Poincaré (1893) directly from the two-dimensional Navier–Stokes equations, are

$$\langle \zeta(t,x,y)\rangle = \langle \zeta_0(x,y)\rangle = \Gamma_0 \ , \tag{6.2.7}$$

$$\langle x\,\zeta(t,x,y)\rangle = \langle x\ \zeta_0(x,y)\rangle = C_1 \ , \tag{6.2.8}$$

$$\langle y\,\zeta(t,x,y)\rangle = \langle y\ \zeta_0(x,y)\rangle = C_2 \ , \tag{6.2.9}$$

and

$$\langle \sigma^2\zeta(t,x,y)\rangle = 4\nu\Gamma_0 t + D_3 \ , \tag{6.2.10}$$

where the constant D_3 is defined by the initial data,

$$D_3 = \langle r^2\zeta_0(x,y)\rangle \ . \tag{6.2.11}$$

Here, $\langle f\rangle$ denotes the area integral of f over the xy plane. Note that (6.2.7)–(6.2.10) are also valid for solutions of the linear diffusion equation, which is (6.2.1) without the nonlinear convection term.

Using the above invariants, we arrive at the far-field behavior of the stream function in polar coordinates, σ, θ,

$$\begin{aligned}\psi(t,\sigma,\theta) = &-\frac{\Gamma_0}{2\pi}\ln\sigma + \frac{1}{2\pi\sigma}[C_1\cos\theta + C_2\sin\theta] \\ &+\frac{1}{4\pi\sigma^2}[F_3(t)\cos 2\theta + H_3(t)\sin 2\theta] + O(\sigma^{-3}) \ ,\end{aligned} \tag{6.2.12}$$

where $F_3 = \langle (x^2-y^2)\zeta\rangle$ and $H_3 = 2\langle xy\zeta\rangle$. This equation provides the boundary data on the computational domain for a total merging of type I(a).

For a total merging of type I(b), we divide the computational domain $\mathcal{D}$ into M subdomains $\mathcal{D}_i$, $i = 1,\ldots,M$, and write

$$\psi = \sum_{i=1}^{M}\psi_i \ , \tag{6.2.13}$$

where ψ_i is the stream function induced by ζ_i. The latter is the vorticity distribution in $\mathcal{D}_i$, which is equal to ζ in $\mathcal{D}_i$ and to zero elsewhere. Note that $\psi_i = 0$ when $\mathcal{D}_i$ are adjacent to the boundary $\partial\mathcal{D}$, where $\zeta_i = 0$ is an exponentially accurate approximation.

We can then express ψ_i by its power series expansion

$$\begin{aligned}\psi_i = &-\frac{\langle\zeta_i\rangle}{2\pi}\ln\sigma_i + \frac{1}{2\pi\sigma_i}[\langle x_i\zeta_i\rangle_i\cos\theta_i + \langle y_i\zeta_i\rangle\sin\theta_i] \\ &+\frac{1}{4\pi\sigma_i^2}[\langle (x_i^2-y_i^2)\zeta_i\rangle\cos 2\theta_i + 2\langle x_i y_i\zeta_i\rangle\sin 2\theta_i] \\ &+O(\sigma_i^{-3}) \ .\end{aligned} \tag{6.2.14}$$

Here, x_i and y_i are Cartesian coordinates and σ_i, θ_i are the corresponding polar coordinates with the origin located at the center of $\mathcal{D}_i$. Because the vorticity distribution ζ_i is discontinuous on $\partial\mathcal{D}_i$, the invariance conditions (6.2.7)–(6.2.10) are not applicable to ζ_i. The moments, $\langle\zeta_i\rangle$, $\langle x_i\zeta_i\rangle$, etc., in (6.2.14) are, in general, time dependent, and must be evaluated at each instant.

6.2.1 Numerical Simulation of Vortex Merging and the Roll-Up of Thin Shear Layers

In the numerical simulation of merging problems, the computational domain $\mathcal{D}$ is allowed to move so that the dominant part of the vorticity distribution is always located near the center of $\mathcal{D}$. Here, we shall explain how to define the velocity of $\mathcal{D}$ and how to simplify the far-field condition (6.2.12) depending on whether the total strength of the vorticity distribution in (6.2.7) is zero or not.

6.2.1.1 Nonzero Circulation $\langle\zeta\rangle \neq 0$. In this case, we can use the invariances (6.2.8) and (6.2.9) of the first moments to define the center of vorticity ($X_{C.G.}$,$Y_{C.G.}$) as

$$X_{C.G.} = \frac{\langle x\zeta\rangle}{\langle\zeta\rangle} = \frac{C_1}{\Gamma_0} \quad \text{and} \quad Y_{C.G.} = \frac{\langle y\zeta\rangle}{\langle\zeta\rangle} = \frac{C_2}{\Gamma_0} \tag{6.2.15}$$

and we conclude that the center of vorticity is stationary. By locating the origin at the center of vorticity from the outset, we obtain $C_1 = C_2 = 0$. The computational domain $\mathcal{D}$ will remain stationary and centered at the origin. The boundary values of ψ on $\partial\mathcal{D}$ are then given by its far-field behavior (6.2.12) in the simplified form,

$$\psi(t,\sigma,\theta) = -\frac{\Gamma_0}{2\pi}\ln\sigma + \frac{1}{4\pi\sigma^2}[F_3(t)\cos 2\theta + H_3(t)\sin 2\theta] + O(\sigma^{-3}) \ . \tag{6.2.16}$$

Thus in the far field, we see a point vortex of constant strength Γ_0 and two quadrupoles of temporally varying strength but no doublet. The point vortex is located at the origin, that is, at the center of vorticity.

6.2.1.2 Zero Circulation $\langle\zeta\rangle = 0$ but $\langle x\zeta\rangle \neq 0$. With $\Gamma_0 = 0$, we cannot define the center of vorticity by (6.2.15). Instead, we shall define the center of an optimum doublet in terms of the far-field behavior (6.2.12), which now becomes

$$\begin{aligned}\psi(t,\sigma,\theta) &= \frac{1}{2\pi\sigma}\left[C_1\cos\theta + C_2\sin\theta\right] \\ &\quad + \frac{1}{4\pi\sigma^2}[F_3(t)\cos 2\theta + H_3(t)\sin 2\theta] + O(\sigma^{-3}) \ .\end{aligned} \tag{6.2.17}$$

The first term in (6.2.17) represents a doublet of constant strength $|\boldsymbol{C}| = \sqrt{C_1^2 + C_2^2}$ oriented along the direction $C_1\hat{\imath} + C_2\hat{\jmath}$ and the second term again represents two quadrupoles of temporally varying strength located at the origin. If both C_1 and C_2 are nonzero, we can represent the leading two terms by a doublet of the same strength and orientation but moving with the velocity, $(\dot{X}_D(t), \dot{Y}_D(t))$, such that

$$\psi(t, \bar{\sigma}, \bar{\theta}) = \frac{1}{2\pi\bar{\sigma}}[C_1 \cos\bar{\theta} + C_2 \sin\bar{\theta}] + O(\bar{\sigma}^{-3}) \tag{6.2.18}$$

$$\dot{X}_D = \frac{C_1\dot{F}_3 + C_2\dot{H}_3}{2|\boldsymbol{C}|^2}, \qquad \dot{Y}_D = \frac{-C_2\dot{F}_3 + C_1\dot{H}_3}{2|\boldsymbol{C}|^2}. \tag{6.2.19}$$

Here, $\bar{\sigma}$ and $\bar{\theta}$ denote the polar coordinates relative to the center $(X_D(t), Y_D(t))$. From (6.2.10), we see that the polar moment is conserved since $\Gamma_0 = 0$.

To get additional information for this case, we express the two-dimensional space $\mathbb{R}^2$ as the union of $\mathcal{S}^+$ and $\mathcal{S}^-$ in which ζ is nonnegative and nonpositive, respectively. That is

$$\mathbb{R}^2 = \mathcal{S}^+ \cup \mathcal{S}^- \tag{6.2.20}$$

$$\text{with} \quad \zeta \geq 0,\ \zeta \leq 0 \text{ and } \zeta = 0 \quad \text{for} \quad \boldsymbol{x} \in \mathcal{S}^+,\ \mathcal{S}^- \text{ and } \mathcal{S}^+ \cap \mathcal{S}^-. \tag{6.2.21}$$

We denote the nonnegative and nonpositive vorticity distribution by $\zeta^\pm$, that is,

$$\begin{aligned} \zeta^\pm &= \zeta\,, \quad \boldsymbol{x} \in \mathcal{S}^\pm\,, \\ \zeta^\pm &= 0\,, \quad \boldsymbol{x} \in \mathcal{S}^\mp\,, \end{aligned} \tag{6.2.22}$$

and then study the moments of $\zeta^\pm$. The condition $\langle\zeta\rangle = 0$ becomes

$$\langle\zeta^+\rangle = -\langle\zeta^-\rangle = \Gamma^+(t)\,, \tag{6.2.23}$$

which is in general not conserved. Its rate of change is given by

$$\langle\zeta^+\rangle_t = \int_{\partial\mathcal{S}^+} \hat{n}\cdot[-\boldsymbol{V}\zeta + \nu\nabla\zeta]\,ds\,, \tag{6.2.24}$$

where $\hat{n}$ denotes the outward unit normal vector and s the arclength. From the definitions (6.2.20)–(6.2.23), we have $\zeta = 0$ and $\hat{n}\cdot\nabla\zeta \leq 0$ along $\partial\mathcal{S}^+$ and hence the total strength of positive vorticity cannot increase

$$\frac{d\Gamma^+}{dt} \leq 0\,. \tag{6.2.25}$$

The equality sign holds only if

$$\hat{n}\cdot\nabla\zeta = 0 \text{ along } \partial\mathcal{S}^+\,. \tag{6.2.26}$$

6.2.1.3 Antisymmetric Vorticity Field. Now we consider the special case that the vorticity field is antisymmetric with respect to an axis, say the y-axis. In addition to $\langle\zeta\rangle = 0$, we have

$$\zeta(t, x, y) = -\zeta(t, -x, y) \ , \quad \zeta(t, 0, y) = 0 \ . \tag{6.2.27}$$

Let $\langle\ \rangle^+$ and $\langle\ \rangle^-$ denote the area integrals over the right and left half plane, respectively. Then (6.2.7)–(6.2.9) and (6.2.26) yield

$$\langle\zeta\rangle^+ = -\langle\zeta\rangle^- \neq 0 \ , \quad \langle x\zeta\rangle^+ = \langle x\zeta\rangle^- = C_1/2 \ , \quad \langle y\zeta\rangle^+ = -\langle y\zeta\rangle^- \ , C_2 = 0 \ . \tag{6.2.28}$$

Note that we have assumed that $\langle\zeta\rangle^+ \neq 0$, but that ζ may change signs in the half plane $x > 0$. Because of the antisymmetry, we need to analyze only the flow field on a half plane say the right half, $x \geq 0$. The computational domain $\mathcal{D}$ will be bounded on its left by the y-axis with the boundary condition,

$$\zeta(t, 0, y) = 0 \ , \quad \psi(t, 0, y) = 0 \ . \tag{6.2.29}$$

To contain the vorticity field away from the remaining boundary of $\mathcal{D}$, we move $\mathcal{D}$ with a velocity $\dot{Y}^+$ parallel to the y-axis. Specifically, we choose the domain $\mathcal{D}$ to be $0 \leq x \leq D$ and $|y - Y^+| \leq D$ with $D \gg L$, and identify Y^+ as the y-coordinate of the center of vorticity on the right half plane (also that on the left half plane), that is,

$$Y^+(t) = Y^-(t) = \langle y\zeta\rangle^+ \ / \ \langle\zeta\rangle^+ \ . \tag{6.2.30}$$

Note that this definition is meaningful only for an antisymmetric vorticity field described by (6.2.27) with $\langle\zeta\rangle^+ \neq 0$. Equation (6.2.30) can be considered as the equation for the ordinate of the center of the vorticity field that behaves as a doublet in the far field. But this equation differs from (6.2.19) for the center of an optimum doublet. The latter yields $Y_D(t) = \langle yx\zeta\rangle^+/C_1$. The difference between these two definitions will be elaborated later in Subsect. 6.2.3.

For the merging of a vortex pair of opposite strength, we assume in addition to the antisymmetry (6.2.27) that the vorticity on either side of the y-axis does not change sign. We assume that the vorticity on the right side of the y-axis is positive, that is, $\zeta \geq 0$ for $x \geq 0$. Noting that $\zeta_x(t, 0, y) \geq 0$ we conclude from (6.2.25) that the total strength of positive (negative) vorticity has to decrease (increase)

$$\Gamma_t^+ = \langle\zeta\rangle_t^+ < 0 \ . \tag{6.2.31}$$

It follows that the x-coordinate of the center of positive vorticity has to move away from the y-axis since

$$\dot{X}^+ = \left[\frac{\langle x\zeta\rangle^+}{\langle\zeta\rangle^+}\right]_t = -\frac{C_1\Gamma_t^+}{2(\Gamma^+)^2} > 0 \ . \tag{6.2.32}$$

If in addition there is no significant diffusion of vorticity across the y-axis, for example, when the vorticity is negligible near the line $x = 0$, we have

$$\zeta_x(t, 0, y) = 0 \quad \text{in addition to} \quad \zeta(t, 0, y) = 0 \,. \tag{6.2.33}$$

This is true in the model problem for the roll-up of a thin trailing vortical layer behind an airplane wing of high aspect ratio. This model problem will be discussed in detail below. Under (6.2.33) we obtain from the vorticity evolution equation that $\langle\zeta\rangle_t^+ = 0$ or

$$\langle\zeta(t, x, y)\rangle^+ = \langle\zeta_0(t, x, y)\rangle^+ = \Gamma^+(0) = \text{ constant} \tag{6.2.34}$$

and then conclude that the center of vorticity on the right half plane is stationary with

$$X^+ = \langle x\zeta\rangle^+ / \langle\zeta\rangle^+ = C_1/\Gamma^+(0) = \text{ constant.} \tag{6.2.35}$$

We now apply the far-field behavior and the equation of the center of a vortical field to define the approximate boundary conditions and the motion of the center of the computational domain in the numerical schemes formulated in Sect. 6.1 for global mergings of type I(a) and I(b). The corresponding computational codes are then employed to study the merging of several viscous vortices in Subsubsect. 6.2.1.4 and the roll-up of a thin trailing vortical layer in Subsubsect. 6.2.1.5.

6.2.1.4 Merging of Several Vortices. Numerical studies of the merging of several viscous vortices, which were initially far apart from each other, were carried out by Lo and Ting (1975, 1976). The numerical results demonstrate the merging of the vortices to a single one with total strength Γ_0, which is the sum of the strengths of the initial vortices, provided that $\Gamma_0 \neq 0$. The vortices merge to a limiting doublet configuration if $\Gamma_0 = 0$. The results also show that the asymptotic solutions for a collection of Lamb vortices remain good approximations even when their vortical cores overlap each other. Furthermore, in the long-time limit these extended asymptotic solutions yield the correct solutions for the final stages of merging. Lo and Ting employed the numerical scheme of Wu and Thompson (1973). At each time step the Poisson equation (6.2.2) for the stream function was solved by numerical evaluation of the Poisson integral. Thus a special construction of boundary data for ψ was not needed, but the number of computational steps per time step was of order N^2 instead of $N \ln N$. This scheme is very inefficient. Nevertheless, the numerical results in Lo and Ting (1975) are useful in verifying the accuracy of the more efficient schemes for global merging of type I(a) and I(b), which employ approximate boundary conditions for ψ in conjunction with a fast Poisson solver.

When the vortices are not too far apart from each other, we say that the merging of vortices commences. To be more precise, we apply the criterion (5.1.11) for the practical region of validity of the asymptotic theory and set the lower bound for the ratio of the distance d_{ij} between two vortices to the sum of their core radii $\delta_i + \delta_j$ equal to 2. Let $t = t_c$ be the instant when $d_{ij}/(\delta_i + \delta_j) = 2$. Using the asymptotic solution at $t = t_c$ as the initial datum, a numerical solution of the Navier–Stokes equations is constructed to simulate the flow field for $t > t_c$. To illustrate the process of merging, we quote several numerical examples dealing with the merging of vortices of equal strength and core size. Additional examples can be found in Lo and Ting (1975) and Liu and Hsu (1984).

The merging of n identical vortices was studied for $n = 2$, 3, and 4. The vortex centers are distributed symmetrically on a circle of radius ρ around the origin. The initial vorticity distribution for each vortex is that of a Lamb vortex with core size δ_0. If we choose δ_0 as the unit length scale and $\delta_0^2/4\nu$ as the time scale, then the numerical solution of the Navier–Stokes equations begins at $t = t_c = 1$ with $\rho = 2$ and $\mathrm{Re} = \Gamma/\nu = 100$.

Because of the diffusion of vorticity, the vortical cores grow and overlap more and more as time increases. The n points of locally maximal vorticity move gradually away from the vortex centers and spiral inward. Eventually, say at $t = t_m$, the n points of local maximum coincide at the origin. We say that the n vortices merge to a single one and lose their individual identity for $t \geq t_m$ and note that the origin is the location of the center of vorticity for all $t \geq 0$ according to (6.2.15).

Shown in Fig. 6.3 are the trajectories of the points of maximum vorticity given by the numerical solutions for $n = 2$, 3, and 4. The dotted lines are the corresponding trajectories defined by the *extended asymptotic solution*, which will be explained in Subsect. 6.2.2. Shown in Fig. 6.4 is the decay of the maximum vorticity given by these two solutions. We see from Figs. 6.3 and 6.4 that the asymptotic solution is in good agreement with the numerical solution even when the core size $\delta(t) = [4\nu(t+1)]^{1/2}$ is of the order of the distance between the adjacent vortex centers. Thus, the practical region of validity of the asymptotic solution can be extended from $\delta/\ell \sim 1/4$ to $\delta/\ell \sim 1$, where ℓ denotes the distance between two adjacent vortex centers. In contrast, under the inviscid theory for n point vortices, the vortex locations will be rotating along the circle of radius ρ forever with zero core size.

The contour lines of constant vorticity for two and three vortices, $n = 2, 3$ are shown in Figs. 6.5 and 6.6, respectively. At each instant, four contour lines at 0.995, $e^{-1/2}$, e^{-1}, and $e^{-3/2}$ times the maximum vorticity are shown. The initial contour lines of the n nonoverlapping Lamb vortices are displayed in Figs. 6.5a and 6.6a. The successive merging of the contour lines of n vortices is shown in Figs. 6.5b and c and 6.6b and c. For $t \geq t_m$ with t_m equal to 5.5 and 4.6 for $n = 2$ and 3, respectively, the n local maximum vorticity points converge to the origin, and the character of three distinct vortices disappears. Thus ends the first stage of merging. Figures 6.5f–h and 6.6f–h then show the gradual circularization of the contour lines beginning from the innermost and continuing toward the outermost ones as t increases. This is the second stage

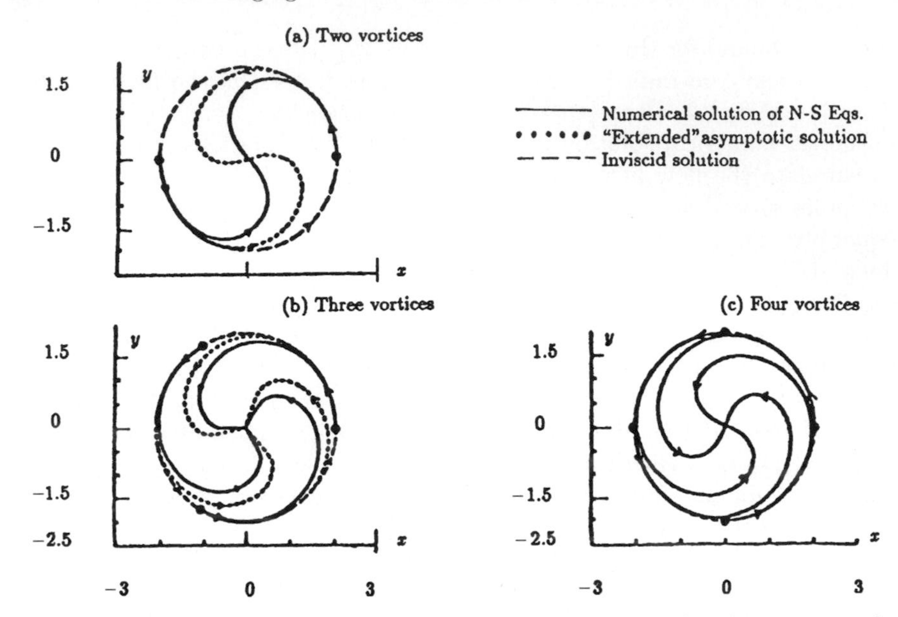

Fig. 6.3. Trajectories of the location of maximum vorticity for the merging of n vortices. Adapted from Ting (1986) and Ting and Liu (1986)

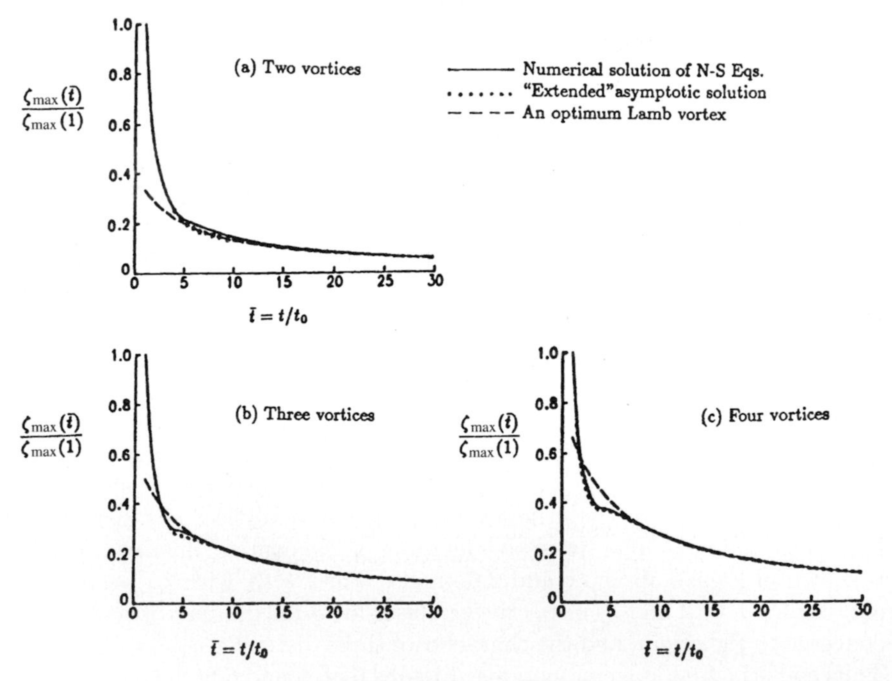

Fig. 6.4. The decay of maximum vorticity, for the merging of n vortices. Adapted from Ting (1986) and Ting and Liu (1986)

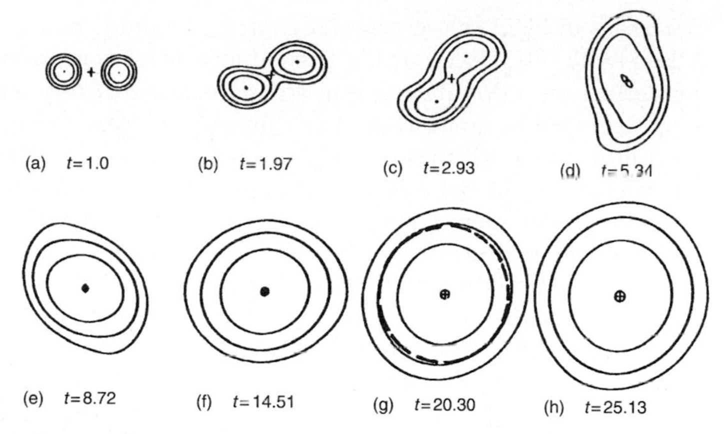

Fig. 6.5. Contour lines of constant vorticity $\zeta/\zeta_{\max}$ in the merging of 2 vortices: — numerical solution of Navier–Stokes equations, – – optimum Lamb vortex. Adapted from Ting (1986) and Ting and Liu (1986)

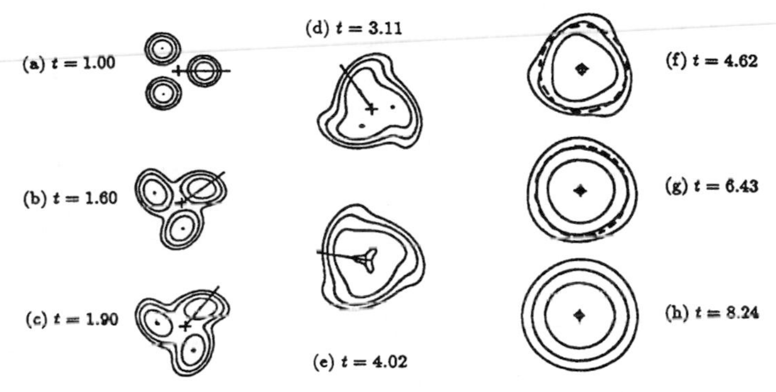

Fig. 6.6. Contour lines of constant vorticity $\zeta/\zeta_{\max}$ in the merging of 3 vortices: — numerical solution of Navier–Stokes equations, – – optimum Lamb vortex. Adapted from Ting (1986) and Ting and Liu (1986)

of merging. In Figs. 6.5h and 6.6h for $n = 2, 3$, all four contour lines are nearly circular and coincide with those of an optimum Lamb vortex with unit strength. Thus, we come to the final stage of merging of n vortex to a single one. Note also that in Fig. 6.4, the decay of maximum vorticity given by the optimum Lamb vortex and that by the extended asymptotic solution are in good agreement with the numerical solution for $t > t_m$. The meaning of the optimum Lamb vortex and the rule of merging will be explained in Subsect. 6.2.2. We shall also explain why the extended asymptotic solution gives a good description of not only the early stage of merging but the final stage of the recircularization of the constant vorticity lines as well.

6.2.1.5 Roll-Up of a Thin Trailing Vortical Layer. The evolution and control of thin vortex wakes trailing airplane wings are of great practical interest for reasons of flight safety, especially during landing and takeoff operations (Olsen et al., 1971). The steady three-dimensional flow downstream of a wing of high aspect ratio submerged in a uniform stream of velocity $U_\infty \hat{k}$ is usually approximated by an unsteady two-dimensional cross flow in the yz plane moving downstream with velocity $U_\infty \hat{\imath}$ relative to the wing. Here x, y, and z are coordinates in the spanwise, vertical, and streamwise direction, respectively. The approximation requires that the streamwise variation of the flow field is negligible and hence is valid only far downstream. The initial conditions for the unsteady cross flow are obtained from the vorticity distribution behind the trailing edge of the wing. Such distributions either are obtained from experimental data or are derived from an idealization based on the spanwise load or lift distribution.

It should be noted that matched asymptotic solutions for the roll-up of an inviscid vortex sheet were presented by Guiraud and Zeytounian (1977, 1980, 1982). Also, a viscous core structure for a tightly rolled-up vortex sheet was presented by Hall (1961). Later, the structure of decaying thin vortical layers was matched to the outer solution of an inviscid vortex sheet by Guirand and Zeytounian (1982) and by Moore (1975).

Here, we will analyze trailing wakes of finite thickness sufficiently far behind the wing so that the shear layer structure can be resolved with sufficient accuracy in the computations. Before describing the numerical solutions, we point out some salient features of this problem. Since the flow field is symmetric with respect to the vertical plane passing through the centerline of the airplane, the yz plane, the vortical field is an odd function of x. In the corresponding unsteady cross flow, the vorticity field $\zeta(t, x, y)$ is antisymmetric with respect to the y-axis. Near the y-axis, the vorticity is weak initially and becomes weaker as t increases (or in going downstream) because the vortical layer is rolling away from the y-axis. The vorticity field fulfills (6.2.33), that is, $\zeta_x = 0$ and $\zeta = 0$ on $x = 0$. We obtain from (6.2.34)–(6.2.35) that the total strength and the center of vorticity on the right half of the xy plane is stationary, that is,

$$\Gamma^+(t) = \langle \zeta \rangle^+ = \langle \zeta_0 \rangle^+ = \Gamma_0 \tag{6.2.36}$$

$$X^+ = \frac{\langle x\zeta \rangle^+}{\langle \zeta \rangle^+} = \frac{\langle x\zeta_0 \rangle^+}{\Gamma_0} \,, \tag{6.2.37}$$

where $\langle \; \rangle^+$ denotes the area integral over the half plane $x \geq 0$.

From the classical theory of wings of finite span (see, for example (von Kármán and Burgers, 1963)), the linear strength $\gamma(x)$ of the vortical layer at the initial station $z = z_0$ can be related to the load or lift per unit span, $\Lambda(x)$ as follows,

$$\gamma(x) = \int_{-\infty}^{\infty} \zeta_0 dy \,, \quad \text{with} \quad \gamma(x) = 0 \quad \text{for } x > S \,, \tag{6.2.38}$$

$$\Lambda(x) = \rho U_\infty \int_x^S \gamma(x')dx' \,, \quad \text{for } 0 \leq x \leq S \tag{6.2.39}$$

and

$$\Lambda(0) = \rho U_\infty \Gamma_0 \,, \tag{6.2.40}$$

where S denotes the half span. Note that Γ_0 is equal to the circulation around the root section of the wing ($x = 0$). The special case of an elliptic load distribution is of great importance because it yields the minimum induced drag for a given span and total lift. In this case, Λ and γ are given by

$$\Lambda(x) = \Lambda(0)\sqrt{1 - x^2/S^2} \,, \tag{6.2.41}$$

$$\gamma(x) = \Gamma_0 z H(S^2 - x^2)/\sqrt{1 - x^2/S^2} \,, \tag{6.2.42}$$

where H denotes the Heaviside function. From (6.2.37), (6.2.41), and (6.2.42), we obtain the stationary spanwise coordinate of the center of vorticity with $x \geq 0$,

$$X^+ = \frac{\pi}{4} S \,, \tag{6.2.43}$$

for $z \geq z_0$. This is a well-known result for an inviscid vortex sheet induced by an elliptic loading. Here we show that it is also true for a viscous thin vortical layer.

We now apply the numerical schemes presented in the preceding sections for global mergings to simulate the vortex wakes evolving from specified spanwise load distributions. The initial vorticity distribution is generated by placing a series of Lamb vortices along the computational grid line corresponding to the wing trailing edge. The strengths of the Lamb vortices are determined from the change in the level of the load distribution between grid points. The core radius of each lamb vortex is equal to one half the thickness of the wake. The initial thin wake is then replaced by the sum of the Lamb vortices.

The first case considered is the above elliptic load distribution. Calculations for this load distribution with a wake thickness of 5% of the semispan were made by Weston and Liu (1982) using the numerical scheme given in Subsect. 6.2.2 for a global merging of type I(a). The initial vorticity field is illustrated in Fig. 6.7 using contours of constant vorticity. These calculations were performed at a Reynolds number (Γ_0/ν) of 20,000. Because of the symmetry of the flow field with respect to the line $x = 0$, it was sufficient to consider only the right half-plane, $x \geq 0$. Since the linear extension of the vorticity distribution is about the half span S, the size D of the computational domain, $\mathcal{D}$, for global merging of type I(a) should be much larger than S. In Weston and Liu (1982), the size of $\mathcal{D}$ is chosen to be $4S$ and the center of the computational domain moves with the center of vorticity with respect to the half space $x \geq 0$. Figures 6.7a–d show contour lines of constant vorticity in several cross-sectional planes downstream from the wing. The vortex

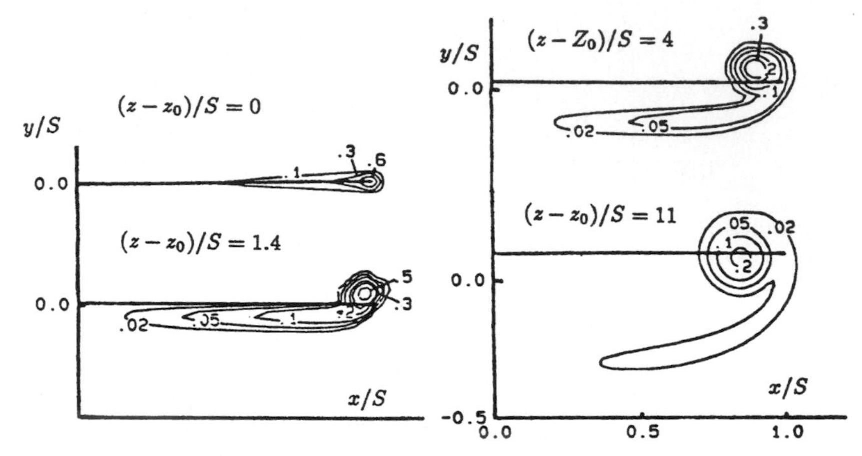

Fig. 6.7. Vortex roll-up for elliptic span load at $Re = 20,000$: $(z - z_0)/S =$ (a) 0, (b) 1.4, (c) 4, (d) 11. Adapted from Weston and Liu (1982)

sheet rapidly rolls up into a single, nearly circular vortex with an eventual lateral position near $x/S = \pi/4$ as predicted by the analytical result (6.2.43). The lack of smoothness in the contour lines for the tip region is observable. This problem is a result of a computational grid that is too coarse to resolve the flow, but could not be finer in this computation, because of the restricted computer memory available at the time the calculation was performed. Approximately 300 grid points across the semispan are required to properly resolve the flow, while only 50 grid points across the semispan were used to fit the entire computational domain into the computer memory (nonvirtual machine). This difficulty was subsequently eliminated using the scheme given in Subsect. 6.2.3 for global merging of type I(b). Here, the computational domain can be reduced to a thin rectangle containing the vorticity distribution in a region $\mathcal{D}$ of width S and thickness $S/20$. Consequently, 200 grid points could be placed on the semispan in that calculation.

A more severe initial vorticity distribution is obtained with an initial wake thickness of 1.0% of S and a Reynolds number of 40,000 requiring initially as much as 500 grid points across the semispan. The resulting initial vorticity distribution is shown in Fig. 6.8a. Figures 6.8b–d illustrate the subsequent roll-up of the vorticity. The division of the computational domain into subdomains is indicated in each figure, with the size of each subdomain chosen to be 10% of the semispan. The extent of the domain boundary is maintained at a size that keeps at least two subdomains between any region of significant vorticity and the closest boundary except near the y-axis. The vortex development at the tip is more rapid than in the previous case, and results, as expected, in a smaller vortex with a higher level of vorticity. The lateral position of the main vortex again approaches $x = \pi/4$.

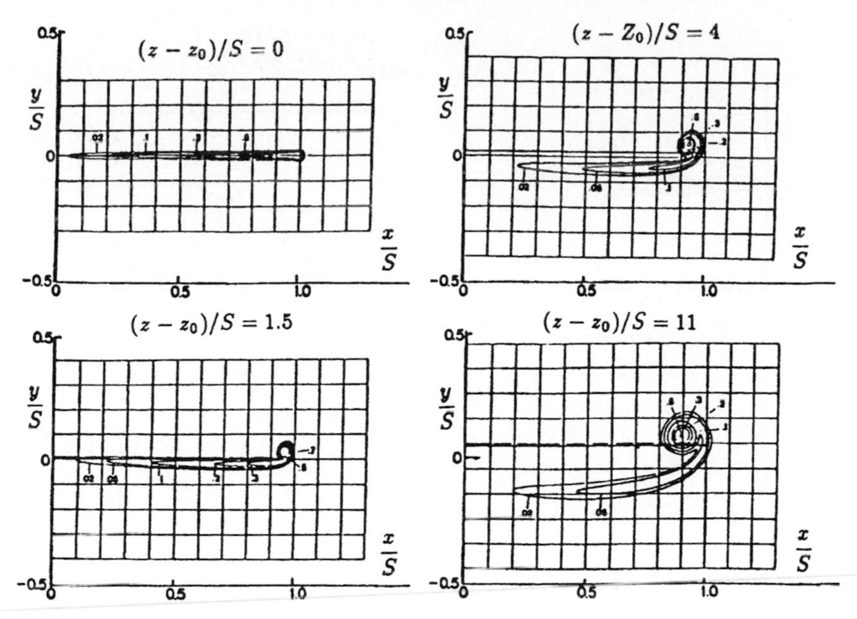

Fig. 6.8. Vortex roll-up for elliptic span load at $Re = 40{,}000$: $(z - z_0)/S =$ (a) 0, (b) 1.5, (c) 4, (d) 11. Adapted from Weston et al. (1986)

Another example presented by Weston and Liu (1982) is for a load distribution similar to that of a transport aircraft using flaps on landing. Note that the appearance of negative values of vorticity is associated with the reduction in lift around the fuselage-wing juncture. Figure 6.9a displays the initial vorticity distribution for a wake thickness of 1.5% at a Reynolds number of 20,000. This example requires 500 grid points across the semispan to properly resolve the flow. Figures 6.9b–d show the subsequent evolution of this vortex system in time or downstream distance and demonstrate the ultimate merging of the vortices from the wing tip and flap tip. Experimental investigations were carried out using a 3% scale model of a B-747 aircraft with the inboard flaps deflected to give a spanwise load distribution similar to that used in the numerical model. The visualized vortex positions are qualitatively similar to the calculated ones shown in Figs. 6.9b and 6.9c at $16S$ and $40S$ downstream of the initial station. The comparison of the numerical simulation with the experimental vortex wake can be found in Weston and Liu (1982) and Liu et al. (1985).

6.2.2 Rules of Merging of Vortices

When we apply the Poincaré relations (6.2.7)–(6.2.10) to two solutions of the initial value problem (6.2.1)–(6.2.6) that have different initial data, we arrive at the following three statements:

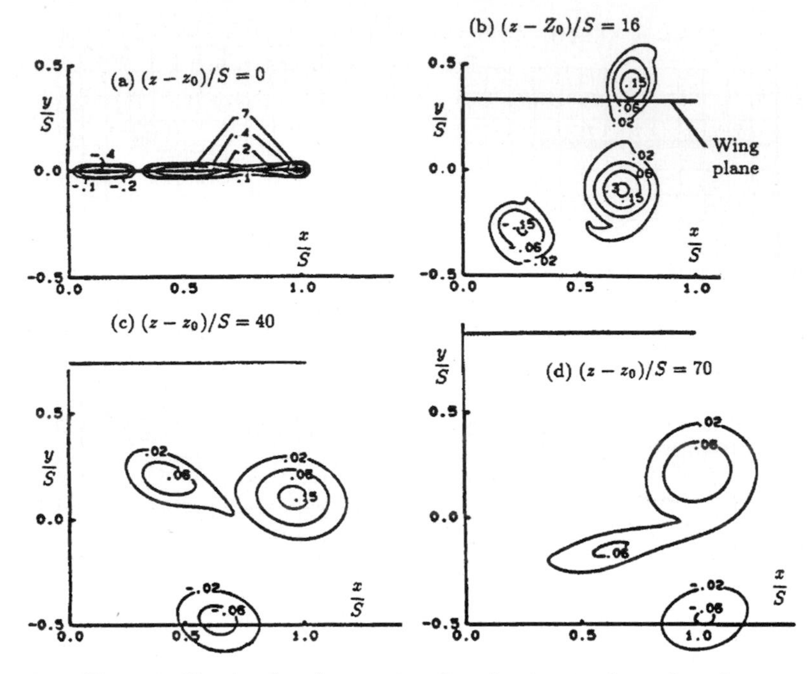

Fig. 6.9. Numerically simulated vorticity distribution in the wake of a transport aircraft. Adapted from Weston and Liu (1982)

If there are two initial data having the same total strength and first moments, the corresponding solutions retain the same strength and first moments for $t \geq 0$.

(A)

If in the above case the total strength is nonzero, then the solutions also have the same center of vorticity for $t \geq 0$.

(B)

If two initial data have the same total strength and polar moment, then the two solutions retain the same total strength and match in their polar moment for $t \geq 0$.

(C)

It was explained in Sect. 3.2 that, on a long time scale, one can approximate the solution of our initial value problem, (6.2.1)–(6.2.6), by a similarity solution, $\Gamma\zeta^*$, that is,

$$\zeta(\bar{r}, t) = \Gamma\zeta^*(\bar{r}, t + t_0)\,\{1 + O[(t + t_0)^{-1}]\}\,, \tag{6.2.44}$$

for any $t_0 > 0$, where

$$\zeta^*(\bar{r}, t) = \frac{1}{4\pi\nu t}\, e^{-\bar{r}^2/(4\bar{\nu}t)}\,. \tag{6.2.45}$$

Here, $\zeta^*(\bar{r}, t)$ represents a Lamb vortex of unit strength created at the instant $t = 0$ with zero initial core radius. The similarity solution $\Gamma\zeta^*(\bar{r}, t + t_0)$ represents a Lamb vortex with the same total strength Γ as that of the actual flow field. Its core size is

$$\delta_S(t) = [4\nu(t + t_0)]^{1/2} \tag{6.2.46}$$

Since $\delta_S = 0$ at $t = -t_0$, we say that the Lamb vortex in (6.2.44) was created at $t = -t_0$ and that its initial "age" (at the instant $t = 0$) is t_0. This initial age is a free parameter in the similarity solution. We define an optimum time shift $t_0 = t^*$ by the condition that ζ approaches the corresponding similarity solution in the fastest possible manner, that is,

$$\zeta(\bar{r}, t) = \Gamma\, \zeta^*(\bar{r}, t + t^*)\, \{1 + O[(t + t_0)^{-2}]\} \, . \tag{6.2.47}$$

This is achieved when the coefficient C_1 of the second term in the series representation (3.2.59) of ζ vanishes; see also (3.2.64). The condition $C_1 = 0$ yields

$$(4\nu t^*)\Gamma = 2\pi \int_0^\infty r^2 \zeta_0\, r\, dr \, , \tag{6.2.48}$$

where $\Gamma = \langle\zeta_0\rangle$. Recall that t^* should be positive. This is certainly true when ζ_0 does not change sign.

The right-hand side of (6.2.48) represents the polar moment of the initial vorticity distribution. The left-hand side is equal to $\Gamma\delta_0^2$, which is the polar moment of $\Gamma\zeta^*$ at $t = 0$. Hence, (6.2.48) is equivalent to the condition of matching the initial polar moment of the actual distribution with that of the leading-order approximation,

$$\langle r^2 \Gamma\zeta^*\rangle|_{t=0} = \Gamma\delta_S^2(0) = \langle r^2\zeta_0\rangle \, . \tag{6.2.49}$$

It follows from statement (C) that under (6.2.49) the polar moment of the optimum similarity solution matches that of the exact solution (3.2.47) for all $t \geq 0$.

We should mention a relevant result obtained by Kleinstein and Ting (1971) regarding the solution of the linear diffusion equation $\zeta_t = \nu\, \Delta\zeta$, with nonaxisymmetric initial data $\zeta_0(x, y)$. The solution soon loses its asymmetry and approaches the optimum axisymmetric similarity solution $\Gamma\zeta^*$ in the sense of (6.2.47). Thus, $\Gamma\zeta^*$ matches the total strength and polar moment of $\zeta_0(x, y)$ and the center of $\Gamma\zeta^*$ is located at the center of gravity of $\zeta_0(x, y)$. In the next section we shall observe the same behavior for the nonlinear initial value problem of the merging of several vortices to a single one.

We now summarize the properties of an optimum similarity solution (or a Lamb vortex with the optimum initial core size):

A similarity solution, $\Gamma\zeta^*(\bar{r}, t + t_0)$, is specified by two parameters, its strength Γ and its initial age t_0 (or its initial core size $\sqrt{4\nu t_0}$) with $t_0 > 0$. (D)

All the similarity solutions preserve the total strength and the first moments (the location of the center of vorticity) of the exact solution.

(E)

The optimum similarity solution, with $t_0 = t^*$, in the sense of (6.2.47), has the same polar moment as the initial data at time $t = 0$ and hence has the same polar moment as the exact solution for all times.

(F)

By combining statements (A,B,C) on two vorticity fields with different initial data, and (D,E,F) on an optimum single Lamb vortex, we come up with the rule of merging of n vortices:

In the long-time limit, n initially nonoverlapping vortices merge to an optimum Lamb vortex. Its total strength is given by the initial value of the circulation around the n vortices, its center remains at the initial center of vorticity and its optimum age, t^*, is defined by matching of its polar moment with that of the n vortices.

(G)

Tests of the rule of merging can be found in Figs. 6.4 and 6.5. They show that the numerical solutions of Navier–Stokes equations are in good agreement with the solutions predicted by the corresponding optimum Lamb vortex for $t > t_m$ and the agreement improves as t increases.

Next, we will express an extended asymptotic solution, which is an asymptotic solution used beyond its region of validity, as sum of Lamb vortices and employ the rule of merging to explain why the solution is a good approximation not only to the early stage but also to the final stage of merging.

6.2.2.1 Extended Asymptotic Solutions. Let us consider an initial vorticity field which can be represented approximately by a sum of n Lamb vortices. Let their strengths, initial ages, and the coordinates of their centers be denoted by Γ_k, t_k^* and $X_k(0), Y_k(0)$ for $k = 1, \ldots, n$. If these n vortices are far apart from each other, the asymptotic solution of the flow field is given by the sum of the Lamb vortices for $t \geq 0$. The vorticity field is

$$\zeta(x, y, t) = \sum_{k=1}^{n} \Gamma_k \zeta^*(x - X_k, y - Y_k, t + t_k^*) = \sum_{i=1}^{n} \frac{\Gamma_k}{\pi \delta_k^2(t)} e^{-r_k^2/\delta_k^2(t)} \tag{6.2.50}$$

where

$$r_k^2 = (x - X_k)^2 + (y - Y_k)^2 \tag{6.2.51}$$

and

$$\delta_k^2(t) = 4\nu(t + t_k^*) \; . \tag{6.2.52}$$

The corresponding velocity field is

$$\begin{aligned} \boldsymbol{V}(x,y,t) &= \sum_{k=1}^{n} \boldsymbol{V}_S(x - X_k, y - Y_k, t + t_k^*) \\ &= \sum_{k=1}^{n} \frac{\Gamma_k}{2\pi r_k^2} \left[-(y - Y_k)\hat{\imath} + (x - X_k)\hat{\jmath} \right] \left[1 - e^{-r_k^2/\delta_k^2(t)} \right] . \end{aligned} \tag{6.2.53}$$

The velocity of the kth vortex center is

$$\dot{X}_k(t)\hat{\imath} + \dot{Y}_k(t)\hat{\jmath} = \boldsymbol{V}(X_k, Y_k, t) \tag{6.2.54}$$

for $k = 1, \ldots, n$. We note that the vorticity and velocity defined by (6.2.50)–(6.2.53) are uniformly valid in the xy plane and agree with the classical inviscid solution for n point vortices in the region away from the vortex centers.

The asymptotic solution (6.2.50)–(6.2.53) remains valid so long as each core size δ_i is much smaller than the distance, d_{ij}, to an adjacent vortex center, that is,

$$\delta_i / d_{ij} \leq \alpha \ll 1 \qquad \text{for } i, j = 1 \ldots n \text{ and } j \neq i\,. \tag{6.2.55}$$

In practice we choose $\alpha = 1/4$ as noted in Sect. 5.1. When condition (6.2.55) is no longer satisfied, the ith and jth vortical cores are overlapping each other and the merging of vortices commences. When we continue to use the asymptotic solution in this case, we call it the extended asymptotic solution. We then study its accuracy by comparison with the numerical (finite difference) solution of the merging process described in Subsect. 6.2.1.

Numerical results for the merging of n identical vortices symmetrically located on a circle centered at the origin were described in Subsect. 6.2.1 for $n = 2, 3, 4$. The corresponding results given by the extended asymptotic solutions were also included in Figs. 6.3 and 6.4.

As shown in Fig. 6.3, the trajectories of n points of maximum vorticity, given by the extended asymptotic solution, spiral inward and end at the origin at an instant t_m, which depends on n. To illustrate this process, we consider $n = 2$. The points of maximum vorticity are located along the radial line joining the two vortex centers and hence may be found by computing the roots of

$$\zeta_r(t, r, 0) = \frac{2\Gamma}{\pi\delta^4} \left[(a - r)\, e^{-(r-a)^2/\delta^2} - (r + a)\, e^{-(r+a)^2/\delta^2} \right] = 0\,. \tag{6.2.56}$$

For $t < t_m$ there are two roots, $r = 0$ for the local minimum, and $r^*(t) < a$ for the maximum with $r^*(t_m) = 0$. For $t > t_m$, there is only one root, $r = 0$, where ζ is the maximum. The instant t_m is defined by the condition $r^*(t_m) = 0$ or $\zeta_{rr} = 0$ at $t = t_m$ and $r = 0$. The result for the case in Fig. 6.3a is $t_m = a^2/(2\nu) = 8$. This value is greater than the value 5.5 predicted by the finite difference solution but the difference gets smaller for larger n.

The decay of the maximum vorticity given by the extended asymptotic solution is also shown in Fig. 6.4. The difference between the extended asymptotic solution and the finite difference solution is noticeable only during the second stage of merging, that is, for the period from the disappearance of n local maxima $t \sim t_m$ to the recircularization of contour lines around the origin ($t \sim t_f$). The difference during this period is smaller for larger values of n. For $t > t_f$, the extended asymptotic solution is again in good agreement with the numerical solution.

In Figs. 6.5 and 6.6 we see the merging of $n = 2, 3$ initially distinct vortices to a single one with nearly circular contours of constant vorticity. The $|\Omega|$-contours gradually become circularized beginning with the innermost line and progressing to the outer ones as time goes on. The dotted lines are the corresponding contour lines predicted by an optimum Lamb vortex defined by the rule of merging (G). The contour lines of the optimum Lamb vortex and those given by the numerical solution become indistinguishable in the last time frame in Figs. 6.5 and 6.6. This agreement was pointed out right after (G). We note here that the contour lines given by the extended asymptotic solutions also become indistinguishable with those by the numerical solutions. The fact that the extended asymptotic solution is valid in the early stage of merging is to be expected because it is within the practical region of validity of the asymptotic solution. The extended solution becomes accurate again for large $t > t_m$; this can be explained by the fact that the solution obeys the linear diffusion equation and hence conserves the total strength and first moments and matches the polar moment of the numerical solution. When the latter approaches that of an optimum single Lamb vortex for $t > t_f$, the extended asymptotic solution converges to the same limit.

The fact that the extended asymptotic solution remains quite accurate for all t except for the finite period of the second stage of merging, $t_c < t < t_f$, prompts us to recast the sum of Lamb vortices as an *approximate solution* to the Navier–Stokes equations. In this process, we identify the source of errors and derive an improvement. This is the topic of Subsect. 6.2.3, in which we point out to what extent the approximate solution compensates for the deficiencies of the extended asymptotic solution.

Finally, we mention another application of the extended asymptotic solution and the rule of merging. They were employed by Ting (1986) and Ting and Liu (1986) to simulate the roll-up and decay of a thin trailing vortical layer behind an aircraft wing. An estimate of the accuracy of the simulation was obtained by comparing the results with those of the numerical solutions presented in Subsubsect. 6.2.1.5. Recall that in that subsection the initial data for the numerical simulation of a trailing vortical layer were a collection of Lamb vortices that might overlap the adjacent ones. The same collection is now used to initialize the extended asymptotic solution (6.2.50)–(6.2.54). The initial data used by Weston et al. (1986) are a collection of 100 vortices equally spaced along the semispan S with initial core sizes equal to 2.5%

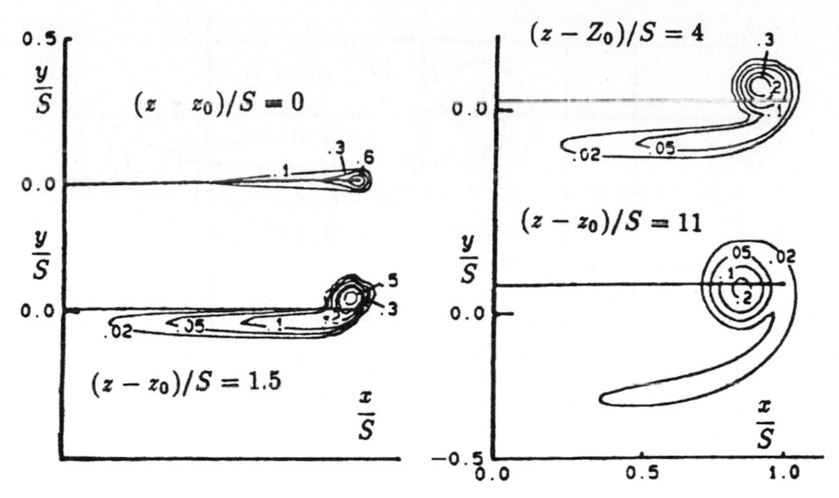

Fig. 6.10. Contour lines of constant vorticity $\zeta/\zeta_{\max}$ in the roll-up and decay of a trailing vortex sheet – extended asymptotic solution: $z - z_0 =$ (a) 0, (b) 1.5, (c) 4, (d) 11. Adapted from Ting (1986)

of the semispan. The Reynolds number is 20,000 based on the circulation Γ_0 around the root section, $x = 0$. Figure 6.10 shows the contour lines at distances of $1.4S$, $4S$, and $11S$ downstream from the wing. They are similar to those of numerical solutions to the Navier–Stokes equations shown in Fig. 6.7. The contour lines of high vorticity show the core structure of an "eye" known as the tip vortex. At the station $z = 11S$, more than half of the initial vortices, accounting for 86% of Γ_0 are packed in and around the eye. The contour lines of constant vorticity near the *eye* become more wavy for larger z/S. This is due to the fact that near the eye many vortices have been closely packed together and subjected to intense merging.

To improve the solution, we employ the rule of merging (G) to allow two Lamb vortices, say the ith and mth, to merge into a single one when the ratio $\Lambda_{km} = d_{km}^2/(\delta_k^2 + \delta_m^2)$ is less than a given critical value. Here d_{km} stands for the distance between the vortex centers. We represent the initial vorticity distribution by 160 Lamb vortices equally spaced along the semispan with core radii equal to 2.5% semispan. We apply the rule of merging (G) to the kth and mth vortices when the ratio Λ_{km} is less than the critical value 1/32.

The results of this modification are as follows: At the station $z = 11S$, there are 94 Lamb vortices left and the contour lines are shown in Fig. 6.11. Also shown in the insert are the contour lines in the center of the eye enlarged five times. We see that these contour lines are in much better agreement with those given by the numerical solutions in Fig. 6.7d than were those in Fig. 6.10d. The rule of merging allows vortices to merge into the eye while those away from the eye spread apart from each other. As a consequence,

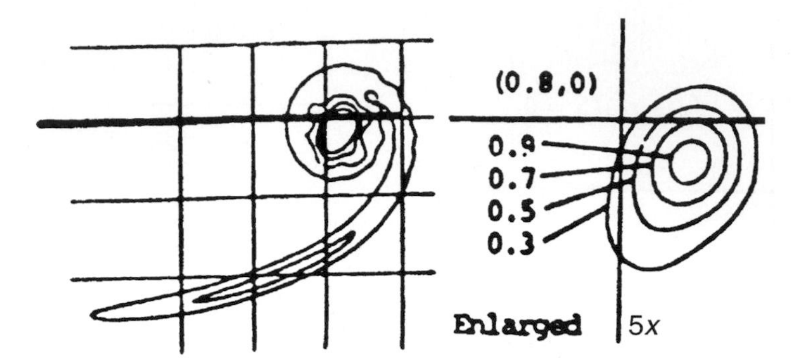

Fig. 6.11. Contour lines of constant vorticity at station $z = 11S$ — initial 160 Lamb vortices merged to 96 vortices. Adapted from Ting and Liu (1986)

the waviness of the contour lines is reduced in comparison with the extended solution without merging.

As mentioned before, we shall improve the extended asymptotic solution by introducing a new criterion for the velocities of the vortex centers. The velocity of the vortex centers are adjusted so that the solution fulfills the Navier–Stokes equations approximately under a minimum principle. The new solution to be described next will be called the *approximate solution.*

6.2.3 Approximate Solution of Navier–Stokes Equations Using Superposition of Lamb Vortices

To prepare for the construction of our *approximate solution,* we use $\zeta^*(x, y, \delta(t))$ to denote the vorticity distribution of a Lamb vortex of unit strength and centered at the origin, that is,

$$\zeta^*(x, y, \delta(t)\,) = \frac{1}{\pi\delta^2(t)}\, e^{-r^2/\delta^2(t)} \tag{6.2.57}$$

with

$$\delta^2(t) = 4\nu t \quad \text{and} \quad r^2 = x^2 + y^2\,. \tag{6.2.58}$$

The Lamb vortex is created at $t = 0$ since the core size δ vanishes at that time. We denote the corresponding stream function by $\psi^*(x, y, \delta)$ and obtain the induced velocity

$$\boldsymbol{V}^*(x, y, \delta) = \psi^*_y\hat{\imath} - \psi^*_x\hat{\jmath} = \frac{1}{2\pi r^2}[-y\hat{\imath} + x\hat{\jmath}]\,[1 - e^{-r^2/\delta^2(t)}]\,. \tag{6.2.59}$$

We then construct an *approximate solution* to the initial value problem, (6.2.1)–(6.2.6), by linear superposition of n Lamb vortices as follows:

$$\zeta(t, x, y) \;= \sum_{k=1}^{n} \Gamma_k\, \zeta^*_k \tag{6.2.60}$$

$$\psi(t,x,y) = \sum_{k=1}^{n} \Gamma_k \, \psi_k^* \,, \tag{6.2.61}$$

$$\zeta_k^* = \zeta^*(x - X_k(\tau) \,,\, y - Y_k(\tau) \,,\, \delta_k(\bar{t} + t_k^*)) \tag{6.2.62}$$

$$\psi_k^* = \psi^*(x - X_k(\tau) \,,\, y - Y_k(\tau) \,,\, \delta_k(\bar{t} + t_k^*)\,) \,. \tag{6.2.63}$$

The formal introduction of two time variables t, τ will be explained shortly. The strength Γ_k, initial vortex center $X_k(0)$, $Y_k(0)$, and age t_k^* for $k = 1 \ldots n$ are chosen to fit the initial data $\zeta_0(x, y)$. The representation of the approximate solution, (6.2.62)–(6.2.63), by a superposition of n Lamb vortices is identical to that for the extended asymptotic solution, (6.2.50)–(6.2.53). They differ only in the velocities of the vortex centers. For the latter, the velocities of the centers are prescribed by the asymptotic theory for nonoverlapping vortices, (6.2.54) and are in error for overlapping vortices. Now the velocities $\dot{X}_k(\tau)$ and $\dot{Y}_k(\tau)$ are treated as $2n$ unknowns to be defined such that (6.2.60)–(6.2.63) give the "best" approximate solution to the Navier–Stokes equations; that is, they are derived from a minimum principle. This is described in the following.

Note that the dependence of the solution (6.2.60)–(6.2.63) on t appears indirectly through the changing core sizes and the moving vortex centers. To show this difference explicitly, we have denoted the time variable t in the core size δ_k and the position of the center (X_k, Y_k) by $\bar{t}$ and τ, respectively. Thus, the time derivative of ζ_k^* is written as

$$\partial_t \zeta_k^* = (\partial_{\bar{t}} + \partial_\tau)\zeta_k^* = \left[\partial_{\bar{t}} - (\dot{X}_k(\tau)\partial_x + \dot{Y}_k(\tau)\partial_y)\right] \zeta_k^* \,. \tag{6.2.64}$$

We note that the vorticity ζ_k^*, stream function ψ_k^*, and the velocity $\boldsymbol{V}_k^*$ fulfill the following equations:

$$\partial_{\bar{t}} \zeta_k^* = \nu \Delta \zeta_k^* \,, \tag{6.2.65}$$

$$(\boldsymbol{V}_k^* \cdot \nabla)\zeta_k^* = 0 \,, \tag{6.2.66}$$

$$\Delta \psi_k^* = -\zeta_k^* \tag{6.2.67}$$

$$\boldsymbol{V}_k^* = (\psi_k^*)_y \hat{\imath} - (\psi_k^*)_x \hat{\jmath} \,. \tag{6.2.68}$$

We observe that the stream function ψ in (6.2.63) fulfills the Poisson equation (6.2.2) with an inhomogeneous term given by ζ in (6.2.60) and that the velocity $\boldsymbol{V}$ defined by (6.2.3) yields

$$\boldsymbol{V}(t,x,y) = \sum_{k=1}^{n} \boldsymbol{\Gamma}_k \, V_k^* \,. \tag{6.2.69}$$

The far-field conditions (6.2.5) and (6.2.6) are also fulfilled by ζ and $\boldsymbol{V}$ of (6.2.60) and (6.2.69). Only the nonlinear vorticity diffusion equation, (6.2.1), remains to be checked. By using (6.2.64), (6.2.1) becomes

$$\sum_k [\boldsymbol{V}(t,x,y) - \dot{X}_k(\tau)\hat{\imath} - \dot{Y}_k(\tau)\hat{\jmath}] \cdot \nabla(\Gamma_k \zeta_k^*) = \sum_k \Gamma_k [\nu \Delta \zeta_k^* - \partial_{\bar{t}} \zeta_k^*] \,. \quad (6.2.70)$$

Because of (6.2.65) and (6.2.66), the right-hand side of (6.2.70) vanishes, while the left-hand side reduces to

$$F(t,x,y) = \sum_k \Gamma_k \left[\left(\sum_{\ell \neq k} \Gamma_\ell \boldsymbol{V}_\ell^* \right) - \dot{X}_k(t)\hat{\imath} - \dot{Y}_k(t)\hat{\jmath} \right] \cdot \nabla \zeta_k^* \qquad (6.2.71)$$

and (6.2.70) becomes

$$F(t,x,y) = 0 \,. \qquad (6.2.72)$$

The function F vanishes in the domain of zero vorticity but it cannot vanish for all x, y in the domain where $\zeta \neq 0$. To fulfill (6.2.72) approximately, we seek $\dot{X}_k(t)$, $\dot{Y}_k(t)$, $k = 1 \dots n$, such that the function

$$H(t, \dot{X}_1 \dots \dot{X}_n \,,\, \dot{Y}_1 \dots \dot{Y}_n) = \langle F^2(t,x,y) \rangle = \min \qquad (6.2.73)$$

for $t \geq 0$, where $\langle \quad \rangle$ again denotes the area integral over the xy plane. In this sense, we say that ζ and ψ in (6.2.60) and (6.2.61) are the "best" approximate solutions of the Navier–Stokes equations.[2] At each instant t, $H(t, \dot{x}_1, \dots \dot{y}_n)$ is minimized by $\dot{X}_1 \dots \dot{Y}_n$, if and only if the following $2n$ linear equations for $\dot{X}_m$ and $\dot{Y}_m$ are fulfilled:

$$\sum_m (a_{km} \dot{X}_m + e_{km} \dot{Y}_m) = C_k \quad \text{and} \quad \sum_m (b_{km} \dot{Y}_m + e_{mk} \dot{X}_m) = -D_k \,, \qquad (6.2.74)$$

for $k = 1 \dots n$ with

$$a_{km} = \Gamma_k \Gamma_m \langle \partial_x \zeta_k^* \, \partial_x \zeta_m^* \rangle \,,$$

$$e_{km} = \Gamma_k \Gamma_m \langle \partial_x \zeta_k^* \, \partial_y \zeta_m^* \rangle \,,$$

$$C_k = \sum_m c_{km} \,, \quad D_k = \sum_m d_{km} \,, \qquad (6.2.75)$$

$$c_{km} = \Gamma_k \Gamma_m \left\langle \partial_x \zeta_k^* \left(\partial_x \zeta_m^* \sum_{\ell \neq m} \Gamma_\ell \partial_y \psi_\ell^* - \partial_y \zeta_m^* \sum_{\ell \neq m} \Gamma_\ell \partial_x \psi_\ell^* \right) \right\rangle \,,$$

[2] We note that the sum of Lamb vortices in (6.2.62) fulfills the conservation of total strength (6.2.7). Because of the special initial configurations of Lamb vortices for the examples studied in Book-I and in this section, the approximate solutions minimizing H happen to fulfill the three auxiliary conditions, the two invariants of the first moments of vorticity, (6.2.8) and (6.2.9), and the Poincaré relationship (6.2.10). In general, we have to add three Lagrange multipliers to account for three auxiliary conditions. This general formulation was presented in Ting and Bauer (1993).

and b_{km} and d_{km} are the same as a_{km} and c_{km} respectively with ∂_x interchanged with ∂_y. These coefficients are elementary functions of X_k, X_m, Y_k, Y_m, δ_m, and δ_k, listed in Appendix D. In particular, we have $a_{kk} = b_{kk} = 2\Gamma_k^2/(\pi\delta_k^4)$ and $e_{kk} = 0$. Since ζ^* decays exponentially on the length scale δ, all the nondiagonal elements, $a_{km} \cdots e_{km}$, contain a factor $\exp(-\Lambda_{km}^2)$, where $\Lambda_{km}^2 = |\boldsymbol{X}_k - \boldsymbol{X}_m|^2/(\delta_k^2 + \delta_m^2)$. The n pairs of equations can be rearranged as

$$GZ = J \, . \tag{6.2.76}$$

Here, Z and J denote respectively the column matrices of $\{\dot{X}_1, \dot{Y}_1, \cdots, \dot{X}_n, \dot{Y}_n\}$, and G denotes the corresponding $2n \times 2n$ matrix. Because the minimum principle is applied to the square of a linear function in the $\{\dot{X}_1, ..., \dot{Y}_1, ...\}$, G is symmetric and positive definite. Since the nondiagonal elements vanish as $\exp(-\Lambda_{km}^2)$, G is dominated by its diagonal elements. Equation (6.2.76) can be solved readily for $\dot{X}_k$ and $\dot{Y}_k$, $k = 1 \cdots n$, the velocities of the n Lamb vortices in the representation of the approximate solution, (6.2.60)–(6.2.63).

The ratio Λ_{km}^2, in the exponential factor, has been used to characterize the interaction between the kth and mth vortices. When $\Lambda_{km}^2 \gg 1$, these two vortices are isolated from each other. When $\Lambda_{km}^2 \ll 1$, they have merged to a single one. For a kth vortex, that is far apart from the others such that $\Lambda_{km}^2 \gg 1$, for $m \neq k$, both the $(2k-1)$st and $2k$th equation of (6.2.76) are decoupled from the remaining $2n-2$ equations and yield the asymptotic classical result for the velocity of the kth vortex center given in (6.2.54). The nondiagonal elements of G and the c_{km}, d_{km} for $k \neq m$ in J account for the effects of interactions between overlapping adjacent vortical cores. Thus we observe the following:

- For a vortex that is far apart from the others, the velocity of its center given by the approximate solution reduces to that by the asymptotic solution.
- For overlapping vortices, the velocities of the vortex centers are determined by minimizing the effect of the nonlinear interaction terms in (6.2.71).

The above observations also explain why the simulation of a thin trailing vortical layer in Subsect. 6.2.2 by the extended asymptotic solution, together with the rule of merging, agrees quite well with the finite difference solution even in the later stage of the roll-up process. The reason is that the core of a vortex will overlap at most with one or two adjacent vortices at a time, while the velocity of a vortex center is the resultant of the velocities induced by all the vortices (over 100 of them) used in the modeling of the trailing vortex layer.

Since the probability of a simultaneous merging of more than two vortices is much less than that of only two, the merging of two vortices is clearly the simplest situation, but nonetheless the fundamental building block in the evolution of a vortical field simulated by many vortices. We shall test the accuracy of our approximate solution in the following studies of the merging of two vortices of the same strength and of opposite strength.

6.2.3.1 Merging of Two Identical Vortices. In Fig. 6.3a we compare the trajectories of the locations of maximum vorticity for two identical vortices with strength $\Gamma = 25$ and Re $= 100$, which were initially ($t = 1$) centered at (2,0) and $(-2, 0)$, with core size $\delta_0 = 1$. With the initial value of the parameter $\Lambda^2 = 8$, the two vortices are considered to be far apart from each other and each one is represented by a Lamb vortex. Under the inviscid theory, the two point vortices will spin around the origin at constant angular velocity forever. The trajectories of the two points of maximum vorticity given by the extended asymptotic solution spiral inward, and combine to a single one at the origin to end the first stage of merging. These trajectories agree only qualitatively with the corresponding finite difference solutions described in Subsect. 6.2.1.

Figure 6.3a is reproduced in Fig. 6.12 (left) with the addition of the trajectories given by the approximate solution. These additional curves are much closer to those of the finite difference solution than those of the extended asymptotic solution. Figure 6.5 showed the contour lines of $\zeta/\zeta_{\max} = 1/\sqrt{e}$, $1/e$, and $1/\sqrt{e^3}$ of the finite difference solution. The heavy dotted lines show the lines based on the optimum single vortex. Hence, the merging to a single vortex is nearly completed by $t = 25$.

Shown in Fig. 6.12 (right) is the rotation of the major axis of vorticity during the merging process. As expected, the inviscid solution yields a straight line while the approximate solution is in considerably better agreement with the finite difference solution than the extended asymptotic solution. It should be noted that in the final phase of merging, $t > 20$, the contour lines are approaching circles and then it becomes rather difficult (or less accurate) to determine the orientations of the major and minor axes. However, the contour lines are no longer sensitive to the orientations of the axes that account only for the small deviations of the contour lines from circles. At that stage, the mean radii of the contour lines given by all three solutions, the numerical,

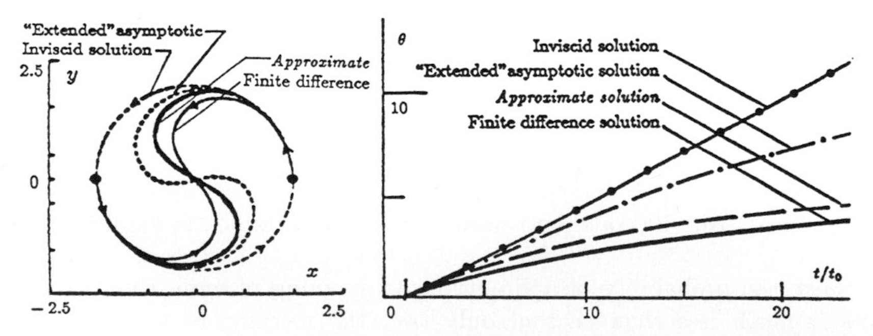

Fig. 6.12. *Left*: trajectories of the locations of the maximum vorticity in the merging of two identical vortices. *Right*: rotation of the major axis of vorticity distribution during the merging of two identical vortices. Adapted from Ting and Liu (1986)

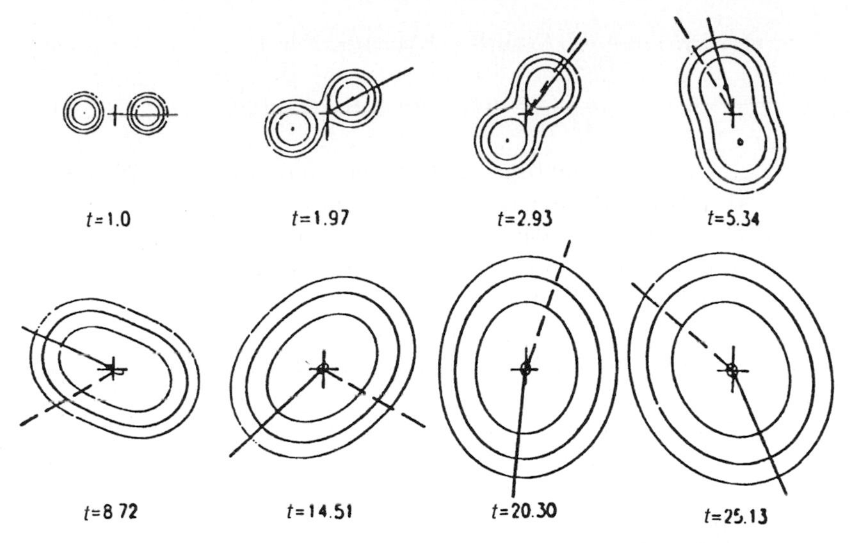

Fig. 6.13. Contour lines of constant vorticity in the merging of two vortices – approximate solution based on a minimum principle. Adapted from Ting and Liu (1986)

extended asymptotic, and approximate solutions, are in good agreement with those predicted by the optimum Lamb vortex.

Figure 6.13 shows the contour lines of the *approximate solution* based upon the minimum principle. The contours agree with those in Fig. 6.5 in size and in the orientation of the principal axes, but they differ in details. In particular, the numerical results are not symmetric with respect to the principal axes while the approximate results are. Also shown in dashed lines are the corresponding axes based on the extended asymptotic solution. We see that the contour lines of the latter will be completely out of phase with the finite difference solution.

6.2.3.2 Evolution of a Viscous Doublet. Here we consider the case that the total vorticity field vanishes at $t = 0$, and hence $\langle\zeta\rangle = \Gamma_0 = 0$ for all $t > 0$. The first moment, which is conserved, is given by its initial data. We choose to specify the two first moments via the components of the vector,

$$\langle \mathbf{x} \times \hat{k}\zeta\rangle = \hat{\imath}\langle y\zeta\rangle - \hat{\jmath}\langle x\zeta\rangle = \mathbf{E} \neq 0\ , \tag{6.2.77}$$

so that in the far field, the flow behaves like that of an inviscid doublet formed by a source sink pair oriented in the direction of $\mathbf{E}$. We call such a flow field a viscous doublet with strength $\mathbf{E}$. With $\Gamma_0 = 0$, we see from (2.4) that the polar moment is conserved, that is,

$$I(t) = \langle\sigma^2\zeta\rangle = I(0)\ . \tag{6.2.78}$$

Now the center of vorticity is not defined. Instead, we shall define the center of a viscous doublet in terms of its far-field behavior

$$\psi(t,\sigma,\theta) = \frac{1}{2\pi\sigma}\left[E_1 \sin\theta - E_2 \cos\theta\right] + \frac{1}{4\pi\sigma^2}[F_3(t)\cos 2\theta + H_3(t)\sin 2\theta] + O(\sigma^{-3}) \ . \quad (6.2.79)$$

This is (6.2.17). The second term on the right-hand side of (6.2.79) represents two inviscid quadrupoles located at the origin of strengths $F_3(t) = \langle (x^2 - y^2)\zeta \rangle$ and $H_3(t) = \langle 2xy\zeta \rangle$. We can absorb the quadrupoles in (6.2.79) into a doublet of the same strength and orientation but moving with the velocity, $\hat{\imath}\dot{X}_D(t) + \hat{\jmath}\dot{Y}_D(t)$, and (6.2.79) becomes

$$\psi(t,\bar{\sigma},\bar{\theta}) = \frac{1}{2\pi\bar{\sigma}}[E_1 \sin\bar{\theta} - E_2 \cos\bar{\theta}] + O(\bar{\sigma}^{-3}) \ , \quad (6.2.80)$$

with

$$\dot{X}_D = \frac{E_2\dot{F}_3 - E_1\dot{H}_3}{2|\mathbf{E}|^2} \ , \qquad \dot{Y}_D = \frac{E_1\dot{F}_3 + E_2\dot{H}_3}{2|\mathbf{E}|^2} \ . \quad (6.2.81)$$

Here, $\bar{\sigma}$ and $\bar{\theta}$ denote the polar coordinates relative to the center $X_D(t)$, $Y_D(t)$. With the velocity of the doublet center defined by (6.2.81), we have

The far-field of a viscous doublet behaves like an inviscid doublet of constant strength $\mathbf{E}$ *moving with the doublet center plus singularities of orders higher than the quadrupoles. The velocity of the center defined by (6.2.81) depends on the second moments of vorticity and is in general not parallel to the doublet strength* $\mathbf{E}$.

Noting the long-time behavior of a two-dimensional viscous doublet in Subsect. 6.1.3, in particular (6.1.23) for the velocity of the viscous doublet, we get the long-time behavior of the displacement of the center of the doublet,

$$\hat{\imath}X_D + \hat{\jmath}Y_D = \mathbf{C_0}\ln t + \mathbf{C_1} + \mathbf{C_2}t^{-1/2} + O(t^{-1}) \ . \quad (6.2.82)$$

Now we consider the special case that the vorticity field is antisymmetric with respect to an axis, say the y-axis. In addition to $\langle\zeta\rangle = 0$, we have

$$\zeta(t,x,y) = -\zeta(t,-x,y) \ , \quad \zeta(t,0,y) = 0 \ . \quad (6.2.83)$$

Let $\langle\ \rangle^+$ and $\langle\ \rangle^-$ denote the area integrals over the right and left half plane, respectively. Although $\langle\zeta\rangle = \Gamma_0 = 0$, we have, in general,

$$\langle\zeta\rangle^+ = -\langle\zeta\rangle^- \neq 0 \ . \quad (6.2.84)$$

From (6.2.77) and (6.2.83), we have

$$\langle x\zeta\rangle^+ = \langle x\zeta\rangle^- = -E_2/2 \ , \ \langle y\zeta\rangle^+ = -\langle y\zeta\rangle^- \ , \ E_1 = 0 \quad \text{with } \mathbf{E} = E_2\hat{\jmath} \ . \quad (6.2.85)$$

Because of the antisymmetry, we need to analyze only the flow field on a half plane, say the left half, $x \leq 0$. The computational domain $\mathcal{D}$ will be bounded on its right by the y-axis with the boundary condition,

$$\zeta(t,0,y) = 0 \quad \text{and} \quad \psi(t,0,y) = 0 \ . \tag{6.2.86}$$

Using (6.2.79) and (6.2.83), we have

$$F_3 = \langle (x^2 - y^2)\zeta \rangle = 0 \quad \text{and} \quad \langle xy\zeta \rangle^+ = \langle xy\zeta \rangle^- = H_3/4 \ . \tag{6.2.87}$$

The equation of the doublet center, (6.2.81), becomes

$$\dot{X}_D = 0 \ , \ \ \dot{Y}_D 0 = \frac{\dot{H}_3}{E_2} \quad \text{or} \quad X_D = 0 \ , \ \ Y_D = \frac{\langle xy\zeta \rangle}{E_2} = -\frac{\langle xy\zeta \rangle^-}{\langle x\zeta \rangle^-} \ , \tag{6.2.88}$$

when the initial doublet center is placed at the origin. Thus

> *When the vorticity is antisymmetric with respect to a line, say the y-axis, the doublet strength* **E** *is parallel to the y-axis and the doublet center is moving along it.*

When we assume in addition to the antisymmetry (6.2.84) that the initial vorticity on each side of the y-axis has the same sign, say the vorticity on the left side of the y-axis is positive, that is, $\zeta \geq 0$ for $x \leq 0$. Noting that $\zeta_x(t,0,y) \leq 0$ and using condition (6.2.86), the rate of change of the integral of vorticity over the left half plane yields

$$\dot{\Gamma}^-(t) = \langle \zeta_t \rangle^- = -\dot{\Gamma}^+(t) = \nu \int_{-\infty}^{\infty} \zeta_x(t,0,y)dy < 0 \ , \tag{6.2.89}$$

where $\Gamma^+ = \langle \zeta \rangle^+$. Equation (6.2.89) means that the absolute value of the strength of these two vorticity fields on the left side of the y-axis and that on the right side decrease as t increases. Thus the evolution of a viscous doublet simulates the gradual cancellation of two vortices of opposite strength.

To demonstrate this merging process, we study the motion and diffusion of two vortices of opposite strength $\pm\Gamma_0$ created at $t = 0$ at a distance 2ℓ apart. We can choose the coordinate axes such that at $t = 0$, there are a positive and a negative vortex with zero core radius located at $(-\ell, 0)$ and $(\ell, 0)$, respectively. As t increases, the strength of the vorticity on the left half plane, Γ^+, decreases from the initial value, Γ_0, but the total strength of the first x-moment remains constant. The strength of the doublet is

$$\mathbf{E} = E_2 \hat{\jmath} \ , \quad \text{with} \quad E_2 = -\langle x\zeta \rangle = -2\langle x\zeta \rangle^- = 2\Gamma_0 \ell \ . \tag{6.2.90}$$

Under the inviscid theory, both vortices remain point vortices and move with constant vertical velocity $\Gamma_0/(4\pi\ell) = E_2/(8\pi\ell^2)$. The vertical displacement of the centers is

$$Y_{\text{inv}} = \frac{E_2 \, t}{8\pi\ell^2} \quad \text{or} \quad \tilde{Y}_{\text{inv}} = \frac{\tilde{t}}{4\pi} \ , \tag{6.2.91}$$

with no change in their x-coordinates, that is, $X^{\pm}_{\text{inv}}(t) \equiv X^{\pm}_{\text{inv}}(0)$. Here $\tilde{Y}$ and $\tilde{t}$ denote the variables,

$$\tilde{Y} = 8\nu Y/E_2 \qquad \text{and} \qquad \tilde{t} = 4\nu t/\ell^2 \ , \tag{6.2.92}$$

The vorticity fields given by the extended asymptotic solution and by the approximate solution are represented by superposition of two Lamb vortices of strengths $\pm\Gamma_0$ with the same core radius $\delta(t) = \sqrt{4\nu t}$. They differ only in the vertical velocity of the vortex centers, $\dot{Y}_D$, while the x-coordinates of the vortex centers remain stationary. In the coordinate system moving with the vortex pair, the vorticity fields given by these two solutions are identical. They give the same lateral motion of the *point of maximum* ζ, from the vortex center $-\ell$ away from the y-axis. The motion is induced by the cancellation of vorticity in the half plane $x < 0$ by its negative image in $x > 0$, and the cancellation in turn is due to diffusion of vorticity across the y-axis.

In terms of the dimensionless variables, $\tilde{Y}$ and $\tilde{t}$, the vertical velocity $\dot{Y}_D$ given by the extended asymptotic solution becomes

$$\frac{d}{d\tilde{t}}\tilde{Y}_{\text{ext}} = \frac{1 - e^{-4/\tilde{t}}}{4\pi} \approx \frac{1}{\pi}\tilde{t}^{-1} \quad \text{for } \tilde{t} \gg 1 \ . \tag{6.2.93}$$

The vertical displacement of the doublet center is

$$\begin{aligned} \tilde{Y}_{\text{ext}} &= \frac{1}{4\pi}\int_0^{\tilde{t}} [1 - e^{-4/\xi}]d\xi = \frac{\tilde{t}}{4\pi}[1 - e^{-4/\tilde{t}}] + \frac{\mathcal{E}_1(4/\tilde{t})}{\pi} \\ &\approx \frac{1}{\pi}\ln\tilde{t} - \frac{1}{\pi}(2\ln 2 + \gamma - 1) + \frac{2}{\pi}\tilde{t}^{-1} \quad \text{for } \tilde{t} \gg 1 \ , \end{aligned} \tag{6.2.94}$$

where $\mathcal{E}_1$ denotes the exponential integral and $\gamma \approx 0.5772$ is the Euler number (Magnus et al., 1966).

Using (6.2.76), the approximate solution yields $\dot{X}_j = 0$, or $X_j = (-1)^j\ell, j = 1, 2$; therefore, the only nontrivial constraint for this antisymmetric vorticity field, the constraint on x-component of the first moment of vorticity is fulfilled. This is another case in which the approximate solution of (6.2.76), which minimizes H, also fulfills the Poincaré constraints. The vertical velocity given by the approximate solution is defined by the third equation of (6.2.76) with $N = 2$. In terms of the dimensionless variables, the vertical velocity is

$$\frac{d}{d\tilde{t}}\tilde{Y}_{\text{app}} = \frac{1 - 4e^{-2/\tilde{t}} + 3e^{-8/3\tilde{t}}}{4\pi[1 - e^{-2/\tilde{t}}]} \approx \frac{1}{3\pi}\tilde{t}^{-1} - \frac{5}{27\pi}\tilde{t}^{-2} \quad \text{for } \tilde{t} \gg 1 \ . \tag{6.2.95}$$

We integrate the equation from $\tilde{t} = 0$ with $\tilde{Y} = 0$ to a large $\tilde{t}$, say $\tilde{t} = 50$, and then determine the constant in the long-time behavior of $\tilde{Y}$. The result is

$$\tilde{Y}_{\text{app}} \approx \frac{1}{3\pi}\ln\tilde{t} + 0.022428 + \frac{5}{27\pi}\tilde{t}^{-1} \ . \tag{6.2.96}$$

Note that both solutions, $\tilde{Y}_{\text{ext}}$ and $\tilde{Y}_{\text{app}}$, reduce to the $\tilde{Y}_{\text{inv}}$ for small $\tilde{t}$. For long time, $\tilde{t} \gg 1$, the vertical displacements given by these two solutions, are consistent with the formula (6.2.82) obtained from dimensional analysis,

$$\tilde{Y}_D \approx C_0 \ln \tilde{t} + C_1 + C_2 \tilde{t}^{-1/2} + C_3 \tilde{t}^{-1} , \tag{6.2.97}$$

with $C_2 = 0$. Because of the difference between the coefficient of the leading-order term, $\ln \tilde{t}$ in the long-time solution (6.2.96) and in (6.2.94), the vertical displacement $\tilde{Y}_{\text{app}}$ is smaller by a factor of 3 than $\tilde{Y}_{\text{ext}}$. The latter in turn is much smaller than $\tilde{Y}_{\text{inv}}$ by the order $\ln \tilde{t}\ /\tilde{t}$.

In long time, the vertical displacement $\tilde{Y}_{\text{ext}}$, $\tilde{Y}_{\text{app}}$, or $\tilde{D}$ is $O(\ln \tilde{t})$ and does not approach a stationary value. But relative to the diffusion length, which increases as $\sqrt{\tilde{t}}$, the displacement of the doublet center is negligible. Hence, the doublet center appears to be stationary. This is not true for the inviscid theory, where $\tilde{Y}_{\text{inv}} \approx \tilde{t}$ increases faster than the diffusion length of the viscous problem by $O(\sqrt{\tilde{t}})$.

Note that the vorticity distributions in both the extended asymptotic and the approximate solution are symmetric with respect to the horizontal axis joining the two Lamb vortex centers. Therefore, the vertical coordinates of the point of maximum (minimum) vorticity, $Y_{\max}$ ($Y_{\min}$), and the two centers of the Lamb vortices on the left and right sides of the y-axis, coincide with that of the center of the viscous doublet, Y_D, that is,

$$Y_{\max} = Y_{\min} = Y_1 = Y_2 = Y_D . \tag{6.2.98}$$

Note that the relationship between the scale variables $\tilde{Y}_{\text{ext}}$ and $\tilde{t}$, and between $\tilde{Y}_{\text{app}}$ and $\tilde{t}$, is independent of the Reynolds number of the doublet,

$$\text{Re} = |\mathbf{E}|/2\nu\ell . \tag{6.2.99}$$

We recall that the extended asymptotic solution, which is a linear superpositions of Lamb vortices, fulfills the Navier–Stokes equation if the nonlinear convection terms coming from two different vortices are omitted, while the corresponding approximate solution, which minimizes the error due to those nonlinear convection terms, is expected to improve the extended solution. To show the improvement, we compare these two solutions with the numerical solution of the Navier–Stokes equation. The latter is Reynolds number dependent.

We use the finite difference solution of the Navier–Stokes equations, (3.2.1) and (3.2.2), to simulate the motion and diffusion of a viscous doublet that is initially a vortex pair. At $t = 0$, there are two vortices located at $(-\ell, 0)$ and $(\ell, 0)$ with strengths $\Gamma_0 > 0$ and $-\Gamma_0$ and zero radius, $\delta = 0$. Since we cannot start the numerical solution from $t = 0$, we use the asymptotic solution for small t or rather $\tilde{t} \ll 1$ to define the vorticity field for $0 < \tilde{t} \leq \tau_0 \ll 1$. We choose a small τ_0 so that the accuracy of the asymptotic solution is well within that of the numerical solution. We use the asymptotic solution at $\tilde{t} = \tau_0$ as the initial data and carry out the numerical solution for $\tilde{t} \geq \tau_0$. We carried out numerical solutions for $\tilde{t} > \tau_0$, for $\tau_0 = (1/4)^2$ and $(3/8)^2$, and found that the difference in the two numerical solutions is insignificant. Therefore, in the numerical results reported here, we use $\tau_0 = (1/4)^2$. Note that the scaled time $\tilde{t}$ defined by (6.2.92) is equal to the square of the ratio of the effective radius to ℓ. At τ_0, we have $t = t_0 = \ell^2\ \tau_0/\sqrt{4\nu}$ or $\delta(t_0)/\ell = 1/4$, that is, the core radius is one eighth of the distance 2ℓ between the two vortex centers. The effect of the vorticity field of one Lamb vortex on the other one is extremely small, $O(e^{-64})$.

In the numerical solution of the Navier–Stokes equations, (3.2.1) and (3.2.2), we use ℓ as the length scale, Γ_0/ℓ as the velocity scale, and ℓ^2/Γ_0 as the time scale. The Reynolds number $\mathrm{Re} = \Gamma_0/\nu$ then appears in (6.2.1) and the solution depends on Re. Numerical solutions are constructed for Re = 100 and 1,000. Figure 6.14 shows five contour lines of constant vorticity, $\zeta/\zeta_{\max} = 2.5/e$, $2/e$, $1.5/e$, $1/e$, and $0.5/e$, with Re = 1,000 at an early stage $\tilde{t} = 0.5292$ and at a later stage $\tilde{t} = 2.558$. The origin of the coordinate system is moving with the doublet center $(0, Y_D)$ where Y_D is defined by (6.2.93). The ordinate of the point of maximum vorticity shown in Fig. 6.14 denotes the difference between the vertical displacement of the point and the doublet center, $Y_{\max} - Y_D$, which increases with time. This difference arises because the numerical solution defines the convection of the vorticity at each grid point while both the approximate solution and extended asymptotic solution assume that the vorticity field moves as a whole with the doublet center. In the numerical solution, the vorticity field near the horizontal line joining the points of maximum and minimum vorticity drifts faster than that far below the line and the field does not have a horizontal line of symmetry. The region of low vorticity below the horizontal axis through the doublet center grows larger and larger.

Numerical simulations of the evolution of a viscous doublet are carried out from $t = t_0$ until $t = t_f$ when the maximum vorticity is <1% of the value

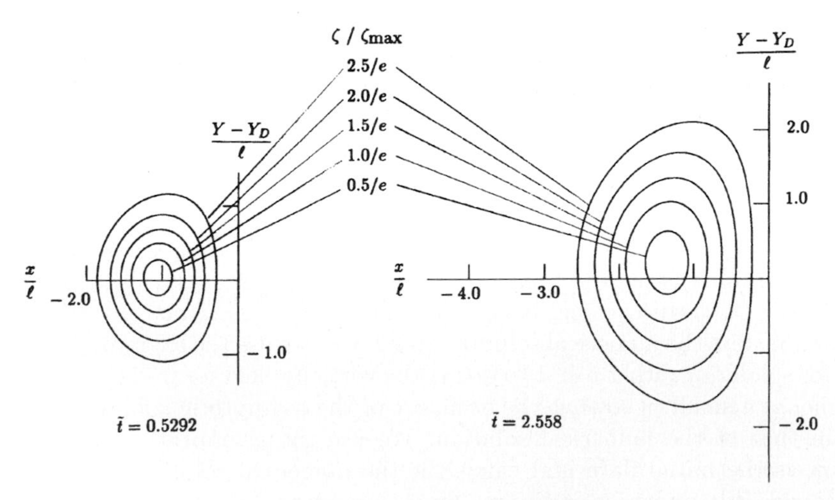

Fig. 6.14. Contour lines of constant vorticity of a two-dimensional viscous doublet of strength $E_2\hat{\jmath}$. Adapted from Ting and Bauer (1993)

at t_0. The invariance of the first moment of vorticity has been employed to check the accuracy of the solution for the computations. To match with the long-time behavior (6.2.97) of the vertical displacement of the doublet, we assume that the coefficient of the $\ln \tilde{t}$ term is equal to the value $1/3\pi$ given by the approximate solution and hence is independent of Re. We write the scaled displacement as

$$\tilde{Y}_D \approx \frac{1}{3\pi} \ln \tilde{t} + C_1 + C_2 \tilde{t}^{-1/2} . \quad (6.2.100)$$

For each Re, the data for the vertical displacement of the doublet, Y_D, near the end of computation, t_f, are used to determine the coefficients C_1 and C_2 in (6.2.100). The same procedure is employed to determine the long-time behavior of the scaled vertical displacement of the point of maximum vorticity, $\tilde{Y}_{max}$. The coefficients C_1 and C_2 for $\tilde{Y}_D$ and for $\tilde{Y}_{max}$ depend on Re. These coefficients are listed in Table 6.1.

Table 6.1. Coefficients, C_1 and C_2, of long-time expansions of forward displacements, $\tilde{Y}$'s

	R_e	C_1	C_2
$\tilde{Y}_D$	100	−0.052	0.119
$\tilde{Y}_D$	1,000	−0.293	0.367
$\tilde{Y}_{max}$	100	−0.011	0.083
$\tilde{Y}_{max}$	1,000	−0.288	0.358

Using the coefficients in each row of Table 6.1 in the expression (6.2.100), for the long-time behavior, we see that the result given by the dotted line coincides with the numerical data given by the solid line from $t_f/2$ to t_f. This indicates that t_f is large enough so that the numerical solution has already overlapped with the long-time expression. Consequently,

> *The use of the same coefficient, $1/3\pi$, in the leading term of the approximate solution and the far-field representation (6.2.100) of the numerical solution is justified. This coefficient is independent of* Re *and valid for $\tilde{Y}_D$ and $\tilde{Y}_{max}$. The next two coefficients, C_1 and C_2, for $\tilde{Y}_{max}$ are different from those for $\tilde{Y}_D$ and are* Re *dependent.*

Figure 6.15 shows the variation of the vertical displacement $\tilde{Y}_D$ and that of $\tilde{Y}_{max}$ given by the numerical results in solid lines matched with the far-field expression (6.2.100) $(\cdots)$ for Re = 100, and 1,000 using the coefficients listed in Table 6.1. Since $\tilde{Y}_D$ nearly coincides with $\tilde{Y}_{max}$ for Re = 1,000, it is not shown in the figure. Also shown are three curves given by the inviscid theory, $\tilde{Y}_{inv}$, the extended asymptotic solution, $\tilde{Y}_{ext}$, and the approximate

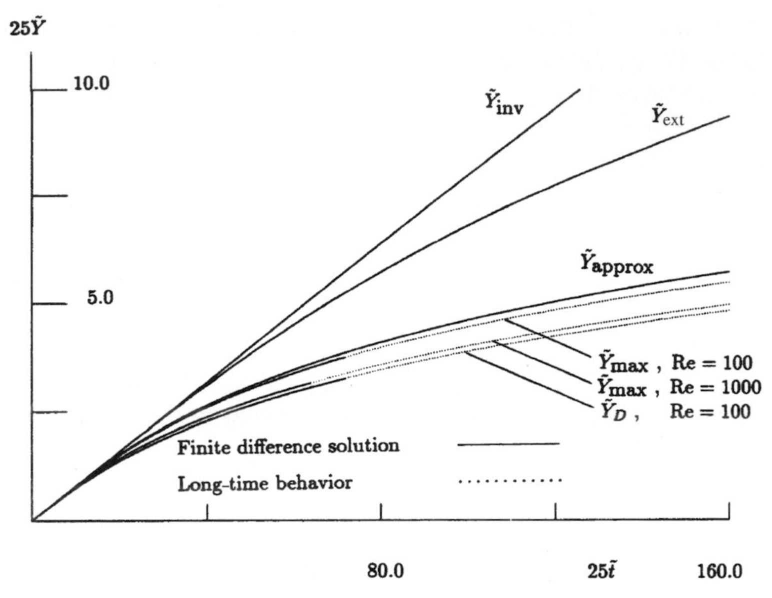

Fig. 6.15. Motion of the point of maximum vorticity in a two-dimensional viscous doublet: The inviscid solution, $\tilde{Y}_{\rm inv}$; the extended asymptotic solution, $\tilde{Y}_{\rm ext}$; the approximate solution, $\tilde{Y}_{\rm app}$; and the finite difference solutions of the doublet center, $\tilde{Y}_D$, and the point of maximum vorticity, $\tilde{Y}_{\rm max}$, for Re $=$ 100 and 1,000. Adapted from Ting and Bauer (1993)

solution, $\tilde{Y}_{\rm app}$. These three curves are independent of the Reynolds number. As expected, for small $\tilde{t}$, say $\tilde{t} < 0.5$, all these curves $\tilde{Y}_{\rm ext}$, $\tilde{Y}_{\rm app}$ and the numerical solutions of $\tilde{Y}_{\rm max}$ and $\tilde{Y}_{\rm D}$ at Re $=$ 100 and 1,000, converge to the straight line, $\tilde{Y}_{\rm inv}$, given by the inviscid theory. For large $\tilde{t}$ the $\tilde{Y}_{\rm app}$ is much closer to the numerical solution $\tilde{Y}_{\rm max}$ for Re $=$ 100 than $\tilde{Y}_{\rm ext}$ while $\tilde{Y}_{\rm inv}$ is way off. We recall that in the approximate solution there is no difference between $\tilde{Y}_{\rm max}$ and $\tilde{Y}_{\rm D}$. The approximate solution is close to the numerical result for $\tilde{Y}_{\rm max}$, because these two solutions emphasize the local field near the point of maximum vorticity, whereas the numerical result $\tilde{Y}_{\rm D}$ includes the global effect of the vorticity field.

6.3 Merging or Intersection of Vortex Filaments

To illustrate how to implement the numerical schemes formulated in Sect. 6.1, that is, how to choose the appropriate computational domain and how to generate the boundary data for different types of initial vorticity distributions, we consider first the axisymmetric problems in Subsect. 6.3.1 to illustrate global mergings of type I(a) and I(b); type II (local) merging will not happen due to axisymmetry. In Subsect. 6.3.2, we present numerical examples and relevant experimental results of global merging of two rings to a single ring or filament. The global mergings are of type I(a). We note that initial data leading to mergings of type I(b) are hard to create and are rarely observed. Also we recall that we recommend that problems of type II, that is, the local merging of two filaments where they intersect and finally reconnect, be analyzed by the extended vortex element method outlined in Sect. 9.1.

6.3.1 Merging of Coaxial Vortex Rings

We consider axisymmetric incompressible flows with zero circumferential velocity and repeat some relevant equations, (2.3.14)–(2.3.17), in Sect. 2.3. Let the axis of symmetry be the z-axis, and σ and θ denote the radial and circumferential coordinates. Since the vortex lines are coaxial circles, the vorticity vector has only a circumferential component, ϖ, that is,

$$\Omega = \hat{\theta}\varpi(t,\sigma,z) = [-\sin\theta\,\hat{\imath} + \cos\theta\,\hat{\jmath}]\varpi(t,\sigma,z) . \tag{6.3.1}$$

The vector potential $\mathbf{A}$ is also a circumferential vector,

$$\mathbf{A} = \hat{\theta}\,[\psi(t,\sigma,z)/\sigma] . \tag{6.3.2}$$

Here ψ represents the stream function in a meridian plane, σz with $\sigma \geq 0$, and is related to the circumferential vorticity ϖ by the axisymmetric Poisson equation,

$$\psi_{\sigma\sigma} + \psi_{zz} - \psi_\sigma/\sigma = -\sigma\varpi . \tag{6.3.3}$$

The vorticity, ϖ, in turn is governed by the evolution equation,

$$\varpi_t - \psi_z\left(\frac{\varpi}{\sigma}\right)_\sigma + \frac{\psi_\sigma}{\sigma}\varpi_z = \nu\left[\varpi_{\sigma\sigma} + \varpi_{zz} + \left(\frac{\varpi}{\sigma}\right)_\sigma\right] . \tag{6.3.4}$$

The three time invariants of first moments of vorticity, (2.3.7), reduce to one nontrivial invariant of the axial component of $< \mathbf{x} \times \Omega >$. It is

$$2\pi \int_{-\infty}^{\infty}\int_0^{\infty} \sigma^2\varpi(t,\sigma,z)\,d\sigma dz = E_3 = |\mathbf{E}| . \tag{6.3.5}$$

In the far field, the flow behaves as a doublet with constant strength E_3 oriented along the z-axis.

Because of axisymmetry and the circular vorticity field, (6.3.1), only two of the five linear combinations of second moments of vorticity in (2.3.8) and (2.3.9) are nontrivial. They are

$$F_i(t) = 0 , \quad i = 1,2,3 ,$$
$$H_3(t) = 0 , \quad H_1(t) = -H_2(t) \neq 0 . \tag{6.3.6}$$

The instantaneous position and velocity of the doublet center are defined by Ting and Bauer (1993):

$$\boldsymbol{X}(t) = -\frac{\hat{\imath}F_2(t) + \hat{\jmath}F_1(t) + \hat{k}\left[H_1(t) - H_2(t)\right]/3}{E_3} \tag{6.3.7}$$

$$\dot{\boldsymbol{X}}(t) = -\frac{\hat{\imath}\dot{F}_2(t) + \hat{\jmath}\dot{F}_1(t) + \hat{k}[\dot{H}_1(t) - \dot{H}_2(t)]/3}{E_3} . \tag{6.3.8}$$

Since the velocity of the doublet center is parallel to the z-axis, we have

$$\dot{\mathbf{X}}(t) = \dot{Z}_D(t)\hat{k} = \langle z \ \mathbf{x} \times \Omega \rangle \hat{k}/|\mathbf{E}| . \tag{6.3.9}$$

The long-time representation of the displacement of the doublet center (Ting and Bauer, 1993),

$$\boldsymbol{X} = \boldsymbol{C}_0 + \boldsymbol{C}_1 \tilde{t}^{-1/2} + \boldsymbol{C}_2 \tilde{t}^{-3/2} + \cdots \tag{6.3.10}$$

reduces to one along the z-axis,

$$Z_D(t) \approx C_0 + C_1 \tilde{t}^{-1/2} + C_2 \tilde{t}^{-3/2} , \tag{6.3.11}$$

where $\tilde{t} = 4\nu t/\ell^2$. With the origin moving with the doublet center, the far-field representation,

$$\Phi(t, \boldsymbol{x}) = \Phi^{(1)}(\boldsymbol{x}) + \Phi^{(2)}(t, \boldsymbol{x}) + O(\bar{r}^{-4}) \tag{6.3.12}$$

with $\Phi^{(1)}$ and $\Phi^{(2)}$ given by (2.3.44) and (2.3.45), respectively, becomes

$$\Phi(t, \mathbf{x}) = \frac{|\mathbf{E}|}{8\pi} \partial_z \frac{1}{\bar{r}} + O(\bar{r}^{-4}) , \tag{6.3.13}$$

where $\bar{r} = |\mathbf{x} - Z_D \hat{k}|$. Because of axial symmetry, $F_3 = 0$ and $H_3 = 0$ and the quadrupole term vanishes (Ting and Bauer, 1993).

Now we use the finite difference solutions of axisymmetric Navier–Stokes equations, (6.3.3) and (6.3.4), to simulate the motion and diffusion of a viscous doublet which is initially a circular vortex line. At $t = 0$, there is a circular vortex line lying on the plane $z = 0$ with radius ℓ centered at the origin. The initial core radius of the vortex ring is zero, $\delta(0) = 0$, and the initial circulation or the strength of the ring is related to the time invariant doublet strength by

$$\Gamma_0 = \Gamma(0) = |\mathbf{U}|/(2\pi\ell^2) . \tag{6.3.14}$$

Since we cannot start the numerical solution from $t = 0$, we use the asymptotic solution for a small scaled time, $\tilde{t} = 4\nu t/\ell^2 \ll 1$, to define the vorticity field for $0 < \tilde{t} \leq \tau_0 \ll 1$. During this period, the ring moves along the z-axis with constant radius $R(t) = \ell$ and strength Γ_0. The diffusion of the core structure is given by the similarity solution, with the core radius and the axial displacement of the centerline of the ring given by

$$\delta^2(t) = 4\nu t \quad \text{or} \quad \delta^2 t/\ell^2 = \tilde{t} \tag{6.3.15}$$

and

$$Z(t) = \frac{\Gamma_0 t}{4\pi\ell}\left[\ln\frac{8\ell}{\delta(t)} + 0.442\right], \quad \text{or} \quad \tilde{Z}(\tilde{t}) = \frac{\tilde{t}}{4\pi}\left[\ln\frac{8}{\sqrt{\tilde{t}}} + 0.442\right], \quad (6.3.16)$$

where $\tilde{Z} = 4(Z/\ell)(\nu/\Gamma_0)$ denotes the scaled axial displacement. We need to choose a small $\tilde{\tau}_0$ so that the asymptotic solution at $\tilde{t} = \tilde{\tau}_0$ could provide the initial data to begin the numerical solution for $\tilde{t} > \tilde{\tau}_0$. The choice of $\tilde{\tau}_0$ was made by comparing two numerical solutions starting at $\tilde{\tau}_0 = (1/4)^2$ and $\tilde{\tau}_0 = (3/8)^2$. We find that the difference between the two solutions is insignificant, say less than 1%. Thus, we use the asymptotic solution at $\tilde{\tau}_0 = (1/4)^2$ to provide the initial data. In the numerical solution of the Navier–Stokes equations, (6.3.3) and (6.3.4), we use ℓ and $|\mathbf{E}|/(2\pi\ell^2)$ as the length and velocity scales. In the scaled variables, the solution depends only on the Reynolds number, $\mathrm{Re} = \Gamma_0/\nu = |\mathbf{E}|/(2\pi\nu\ell^2)$. The choice of the finite computational domain and the formulae for the approximate boundary data used to simulate the solution in an unbounded domain are described in Sect. 6.1.

Numerical simulations of the evolution of a viscous doublet are carried out for Re = 100, 1,000, respectively from $\tilde{t} = \tilde{t}_0$ until $\tilde{t} = \tilde{t}_f$ when the maximum vorticity is less than 1% of the value at $\tilde{t}_0$. The invariance of the first moment of vorticity was employed to check the accuracy of the solution for the computations.

Figure 6.16 shows five contour lines of constant vorticity, $\varpi/\varpi_{\max} = 2.5/e,\ 2/e,\ 1.5/e,\ 1/e$, and $0.5/e$, with Re = 1,000, at an early stage $\tilde{t} = 0.4125$ and at a late stage $\tilde{t} = 4.495$. The origin of the coordinate system in the σz plane is moving with the doublet center $(0, Z_D)$, where Z_D is defined by (6.3.9). The abscissa $\sigma_{\max}$ of the point of maximum vorticity shown in Fig. 6.16 can be used to denote the radius of the ring centerline. The ordinate denotes the difference between the vertical displacements of the centerline and the doublet center, $Z_{\max} - Z_D$, which is equal to 0 in the asymptotic theory; that is, it is small for small $\tilde{t}$. The difference increases with time and approaches a constant as $\tilde{t} \to \infty$ since both $Z_{\max}$ and Z_D approach finite limits. The difference, $Z_{\max} - Z_D$, and the vertical elongation of the contour lines below the line of maximum vorticity for the axisymmetric case shown in Fig. 6.16 are more pronounced than the two-dimensional case shown in Fig. 6.14. The region of low vorticity below the horizontal plane centerline grows larger and larger as shown in Fig. 6.16.

For each Re, the data near $\tilde{t}_f$ for the vertical displacement of the doublet, $\tilde{Z}_D$, and that of the centerline, $\tilde{Z}_{\max}$, are used to determine the leading two coefficients in the long-time expression (6.3.9) with Z replaced by $\tilde{Z}$. These coefficients are listed in Table 6.2 below.

Using the appropriate coefficients, the expression (6.3.11) in Figs. 6.17 and 6.18 $(\cdots)$, fits the numerical data given by the solid line from $\tilde{t}_f$ backward to $\tilde{t}_f/2$. This indicates that the finite difference solution has already overlapped with the long-time expression. Figure 6.17 compares the vertical

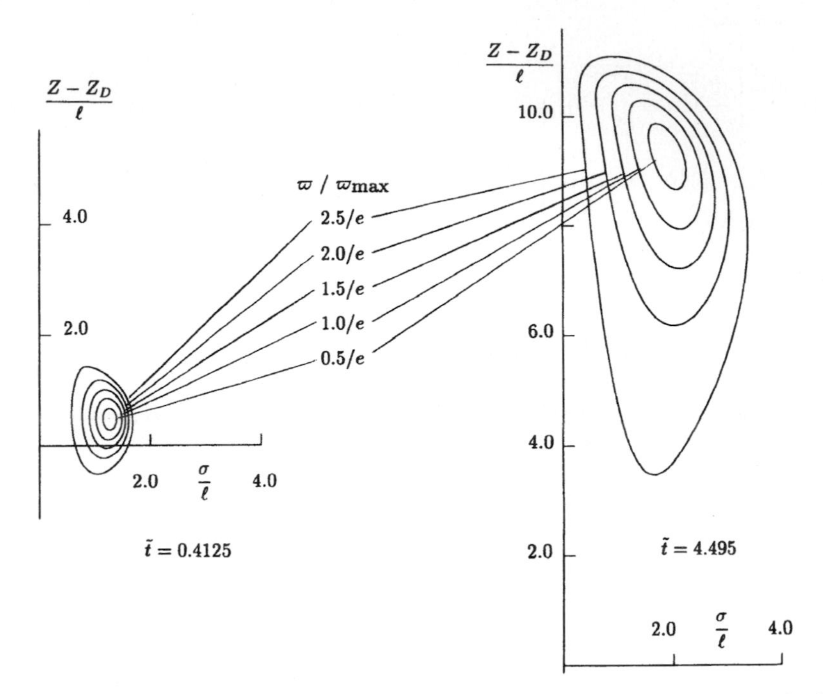

Fig. 6.16. Contour lines of constant vorticity of an axisymmetric viscous doublet of strength $E_3\hat{k}$. Adapted from Ting and Bauer (1993)

Table 6.2. First two coefficients of the long-time expansions of the axial displacements, $\tilde{Z}$

	R_e	C_0	C_1
$\tilde{Z}_D$	100	0.282	−0.180
$\tilde{Z}_D$	1,000	0.219	−0.104
$\tilde{Z}_{max}$	100	0.315	−0.112
$\tilde{Z}_{max}$	1,000	0.296	−0.191

displacement of the centerline $\tilde{Z}_{max}$ with that of the doublet center $\tilde{Z}_D$ for Re = 1,000 and with the asymptotic solution valid for small $\tilde{t}$. Figure 6.18 shows the variation of the vertical displacement Z_{max} and $\tilde{Z}_D$ given by the finite difference solution (– – – –) matched with the far-field expression (· · ·) for Re = 100 and 1,000. In particular, the figure shows the dependence of the solution on Re.

6.3.1.1 Type I(b) Vorticity Concentrated in a Torus. The merging or collision of two coaxial vortex rings with their initial core radii, $\delta_i(0)$, $i = 1, 2$, much smaller than their ring radii, $R_i(0)$, are examples of global merging

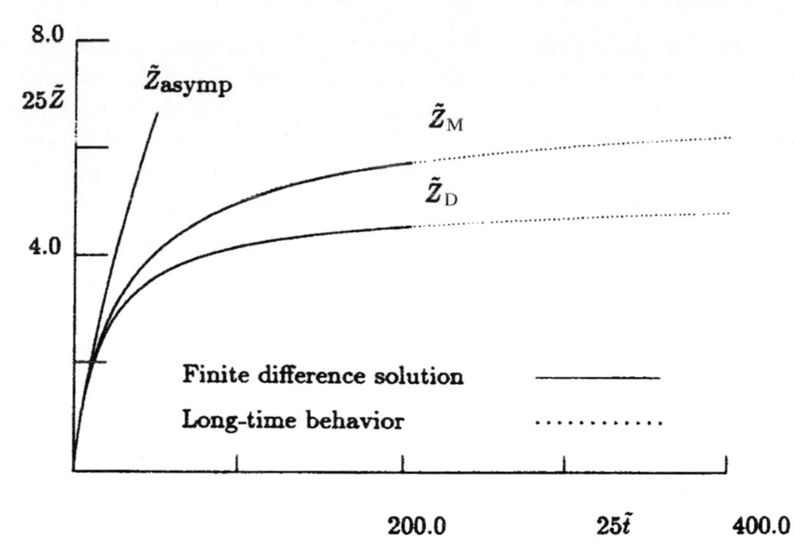

Fig. 6.17. Finite difference solution and long-time behavior of the axial motion of the center of an axisymmetric viscous doublet and of the line of maximum vorticity, for Re = 1,000, and the asymptotic solution for the short time. Adapted from Ting and Bauer (1993)

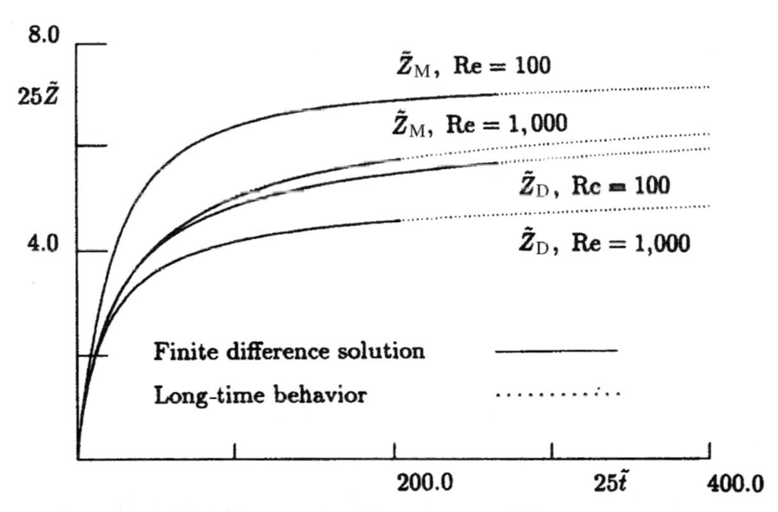

Fig. 6.18. Vertical motion of the center of an axisymmetric viscous doublet, $\tilde{Z}_D$, and the line of maximum vorticity, $\tilde{Z}_{\max}$, for Re = 100 and 1,000. Adapted from Ting and Bauer (1993)

problems of type I(b). The vorticity field associated with both rings can be effectively contained inside a torus whose cross-sectional radius $a = O(\delta_i)$ is much smaller than the radius $R = O(\ell)$ of its centerline. In general, the torus containing its centerline moves with the axial velocity of the center of the vorticity field in the σz plane. The plane $z = Z(t)$ containing the centerline of the torus remains stationary if the two vortex rings are of the same size but of opposite circulation. Equations defining the center of the vorticity in the meridian plane will be provided later. The assumption of $a \ll R$ implies that the vorticity decays exponentially in ρ_+, where ρ_+ is the minimum distance from a reference point $Q(\sigma, z)$ in the flow field to the centerline of the torus. The maximum distance is denoted by ρ_- as shown in Fig. 6.19. Also shown are the minimum and maximum distances, τ_+ and τ_-, from the point $Q(\sigma, z)$ to a circular vortex line of radius σ' in the plane $z = z'$.

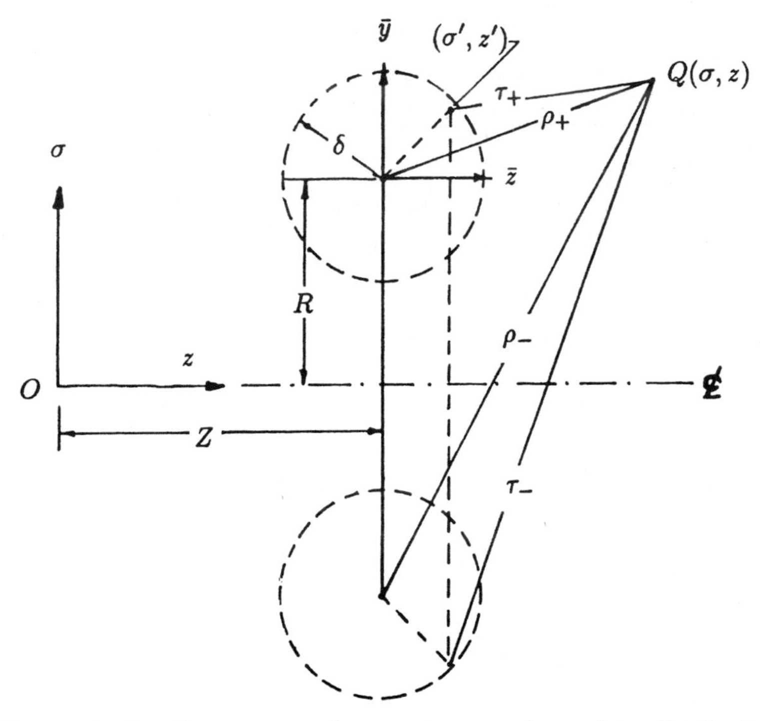

Fig. 6.19. Coordinate system for an axisymmetric vorticity distribution

The computational domain $\mathcal{D}$ in the meridian plane can be a square with side $2ma$ centered at $(R(t), Z(t))$, provided that $R > ma$. Otherwise, the lower boundary, $\sigma = R - ma$, of $\mathcal{D}$ shall be replaced by $\sigma = 0$. To demonstrate the choice of the computational domain and the generation of boundary data suitable for different types of initial vorticity distributions, we discuss a numerical example from Liu and Ting (1982) for the head-on collision of two vortex rings of identical shape but with opposite circulations. The vorticity field is an odd function of z. We set $Z(t) = 0$ and identify $R(t)$ as the radial coordinate of the point of maximum $|\varpi|$. At $t = 0$, we have $R(0)$ equal to the initial radius R_0 of the vortex rings.

Again, we adjust the number m, which characterizes the size of $\mathcal{D}$, so as to achieve a required accuracy. Depending on the order of magnitude of R/a relative to m, that is, $R/a = O(m)$ or $R/a \gg m$, two different expansions for the Poisson integral must be developed to obtain an accurate far-field description. We will explain these expansions explicitly for the one-domain method. The analogous procedures for the *two-domain method* will then become quite obvious.

Size of Computational Domain $ma = O(R)$. In this case we can choose the computational domain $\mathcal{D}$ for the finite difference solution of the Poisson equation to be a rectangle,

$$|z - Z(t)| \leq ma \quad \text{and} \quad R(t) + ma \geq \sigma \geq R(t) - ma \geq 0 \tag{6.3.17}$$

with a sufficiently large m, say $m = 4$, and impose $\varpi = 0$ on $\partial\mathcal{D}$ since the decay length is $\ell_d = O(a)$. In case $R < ma$, we replace the lower boundary by the axis of symmetry, $\sigma = 0$, on which we have $\varpi = 0$ and $\psi = \psi_\sigma = 0$. The area of $\mathcal{D}$ is of the order $4R^2$, while the area of a computational domain for the preceding case of type I(a) was $O(2m^2R^2)$. Thus, the number of grid points in $\mathcal{D}$ is reduced from that for type I(a) by a factor of $m^2/2$. To obtain far-field boundary data for ψ as $\rho_+ \gg a$, we make use of the axisymmetry, and the formula for the stream function induced by a circular vortex line of strength $\omega(t, \sigma', z')d\sigma'\,dz'$, (see Lamb, 1932, p. 237) to obtain the stream function of the vortex ring

$$\psi(t, \sigma, z) = \frac{1}{2\pi} \int_{-\infty}^{\infty} \int_{0}^{\infty} d\sigma'\,dz'\,(\tau_1 + \tau_2)\,G(\lambda)\,\varpi(t, \sigma', z')\,, \tag{6.3.18}$$

where

$$\tau_\pm = \sqrt{(z - z')^2 + (\sigma \mp \sigma')^2}\,, \quad \lambda = \frac{\tau_- - \tau_+}{\tau_- + \tau_+} = \frac{4(R + \bar{y}')(R + \bar{y})}{(\tau_- + \tau_+)^2} \tag{6.3.19}$$

and

$$G(\lambda) = F(\lambda) - E(\lambda)\,. \tag{6.3.20}$$

Here, F and E are the complete elliptic integrals of the first and second kind respectively and we have introduced the coordinates $(\bar{y}, \bar{z})$ relative to (R, Z) by

$$\sigma = \bar{y} + R \quad \text{and} \quad z = \bar{z} + Z\,. \tag{6.3.21}$$

Next we extend the integrals in (6.3.18), which cover the half σz plane with $\sigma \geq 0$, to the entire $\bar{y}\bar{z}$–plane using the fact that ϖ decays exponentially in terms of the scaled radius

$$\rho_+^2/a^2 = (\bar{y}^2 + \bar{z}^2)/a^2\,. \tag{6.3.22}$$

Then we expand $\tau_\pm$, λ and the integral (6.3.18) in powers of $a/\rho_\pm = O(\delta/\ell) = O(1/m)$. The result is again a power series expansion for the stream function of the form

$$2\pi\,\psi(t,\sigma,z) = \langle\varpi\rangle^* a_{00} + \langle\varpi\bar{z}\rangle^*\, a_{10} + \langle\varpi\bar{y}\rangle^* a_{01} + O(\delta^2/\ell^2)\;, \qquad (6.3.23)$$

where $\langle f\rangle^*$ stands for $\int\int_{-\infty}^{\infty} f(\bar{y},\bar{z})\,d\bar{y}\,d\bar{z}$. In particular, $\langle\varpi\rangle^*$ is equal to the strength of the axisymmetric vortical field in $\mathcal{D}$, and in case of the merging of two vortex rings of opposite circulations, we have $\langle\varpi\rangle^* = 0$.

The coefficients in (6.3.23) are

$$a_{00} = (\rho_+ + \rho_-)\,G(\lambda_0) \qquad (6.3.24)$$

$$a_{10} = \left(\frac{\bar{z}}{\rho_+} + \frac{\bar{z}}{\rho_-}\right)[-G(\lambda_0) + 2\lambda_0 G'(\lambda_0)] \qquad (6.3.25)$$

$$a_{01} = -\left(\frac{\bar{y}}{\rho_+} - \frac{\bar{y}+2R}{\rho_-}\right) G(\lambda_0)$$

$$+\left[\frac{\rho_+ + \rho_-}{R} + 2\left(\frac{\bar{y}}{\rho_+} - \frac{\bar{y}+2R}{\rho_-}\right)\right]\lambda_0\, G'(\lambda_0)\;, \qquad (6.3.26)$$

where $\lambda_0 = 4R\sigma/(\rho_+ + \rho_-)^2$. Although only the first two terms in ψ are given in (6.3.23), the original numerical computations of Liu and Ting (1982) included two more terms so that the error of the boundary data was $O(\delta^4/\ell^4)$.

Size of Computational Domain $ma \ll R$. In this case the computational domain $\mathcal{D}$ can again be a square of size $2ma$ centered at (R, Z). With the size of $\mathcal{D}$ being now much smaller than R, the leading-order solution in $\mathcal{D}$ should be a two-dimensional solution since the far-field expansion parameter m^{-1} is independent of δ/R. The far-field behavior of ψ on $\partial\mathcal{D}$ is obtained from (6.3.18)–(6.3.20) with $\delta/\rho_1 = O(1/m)$ and $\Lambda = \rho_1/\rho_2 = O(ma/R)$ as small independent parameters. The first three terms, also given by Liu and Ting (1982), are

$$\frac{2\pi}{\sqrt{\sigma R}}\,\psi(t,\sigma,z) = \langle\varpi\rangle^* b_{00} + \langle\varpi\bar{z}\rangle^* b_{10}$$

$$+\langle\varpi\bar{y}\rangle^* b_{01} + o\left(\frac{\delta^j}{\ell^j}\Lambda^n \log\Lambda\right),\; j+n=2\;, \qquad (6.3.27)$$

where

$$b_{00} = \ln\frac{4}{\Lambda} - 2\;,$$
$$b_{10} = \frac{\bar{z}}{\rho_+^2}\left(1 - \frac{1}{2R}\right) - \frac{\bar{z}}{\rho_-^2}\;, \qquad (6.3.28)$$
$$b_{01} = \frac{\bar{y}}{\rho_+^2}\left(1 - \frac{1}{2R}\right) - \frac{\bar{y}+2R}{\rho_-^2} + \frac{b_{00}}{2R}\;.$$

Note that the leading term containing $\ln(4/\Lambda)$ represents the local two-dimensional result. Again, in the numerical solutions of Liu and Ting (1982), the expansion (6.3.27) was carried out up to terms of $O(\Lambda^n \log \Lambda\, \delta^i/\ell^i)$ with $j+n=4$. It should be mentioned that the integral invariant corresponding to (1.2.23) becomes

$$\langle (R+\bar{y})^2\varpi\rangle = R^2\langle\varpi\rangle^* + 2R\langle\varpi\bar{y}\rangle^* + \langle\varpi\bar{y}^2\rangle^* = \text{constant}. \tag{6.3.29}$$

6.3.1.2 Interaction of Coaxial Vortex Rings. The present scheme was applied to study the interaction of a pair of coaxial vortex rings. A particular example is the interaction of identical vortex rings of opposite strength in a "head-on collision" as sketched in Fig. 6.20 (left). The initial vorticity distribution at $t_0 = 1$ is given by a superposition of two nonoverlapping vortex rings with the core structure of a Lamb vortex,

$$\begin{aligned}\varpi(r,z) = {} & \frac{\Gamma_1}{\pi}\exp\left\{-[(z+Z_0)^2+(r-R_0)^2]\right\} \\ & + \frac{\Gamma_2}{\pi}\exp\left\{-[(z-Z_0)^2+(r-R_0)^2]\right\}.\end{aligned} \tag{6.3.30}$$

The length and the time scales are chosen so that the initial core size $\delta_0 = \sqrt{4\nu} = 1$. We set $\Gamma_1 = -\Gamma_2 = 16\pi$, $R_0 = 20$, and $Z_0 = 2$. The two vortex rings are of opposite strength, and their centers are located in the σz plane at $(20, \pm 2)$. The initial distance between the centerlines of the rings is equal to 4, that is, $4\delta_0$, so that they are considered to be nonoverlapping. As t increases, $R(t)$ increases. Consequently, the condition of $ma = O(m\delta) \ll R$ is realized for $t \geq 1$. Recall that $(R, Z_\perp)$ denote the locations of the maximum $|\varpi|$ associated with the two rings.

The computed trajectory of the center of the vortex ring in the left half plane $z < 0$, that is, $R(t)$ versus $-Z(t)$, is shown in Fig. 6.20 (right) along with the predictions of inviscid theory and extended asymptotic analysis. The variations of the maximum vorticity, $|\varpi|_{\max}$, predicted by these three solutions are shown in Fig. 6.21 (left) while the corresponding circulations or integrals of vorticity over the left half plane, $\Gamma^-(t)$, are plotted versus the axial displacement, $-Z(t)$, in Fig. 6.21 (right). The ratio of the circulation to its initial value, $\Gamma^-(t)/\Gamma_2$, serves as a measure of the extent of vorticity cancellation as the two vortex rings overlap each other.

For an inviscid slender vortex ring, we represent the core structure by a Lamb vortex with the core radius defined by the formula $\delta_{\text{inv}}(t) = \delta_0\sqrt{R_0/R(t)}$ to account for the stretching of the ring. The inviscid theory for the head-on collision of two vortex rings with small vortical cores predicts that (1) the circulation is conserved, $\Gamma^-(t)/\Gamma_2 = 1$, (2) the core radius $R(t)$ increases with t, (3) the distance between the rings, $2Z(t)$, decreases to 0, and (4) the maximum $|\varpi(t)|$ increases as the core size decreases because of stretching. The prediction on $|\varpi(t)|_{\max}$ is incorrect for $t > 1$. The first three predictions become erroneous for a larger t, certainly for $t > 3.5$, when

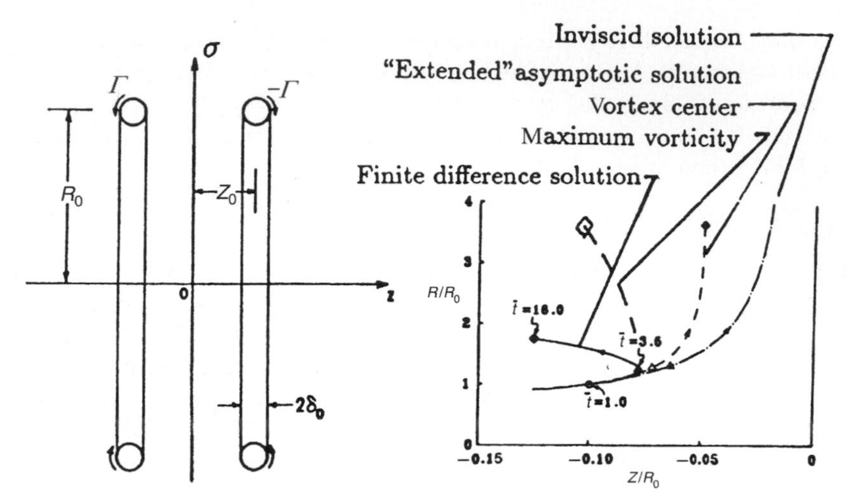

Fig. 6.20. *Left*: initial geometry for two coaxial vortex rings. *Right*: path of the colliding vortex ring. Adapted from Liu and Ting (1982)

the cancellation of vorticity in the overlapping region becomes substantial. This we see from Fig. 6.21 (right). At $t = 3.5$, the circulation given by the numerical solution is already reduced to 75% of its initial value.

For a slender vortex ring given by the extended asymptotic solution, the core structure is that of a Lamb vortex, with its core radius $\delta_{\text{asy}}(t) = \delta_0\sqrt{tR_0/[t_0R(t)]}$ to account for diffusion and stretching. The effect of diffusion dominates that of stretching so long as $R(t)/t$ decreases as t increases. This is the case for our example. As the core size decreases the maximum vorticity, $|\varpi|_{\max}$ increases as $1/\delta_{\text{asy}}$. As shown in Fig. 6.21 (left), the max-

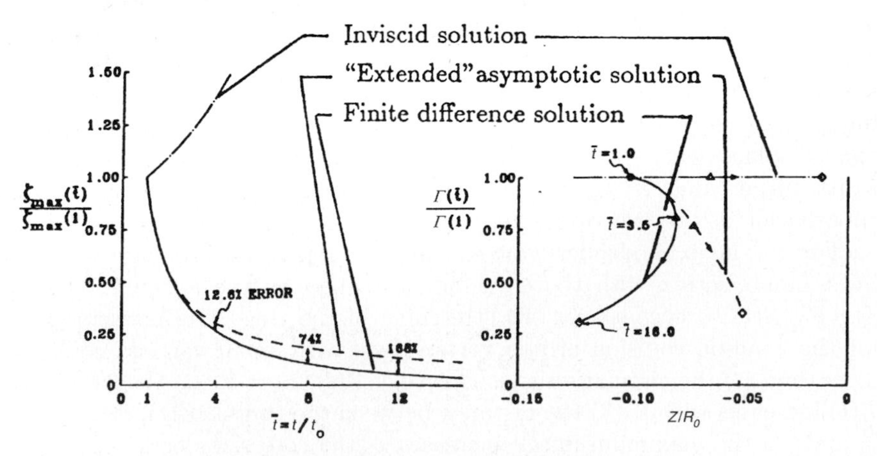

Fig. 6.21. *Left*: decay of the maximum vorticity. *Right*: decay of the circulation and the half σz plane. Adapted from Liu and Ting (1982)

imum begins to deviate from the numerical solution when $t/t_0 = 2$. The relative deviations are 12.6, 74, 168, and 233% at $\bar{t} = t/t_0 = 4, 8, 12$, and 16, respectively. The deviations represent the cumulative effects of the nonlinear convection terms that are not accounted for in the extended asymptotic solution.

In the early stage, $t \sim 1$, the rings are sufficiently far apart relative to their core size; hence, the trajectories given by the three solutions in Fig. 6.20 (right) are in good agreement. As t increases, the rings are approaching each other while the ring radius is increasing, that is, $\dot{Z}(t) < 0$ and $\dot{R}(t) > 0$. The trajectory given by the finite difference solution deviates from the other two at about $t = 3.5$, when the distance between the two rings $2Z(t)$ reaches its minimum and begins to increase. Recall that $\pm Z(t)$ denote the axial coordinates of the two maxima of $|\varpi|$. The reversal of the axial velocity $\dot{Z}(t)$ is caused by the cancellation of vorticity in the overlapping vortical cores and by the interaction of nonlinear convection terms. At $t = 3.5$, the core radius predicted by the extended asymptotic solution is already about 88% of the semidistance between the core centers. Because of the overlap of the vortical cores, the trajectory of the point of maximum vorticity, $(R(t), -Z_{\mathrm{asy}}(t))$, differs from that of the center of the ring, $(R(t), Z_{-}(t))$, in their z-coordinates and their difference increases with t. These two trajectories are shown in Fig. 6.20 (right). Note that the reversal of the trajectory of the point of maximum vorticity brings the trajectory to better agreement with that of the numerical solution at least qualitatively for $t > 3.5$. From Fig. 6.21 (right), we see that the deviation of the circulation given by the extended asymptotic solution from that of the numerical solution is within 10% even when $t = 16$. This is not surprising because the rate of the decrease of circulation is measured by the line integral of vorticity gradient along the positive σ-axis $(z = 0, \sigma \geq 0)$. There, the effect of the nonlinear convection terms on the vorticity gradient is weak initially $(t \sim 1)$, reaches a maximum, and then decreases as the points of maximum vorticity begin to move away from each other while the vorticity in the overlapping region vanishes because of cancellation. The deficiency of the extended asymptotic solution for large t lies in the inaccurate prediction of the radial velocity, $\dot{R}(t)$. An improvement for large t that accounts for the nonlinear interaction terms and corrects the radial velocity is needed. We next discuss problems without axisymmetry.

6.3.2 Numerical Modeling of Merging of Vortex Filaments

The numerical solutions of the Navier–Stokes equations modeling the merging of vortex filaments were first obtained by Chamberlain and Liu (1985) and by Chamberlain and Weston (1984). They considered the self-merging of an elliptical vortex ring with a core size comparable to the minimum radius of curvature of its centerline. They also simulated the oblique collision and merging of two and four vortex rings respectively. These merging problems were treated as problems of type I(a) with sizes of the computational domain $\mathcal{D}$ being much larger than those of the vortical regions.

Initially there are two vortex rings with equal strengths, core radii and toroidal (ring) radii. The planes containing their centerlines (ring circles) intersect each other at an angle of 45°. The angle is bisected by the zx plane and the centers of the ring circles lie on the y axis at ± 1.5 units, where a unit of length represents the toroidal radius of each ring. The form of the initial vorticity distribution, $|\Upsilon|$, in a cross section (the core) of a ring is Gaussian, and the core radii of the rings are 0.5. The initial surfaces of constant vector potential magnitude $|\mathbf{A}|$ and constant vorticity magnitude, $|\Omega|$, are shown in Figs. 6.22a and b, respectively. The constant $|\Omega|$ surfaces show the distinct features of two vortex rings with little interaction, while the constant $|\mathbf{A}|$ surfaces clearly indicate the effect of interaction. This demonstrates the fact that the vorticity decays exponentially while its induced velocity and vector potential fields decay much slower in inverse powers of the distance.

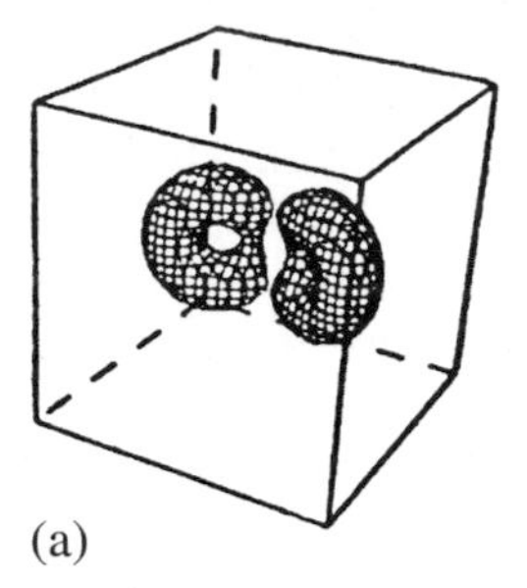

(a)

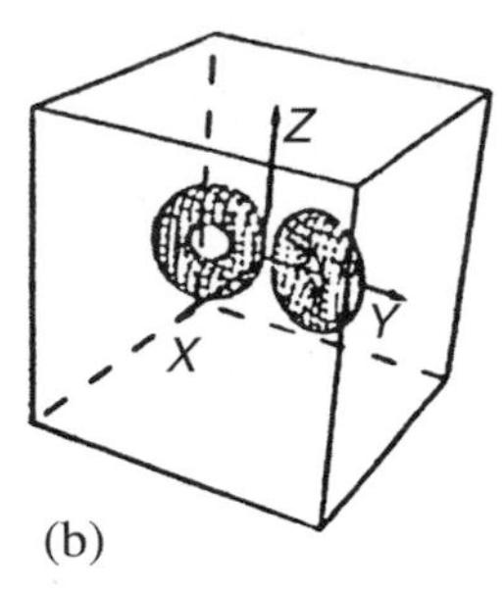

(b)

Fig. 6.22. Initial isosurfaces of (**a**) vector potential $|\mathbf{A}|$ and (**b**) vorticity $|\Omega|$. Adapted from Chamberlain and Liu (1985)

Perspective views of surfaces of constant magnitude of vector potential, $|\mathbf{A}|$, at several stages are shown in Figs. 6.23a–d. (The corresponding top, side, and front views are presented in Chamberlain and Liu (1985).) The computational domain is a cube centered about the origin with sides $8\times8\times8$, while the initial vorticity distribution is contained effectively in a box of sides $4 \times 3 \times 2$. Figures 6.23a–d show the subsequent merging of the two into a single distorted oblong ring, and the exchange of the roles of the major and minor axes of the ring. This behavior is in good qualitative agreement with experimental visualizations by Fohl and Turner (1975) and Oshima and Asaka (1975, 1977).

To achieve an accurate resolution of the merging of constant $|\Omega|$ surfaces, the above global merging problems were reconsidered by Ishii et al. (1989) as problems of type I(b) (instead of as type I(a)) because the decay length δ is much smaller than a typical size of the vorticity distribution, ℓ, by a factor of at least 4. The authors obtained the evolution of the isosurfaces of constant vorticity magnitude for the self-merging of a 4:1 elliptical vortex

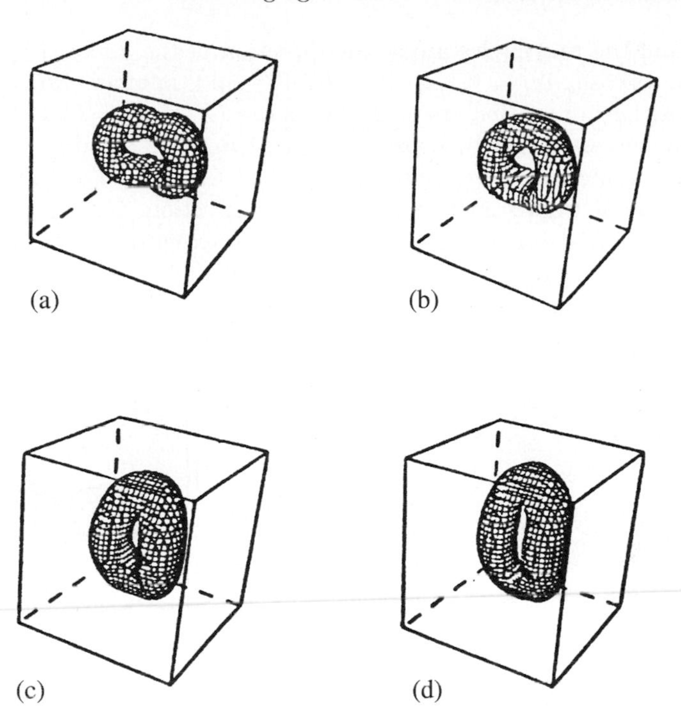

Fig. 6.23. Isosurface of vector potential magnitude at $t|\Omega_0|_{\max}$ = (**a**) 10.4, (**b**) 20.2, (**c**) 44.9, and (**d**) 60.1. Adapted from Chamberlain and Liu (1985)

ring and for the global merging of two vortex rings. The calculation used a cubic grid of 131^3 points, a second-order central difference scheme in space, and the explicit Euler method for the time evolution with $\Delta t = 0.0125$.

A total of 26^3 subdomains, with 5^3 grid points per subdomain, was employed to evaluate the values of $\mathbf{A}$ at the boundaries using a series representation of the Poisson integral. However, by taking advantage of the fact that the length and time scales of $\mathbf{A}$ on the boundary are much larger than those within the vortical region, these computations were simplified considerably. The vector potential was computed at only 6,700 points on the boundary (instead of all 67,604 points), while the values of $\mathbf{A}$ at the intermediate positions were obtained by cubic interpolation. Furthermore, the boundary data were updated only every five time steps.

Here we report the descriptions of the numerical results by Ishii et al. (1989, 1993) for the merging of two vortex rings for two different Reynolds numbers $\Gamma_0/\nu = 628$ and $\Gamma_0/\nu = 62.8$. The initial vorticity distribution represents a pair of circular vortex rings of identical shape. Their centerlines are coplanar, lying in the xy plane. The rings have the same centerline radius $R_0 = 4.0$, circulation $\Gamma_0 = 20\pi$, and a Gaussian vorticity distribution with core radius $\delta_0 = 1.0$. The direction of the circulation is chosen so that the velocities of the ring circles are in the direction of the z-axis. In case 1, the kinematic viscosity is $\nu = 0.1$, leading to a Reynolds number of $Re = 628$. The initial positions of the centers of the rings are $(X, Y, Z) = (\pm 4.0, \pm 4.0, 0.0)$. In case 2, Ishii et al. (1989) choose $\nu = 1$, and $Re = 62.8$ and the initial positions of the centers of the rings to be $(\pm 4.25, \pm 4.25, 0.0)$. The computational domain extends from -20.0 to 20.0 in each direction. Case 2 is of interest because the initial configuration simulates the experiments of Oshima and Izutsu (1988). The three views (top, front, and side) of the surfaces of constant vorticity magnitude are presented by Ishii et al. (1989). The perspective views of surfaces of constant $|\Omega|$ for case 2 are shown in Fig. 6.24.

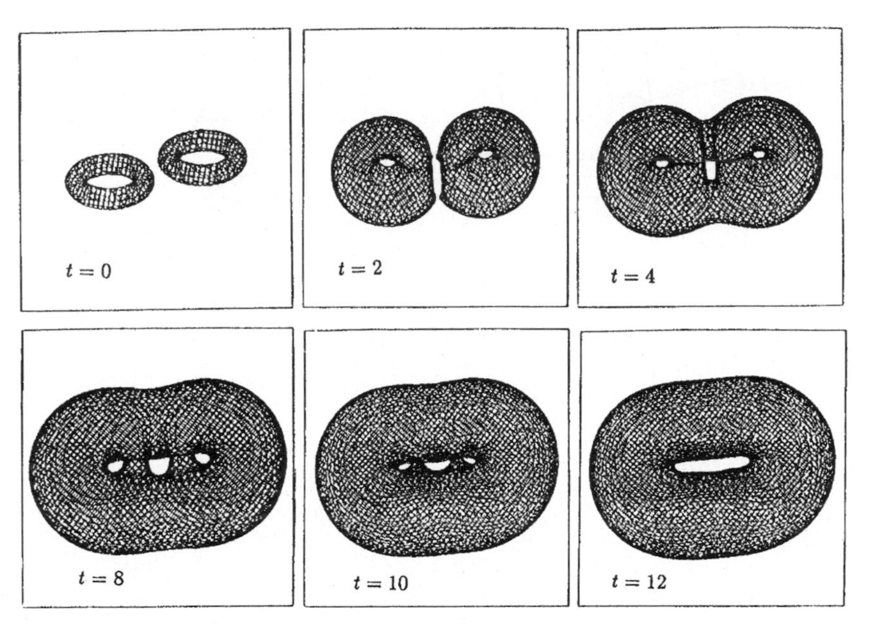

Fig. 6.24. Merging of two vortex rings, perspective views of surfaces of constant $|\Omega|$. Adapted from Ishii et al. (1989)

From the initial data, we see that the y- and z-components of vorticity, ω_2 and ω_3, are odd functions of x while ω_1 is an even function. In the yz plane, we have $\omega_2 = \omega_3 = 0$, that is, $\Omega = \omega_1 \hat{\imath}$ on $x = 0$. Therefore, a vortex line or an isosurface crossing over the yz plane has to be orthogonal to the plane. The vorticity vector in the segments of these two rings closest to each other are dominated by the components in the yz plane and are of opposite sense. The decrease of vorticity magnitude due to the mutual vorticity cancellation of the vorticity fields of these two rings are most enhanced in the region close or common to these two segments. As time increases, the rings turn toward and approach each other while their core size increases. The mean forward motion of the rings is accounted for by the translation of the computational domain.

The two initially disjoint isosurfaces of vorticity magnitude (at 30% of the maximum $|\Omega|$) begin to join across the yz plane where $|\Omega| = |\omega_1|$. We see the formation of a third hole by the isosurface in addition to the original two. Part of the vortex lines in the isosurface coming from the far side of one ring, say from $x > X$, crosses over the yz plane at a right angle and is continued or connected to the corresponding part on the $x < 0$ side. The remaining part of the vortex lines in the isosurface remains on the same side $x > 0$, is continued to the segment separating the third hole and the hole on the $x > 0$ side, and sustains the shape of a ring on the $x > 0$ side. The vorticity magnitude in the third hole is lower than the value of the isosurface because of the cancellation of the vorticity belonging to the overlapping segments of the two rings. As time increases the percentage of the vortex lines in the isosurface crossing over the yz plane increases. Eventually, all the vortex lines in the isosurface cross over the plane from one ring to the other, and the three holes formed by the isosurface merge to a single one. Thus, the isosurface with $|\Omega| = 30\%\ |\Omega|_{\max}$ gives the impression of the merging of two rings to a single ring or filament. This is the same type of behavior as those seen in the experimental vortex ring pairs by Oshima and Izutsu (1988). The corresponding experimental results are reproduced in Fig. 6.25.

To understand the essential mechanism of the reconnection process, we present a simple model that gives qualitative simulation of the merging process and helps identify the criteria for reconnection. See Ting and Bauer (1993). The model emerged from the studies of the merging and cancellations of vorticity in two-dimensional flow in Sect. 6.2. We note the following:

> In the vorticity evolution equation, (2.3.2), the temporal variation of vorticity is driven by viscous diffusion and by nonlinear convection.
>
> In the inviscid theory, that is, dropping the viscous terms, the circulation around a vortex tube is conserved. The vorticity inside the tube can intensify or diminish as the tube contracts or expands, but the cancellation of vorticity in one tube or with that in another tube will not occur.
>
> If the nonlinear convection terms are neglected, the flow is viscous while the governing equation is linear. Thus, linear superposition of vorticity fields, initially non-overlapping, would remain valid through their merging stage and beyond.

From the numerical and experimental investigations of the merging of two vortex rings as shown in Figs. 6.24 and 6.25, we see that the merging is local and takes place in a short duration; the two rings then reconnect to a single one. Before proposing a simple model simulating the phenomena of merging and reconnection, we first recall the following from the asymptotic theory of slender filaments in Chap. 3.

> The strength of a filament, circulation Γ, remains constant, that is, constant along its centerline $\mathcal{C}$, and time invariant. During the

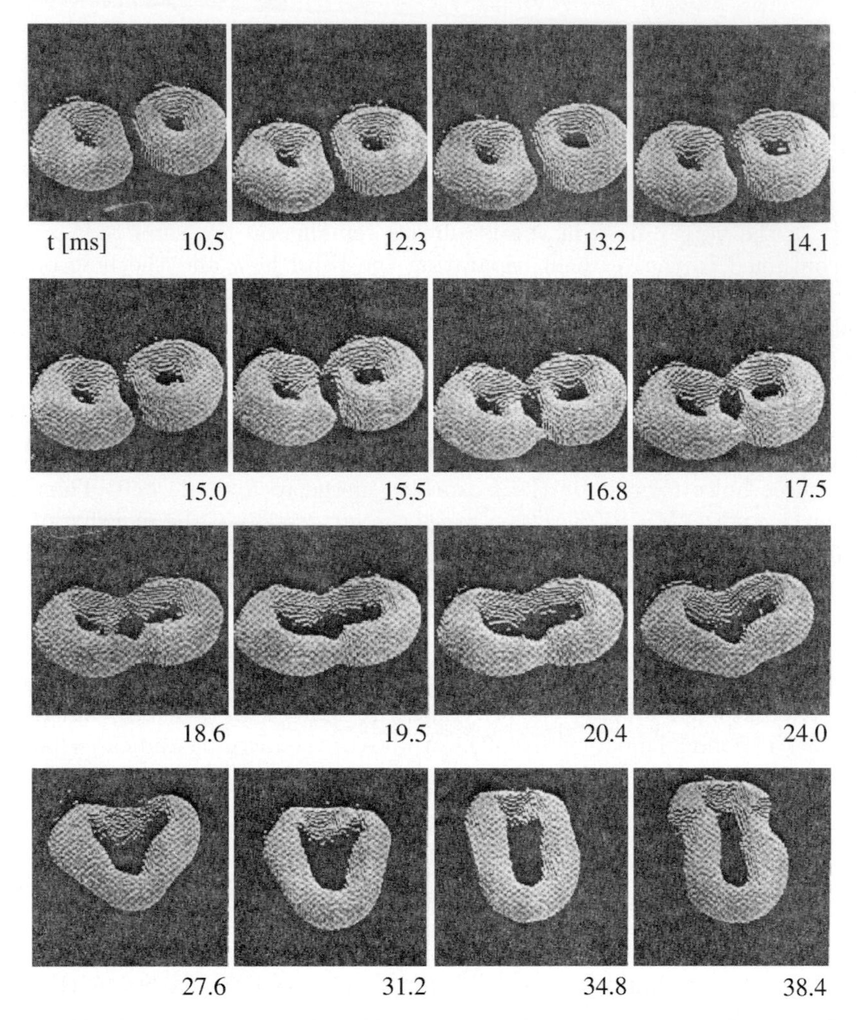

Fig. 6.25. Experimental results of the merging of two vortex rings. Adapted from Oshima and Izutsu (1988)

merging process, the asymptotic theory remains applicable to the segments outside the local merging region(s). Hence, in the post-merging period, only segments of the same circulation can be reconnected.

For a slender filament with large swirling flow and $O(1)$ axial flow, the leading-order core structure, cannot have axial variation. The vorticity field $\boldsymbol{\Omega}$ in the core has only an axial component that is, $\boldsymbol{\Omega} = |\boldsymbol{\Omega}|\tau$, where τ is the unit tangent to $\mathcal{C}$, and the vorticity magnitude $|\boldsymbol{\Omega}|$ is independent of the axial variable s along $\mathcal{C}$.

In our simple model, we identify *the superposition of the vorticity fields associated with the filaments as the essential mechanism in the merging process.*

We simulate the vorticity field in the merging region by the superposition of the asymptotic solutions for the two filaments as if they were nonoverlapping and "frozen" in time. This model is also consistent with the fact that, away from the local merging region(s), the filaments remain slender and hence retain their original structure, and that the period of merging is small relative to the diffusion time scale. We consider two slender vortex filaments of the same circulation, $\Gamma_1 = \Gamma_2 = \Gamma$, having only large swirling flow. Let the vorticity field be represented by a similarity solution, and the maximum vorticity vector be denoted by Ω_m. Then the centerline $\mathcal{C}$ is the line of maximum vorticity magnitude and is the vortex line in the scale ℓ with strength Γ_T. We define the angle of intersection of two filaments by the angle α between $\mathcal{C}_1$ and $\mathcal{C}_2$ prior to the merging stage, or the angle formed by the continuation of the vortex segments through the merging region (Fig. 6.26).

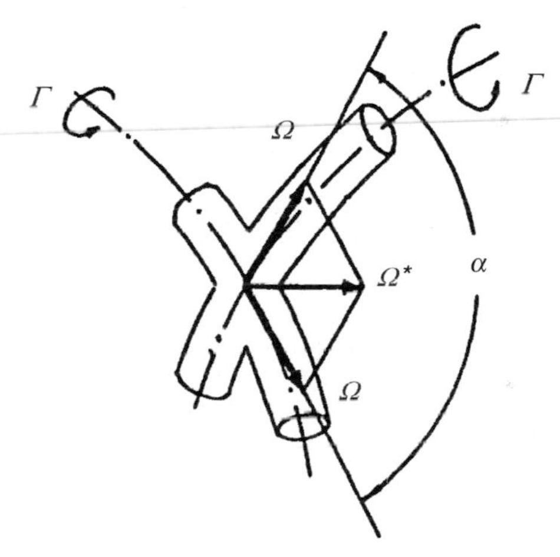

Fig. 6.26. Superposition of vorticity vectors in a local merging region. Adapted from Ting and Bauer (1993)

Since the two filaments have the same core structure, they have the same maximum vorticity magnitude $|\boldsymbol{\Omega}|_m$. Thus the magnitude of the resultant of the two vorticity vectors, given by the law of cosines, is

$$|\boldsymbol{\Omega}^*|_m = |\boldsymbol{\Omega}|_m \sqrt{2(1+\cos\alpha)} \quad \text{also} \quad \Gamma^* = 2\Gamma\sqrt{2(1+2\cos\alpha)}\,, \qquad (6.3.31)$$

and the resultant vector is along the bisector of the angle α. We see that the maximum vorticity and the total strength in the merging region are equal to that of each filament outside of the merging region only when the two filaments intersect at angle $\alpha = 2\pi/3$ and then the two segments in the merging region will become or reconnect to a single one of total strength Γ in the direction of the bisector of the angle α.

This statement seems to be consistent with the numerical results shown in Fig. 6.24, where the two filaments intersect at an angle $\approx 2\pi/3$ and reconnect by a horizontal segment in the direction of the bisector.

With the angle of intersection α as the key parameter for vorticity cancellation and reconnection of filaments, conjectures for the cases, $\alpha >$ or $<$ $2\pi/3$, that is, $\Gamma^* <$ or $> \Gamma$ can be found in Ting and Bauer (1993). We present this simple model and the conjectures with the hope that they will suggest more test cases for numerical and experimental investigations and be employed for the interpretation of the data leading to confirmation, modification, or disapproval of the conjectures.

7. Flow Generated

Motivation

Thus far we have considered only incompressible flows in detail, which are characterized by a divergence constraint for the velocity field, $\nabla \cdot \boldsymbol{v} = 0$. Here we acknowledge that this is merely an idealization justified through an asymptotic limit of the compressible flow equations for vanishing Mach number $\mathrm{M} = u_{\mathrm{ref}}/c_{\mathrm{ref}} \to 0$. Here u_{ref} is a characteristic flow velocity and c_{ref} a typical value of the speed of sound of the fluid. Thus, in low Mach number flows with $\mathrm{M} \ll 1$, sound waves propagate much faster than the fluid itself.

Multiple Space Versus Multiple Time Scales. As we shall see in the subsequent sections, this limit is singular and, depending on initial and boundary data, very different asymptotic regimes need to be distinguished. Here we walk through an intuitive discussion of how these different regimes may arise. Consider to this end Fig. 7.1. We distinguish *acoustic* perturbations, spreading at the speed of sound, $c \approx c_{\mathrm{ref}}$, from *advective* fluctuations of entropy, passive scalars, and vorticity, which are carried by the flow at speeds comparable to a typical flow velocity, u_{ref}. If a common source characterized

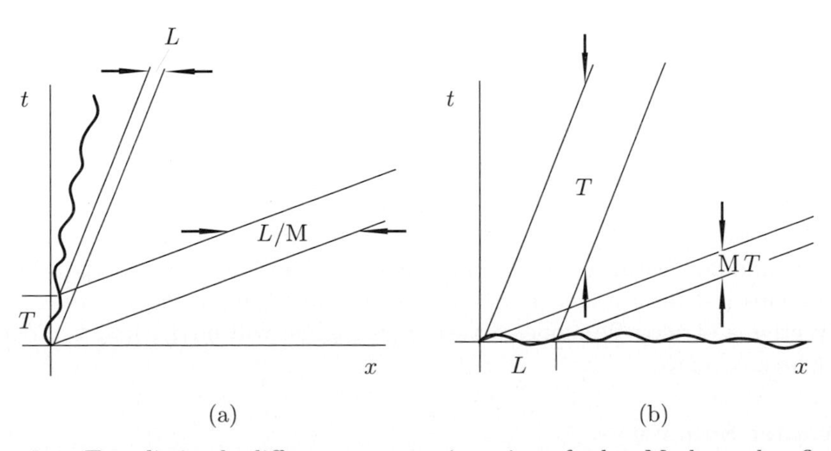

Fig. 7.1. Two distinctly different asymptotic regimes for low Mach number flows: (**a**) single time scale – multiple space scales, (**b**) multiple time scales – single space scale

by a typical time scale T, say, is driving both acoustic and advective disturbances, then the discrepancy of the signal speeds leads to spacial scale separation. Advective perturbations will acquire a characteristic length of $L \approx u_{\text{ref}} T$, whereas acoustic modes will develop much larger wavelengths of order $c_{\text{ref}} T \sim L/\mathrm{M}$. This situation is sketched in Fig. 7.1a.

In contrast, the regime sketched in Fig. 7.1b may result from initial data with a given characteristic length scale L, say, which generate acoustic as well as advective perturbations. In this setting, the signal speed differences induce a separation of time scales with the spatial scales fixed. Advective disturbances evolve on a characteristic time scale $T \approx L/u_{\text{ref}}$, whereas the acoustic modes have a short oscillation time scale of order $L/c_{\text{ref}} \approx \mathrm{M}T$.

Localized Disturbances Versus Far-Field Sound. The single time–multiple space scale regime from Fig. 7.1a has two subregimes. The first involves a spatially localized, nearly incompressible vortical flow radiating weak, long-wave acoustic waves into its far field. Typical pressure variations correspond to those arising in incompressible flows. They are of the order of ρu_{ref}^2 so that $\delta p/p_{\text{ref}} = \mathrm{O}(\mathrm{M}^2)$ as $\mathrm{M} \to 0$. *Acoustic* pressure amplitudes, however, are by one order smaller in terms of the Mach number so that $\delta p_{\text{ac}}/p_{\text{ref}} = \mathrm{O}(\mathrm{M}^3)$ as $\mathrm{M} \to 0$ in this regime. Acoustics-induced flow velocities are of the order $O(\mathrm{M}^2) u_{\text{ref}}$ only.

In the second subregime, acoustic waves instead have a leading-order effect on velocities. To estimate the order of magnitude of pressure fluctuations in this case, we consider the compatibility condition for sound wave characteristics in one-dimensional flows, that is,

$$dp \pm \rho c \, du = 0 \,, \tag{7.0.1}$$

for characteristic curves $x_\pm : \mathbb{R}^+ \to \mathbb{R};\ t \to x_\pm(t)$ with

$$\frac{dx_\pm}{dt} = u \pm c \,. \tag{7.0.2}$$

Nondimensionalizing (7.0.1), we find that

$$\frac{dp}{p_{\text{ref}}} = \mathrm{O}(\mathrm{M}) \,. \tag{7.0.3}$$

This is to be contrasted with the much weaker pressure perturbations of order $\mathrm{O}(\mathrm{M}^3)$ in the first subregime. At the same time, we allow the size of the domain featuring small-scale advective motion to be comparable to the acoustic length scale, in which case a mutual interaction between the advective and acoustic modes arises. This regime will be discussed in detail in Chap. 8 below.

Chapter Summary

In the rest of this chapter we consider various examples of sound generation by vortical flows, which belong to the first subregime.

In Sect. 7.1, we identify the solution for an incompressible vortical flow as the leading solution of a low Mach number flow and include in its far-field behavior the effect of density fluctuations induced by acoustic pressure waves. We apply the method of matched asymptotics to obtain the outer acoustic field induced by the vortical flow. The leading acoustic field is composed of five quadrupoles and a monopole. The strengths of the quadrupoles are defined by the retarded values of five linear combinations of the second moments of vorticity while the strength of the monopole is defined by the retarded value of the energy dissipation rate of the inner vortical flow. Only in the absence of sizeable energy dissipation and of energy sources do we recover Lighthill's result stating that the leading-order acoustic modes should be acoustic quadrupoles.

In Sect. 7.2 we study the vortical flows in the presence of a rigid body. We obtain explicit results in case of an obstacle having the shape of a sphere. For this case, we obtain the additional contributions to the flow field and the sound generation using the method of images. Subsection 7.2.1 recounts the analysis of Knio and Ting (1997), which extends the analysis of Weiss (1944) for a potential flow outside a sphere to that of a rotational flow. With the sphere replaced by appropriate images inside, we present formulae for the image velocity, $\boldsymbol{v}^*$ in $\mathcal{R}^3$, and the corresponding image potential, Φ^*. In Subsect. 7.2.2, we derive the far-field representation of the image potential and apply the matching procedure described in Subsect. 7.1.1 and Subsect. 7.1.2 to obtain the additional far-field sound due to the presence of the sphere.

We specialize the results for the special case of closed-loop slender vortex filaments interacting with the sphere in Subsect. 7.2.3. Finally, in Subsect. 7.2.4, we reveal the influence of viscosity and of a nonsymmetric filament–sphere interactions on the generated sound modes through several explicit computational examples. These simulations make use of the improved thin-tube numerical methods from Chap. 5.

7.1 Sound Generation by an Unsteady Vortical Flow

In this section, we study the acoustic field induced by an unsteady viscous vortical flow at low Mach number. We define the reference Mach number M by

$$M = U/C \ll 1 \, , \tag{7.1.1}$$

where C denotes the ambient speed of sound. In the vortical flow field scaled by length ℓ and velocity U, compressibility effects are of the order M^2. Therefore, the leading-order solution of the vortical flow is an incompressible viscous flow as described in Sect. 2.1 and its far-field behavior is defined by (2.3.24), (2.3.44), and (2.3.45). In the far field, the leading unsteady term of the velocity decays as $O(\ell^4/r^4)$. When r becomes much larger than ℓ by an order $O(M^{-1})$, the scaled velocity contributions decrease to $O(M^4)$. At such distances, effects of compressibility are no longer negligible. This is a typical problem amenable to matched asymptotic analysis. We consider the flow field of length scale ℓ to be the inner region, the acoustic field of length scale

$$L_a = \ell/M \tag{7.1.2}$$

to be the outer region, and we use the Mach number M as a small parameter in an asymptotic expansion and for the matching of inner and outer solutions.

Before describing the analysis, we note that sound generation due to turbulence has been studied by several authors. In pioneering work, Lighthill (1952) found that the leading-order acoustic field should be dominated by quadrupoles. Later on, Ribner (1962) pointed out that there could be an acoustic monopole due to fluid dilatation. Crow (1970a) carried out an asymptotic analysis for an inviscid *isentropic* flow and concluded that the leading-order acoustic field could be composed of only quadrupoles. Extension of Crow's analysis to sound generation by a vortical flow induced by a vorticity distribution of bounded support, in particular for vortex rings, was reported by Möhring (1978). Viscous effects were taken into account by Obermeier (1985). A different derivation of the results, essentially equivalent to those given by Möhring and Obermeier, was obtained by Ting and Miksis (1990) albeit the recent derivation is valid under the more general condition of a vorticity distribution which decays rapidly in r. Here we will highlight the analysis of Ting and Miksis (1990), because it is a follow-up of the analyses in the preceding sections for the far-field behavior of an incompressible viscous vortical flow.

Let the fluid be an ideal gas so that the ambient speed of sound is related to the thermodynamic state variables of the gas by

$$C^2 = \gamma P_0/\rho_0, \tag{7.1.3}$$

where γ denotes the ratio of specific heats and P_0 and ρ_0 stand for the ambient pressure and density. Generally, the gas is assumed to be at rest at infinity such that

$$|\boldsymbol{v}| \to 0, \quad p \to P_0 \quad \text{and} \quad \rho \to \rho_0 \qquad \text{as} \quad r \to \infty. \tag{7.1.4}$$

In Subsect. 7.1.1 we introduce expansion schemes similar to those for an inviscid turbulent flow Crow (1970a) and obtain the governing equations for the inner viscous vortical flow and for the outer acoustic field. In Subsect. 7.1.2, we study the inner flow field and identify its leading solution with that of an incompressible vortical flow as analyzed in the preceding sections. Using the far-field behavior of the vortical flow, we show that the leading unsteady terms, which all contribute to the acoustic field, are five quadrupoles defined by (2.3.45). We then study the pressure variation $(p - P_0)$ and relate the total pressure fluctuation, $\langle p - P_0\rangle_t$, to the rate of energy dissipation, $\langle\Theta\rangle$, where Θ is the dissipation function of the incompressible viscous flow as defined in (7.1.11). From the leading energy equation, we obtain the total density fluctuation, $\langle \rho \rangle_t$, which in turn yields a measure of the global dilatation effect of the next order inner solution and defines the strength of an unsteady source, or monopole, in the far field. The source strength is two orders smaller, $O(M^2)$, than that of the quadrupoles of the leading-order incompressible flow, but it decays two orders slower; $O(r^{-2})$ instead of $O(r^{-4})$. Therefore, the corresponding acoustic source and quadrupole terms are of the same order on the acoustic length scale L_a. In Subsect. 7.1.3 we show that the leading solution of the acoustic field is composed of five quadrupoles and one monopole and that their strengths are related to the retarded values of the five time-dependent linear combinations of the second moments of vorticity and to the retarded value of the rate of energy dissipation, respectively. In Subsect. 7.1.4, we apply our results on the sound generation due to unsteady vortical flows to turbulent flows and show that the strengths of the quadrupoles are defined by the moments of the mean vorticity field while the effect of turbulence appears explicitly only in the strength of the monopole. We note that only when the rate of energy dissipation is negligible, the acoustic field will be dominated by the quadrupoles in agreement with Lighthill (1952) and Crow (1970a).

7.1.1 Expansion Schemes and the Governing Equations

As mentioned in Chap. 2, in a vortical flow field the position vector $\boldsymbol{x}$ and velocity $\boldsymbol{v}$ are scaled respectively by the reference length ℓ and velocity U. Time t and vorticity $\boldsymbol{\Omega}$ are then scaled by ℓ/U and U/ℓ. For low-speed flow, the pressure variation is of the order of $\rho_0 U^2$. Therefore, the pressure variation scaled by $\rho_0 C^2$ and the density variation scaled by ρ_0 are of the order $O(M^2)$. We expand the scaled variables in the vortical flow field, the inner region, in power series in M^2 as follows:

$$\begin{aligned} \boldsymbol{v}/U &= \boldsymbol{v}^{(0)} + M^2 \boldsymbol{v}^{(1)} + \cdots, \\ \frac{p - P_0}{\rho_0 C^2} &= M^2 p^{(1)} + \cdots, \\ \frac{\rho - \rho_0}{\rho_0} &= M^2 \rho^{(1)} + \cdots \end{aligned} \tag{7.1.5}$$

This is the well-known expansion of Janzen and Rayleigh (see Sears, 1954; Majda, 1984). Since we shall not carry out the matching procedure to higher order, we can only rule out the terms linear in M but not all higher odd powers of M in the expansions in (7.1.5). Therefore, we may have to admit M^3 terms in the expansions in (7.1.5) if it is necessary for the matching of higher order terms.

In the acoustic field (the outer region), we scale $\boldsymbol{x}$ by the reference length $L_a = \ell/M$. We define the outer space variable

$$\tilde{\boldsymbol{x}} = M\boldsymbol{x}/\ell \tag{7.1.6}$$

and add a tilde to all variables in the outer region to indicate that they are functions of $\tilde{\boldsymbol{x}}$. The scaled variables are now expanded in power series of M as follows:

$$\begin{aligned}
\frac{\boldsymbol{v}}{U} &= \epsilon M^{-1}[\tilde{\boldsymbol{v}}^{(0)} + M\tilde{\boldsymbol{v}}^{(1)} + M^2\tilde{\boldsymbol{v}}^{(2)} \cdots], \\
\frac{p - P_0}{\rho_0 C^2} &= \epsilon[\tilde{p}^{(0)} + M\tilde{p}^{(1)} + M^2\tilde{p}^{(2)} + \cdots], \\
\frac{\rho - \rho_0}{\rho_0} &= \epsilon[\tilde{\rho}^{(0)} + M\tilde{\rho}^{(1)} + \tilde{\rho}^{(2)} + \cdots].
\end{aligned} \tag{7.1.7}$$

Later, ϵ shall be identified as M^3 to match with the inner solution.

After substituting the series (7.1.5) into the compressible Navier–Stokes equations and equating the coefficients of like powers of M, we obtain the governing equations for the inner solution. Since we shall deal with only the first two terms in the series (7.1.5), we shall drop the superscript (0) and replace the superscript (1) by a prime, for example, $p^{(1)}$ is replaced by p'. The leading continuity and momentum equations are the incompressible Navier–Stokes equations for $\mathbf{v}$ and p'. They are

$$\nabla \cdot \boldsymbol{v} = 0 \tag{7.1.8}$$

and

$$\boldsymbol{v}_t + \boldsymbol{v} \cdot \nabla \boldsymbol{v} = -\nabla p' + \frac{1}{Re}\Delta \boldsymbol{v}. \tag{7.1.9}$$

The leading energy equation becomes an equation for the density fluctuation ρ',

$$(p' - \rho')_t + \nabla \cdot [(p' - \rho')\boldsymbol{v}] = \frac{1}{Re\ Pr}\Delta(\gamma p' - \rho') + (\gamma - 1)\Theta, \tag{7.1.10}$$

where Pr is the Prandtl number and Θ is the dissipation function,

$$\Theta = \frac{1}{2\,Re}\sum_{j,k=1}^{3}\left(\frac{\partial v_j}{\partial x_k} + \frac{\partial v_k}{\partial x_j}\right)^2. \tag{7.1.11}$$

The next-order continuity equation shows the density variations acting as sources for the next order velocity field,

$$\nabla \cdot \mathbf{v}' = -\rho'_t - \nabla \cdot (\rho' \mathbf{v}). \tag{7.1.12}$$

Similarly, we obtain the governing equation for the first two terms of the outer solution (7.1.8). They are

$$\begin{aligned}
\tilde{\rho}_t^{(i)} &= -\nabla \cdot \tilde{\boldsymbol{v}}^{(i)}, \\
\tilde{\mathbf{v}}_t^{(i)} &= -\nabla \tilde{p}^{(i)}, \\
\tilde{p}^{(i)} &= \tilde{\rho}^{(i)},
\end{aligned} \tag{7.1.13}$$

for $i = 0, 1$, so long as $\epsilon = o(M)$. They can be reduced to a simple wave equation for $\tilde{p}^{(i)}$ or the corresponding acoustic potential $\tilde{\Phi}^{(i)}$.

7.1.2 The Vortical Flow Field

The leading inner solution of (7.1.8) and (7.1.9), subject to the initial condition (2.1.14a) and the far-field conditions of (2.1.18) and (2.1.19), is the solution of an incompressible viscous vortical flow studied in Sects. 2.1 and 2.3. We shall summarize the relevant results derived in Sect. 2.3 on the far-field behavior of the vortical field and then study the pressure and density fluctuations and the global dilatation of the inner region.

It was shown in Sect. 2.3 that the far-field velocity can be expressed as the gradient of a scalar potential, $\Phi(t, \boldsymbol{x})$, that is

$$\boldsymbol{v} = \sum_{n=1,2\ldots} \nabla \Phi^{(n)} \tag{7.1.14}$$

with $\Phi^{(n)} = O(r^{-(n+1)})$. In particular, from (2.3.44) and (2.3.45) we have,

$$\Phi^{(1)}(\boldsymbol{x}) = \frac{1}{8\pi}\boldsymbol{E} \cdot \left[\nabla \frac{1}{r}\right] \tag{7.1.15}$$

and

$$\begin{aligned}\Phi^{(2)}(t, \boldsymbol{x}) = & \frac{1}{8\pi}[\partial_2\partial_3 F_1 + \partial_3\partial_1 F_2 + \partial_1\partial_2 F_3]\left[\frac{1}{r}\right] \\ & + \frac{1}{36\pi}[\partial_1{}^2(H_2 - H_3) + \partial_2{}^2(H_3 - H_1) + \partial_3{}^2(H_1 - H_2)]\left[\frac{1}{r}\right] .\end{aligned} \tag{7.1.16}$$

We see from (7.1.15) that $\Phi^{(1)}$ represents a doublet with constant strength and orientation defined by the initial first moment vector $\boldsymbol{E}/2$, (2.3.7). From (7.1.16) we see that $\Phi^{(2)}$ represents quadrupoles with strengths given by the second moments of vorticity $\boldsymbol{F}(t)$ and $\boldsymbol{H}(t)$ defined by (2.3.8) and (2.3.9), respectively. Only the unsteady terms, that is, the quadrupoles, can generate far-field sound and the corresponding velocity potential is small of order $O(M^3)$ when $r = O(M^{-1})$.

We note that the next-order velocity field $M^2\boldsymbol{v}'$ which obeys the continuity equation (7.1.12) is no longer divergence free due to the density fluctuation, $M^2\rho'_t$, which induces an acoustic source of strength $O(M^3)$. To study the effect of the density fluctuations, we express $\boldsymbol{v}'$ as the sum of an irrotational term and a divergence-free term, that is,

$$\boldsymbol{v}' = \nabla\phi + \nabla \times \boldsymbol{A}'. \tag{7.1.17}$$

Equation (7.1.12) becomes

$$\triangle\phi = -\rho'_t - \nabla \cdot (\rho'\boldsymbol{v}). \tag{7.1.18}$$

Thus the effect of ρ'_t is contained in the irrotational term, $\nabla\phi$. By repeating what we did for the vector velocity potential $\boldsymbol{A}$ in Sect. 2.3, we can show that the far-field behavior of the divergencer-free flow, $\nabla\times\boldsymbol{A}'$, cannot have a source term, and hence the acoustic potential induced by $M^2\boldsymbol{A}'$ will be at most $O(M^4)$, that is, at least one order higher than that induced by the unsteady quadrupoles in $\boldsymbol{A}$. In Subsect. 7.1.4, we shall show that the irrotational term in (7.1.17) does induce an acoustic source and that its strength is defined by the retarded value of the total density fluctuation, which is related to the rate of energy dissipation of the incompressible flow field.

As we explained before, the far-field expansion of the vector potential $\boldsymbol{A}$, (2.3.21)–(2.3.22), remains valid even if the next higher order term for $n = 3$ is included so long as the point $\boldsymbol{x}$ lies in the inner solution, that is,

$$r = o(M^{-1}) \quad \text{but} \quad r \gg 1\,, \qquad \text{say} \qquad r = O(M^{-1/4})\,. \tag{7.1.19}$$

The far-field expansion can therefore be employed to prescribe appropriate boundary data for a numerical solution of the vector Poisson equation (2.1.13) in a finite computational domain $\mathcal{D}$ disregarding effects of compressibility, provided that its size fulfills condition (7.1.19). Thus there is no feedback from the outer acoustic field to the inner flow field. Once a numerical solution of the inner field is obtained, we can evaluate the five linear combinations of second moments of vorticity and the rate of energy dissipation. The next step is to relate the rate of energy dissipation to the total pressure and density fluctuations and then to the global dilation of the vortical flow field.

Using the far-field behavior of $\boldsymbol{v}$, we obtain from (7.1.9), $\nabla p' = -\boldsymbol{v}_t + O(r^{-6})$, and hence the far-field behavior of p' is

$$p'(t,\boldsymbol{x}) = -\Phi_t^{(2)}(t,\boldsymbol{x}) + O(r^{-4})\,. \tag{7.1.20}$$

The leading term in p' represents quadrupoles and is $O(r^{-3})$. To determine p' in the near field, we have to solve p' from (7.1.9), which reduces to a Poisson equation,

$$\triangle p' = -\nabla\cdot\big[\boldsymbol{v}\cdot\nabla\boldsymbol{v}\big] = -\sum_{i,j=1,2,3} \partial_i\partial_j[v_iv_j]\,. \tag{7.1.21}$$

The solution is expressed as the sum of the Poisson integrals of the terms on the right-hand side of (7.1.21),

$$p'(t,\boldsymbol{x}) = \sum_{i,j=1,2,3} p'_{ij} \quad \text{and} \quad p'_{ij}(t,\boldsymbol{x}) = \partial_i\partial_j\, q_{ij}(t,\boldsymbol{x})\,, \tag{7.1.22a}$$

where q_{ij} denotes the Poisson integral of the inhomogeneous term $-v_iv_j$,

$$q_{ij}(t,\boldsymbol{x}) = \frac{1}{4\pi}\int\!\!\int\!\!\int_{-\infty}^{\infty} \frac{v_iv_j}{|\boldsymbol{x}-\boldsymbol{x}'|}d\boldsymbol{x}'\,. \tag{7.1.22b}$$

For the later process of matching of the inner and outer solutions we need a particular notion of the volume integral of p'. Since p' is of order $O(r^{-3})$ in the far field, its volume integral over all of space is, in general, divergent. By making use of the far-field behavior of p' or rather p'_{ij}, we show that as

$R \to \infty$ the limit of the volume integral of p' over a sphere of radius R exists, and then define the improper volume integral of p' by the limit

$$\langle p' \rangle = \lim_{R\to\infty} \int\int\int_{\mathcal{S}} p'(t, \boldsymbol{x})\, d\boldsymbol{x} \ , \tag{7.1.23}$$

where $\mathcal{S}$ denotes the sphere $|\boldsymbol{x}| \le R$.

Let us evaluate one of the volume integrals of the p'_{ij}'s over $\mathcal{S}$, say p'_{11}. It can be reduced to an area integral,

$$\int\int\int_{\mathcal{S}} p'_{11} d\boldsymbol{x} = \int\int_{\mathcal{B}} dx_2 dx_3 \left[\partial_1 \, q_{11}\right]_{-x_1^*}^{+x_1^*} \ , \tag{7.1.24}$$

where $\mathcal{B}$ denotes the circular disc $\sigma = \sqrt{x_2^2 + x_3^2} \le R$ and $x_1^* = \sqrt{R^2 - \sigma^2}$. Note that the upper and lower points, $\boldsymbol{x}^{\pm} = (\pm x_1^*, x_2, x_3)$, lie on the spherical surface $\partial\mathcal{S}$, that is $|\boldsymbol{x}^{\pm}| = R$. From (7.1.22b), we obtain the leading term of q_{ij} in the far field by means of a power series expansion in $\boldsymbol{x}'/r$,

$$q_{ij}(t, \boldsymbol{x}) = \frac{\langle v_i v_j \rangle}{4\pi r} + O(r^{-2}) \ .$$

Equation (7.1.24) becomes

$$\begin{aligned} \int\int\int_{\mathcal{S}} p'_{11}\, dx &= -\frac{\langle v_1 v_1 \rangle}{2\pi} \int\int_{\mathcal{B}} \frac{\sqrt{R^2 - \sigma^2}\; dx_2 dx_3}{R^3} + O(R^{-1}) \\ &= -\frac{1}{3}\langle v_1 v_1 \rangle + O(R^{-1}) \ . \end{aligned} \tag{7.1.25}$$

By repeating the steps in arriving at (7.1.25) for all of the p'_{ij}'s, we obtain the volume integral of p'

$$\int\int\int_{\mathcal{S}} p'\, d\mathbf{x} = -\frac{1}{3}\langle \mathbf{v}\cdot\mathbf{v} \rangle + O(R^{-1})$$

and (7.1.23) then yields

$$\langle p' \rangle = -\frac{1}{3}\langle |\mathbf{v}|^2 \rangle \ . \tag{7.1.26}$$

Since the rate of decrease of the total kinetic energy is equal to the rate of energy dissipation, (7.1.11), we have

$$\langle p' \rangle_t = \frac{2}{3}\langle \Theta \rangle. \tag{7.1.27}$$

The density fluctuation, ρ'_t, is defined by the energy equation (7.1.10). In the far field, we get

$$\rho'_t = p'_t + O(r^{-5}), \tag{7.1.28}$$

which yields the isentropic relationship when $O(r^{-5})$ terms are neglected. As we shall see later, the fluid dilatation due to viscous dissipation induces a source term in the far field whose strength depends only on the volume integral of the density fluctuation. It is obtained by evaluating the volume integrals of both sides of (7.1.10) in the sense of (7.1.23) and by making use of the far-field behaviors of $\boldsymbol{v}$ and p' and then (7.1.27). The result is

$$\langle \rho' \rangle_t = \langle p' \rangle_t - [\gamma - 1]\langle \Theta \rangle = \frac{5 - 3\gamma}{3}\langle \Theta \rangle . \qquad (7.1.29)$$

Now we are ready to study the next order velocity field induced by the density fluctuation, that is, the velocity field associated with the scalar potential ϕ, which obeys the Poisson equation (7.1.18) and the far-field condition,

$$\nabla \phi = 0 \qquad \text{as} \quad r \to \infty. \qquad (7.1.30)$$

We call the solution of (7.1.18) and (7.1.30) Problem I. However, to describe the acoustic field, we merely need the far-field behavior of ϕ and consider a simpler problem, which is to solve (7.1.18) in the far field outside a large sphere $\mathcal{S}$ of radius $r = R \gg \ell$. This is called Problem II, for which we replace the inhomogeneous part of (7.1.18) by its far-field representation. On account of (2.2.12), (7.1.28), and (7.1.20), we reduce (7.1.18) to

$$\triangle \phi = \Phi_{tt}^{(2)} + O(r^{-4}) , \qquad \text{for} \quad r \geq R . \qquad (7.1.31)$$

Solutions are subject to the condition (7.1.30) at infinity and to an appropriate boundary condition on the spherical surface $\partial\mathcal{S} : r = R$ consistent with the leading far-field solution of Problem I.

We solve Problem II in two steps. First, we construct a particular solution $\eta_0(t, \mathbf{x})$ satisfying the above Poission equation without the $O(r^{-4})$ term and the condition (7.1.30). Second, we derive an appropriate Neumann condition on $\partial\mathcal{S}$ from the leading term of the normal velocity of a solution of Problem I. The leading term represents the average mass flux density through $\partial\mathcal{S}$, namely $m/(4\pi R^2)$. Next we construct the homogeneous solution, ξ, of (7.1.31) outside of $\mathcal{S}$ so that $\xi + \eta_0$ satisfies the Neumann condition.

Since the total flux of η_0 across $\partial\mathcal{S}$ turns out to be zero, it follows that ξ solves the inhomogeneous equation $\triangle\xi = m\delta(\boldsymbol{x})$ for all $\boldsymbol{x}$. The source strength m is related to the volume integral of the right-hand side of (7.1.18) over $\mathcal{S}$, that is, to the total density fluctuation. Thus, the contributions, ξ and η_0, of the leading far-field potential ϕ reflect the overall fluid dilatation and the quadrupole pressure field, respectively.

By carrying out the aforementioned steps, we obtain the following results:

$$\phi = \xi + \eta_0 + O(r^{-2}) , \qquad r \geq R , \qquad (7.1.32)$$

where

$$\eta_0 = -[r^3 \Phi_{tt}^{(2)}]/[6r] \qquad (7.1.33)$$

and

$$\xi(t, \boldsymbol{x}) = -\frac{m(t, R)}{4\pi r} \left[1 + O\left(\frac{R}{r}\right)\right] . \tag{7.1.34}$$

Here, $m(t, R)$ is the total flux of ξ through $\partial\mathcal{S}$,

$$m(t, R) = \int\int_{\partial\mathcal{S}} \phi_r(t, \boldsymbol{x}')\, d\Sigma + O(R^{-1}), \tag{7.1.35}$$

where $d\Sigma$ denotes the surface area element on $\partial\mathcal{S}$. The derivation of (7.1.31)–(7.1.35) is given in Ting and Miksis (1990).

Using the divergence theorem, (7.1.18) and (7.1.29), we convert (7.1.35) to

$$\begin{aligned} m(t, R) &= -\int\int\int_{\mathcal{S}} \rho_t'(t, \boldsymbol{x}) d\boldsymbol{x} + O(1/R) \\ &= \frac{3\gamma - 5}{3} \langle\Theta\rangle + O(1/R). \end{aligned} \tag{7.1.36}$$

From (7.1.32), (7.1.34), and (7.1.36), we get the far-field behavior of ϕ. It is

$$\phi(t, \boldsymbol{x}) = -\frac{\bar{m}(t)}{4\pi r} + \eta_0 + O\left(\frac{R}{r^2}\right) \tag{7.1.37}$$

where

$$\bar{m} = \frac{3\gamma - 5}{3} \langle\Theta\rangle = m(t, R)|_{R\to\infty} . \tag{7.1.38}$$

In deriving (7.1.37)–(7.1.38), we have assumed that $r \gg R \gg \ell = 1$. For example, if $r/\ell = R^2/\ell^2 \gg 1$, then the last term in (7.1.37) is $O(r^{-3/2})$. This is stronger than $O(r^{-1})$, which is all we need; the last term should be much smaller than the first two terms, both of which are $O(r^{-1})$.

Finally, we obtain the far-field behavior of the vortical flow, including the second-order solution, $O(M^2)$,

$$\boldsymbol{v} = \nabla\Phi + M^2\nabla\phi + O(M^2 r^{-3}) , \tag{7.1.39a}$$

with

$$\Phi = \Phi^{(1)}(\mathbf{x}) + \Phi^{(2)}(t, \mathbf{x}) + O(r^{-4}) , \tag{7.1.39b}$$

$$M^2\phi = -\frac{M^2\bar{m}(t)}{4\pi r} + M^2\eta_0(t, \mathbf{x}) + O\left(\frac{M^2}{r^2}\right) \tag{7.1.39c}$$

and

$$\eta_0 = -r^3\Phi_{tt}^{(2)}/(6r) = O(r^{-1}) . \tag{7.1.39d}$$

In terms of the outer variable $\tilde{r}$, see (7.1.6), the orders of magnitude of the first two terms in Φ and $M^2\phi$ are

$$\Phi^{(1)} = O(M^2\tilde{r}^{-2}) = O(M^2) , \qquad \Phi^{(2)} = O(M^3)$$

and

$$M^2 r^{-1} = O(M^3) , \qquad M^2 \eta_0 = O(M^3) .$$

The first term of Φ representing a doublet is $O(M^2)$ but is independent of t and hence does not contribute to the acoustic field. The second term of Φ and the first two terms of $M^2\phi$ are of the same order and generate an acoustic pressure field of $O(M^3)$. Thus ϵ in (7.1.8) can be equated to M^3. In the next section, we shall match the inner and outer solutions and see that the terms $M^2\Phi^2$ and $M^2\eta_0$ are matched with the first two terms of acoustic quadrupoles, while the term $M^2\bar{m}/(4\pi r)$ is matched with the leading term of an acoustic monopole.

7.1.3 The Acoustic Field

To prepare for the matching of the outer solution, governed by the simple wave equation, with the inner solution (7.1.39a–d), we find the behavior of an acoustic monopole and quadrupole for $\tilde{r} \ll 1$.

For an acoustic monopole of strength $M^3\bar{m}(t)$ at the origin, the acoustic potential behaves as

$$\tilde{\phi}(\tau, \tilde{\boldsymbol{x}}) = -\frac{M^3\bar{m}(\tau)}{4\pi\tilde{r}} = -\frac{M^3\bar{m}(t)}{4\pi\tilde{r}} + \frac{M^3\bar{m}'(t)}{4\pi} + O(M^3\tilde{r}) \quad \text{for} \quad \tilde{r} \ll 1 . \tag{7.1.40}$$

Here $\tau = t - \tilde{r}$ is the retarded time. In terms of the inner variable r, the right-hand side becomes a power series in M. The first term, $O(M^2)$, matches with the potential source in the second-order inner solution $M^2\phi$. The second term, $O(M^3)$, has no contribution to the velocity field since it is independent of $\boldsymbol{x}$.

For a quadrupole at the origin, of strength $M^3 q_{ij}(t)$ and orientation ij, the acoustic potential behaves as

$$\begin{aligned}\frac{M^3}{4\pi}\tilde{\partial}^2_{ij}\left[\frac{q_{ij}(\tau)}{\tilde{r}}\right] &= \frac{M^3 q_{ij}(t)}{4\pi}\tilde{\partial}^2_{ij}\frac{1}{\tilde{r}} \\ &\quad + \frac{M^3 q''_{ij}(t)}{24\pi}\left[\frac{2\delta_{ij}}{\tilde{r}} - \tilde{r}^2\tilde{\partial}^2_{ij}\frac{1}{\tilde{r}}\right] + O(M^3\tilde{r}^0) ,\end{aligned} \tag{7.1.41a}$$

where $\tilde{\partial}_i$ stands for partial derivative with respect to $\tilde{x}_i$. In terms of the inner variable r, (7.1.41a) becomes

$$\begin{aligned}\frac{M^3}{4\pi}\tilde{\partial}^2_{ij}\left[\frac{q_{ij}(\tau)}{\tilde{r}}\right] &= \frac{q_{ij}(t)}{4\pi}\partial^2_{ij}\frac{1}{r} \\ &\quad + \frac{M^2 q''_{ij}(t)}{24\pi}\left[\frac{2\delta_{ij}}{r} - r^2\partial^2_{ij}\frac{1}{r}\right] + O(M^3) .\end{aligned} \tag{7.1.41b}$$

The first term on the right-hand side, which is $O(1)$, matches with the potential of the quadrupole of orientation ij in the leading unsteady inner solution

$\Phi^{(2)}$. The second term, which is $O(M^2)$, will match with a term in $M^2\phi$ for large r.

We note once more that the first term of Φ in (7.1.39a–d) represents a doublet and its strength is independent of time. Therefore, the corresponding acoustic potential $M^2\tilde{\Phi}^{(1)}(\tilde{\boldsymbol{x}})$ is an analytic continuation of the inner solution $\Phi^{(1)}(\boldsymbol{x})$ with $\boldsymbol{x}$ replaced by $M^{-1}\tilde{\boldsymbol{x}}$, and it does not generate sound because $\tilde{\Phi}_t^{(1)} = 0$.

To match with the quadrupole of the inner solution in $\Phi^{(2)}(t,\boldsymbol{x})$, of (7.1.16), we take note of (7.1.41a,b) and define the corresponding acoustic potentials by replacing the variable t in the strengths $\boldsymbol{F}(t)$ and $\boldsymbol{H}(t)$ by the retarded time τ and the inner variable $\boldsymbol{x}$ in $\Phi^{(2)}$ by the outer variable $\tilde{\boldsymbol{x}}/M$, that is,

$$\tilde{\Phi}(\tau,\tilde{\boldsymbol{x}}) = \Phi^{(2)}(\tau,\tilde{\boldsymbol{x}}/M). \tag{7.1.42a}$$

Using (7.1.41a) again and observing that $q_{11}+q_{22}+q_{33} = 0$, we arrive at the behavior of $\tilde{\Phi}$ for $\tilde{r} \ll 1$,

$$\tilde{\Phi}(\tau,\tilde{\mathrm{x}}) = \tilde{\Phi}(t,\tilde{\mathrm{x}}) - \frac{1}{6}\tilde{r}^2\tilde{\Phi}_{tt}(t,\tilde{\mathrm{x}}) + O(\tilde{r}^0). \tag{7.1.42b}$$

From (7.1.16), (7.1.42a), and (7.1.39b–d), we see that the first term on the right-hand side of the above equation is $O(\tilde{r}^{-3})$ and is equal to the first unsteady term $\Phi^{(2)}$ in Φ, whereas the second term is $O(\tilde{r}^{-1})$ and is equal to the second term $M^2\eta_0$ in $M^2\phi$. Since the first term of the acoustic monopole for $\tilde{r} \ll 1$ in (7.1.40) matches with the corresponding term in the inner solution, the first term of $M^2\phi$ in (7.1.39c), we conclude that

- *the leading acoustic pressure is of order $O(M^3)$ and is composed of the contributions from the acoustic quadrupoles $\tilde{\Phi}$ and a monopole $\tilde{\phi}$. It is*

$$\frac{p-P_0}{\rho_0 U^2} \approx M^3\tilde{p}^{(0)} = \tilde{\Phi}_t + \tilde{\phi}_t\,. \tag{7.1.43}$$

The strengths of the acoustic quadrupoles in the first term on the right-hand side are related through (7.1.16) and (7.1.39a–d) to the retarded values of the five linear combinations of second moments of vorticity, F_1, F_2, F_3, H_1, and H_2, which are defined by (2.3.8) and (2.3.9). The strength of the monopole $\bar{m}$ in the second term is related to the retarded value of the rate of energy dissipation by (7.1.38). This defines the leading acoustic field in terms of the far-field behavior of the viscous vortical flow.

7.1.4 Sound Generation due to Turbulence

It is well known that a turbulent flow is a special class of unsteady flow for which the initial vorticity or velocity is not available or not precisely specified but the flow field has distinct length scales: a length scale ℓ for the mean flow and much smaller multiple scales for the eddies. If the mean flow is unsteady, the initial data for the mean flow, say the mean vorticity distribution in the length scale ℓ, have to be prescribed. Nevertheless, the results obtained above for an unsteady flow remain valid for a turbulent flow. The question is how to obtain the corresponding results for the mean flow alone from those for the resultant flow, which is the sum of the mean flow and the turbulent eddies. Let us use bar and check to denote the mean and the turbulent part respectively. For example, we write for the velocity and vorticity,

$$\boldsymbol{v} = \bar{\boldsymbol{v}} + \check{\boldsymbol{v}} \qquad \text{and} \qquad \boldsymbol{\Omega} = \bar{\boldsymbol{\Omega}} + \check{\boldsymbol{\Omega}} \,. \tag{7.1.44}$$

The volume integral of the pressure fluctuation is equal to that of the mean pressure fluctuation, that is, $\langle p_t' \rangle = \langle \bar{p}_t' \rangle$. The same holds for the density fluctuation. For the rate of energy dissipation, we have

$$\langle \Theta \rangle = \langle \Theta_0 \rangle + \langle \Theta_1 \rangle \tag{7.1.45}$$

with

$$\Theta_0 = \frac{1}{Re} \sum_{j,k=1}^{3} \left(\frac{\partial \bar{v}_j}{\partial x_k} + \frac{\partial \bar{v}_k}{\partial x_j} \right)^2 \qquad \text{and} \qquad \Theta_1 = \frac{1}{Re} \sum_{j,k=1}^{3} \left(\frac{\partial \check{v}_j}{\partial x_k} + \frac{\partial \check{v}_k}{\partial x_j} \right)^2 . \tag{7.1.46}$$

The relationships (2.3.8) and (2.3.9) between the strengths of the quadrupoles and the second moments of vorticity remain valid. Since the latter are linear in Ω, the moments of vorticity are equal to the moments of the mean vorticity. Thus we replace ω_i by $\bar{\omega}_i$ in (2.3.8) and (2.3.9). The diffusion equation for the mean vorticity is obtainable from the momentum equation for the mean velocity when model equations relating the Reynolds stresses to the velocity gradient are introduced (see, e.g., Schlichting, 1978).

The strength of the source $\bar{m}$ is related by (7.1.38) to the rate of energy dissipation in which the effect of turbulent eddies appears directly as the second term on the right-hand side of (7.1.45). An additional turbulence modeling equation is needed to determine the second term, $\langle \Theta_1 \rangle$. Of course, when the rate of energy dissipation of the eddies and that of the average flow are negligible, the acoustic field induced by a turbulent field will be dominated by acoustic quadrupoles induced by the mean flow.

7.1.5 Sound Generation by Vortex Filaments

We now collect the formulae presented in Subsect. 7.1.3 for the sound generation induced by a vortical field. The leading-order acoustic field is composed of an acoustic monopole (7.1.40) and quadrupoles (7.1.42b).

The strength of the acoustic quadrupoles is related through (7.1.16) and (7.1.39a)–(7.1.39d) to the retarded values of the five linear combinations of second moments, F_1, F_2, F_3, H_1, and H_2, which are defined by (2.3.8) and

(2.3.9). These five moments for a slender filament with centerline $\mathcal{C}: \boldsymbol{X}(t,s)$ are

$$F_i(t) = \Gamma \boldsymbol{\imath} \cdot \int_{\mathcal{C}} [X_j^2 - X_k^2]\, d\boldsymbol{X} + O(\epsilon^2) \tag{7.1.47}$$

and

$$H_i(t) = \Gamma \int_{\mathcal{C}} [2X_j X_k \boldsymbol{\imath} - X_k X_i \boldsymbol{\jmath} - X_i X_j \hat{k}] \cdot d\boldsymbol{X} + O(\epsilon^2) \,, \tag{7.1.48}$$

for $i = 1, 2, 3$, and i, j, k in cyclic order. We also note the identity, $H_1 + H_2 + H_3 = 0$. Here we used the fact that the core size is $O(\epsilon\ell)$ and hence the contributions of the second moments relative to $\mathcal{C}$ are $O(\epsilon^2)$. Therefore, the strengths of the quadrupoles depend explicitly only on the motion of the filament centerline(s).

From (7.1.40), one observes that the strength of the monopole is related to the retarded value of the strength of the total volume dilatation of the vortical field, or the strength of the equivalent source term in the far-field representation of the vortical flow. The strength in turn is related to the rate of energy dissipation Θ defined by (7.1.38). For a slender filament with leading-order core structure independent of the polar coordinate θ and the axial parameter s, the strength of the monopole is the retarded value of

$$\frac{3\gamma - 5}{3} \langle \Theta \rangle = \frac{3\gamma - 5}{3Re} \langle |\Omega|^2 \rangle \,, \tag{7.1.49}$$

with

$$\langle |\Omega|^2 \rangle = 2\pi S^{(0)}(t) \int_0^\infty \left[\left(\zeta^{(0)} \right)^2 + \left(w_{\bar{r}}^{(0)} \right)^2 \right] \bar{r} d\bar{r} \,. \tag{7.1.50}$$

The strength of the monopole depends only on the core structure and the length of the filament. From (7.1.40), (7.1.42b), and (7.1.47) to (7.1.50), we conclude that

- *The motion of the filament(s) creates quadrupole noise while the evolution of the core structure generates monopole noise.*

When the filament has a similar core structure, (3.3.184), (3.3.185), and (7.1.50) reduce to

$$\langle |\Omega|^2 \rangle = \frac{\Gamma^2 S^{(0)}(t)}{\sqrt{2}\bar{\delta}^2(t)} + \frac{m(0)^2 S_0^4}{\sqrt{2}[S^{(0)}]^3} \,. \tag{7.1.51}$$

We see that the contribution of the core structure to the acoustic monopole is defined by three parameters: the circulation Γ, the initial axial flux $m(0)$, and the core size $\delta(0) = \bar{\delta}(0)\epsilon$. It was noted before that accurate initial data for the vortical core other than the above three global quantities are very hard to obtain. We could assume that the core structure can be approximated by a similarity solution, and then use acoustic measurements of the strength of the monopole to determine the initial core size. The difference between the theoretical predictions and the experimental data on the motion of the filament and the acoustic field would give a measure of the error in the initial axial mass flux and the deviation of the true core structure from the assumed similarity solution.

7.2 Vortical Flow Outside a Sphere and Sound Generation

In the preceding sections of this chapter, we considered unsteady vortical flows in free space. In this section, we shall study the vortical flows in the presence of a body. To show analytically the difference of the flow field from that in free space, we consider the body to be a sphere, for which the additional contributions to the flow field and the sound generation are obtained by the method of images.

In Subsect. 7.2.1 we describe the analysis of Knio and Ting (1997), which is an extension of the analysis of Weiss (1944) for a potential flow outside a sphere to that of a rotational flow. With the sphere replaced by appropriate images inside, we present formulae for the image velocity, $\boldsymbol{v}^*$ in $\mathcal{R}^3$, and the corresponding image potential, Φ^*. In Subsect. 7.2.2, we derive the far-field representation of the image potential and apply the matching procedure described in Subsects. 7.1.1 and 7.1.2 to obtain the additional far-field sound due to the presence of the sphere.

7.2.1 The Image of a Rotational Flow due to the Presence of a Sphere

We study an incompressible inviscid velocity field $\boldsymbol{V}(\boldsymbol{x})$ induced by a vorticity field $\boldsymbol{\Omega}(\boldsymbol{x})$ outside of a rigid sphere, $\mathcal{S}$, $r = |\boldsymbol{x}| = a$. The independent variable t shall be shown explicitly later when we discuss the acoustic field. The velocity $\boldsymbol{V}$ is expressed as the sum of an image velocity $\boldsymbol{v}^*$ and the velocity $\boldsymbol{v}$ in $\mathcal{R}^3$. The latter is induced by the same vorticity field $\boldsymbol{\Omega}$ outside the sphere $\mathcal{S}$ and the extension, $\boldsymbol{\Omega} = 0$ inside. That is

$$\boldsymbol{V}(\boldsymbol{x}) = \boldsymbol{v}(\boldsymbol{x}) + \boldsymbol{v}^*(\boldsymbol{x}) , \tag{7.2.1}$$

in $\mathcal{R}^3$ with

$$\nabla \times \boldsymbol{v} = \boldsymbol{\Omega} , \qquad \nabla \times \boldsymbol{v}^* = 0 , \tag{7.2.2}$$

and zero radial velocity,

$$\hat{\boldsymbol{x}} \cdot \boldsymbol{V} = 0 , \qquad \text{on } \mathcal{S},\ |\boldsymbol{x}| = a . \tag{7.2.3}$$

In addition to the assumption that $\boldsymbol{\Omega}$ is of bounded support or decays exponentially in $|\boldsymbol{x}|$, we have

$$\nabla \times \boldsymbol{v} = \boldsymbol{\Omega} = 0\ , \quad \text{for} \quad |\boldsymbol{x}| \le a\ . \tag{7.2.4}$$

The velocity $\boldsymbol{v}$ is expressed as $\nabla \times \boldsymbol{A}$ with the vector potential $\boldsymbol{A}$ related to the vorticity field $\boldsymbol{\Omega}$ by the Poisson integral (2.3.4).

We now quote the formula of Weiss (1944) for the image of a potential flow induced by a rigid sphere, $\mathcal{S}$, and recast it to a form applicable to a rotational flow outside $\mathcal{S}$.

Let $\Phi_0(\boldsymbol{x})$ be a velocity potential in $\mathcal{R}^3$ with singularities outside the sphere $\mathcal{S}$. Then the potential solution Φ due to the presence of the sphere, $\mathcal{S}$, can be written as

$$\Phi(\boldsymbol{x}) = \Phi_0(\boldsymbol{x}) + \Phi^*(\boldsymbol{x})\ , \tag{7.2.5}$$

with $\boldsymbol{v} = \nabla\Phi_0$ and $\boldsymbol{v}^* = \nabla\Phi^*$. The formula relating the image potential, Φ^*, to Φ_0 is

$$\Phi^*(\boldsymbol{x}) = \frac{a}{r}\Phi_0\left(\frac{a^2\boldsymbol{x}}{r^2}\right) - \frac{2}{ar}\int_0^a \lambda\Phi_0\left(\frac{\lambda^2\boldsymbol{x}}{r^2}\right) d\lambda\ , \tag{7.2.6}$$

or

$$\Phi^*(r,\theta,\phi) = \frac{a}{r}\Phi_0(\bar{r},\theta,\phi) - \frac{1}{a}\int_0^{a^2/r} \Phi_0(r',\theta,\phi)dr'\ , \tag{7.2.7}$$

where (r,θ,ϕ) denote the spherical coordinates of $\boldsymbol{x}$ and $\bar{r} = a^2/r\langle a \text{ for } r\rangle a$, Φ^* fulfills the requirements:

{1} Φ^* is regular and harmonic for $r \geq a$.
{2} $\Phi^* = O(r^{-2})$ for large r.
{3} Φ^* satisfies the boundary condition, (7.2.4), that is, $\partial_r(\Phi^* + \Phi_0) = 0$ on $\mathcal{S}$.

The radial derivative of (7.2.7) yields

$$\partial_r\Phi^*(r,\theta,\phi) = -[a/r]^3\partial_{\bar{r}}\phi_0(\bar{r},\theta,\phi) \quad \text{or} \quad u^*(P) = -[a/r]^3u(Q)\ , \tag{7.2.8}$$

where u^* and u denote respectively the radial component of $\boldsymbol{v}^*$ and $\boldsymbol{v}$, and point $Q(\bar{r},\theta,\phi)$ inside $\mathcal{S}$ is the mirror image of point $P(r,\theta,\phi)$ outside. Integrating (7.2.8) with respect to r using {2}, we recover the formula given by Keller and Rubinow (1971)

$$\Phi^*(r,\theta,\phi) = \frac{1}{a}\int_0^{\bar{r}} \partial_{r'}\Phi_0(r',\theta,\phi)\ r'dr'\ . \tag{7.2.9}$$

Equation (7.2.8) and its equivalent (7.2.9) say that

{4} The image potential Φ^* and radial velocity u^* outside the sphere depend only on the radial velocity $u = \partial_r\Phi_0$ inside.

Now we need to show that (7.2.9) is applicable to a rotational flow outside $\mathcal{S}$. Because of (7.2.2), the image velocity is irrotational and hence there is an image potential Φ^* such that $\boldsymbol{v}^* = \nabla\Phi^*$. Because of (7.2.4), the velocity field $\boldsymbol{v}$ is irrotational inside $\mathcal{S}$ and hence there exists a velocity potential Φ_0 such that $\nabla\Phi_0 = \boldsymbol{v}$. Thus (7.2.8) and hence (7.2.9) are applicable to a rotational flow outside $\mathcal{S}$.

With $\boldsymbol{v} = \nabla \times \mathbf{A}$ and the Poisson integral (2.3.4), we relate the radial component of $\boldsymbol{v}$ to the vorticity field $\boldsymbol{\Omega}$,

$$u(r,\theta,\phi) = \frac{1}{4\pi}\iiint_{R\geq a} \frac{\hat{r}\cdot\boldsymbol{y}\times\boldsymbol{\Upsilon}(\boldsymbol{y})}{|\boldsymbol{x}-\boldsymbol{y}|^3}\, d\mathcal{V}\ , \tag{7.2.10}$$

where $\hat{\boldsymbol{y}}$ denotes the unit vector of $\boldsymbol{y}$ and $\boldsymbol{\Upsilon}$ denotes the transverse vorticity vector, the vector normal to $\boldsymbol{y}$,

$$\boldsymbol{\Upsilon}(\boldsymbol{y}) = \boldsymbol{\Omega} - (\boldsymbol{\Omega}\cdot\hat{\boldsymbol{y}})\hat{\boldsymbol{y}}\ . \tag{7.2.11}$$

In (7.2.10), the dependence of u on r appears implicitly via the distance,

$$|\boldsymbol{x}-\boldsymbol{y}| = [R^2 - 2rR\mu + r^2]^{1/2}\ , \tag{7.2.12}$$

where $R = |\boldsymbol{y}|$, $\mu(\theta,\phi) = \cos\beta = \hat{r}(\theta,\phi)\cdot\hat{\boldsymbol{y}}$, and β denotes the angle between the position vectors $\boldsymbol{x}$ and $\boldsymbol{y}$.

Using (7.2.10) for u inside $\mathcal{S}$, (7.2.9) yields the image potential outside $\mathcal{S}$,[1]

$$\Phi^*(r,\theta,\phi) = \iiint_{R\geq a}\left[\frac{G(\lambda,\mu)\ \hat{r}\times\hat{\boldsymbol{y}}}{4\pi a}\right]\cdot\boldsymbol{\Upsilon}\, d\mathcal{V}\ , \tag{7.2.13}$$

where

$$G(\lambda,\mu) = \int_0^\lambda \frac{\lambda'\, d\lambda'}{Z^3(\lambda',\mu)} = \frac{\lambda^2}{Z[Z+1-\lambda\mu]}\ , \tag{7.2.14}$$

and

$$Z(\lambda,\mu) = [1 - 2\lambda\mu + \lambda^2]^{1/2} \quad \text{and} \quad \lambda = \bar{r}/R = a^2/(rR)\ . \tag{7.2.15}$$

Note that G is an elementary function of λ and μ, which depends only on $R = |\boldsymbol{y}|$, $r = |\boldsymbol{x}|$ and the angle β between these two position vectors. Thus, the image potential outside $\mathcal{S}$ is a volume integral over the support of the vorticity field with integrand regular. From (7.2.13), we observe that

{5} The image potential Φ^* is independent of the radial vorticity component. In the integrand of the volume integral in (7.2.13), the transverse component of the vector inside the square bracket can be interpreted as the image potential of a vortex element of unit strength induced by the sphere.

[1] It is obvious that $\Phi^* \equiv 0$ when the radius $a = 0$ and we recover the vortical flow in free space, as expected.

To obtain the resultant velocity $\boldsymbol{v}+\boldsymbol{v}^*$, with the vortical flow $\boldsymbol{v}$ usually expressed in Cartesian components, we present formulae for the image velocity also in Cartesian components, $v_j^* = \partial_{x_j}\Phi^*$, for $j = 1, 2, 3$. They are

$$v_j^* = \frac{-1}{4\pi a}\iiint_{R\geq a} d\mathcal{V}\ \{G\ \partial_{x_j}\hat{r} + [\partial_\lambda G\ \partial_{x_j}\lambda + \partial_\mu G\ \partial_{x_j}\mu]\ \hat{r}\}\cdot[\hat{\boldsymbol{y}}\times\boldsymbol{\Omega}(\boldsymbol{y})]\ , \tag{7.2.16}$$

where

$$\partial_\lambda G = \lambda/Z^3(\lambda,\mu)\ , \qquad \partial_{x_j}\lambda = -a^2 x_j/(Rr^3)\ , \tag{7.2.17a}$$

$$\partial_\mu G = \frac{\lambda^3[Z^2+2Z+1-\lambda\mu]}{Z[Z^2+Z(1-\lambda\mu)]^2}\ , \tag{7.2.17b}$$

$$\partial_{x_j}\hat{r} = \partial_{x_j}\left(\frac{\boldsymbol{x}}{r}\right) = \frac{\hat{j}}{r} - \frac{x_j\boldsymbol{x}}{r^3}\ , \quad \text{and} \quad \partial_{x_j}\mu = \hat{\boldsymbol{y}}\cdot\partial_{x_j}\hat{r}\ . \tag{7.2.17c}$$

7.2.2 Sound Generation due to the Presence of a Sphere

To obtain the far-field sound induced by the image potential, we need the far-field behavior of Φ^*, that is, where $r/a \gg 1$ and show its dependence on time t. With $\lambda = a^2/(rR) = O(r^{-1})$ and $\mu = \hat{r}\cdot\hat{\boldsymbol{y}}$, we have in the far field

$$Z = 1 - \lambda\mu + O(r^{-2})\ , \qquad G = \lambda^2[1 + 2\lambda\mu + O(r^{-2})]/2\ , \tag{7.2.18}$$

and the far-field image potential

$$\begin{aligned}\Phi^*(t,\boldsymbol{x}) &= \frac{1}{8\pi r^2}\iiint_{R\geq a} d\mathcal{V}\ \hat{r}\cdot\boldsymbol{y}\times\boldsymbol{\Omega}(t,\boldsymbol{y})\left[\frac{a}{R}\right]^3 \\ &\quad + \frac{1}{4\pi r^3}\iiint_{R\geq a} d\mathcal{V}\ \hat{r}\cdot\boldsymbol{y}\times\boldsymbol{\Omega}(t,\boldsymbol{y})[\hat{r}\cdot\boldsymbol{y}]\left[\frac{a}{R}\right]^5 + O(r^{-4}) \\ &= \frac{-1}{4\pi}\sum_{i=1}^{3}\partial_{x_i}\left[\frac{D*_i(t)}{r}\right] + \frac{1}{12\pi}\sum_{i=1}^{3}\sum_{l=1}^{3}\partial_{x_i}\partial_{x_l}\left[\frac{q_{il}^*(t)}{r}\right] + O(r^{-4})\ .\end{aligned} \tag{7.2.19}$$

The first term represents doublets with strengths,

$$D_i^*(t) = \frac{1}{2}\iiint_{R\geq a} d\mathcal{V}\ [y_j\omega_k - y_k\omega_j]\left(\frac{a}{R}\right)^3\ , \tag{7.2.20}$$

for $i = 1, 2, 3$, and i, j, k in cyclic order. Here y_i and ω_i denote the ith component of $\boldsymbol{y}$ and $\boldsymbol{\Omega}(t,\boldsymbol{y})$, respectively. The strength $D_i^*(t)$ is a first moment of vorticity weighted by $[a/R]^3$.

The second term in (7.2.20) represents quadrupoles with components,

$$q_{il}^* = \iiint_{R\geq a} d\mathcal{V}\ [y_j\omega_k - y_k\omega_j]y_l\left(\frac{a}{R}\right)^5\ , \tag{7.2.21}$$

for $i, l = 1, 2, 3$, and i, j, k in cyclic order. The strength $q^*_{il}(t)$ is a second moment of $\boldsymbol{\Omega}$ weighted by $(a/R)^5$.

The dependence of the doublet and quadrupole strengths, D^*_i and q^*_{il}, on t is defined by the solution of the vorticity evolution equation outside the sphere with convective velocity $\boldsymbol{v} + \boldsymbol{v}^*$.

By matching the far-field solution of the image potential in $\boldsymbol{x}$ to the outer solution of the acoustic field in the stretched variable $\tilde{r} = Mr$ or $\tilde{\boldsymbol{x}} = M\boldsymbol{x}$, we obtain the acoustic pressure induced by the image potential Φ^*,

$$\begin{aligned}\tilde{p}^* &= -\rho\, \partial_t \tilde{\Phi}^*(t, \tilde{\boldsymbol{x}}) \\ &= \left[\frac{\rho}{4\pi}\right] \left\{ M^2 \sum_{i=1}^{3} \partial_{\tilde{x}_i} \left[\frac{\dot{D}^*_i(\tau)}{\tilde{r}}\right] \right. \\ &\quad \left. - \frac{M^3}{3} \sum_{i=1}^{3} \sum_{l=1}^{3} \partial^2_{\tilde{x}_i \tilde{x}_l} \left[\frac{\dot{q}^*_{il}(\tau)}{\tilde{r}}\right] + O(M^4) \right\} , \end{aligned} \tag{7.2.22}$$

where $\tau = t - r/C = t - \tilde{r}/(MC)$ denotes the retarded time, ρ is the ambient density, and $M \ll 1$ is a typical Mach number of the vortical flow. The leading term in (7.2.22) is $O(M^2)$ representing dipoles and the second one in (7.2.22) is $O(M^3)$ representing quadrupoles.

In the absence of the sphere, the *resultant* far-field doublet strength of a vortical flow is given by the first moment of vorticity in $\mathcal{R}^3$. The first moment was shown to be time-invariant by Moreau (1948a), using the first moment of the vorticity evolution equation. Thus the far-field doublet strength is independent of t and the corresponding dipole sound vanishes. The leading term in the acoustic field is $O(M^3)$ representing quadrupoles (see Sect. 7.1). Now in the presence of a rigid sphere, Moreau's theorem is no longer applicable. The first moment of vorticity $\boldsymbol{\Omega}$ is time-dependent because the vorticity field is now convected by the image velocity $\boldsymbol{v}^*$ in addition to $\boldsymbol{v}$. Furthermore, the contribution of the image potential to the doublet is defined by the weighted vorticity first moment (7.2.20). Both the unsteady vector potential $\boldsymbol{A}$ induced by the vorticity field and the image potential Φ^* contribute to dipole sound. It follows that

{6} the acoustic field induced by a vortical flow in the presence of the sphere is one order stronger than that in absence of the sphere.

We now point out the effect of a shift of origin by $O(\ell)$ on the far-field sound. The effect appears in the change of $r = |\boldsymbol{x}|$ by $O(\ell/r)$ and in the strengths of the dipoles and quadrupoles.

For a vortical flow in $\mathcal{R}^3$, the dipole strength, that is, the first moment of vorticity, is independent of t and also of the shift of origin. This is true because the volume integral of vorticity is zero. A shift of origin will change the strengths of quadrupoles, that is, the second moments of vorticity, by $O(1)$. But these changes are independent of t, because the first moments are time-invariant. Thus, a shift of origin will change the acoustic pressure prediction by only $O(\ell/r) = O(M\ell/\tilde{r})$.

For a vortical flow outside a rigid body, for example, a sphere, the image solution (7.2.13) and hence the convective velocity depend on the choice of the origin. Consequently, a shift of origin by $O(\ell)$ will change the strengths of the far-field dipoles and quadrupoles of the vortical field plus the image potential flow by $O(1)$. Thus for the computation of the strengths, the origin has to coincide with that for the vortical flow, that is, the center of the sphere.

We also note that the above representation of the acoustic dipole in terms of the first moment of vorticity provides an attractive alternative to the classical results of Powell (1964), in which the acoustic dipole is expressed in terms of an area integral over the surface distribution of Bernoulli pressure. In particular, the present formulation avoids the need to perform delicate and time-consuming determination of the Bernoulli pressure, which generally requires inverting a pressure Poisson equation or integrating the momentum equation along solid surfaces.

In the next section we shall apply the formulae for the image velocity, image potential and acoustic field, (7.2.10)–(7.2.22), to the special case that the vorticity is highly concentrated in the core of a slender vortex filament.

7.2.3 A Slender Vortex Filament Outside a Sphere

Consider the vector potential $\boldsymbol{A}$ induced by a slender vortex filament of strength or circulation Γ. Let $\boldsymbol{X}(s)$ denote the position vector of the centerline $\mathcal{C}$ of a slender vortex filament of length L and s denote the arclength of $\mathcal{C}$ (at instant t). For a point $\boldsymbol{x}$ at a distance from $\mathcal{C}$ much larger than the core size, δ, the leading-order velocity term is given by the line Biot–Savart integral,

$$\boldsymbol{v}(t, \boldsymbol{x}) = -\frac{\Gamma}{4\pi} \int_{\mathcal{C}} \frac{(\boldsymbol{x} - \boldsymbol{X}) \times \hat{\tau}\, ds}{|\boldsymbol{x} - \boldsymbol{X}|^3}\,, \tag{7.2.23}$$

where $|\boldsymbol{x} - \boldsymbol{X}| = O(1) \gg \delta$, $d\boldsymbol{X}(s) = \hat{\tau}(s)\, ds$ and $\hat{\tau}(s)$ denotes the unit tangent vector of $\mathcal{C}$ at $\boldsymbol{X}(s)$. The volume integral over the support of vorticity is reduced to the above line integral along $\mathcal{C}$ by the following approximation rule:

{7} Replace $\boldsymbol{y}$ by $\boldsymbol{X}$, the vortex volume element $\boldsymbol{\Omega}(\boldsymbol{y})\, d\mathcal{V}$ by the vortex line element $\Gamma d\boldsymbol{X}(s)$.

We assume that the filament lies outside of the sphere $r = a$, with $a = O(1)$, and that the distance between the filament and the sphere is much larger than the typical vortical core size δ and the boundary layer thickness around the sphere. Thus, we have

$$|\boldsymbol{X}(s)|\rangle a \quad \text{for} \quad 0 \leq s \leq L\,. \tag{7.2.24}$$

This condition is consistent with the assumption that the support of $\boldsymbol{\Omega}$ lies outside the sphere.

By omitting the boundary layer, that is, assuming the Reynolds number Re $\gg 1$, the formulae derived in the preceding section for the *image* velocity, potential and far-field sound are applicable. Using the rule of approximations {7}, we identify $\hat{\boldsymbol{y}}$ as $\hat{\boldsymbol{X}}$, $R = |\boldsymbol{y}|$ as $R = |\boldsymbol{X}|$ and $\mu = \cos\beta = \hat{r} \cdot \hat{\boldsymbol{X}}$ and keep the same definition for $\bar{r} = A^2/r$, and the definitions of λ and $Z(\lambda, \mu)$ by (7.2.15). Equation (7.2.13) for the *image* potential becomes

$$\Phi^*(\boldsymbol{x}) = \int_C \left[\frac{G(\lambda, \mu)\hat{r} \times \hat{\boldsymbol{X}}(s)}{4\pi a} \right] \cdot \hat{\tau}(s)\, \Gamma ds \ . \tag{7.2.25}$$

Thus, the *image* potential $\Phi^*(\boldsymbol{x})$ outside the sphere is a line integral along $\mathcal{C}$ with integrand regular. From (7.2.25), we arrive at the following observation corresponding to {7}:

{8} In the integrand of the line integral for the *image* potential Φ^*, (7.2.25), the tangential component of the vector inside the square bracket can be interpreted as the *image* potential of a tangential vortex element of unit strength induced by the sphere.

Formula (7.2.16) for the components of the *image* velocity, $\boldsymbol{v}^*$, becomes

$$v_j^* = \frac{-\Gamma}{4\pi a} \int_C \{G\, \partial_{x_j} \hat{r} + [\partial_\lambda G\, \partial_{x_j} \lambda + \partial_\mu G\, \partial_{x_j} \mu]\, \hat{r}\} \cdot [\hat{\tau}(s) \times \hat{\boldsymbol{X}}(s)]\, ds \ , \tag{7.2.26}$$

for $j = 1, 2, 3$. The three partial derivatives in the integrand are defined by (7.2.17a), (7.2.17b), and (7.2.17c) with $\boldsymbol{y}$ replaced by $\boldsymbol{X}$. The *image* velocity is regular outside the sphere, in particular on the centerline $\mathcal{C}$; therefore, the velocity of the filament in the presence of the sphere is

$$\dot{\boldsymbol{X}}(s_0) = \mathbf{q}(s_0) + \boldsymbol{v}^*(\boldsymbol{X}(s_0)) \ , \tag{7.2.27}$$

where $\mathbf{q}$ denotes the velocity induced by the filament alone. The latter depends on the core structure and the finite part of the Biot–Savart integral. The contribution of the *image* potential to the far-field pressure due to the presence of the sphere can be found from (7.2.20) and (7.2.21) using the following expressions for the strengths of doublets and quadrupoles,

$$D_i^*(t) = \frac{\Gamma}{2} \int_C ds\ [X_j \tau_k - X_k \tau_j] \left(\frac{a}{|\boldsymbol{X}|} \right)^3 , \tag{7.2.28}$$

and

$$q_{il}^* = \Gamma \int_C ds\ [X_j \tau_k - X_k \tau_j] X_l \left(\frac{a}{|\boldsymbol{X}|} \right)^5 . \tag{7.2.29}$$

In 1956, Lighthill (1956) constructed the image of a given vortex element, $\boldsymbol{J}$, located at $\boldsymbol{X}$ outside of the sphere, for example, a segment of a vortex line

$\mathcal{C}$, with $\boldsymbol{J} = \Gamma d\boldsymbol{X}(s)$ and $|\boldsymbol{X}|\rangle a$, so that the resultant velocity $\boldsymbol{V}$ induced by the vortex element $\boldsymbol{J}$ outside and the image vorticity $\boldsymbol{W}$ inside the sphere has zero radial component on the sphere. It was duly noted that a radial vorticity vector does not induce radial velocity on the sphere. The radial velocity induced by the transverse vortex element outside of the sphere can be cancelled only by an image transverse element scaled by $-a/R$ located at the inverse point $a^2\boldsymbol{X}/R^2$ with $R = |\boldsymbol{X}|$. To make the image vorticity field inside the sphere divergence free, an image radial vortex element at the inverse point and a radial line-vortex extending from the inverse point to the origin are needed. The major effort of Lighthill (1956) is devoted to applying the divergence-free condition to a distributed vorticity field and then interpreting the result for the vortex element by physical reasoning.

When the vorticity field $\boldsymbol{\Omega}$ outside the sphere is a vortex line $\mathcal{C}$, the image vorticity field $\boldsymbol{W}$ inside the sphere is composed of a line distribution of vorticity along the image of $\mathcal{C}$ plus a conical vortex surface with the origin as its vertex and the image of $\mathcal{C}$ as the baseline. However, the velocity $\boldsymbol{V}$ induced by the combined vorticity field $\boldsymbol{\Omega} + \boldsymbol{W}$ in $\mathcal{R}^3$ can be expressed as a line integral along $\mathcal{C}$ and the algebraic complexity of the integrand is not increased by a factor more than three over that of the Biot–Savart integral for $\boldsymbol{\Omega}$ alone. The formulae for the three Cartesian components of $\boldsymbol{V}$ in Lighthill (1956) are in agreement with the formula (7.2.16) given above. The formula coming from (7.2.13) shows that the additional or *image* velocity $\boldsymbol{v}^*$ due to the sphere depends only on the transverse vorticity field outside of the sphere. The derivation presented here is straightforward because it makes use of the fact that the *image* velocity $\boldsymbol{v}^*$ outside of the sphere is irrotational and Weiss's theorem (Weiss, 1944) relates the radial component u^* of $\boldsymbol{v}^*$ directly to that of $\boldsymbol{v}$ inside the sphere, where $\boldsymbol{v}$ is irrotational. From the *image* radial component u^*, we obtain the *image* potential and the corresponding acoustic potential and pressure.

In hindsight, we could say that the determination of the relatively complex radial image vorticity field, making the vorticity field $\boldsymbol{W}$ divergence free, is not needed to define the additional velocity $\boldsymbol{v}^*$. It is needed (Lighthill, 1956) when we apply the Biot–Savart law to define the velocity induced by the vorticity field or the line Biot–Savart integral (7.2.23) for a vortex line.

We also note that we can use Lighthill's formula for the velocity $\boldsymbol{V}$ induced by the vorticity field $\boldsymbol{\Omega}+\boldsymbol{W}$ in $\mathcal{R}^3$, with the rigid sphere replaced by the *image* $\boldsymbol{W}$ inside the sphere. We can then derive the corresponding acoustic potential, $\tilde{\Phi}$ at instant t, and find that the leading terms are dipoles with strengths given by the first moments of $\boldsymbol{\Omega} + \boldsymbol{W}$ over $\mathcal{R}^3$ and are equivalent to those defined above. The vorticity evolution equation with convective velocity $\boldsymbol{V}$ defines the temporal variation of the vorticity field $\boldsymbol{\Omega}$ outside of the sphere while the image field $\boldsymbol{W}$ inside the sphere at each instant is given by Lighthill's formula. Thus we cannot apply Moreau's theorem to the first moments $\boldsymbol{\Omega}+\boldsymbol{W}$

in Re^3, and conclude that the dipole strengths are time-invariant producing zero acoustic pressure.

7.2.4 Computational Examples

This section provides computational results that illustrate the theoretical development outlined above. Once again we focus our attention to the case of a slender vortex ring passing over a rigid stationary sphere. We start with an outline for the vortex flow and sound models and then consider several scenarios to investigate the relationships between three-dimensional slender filament dynamics and radiated noise.

7.2.4.1 Simulation Scheme. Simulation of the motion of the vortex filament is performed using a Lagrangian vortex scheme that generalizes the thin-tube model of Chap. 5. The generalization is based on the two key observations that (1) the corrected thin-tube model for filaments in free-space provides a velocity that is consistent with that predicted by the asymptotic theory, and (2) in the presence of the sphere, the filament velocity can be expressed as the sum of two velocity vectors, the first corresponding to the free-space filament velocity whereas the second is given by the image of the filament, as expressed in (7.2.26). Thus, to extend the computations to account for the presence of the sphere, one replaces the free-space equation of motion for the Lagrangian particles with

$$\frac{\partial \chi_i}{\partial t} = \boldsymbol{v}^{\mathrm{ttm}}(\chi_i(t), t) + \boldsymbol{v}^{\mathrm{im}}(\chi_i(t), t) \,. \tag{7.2.30}$$

The image velocity is obtained by numerically evaluating the integral in (7.2.26) using the same quadrature used for the evaluation of the self-induced velocity. With this modification, application of the slender filament scheme to simulate the passage of a vortex ring over a sphere can be summarized as follows. (a) The corrected slender filament scheme is used to evaluate the self-induced and image velocities at the Lagrangian particles that represent the filament centerline (Klein and Knio, 1995; Klein et al., 1996). (b) The second-order Adams–Bashforth scheme is used to update the Lagrangian position vector of the vortex elements, that is, by integrating (7.2.30). (c) Based on the new particle positions, the new value of the filament arclength is computed. (d) The Crank–Nicolson scheme is used to advance the stretched variable ς, and thus update the core structure coefficients C_v and C_w, and consequently the numerical core radius (Subsect. 5.2.3). Steps (a–d) are repeated to advance the solution in time.

7.2.4.2 Far-Field Sound. Combining the results of Sects. 7.1 and 7.2, the far-field acoustic pressure due to the filament sphere interaction is expressed as follows:

$$p_a(\boldsymbol{x},t) = \frac{1}{4\pi r^2}\dot{D}_i(t_r)\hat{x}_i + \frac{M}{4\pi r}\ddot{D}_i(t_r)\hat{x}_i + \frac{1}{4\pi r^3}\dot{Q}_{il}(t_r)\hat{x}_i\hat{x}_l + \frac{M}{4\pi r^2}\ddot{Q}_{il}(t_r)\hat{x}_i\hat{x}_l + \frac{M^2}{12\pi r}\dddot{Q}_{il}(t_r)\hat{x}_i\hat{x}_l \tag{7.2.31}$$

where $\boldsymbol{x}$ is the observer location, $r = |\boldsymbol{x}|$, $\hat{\boldsymbol{x}} = \boldsymbol{x}/r$, and $t_r = t - Mr$ is the retarded time. The dipole strengths are given by

$$D_i(t) = d_i(t) - d_i^{\text{im}}(t)\ , \tag{7.2.32}$$

where

$$d_i(t) = \frac{\Gamma}{2}\int_{\mathcal{L}(t)} [X_j\tau_k - X_k\tau_j]\ ds \tag{7.2.33}$$

and

$$d_i^{\text{im}}(t) = \frac{\Gamma}{2}\int_{\mathcal{L}(t)} [X_j\tau_k - X_k\tau_j]\left(\frac{a}{|\boldsymbol{x}|}\right)^3 ds \tag{7.2.34}$$

for $i = 1,2,3$, and i,j,k in cyclic order. The quadrupole strengths are expressed as

$$Q_{il}(t) = q_{il}(t) - q_{il}^{\text{im}}(t)\ , \tag{7.2.35}$$

where

$$q_{il}(t) = \Gamma\int_{\mathcal{L}(t)} [X_j\tau_k - X_k\tau_j]\ X_l\, ds \tag{7.2.36}$$

are

$$q_{il}^{\text{im}} - \Gamma\int_{\mathcal{L}(t)} [X_j\tau_k - X_k\tau_j]\ X_l\left(\frac{a}{|\boldsymbol{X}|}\right)^5 ds \tag{7.2.37}$$

for $i,l = 1,2,3$, and i,j,k in cyclic order. In the acoustic far field, $Mr \gg 1$, only the second and last terms on the right-hand side of Eq. (7.2.31) survive, and one obtains the following far-field acoustic pressure expression:

$$p_a^F(\boldsymbol{x},t) = \frac{M}{4\pi r}\ddot{D}_i(t_r)\hat{x}_i + \frac{M^2}{12\pi r}\dddot{Q}_{il}(t_r)\hat{x}_i\hat{x}_l\ . \tag{7.2.38}$$

Thus, the far-field acoustic pressure is governed by $O(M)$ dipoles and $O(M^2)$ quadrupoles. This behavior will be exploited in the analysis of the computations.

It is interesting to indicate how the present representations of the acoustic pressure in (7.2.31) and (7.2.38) relate to classical results of the aerodynamic theory of sound. We first note that the leading-order dipole term, expressed in terms of the first moments of vorticity, reflects the changes in the impulse of the slender filament and of its image within the sphere, that is, in the total impulse of the flow (Wells, 1996). Thus, the present three-dimensional expressions, which generalize the axisymmetric results of Miyazaki and Kambe (1986), coincide with the classical results for sound generation in the presence of a compact rigid body (Lighthill, 1978; Howe, 1975;

Curle, 1955). It is also interesting to point out that the leading dipole term in (7.2.38) has exactly the same form as that predicted by Obermeier (1980) for vortical flow outside a rigid sphere. The contributions of the present analysis are that, for the case of a slender vortex interacting with a rigid sphere, it (1) provides simple explicit formulae for the acoustic source terms and (2) relates the evolution of the acoustic source terms to the *correct* leading-order behavior of the slender filament motion. The present expressions of the far-field pressure account for dipole and quadrupole emission and may be easily extended, if so desired, to account for octopoles and higher order contributions as well. We also point out that the above explicit representation of the acoustic dipole in terms of the first moments of vorticity is computationally more attractive than the well-known expressions of Powell (1960, 1964) and Curle (1955), since they do not involve integrals of velocity derivatives over the solid surface.

As noted earlier, in the absence of the sphere the contributions of the image of the filament, d_i^{im} and q_{il}^{im}, drop out, and the first and second moments of vorticity reduce to $D_i = d_i$ and $Q_{il} = q_{il}$, respectively. Since in this case d_i is time invariant (Moreau, 1948a), the dipole contribution to the acoustic field vanishes identically, and the acoustic pressure is dominated by the effect of the quadrupoles. In particular, in the absence of the sphere, (7.2.38) simplifies to

$$p_a^F(\boldsymbol{x}, t) = \frac{M^2}{12\pi r} \dddot{Q}_{il}(t_r)\hat{x}_i\hat{x}_l \tag{7.2.39}$$

and one exactly recovers Möhring's formula (Möhring, 1978) (see also Powell, 1995a, 1995b) for the far-field acoustic pressure due to vortex sound at low Mach number. Note that the dependence of the far-field acoustic pressure on the distance of the observer can be absorbed by defining a reduced acoustic pressure $p^R \equiv 4\pi r p_a^F$. Using this definition, the far-field sound predictions can be interpreted using p^R in lieu of p_a^F. This reduction will be adopted in the discussion below, where we refer to both p_a^F and p^R simply as far-field pressure.

In the implementation of (7.2.31) and (7.2.38), the first moments d_i and d_i^{im} and the second moments q_{il} and q_{il}^{im} are determined by numerically evaluating the integrals in (7.2.33)–(7.2.34) and (7.2.36)–(7.2.37) using the same spectral collocation approximation that is employed for computing the line Biot–Savart integral. The first and second moments are stored during the computations; second-order centered differences are then used to estimate the first and second time derivatives of the first moments, and the first, second, and third time derivatives of the second moments of vorticity.

7.2.4.3 Setup. Consistent with the approximation used in the formulation of the model, we restrict our attention to interaction events in which the filament remains at all times well separated from the surface of the sphere, by a distance of at least a few core radii. In addition, the chatacteristic Mach number is assumed to be so small that the motion surrounding the vortex

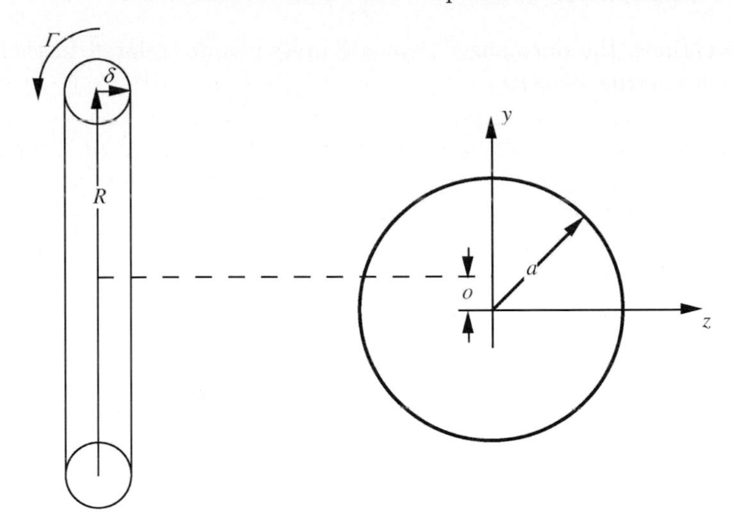

Fig. 7.2. Schematic illustration of the initial flow configuration. Adapted from Knio et al. (1998)

and the sphere can be treated as essentially incompressible. Furthermore, the Reynolds number is assumed to be large enough so that the thickness of the boundary layer on the surface of the sphere is much smaller than the sphere radius. Consequently, viscous effects near the surface of the sphere (Pedrizetti, 1992), and their potential contribution to sound emission (Howe, 1996), are ignored.

As illustrated in Fig. 7.2, the physical setup presently considered consists of a vortex ring of radius $\tilde{R}$, circulation $\tilde{\Gamma}$, and core size $\tilde{\delta}$ initially located "upstream" of a sphere of radius $\tilde{a}$. A Cartesian (x, y, z) coordinate system is used whose origin coincides with the center of the sphere. The axis of the vortex ring is assumed to point along the z-direction, and to intersect the y-axis at an offset distance o from the origin.

For brevity, we illustrate only three-dimensional interactions corresponding to two sets of initial conditions. In the first case, the offset $o = 0$, but the centerlines of the vortex rings are initially perturbed using helical waves, and the computations are used to analyze the effect of perturbation amplitude and wavenumber on the dynamics of the filament and the radiated noise. In the second case, we consider asymmetric initial configurations corresponding to circular vortex rings whose axis does not pass through the center of the sphere, that is, $o \neq 0$.

To interpret the results of the computations, the following scaling parameters are used. We choose the sphere radius $\tilde{a}$ as reference lengthscale, that is, $\tilde{L}_{\text{ref}} = \tilde{a}$, and use the circulation of the filament in defining the following reference velocity $\tilde{U}_{\text{ref}} = \tilde{\Gamma}/\tilde{a}$. With these definitions, the normalized circulation of the filament $\Gamma = 1$, and the Reynolds number $\text{Re} = \tilde{\Gamma}/\tilde{\nu}$. Following these conventions, the normalized acoustic pressures are related to their dimensional counterparts using

$$p = \frac{\tilde{p}}{\tilde{\rho}_0 \tilde{\Gamma}^2 / \tilde{a}^2} \tag{7.2.40}$$

and the Mach number $M = \tilde{\Gamma}/\tilde{a}\tilde{c}_0$. Here, $\tilde{\rho}_0$ and $\tilde{c}_0$ are the dimensional density and speed of sound of the undisturbed medium, respectively. Meanwhile, the normalized viscosity parameter used in the description of the core structure evolution is given by $K = \delta^{-2}\tilde{\nu}/\tilde{\Gamma} = 1/\mathrm{Re}\delta^2$, where $\delta = \tilde{\delta}/\tilde{a}$ is the normalized physical core radius.

In all of the computations presented below, the core function used to regularize the vorticity field is $f(r) = \mathrm{sech}^2(r^3)$, with corresponding velocity kernel $\kappa(r) = \tanh(r^3)$ (Beale and Majda, 1985). For this choice of core smoothing function, the numerical core structure coefficient $C^{\mathrm{ttm}} = -0.4202$ (Klein and Knio, 1995). As discussed in Chap. 5, once the core smoothing function and the corresponding value of C^{ttm} are specified, the initial core structure of the filament can be described in terms of the initial value of the numerical core radius, σ, only.

7.2.4.4 Passage of a Perturbed Ring Over the Sphere. We now turn our attention to three-dimensional filament/sphere interactions and focus, in this section, on the passage of nonaxisymmetric rings over the rigid sphere. Specifically, we generalize the setup of the previous section by imposing at the start of the computations a single-mode helical perturbation to the filament centerline (Knio and Juvé, 1996). The perturbation is specified in terms of its wavenumber k and (dimensionless) "amplitude" ϵ', and is applied in such a way that the filament centerline is initially described by

$$\begin{aligned} \chi_1(\theta') &= R(1+\epsilon')\sin(k\theta')\cos(\theta') \\ \chi_2(\theta') &= R(1+\epsilon')\sin(k\theta')\sin(\theta') \\ \chi_3(\theta') &= \epsilon' R\cos(k\theta') + z_i \ , \end{aligned} \tag{7.2.41}$$

where θ' is the azimuthal angle, $0 \le \theta' \le 2\pi$, and z_i is the original undisturbed position of the center of the ring. In the computations, we consider perturbations with three different wavenumbers $k = 1$, 2, and 8, and two different amplitudes $\epsilon' = 0.01$ and 0.1, and restrict the analysis to slender rings with initial core radius $\sigma_0 = 0.06$, and unperturbed mean radius $R_0 = 1.5$.

As the ring propagates under its own self-induced velocity, the helical perturbations travel around its circumference. Since the perturbations have small (but finite) amplitudes, their evolution does not significantly alter the broad features of the propagation of the slender ring nor its passage over the sphere. Thus, the objective of the present computations is to determine whether the evolution of the helical waves affects the sound emission, and if so, to quantify the associated effects as a function of the properties of the waves. In particular, because of the moderate amplitudes, the spinning

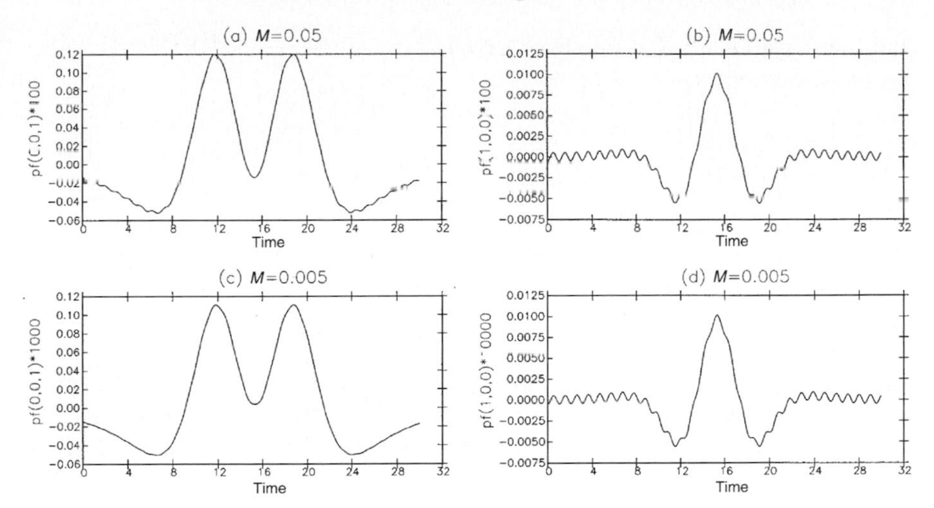

Fig. 7.3. Evolution of the far-field pressure p^R during the passage of a three-dimensional vortex ring over a rigid sphere. The initial ring radius $R = 1.5$, and core radius $\sigma = 0.06$. The centerline of the ring is initially perturbed with a helical mode with $k = 8$ and $\epsilon' = 0.01$. The plots show the far-field emission along the z-axis (a,c) and the x-axis (b,d). The value of the Mach number M is indicated. Adapted from Knio et al. (1998)

of the waves occurs at a frequency that is in large part determined by the wave number (e.g., Klein and Majda, 1991b). Thus, an interesting issue is the investigation of waves whose frequencies are higher than or comparable to the "frequency" of the passage event itself. This motivates our selection of the different wavenumbers specified above, as will be evident shortly.

Figure 7.3 shows the evolution of the far-field pressure p^R for a vortex ring perturbed using a helical wave with $k = 8$ and $\epsilon' = 0.01$. The figure shows that the present perturbation has a weak effect on the far-field pressure amplitude. The effect of the helical wave is more pronounced in Figs. 7.3b and d, which depict the evolution of the far-field pressure along the x-axis for $M = 0.05$ and $M = 0.005$, respectively. These plots clearly reflect the spinning of the wave, whose characteristic period is substantially smaller than the interaction time between the ring and the sphere. Furthermore, comparison of Figs. 7.3b and 7.3d indicates that the effect of the waves on the acoustic pressure along the x-axis is proportional to the square of the Mach number, thus suggesting that the present perturbation affects the evolution of the quadrupoles only. Further examination of the results (not shown) indicates that this is in fact the case. Specifically, the computations show that the dipole vector remains aligned with the ring axis and that its amplitude does not appear to be affected by the evolution of the perturbation. Thus, one would expect that for the present small-amplitude perturbation, the far-field emission along the z-axis remains dominated by the contribution of the dipoles. Not surprisingly, the pressure signals of Figs. 7.3a and c show only small departures from the axisymmetric

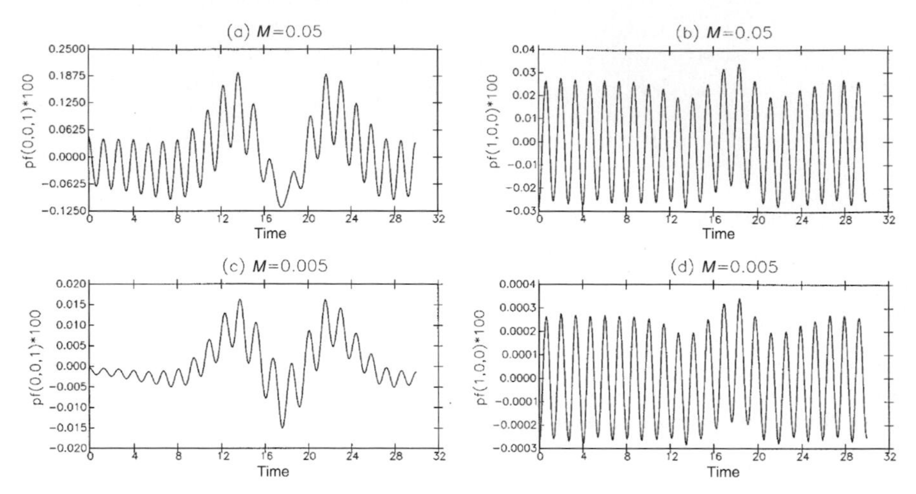

Fig. 7.4. Evolution of the far-field pressure p^R during the passage of a three-dimensional vortex ring over a rigid sphere. The initial ring radius $R = 1.5$, and core radius $\sigma = 0.06$. The centerline of the ring is initially perturbed with a helical mode with $k = 8$ and $\epsilon' = 0.1$. The plots show the far-field emission along the z-axis (a,c) and the x-axis (b,d). The value of the Mach number M is indicated. Adapted from Knio et al. (1998)

prediction, which consist of weak ondulations at the same frequency of the helical wave. The ondulations are easier to detect prior to and following the passage of the ring over the sphere and, as expected, become more pronounced as M increases.

The situation described above is dramatically altered as the amplitude of the perturbation is increased, as illustrated in Fig. 7.4. The latter shows that for a helical perturbation with $k = 8$ and $\epsilon' = 0.1$, the *quadrupolar* emission associated with the helical wave can dominate the sound emission. Specifically, the pressure signal along the x-axis is overwhelmed by the effect of the perturbation (Figs. 7.4b and d). In addition, for the present case, the amplitude of the quadrupole sound becomes comparable to that of the dipole sound. As shown in Figs. 7.4a and c, the "underlying" dipole emission due to the interaction of the filament with the sphere can be hidden by the effect of the perturbation. Naturally, the observed "competition" between dipole and quadrupole sound is dependent on the Mach number. The role of the quadrupoles becomes relatively more important as M increases but nonetheless may not be ignored even at low Mach numbers.

We now consider a helical perturbation with larger pitch, starting with waves having $k = 2$ and amplitudes $\epsilon' = 0.01$ and 0.1. Results of the computations are plotted in Fig. 7.5, which shows, for both wave amplitudes, the evolution of p^R along (a) the z-axis and (b) the x-axis. The results show that for the present wavenumber, the characteristic period of the perturbation is comparable to the passage timescale. In these situations, we find that for ob-

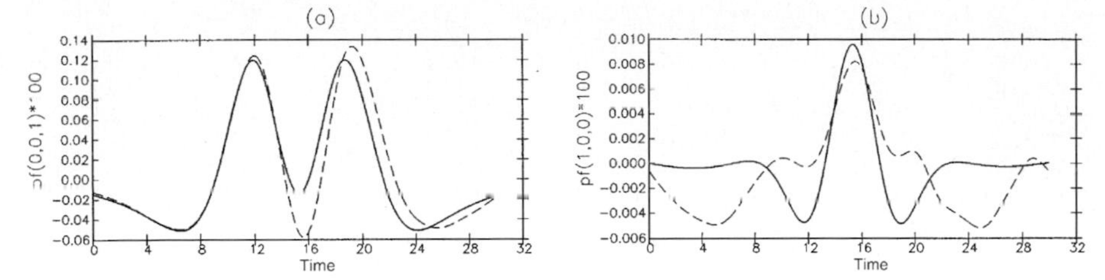

Fig. 7.5. Evolution of the far-field pressure p^R during the passage of a three-dimensional vortex ring over a rigid sphere. The initial ring radius $R = 1.5$, and core radius $\sigma = 0.06$. The plots show the far-field emission along (a) the z-axis and (b) the x-axis. The centerline of the ring is initially perturbed with a helical mode with $k = 2$: ——, $\epsilon' = 0.01$; – – –, $\epsilon' = 0.1$. The Mach number $M = 0.05$. Adapted from Knio et al. (1998)

server locations that are significantly affected by the dipole emission, the role of the perturbation is restricted to a moderate modulation of the acoustic pressure signal (Fig. 7.5a). On the other hand, for emission directions for which quadrupole sound is dominant (Fig. 7.5b), the perturbation can still significantly affect the acoustic far field.

Results obtained using mode 1 helical perturbations reveal many similarities to those obtained with $k = 2$, and also some unexpected differences. The evolution of the far-field acoustic pressure p^R is shown in Fig. 7.6 for a perturbation having $k = 1$ and $\epsilon' = 1$. The pressure signals show that, similar to the effect of mode 2 perturbations, the evolution of the helical wave leads to a modulation of sound emission along both the z- and x-axes. However, unlike the mode 2 case, it is now observed that the acoustic pressure along the x-axis is comparable in amplitude to the emission along the z-axis. Furthermore, examination of the results in Figs. 7.6c and d indicates that the acoustic pressure along the x-axis no longer follows the $O(M^2)$ dependence that one would expect for quadrupole emission. Detailed examination of the results reveals that the origin of this phenomenon is due to asymmetry of the perturbation. As a result of this asymmetry, the ring is initially slightly tilted toward the x-direction, and its interaction with the sphere occurs in such a way that the dipole vector is gradually displaced from its initial position. As shown in Fig. 7.7, the dipole vector initially has a major component along the z-direction and admits a small but finite component along the x-direction. During the passage, all the components of the dipole vector change in a time-dependent fashion and, unlike the axisymmetric case, the ring does not return to its initial state. In particular, the results in Fig. 7.7 clearly show that during the interaction a net (finite) momentum is transferred to the fluid. These observations enable us to relate the behavior of the acoustic pressure signals in Figs. 7.6b and d to the time-dependent directivity of the dipole emission.

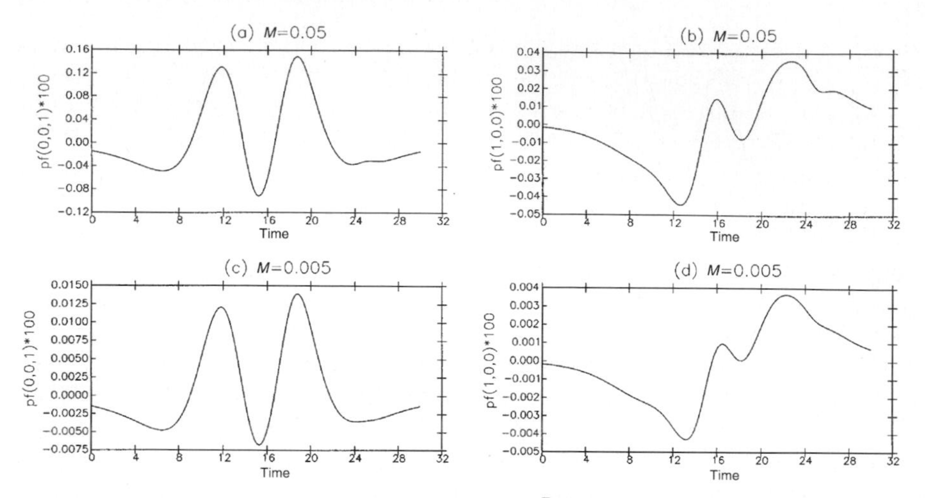

Fig. 7.6. Evolution of the far-field pressure p^R during the passage of a three-dimensional vortex ring over a rigid sphere. The initial ring radius $R = 1.5$, and core radius $\sigma = 0.06$. The centerline of the ring is initially perturbed with a helical mode with $k = 1$ and $\epsilon' = 0.1$. The plots show the far-field emission along the z-axis (a,c) and the x-axis (b,d). The value of the Mach number M is indicated. Adapted from Knio et al. (1998)

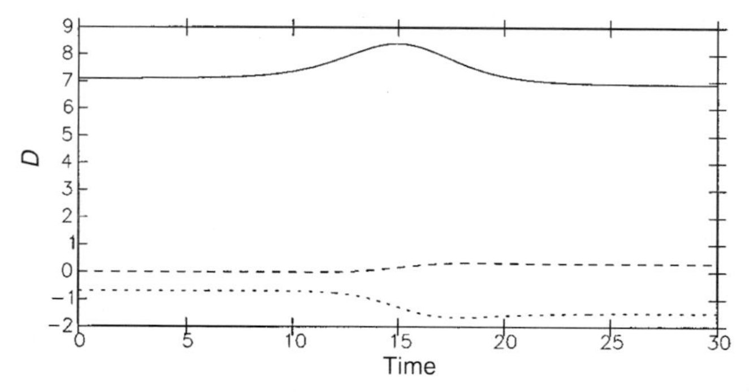

Fig. 7.7. Evolution of the dipole vector for the ring of Fig. 7.6: ----, D_1; −−−, D_2; ——, D_3. Adapted from Knio et al. (1998)

7.2.4.5 Effects of Offset and Viscosity. We consider in this section the noncoaxial passage of an initially circular slender ring. Specifically, we focus on a single initial condition consisting of a vortex ring of radius $R_0 = 1.5$ that is initially centered at $(0, 0.4, -4)$. With respect to Fig. 7.2, the axis of the ring is displaced in the y-direction by an offset $o = 0.4$. We assume that the ring has a self-similar Gaussian core structure (Klein and Ting, 1995; Klein et al., 1996), with numerical core size $\sigma = 0.06$. We compare results of an inviscid computation to predictions obtained using the same initial conditions but with diffusivity $K = 0.1$ and $K = 0.3$ which, based on the present normalization convention, correspond to Reynolds numbers $\mathrm{Re} \equiv \tilde{\Gamma}/\tilde{\nu} = 3659$ and 1220, respectively.

In the computation of filaments with viscous cores, we still rely on expressions (7.2.31) and (7.2.38) for the prediction of sound emission. Thus, the formulation is not altered to account for the effects of monopole radiation (Kambe, 1986; Ting and Miksis, 1990). This simplification facilitates the analysis of the acoustic far-field and enables straightforward comparison of the different cases. For the present setup, it also constitutes a reasonable approximation since the vortices are mildly stretched and the Reynolds number is high (Iafrati and Riccardi, 1996).

Figure 7.8 shows the projection of the ring centerline on the y–z plane for a simulation with $K = 0$. The figure shows that as the ring passes over the sphere its centerline is deformed such that elements that come closer to the sphere surface acquire a larger velocity than those further away. As a result, the centerline of the vortex ring is no longer axisymmetric, but symmetry with respect to the $y - z$ plane is maintained. The results also indicate that the trajectory of the center of the ring is deflected during the interaction. Initially, the ring propagates along the z-axis and, once the passage over the sphere is completed, the self-induced velocity is tilted toward the y-axis. The

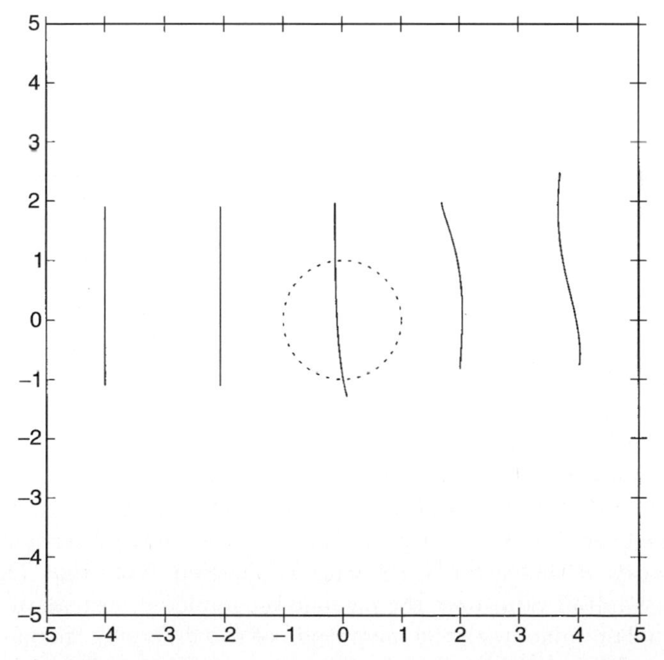

Fig. 7.8. Projections of the vortex ring on the $y - z$ plane at different instances of time. The surface of the sphere is indicated using a dashed line. The results are obtained using an inviscid computation of a slender ring initially having $R = 1.5$, $\sigma = 0.06$, and located at $(-4, 0, 0.4)$. Adapted from Knio et al. (1998)

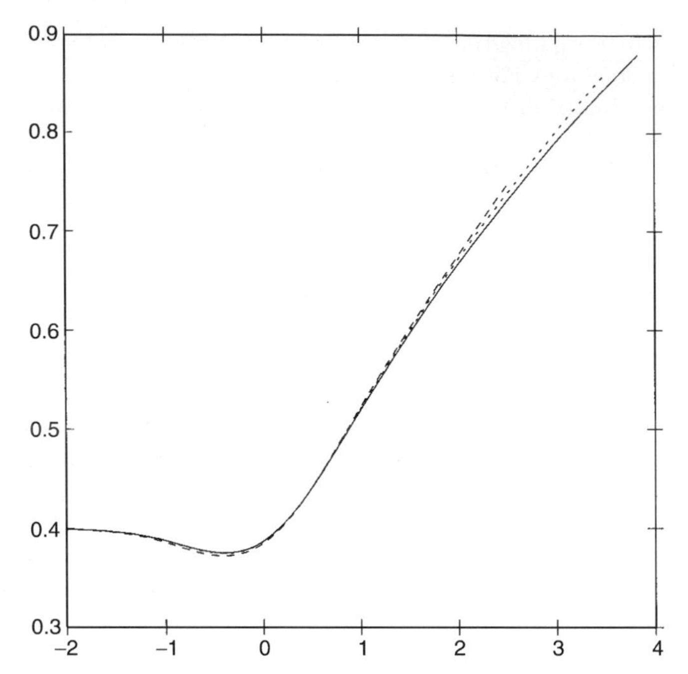

Fig. 7.9. Trajectories of the center of the vortex ring for an asymmetric passage with $o = 0.4$: ——, $K = 0$ (inviscid); - - - -, $K = 0.1$; − − −, $K = 0.3$. Adapted from Knio et al. (1998)

deflection of the trajectory of the center of the ring is further examined in Fig. 7.9 for all three cases considered. The results show that for $K = 0$, the deflection of the trajectory is roughly 15°. As the normalized viscosity is increased to $K = 0.3$, the deflection angle increases to approximately 17°; that is, it changes by more than 10%. Thus, the effect of viscosity is noticeable, but moderate.

The symmetry of the flow with respect to the $y - z$ plane enables us to easily interpret the evolution of the dipole vector, whose component along the x-axis vanishes identically. The evolutions of the z- and y-components of the dipole vector are shown in Figs. 7.10a and b, respectively; plotted are curves obtained with $K = 0$, 0.1, and 0.3. Figure 7.10a shows that the z-component of the dipole vector increases as the filament is stretched during the passage over the sphere, and then decreases as the arclength shrinks and the trajectory of the center of the ring is deflected. Note that D_3 does not return to its initial value once the passage is completed; instead, it assumes a slightly smaller value, with the magnitude of the difference increasing as the Reynolds number decreases. Meanwhile, the overall trend in the evolution of D_2 is generally increasing with time. The evolution curves of D_2 exhibit a slight dip at the start of the passage and a more pronounced overshoot toward the end of the interaction, the magnitude of which increases as Re decreases.

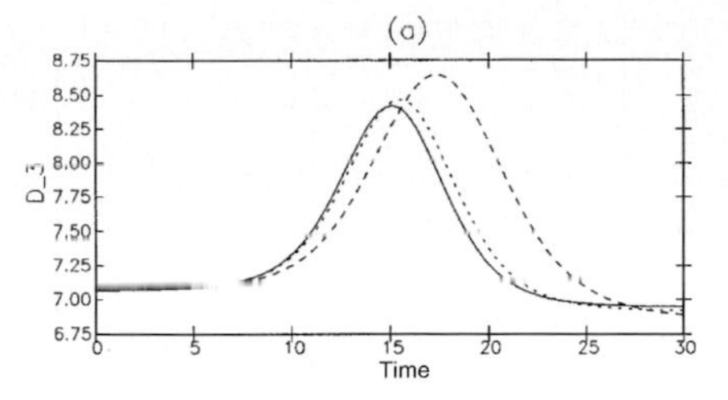

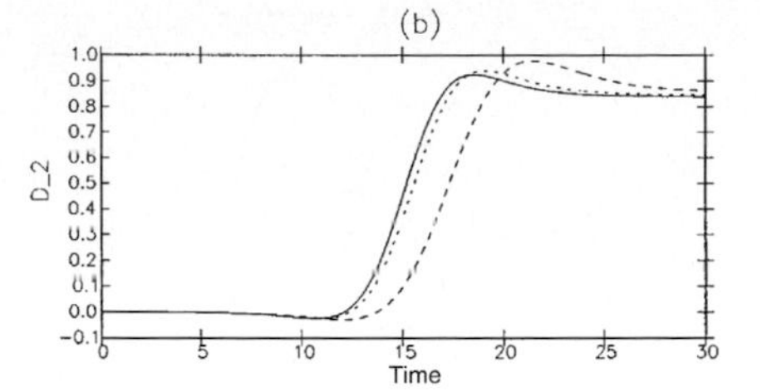

Fig. 7.10. Evolution of (**a**) D_3 and (**b**) D_2 for asymmetric passage with $o = 0.4$: ——, $K = 0$ (inviscid); ----, $K = 0.1$; – – –, $K = 0.3$. Adapted from Knio et al. (1998)

These observations suggest that the evolution of the dipole is governed by two, at times competing, effects: the stretching of the ring centerline and the deflection of its trajectory. It is interesting to note the similarity between the evolution of the dipole vector and earlier observations of the motion of the ring, and that, consistent with these observations, the amplitude of the changes in the dipole vector are noticeably more pronounced as viscosity increases.

Following the discussion of the previous section, computed results on the dynamics of the filament lead us to expect a number of trends in the far-field sound. In particular, because of the tilting of the dipole vector toward the y-axis, one would expect that emission along this direction would now be dominated by dipole sound. However, the effects of the (small) viscosity are hard to guess. On one hand, as discussed above, viscous effects tend to increase the variations of the vorticity moments. On the other hand, viscous effects tend to slow down the time scales over which these variations occur. This dual role makes it difficult to predict its net effect on sound generation.

To examine this issue, we plot in Fig. 7.11 the evolution of the far-field acoustic pressure along the x, y, and z directions, for all values of K considered and Mach numbers $M = 0.05$ and 0.005. The results show that, as expected, the acoustic pressure signals along the y and z directions are dominated by dipole sound, while emission along the x-axis is due to quadrupoles only. The pressure curves in Fig. 7.11a–d indicate that viscous effects lead to a slight increase in the amplitude of the dipole sound, and result in more significant variations than one would anticipate on the basis of the ring dynamics. Meanwhile, the acoustic pressure amplitudes associated with the quadrupoles (Figs. 7.11e and f) are less affected by viscosity. The pressure curves in Figs. 7.11e and f have similar shapes and assume very close values, and the major effect of viscosity is an increase in the characteristic time of the pressure variations. The present observations also underscore the effect of the core structure evolution on the filament motion, and its potentially large impact on sound generation.

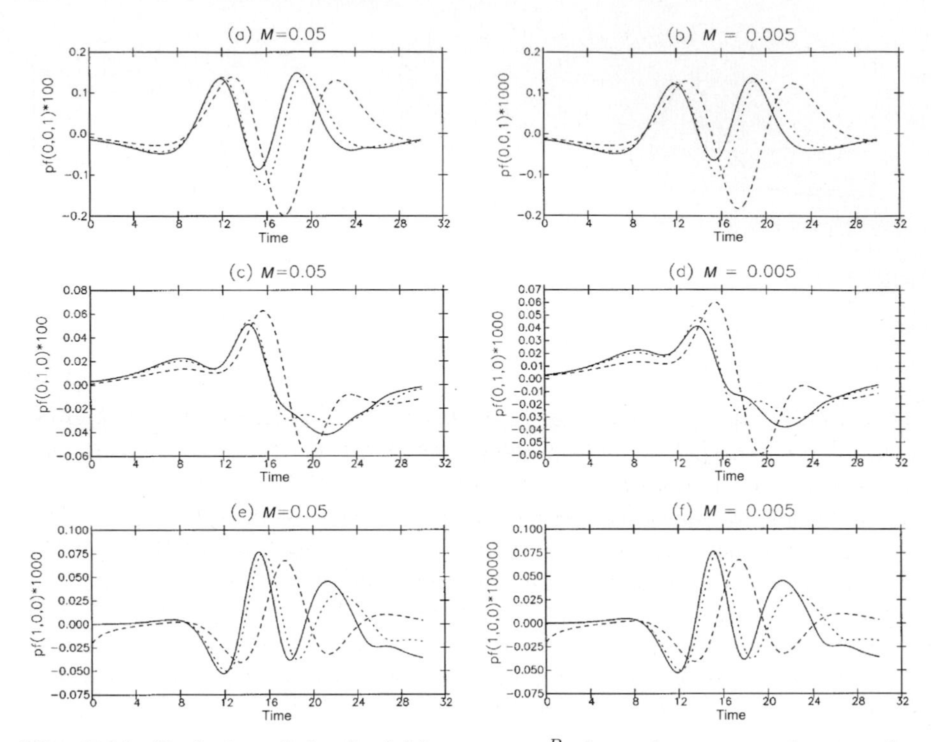

Fig. 7.11. Evolution of the far-field pressure p^R along the x-, y- and z-axes for asymmetric passage with $o = 0.4$: ——, $K = 0$ (inviscid); ----, $K = 0.1$; – – –, $K = 0.3$. The value of the Mach number is indicated. Adapted from Knio et al. (1998)

8. Sound Generated Flow

Motivation

Fluids are primary media of choice in energy conversion devices. Prominent examples are furnace burners, gas turbines, internal combustion engines, and refrigerators. In the interest of thermodynamic efficiency, flow velocities remain controlled and quite low within such devices, with characteristic flow Mach numbers in the range of $M \sim 0.1 \ldots 0.2$ and smaller. In all of the examples, there is conversion between mechanical and internal energy so that large temperature variations must be accounted for. From a theoretical point of view, the relevant flow regime therefore involves small Mach numbers and, due to the temperature effects, density variations of order unity. We refer to such flows below as "quasi-incompressible."

As long as flow domains remain "acoustically compact," a notion to be explained shortly, such flows are well described by an extension of the classical theory of incompressible, constant density flows. The zero divergence constraint for the velocity field from the classical theory is replaced in the extended equations with an inhomogeneous divergence constraint for the energy and/or mass flux densities, thereby allowing for nonmechanical energy exchange between fluid parcels as well as for internal energy conversion, for example, through bulk compression or exothermal chemical reactions.

By the term "acoustically compact domain" we refer to a flow domain that is too small to support low frequency, long wave acoustic modes with time scales comparable to the characteristic advection time of the underlying small scale quasi-incompressible flow. See also the discussion of Fig. 7.1a in the previous chapter. In acoustically noncompact domains, interactions between such long-wave acoustical modes and the local, small-scale quasi-incompressible transport of entropy, vorticity, and active or passive scalars arise. On one hand, these interactions are often undesirable, for example, in the context of thermoacoustic instabilities of combustion devices. On the other hand, the same thermoacoustic mechanism that is responsible for these instabilities can also be exploited as a key component of an energy conversion process, for example, in thermoacoustic refrigerators.

Chapter Summary

In Sect. 8.1 we summarize the multiple scales asymptotic theory of single time scale – multiple-lengthscale, low-Mach-number flows, which provides the mathematical background for the flow regimes discussed above. We emphasize the scale interactions that are responsible not only for thermoacoustic effects but also for the generation of small-scale vorticity through baroclinic torques.

In Sect. 8.2, we present two alternative approaches to the numerical simulation of zero Mach number, variable density flows, one based on a vorticity-stream function formulation, the other on the conservation laws for mass, momentum, and total energy.

Section 8.3 presents the results of an extensive study of thermoacoustic devices that used the theory and numerical simulation techniques introduced in the previous sections. We first describe a simplified mathematical model of a thermoacoustic device focusing, in particular, on the coupling of long-wave acoustic, and small-scale zero-Mach-number computations in an asymptotically correct fashion. Unphysical accumulations of modelling errors that often arise from ad hoc coupling procedures are thus avoided. We then document comparisons of simulations with experimental obervations and move on to an extensive parameter study focusing on the performance of thermoacoustic refrigerators.

8.1 Single Time Scale, Low Mach Number Limits

In Chap. 7 we discussed two essentially different asymptotic low-Mach-number regimes characterized by (multiple time/single space scales) and (single time/multiple space scales), respectively. We then considered a subregime of the second case, in which localized quasi-incompressible flow generates low-amplitude far-field sound. Here we analyze the second, "true" multiscale subregime, in which the spatial extent over which small-scale fluctuations are found is comparable with the long acoustic wave lengths.

In this section we discuss the formal asymptotics for this regime following (Klein, 1995) and we refer the reader to Schochet (2005) and references therein for subsequent developments on the rigorous theory. Subsection 8.1.1 provides the hierarchy of perturbation equations, Subsect. 8.1.2 discusses their specialization to zero-Mach-number and acoustically compact domains, and in Subsect. 8.1.3 we develop the multiple scales theory for the interaction of long-wave acoustics with small-scale, quasi-incompressible, variable density flow.

8.1.1 Expansions in a Single Time and Multiple Spacial Scales

8.1.1.1 Flow Regime and Key Results. Here we analyze the compressible Euler equations in nondimensional form

$$
\begin{aligned}
\rho_t + \nabla \cdot \boldsymbol{m} &= 0 \\
\boldsymbol{m}_t + \nabla \cdot \left(\boldsymbol{m} \circ \boldsymbol{v} + \frac{1}{\mathrm{M}^2} p \, \mathbf{1} \right) &= 0 \\
e_t + \nabla \cdot (\boldsymbol{v} \, [e + p] \,) &= 0
\end{aligned}
\tag{8.1.1}
$$

where $\rho, \boldsymbol{m}$, and e, are the densities of mass, momentum, and total energy per unit volume, $\boldsymbol{v} \equiv \boldsymbol{m}/\rho$ is the flow velocity, p the pressure, $\mathbf{1}$ denotes the unit tensor, and the $\circ$ symbol indicates the tensorial product. The pressure is related to the conserved variables through the equation of state of a perfect gas,

$$
p = (\gamma - 1) \left[e - \mathrm{M}^2 \frac{1}{2} \frac{\boldsymbol{m}^2}{\rho} \right] \quad \text{with} \quad \gamma = \text{const.} \tag{8.1.2}
$$

To obtain (8.1.1) and (8.1.2), we have used the nondimensionalization

$$
\begin{aligned}
\rho &= \rho' / \rho'_{\text{ref}} \\
\boldsymbol{v} &= \boldsymbol{v}' / u'_{\text{ref}} \\
p &= p' / p'_{\text{ref}} \\
\boldsymbol{x} &= \boldsymbol{x}' / \ell'_{\text{ref}} \\
t &= t' u'_{\text{ref}} / \ell'_{\text{ref}}
\end{aligned}
\quad , \tag{8.1.3}
$$

where primes denote dimensional quantities. Notice that we use a reference velocity, u'_{ref}, independent of the characteristic sound speed $(p'_{\text{ref}}/\rho'_{\text{ref}})^{1/2}$. Through this choice we ensure that the nondimensional velocity, $\boldsymbol{v}$, remains well-defined and finite in the limit of vanishing Mach number parameter

$$
\mathrm{M} = \frac{u'_{\text{ref}}}{(p'_{\text{ref}}/\rho'_{\text{ref}})^{1/2}} \to 0 \, . \tag{8.1.4}
$$

We emphasize that M is a global parameter characterizing the normalization, but *not* the local flow Mach number. Furthermore, time is normalized by a characteristic flow time scale, not by a characteristic time for sound wave propagation.

We analyze the behavior of specific solutions of (8.1.1) and (8.1.2) which for $\mathrm{M} \ll 1$ have an expansion

$$
\mathbf{U} = \mathbf{U}^{(0)} + \mathrm{M}\mathbf{U}^{(1)} + \mathrm{M}^2\mathbf{U}^{(2)} \, , \tag{8.1.5}
$$

where

$$
\mathbf{U} = (\rho, \boldsymbol{m}, e, p, \boldsymbol{v}) \tag{8.1.6}
$$

represents the vector of all the unknowns. Each of the expansion functions in (8.1.5) depends on two space variables, that is,

$$
\mathbf{U}^{(j)} = \mathbf{U}^{(j)}(\boldsymbol{x}, \boldsymbol{\xi}, t) \, , \qquad \text{where} \qquad \boldsymbol{\xi} = \mathrm{M}\boldsymbol{x} \, , \tag{8.1.7}
$$

with $\boldsymbol{x}$ resolving small-scale entropy fluctuations and vortical structures, while $\boldsymbol{\xi}$ is an acoustic scale coordinate.

In this section, we explore the consequences of the ansatz in (8.1.5), using (1) identification of terms multiplied by the same powers of the Mach number and (2) the condition of sublinear growth in $\boldsymbol{x}$ of the $\mathbf{U}^{(i)}$ to derive the dependencies of the ansatz functions $\mathbf{U}^{(i)}(\boldsymbol{x}, \boldsymbol{\xi}, t)$ on both $\boldsymbol{x}$ and $\boldsymbol{\xi}$. As usual in multiple-scales asymptotics, we will *assume* that the small-scale averages arising in the derivations of the sublinear growth conditions exist when needed.

The principal results of the asymptotics are as follows: The key singularity of the Euler equations as $\mathrm{M} \to 0$ arises from the pressure gradient term, $\frac{1}{\mathrm{M}^2}\nabla p$, in the momentum equation. In the asymptotic limit we will find that

$$\nabla_{\boldsymbol{x}} p^{(0)} = \nabla_{\boldsymbol{\xi}} p^{(0)} = \nabla_{\boldsymbol{x}} p^{(1)} = 0 . \tag{8.1.8}$$

As the leading-order pressure is homogeneous in space in terms of both $\boldsymbol{x}$ and $\boldsymbol{\xi}$, the same holds for its time derivative, $dp^{(0)}/dt$, and, following (8.1.2), for $\partial e/\partial t$. This, in turn, amounts to a constraint on the divergence of the energy flux, $\nabla \cdot (\boldsymbol{v}\,[e+p])$, which must be constant in space as well. Finally, because in the limit $[e+p]$ becomes constant in space, too, we obtain the divergence constraint on the velocity field,

$$\nabla_{\boldsymbol{x}} \cdot \boldsymbol{v}^{(0)} = -\frac{1}{[e+p]^{(0)}(t)} \frac{d}{dt} \frac{p^{(0)}}{\gamma - 1}(t) = -\left(\frac{1}{\gamma p^{(0)}} \frac{dp^{(0)}}{dt}\right)(t) . \tag{8.1.9}$$

In the special case of a time-independent leading-order pressure, this reduces to the zero divergence constraint of incompressible flows, that is, to $\nabla_{\boldsymbol{x}} \cdot \boldsymbol{v}^{(0)} = 0$.

Importantly, however, incompressibility does *not* imply constant density! In fact, in a setting where $\nabla_{\boldsymbol{x}} \cdot \boldsymbol{v}^{(0)} \equiv 0$, one finds from $(8.1.1)_1$ that

$$\rho_t^{(0)} + \boldsymbol{v}^{(0)} \cdot \nabla_{\boldsymbol{x}} \rho^{(0)} = 0 , \tag{8.1.10}$$

which describes advection of small-scale, leading-order density variations along particle paths. Incidentally, the mass conservation equation does *not* yield the divergence-free condition in the present conservative formulation.

The asymptotic analysis of the second-order momentum and first order energy equations shows that

1. the second-order pressure $p^{(2)}$ reduces to the standard pressure variable for incompressible flows that satisfies the well-known Poisson equation as $\mathrm{M} \to 0$ and
2. the first order pressure $p^{(1)}$ appears as an acoustic wave amplitude which, under certain restrictions, satisfies the linearized, nonconstant coefficient wave equation

$$p_{tt}^{(1)} - \nabla_{\boldsymbol{\xi}} \cdot \left(c_0^2(\boldsymbol{\xi})\, \nabla_{\boldsymbol{\xi}} p^{(1)}\right) = 0 , \tag{8.1.11}$$

where the signal velocity $c_0 = (\gamma P_0/\rho_0)(\boldsymbol{\xi})$ is slowly varying on the time scale considered.

8.1.1.2 Detailed Expansions. Here we describe the hierarchy of perturbation equations that follows from the asymptotic expansion in (8.1.5). Following Klein (1995), we selectively present only those equations that will be relevant in designing a numerical scheme later in Subsect. 8.2.1, but we do not attempt to systematically solve the asymptotic system.

The leading-order continuity equation reads

$$\rho_t^{(0)} + \nabla_{\boldsymbol{x}} \cdot (\boldsymbol{v}^{(0)} \rho^{(0)}) = 0\,. \tag{8.1.12}$$

The leading-, first-, and second-order momentum equations are

$$\begin{aligned} \nabla_{\boldsymbol{x}}\, p^{(0)} &= 0 \\ \nabla_{\boldsymbol{x}}\, p^{(1)} + \nabla_{\boldsymbol{\xi}}\, p^{(0)} &= 0 \\ \boldsymbol{m}_t^{(0)} + \nabla_{\boldsymbol{x}} \cdot (\boldsymbol{m}^{(0)} \circ \boldsymbol{v}^{(0)}) + \nabla_{\boldsymbol{x}}\, p^{(2)} &= -\nabla_{\boldsymbol{\xi}}\, p^{(1)} \end{aligned} \tag{8.1.13}$$

and the set of energy equations is

$$\begin{aligned} e_t^{(0)} + \nabla_{\boldsymbol{x}} \cdot (\boldsymbol{v}h)^{(0)} &= 0\,, \\ e_t^{(1)} + \nabla_{\boldsymbol{x}} \cdot (\boldsymbol{v}h)^{(1)} &= -\nabla_{\boldsymbol{\xi}} \cdot (\boldsymbol{v}h)^{(0)}\,, \\ e_t^{(2)} + \nabla_{\boldsymbol{x}} \cdot (\boldsymbol{v}h)^{(2)} &= -\nabla_{\boldsymbol{\xi}} \cdot (\boldsymbol{v}h)^{(1)}\,, \end{aligned} \tag{8.1.14}$$

where $h = e + p = \rho H$ is the total enthalpy per unit volume.

The first two equations in (8.1.13) are the first to indicate the change of type of the governing equations upon transition from finite to vanishing Mach number as they do not involve a time derivative. One first concludes from $(8.1.13)_1$ that $p^{(0)} = p^{(0)}(\boldsymbol{\xi}, t)$. Then, by integrating $(8.1.13)_2$ in $\boldsymbol{x}$, one obtains

$$\frac{p^{(1)}(\boldsymbol{x}_1, \boldsymbol{\xi}, t) - p^{(1)}(\boldsymbol{x}_0, \boldsymbol{\xi}, t)}{|\boldsymbol{x}_1 - \boldsymbol{x}_0|} = \boldsymbol{n}_{01} \cdot \nabla_{\boldsymbol{\xi}} p^{(0)}\,, \tag{8.1.15}$$

where $\boldsymbol{n}_{01} = (\boldsymbol{x}_1 - \boldsymbol{x}_0)/|\boldsymbol{x}_1 - \boldsymbol{x}_0|$. The sublinear growth condition for $p^{(1)}$ states that the left-hand side of this equation should vanish as $|\boldsymbol{x}_1 - \boldsymbol{x}_0| \to \infty$ for any unit vector $\boldsymbol{n}_{01}$. As a consequence

$$\nabla_{\boldsymbol{\xi}} p^{(0)} \equiv 0 \qquad \text{and} \qquad \nabla_{\boldsymbol{x}} p^{(1)} \equiv 0\,. \tag{8.1.16}$$

Equivalently, there must be functions $P_0(t)$ and $P_1(\boldsymbol{\xi}, t)$ such that

$$\begin{aligned} p^{(0)} &= P_0(t) \\ p^{(1)} &= P_1(\boldsymbol{\xi}, t)\,. \end{aligned} \tag{8.1.17}$$

Consider next the leading-order energy equation, $(8.1.14)_1$, taking into account $(8.1.17)_1$, and the equation of state, (8.1.2). One finds that

$$\nabla_{\boldsymbol{x}} \cdot \boldsymbol{v}^{(0)} = -\frac{1}{\gamma P_0} \frac{dP_0}{dt}\,. \tag{8.1.18}$$

Given these results, we discuss next two very different, yet equally important regimes and draw conclusions from (8.1.18): The first is the regime of acoustically compact domains that are too small to accomodate sound waves with characteristic frequencies comparable to the inverse of the advection time scale over distances of ℓ'_{ref}. The second is the regime of long-wave acoustics, in which we find multiple scales interactions between advective and acoustic modes.

8.1.2 A Single Spatial Scale: Zero Mach Number Flow in Acoustically Compact Domains

In this regime, we assume the overall system dimensions to be comparable with the reference length, ℓ'_{ref}, from (8.1.3). Then we have a single-length scale solution with $\boldsymbol{x} \in V \subset I\!R^d$, where V is the finite flow domain, and d the number of space dimensions considered. The large-scale variable $\boldsymbol{\xi}$ from the multiple scales expansion becomes void, because for any finite $|\boldsymbol{x}|$, we have $\boldsymbol{\xi} = \mathrm{M}\boldsymbol{x} \to 0$ as $\mathrm{M} \to 0$.

This is the regime addressed in the theories of Janzen (1913) and Rayleigh (1916), and for which the formal asymptotics has been shown to hold rigorously, e.g., by Ebin (1982) for flows in a bounded domain and by Klainerman and Majda (1982) for initial value problems. See also (Schochet, 2005).

Integrating (8.1.18) with respect to $\boldsymbol{x}$ yields

$$\frac{d}{dt} \ln P_0^{1/\gamma} = -\frac{1}{|V|} \oint_{\partial V} \boldsymbol{v}^{(0)} \cdot \boldsymbol{n} \, d\sigma \,, \tag{8.1.19}$$

where $|V|$ denotes the size of the flow domain and ∂V its boundary. If the normal velocity $\boldsymbol{v}^{(0)} \cdot \boldsymbol{n}$ is known via boundary conditions all along the system boundary ∂V, then this equation yields the overall pressure change due to compression/expansion from the boundaries. A typical example for this kind of situation is a closed piston-cylinder system, where the piston motion determines the global pressure rise. If, on the other hand, the pressure is imposed at least on one part of the boundary, then, due to the spatial homogeneity of P_0, (8.1.19) yields an integral constraint on the velocities on ∂V.

In both cases (8.1.18) together with (8.1.19) provides a second equation that exhibits a change of type of the governing equations: Either the flow divergence $\nabla_{\boldsymbol{x}} \cdot \boldsymbol{v}^{(0)}$ is determined at any time by the volume flux across the boundary,

$$\nabla_{\boldsymbol{x}} \cdot \boldsymbol{v}^{(0)} = \frac{1}{V} \oint_{\partial V} \boldsymbol{v}^{(0)} \cdot \boldsymbol{n} \, d\sigma \,, \tag{8.1.20}$$

or it is imposed from outside due to an imposed global pressure change. In that case, the right-hand side of (8.1.18), and thus the local flow divergence, are explicitly prescribed. Of course, combinations of these two extremes are possible, for example, in the case of a high-pressure vessel with a small leakage.

Together with the second-order momentum equation, $(8.1.13)_3$, we obtain, for the present single length scale regime, the *zero-Mach-number, variable-density flow equations*:

$$\begin{aligned} \rho_t^{(0)} + \nabla_{\boldsymbol{x}} \cdot (\rho \boldsymbol{v})^{(0)} &= 0 \\ (\rho \boldsymbol{v})_t^{(0)} + \nabla_{\boldsymbol{x}} \cdot (\rho \boldsymbol{v} \circ \boldsymbol{v})^{(0)} + \nabla_{\boldsymbol{x}} p^{(2)} &= 0 \\ \frac{dP_0}{dt} + \nabla_{\boldsymbol{x}} \cdot \left(\gamma P_0 \boldsymbol{v}^{(0)}\right) &= 0\,. \end{aligned} \tag{8.1.21}$$

The present derivations have been formulated consistently in *conservation form* for mass, momentum, and energy. In this setting, the divergence condition for incompressible flows follows from the *energy equation*, not from mass continuity. In fact, the continuity equation $(8.1.21)_1$ plays quite a different role here as is seen when we rewrite it as

$$\rho_t^{(0)} + \boldsymbol{v}^{(0)} \cdot \nabla_{\boldsymbol{x}} \rho^{(0)} = -\rho^{(0)} \nabla_{\boldsymbol{x}} \cdot \boldsymbol{v}^{(0)}\,, \tag{8.1.22}$$

or, using $(8.1.21)_3$,

$$\frac{\mathcal{D} \ln \rho^{(0)}}{\mathcal{D}t} = \frac{1}{\gamma} \frac{d \ln P_0}{dt} \quad \text{where} \quad \frac{\mathcal{D}}{\mathcal{D}t} = \partial_t + \boldsymbol{v}^{(0)} \cdot \nabla\,. \tag{8.1.23}$$

This is the equation for quasistatic adiabatic compression of mass elements along particle paths, which justifies labelling $p^{(0)}$ the "thermodynamic pressure" as is often done in the literature.

We emphasize that for more general equations of state, when the internal energy per unit volume, $\rho e - \rho \boldsymbol{v}^2/2$, does depend nontrivially on both pressure *and* density, the divergence constraint can be derived only by a suitable combination of the mass and energy equations. Consider the more general equation of state

$$\rho e = (\rho e_i)(\rho, p) + \frac{\mathrm{M}^2}{2} \rho \boldsymbol{v}^2\,. \tag{8.1.24}$$

At zero Mach number, only the internal energy contribution, $(\rho e_i)(\rho, p)$, is relevant. Assuming differentiability, the energy conservation law becomes

$$(\rho e_i)_\rho \left(\rho_t + \boldsymbol{v} \cdot \nabla \rho\right) + (\rho e_i)_p p_t + (\rho e_i + p) \nabla \cdot \boldsymbol{v} = 0\,, \tag{8.1.25}$$

where we have already used the fact that $\nabla p \equiv 0$, which is true independently of the specific form of the equation of state. To arrive at a constraint relating pressure changes to the local flow divergence, we have to replace the first term in (8.1.25) with $-\rho(\rho e_i)_\rho \nabla \cdot \boldsymbol{v}$ using the mass conservation law. Then

$$\nabla \cdot \boldsymbol{v} = -\frac{(\rho e_i)_p}{((\rho e_i) + p - \rho(\rho e_i)_\rho)} \frac{dp}{dt}\,. \tag{8.1.26}$$

See also the discussions of this issue in the context of the construction of numerical techniques for low-Mach-number flows in van der Heul et al. (2003) and Nerinckx et al. (2005).

In Subsect. 8.2.1 we will discuss a numerical scheme for the zero-Mach-number flow equations that conserves mass, momentum, and energy, and represents a direct discretization of the asymptotic limit equations in (8.1.21). The present zero-Mach-number limit equations will also provide the relevant mathematical model for the numerical simulations of thermo-acoustic effects in Sect. 8.3.

8.1.3 Multiple Spatial Scales: Long-Wave Acoustics and Baroclinic Small-Scale Flow

In this regime the system dimensions are large compared to ℓ_{ref} and the acoustic scale variable $\boldsymbol{\xi}$ carries nontrivial information. In other words, we assume that $\text{diam}(V) = O(1/\text{M})$ as $\text{M} \to 0$. In contrast to the setting in the previous section, where we discussed acoustically compact domains, we now consider some fixed $\boldsymbol{\xi}$ and then have $|\boldsymbol{x}| = |\boldsymbol{\xi}/\text{M}| \to \infty$ as $\text{M} \to 0$.

As before, integration of (8.1.18) with respect to $\boldsymbol{x}$ over some $\hat{V} \subset V$ then yields

$$\frac{1}{|\hat{V}|} \oint_{\partial \hat{V}} \boldsymbol{v}^{(0)} \cdot \boldsymbol{n} \, d\sigma = -\frac{1}{\gamma P_0} \frac{dP_0}{dt} . \tag{8.1.27}$$

Let us now parameterize $\hat{V}$ in terms of a finite size, Mach number independent domain $\hat{\mathcal{V}}$, such that for all $\boldsymbol{x} \in \hat{V}$ there is a $\boldsymbol{\xi} \in \hat{\mathcal{V}}$ independent of M with $\boldsymbol{x}(\boldsymbol{\xi}, \text{M}) = \boldsymbol{\xi}/\text{M}$. Then we will show that the l.h.s. of (8.1.27) vanishes as $\text{M} \to 0$ so that $dP_0/dt \equiv 0$, and $P_0 \equiv P_\infty = \text{const}$. To this end, we first observe that

$$\left| \frac{1}{|\hat{V}|} \oint_{\partial \hat{V}} \boldsymbol{v}^{(0)} \cdot \boldsymbol{n} \, d\sigma \right| \leq c \, \frac{|\partial V|}{|V|} \sup_{\boldsymbol{x} \in \partial V} \left| \boldsymbol{v}^{(0)}(\boldsymbol{x}, \boldsymbol{\xi}, t) \right| . \tag{8.1.28}$$

Using the parameterization for V introduced above, which implies $|\partial V|/|V| = \text{M}\, |\partial \mathcal{V}|/|\mathcal{V}|$ with $|\partial \mathcal{V}|/|\mathcal{V}|$ independent of M, we conclude that

$$c \, \frac{|\partial V|}{|V|} \sup_{\boldsymbol{x} \in \partial V} \left| \boldsymbol{v}^{(0)}(\boldsymbol{x}, \boldsymbol{\xi}, t) \right| = c \, \frac{|\partial \mathcal{V}|}{|\mathcal{V}|} \sup_{\boldsymbol{\xi}^* \in \partial \mathcal{V}} \left| \text{M} \, \boldsymbol{v}^{(0)}(\boldsymbol{\xi}^*/\text{M}, \boldsymbol{\xi}, t) \right| . \tag{8.1.29}$$

Now sublinear growth for $\boldsymbol{v}^{(0)}$ in terms of the fast coordinate, $\boldsymbol{x}$, requires that

$$\lim_{\text{M} \to 0} \left(\text{M} \, \boldsymbol{v}^{(0)}(\boldsymbol{\xi}^*/\text{M}, \boldsymbol{\xi}, t) \right) = 0 . \tag{8.1.30}$$

Collecting (8.1.27)–(8.1.30), we conclude that

$$\frac{dP_0}{dt} \equiv 0 \tag{8.1.31}$$

so that P_0 does not vary on the time scale considered and as a consequence (8.1.18) yields the small-scale divergence condition

$$\nabla_{\boldsymbol{x}} \cdot \boldsymbol{v}^{(0)} = 0\,. \tag{8.1.32}$$

Following (8.1.22), this also implies

$$\rho_t^{(0)} + \boldsymbol{v}^{(0)} \cdot \nabla_{\boldsymbol{x}}\, \rho^{(0)} = 0 \tag{8.1.33}$$

such that the leading-order density is just advected along the (small-scale) particle paths.

As an intermediate result we collect here the leading-order mass and energy balances and the second-order momentum equation with the relations from (8.1.32) and (8.1.33) inserted, which yields

$$\begin{aligned} \rho_t^{(0)} + \nabla_{\boldsymbol{x}} \cdot (\rho\boldsymbol{v})^{(0)} &= 0 \\ (\rho\boldsymbol{v})_t^{(0)} + \nabla_{\boldsymbol{x}} \cdot (\rho\boldsymbol{v} \circ \boldsymbol{v})^{(0)} + \nabla_{\boldsymbol{x}} p^{(2)} &= -\nabla_{\boldsymbol{\xi}} P_1 \\ \nabla_{\boldsymbol{x}} \cdot \boldsymbol{v}^{(0)} &= 0\,. \end{aligned} \tag{8.1.34}$$

These equations are closed except for the yet unknown acoustic pressure gradient $\nabla_{\boldsymbol{\xi}} P_1$ in the momentum equation. To close the sytem, we have to derive appropriate long-wave acoustic evolution equations through systematic use of sublinear growth conditions.

To this end we proceed as in (8.1.27)–(8.1.32) and integrate $(8.1.34)_2$ in $\boldsymbol{x}$ over some $\hat{V} \subset V$, parameterize $\hat{V}$ by some finite size domain $\hat{\mathcal{V}}$ so that $\mathrm{M}\boldsymbol{x} \in \hat{\mathcal{V}}$, let $\mathrm{M} \to 0$, and obtain

$$\overline{(\rho\boldsymbol{v})_t^{(0)}} + \nabla_{\boldsymbol{\xi}} P_1 = 0\,. \tag{8.1.35}$$

Here the overbar denotes the small-scale average, that is,

$$\overline{(\rho\boldsymbol{v})^{(0)}}(\boldsymbol{\xi}, t) = \lim_{\mathrm{M}\to 0} \frac{1}{|\hat{\mathcal{V}}|} \int_{\hat{\mathcal{V}}} (\rho\boldsymbol{v})^{(0)}(\boldsymbol{\xi}^*/\mathrm{M}, \boldsymbol{\xi}, t)\, d\boldsymbol{\xi}^*\,. \tag{8.1.36}$$

It is generally *assumed* in formal multiple scales asymptotics that such limits are independent of the particular choice of $\hat{\mathcal{V}}$ as long as $\boldsymbol{\xi} \in \hat{\mathcal{V}}$.

To see the physics of (8.1.35), we combine it with the analogous $\boldsymbol{x}$-average of the first order energy equation in $(8.1.14)_2$. Taking into account that

$$e^{(0)} = \frac{1}{\gamma - 1} P_0\,, \tag{8.1.37}$$

we find that

$$P_{1,t} + \nabla_{\boldsymbol{\xi}} \cdot \left(\gamma P_0\, \overline{\boldsymbol{v}^{(0)}}\right) = 0\,. \tag{8.1.38}$$

Consider for a moment some flow field that features order $O(1)$ variation of the density only on the acoustic scale so that

$$\rho^{(0)} \equiv \overline{\rho}_0(\boldsymbol{\xi}, t)\,. \tag{8.1.39}$$

As the sublinear growth condition derived from equation $(8.1.34)_1$ via small-scale averaging yields

$$\overline{\rho^{(0)}}_t \equiv 0\,, \tag{8.1.40}$$

we conclude that $\rho^{(0)}$ can vary only on time scales much longer than the reference time $t_{\text{ref}} = \ell_{\text{ref}}/u_{\text{ref}}$. In fact, since the density distribution is mainly advected by the flow, leading-order changes of $\overline{\rho}_0$ can occur only when mass elements have passed $O(\ell_{\text{ref}}/\text{M})$ distances, that is, over times of order t_{ref}/M. Using (8.1.39) and (8.1.40), we find

$$\nabla_{\boldsymbol{\xi}} \cdot \left(\gamma P_0 \overline{\boldsymbol{v}^{(0)}}\right) = \nabla_{\boldsymbol{\xi}} \cdot \left(\frac{\gamma P_0}{\rho_0}\ \overline{(\rho \boldsymbol{v})^{(0)}}\right) \tag{8.1.41}$$

for the second term in (8.1.38). Together with (8.1.38), this is the system of nonconstant coefficient linearized acoustics

$$\begin{aligned} \overline{\boldsymbol{m}^{(0)}}_t + \nabla_{\boldsymbol{\xi}} P_1 &= 0 \\ P_{1,t} + \nabla_{\boldsymbol{\xi}} \cdot \left(\overline{c}^2\, \overline{\boldsymbol{m}^{(0)}}\right) &= 0\,, \end{aligned} \tag{8.1.42}$$

where

$$\overline{c}^2(\boldsymbol{\xi}) = \frac{\gamma P_0}{\rho_0(\boldsymbol{\xi})} \qquad \text{and} \qquad \boldsymbol{m}^{(0)} = \rho^{(0)} \boldsymbol{v}^{(0)}\,. \tag{8.1.43}$$

This description of the long-wave acoustic modes is valid if $\rho^{(0)} \equiv \rho_0(\boldsymbol{\xi})$ and may be transformed into the linear, nonconstant coefficient wave equation for P_1

$$P_{1,tt} - \nabla_{\boldsymbol{\xi}} \cdot \left(\overline{c}^2(\boldsymbol{\xi})\ \ \nabla_{\boldsymbol{\xi}} P_1\right) = 0\,. \tag{8.1.44}$$

Notice that we do *not* recover this system of linearized acoustics if we allow for $O(1)$-variations of the leading-order density on the small reference length scale. Instead, with the separation of $\rho^{(0)}$ and $\boldsymbol{v}^{(0)}$ into $\boldsymbol{x}$-scale fluctuations and $\boldsymbol{\xi}$-scale averages according to

$$\begin{aligned} \rho^{(0)} &= \widetilde{\rho^{(0)}}(\boldsymbol{x}, \boldsymbol{\xi}, t) + \overline{\rho^{(0)}}(\boldsymbol{\xi}, t) \\ \boldsymbol{v}^{(0)} &= \widetilde{\boldsymbol{v}^{(0)}}(\boldsymbol{x}, \boldsymbol{\xi}, t) + \overline{\boldsymbol{v}^{(0)}}(\boldsymbol{\xi}, t)\,, \end{aligned} \tag{8.1.45}$$

one has

$$\nabla_{\boldsymbol{\xi}} \cdot \left(\gamma P_0 \overline{\boldsymbol{v}^{(0)}}\right) = \nabla_{\boldsymbol{\xi}} \cdot \left(\overline{c}^2\ \overline{\boldsymbol{m}^{(0)}}\right) - \nabla_{\boldsymbol{\xi}} \cdot \left(\overline{c}^2\ \overline{\widetilde{\rho^{(0)}}\widetilde{\boldsymbol{v}^{(0)}}}\right)\,. \tag{8.1.46}$$

In this situation, we find a modified acoustic system that couples to small-scale fluctuations through the accumulation of small-scale density–velocity correlations in the energy flux,

$$\begin{aligned} \overline{\boldsymbol{m}^{(0)}}_t + \nabla_{\boldsymbol{\xi}} P_1 &= 0 \\ P_{1,t} + \nabla_{\boldsymbol{\xi}} \cdot \left(\overline{c}^2 \, \overline{\boldsymbol{m}^{(0)}} \right) &= \nabla_{\boldsymbol{\xi}} \cdot \left(\overline{c}^2 \, \overline{\widetilde{\rho^{(0)}} \widetilde{\boldsymbol{v}^{(0)}}} \right) . \end{aligned} \tag{8.1.47}$$

Instead of writing the long-wave evolution equations as a system for $\left(\overline{\boldsymbol{m}^{(0)}}, P_1\right)$, one may as well use $\left(\overline{\boldsymbol{v}^{(0)}}, P_1\right)$ as the primary variables. This has been done in Klein (1995) and Munz et al. (2003) and leads to the alternative system

$$\begin{aligned} \overline{\boldsymbol{v}^{(0)}}_t + \frac{1}{\overline{\rho^{(0)}}} \nabla_{\boldsymbol{\xi}} P_1 &= -\frac{1}{\overline{\rho^{(0)}}} \left(\overline{\widetilde{\rho^{(0)}} \widetilde{\boldsymbol{v}^{(0)}}} \right)_t \\ P_{1,t} + \gamma P_0 \nabla_{\boldsymbol{\xi}} \cdot \overline{\boldsymbol{v}^{(0)}} &= 0 . \end{aligned} \tag{8.1.48}$$

We notice that, in contrast to (8.1.47), this system is *not* in conservation form.

8.1.3.1 Single Time/Multiple Space Scale Weakly Compressible Flow. Here we summarize the multiple scales equations derived in the previous section. To streamline the notation, we drop here the $(\ )^{(0)}$ superscript on the averages and fluctuations of the leading-order density and momentum, $\rho, \boldsymbol{m}$ as well as on the leading-order velocity, $\boldsymbol{v}$. Thus we have

$$\begin{aligned} \rho^{(0)}(\boldsymbol{x}, \boldsymbol{\xi}, t) &= \widetilde{\rho}(\boldsymbol{x}, \boldsymbol{\xi}, t) + \overline{\rho}(\boldsymbol{\xi}, t) \\ \boldsymbol{m}^{(0)}(\boldsymbol{x}, \boldsymbol{\xi}, t) &= \widetilde{\boldsymbol{m}}(\boldsymbol{x}, \boldsymbol{\xi}, t) + \overline{\boldsymbol{m}}(\boldsymbol{\xi}, t) . \end{aligned} \tag{8.1.49}$$

In addition we define an analogous decomposition of the velocity field through

$$\boldsymbol{v} = \frac{\widetilde{\boldsymbol{m}} + \overline{\boldsymbol{m}}}{\widetilde{\rho} + \overline{\rho}} = \widetilde{\boldsymbol{v}} + \overline{\boldsymbol{v}} . \tag{8.1.50}$$

Unique decompositions of $\rho^{(0)}$, $\boldsymbol{m}^{(0)}$, and $\boldsymbol{v}$ into $\boldsymbol{x}$-averages and fluctuations are achieved by requiring that

$$\overline{\widetilde{\rho}} = 0 , \qquad \overline{\widetilde{\boldsymbol{m}}} = \overline{\widetilde{\boldsymbol{v}}} = 0 . \tag{8.1.51}$$

Subtracting the long-wave mass and momentum balances in (8.1.40) and $(8.1.47)_1$ from the respective equations in (8.1.34), we obtain evolution equations for the small-scale components of the flow variables

$$\begin{aligned} \widetilde{\rho}_t + \nabla_{\boldsymbol{x}} \cdot \widetilde{\boldsymbol{m}} &= 0 \\ \widetilde{\boldsymbol{m}}_t + \nabla_{\boldsymbol{x}} \cdot \left((\widetilde{\rho} + \overline{\rho}) \, \boldsymbol{v} \circ \boldsymbol{v} \right) + \nabla_{\boldsymbol{x}} p^{(2)} &= 0 \\ \nabla_{\boldsymbol{x}} \cdot \boldsymbol{v} &= 0 . \end{aligned} \tag{8.1.52}$$

The long-wave components evolve according to (8.1.40) and (8.1.47), that is,

$$\overline{\rho}_t = 0$$
$$\overline{\boldsymbol{m}}_t + \nabla_{\boldsymbol{\xi}} P_1 = 0 \tag{8.1.53}$$
$$P_{1,t} + \nabla_{\boldsymbol{\xi}} \cdot \left(\overline{c}^2\, \overline{\boldsymbol{m}}\right) = \nabla_{\boldsymbol{\xi}} \cdot \left(\overline{c}^2\, \overline{\widetilde{\rho \boldsymbol{v}}}\right) ,$$

where

$$\overline{c}^2 = \frac{\gamma P_0}{\overline{\rho}} . \tag{8.1.54}$$

This completes the separation of the asymptotic multiple scales equations into small-scale and long-wave components that are coupled through (1) the net large-scale energy flux divergence that results from density and velocity correlations $\overline{\widetilde{\rho \boldsymbol{v}}}$ in $(8.1.53)_3$, and (2) the appearance of $\boldsymbol{v} \neq \widetilde{\boldsymbol{m}}/\widetilde{\rho}$ in the small-scale advective momentum and energy fluxes in $(8.1.52)_{2,3}$.

8.1.4 Localized Small-Scale Flow and Multiple Time Scales: Thermoacoustics

In the previous section, we discussed situations in which long-wave modes and small-scale structures coexist essentially everywhere within the flow domain. Important applications differ from this pattern in that they involve the interaction of long-wave acoustics with a localized small-scale structure. An interesting example is thermoacoustic refrigeration, the mechanics of which we will now discuss briefly.

8.1.4.1 Thermoacoustic Refrigerator. Consider a long, straight resonance tube, closed at one end and equipped with a loudspeaker at the other. At a distance of about one quarter of the length of the tube away from the closed end, there are three narrowly spaced stacks of heat exchanger plates, as sketched in Fig. 8.1 (top, lower graph). The two outer stacks are in contact with the environment and with the object to be cooled, the middle stack is isolated from the environment and from the neighboring stacks and functions as a medium for the transport of heat between the cold and the hot heat exchanger stacks.

Assume that a standing acoustic wave with a wavelength equal to four times the tube length is being maintained by the loudspeaker (Fig. 8.1, top, upper graph), and consider the motion of fluid parcels within the tube (Fig. 8.1, bottom left, upper graph). Because of velocity perturbations induced by the pressure wave, a parcel will be displaced in an oscillatory fashion and it will undergo pressure and temperature fluctuations. For example, a parcel near the closed end of the tube moves toward the closed end as the medium reaches maximum compression. When the gas expands in the opposite phase of the cycle, its pressure and temperature drop and it moves away from the

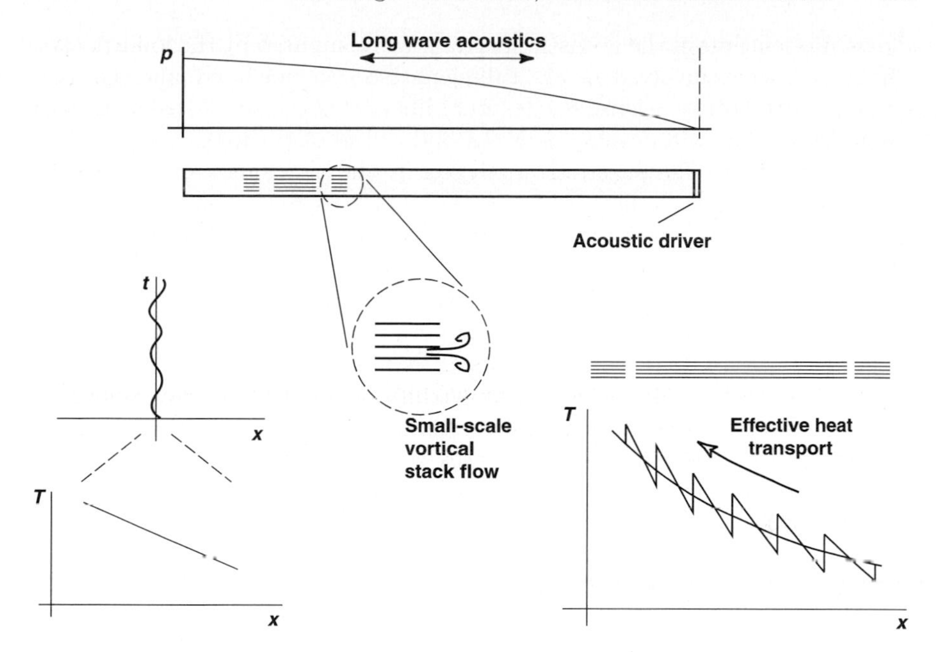

Fig. 8.1. Conceptual sketches illustrating the mechanism enabling thermoacoustic refrigeration. *Top left*: Quarter wavelength acoustic standing wave and resonance tube with heat exchanger stacks. *Bottom left*: A parcel's motion between the plates in a space–time diagram and the temperature gradient it traces because of the cyclic adiabatic compression and expansion. *Bottom right*: Mechanism of cold-to-hot heat transfer along the middle stack. Adapted from Klein (2005)

closed end. Thus, fluid particles generally trace a local temperature gradient by this oscillatory motion (Fig. 8.1, bottom left, lower graph).

If the stack has thin, closely spaced plates, then axial conduction along the plates is minute, while heat exchange between the gas and the plates is intense. Initially, the plates are in thermal equilibrium with the gas, which in turn has constant temperature. A parcel of the gas will, when moving toward the closed end of the tube, become hotter than the plate locally and it will release heat to the plate. As it moves to the left and expands, it becomes colder than the plate and receives heat. The net effect of one oscillatory cycle is to move heat along the plate. Over many cycles, unless the plate is heated, its left end and the surrounding gas are cooled; ultimately, the plates will acquire a temperature distribution that at every point has a gradient similar to that traced by the parcels during their acoustic motion. If heat is added to the right end at some temperature lower than ambient but higher than the plate's long-time equilibrium temperature, a persistent heat transport from this low-temperature reservoir to the ambient is established.

The flow field within such a device is a perfect example of a *single time scale/multiple space scale* low-Mach-number flow. Its characteristic time is the oscillation time of the acoustic standing wave imposed by the loudspeaker. The spatial scales involved are the tube length on the one hand, and the stack spacing on the other. Figure 8.1 (center) illustrates the small-scale, strongly vortical flow that will develop near the end of the plate stack.

In contrast to the situation analyzed in the previous section, the appearance of small-scale flow structures is restricted to the immediate vicinity of the stack of plates. In comparison with the acoustic wavelength (the tube length) this region is small, and we have a multiple scales regime with localized small-scale flow. In terms of the techniques of asymptotic analysis, this implies matched asymptotic expansions in space as the appropriate means of analysis. This will be discussed shortly.

However, the thermoacoustics mechanism explained in the context of Fig. 8.1 requires very many acoustic cycles to produce a sizeable result. A full asymptotic description that faithfully represents these cumulative effects will have to account for multiple time scales as well. The relevant theory of weakly nonlinear (resonant) acoustics was developed in the 1980s and 1990s (see Hunter et al., 1986; Majda et al., 1988; Klein and Peters, 1988; Joly et. al., 1993; Schochet, 1994a, 1994b; Meister, 1999). For more recent developments, see Abgrall and Guillard (2005) and Schochet (2005) and references therein. An analysis of weakly nonlinear effects in the context of thermoacoustics is found in Karpov and Prosperetti (2002).

8.1.4.2 Matched Asymptotics Expansions in Space. Here we discuss how short-time acoustic motions drive the small-scale flow within a thermoacoustic stack using techniques of matched asymptotic expansions. Outside of the plate stack there is no small-scale flow and the flow will be essentially one-dimensional. Thus we expand the solution in the form

$$\mathbf{U}(x,t;\mathrm{M}) = \begin{pmatrix} \rho \\ u \\ p \end{pmatrix}(x,t;\mathrm{M}) = \mathbf{U}^{(0)} + \mathrm{M}\mathbf{U}^{(1)} + \mathrm{M}^2\mathbf{U}^{(1)} + \cdots \qquad (8.1.55)$$

where the expansion functions

$$\mathbf{U}^{(i)} = \begin{pmatrix} R \\ U \\ P \end{pmatrix}^{(i)} \qquad (8.1.56)$$

have the signature

$$\mathbf{U}^{(i)}(\xi,t) \qquad \text{with} \qquad \xi = \mathrm{M}x\,. \qquad (8.1.57)$$

It will be sufficient to consider the Euler equations in the region outside of the plate stack.

From the analysis in Subsect. 8.1.3 we conclude that, for homentropic initial data in the tube, the leading-order thermodynamic variables will be constant in time so that

$$P^{(0)}(\xi, t) \equiv P_0 = \text{const} \qquad \text{and} \qquad R^{(0)}(\xi, t) \equiv R_0 = \text{const}\,. \tag{8.1.58}$$

Then, on the time scale of a single acoustic cycle, the flow in the tube will be described by the one-dimensional linear acoustic equations,

$$\begin{aligned} R_0\, U_t^{(0)} + P_\xi^{(1)} &= 0 \\ P_t^{(1)} + c_0^2 R_0\, U_\xi^{(0)} &= 0\,. \end{aligned} \tag{8.1.59}$$

The boundary data require zero velocity at the closed end and an oscillatory motion about a fixed central position at the end of the acoustic driver, that is,

$$U(0, t) = 0\,, \qquad U(L, t) = U_d(t)\,, \tag{8.1.60}$$

with $U_d(t)$ the prescribed velocity of the acoustic driver.

To begin with, the linear acoustic equations in (8.1.59) hold only in the subintervals $0 \le \xi < \xi_{\text{st}}$ and $\xi_{\text{st}} < \xi \le L$, where ξ_{st} is the position of the stack of plates. Conditions connecting the linear acoustic solutions on both sides of the stack have to be derived from an analysis of the local small-scale flow around the stack plates.

Leading-Order Small-Scale Flow Around the Stack. The stack length in the direction of the tube axis is assumed to be comparable to our reference scale. There are (at least) two options for the scaling of the plate separation distance (in the direction normal to the tube axis): one option is to not assume an asymptotic scaling of the plate separation with the Mach number, that is, to assume it to be comparable to the plate length as $\mathrm{M} \to 0$. An alternative option is to let the plate separation distance be of order $O(\mathrm{M})$ as $\mathrm{M} \to 0$.

Here we intend to prepare for the numerical simulations in Sect. 8.3, for which the first option is relevant. Then, the analysis of Subsect. 8.1.3, specialized to include only the smaller of the two length scales, immediately yields the constant density, incompressible Navier–Stokes equations as the leading-order flow model. That is,

$$\begin{aligned} \boldsymbol{u}_t^{(0)} + \nabla_{\boldsymbol{x}} \cdot (\boldsymbol{u}^{(0)} \circ \boldsymbol{u}^{(0)}) + \frac{1}{\rho_0} \nabla_{\boldsymbol{x}} p^{(2)} &= -\nabla_{\boldsymbol{x}} \cdot \boldsymbol{\tau}[\boldsymbol{u}^{(0)}] \\ \nabla_{\boldsymbol{x}} \cdot \boldsymbol{u}^{(0)} &= 0\,. \end{aligned} \tag{8.1.61}$$

Here $\boldsymbol{\tau}[\boldsymbol{u}^{(0)}]$ represents the equation of state for the stress tensor evaluated with the leading-order flow velocity.

We conclude first that the small-scale flow field in the regime considered here is a fully nonlinear Navier–Stokes flow. However, it is also a constant density, that is, constant temperature, flow, so that any thermal effects can show up at most at first order in the Mach number. To describe these effects, one will have to consider the next-order perturbation equations both for the external, long-wave motion and for the flow around the plate stack.

Before doing so, we can derive the still missing connection conditions for the partial acoustic solutions outside the stack. As we have seen in the earlier multiple scales analysis, the first order pressure, $p^{(1)}$, has no variation on the small-scale (see (8.1.17)). We conclude that $p^{(1)}$ must be constant across the plate stack. This is the first matching condition, that is,

$$\lim_{\xi \nearrow \xi_{\mathrm{st}}} P^{(1)}(\xi, t) = \lim_{\xi \searrow \xi_{\mathrm{st}}} P^{(1)}(\xi, t) , \tag{8.1.62}$$

so that the first order pressure will be continuous across the stack.

Far upstream and downstream of the stack, the flow will match into the one-dimensional flow of the long-wave acoustic modes. As the small-scale flow is divergence free, these cross-sectional average velocities (or mass fluxes) will not change across the plate stack, and we obtain

$$\lim_{\xi \nearrow \xi_{\mathrm{st}}} U^{(0)}(\xi, t) = \lim_{\xi \searrow \xi_{\mathrm{st}}} U^{(0)}(\xi, t) . \tag{8.1.63}$$

With both $P^{(1)}$ and $U^{(0)}$ being continuous across the stack and the linear acoustic equations (8.1.59) being a first order hyperbolic system, we conclude that the long-wave equations simply hold within the entire tube and that the stack can at most influence long-wave motion at higher orders in the Mach number.

In turn, we can now draw conclusions regarding the boundary conditions for the small-scale flow. As the long-wave acoustics evolves independently of the flow within the stack of plates at leading order, the mean mass flux through the stack is prescribed as a function of time. The flow outside the stack is assumed to be homentropic and due only to the motion of acoustic waves. As a consequence, it will be irrotational and at large distances it will be one-dimensional. This defines the matching conditions for the small-scale flow:

$$u(x, y, t) \to \pm U(\xi_{\mathrm{st}}, t) , \quad v(x, y, t) \to 0 \qquad \text{as} \qquad x \to \pm\infty . \tag{8.1.64}$$

8.1.4.3 Temperature Evolution. From the compressible Navier–Stokes equations in (2.1.2)–(2.1.6) we can derive the following reaction–diffusion equation for temperature

$$T_t + \boldsymbol{u} \cdot \nabla_{\boldsymbol{x}} T = (\gamma - 1) \left(\frac{1}{\rho} \nabla \cdot (\lambda \nabla T) - T \nabla \cdot \boldsymbol{u} \right) + \mathrm{M}^2 \mathrm{Diss} , \tag{8.1.65}$$

where "Diss" denotes viscous dissipation.

According to the previous derivations, the leading-order temperature, $T^{(0)} = p_0/\rho_0$, is constant so that the leading-order version of (8.1.65) is solved trivially. The first order temperature equation reads

$$T^{(1)}_t + \boldsymbol{u}^{(0)} \cdot \nabla_{\boldsymbol{x}} T^{(1)} = (\gamma - 1) \left(\frac{1}{\rho} \nabla \cdot (\lambda \nabla T^{(1)}) - T^{(0)} \nabla \cdot \boldsymbol{u}^{(1)} \right) . \tag{8.1.66}$$

This is a linear advection–diffusion equation with the first order flow divergence entering as a source term, and this remains to be determined.

Just as for the flow outside of the stack, the flow divergence is related to the temporal variation of the first order pressure. In fact, the analogue of $(8.1.59)_2$ for the small-scale flow within the plate stack reads

$$\nabla \cdot \boldsymbol{u}^{(1)} = -\frac{1}{\gamma P_0} p_t^{(1)} . \tag{8.1.67}$$

As we have seen above, the first order pressure in the plate stack matches that of the acoustic mode outside of the stack, evaluated at the (large-scale) stack location. Therefore, the first order flow divergence is simply imposed by the acoustics, and we have

$$p_t^{(1)}(\boldsymbol{x}, t) = \frac{d}{dt} P^{(1)}(t, \xi_{\mathrm{st}}) , \tag{8.1.68}$$

and the first-order temperature equation becomes

$$T_t^{(1)} + \boldsymbol{u}^{(0)} \cdot \nabla_{\boldsymbol{x}} T^{(1)} = \frac{\gamma - 1}{\rho_0} \left(\nabla \cdot (\lambda \nabla T^{(1)}) + \frac{1}{\gamma} \frac{d}{dt} P^{(1)}(t, \xi_{\mathrm{st}}) \right) . \tag{8.1.69}$$

To determine the time evolution of the local temperature distribution within the plate stack, this equation is to be solved as a coupled problem together with the temperature evolution in the plates themselves. The key effects discussed qualitatively in the context of Fig. 8.1, namely the oscillatory temperature change due to adiabatic compression and the oscillatory motion between the plates, is captured in (8.1.69) via the advection terms and the source term induced by the acoustic pressure variation.

This completes the summary of the short-time problem governing thermoacoustic effects. In Sect. 8.3 we describe the results of numerical simulations that explore these effects, but are based on a slightly more general approach than that of evaluating the asymptotic limit equations just derived. The numerical strategies used in these simulations are described in the next section. The philosophy is to consider the pressure variations in the plate stack not as small perturbations, but to assume that they are $O(1)$. Then, the appropriate limit equations are the zero-Mach-number, variable density Navier–Stokes equations. In this fashion the simulations include weakly nonlinear effects, which become noticeable only after a large number of acoustic oscillations. These effects can be captured via asymptotics only by employing a multiple-time scale analysis as mentioned earlier, and we refer the reader to the published literature.

8.2 Computational Schemes for Low Mach Number Flows

The variety of low-Mach-number flow solvers is, of course, too broad to be covered with reasonable completeness in the present volume. For a comprehensive overview the reader is referred to the vast amount of advanced textbooks on computational fluid dynamics, such as Wesseling (1991), Kröener (1996) and LeVeque (2002). In this section we describe a finite volume projection method (Subsect. 8.2.1) and a finite difference method based on the vorticity-stream function formulation of the flow equations (Subsect. 8.2.2). These schemes have been developed and used successfully in applications by two of the authors.

8.2.1 A Zero Mach Number Godunov-Type Scheme

The zero-Mach-number inviscid flow solver of Schneider et al. (1999) procedes in three steps, which we summarize in the following subsections. An explicit predictor step determines the major effects of advection. Then, a first projection guarantees that the advective fluxes comply with the zero-Mach-number divergence constraint. The major difference between the auxiliary equations solved in the predictor step and the zero-Mach-number Euler equations is the pressure contribution to the momentum flux, and this is accounted for in a second projection step. This step is designed to ensure that the cell-centered velocities at the new time level also satisfy the zero-Mach-number divergence constraint.

Closely related work on flows at low but nonzero Mach number was pursued in close cooperation with A. Meister, C.-D. Munz, Th. Sonar, and coworkers. Both conservative and nonconservative discretizations have been considered in Klein et al. (2001), Munz et al. (2005), Park and Munz (2005) and Roller et al. (2005). For recent developments in theory and numerics of low Mach number flows, see Abgrall and Guillard (2005) and references therein. The scheme to be discussed here, on one hand, is an extension of a Godunov-type scheme for compressible flows. On the other hand, for zero Mach number as considered here, it is a particular version of a project method derived directly from the conservation laws for mass, momentum, and total energy. In the context of projection methods, cf. Chorin (1968a, 1968b), Bell and Marcus (1992), Almgren et al. (1996), Puckett et al. (1997), Almgren et al. (1998), Brown et al. (2001), Minion (2004), and references therein.

8.2.1.1 Explicit Predictor for Advection (Ma = 0).

An Auxiliary, Nonsingular Hyperbolic System. Given cell averages of mass, momentum, and energy at time level t^n, the following auxiliary hyperbolic system is solved over one-time step using an explicit standard second-order Godunov-type scheme to obtain explicit predictions for the fluxes of mass, momentum, and energy:

$$\begin{aligned} \rho_t + \nabla \cdot (\rho \boldsymbol{v}) &= 0\,, \\ (\rho \boldsymbol{v})_t + \nabla \cdot (\rho \boldsymbol{v} \circ \boldsymbol{v} + p\mathbf{1}) &= 0\,, \\ (\rho e)_t + \nabla \cdot (\boldsymbol{v}\,[\rho e + \pi]) &= 0\,, \end{aligned} \tag{8.2.1}$$

$$\rho e = \frac{p}{\gamma - 1}, \tag{8.2.2}$$

$$\pi = P_0(t). \tag{8.2.3}$$

This system features the same convective nonlinearities as the compressible Euler equations but its signal speeds are of order $O(1)$. In fact, the characteristic analysis in Klein (1995) for one space dimension yields the eigenvalues of the flux Jacobian,

$$\lambda_0^* = u\,, \quad \lambda_\pm^* = u \pm c^* \tag{8.2.4}$$

where

$$c^* = \sqrt{c^2 + \frac{\pi - p}{\rho}}\,, \quad c^2 = \frac{\gamma p}{\rho}\,. \tag{8.2.5}$$

This is in contrast to the corresponding eigenvalues of the Euler equations in (8.1.1), (8.1.2) which diverge as $\mathrm{M} \to 0$,

$$\lambda_0 = u\,, \quad \lambda_\pm = u \pm \frac{1}{\mathrm{Ma}}\,c\,. \tag{8.2.6}$$

As a consequence, the explicit convective flux predictor step uses Courant numbers with respect to advection of order $\Delta t\, u_{\mathrm{ref}}/\Delta x = O(1)$ for $\mathrm{M} = 0$ without violating the CFL stability criterion.

Following the standard MUSCL-strategy (van Leer, 1979), we obtain a second-order scheme for the auxiliary system by computing a second-order, cell-centered, discrete flux divergence from numerical fluxes that are centered on the grid cell interfaces. These fluxes are also "time-centered" in that they approximate the exact fluxes of the auxiliary system at time levels $t^{n+\frac{1}{2}} = \frac{1}{2}\left(t^n + t^{n+1}\right)$ to first order. All finite volume results presented in this chapter were obtained using the HLLE-numerical flux (Einfeld, 1988), adapted to the auxiliary system.

Accuracy Estimate for the Predictor Step. The fluxes obtained in the predictor step correspond to a second-order accurate scheme for the auxiliary system. Here we will argue that they also yield first-order accurate updates for density and energy under the zero-Mach-number Euler equations.

The fluxes of the auxiliary system differ from those of the zero-Mach-number Euler equations from (8.1.21) by

$$\boldsymbol{F}_{\mathrm{Euler}} - \boldsymbol{F}_{\mathrm{aux}} = \begin{pmatrix} \rho \boldsymbol{v} \\ \rho \boldsymbol{v} \circ \boldsymbol{v} + \mathbf{1}\,p^{(2)} \\ [\rho e + p]\,\boldsymbol{v} \end{pmatrix} - \begin{pmatrix} \rho \boldsymbol{v} \\ \rho \boldsymbol{v} \circ \boldsymbol{v} + \mathbf{1}\,p \\ [\rho e + \pi]\,\boldsymbol{v} \end{pmatrix} = \begin{pmatrix} 0 \\ \mathbf{1}\left(p^{(2)} - p\right) \\ [p - \pi]\,\boldsymbol{v} \end{pmatrix}. \tag{8.2.7}$$

Suppose that the data computed at some time level t^n are consistent with a solution of the zero-Mach-number Euler equations. Then we have $p^n = \pi^n = P_0(t^n)$ according to (8.1.17) and (8.2.3), and the differences in both the mass and energy fluxes in (8.2.7) vanish. That is,

$$(\boldsymbol{F}_{\text{Euler}} - \boldsymbol{F}_{\text{aux}})^n = \begin{pmatrix} 0 \\ \mathbf{1}\left(p^{(2)} - p\right)^n \\ 0 \end{pmatrix}, \tag{8.2.8}$$

and

$$\left(\frac{\partial \rho}{\partial t}\right)_{\text{Euler}} = \left(\frac{\partial \rho}{\partial t}\right)_{\text{aux}}, \qquad \left(\frac{\partial \rho e}{\partial t}\right)_{\text{Euler}} = \left(\frac{\partial \rho e}{\partial t}\right)_{\text{aux}} \qquad \text{at} \qquad t = t^n. \tag{8.2.9}$$

We conclude that a first-order time accurate update to the half time level $t^{n+\frac{1}{2}}$ for ρ and ρe derived from the auxiliary system is also a first-order update for the Euler equations. Therefore, the approximated values for ρ and ρe at the grid cell faces obtained in the explicit predictor will be sufficiently accurate for our purpose of constructing a second-order accurate scheme. The momentum prediction, however, is inaccurate (zeroth order) because of order $O(1)$ differences between p and $p^{(2)}$.

From here on we will use the notation $\phi^{*,n+\frac{1}{2}}, \phi^{*,n+1}$ to abbreviate values of some state variable, ϕ, corresponding to data as obtained in the predictor step at time levels $\frac{1}{2}\left(t^n + t^{n+1}\right)$ and t^{n+1}, respectively.

8.2.1.2 First Projection: Energy Fluxes and the Advection Velocity Divergence Constraint. After the predictor step, the mass fluxes, $\rho \boldsymbol{v}$, at the grid cell interfaces need to be corrected to account for the misrepresented pressure gradient effect involving ∇p instead of $\nabla p^{(2)}$ in the momentum equation. Low-Mach-number theory states that the second-order pressure $p^{(2)}$ may be considered as a Lagrangian multiplier that guarantees compliance with the zero divergence constraint for the velocity field. We have seen above that this constraint is a consequence of the energy equation in the present conservative formulation. (Recall, however, the earlier consideration for more complex equations in the context of (8.1.26)!) The determining equation for the second-order pressure, $p^{(2)}$, and for the desired correction of $\rho \boldsymbol{v}$, is thus the energy equation. The resulting "first projection" will be analogous to a "MAC-projection" (Almgren et al., 1998), and it involves the solution of a discrete Poisson-type equation for cell-centered pressure variables.

For a grid cell C of our computational mesh energy conservation over a time step $\Delta t = t^{n+1} - t^n$ implies

$$(\rho e)_C^{n+1} - (\rho e)_C^n = -\frac{\Delta t}{|C|} \int_{\partial C} \left([\rho e + p]\boldsymbol{v}\right)^{n+\frac{1}{2}} \cdot \boldsymbol{n}\, d\sigma. \tag{8.2.10}$$

As discussed in the context of (8.2.9), sufficiently accurate data for $(\rho e)^{n+\frac{1}{2}}$ and—via the zero-Mach-number version of the equation of state (8.1.2)—for $p^{n+\frac{1}{2}}$ are available from the explicit predictor step so that we let

$$(\rho, \rho e, p)^{n+\frac{1}{2}} = (\rho, \rho e, p)^{*,n+\frac{1}{2}}. \tag{8.2.11}$$

Without affecting accuracy, we may either extract these data directly during the flux computation or calculate them from the cell averages available before and after the predictor by spatiotemporal interpolation.

To determine $\boldsymbol{v}^{n+\frac{1}{2}}$ in (8.2.10), we have to correct for the discrepancy between the mass fluxes $(\rho\boldsymbol{v})^{*,n+\frac{1}{2}}$, as predicted by the auxiliary system, and $(\rho\boldsymbol{v})^{n+\frac{1}{2}}$ as it would be predicted by the Euler system, (8.2.8). As observed earlier, $\rho^{*,n+\frac{1}{2}}$ approximates $\rho^{n+\frac{1}{2}}$ with sufficient accuracy so that we may write this correction as

$$\boldsymbol{v}^{n+\frac{1}{2}} = \boldsymbol{v}^{*,n+\frac{1}{2}} - \frac{\Delta t}{2} \frac{1}{\rho^{*,n+\frac{1}{2}}} \nabla (p^{(2)} - p)^{n+\frac{1}{2}} . \tag{8.2.12}$$

A discrete approximation to $p^{(2)}$ remains to be determined.

The left-hand side of (8.2.10) is known for $\mathrm{M} = 0$ after evaluation of (8.1.2) for $\mathrm{M} = 0$ and (8.1.19),

$$(\rho e)_C^{n+1} - (\rho e)_C^n = \frac{1}{\gamma - 1} \left(P_0(t^{n+1}) - P_0(t^n) \right) . \tag{8.2.13}$$

Collecting the results from (8.2.10) to (8.2.13), we find

$$\frac{1}{|C|} \int_{\partial C} h^{*,n+\frac{1}{2}} \, \nabla p_*^{(2)} \cdot \boldsymbol{n} \, d\sigma = \text{r.h.s.} , \tag{8.2.14}$$

where

$$\text{r.h.s.} = \frac{2}{\Delta t} \left(\frac{1}{\gamma - 1} \frac{P_0(t^{n+1}) - P_0(t^n)}{\Delta t} + \frac{1}{|C|} \int_{\partial C} (\rho h \boldsymbol{v})^{*,n+\frac{1}{2}} \cdot \boldsymbol{n} \, d\sigma \right) , \tag{8.2.15}$$

$$p_*^{(2)} = (p^{(2)} - p)^{n+\frac{1}{2}} , \tag{8.2.16}$$

and

$$h = \frac{\rho e + p}{\rho} . \tag{8.2.17}$$

We simplify (8.2.14) and (8.2.15) by taking into account that the energy update over Δt from the predictor step reads

$$\frac{(\rho e)_C^{*,n+1} - (\rho e)_C^n}{\Delta t} = - \frac{1}{|C|} \int_{\partial C} (\rho h \boldsymbol{v})^{*,n+\frac{1}{2}} \cdot \boldsymbol{n} \, d\sigma . \tag{8.2.18}$$

Then, using (8.2.13), we may then replace (8.2.14) with

$$\frac{(\Delta t)^2}{2|C|} \int_{\partial C} h^{*,n+\frac{1}{2}} \, \nabla p_*^{(2)} \cdot \boldsymbol{n} \, d\sigma = \frac{P_0(t^{n+1})}{\gamma - 1} - (\rho e)_C^{*,n+1} . \tag{8.2.19}$$

This is a Poisson-type equation for the discrete degrees of freedom of $p_*^{(2)}$. Once it is solved and approximations to $\nabla p_*^{(2)}$ are available at the grid cell interfaces, we correct all velocity-related components of the flux tensor according to

$$\begin{pmatrix} \rho \boldsymbol{v} \\ \rho \boldsymbol{v} \circ \boldsymbol{v} \\ h \boldsymbol{v} \end{pmatrix}^{n+\frac{1}{2}} = \begin{pmatrix} \rho \boldsymbol{v} \\ \rho \boldsymbol{v} \circ \boldsymbol{v} \\ h \boldsymbol{v} \end{pmatrix}^{*,n+\frac{1}{2}} - \frac{\Delta t}{2} \begin{pmatrix} \nabla p_*^{(2)} \\ \nabla p_*^{(2)} \circ \boldsymbol{v} + \boldsymbol{v} \circ \nabla p_*^{(2)} \\ h^* \, \nabla p_*^{(2)} \end{pmatrix}^{n+\frac{1}{2}} . \tag{8.2.20}$$

This completes the description of the first projection.

In their implementation, Schneider et al. (1999) use a standard cell-centered discrete representation of $p^{(2)}$ and the midpoint rule with second-order centered finite differences for $\nabla p^{(2)} \cdot \boldsymbol{n}$ to evaluate the integrals in (8.2.19). This leads to the standard 5- (7-) point stencil for the left-hand side in (8.2.19) in two (three) dimensions, and corresponds with pointwise evaluation of the velocity corrections from (8.2.20) at the cell face centers. Of course, alternative discretizations/quadrature rules can be thought of. This is work in progress.

8.2.1.3 Second Projection: Pressure-Related Momentum Flux. As a result of the second projection step we will be in the position to evaluate the pressure integral in the momentum balance for the finite volume cells, that is, the last term in

$$(\rho \boldsymbol{v})_C^{n+1} - (\rho \boldsymbol{v})_C^{n} = -\frac{\Delta t}{|C|} \left(\int_{\partial C} (\rho \boldsymbol{v} \circ \boldsymbol{v})^{n+\frac{1}{2}} \cdot \boldsymbol{n} \, d\sigma + \int_{\partial C} p^{(2),n+\frac{1}{2}} \boldsymbol{n} \, d\sigma \right) . \tag{8.2.21}$$

To evaluate this integral, we need a representation of the second-order pressure $p^{(2)}$ on the grid cell interfaces and not, as in the first projection, its normal derivative $\nabla p^{(2)} \cdot \boldsymbol{n}$. One option for calculating pressure values on the cell interfaces would be the interpolation of the (cell-centered) $p^{(2)}$ data obtained in the first projection step. Yet, unless careful steps are taken, this can lead to pressure–velocity decoupling and the occurence of spurious high-frequency oscillations (see Rhie and Chow, 1983).

Here we follow Bell and Marcus (1992), Almgren et al. (1996), and Schneider et al. (1999), and introduce a second projection step designed to explicitly control the divergence of the cell-centered velocities through the pressure integral in (8.2.21). Like Bell and Marcus (1992) and Schneider et al. (1999), we introduce node-centered pressure variables (see Fig. 8.2). Pressure integrals over the grid cell interfaces, that is, the pressure contributions to the momentum fluxes, are then computed by the trapezoidal rule. The determining equations for these node-centered pressures are the zero-Mach-number divergence constraints formulated for the dual cells of our grid, that is, for node-centered control volumes such as $\overline{C}$ in Fig. 8.2.

Bell and Marcus (1992) and Schneider et al. (1999) employ the standard trapezoidal rule to evaluate the node-centered divergences. The resulting discrete pressure Poisson problem allows for a checkerboard decoupling of the

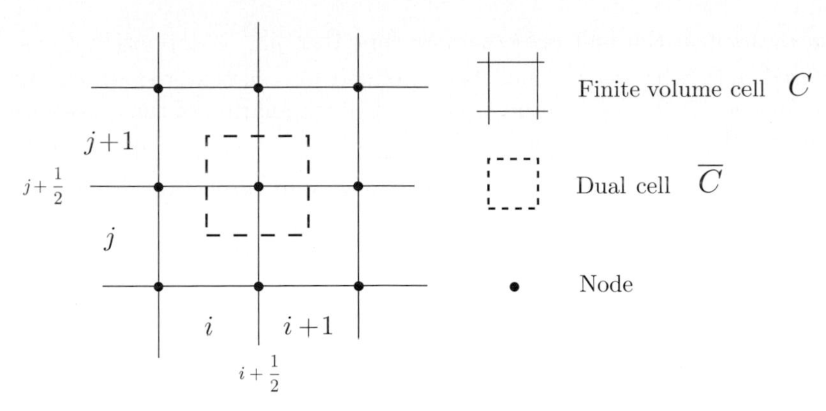

Fig. 8.2. Pressures are associated with cell vertices $(x_{i+\frac{1}{2}}, y_{j+\frac{1}{2}})$ in the second projection

nodal pressures. Careful choices of the Poisson solver and the initial guesses for the pressure solves allow one to control this mode, (Bell and Marcus, 1992; Schneider et al., 1999), and the results presented below all rely on this scheme.

The potential unphysical checkerboard mode, and the fact that it is very difficult to control for problems with realistic boundary conditions or in grid-adaptive computations, have led to the development of approximate projection methods (Almgren et al., 1996). These operate with a stable and computationally less sensitive Poisson stencil borrowed from finite element schemes, yet at the cost of controlling the discrete divergence only up to the truncation error of the scheme instead of up to machine accuracy. The development of a scheme that combines stability with exact projection (divergence control up to machine accuracy) is work in progress (Vater, 2005).

The determining equation for the node-based pressures is again a zero-Mach-number divergence constraint for some energy fluxes. This time it is written as an integral energy balance for the dual cells, $\overline{C}$, shown in Fig. 8.2,

$$(\rho e)_{\overline{C}}^{n+1} - (\rho e)_{\overline{C}}^{n} = -\frac{\Delta t}{|\overline{C}|} \int_{\partial \overline{C}} \frac{1}{2} \left((\rho h \boldsymbol{v})^n + (\rho h \boldsymbol{v})^{n+1} \right) \cdot \boldsymbol{n} \, d\sigma \, . \qquad (8.2.22)$$

The data for $h^{n+1} = \frac{\gamma}{\gamma-1} P_0(t^{n+1})/\rho^{n+1}$ are available from the previous calculations: $P_0(t^{n+1})$ is known, because $\mathrm{M} = 0$ as discussed earlier, and local approximations to ρ^{n+1} are known, because the density update has been completed already after the first predictor step. Local distributions of $(\rho \boldsymbol{v})^{n+1}$ on $\partial \overline{C}$ are constructed by using

$$(\rho \boldsymbol{v})^{n+1} = (\rho \boldsymbol{v})^{*,n+1} - \Delta t \, \nabla p_{**}^{(2)} \qquad (8.2.23)$$

on each of the finite volume cells having a nonzero intersection with the dual cell considered. Here $p_{**}^{(2)}$ is the pressure correction field. It is analogous to $p_{*}^{(2)}$

from (8.2.16) in the first projection, except that $p_{**}^{(2)}$ is supported by node-centered data whereas $p_{*}^{(2)}$ had been defined by cell-centered values. Local data for $(\rho \boldsymbol{v})^{*,n+1}$ on $\partial \overline{C}$ result from linear interpolation of the cell-averaged momenta after the first projection.

Collecting from the above, we arrive at the second projection equation

$$\frac{1}{|\overline{C}|} \int_{\partial \overline{C}} h^{n+1} \, \nabla p_{**}^{(2),n+\frac{1}{2}} \cdot \boldsymbol{n} \, d\sigma = \text{r.h.s.} \,, \tag{8.2.24}$$

where

$$r.h.s. = \frac{(\rho e)_{\overline{C}}^{n+1} - (\rho e)_{\overline{C}}^{n}}{(\Delta t)^2/2} + \frac{2}{|\overline{C}| \Delta t} \int_{\partial \overline{C}} \frac{1}{2} \left((\rho h \boldsymbol{v})^n + h^{n+1} (\rho \boldsymbol{v})^{*,n+1} \right) \cdot \boldsymbol{n} \, d\sigma \,. \tag{8.2.25}$$

All terms on the right are readily evaluated once interpolation schemes providing h^n, h^{n+1} and $(\rho v)^n, (\rho v)^{*,n+1}$ on the dual cell boundaries from cell-centered/cell-averaged data have been selected.

This completes the description of the fully conservative zero-Mach-number Euler solver from Schneider et al. (1999) except for the influence of gravity, which will be important in the test cases discussed below. For Froude numbers $\text{Fr} = u_{\text{ref}}/\sqrt{g \ell_{\text{ref}}} = O(1)$ as $\text{M} \to 0$ the gravity source term in the vertical momentum balance is nonstiff. A straightforward discretization of the gravity source term in this situation procedes by simply replacing $-\Delta t \ \nabla p_{\circ}^{(2)}$ with $-\Delta t \ \left(\nabla p_{\circ}^{(2)} + \rho_{\circ}^{n+\frac{1}{2}} g \boldsymbol{k} \right)$, where $p_{\circ}^{(2)}$ replaces $p_{*}^{(2)}$ and $p_{**}^{(2)}$ in (8.2.20) and (8.2.23), respectively. In both cases it is sufficient to use a first-order accurate prediction of the density $\rho_{\circ}^{n+\frac{1}{2}}$ at time level $t^{n+\frac{1}{2}}$ to obtain a second-order scheme.

Falling Droplets. To demonstrate the performance of this scheme, we discuss here briefly some results for the motion of a two-dimensional "droplet" under the influence of gravity as reported originally in Schneider et al. (1999).

The scheme has been designed specifically to handle zero-Mach-number flows *with variable density.* A particularly hard test for such a scheme is the motion of large amplitude density discontinuities. Consider, for example, the vertical acceleration of a droplet of density $\rho_{\text{drop}} = 1,000$ in an environment of density $\rho_{\text{amb}} = 1$, and its subsequent collision with the surface of a bulk mass of equally high density. This problem has been proposed originally by Puckett et al. (1997) to test an interface tracking method. The situation is shown in Fig. 8.3, which exhibits a series of density contour plots as obtained from a two-dimensional simulation on a cartesian, equally spaced grid with 64×128 grid cells.

The initial data for the test case read

$$p(x, y, 0) = 1 \,, \qquad \boldsymbol{v}(x, y, 0) = \boldsymbol{0} \,, \tag{8.2.26}$$

$$\rho(x,y,0) = \begin{cases} 1000 & \text{if} \quad (0 \le y \le 1) \quad \text{or} \quad (0 \le r \le 0.2) \\ 1 & \text{if} \quad (1 < y \le 2) \quad \text{and} \quad (0.2 < r)\,, \end{cases} \tag{8.2.27}$$

where

$$r = \sqrt{(x-0.5)^2 + (y-1.75)^2}\,. \tag{8.2.28}$$

The density interface is captured throughout the computation over merely a few grid cells as expected from a (modified) Godunov-type scheme. The collision of the two masses of heavy fluid is represented robustly and without spurious overshoots.

A critical issue for the present scheme, in particular on a grid with equal spacing in the two directions, is the potential occurence of checkerboard modes in the pressure field. For the falling droplet example, contours of the grid cell interface pressure at time $t = 1.125$ are given in Fig. 8.4 (left). Even in this severe test with large, discontinuous density changes, no spurious pressure oscillations occur.

For an extension of the scheme to the zero-Mach-number Navier–Stokes equations and further tests, the reader is referred to the original reference (Schneider et al., 1999), and to Sect. 8.3 below, where we discuss some thermoacoustic applications.

8.2.2 Vorticity-Based Formulation

The equations of motion governing zero-Mach-number compressible flow may also be recast in vorticity based form. Several variants have appeared in the literature, including velocity–vorticity formulations (e.g., Daube, 1992) as well as vorticity–stream function formulations. In this section, a vorticity–stream function formulation is presented whose development has been motivated by applications to thermoacoustic refrigerators.

As illustrated in Fig. 8.1, and further discussed in Sect. 8.3 below, in their simplest form thermoacoustic refrigerators may be idealized as consisting of a straight resonance tube housing a stack of parallel plates, heat exchangers, and an acoustic source. These devices have been the focus of considerable recent research (see, e.g., Mironov et al., 2002; Karpov and Prosperetti, 2002; Ishikawa and Mee, 2001; Tijani et al., 2002a, 2002b; Hamilton et al., 2002) as they offer a variety of advantages, including minimal use of mechanical moving parts, amenability to miniaturization, and the elimination of harmful refrigerants.

Thermoacoustic refrigerators exploit the well-known “thermoacoustic effect” (Rayleigh, 1945; Swift, 1988) in order to induce a temperature difference across the stack, and to induce a mean heat flux from one end of the stack plate to the other. Heat exchangers are then used to transfer energy between the thermoacoustic refrigerator and hot or cold reservoirs.

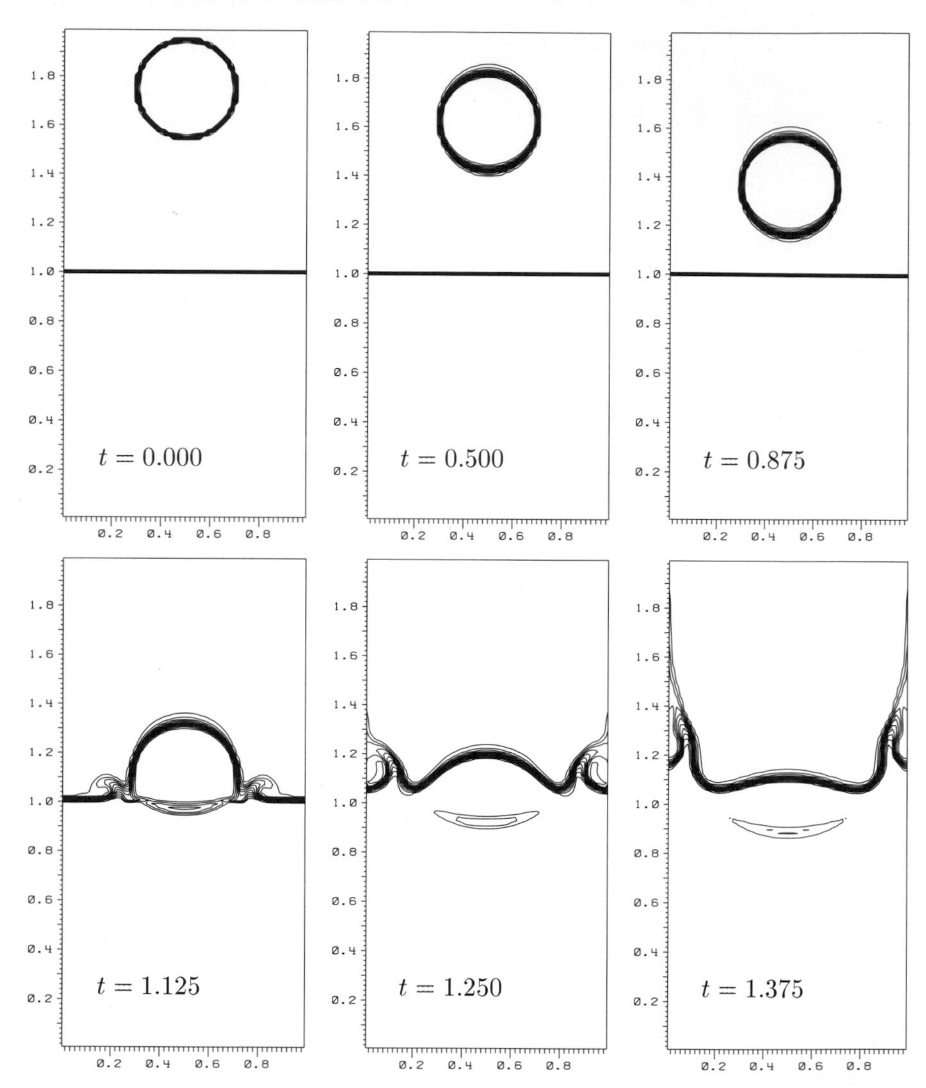

Fig. 8.3. Two-dimensional falling droplet in two space dimensions at Froude number Fr = 1: shown are 10 contours of density in $[1, 1000]$. Adapted from Schneider et al. (1999)

The mechanics of the mean energy transport between the stack plates is relatively well understood, and is in most cases adequately described using the well-known linearized theory (Swift, 1988). In contrast, the behavior of thermoacoustic heat exchangers is far more challenging, in large part due to the presence of concentrated eddy structures in the neighborhood of plate ends, (see Worlikar and Knio, 1996; Worlikar et al., 1998; Wetzel and Herman, 1999, 2000; Besnoin, 2001; Besnoin and Knio, 2001; Dufford, 2001;

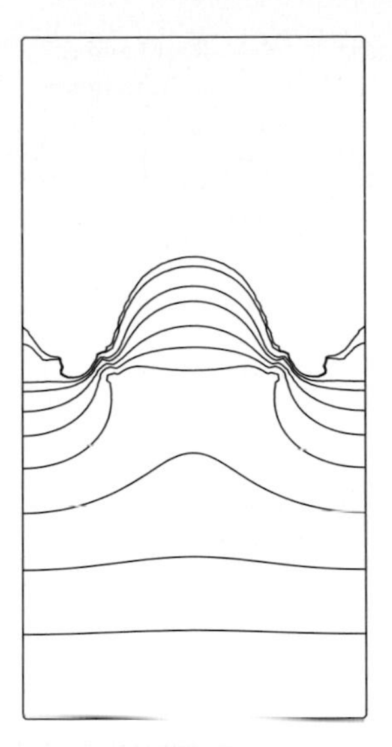

Fig. 8.4. As in Fig. 8.3 shown are 10 contour lines of the cell interface pressure at $-999, -990, -900, -800, -600, -400, -200, 0, 200$, and 400 (left) and 10 contour lines of the density in $[1, 1000]$ (right) at time $t = 1.125$. Adapted from Schneider et al. (1999)

Blanc-Benon, 2003). These vortices can have a dramatic impact on the structure of velocity and temperature distributions, as well as mean heat transfer properties. The complexity of the local flow field provides a significant challenge to simplified modeling studies (Mozurkewich, 1998; Karpov and Prosperetti, 2000, 2002; Poese and Garrett, 2000), as heat and momentum correlations for the local (oscillatory) vortical flow are generally not available.

To perform a detailed investigation of the flow features around, and heat transfer behavior of, thermoacoustic stack and heat exchangers, it is essential to concentrate computational resources in these regions. While multiple alternatives are possible, an attractive approach consists of exploiting the inherent features (and associated scale disparities) of the flow field, which may be viewed as consisting of an acoustically compact zero-Mach-number vortical flow in the neighborhood of the stack and heat exchangers, and an essentially irrotational one-dimensional flow in the remainder of the resonance tube. As discussed in Subsect. 8.1.4, this suggests a domain decomposition approach that is based on combining a simplified one-dimensional representation of resonant tube acoustics, with a resolved two-dimensional representation of the flow around the stack and heat exchangers (Worlikar, 1997; Worlikar and Knio, 1999). Specifically, as illustrated in Figs. 8.5 and 8.6, the detailed multidimensional computations are restricted to a single period of a stack/heat exchanger configuration that is composed of a large number of identical, equispaced plates. The inlet/outlet surfaces limit the computational domain in

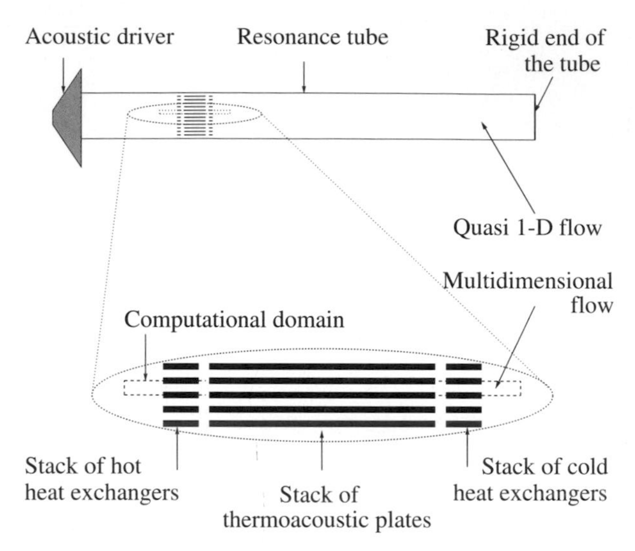

Fig. 8.5. Schematic illustration of the thermoacoustic refrigerator (*top*) with a magnified view of the computational domain (*bottom*). Adapted from Besnoin and Knio (2004)

the "streamwise" direction (Fig. 8.6). On these surfaces, the solution matches the wave representation in the "outer region," which is assumed to be dominated by a quasi-one-dimensional, idealized acoustic standing wave.

Within the computational domain shown in Fig. 8.6, a compressible, acoustically compact zero-Mach-number flow description is used (see Majda and Sethian, 1985; Worlikar and Knio, 1996, 1999; Worlikar et al., 1998; Besnoin and Knio, 2001). Assuming that (1) the Mach number is small and the flow is two-dimensional, (2) the stack and heat exchangers are much shorter than the acoustic wavelength, and (3) the gas is Newtonian and has constant dynamic viscosity and thermal conductivity, the equations

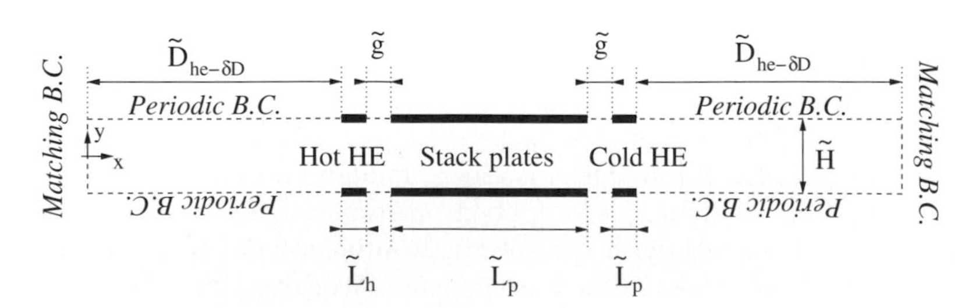

Fig. 8.6. Schematic illustration of the computational domain, showing the dimensions of the stack and heat exchanger plates, the gap g between the stack and heat exchangers, the channel height H and the location of the matching boundaries. Adapted from Besnoin and Knio (2004)

governing the motion of the gas are expressed in the following vorticity-stream function-potential form:

$$\frac{\partial \zeta}{\partial t} = -\nabla \times (\zeta \boldsymbol{k} \times \boldsymbol{u}) - \frac{\nabla \rho}{\rho} \times \frac{D\boldsymbol{u}}{Dt} + \frac{1}{\rho \,\mathrm{Re}} \nabla^2 \zeta \tag{8.2.29}$$

$$\boldsymbol{u} = \nabla \phi + \nabla \times (\psi \boldsymbol{k}) \tag{8.2.30}$$

$$\nabla^2 \phi = \frac{1}{P}\left[-\frac{1}{\gamma}\frac{dP}{dt} + \frac{1}{\mathrm{Pe}}\nabla^2 T + \frac{\mathrm{Ec}}{\mathrm{Re}}\Phi\right] \tag{8.2.31}$$

$$\nabla^2 \psi = -\zeta \tag{8.2.32}$$

$$\rho\frac{DT}{Dt} - \frac{\gamma - 1}{\gamma}\frac{DP}{Dt} = -\frac{1}{\mathrm{Pe}}\nabla \cdot \boldsymbol{q} + \frac{\mathrm{Ec}}{\mathrm{Re}}\Phi \tag{8.2.33}$$

$$\frac{V}{\gamma}\frac{dP}{dt} = -P\int_{\partial D} \boldsymbol{u} \cdot \boldsymbol{n} dA + \frac{1}{\mathrm{Pe}}\int_{\partial D} \nabla T \cdot \boldsymbol{n} dA + \frac{\mathrm{Ec}}{\mathrm{Re}}\int_D \Phi dV \tag{8.2.34}$$

$$P = \rho T\,, \tag{8.2.35}$$

where $\boldsymbol{u} = (u, v)$ is the velocity vector, ζ the vorticity, ϕ the velocity potential, ψ the stream function, $\boldsymbol{k}$ is the right-handed normal to the plane of motion, P the thermodynamic pressure, T the temperature, ρ the density, $D/Dt = \partial/\partial t + \boldsymbol{u} \cdot \nabla$ the material derivative, γ the specific heat ratio, Φ is the viscous dissipation term, D denotes the computational domain, ∂D its boundary, and $\boldsymbol{n}$ the outer normal at ∂D. Variables are normalized with respect to the appropriate combination of the angular frequency of the acoustic standing wave, $\tilde{\Omega}$, the plate separation distance, $\tilde{H}$, the mean pressure, $\tilde{P}_m$, and the mean temperature $\tilde{T}_m$. We use tildes to identify dimensional quantities, and the subscript "m" to denote values at the mean gas pressure and temperature. The normalization process leads to the definition of the following dimensionless parameters: the Reynolds number, $\mathrm{Re} \equiv \tilde{\Omega}\tilde{H}^2/\tilde{\nu}$, the Peclet number $\mathrm{Pe} \equiv \tilde{\Omega}\tilde{H}^2/\tilde{\alpha}_m$, and the Eckert number $\mathrm{Ec} \equiv \tilde{\Omega}^2\tilde{H}^2/(\tilde{C}_p\tilde{T}_m)$.

In the applications discussed below, a simplified thermal model for the heat exchangers shown in Fig. 8.5 is used. Specifically, the heat exchangers are treated as isothermal with hot temperature, T_h, and cold temperature, T_c. On the other hand, the temperature distribution within the stack plates is determined by solving the unsteady heat conduction equation:

$$\frac{\partial T_s}{\partial t} = \frac{1}{\mathrm{Pe}_s}\nabla^2 T_s, \tag{8.2.36}$$

where $\mathrm{Pe}_s \equiv \tilde{\Omega}\tilde{H}^2/\tilde{\alpha}_s$ is the Peclet number for the solid. The thermal model for the stack plate and heat exchangers is coupled to the low-Mach-number flow model by imposing continuity of the temperature and the heat flux at the gas–solid interfaces.

The applications discussed in Sect. 8.3 below are based on the assumption that the thermoacoustic stack is composed of a large number of identical, equi-spaced plates, and that the hot and cold heat exchangers have the same number of plates and the same plate spacing as the stack. We model this setting by restricting the simulations to a single period and imposing periodicity boundary conditions in the direction normal to the acoustic axis. The inlet/outlet surfaces limit the computational domain in the streamwise direction. On these surfaces, the solution matches the wave representation in the outer region, which is assumed to be dominated by a quasi-one-dimensional, idealized acoustic standing wave. In the following, we refer to these computational surfaces as matching boundaries.

To represent the action of the acoustic standing wave, one can impose oscillating velocity conditions at the ends of the computational domain that match with the idealized one-dimensional (linear) standing wave solution at the same location. In this approach, the oscillations in the thermodynamic pressure are obtained from the solution of (8.2.34) (Worlikar et al., 1998). Alternatively, the action of the standing wave can be represented by prescribing oscillations of the thermodynamic pressure, P, and the mean flow velocity. In the latter case, the integral constraint in (8.2.34) is used to compute the oscillating velocities at the ends of the computational domain. The suitability of both approaches is examined in the tests provided below.

Numerical simulation of the governing equations is based on a finite difference methodology. Spatial derivatives are approximated using second-order centered differences and time integration is based on the third-order Adams-Bashforth scheme (Worlikar and Knio, 1996; Worlikar et al., 1998). Reconstruction of the velocity involves the solution of two elliptic equations that are inverted using fast Poisson solvers. The fast solvers are based on domain-decomposition/boundary Green's functions techniques whose construction is discussed in Worlikar and Knio (1996) and Worlikar et al. (1998). For additional details regarding the numerical construction and implementation details, see Worlikar et al. (1998), Worlikar (1997) and Besnoin (2001).

One of the key features of the present model is that it enables us to isolate a small neighborhood of the stack and heat exchangers. Within this region, which is assumed to be "acoustically compact," the model ignores elastic effects but retains bulk compressibility and convective nonlinearities in a stratified medium. This enables us to avoid sonic CFL limitations and thus efficiently analyze the details of the acoustically driven flow.

8.3 Application to Thermoacoustic Refrigeration

This section outlines applications of the zero-Mach-number compressible flow formulations of Subsects. 8.2.1 and 8.2.2 to simplified models of thermacoustic refrigerators. We start in Subsect. 8.3.1 with an analysis of the properties of the numerical models and of modeling approximations. In Subsects. 8.3.2 and 8.3.3, highlights of validation studies are presented; experimental measurements performed on so-called thermoacoustic couples are used for this purpose. Finally, results are presented in Subsect. 8.3.4 from a detailed study of the thermal performance of the simplified thermacoustic heat exchangers.

8.3.1 Convergence Properties and Modeling Approximations

In this section, we highlight results of a numerical study of the convergence properties of the two numerical schemes introduced in Subsects. 8.2.1 and 8.2.2, and of the effect of modeling approximations. For the purpose of computational efficiency, a reduced version of the model provided in Subsect. 8.2.2 is first derived. The reduction is based on using a lumped-parameter description for the temperature distribution within the stack, which is obtained by averaging the energy equation (8.2.36) over the cross section of the stack plate. Thus, an unsteady, quasi-one-dimensional description of the temperature along the stack plates, $T_s(x,t)$, is obtained, which is governed by (Schneider et al., 1999; Besnoin and Knio, 2001):

$$\frac{\partial T_s}{\partial t} = \frac{1}{\mathrm{Pe}_s}\left(\frac{\partial^2 T_s}{\partial x^2} + \frac{2\kappa}{\delta_p}\frac{\partial T}{\partial y}\right), \tag{8.3.1}$$

where κ is the ratio of the gas thermal conductivity to the plate thermal conductivity, and δ_p is the "nominal" plate thickness. Note that the present thin-plate formulation resembles that used by Cao et al. (1996), but differs from it in that the plate temperature distribution is computed rather than prescribed.

In the analysis below, we primarily focus on three specific configurations and operating conditions as summarized in Table 8.1. As indicated in the table, the flow conditions are identified using (a) the drive ratio, Dr, defined as the ratio of the acoustic pressure amplitude to the mean thermodynamic pressure; (b) the Reynolds number, $\mathrm{Re} \equiv \tilde{\Omega}\tilde{H}^2/\tilde{\nu}$, where $\tilde{\nu}$ is the kinematic viscosity of the gas at the mean pressure and temperature; (c) the particle displacement parameter, $R_p \equiv \tilde{u}_a/\tilde{\Omega}\tilde{H}$, where $\tilde{u}_a$ is the acoustic pressure amplitude at the stack location; (d) the ratio H/δ_k of the stack centerline spacing to the thermal penetration depth; (e) the length of the hot and cold heat exchanger, $\tilde{L}_{h/c}$; (f) the gap, $\tilde{g}$, between stack and heat exchanger plates; (g) the ratio, $\tilde{T}_h \;/\; \tilde{T}_m$, of the hot heat exchanger temperature to the mean temperature; and (h) the ratio, $\tilde{T}_c \;/\; \tilde{T}_m$ of the cold heat exchanger

Table 8.1. Configuration and flow parameters. Values of $\tilde{L}_{h/c}$, and $\tilde{g}$ are in millimeters

Case	Dr	Re	R_p	H/δ_k	$\tilde{L}_{h/c}$	$\tilde{g}$	$\tilde{T}_h \;/\; \tilde{T}_m$	$\tilde{T}_c \;/\; \tilde{T}_m$
1	2 %	200	1.95	8.20	1.06	0.53	1.03	0.97
2	8 %	28.43	7.80	3.09	4.59	1.41	1.05	0.95
3	8 %	28.43	7.80	3.09	1.24	3.71	1.05	0.95

temperature to the mean temperature. (The mean temperature refers to the average temperature within the resonance tube when the fluid is at rest.) The center of the stack is assumed to be located midway between the velocity node and the velocity antinode, that is, at a nondimensional wavenumber $kx = 3\pi/4$. The stack plates are 3.18 cm long with a thermal conductivity $\tilde{k}_s = 0.482$ W/m·°K. We also assume that the resonance tube is filled with helium, $\tilde{T}_m = 300$°K, $\tilde{P}_m = 10^5$ N/m^2, the frequency of the standing wave $\tilde{f} = 707$ Hz and the acoustic wavelength $\tilde{\lambda} = 1.44$ m. In the analysis below, the computations are started from a state of rest and carried out until stationary oscillating conditions are reached.

8.3.1.1 Convergence. The validity of the computations was first examined by contrasting the results of the vorticity-based model (Subsect. 8.2.2) and the conservative primitive-variable scheme (Subsect. 8.2.1). A sample of this exercise is provided in Fig. 8.7, which shows the evolution of the instantaneous heat flux into the cold heat exchanger, computed using the vorticity-based and primitive-variable codes (Table 8.1, case 1). The instantaneous heat flux into the cold heat exchanger is defined as

$$\tilde{Q}_{f\to c} = \int_{\mathcal{A}} -\tilde{k}\nabla\tilde{T}\cdot \boldsymbol{n}dA, \tag{8.3.2}$$

where $\mathcal{A}$ denotes the area (perimeter) of the heat exchanger, while $\boldsymbol{n}$ is the unit normal pointing into the heat exchanger solid surface (that is, away from the fluid). We also define the mean heat flux, $\tilde{Q}_{m,f\to c}$, according to

$$\tilde{Q}_{m,f\to c} = \frac{\tilde{\Omega}}{2\pi}\int_0^{\tilde{\Omega}/2\pi} \tilde{Q}_{f\to c}dt. \tag{8.3.3}$$

In other words, $\tilde{Q}_{m,f\to c}$ is the heat flux averaged over one acoustic cycle; accordingly, $\tilde{Q}_{m,f\to c}$ is also referred to as the "cooling load." The comparison in Fig. 8.7 shows that the results of the two models are in good agreement. Close agreement is also obtained for other cases considered, which provides strong support for the validity of both schemes.

In addition to cross-verification, the effect of spatial resolution was analyzed by systematic refinement of the grid size in the vorticity-based computations. A sample of the exercise is shown in Fig. 8.8, which depicts the dependence of the mean heat flux from the cold heat exchanger on the spatial grid size (Table 8.1, case 2). The results indicate that, as the grid is refined, the predictions become insensitive to the mesh size value. In particular, Fig. 8.8 indicates that the predictions become essentially independent of the mesh size when the resolution is such that there are six or more points within a viscous penetration depth.

8.3.1.2 Convergence Acceleration Scheme. A major difficulty facing direct simulations of thermoacoustic devices is that a large number of acoustic cycles (typically on order of several hundreds) is required to reach steady

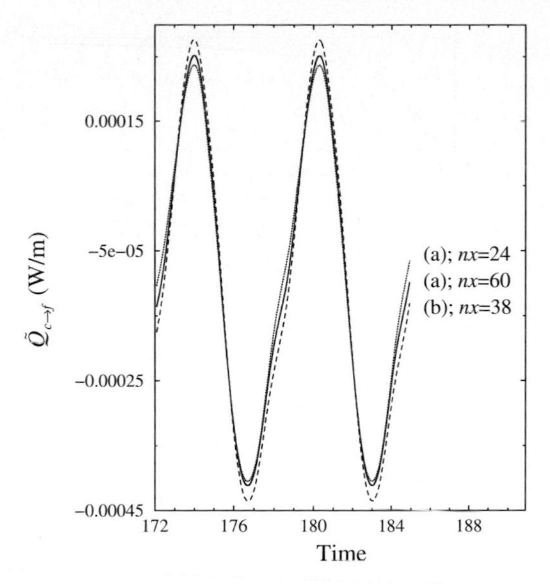

Fig. 8.7. Evolution of the normalized heat flux $Q_{f\to c}$ into the cold heat exchanger. The curves are generated using (a) the vorticity-based scheme with $nx = 60$ and also 24 grid cells in the cross-stream direction, and (b) the finite-volume scheme with $nx = 38$ grid cells (Table 8.1, case 1). Adapted from Besnoin and Knio (2001)

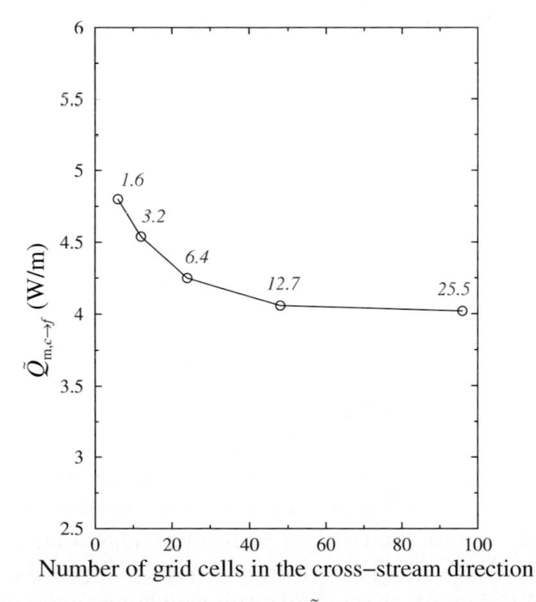

Fig. 8.8. Dependence of the mean heat flux $\tilde{Q}_{m,f\to c}$ on the numerical discretization (Table 8.1, case 2), also showing the resolution of each viscous boundary-layer. The figures correspond to the number of grid cells in the cross-stream direction per viscous penetration depth at the corresponding discretization level. Adapted from Besnoin and Knio (2001)

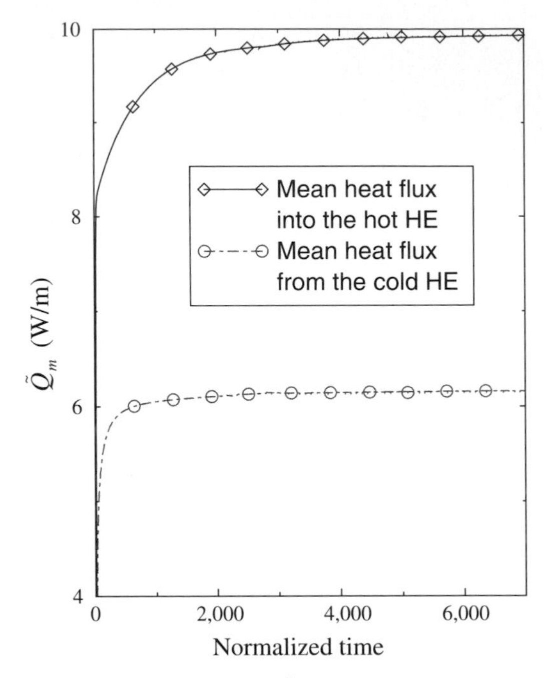

Fig. 8.9. Evolution of the mean heat flux $\tilde{Q}_m$ into the hot heat exchanger (HE) and from the cold heat exchanger. The curves are generated using background pressure oscillations as the representation of the standing wave (Table 8.1, case 3). Adapted from Besnoin and Knio (2001)

state. Since these simulations must resolve the thin structure of oscillating Stokes layers, and since a good approximation of the steady-state field is not available a priori, a large computational overhead is consequently required. To reduce this overhead, a scheme of successive mesh refinement is implemented. Specifically, extended computations are first performed on a coarse computational mesh. Once steady-state conditions are reached on the coarse computational mesh, the solution is interpolated on a finer computational grid. This process is repeated until steady-state conditions are reached on a grid that is fine enough for quantities to be accurately determined.

The advantage of the present approach stems from the fact that, while hundreds of cycles are needed to reach steady conditions on the coarse grid (Fig. 8.9), computations on the refined grids only need to be carried out for a few acoustic cycles (not shown). An additional advantage of this approach is that it automatically provides us with useful information regarding the convergence of the computations. Thus, the convergence acceleration scheme is used both as means for efficiently obtaining steady-state results and for verifying the accuracy of the predictions.

8.3.1.3 Acoustic Wave Representation. As mentioned earlier, we have explored two approaches for representing the action of the acoustic stand-

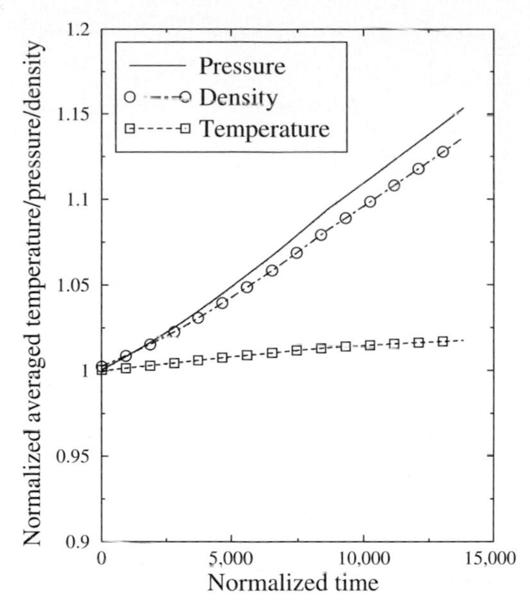

Fig. 8.10. Temporal evolution of the normalized mean pressure, density, and temperature in the computational domain. The velocities at the inlet and outlet of the computational domain are imposed; the background pressure is computed (Table 8.1, case 3). Adapted from Besnoin and Knio (2001)

ing wave. Initially, we adapted the procedure used in Worlikar et al. (1998) and Worlikar and Knio (1999), which is based on using the standing wave solution at the matching surfaces as oscillating velocity boundary condition. This approach proved suitable in earlier stack computations (see also Cao et al., 1996), which were primarily limited to a regime with low/moderate drive ratios. Unfortunately, at high drive ratios the corresponding simulations suffer from a weak but sustained drift in the mean stack conditions. An example of this drift is shown in Fig. 8.10, which shows the evolution of the normalized mean pressure, density, and temperature in the computational domain. The results indicate that in this case the average pressure, density, and temperature in the domain increase steadily with time and that, consequently, steady-state conditions are not reached. Note that the same drift has also been observed with the finite-volume computations (not shown), and thus does not appear to be due to specific details of the physical model or numerical discretization.

To overcome the difficulties outlined above, an alternative procedure is developed for modeling the effect of the acoustic wave. As mentioned earlier, this procedure is based on imposing the oscillations of the thermodynamic pressure and mean volume flux, and computing the velocity at the matching surfaces using the integral divergence constraint in (8.2.34). An immediate advantage of this approach is that the mean pressure remains fixed and,

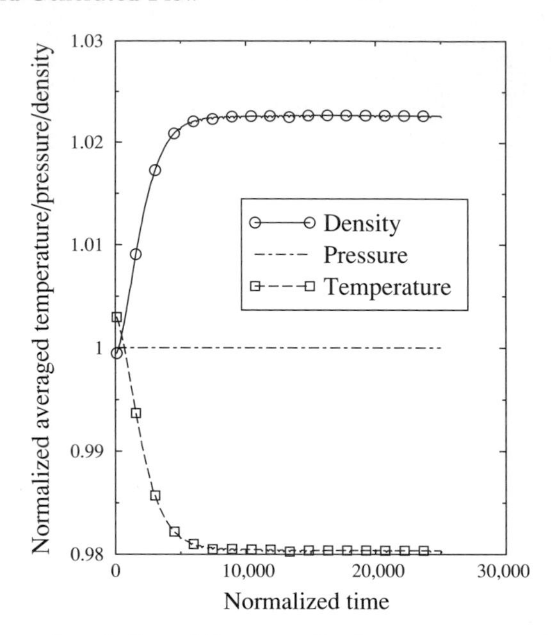

Fig. 8.11. Temporal evolution of the normalized mean pressure, density, and temperature in the computational domain. The background pressure is imposed and inlet and outlet velocities are computed (Table 8.1, case 3). Adapted from Besnoin and Knio (2001)

consequently, is no longer susceptible to drift. In addition, the drive ratio is explicitly imposed and exactly recovered.

We examine the suitability of this approach by plotting the evolution of the mean pressure, temperature, and density within the computational domain. The results are shown in Fig. 8.11 for the same configuration and operating parameters considered earlier. As expected, the mean (thermodynamic) pressure maintains its initial value, whereas the mean temperature and density curves follow an initial transient adjustment period and then settle at a constant value. Since the amplitude of pressure oscillations is fixed, a well-defined stationary regime is obviously reached.

To further examine the predictions, we show in Fig. 8.12 the evolution of the computed velocity at the matching surfaces during one acoustic cycle. The plot is generated when steady conditions are reached, and depicts the velocities obtained by imposing either (a) oscillations in the thermodynamic pressure and mean velocity oscillations are deriving the velocity at the matching surfaces from the integral constraint or (b) sinusoidal signals at the matching surfaces that coincide with the linear standing wave solution at the corresponding locations. The comparison shows that at steady state the derived velocities at the matching surface remain very close to the linear prediction, with very small amplitude differences between the two signals.

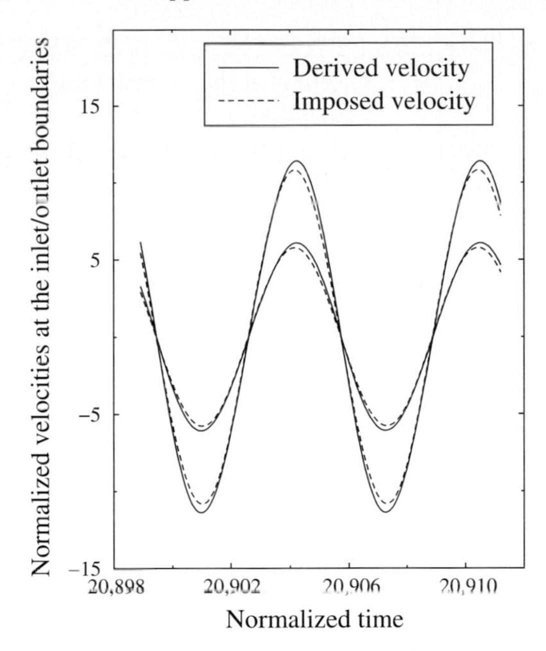

Fig. 8.12. Temporal evolution of the normalized velocities at the inlet and outlet boundaries of the computational domain. Curves are generated when the background pressure (*solid lines*) is imposed and when the inlet and outlet velocities are imposed (*dashed lines*) (Table 8.1, case 3). Adapted from Besnoin and Knio (2001)

This suggests that the drift observed earlier is due to imposition of prescribed sinusoidal velocity oscillations, which may not be suitable at high drive ratios.

8.3.1.4 Domain Size and Thermal Boundary Conditions. In the present model, the computations are restricted to a neighborhood of the stack and heat exchangers so that the size of the computational domain must be specified. This is accomplished by selecting values for the lengths $D_{he-\delta D}$ (Fig. 8.6), which determine the location of the matching surfaces. In this section, a numerical study is performed of the effect of this parameter on the predictions.

We also examine the effect of thermal conditions at the matching surfaces which are imposed using two different approaches. The first approach, which is identical to that used in Worlikar et al. (1998), Worlikar and Knio (1999), and Cao et al. (1996), is based on using vanishing flux conditions at both surfaces, that is, $\partial T/\partial x = 0$. These conditions are motivated by the fact that as one moves away from the stack and heat exchangers the thermoacoustic effect is expected to essentially vanish so that the axial thermal gradients should decrease as well. Meanwhile, the second approach is based on imposing mixed inflow/outflow conditions. The outflow condition is used when the (computed) velocity is in the outward normal direction; it is based on locally

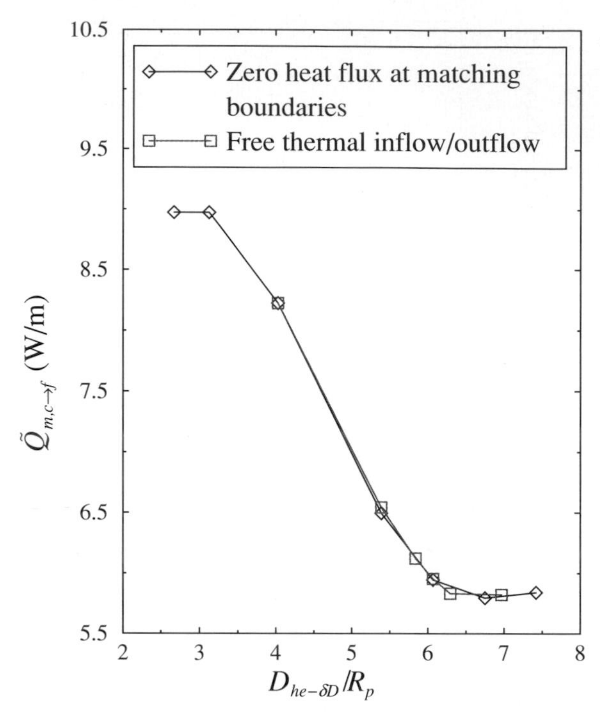

Fig. 8.13. Dependence of the mean heat flux $\tilde{Q}_{m,f\to c}$ on the distance $\tilde{D}_{he-\delta D}$ that separates the hot/cold heat exchanger and the inlet/oulet boundary of the computational domain. Curves are generated using either vanishing flux conditions (diamonds), or inflow/outflow conditions (squares), for temperature and density at the inlet/oulet boundaries of the computational domain (Table 8.1, case 3). Adapted from Besnoin and Knio (2001)

ignoring the diffusive flux from the energy equation and only retaining the effects of transport and bulk compression. The inflow condition is used when the velocity is pointing into the domain. In this case, the boundary temperature is prescribed. To this end, we assume that outside the computational domain the fluid follows an isentropic process with the temperature given by the idealized one-dimensional standing wave solution.

Figure 8.13 shows the dependence of the average cooling load from the cold heat exchanger on the ratio $D_{he-\delta D}/R_p$ of the distance between the matching boundaries and the heat exchangers and the particle displacement amplitude (see Fig. 8.6). Figure 8.13 shows that as $D_{he-\delta D}$ increases the predictions become essentially independent of the selected location of the matching surfaces, and that this occurs when the distance between matching surfaces and the heat exchangers is larger than six times the particle displacement amplitude. The results also indicate that the predictions obtained using both temperature boundary conditions are consistent, at least when the matching surfaces are well separated from the heat exchangers.

It may at first appear that one should conservatively select large ratios $D_{he-\delta D}/R_p \gg 1$. However, this approach should be avoided for two main reasons. First, increasing $D_{he-\delta D}$ results in an unnecessary increase in CPU requirements. Second, for very large values of $D_{he-\delta D}$ the computational domain may no longer remain acoustically compact, as assumed in the formulation of the model. Consequently, the location of the matching surfaces should be carefully selected, with $D_{he-\delta D}$ as small as possible. For the results in Fig. 8.13, the stack is operating at a drive ratio of 8% and a ratio $D_{he-\delta D}/R_p \simeq 6$ is a reasonable selection.

8.3.2 Temperature Difference Across a Thermoacoustic Couple

In this section, the zero-Mach-number thermoacoustic flow model is applied to simulate some of the experimental conditions in the study of Atchley et al. (1990), who focused, in particular, on the steady-state acoustically generated temperature gradients in short thermoacoustic couples (TACs). Briefly, TACs have a similar configuration to thermoacoustic refrigerators (Fig. 8.5), the prime distinction concerning the absence of the heat exchangers. This simplified setting facilitates experimental diagnostics, particularly measurement of flow conditions near the edges of the stack plates. The temperature measurements in Atchley et al. (1990) thus provide a well-established basis for validation of the computations and for analysis of the origin of the limitation of the linear theory (Swift, 1988).

In the computations below, we focus exclusively on a single stack arrangement, which corresponds to TAC # 3 in Atchley et al. (1990). The stack plates are a 0.1905-mm thick, laminate of stainless steel and fiberglass, with an effective thermal conductivity $\tilde{k}_s = 5.76$ W/m·°K. The stack is 6.85 mm long with a half gap $\tilde{y}_0 = \tilde{h}/2 = 0.669$ mm. Thus, the stack has a blockage ratio $h/H = 0.875$ and a plate length parameter of $L/H = 4.48$. The physical dimensions of the stack are approximated in the computations by using a blockage ratio $h/H = 0.837$ and a plate length parameter $L/H = 4.5$.

Table 8.2 summarizes the flow conditions for which the model is run. For cases 1–4, the stack is held fixed at a location midway between the velocity node and antinode (that is, $\tilde{k}\tilde{x} = 3\pi/4$), while the strength of the standing wave is varied. The latter is quantified in terms of the drive ratio, Dr, which is defined as the ratio of the acoustic pressure amplitude to the mean pressure, that is, $Dr \equiv \tilde{P}_0/\tilde{P}_m$. As shown in Table 8.2, Dr increases from 0.28% for case 1 to 2% for case 4. For runs 5–7, the drive ratio is held fixed at 2%, while the position of the stack within the resonance tube is varied. In all cases, the Peclet number Pe $= 69.55$, the solid Peclet number $\text{Pe}_s = 301$, and the ratio of the solid thermal conductivity to the gas thermal conductivity $\kappa = 40.65$.

Table 8.2 also lists three "redundant" dimensionless groups that are nonetheless helpful for interpretation of the results. The first is a dynamic Reynolds number based on the acoustic velocity amplitude at the stack location, $\tilde{u}_a$, and the viscous penetration depth, $\tilde{\delta}_v \equiv \sqrt{2\tilde{\nu}/\tilde{\Omega}}$,

$$Re_a \equiv \sqrt{2}\frac{\tilde{u}_a \tilde{\delta}_v}{\tilde{\nu}} . \tag{8.3.4}$$

As shown in the table, Re_a is small and subcritical in all cases considered (Merkli and Thomann, 1975); thus, transition to turbulence is not expected to occur. The second is a particle displacement parameter, $R_p \equiv \tilde{u}_a/\tilde{\Omega}\tilde{H}$, which reflects the ratio of the characteristic particle displacement within one acoustic cycle to the plate separation distance. It is related to the drive ratio through

$$R_p = \frac{Dr}{\gamma M} \sin(\tilde{k}\tilde{x}) , \tag{8.3.5}$$

where $\tilde{k} = \tilde{\Omega}/\tilde{c}$ is the dimensional wavenumber, $\tilde{c}$ is the speed of sound in the gas, $\tilde{\lambda}$ is the acoustic wavelength ($\tilde{\lambda} = 2\tilde{L}$), $M = \tilde{\Omega}\tilde{H}/\tilde{c}$ is a scaled Mach number, and $\tilde{x}$ is the location of the stack measured from the rigid end of the resonance tube. The third parameter $R_p' \equiv 2R_pH/L$ is a scaled particle displacement parameter, which reflects the ratio of the characteristic particle displacement to the plate length.

8.3.2.1 Linear Theory. Theoretical predictions for the acoustically generated temperature difference across thermoacoustic stacks have been extensively discussed in the literature (see, e.g., Swift, 1988; Atchley et al., 1990; Wheatley et al., 1983). Thus, only a brief description is provided. As in Atchley et al. (1990) and Wheatley et al. (1983), the present approach is based on the following key ingredients: (a) the short stack approximation is invoked, and used to approximate the wave properties using the idealized linear predictions at the stack location, (b) the enthalpy flux through the "channel" between neighboring stack plates can be approximated using the well-known quasi-one-dimensional predictions for an oscillating channel flow (Rott, 1980; Swift, 1988; Panton, 1984), (c) thermal and mechanical energy losses outside the channel are ignored, and (d) at "steady-state" conditions, the mean enthalpy flux within the channel is assumed to be balanced by conduction through the plates. As summarized below, these ingredients can be combined to estimate the steady-state temperature difference across the stack.

Table 8.2. Details of runs

Case	Re	Re_a	$\tilde{k}\tilde{x}$	R_p	R_p'	Dr (%)
1	104.13	4.38	2.35	0.215	0.095	0.28
2	104.13	7.83	2.35	0.384	0.171	0.5
3	104.13	15.69	2.35	0.769	0.342	1.0
4	104.13	31.4	2.35	1.54	0.684	2.0
5	104.13	23.67	2.6	1.16	0.515	2.0
6	104.13	16.46	2.8	0.807	0.358	2.0
7	104.13	8.46	3.0	0.415	0.184	2.0

We first note that expressions for the mean enthalpy flux through the channel, $\bar{H}_2$, are readily available from the detailed review of Swift (1988). Using the short stack approximation, we have (Swift, 1988):

$$\overline{H}_2 = \frac{-\Pi \tilde{\delta}_k \tilde{P}_1^s \langle \tilde{u}_1^s \rangle}{4\,(1+\epsilon_s)(1+\mathrm{Pr})\left[1 - \dfrac{\tilde{\delta}_v}{\tilde{y}_0} + \dfrac{\tilde{\delta}_v^2}{2\tilde{y}_0^2}\right]} \times \left[\frac{d\tilde{T}_m}{d\tilde{x}} \frac{\tilde{\rho}_m \tilde{c}_p \langle \tilde{u}_1^s \rangle}{\tilde{\Omega} \tilde{P}_1^s} \left(\frac{1+\sqrt{\mathrm{Pr}}+\mathrm{Pr}+\mathrm{Pr}\epsilon_s}{1+\sqrt{\mathrm{Pr}}}\right) - \left(1+\sqrt{\mathrm{Pr}} - \frac{\tilde{\delta}_v}{\tilde{y}_0}\right)\right] - \Pi \tilde{y}_0 \tilde{k} \frac{dT_m}{dx}, \tag{8.3.6}$$

where $\tilde{d}$ is the plate thickness, Pr the Prandtl number, $\tilde{\delta}_k = \tilde{\delta}_v/\sqrt{Pr}$ the thermal penetration depth, $\epsilon_s = \tilde{\rho}_m \tilde{c}_p \tilde{\delta}_k / \tilde{\rho} \tilde{c}_{p,s} \tilde{d}$ the ratio of the heat capacity of the gas to the heat capacity of the plate, $\tilde{P}_1^s$ the pressure amplitude at the stack location, $\langle \tilde{u}_1^s \rangle$ the amplitude of the gap-averaged mean velocity at the stack location, and Π the plate perimeter.

Next, we equate the mean conduction heat flux within the solid to the mean enthalpy flux within the channel (Wheatley et al., 1983; Swift, 1988; Atchley et al., 1990). This leads to the following expression for the steady-state temperature difference across the stack (Worlikar et al., 1998):

$$\Delta T = \frac{\alpha_1 \tilde{L} \dfrac{\tilde{\delta}_k}{8} \dfrac{\tilde{P}_o^2}{\tilde{\rho}_m \tilde{c}} sin\left(\dfrac{2\tilde{x}}{\tilde{\lambda}}\right) \dfrac{\tilde{H}}{\tilde{h}}}{\left[\left(\dfrac{\tilde{h}}{2}\tilde{k} + \dfrac{\tilde{d}}{2}\tilde{k}_s\right) + \left(\dfrac{\tilde{H}}{\tilde{h}}\right)^2 \dfrac{\tilde{\delta}_k}{8\tilde{\Omega}} \dfrac{\tilde{c}_p \tilde{P}_o^2}{\tilde{\rho}_m \tilde{c}^2} \alpha_2\right]} \tag{8.3.7}$$

where

$$\alpha_1 \equiv \frac{\left(1+\sqrt{\mathrm{Pr}} - \tilde{\delta}_v/\tilde{y}_0\right)}{(1+\epsilon_s)(1+\mathrm{Pr})\left[1 - \dfrac{\tilde{\delta}_v}{\tilde{y}_0} + \dfrac{\tilde{\delta}_v^2}{2\tilde{y}_0^2}\right]} \tag{8.3.8}$$

and

$$\alpha_2 \equiv \frac{\left[1-\cos\left(\dfrac{2\tilde{x}}{\tilde{\lambda}}\right)\right]\left(1+\sqrt{\mathrm{Pr}}+\mathrm{Pr}+\mathrm{Pr}\epsilon_s\right)}{(1+\epsilon_s)(1+\mathrm{Pr})\left(1 - \dfrac{\tilde{\delta}_v}{\tilde{y}_0} + \dfrac{\tilde{\delta}_v^2}{2\tilde{y}_0^2}\right)\left(1+\sqrt{\mathrm{Pr}}\right)}. \tag{8.3.9}$$

Linear theory predictions of the same nature as in (8.3.7) have been tested against experimental data (Wheatley et al., 1983; Atchley et al., 1990). These tests have shown a good agreement between theoretical predictions and experimental measurements at low drive ratio; however, the agreement deteriorates as the drive ratio increases. Thus, the analysis below examines the origin of the discrepancy. In particular, the results of the computations are compared to measurements in Atchley et al. (1990) and the theoretical predictions in (8.3.7). For the conditions of the experiments, we use $\tilde{\Omega} = 4448$ rad/s, $\tilde{\rho}_m = 0.1813$ kg/m^3, $\tilde{\nu} = 10^{-4}$ m^2/s, $\tilde{\lambda} = 2.74$ m, $\tilde{c} = 1160$ m/s, $\epsilon_s = 7.26 \times 10^{-3}$, and $\tilde{k} = 0.1418$ W/m·°K.

8.3.2.2 The Steady-State Temperature Difference. The computations are started with the gas at rest, and with a uniform temperature distribution in the gas and the plates. They are carried out until ΔT, the temperature difference across the stack averaged over one acoustic cycle, approaches a steady-state value. The results (not shown) indicate that ΔT increases rapidly in the early stages of the simulation, where its growth rate is highest. To limit CPU requirements, the present computations are stopped when the rate of change of ΔT drops to less than 2% of its peak value. We then take advantage of the exponential-like decay in the rate of change of ΔT at large times by extrapolating the results to obtain the steady-state value. To this end, the time derivative of ΔT is fitted with a decaying exponential, that is, approximated as

$$\frac{d(\Delta T)}{dt} \simeq a_0 \exp\left(-\frac{t - t_f}{\tau}\right) , \qquad (8.3.10)$$

where t_f is the time at which the computations are stopped, while a_0 and τ are estimated using the large-time behavior of the computed results. Then, the steady-state temperature difference is estimated as

$$\Delta T_p \simeq \Delta T(t_f) + a_0 \tau , \qquad (8.3.11)$$

where $\Delta T(t_f)$ is the computed value of the ΔT at t_f. Note that, since the simulations are carried to sufficiently large times, the magnitude of the "correction" in (8.3.11) is small, that is, $a_0 \tau \ll \Delta T(t_f)$.

The predicted steady-state temperature difference across the stack for cases 1–4 (Table 8.2) is plotted in Fig. 8.14, along with the experimental measurements of Atchley et al. (1990) and the theoretical prediction based on (8.3.7). The figure reveals a good agreement between the experimental results and the computations at all the values of the drive ratio considered. The results of the computations are in excellent agreement with the predictions of the theory at low drive ratio, $Dr \leq 1\%$. However, as Dr increases, the predictions of the linear theory deviate significantly from both the experimental measurements and the computations. At moderate drive ratios, the deviation can be substantial, and exceeds 25% at a drive ratio of 0.02.

The dependence of ΔT_{ss} on stack position is plotted in Fig. 8.15, which compares the computed results for cases 4–7 (Table 8.2) to the experimental measurements of Atchley et al. (1990) and the prediction of (8.3.7). In all cases, the drive ratio is held fixed at 2%. As observed earlier, Fig. 8.15 reveals a good agreement between the measurements and the computations. On the other hand, as previously noted in Atchley et al. (1990), the theoretical estimates tend to overpredict ΔT at all stack positions, except in

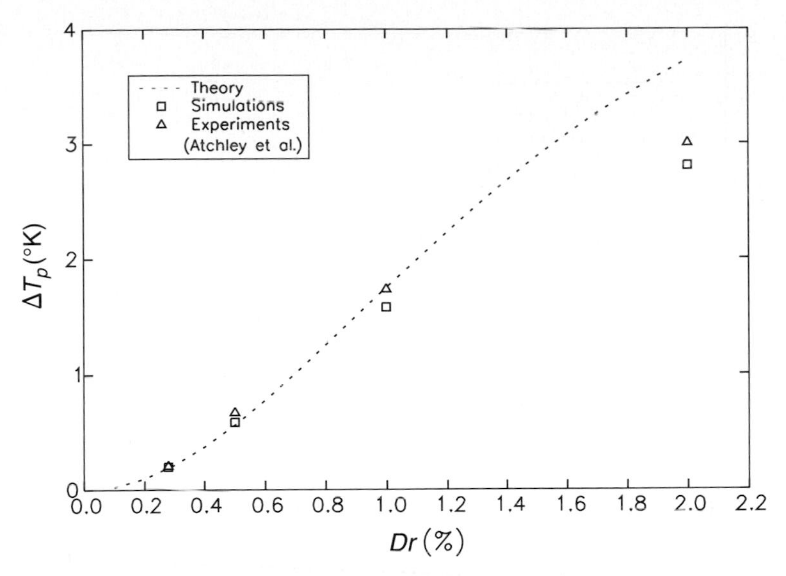

Fig. 8.14. Variation of temperature difference across the stack with drive ratio. The stack is located at $kx = 3\pi/4$. Adapted from Worlikar et al. (1998)

the neighborhood of the velocity node where the thermoacoustic effect naturally vanishes. It is also interesting to note that the computed prediction of the stack position where ΔT peaks, $kx \sim 2.8$, coincides with the corresponding experimental prediction. On the other hand, the simulations and experiments are in slight disagreement with the theory, which predicts that the peak temperature difference occurs at $kx \sim 2.7$.

To analyze the origin of the difference between experimental and computed predictions, on one hand, and predictions of the linear theory, on the other, we examine the distribution of the energy flux density. As noted in Cao et al. (1996), thermal energy transport may be more clearly visualized by plotting the normalized, time-averaged, energy flux density, as defined by

$$\overline{\boldsymbol{H}} = \overline{\rho(T - T_m)\boldsymbol{u}} - \frac{\overline{\nabla T}}{Pe} \tag{8.3.12}$$

where the overbar indicates time-average over one acoustic cycle. Note that $\overline{\boldsymbol{H}}$ is a vector quantity whose magnitude and direction reflect mean energy transport. At the plate surface, the fluid velocity vanishes and the component of $\overline{\boldsymbol{H}}$ normal to the surface indicates the mean heat transfer between the fluid and the plates.

The distribution of the mean energy flux is plotted in Fig. 8.16, which provides, for the same cases just mentioned, enlarged views of the energy flux in the neighborhood of top left plate corner. Figure 8.16 shows that mean heat exchange between the plates and the fluid is concentrated at the ends of the plates. Also the flux through the vertical side of the plates is a

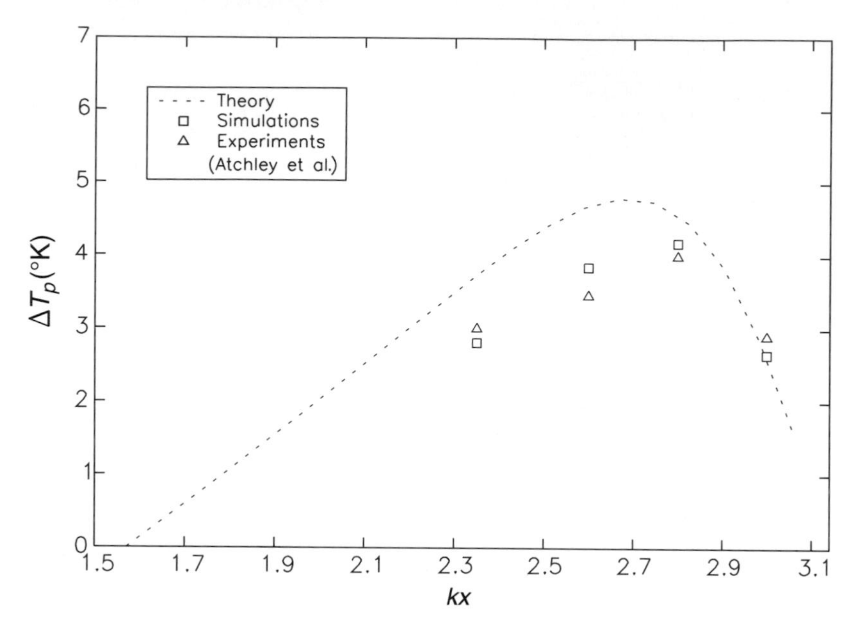

Fig. 8.15. Variation of temperature difference across the stack with position at $Dr = 2\%$. Adapted from Worlikar et al. (1998)

significant portion of the total heat flux leaving or entering the corresponding end. For the cold end shown in Fig. 8.16a ($Dr = 0.5\%$), the heat flux leaving the side of the plate is approximately 26.5% of the total heat flux leaving the cold end. The relative contribution to heat flux from the plate sides increases with drive ratio. At $Dr = 2\%$ (case 4, Fig. 8.16b), the heat flux from the side accounts for roughly 37.6% of the total heat flux leaving the cold end.

Figure 8.16 also provides a deeper insight into the effect of the drive ratio, in particular regarding the path of energy transfer between the hot and cold ends of the stack. At low drive ratio, Fig. 8.16a shows that as they leave the cold end of the stack the energy pathlines rapidly curve around the corner and then follow an essentially straight path toward the hot end. Meanwhile, at higher drive ratio, Fig. 8.16b shows that as they exit the end and side of the plate the energy pathlines first contour a large well-defined recirculation region before they enter into the channel. Thus, for the presently considered conditions, the length of the mean energy path between the ends of the stack significantly increases with increasing drive ratio. This phenomenon, and to some extent the occurrence of a relatively important two-dimensional motion within the gap (see further discussion below), are the major effects of the increase in Dr and appear to be at the origin of the departure between the linear theory and the computations.

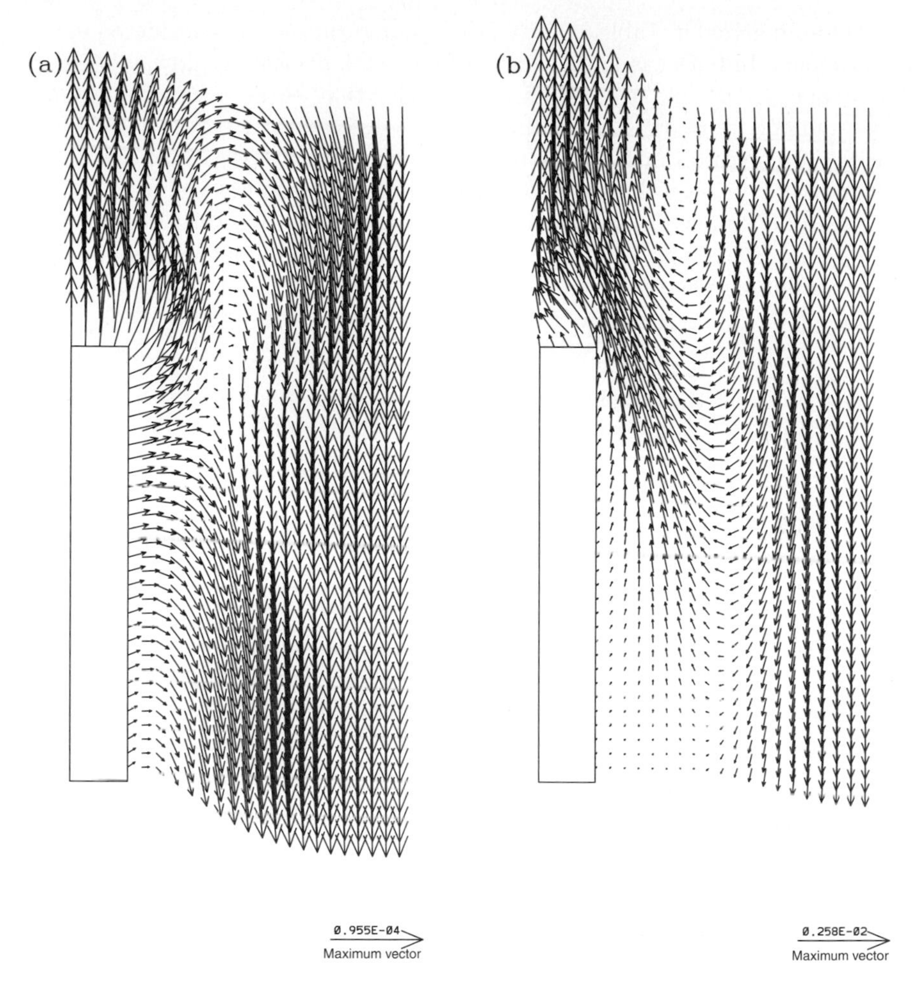

Fig. 8.16. Vector map of the energy flux density near the top left-hand corner of the stack (cold end) for: (**a**) case 2 ($Dr = 0.5\%$), (**b**) case 4 ($Dr = 2\%$). Adapted from Worlikar et al. (1998)

8.3.3 Quantitative Visualization of the Flow Around a Thermoacoustic Couple

In this section, quantitative visualization of the unsteady flow field around the edges of a thermoacoustic couple is performed, and the predictions are contrasted to the experimental PIV measurements of Duffourd (2001). Similar to the experimental setup, thermoacoustic heat exchangers are not present, and attention is focused on stacks operating at low drive ratios. Specifically, results are obtained for two stack configurations that are characterized by disparate ratios of the plate thickness to the viscous penetration depth.

As summarized in Table 8.3, two stack configurations are considered in the experiments. In both cases, the center of the stack is located midway between the velocity node and the velocity antinode, that is, at a nondimensional wave number $kx = 3\pi/4$. In addition to the geometrical parameters of the stack, the computations use as input the thermal properties of the stack plates and of air, as well as the mean pressure and temperature, respectively $P_m = 10^5$ N/m^2 and $T_m = 300°$K. The experimentally observed frequency of the resonant standing wave, $f = 200$ Hz, is also used as input. The amplitude of the standing wave is expressed in terms of the drive ratio, Dr, as reported in Table 8.3.

Table 8.3. Geometric and flow parameters for configurations A and B: d denotes the plate thickness, h is the channel height, and L_p is the plate length

	d (mm)	h (mm)	L_p (mm)	Dr (%)	h/δ_ν	d/δ_ν
Configuration A	1.0	2.0	25.8	0.5	13.	6.5
Configuration B	0.15	1.0	24.0	1.5	6.7	1.0

As can be appreciated from Table 8.3, the two stack configurations are nearly equal in length. However, the thickness and spacing of the stack plates are significantly smaller in configuration A than they are in configuration B. These differences are also highlighted by providing the ratios h/δ_v and d/δ_v, where δ_v is the viscous penetration depth (Swift, 1988). Thus, d is substantially larger than δ_v in configuration A, but is close to δ_v in configuration B. Accordingly, we shall refer to these two stacks as thick-plate and thin-plate configurations, respectively.

In both the experiments and computations, data analysis is performed when a "periodically steady" state is established. This occurs after a large number of acoustic cycles, during which the initial starting transient decays. Once steady state is achieved, instantaneous data on the velocity and vorticity fields are collected at well-defined instants within a single cycle. In both the experiments and computations, 16 equally spaced, instantaneous realizations are obtained.

To perform a meaningful comparison between instantaneous realizations, it is necessary to establish a proper synchronization of the time axes. In the experiments, the phase is measured with respect to t_0, time at which the (imposed) driving voltage of the acoustic speaker peaks within the cycle. In the computations, on the other hand, the "phase" is measured with respect to t_p, time at which the (imposed) acoustic pressure signal peaks within the acoustic cycle. By performing detailed comparison of the computational and experimental results, we find that closest correspondence occurs when t_p and t_0 are separated by $T/16$, where T is the period of the acoustic cycle. This shift, which is identical for both configurations, suggests a small phase difference between the voltage and pressure oscillations in the experiments. However, this could not be independently verified since the experimental pressure signal at the stack location was not recorded during the PIV measurements. On the other hand, apart from this synchronization of time origins, no adjustments have been performed to either the experimental or computational data.

8.3.3.1 Thick-Plate Configuration. Figure 8.17 depicts two instantaneous realizations of the velocity and vorticity fields for configuration A. Shown on the top row are experimental results obtained at $t_0 + T/4$ and $t_0 + 3T/8$; the bottom row depicts computational predictions obtained at $t_p + 3T/16$ and $t_p + 5T/16$. Thus, in both the experiments and computations the two frames are $T/8$ apart. Note that the frame of the computational results has also been restricted to match the 4.2×4.2 mm^2 window of the PIV measurements, which is located near the cold end of the stack plates. In

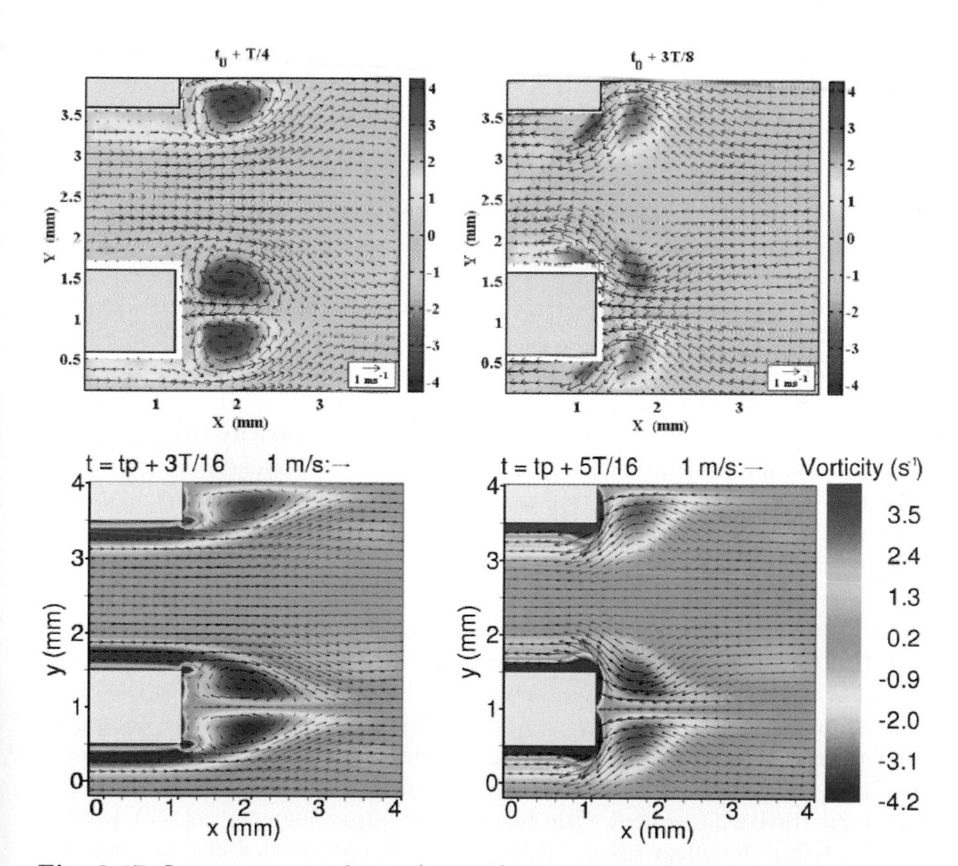

Fig. 8.17. Instantaneous velocity (vectors) and vorticity (contours) fields around the cold end of the stack in configuration A. Top row: experimental PIV measurements; bottom row: computations. The times at which the frames are generated are indicated in the labels. Adapted from Blanc-Benon et al. (2003)

particular, the results are arranged in such a way that the velocity node is located to the left of each frame.

For the present configuration, both the experimental and computational results reveal the presence of concentrated vortices near the edge of the plate. Also evident is the presence of both signs of vorticity, both inside the channel and in the open region. The generation of alternating layers of vorticity within the channel is not surprising, since the ratio of channel height to viscous penetration depth is large and the development of Stokes layers within the channel is accordingly expected. As can be observed in Fig. 8.17, the vorticity distribution outside the channel can also exhibit complex structure, especially when separated vortices are driven back into the channel and impinge on the edge of the plates.

Comparison of the experimental and computational results in Fig. 8.17 reveals a close correspondence between the predictions. In particular, at both phases, very good agreement can be observed between peak vorticity values as well as the size of the recirculating regions.

8.3.3.2 Thin-Plate Configuration. Figure 8.18 provides computed and experimental results obtained for configuration B. Shown in the left column are computed velocity and vorticity distribution at three selected phases within the cycle; plotted in the right columns are contours of the experimental vorticity distribution at the corresponding times. The figure is generated in a similar fashion as Fig. 8.17; that is, the computational test section is restricted so that it matches the experimental window. In the present case, the PIV measurements are performed in a 2.9×2.3 mm^2 window, also located near the cold end of the stack.

The present predictions are in sharp contrast with the observations of configuration A. Specifically, in the present configuration B, the results do not show the formation of well-defined eddies. The vorticity distribution in the wake exhibits the presence of elongated layers that extend well outside the channel. However, significant roll up of these layers is not observed, although evidence of very weak recirculating motion near the plate corners can be detected in the last computational frame.

Note that the top and bottom frames in Fig. 8.18 correspond to t_p and $t_p + T/2$, times at which the acoustic pressure is maximal and minimal, respectively. At these phases, the streamwise extent of the vortical wake is found to be maximal and the direction of the flow in the middle of the channel is opposite to that of the flow along the plate surface. The computed and measured velocities are in good agreement with each other. As noted by Duffourd (Duffourd, 2001), the measured and computed velocities are in good qualitative agreement with, but slightly larger than the velocity profiles derived using the linear theory (Swift, 1988; Arnott et al., 1991) for the same configuration. As discussed in Worlikar and Knio (1996), this discrepancy is most likely due to blockage effects that are ignored by the linear theory.

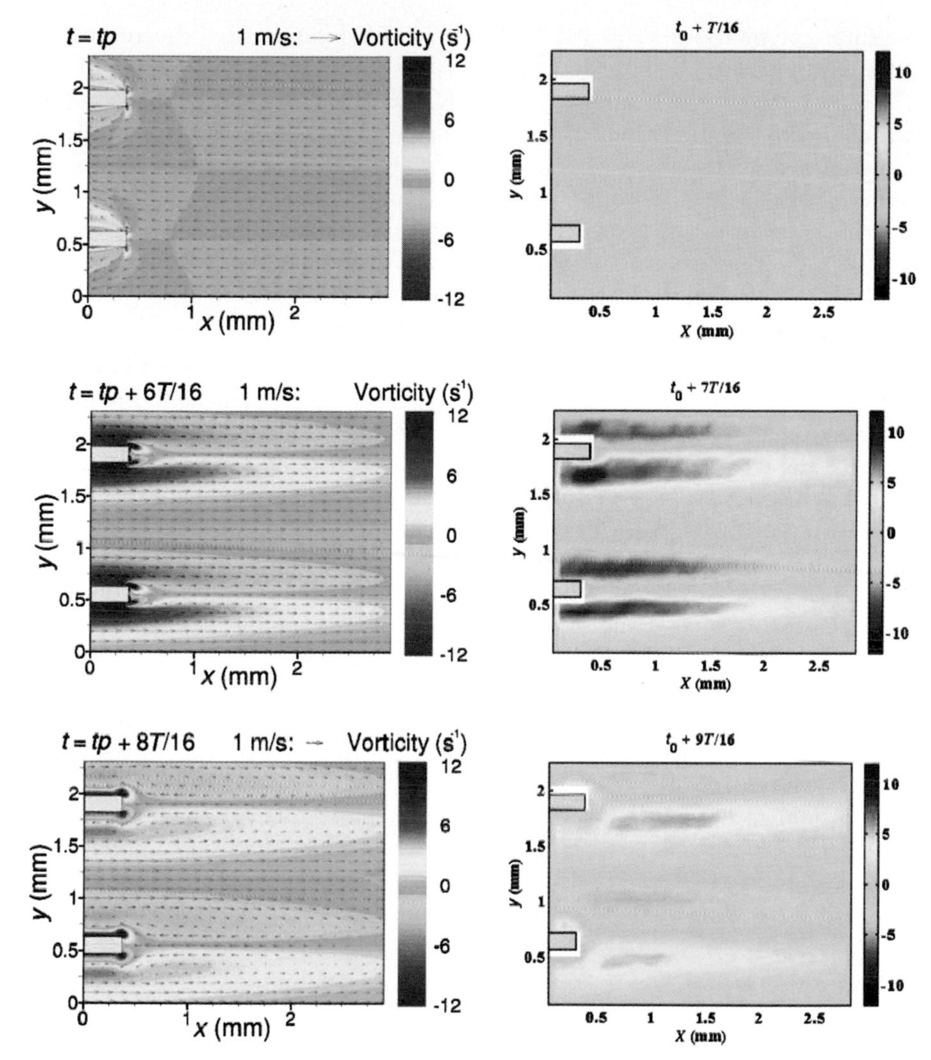

Fig. 8.18. *Left*: Computed velocity (vectors) and vorticity (contours) fields at selected times for configuration B. *Right*: Instantaneous vorticity contours from PIV measurements. The times at which the frames are generated are indicated in the labels. Adapted from Blanc-Benon et al. (2003)

8.3.4 Analysis of Performance

The vorticity-based scheme outlined above is applied to investigate the effect of the stack and heat exchanger configuration on the thermal performance of the device. In all the cases below, the center of the stack is located midway between the velocity node and the velocity antinode, that is, at a nondimensional wavenumber $kx = 3\pi/4$. The stack plates are 3.18 cm long with a thermal conductivity $\tilde{k}_s = 0.482$ W/m·°K. We assume that the resonance tube is filled with helium and use $\tilde{\nu} = 10^{-4}$ m^2/s, $\tilde{k}_f = 0.14$ W/m·°K, $Pr = 0.67$, $\tilde{T}_m = 300$°K, and $\tilde{P}_m = 10^5$ N/m^2. The frequency of the standing wave $\tilde{f} = 707$ Hz and the acoustic wavelength $\tilde{\lambda} = 1.44$ m. A parametric study is performed of the effects of drive ratio ($2\% \leq Dr \leq 8\%$), temperature difference between the hot and cold heat exchangers ($6\ K \leq \Delta\tilde{T} \leq 21\ K$), gap width ($\tilde{g}$) between the heat exchangers and the stack plates, heat exchanger length ($\tilde{L}_c$), thickness of the plates ($\tilde{d}$), and plate spacing ($\tilde{h}$). We start by analyzing the behavior of the cooling load ($\tilde{Q}_{m,f\rightarrow c}$) and then examine the flow structure for selected configurations and operating conditions.

8.3.4.1 Effect of Geometrical and Operating Parameters on Cooling Load. With most configurations, once steady state is reached this cycle-averaged heat flux remains constant; that is, it does not change from one cycle to the next. At high drive ratios however ($Dr \geq 5\%$), in some cases the cycle-averaged heat flux periodically oscillates over several acoustic cycles, as illustrated in Fig. 8.19. In these situations, the cooling load is computed by averaging the heat flux over a suitably large number of cycles. The origin and onset of the oscillations in the cycle-averaged heat flux require detailed analysis that is beyond the scope of the present discussion.

Preliminary analysis of the computations reveals that heat exchanger length, gap width, drive ratio, temperature difference (ΔT), position of the stack in the resonance tube, and plate spacing (h) affect the performance of the device in a complex fashion. The analysis also reveals that three parameters are particularly relevant: the optimal heat exchanger length (L_{opt}), the optimal gap width (g_{opt}), and the optimal gap width and heat exchanger length combination $(L, g)_{\text{opt}}$. L_{opt} denotes the ratio of the heat exchanger length ($\tilde{L}_{h/c}$) to $(2\tilde{R}_p + \tilde{\delta}_k)$ for which the cooling ($\tilde{Q}_{m,f\rightarrow c}$) peaks when the gap width ($\tilde{g}$) is kept constant. On the other hand, g_{opt} denotes the ratio of the gap width ($\tilde{g}$) to $(2\tilde{R}_p + \tilde{\delta}_k)$ for which the cooling peaks when the heat exchanger length ($L_{h/c}$) is kept constant. The combination ($L_{\text{opt}}, g_{\text{opt}}$) corresponds to the optimal combination $(L_c, g)_{\text{opt}}$ for which the cooling peaks when both $L_{h/c}$ and g are varied. Note that the optimal gap (g_{opt}) computed for a given heat exchanger length $L_c \neq L_{\text{opt}}$ is a priori different from the gap width in the optimal combination $(L_c, g)_{\text{opt}}$. The same remark applies to L_{opt}.

The analysis is presented in the following order. A study of the combined effect of heat exchanger length and gap width on the cooling load is first conducted, followed by an analysis of the role that drive ratio and temperature difference play on the performance. Finally, the effects of the plate thickness, stack periodicity, and stack location in the tube are investigated.

8.3.4.2 Combined Effect of Heat Exchanger Length and Gap Width. For all considered values of the drive ratio, temperature difference ($\Delta\tilde{T}$), stack location (kx), stack spacing ($\tilde{H}$), and plate thicknesses ($\tilde{d}$), the computations reveal that the cooling load ($\tilde{Q}_{m,f\rightarrow c}$) peaks at well-defined values of the gap

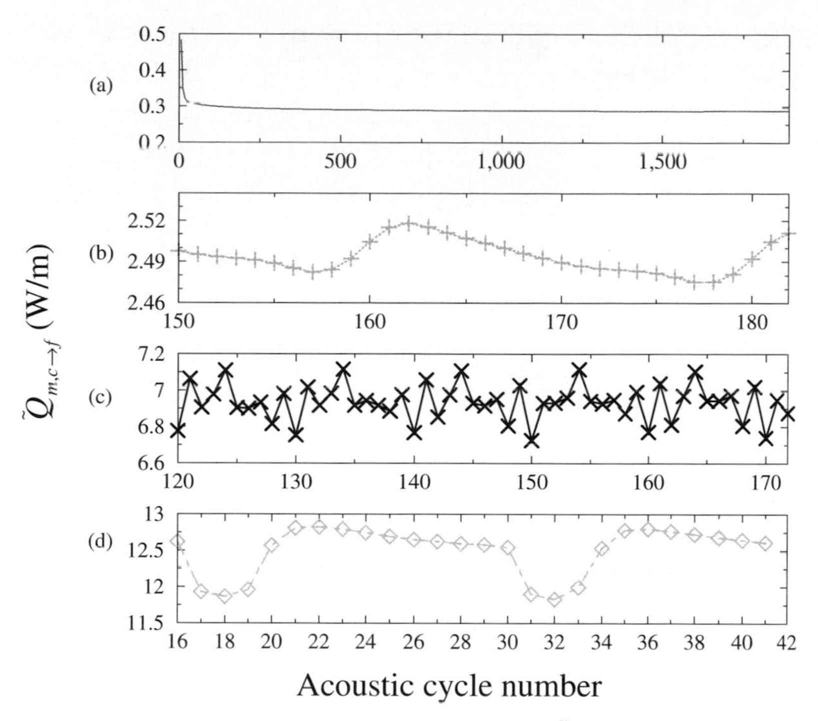

Fig. 8.19. Evolution of the cycle-averaged heat flux $\tilde{Q}_{m,f\to c}$ from the cold heat exchanger: (a) $Dr = 2.5\%$, $\Delta\tilde{T} = 18$ K, $L_{c/h}/Lo = 1.73$, and $g/Lo = 0.59$; (b) $Dr = 5\%$, $\Delta\tilde{T} = 18$ K, $L_{c/h}/Lo = 1.33$, and $g/Lo = 0.66$; (c) $Dr = 6.5\%$, $\Delta\tilde{T} = 18$ K, $L_{c/h}/Lo = 1.03$, and $g/Lo = 0.34$; and (d) $Dr = 8\%$, $\Delta\tilde{T} = 18$ K, $L_{c/h}/Lo = 1.34$, and $g/Lo = 0.33$. Adapted from Besnoin and Knio (2004)

width and heat exchanger length. To illustrate the variation of the cooling load with $\tilde{L}_{h/c}$ and $\tilde{g}$, we plot in Fig. 8.20 the distribution of the mean heat flux $\tilde{Q}_{m,f\to c}$ when $\tilde{L}_{h/c}$ and $\tilde{g}$ are varied, with a stack operating at $Dr = 8\%$ and $\Delta\tilde{T} = 18$K. For the present conditions $(L_c, g)_{\text{opt}} \simeq (1.5, 0.4)$. For larger and smaller gaps, the cooling load drops rapidly, exhibiting with variations of up to 100% in the range considered. Note that the decrease in the cooling load at large values of $\tilde{g}$ is expected. This is the case because the exchange between the stack and the heat exchangers is governed by particle transport and is consequently limited by the particle displacement amplitude. Thus, as the gap becomes large, the thermal coupling between the stack and heat exchangers is diminished, and the cooling load is consequently reduced. As observed in Besnoin and Knio (2001), the reverse effect, namely the drop in the mean heat flux as $\tilde{g}$ becomes small, results from the increase in reverse heat transfer as $\tilde{g}$ decreases.

For the present configuration, the cooling load peaks when $L_{h/c}/(2R_p + \delta_k) \simeq 1.5$. This result strongly contrasts with results obtained in the

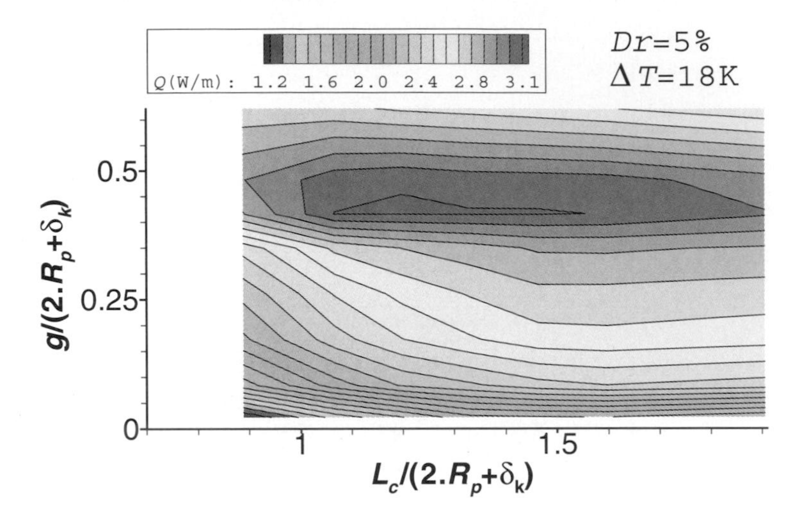

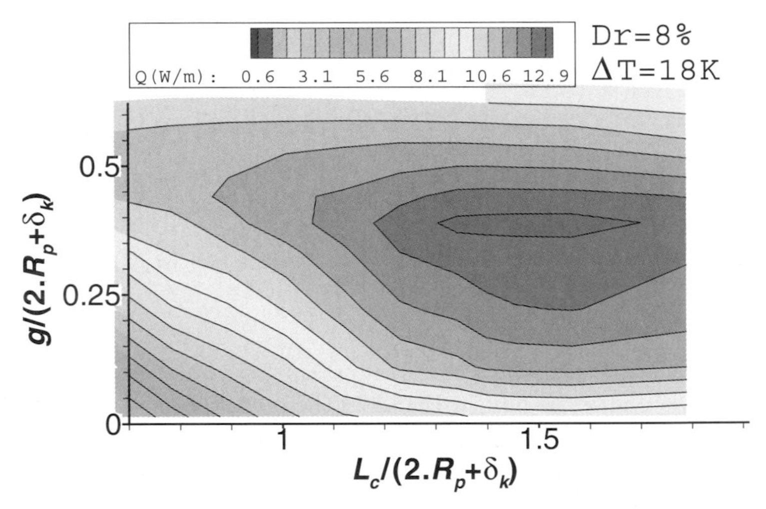

Fig. 8.20. Contours of $\tilde{Q}_{m,f\to c}$ for different values of L and g. The curves are obtained with $\Delta\tilde{T} = 18\text{K}$, $Dr = 5\%$ (*top*), and $Dr = 8\%$ (*bottom*). Adapted from Besnoin and Knio (2004)

thin-plate limit (Besnoin and Knio, 2001), where cooling was found to peak when $L_{c/h}/(2R_p + \delta_k) \simeq 0.3$. The result also differs from Swift's analysis (Swift, 1988), which predicts peak performance when the length of the cold heat exchanger is approximately equal to $(2R_p + \delta_k)$. The analysis below addresses these differences.

8.3.4.3 Effect of Drive Ratio and Temperature Difference Between the Heat Exchangers. To determine the effect of Dr and $\Delta\tilde{T}$ on the cooling load, both parameters are varied in this section. The location of the stack in the resonance tube is kept constant at $kx = 3\pi/4$. Also fixed are the stack plate spacing ($\tilde{H} = 1.4$ mm) and the plate thickness ($\tilde{d} = 0.467$ mm). The temperature difference $\Delta\tilde{T}$ varies between 6 K and 21 K, and the drive ratio Dr is varied between 2% and 8%. Specifically, for $Dr = 2\%$, results are obtained with $\Delta\tilde{T} = 6$ K. For $Dr = 2.5\%$ and 3.5%, results are obtained with $\Delta\tilde{T} = 15$ and 18 K. For $Dr = 3\%$, results are obtained with $\Delta\tilde{T} = 6, 9, 12, 15, 18$ K. For $Dr = 5\%$, results are obtained with $\Delta\tilde{T} = 6, 9, 12, 15, 18, 21$ K. Finally, for $Dr = 8\%$, results are obtained with $\Delta\tilde{T} = 6, 12, 18$ K.

For each configuration, $\tilde{L}_c$ and $\tilde{g}$ are varied and the optimal combination $(L_c, g)_{\text{opt}}$ is identified. Figures 8.20, 8.21, and 8.22 show the distributions of the mean heat flux $\tilde{Q}_{m,f\to c}$ for selected cases. The results indicate that when the operating conditions are varied, the optimal combination $(L_c, g)_{\text{opt}}$ also varies. However, one can differentiate between the effects of the drive ratio and of the temperature difference. As shown in Figs. 8.20 and 8.21, the optimal combination $(L_c, g)_{\text{opt}}$ seems to remain constant when only the drive ratio is varied. In the first case (Fig. 8.21), for all drive ratios considered and a temperature difference $\Delta\tilde{T} = 6$ K, the cooling load peaks at $(L_c, g)_{\text{opt}} \simeq (1.5, 0.1)$, whereas with $\Delta\tilde{T} = 18$ K (Fig. 8.20), the peak occurs at $(L_c, g)_{\text{opt}} \simeq (1.4, 0.4)$ both for $Dr = 5\%$ and 8%. On the other hand, the heat exchanger length ($L_{h/c}$) and gap width (g) for which the peak occurs both vary with $\Delta\tilde{T}$. To isolate the effect of $\Delta\tilde{T}$, we plot in Fig. 8.22 the distribution of the mean heat flux when $\Delta\tilde{T}$ is varied independently. Clearly, the location at which the mean heat flux peaks changes as $\Delta\tilde{T}$ increases. Specifically, the optimal heat exchanger length decreases and the optimal gap increases when $\Delta\tilde{T}$ increases.

8.3.4.4 Optimal Heat Exchanger Length and Optimal Gap Width. The following trends are observed for all cases considered in the analysis. First, as shown in Fig. 8.23, the optimal length L_{opt} varies with the gap width only, and is insensitive to the drive ratio (Dr) and temperature difference ($\Delta\tilde{T}$). When $g/(2R_p + \delta_k) \leq 0.35$, the cooling load systematically peaks for a heat exchanger length equal to about $1.55 \times (2R_p + \delta_k)$. For wider gaps, the optimal length drops rapidly as g increases.

On the other hand, as shown in Fig. 8.24, the optimal gap width varies with $\Delta\tilde{T}$. The results show that in the range $L_c/(2.R_p + \delta_k) \geq 1$, the optimal gap width (g_{opt}) is insensitive to L_c. In this range, the variation of g_{opt} is better depicted by plotting its dependence on $\Delta\tilde{T}$, as shown in Fig. 8.25. Clearly, the optimal gap width increases with $\Delta\tilde{T}$ for all the configurations considered. In addition, for $\Delta\tilde{T} \leq 15$ K, the optimal gap width only varies with $\Delta\tilde{T}$, not with the drive ratio Dr. On the other hand, when $\Delta\tilde{T} > 15$ K, the predictions exhibit a noticeable dependence on the drive ratio. This dependence appears to be linked with the appearance of wavy structures around the egdes of the stack and heat exchanger plates for $Dr \geq 3.5\%$. Further

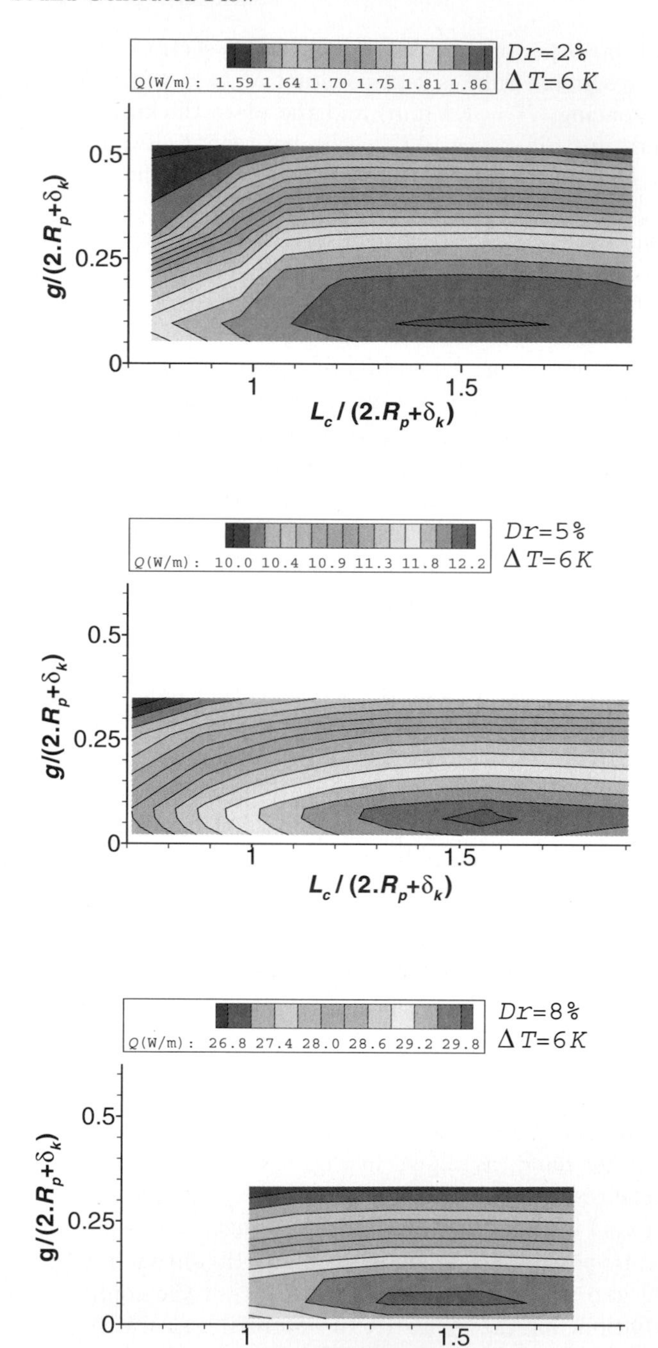

Fig. 8.21. Contours of $\tilde{Q}_{m,f\to c}$ for different values of L and g. The curves are obtained with $\Delta\tilde{T} = 6$ K, and $Dr = 2$ (*top*), 5 (*center*), and 8% (*bottom*). Adapted from Besnoin and Knio (2004)

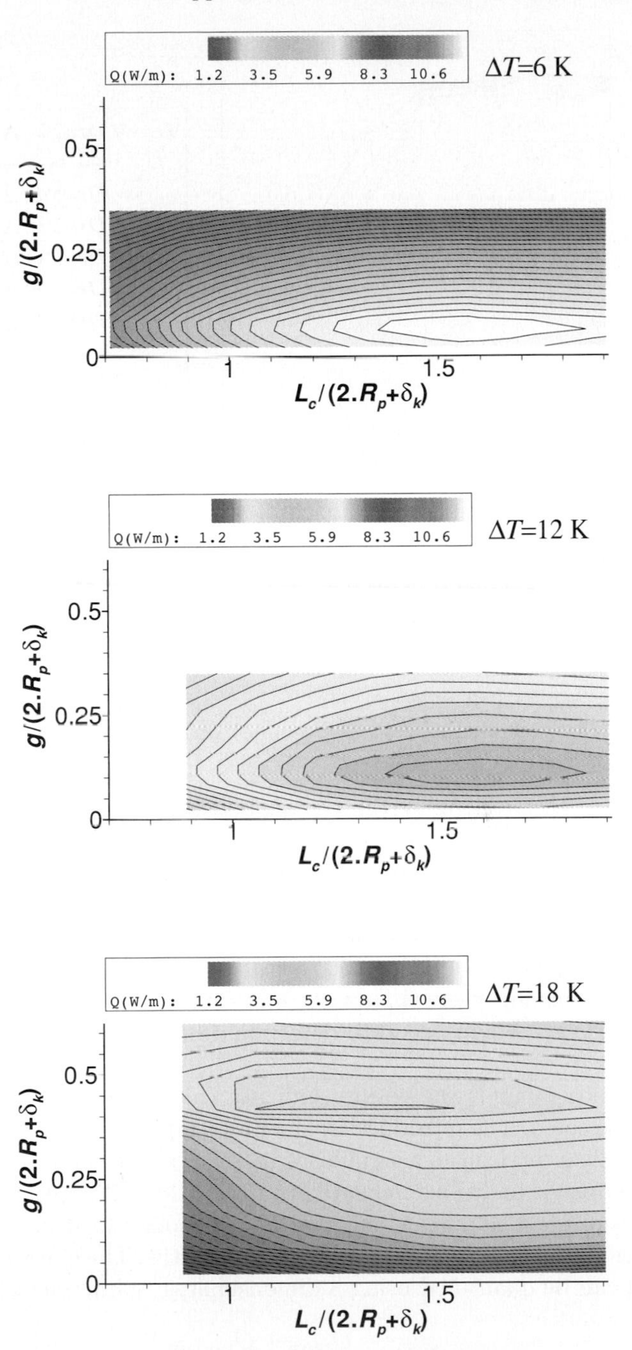

Fig. 8.22. Contours of $\tilde{Q}_{m,f\to c}$ for different values of L and g. The curves are obtained with $\Delta\tilde{T} = 6$ K (*top*), 12 K (*center*), and 18 K (*bottom*). The stack is operating at $Dr = 5\%$. Adapted from Besnoin and Knio (2004)

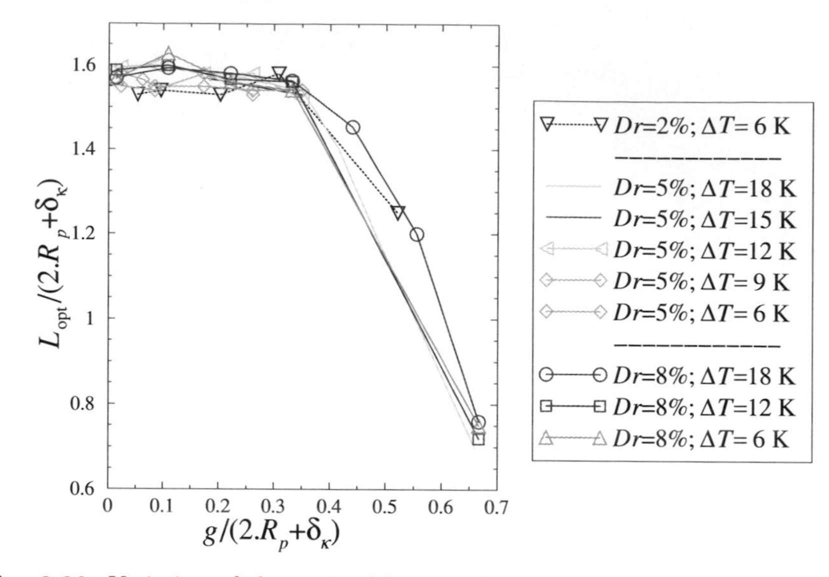

Fig. 8.23. Variation of the optimal heat exchanger length with normalized gap size. L_{opt} is defined as the length for which the flux peaks for a given (fixed) gap width. Adapted from Besnoin and Knio (2004)

analysis of these structures and their effect on heat transfer is provided in the following section.

8.3.4.5 Gain in Cooling After Optimization. The impact of proper selection of L and g is now investigated. For all configurations considered, it is found that the cooling load fluctuates above 90% of its maximal value whenever $0.8 \leq L_{h/c}/(2.R_p + \delta_k) \leq 2$. When the length of the heat exchanger is chosen below this range, the cooling drops sharply. This indicates that, as observed in Besnoin and Knio (2001), in case of uncertainty regarding the optimal heat exchanger length, a wiser solution is to select $L_{h/c}$ in the upper part of the range, where accurate determination of the optimum is far less critical.

On the other hand, the results indicate that the gap width should be accurately selected. The introduction of a finite gap between the stack and the heat exchangers can lead to a significant increase in the cooling load $\tilde{Q}_{m,f\to c}$. For the operating conditions indicated in Figs. 8.20–8.22, optimization of the gap size results in a significant increase in the cooling load in comparison to arrangements having either small or large gap widths. The effect of proper size of the gap can be quantified using a dimensionless "gain," defined according to

$$Gain = \frac{Q_{max} - Q_{no\ gap}}{Q_{no\ gap}}, \tag{8.3.13}$$

where Q_{max} is the optimal heat flux, and $Q_{no\ gap}$ is the optimal heat flux obtained when the heat exchangers are in close proximity with the stack

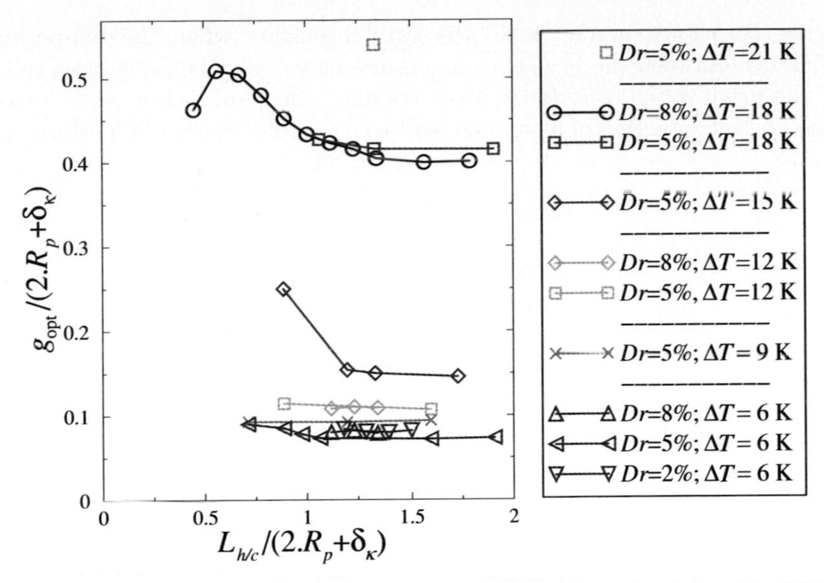

Fig. 8.24. Variation of the optimal gap width with normalized heat exchanger length. g_{opt} is defined as the gap width for which the flux peaks for a given (fixed) heat exchanger length. Adapted from Besnoin and Knio (2004)

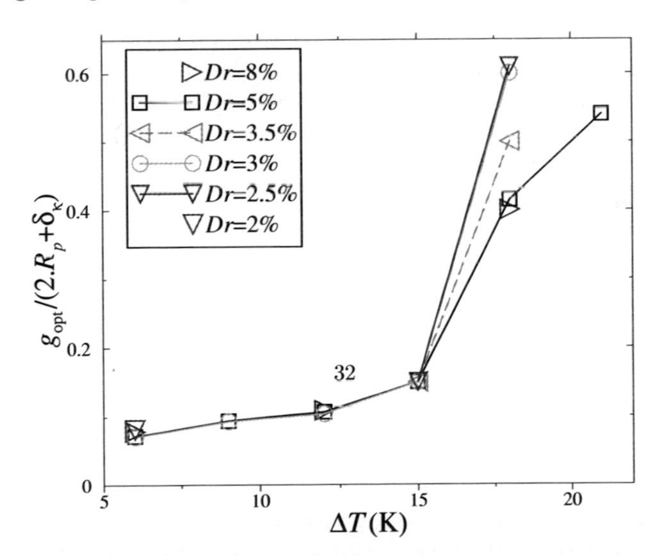

Fig. 8.25. Variation of the optimal gap width with the temperature difference $\Delta\tilde{T}$, when $L_c/(2.R_p + \delta_k) \geq 1$. Adapted from Besnoin and Knio (2004)

plates. As shown in Fig. 8.26, the gain intensifies when the temperature difference between the heat exchangers increases. This trend is observed for all the drive ratios considered. In particular, the results show that in some cases the introduction of a gap can lead to a twofold increase in performance.

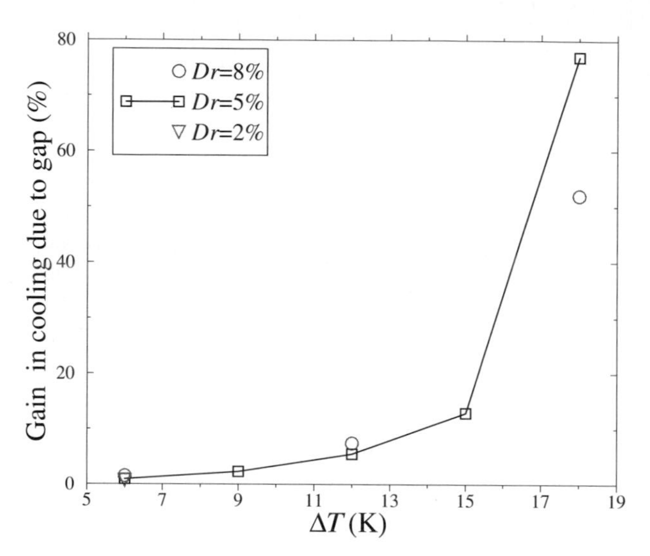

Fig. 8.26. Variation of the gain with the temperature difference $\Delta\tilde{T}$ for a stack located at $kx = 3\pi/4$, for $Dr = 2\%$, 5% and 8%. Adapted from Besnoin and Knio (2004)

8.3.4.6 Effect of Plate Thickness and Stack Period. The effects of the plate thickness and stack period are analyzed in this section. The heat exchanger length, gap width, drive ratio, and temperature difference ($\Delta\tilde{T}$) are kept constant, while several combinations of the stack period ($\tilde{H}$) and plate thickness ($\tilde{d}$) are explored. This enables us to study the effect of blockage ratio (BR) and stack separation ($\tilde{h}$) on the cooling load ($\tilde{Q}_{m,f\to c}$) and on the power density ($\tilde{Q}_{m,f\to c}/\tilde{H}$). Note that the heat exchanger length $\tilde{L}_{h/c} \simeq$ 14.6 mm and the gap width $\tilde{g} \simeq 1.72$ mm approximately correspond to the configuration for which the cooling load peaks when $\Delta T = 15$ K, $Dr = 5\%$.

The results show that the cooling increases rapidly when the stack separation is increased, both when the thickness of the plates is maintained constant (along a horizontal line in Fig. 8.27a) and when the blockage ratio BR is kept constant (Fig. 8.27b). In both cases, the cooling reaches a peak beyond which it varies slightly. The stack period for which the plateau occurs remains essentially constant as the blockage ratio increases. As shown in Fig. 8.27b, for BR $= 66\%$, the plateau occurs for $d \geq 0.46$ mm, which corresponds in this case to $H/\delta_k \geq 4.8$. For BR $= 80\%$, the plateau occurs for $d \geq 0.25$ mm, which also corresponds to $H/\delta_k \geq 4.8$. Similar calculations at other blockage

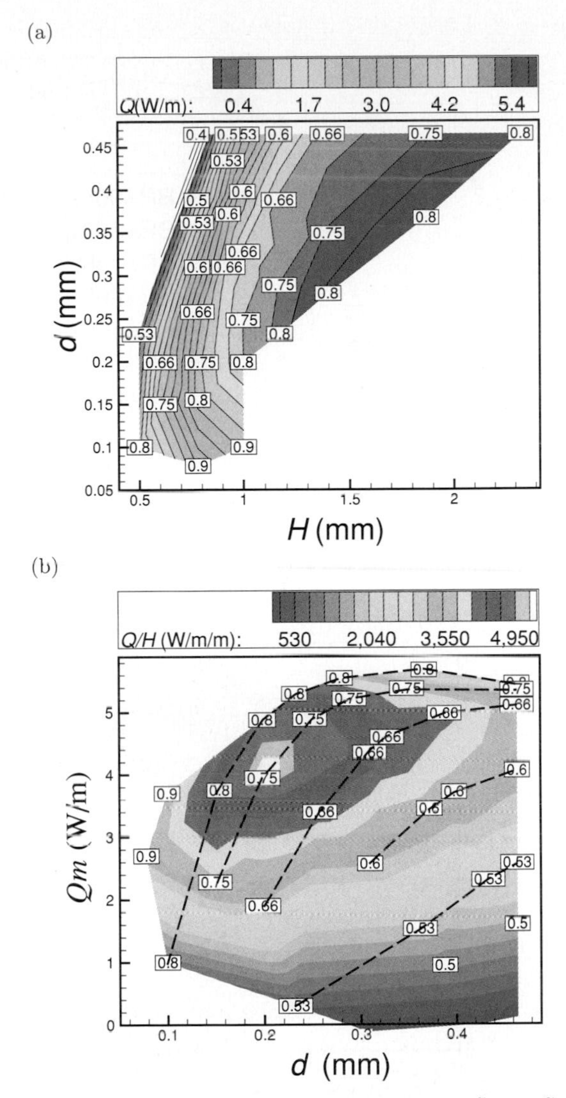

Fig. 8.27. (**a**) Contours of $\tilde{Q}_{m,f\to c}$ for different values of $\tilde{d}$ and $\tilde{H}$. (**b**) Composite plot showing the variation of the cooling load $\tilde{Q}_{m,f\to c}$ (dashed lines) and the power density $\tilde{Q}_{m,f\to c}/\tilde{H}$ (contours) with $\tilde{d}$. In both frames, the boxed labels denote the blockage ratio, BR $\equiv (H-d)/H \equiv h/H$. In all cases, $\Delta\tilde{T} = 15$ K and $Dr = 5\%$, $kx = 3\pi/4$, $\tilde{L}_{h/c} \simeq 14.6$ mm, and $\tilde{g} \simeq 1.72$ mm. Adapted from Besnoin and Knio (2004)

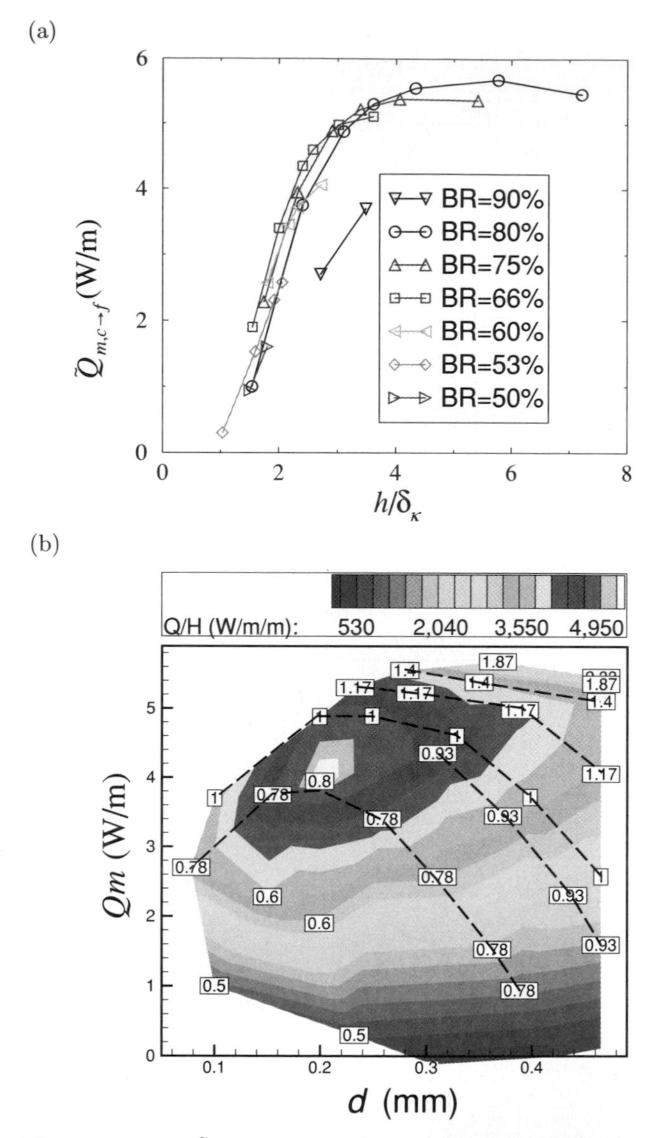

Fig. 8.28. (**a**) Variation of $\tilde{Q}_{m,f\to c}$ with h/δ_k for different values of the blockage ratio, BR. (**b**) Composite plot showing the variation of the cooling load $\tilde{Q}_{m,f\to c}$ (dashed lines) and the power density $\tilde{Q}_{m,f\to c}/\tilde{H}$ with $\tilde{d}$. In all cases, $\Delta\tilde{T} = 15$ K and $Dr = 5\%$, $kx = 3\pi/4$, $\tilde{L}_{h/c} \simeq 14.6$ mm and $\tilde{g} \simeq 1.72$ mm. The boxed labels in (b) denote the stack periodicity $\tilde{H}$ in mm. Adapted from Besnoin and Knio (2004)

ratios lead to the same result. Similarly, as shown in Fig. 8.28a, the cooling reaches a plateau when the ratio h/δ_k exceeds 4.

Figure 8.28a also shows that given a ratio h/δ_k, the amount of heat transfered to one heat exchanger plate remains similar for any combination of plate thickness ($\tilde{d}$) and blockage ratio, as long as BR $\leq 80\%$. When BR $= 90\%$ however, the cooling load is significantly smaller. On the other hand, Fig. 8.28b indicates that, when the stack period ($\tilde{H}$) is maintained constant and $\tilde{d}$ is varied, the cooling load peaks for a well-defined plate thickness. For the present configuration, when $\tilde{d} \leq 0.18$ mm, the cooling load drops sharply. This drop appears to be affected by two distinct phenomena. The first concerns the dynamics of the flow near the edges of the plates. Specifically, as noted in Duffourd (2001) and Blanc-Benon et al. (2003), as the plates are made thinner the flow structure around the plate edges changes from being dominated by the shedding and impingement of concentrated vortices to one that is characterized by elongated vorticity layers. A second effect that occurs when the plate thickness decreases concerns the reduced thermal mass and diffusion timescale within the plates; this leads to amplified temperature oscillations within the plates (not shown). Both of these effects appear to contribute to the drop in cooling load at small values of d.

8.3.4.7 Power Density. Besides variations of the cooling load ($\tilde{Q}_{m,f\to c}$), Figs. 8.27b and 8.28b show contours of the power density ($\tilde{Q}_{m,f\to c}/\tilde{H}$). The results show that an optimal plate thickness exists and that a poor choice of $\tilde{d}$ can significantly affect heat exchanger performance. As shown in Fig. 8.29a, for the present configuration the power density peaks when the plates are separated by a distance $\tilde{h}$ about 2.5 times the thermal penetration depth, and when the thickness of the plates is slightly smaller than the thermal penetration depth. With thinner or thicker plates, the power density drops rapidly.

The high sensitivity of the power density on plate spacing is also illustrated in Fig. 8.29b, which provides results for different values of the blockage ratio. For all considered values of d and BR, the power density peaks when $h/\delta_k \simeq 2.5$–3. As observed in Besnoin and Knio (2001), stacking the plates too closely would result in a considerable loss of efficiency. In case of uncertainty, it appears preferable to select h in the upper part of the uncertainty range. While an overestimate of h/δ_k would result in a substantial reduction in the cooling load compared to the optimal configuration, the drop would not be as large as when h is underestimated. Note also that the power density peaks when BR $= 75\%$, which is consistent with the drop in cooling mentioned above at large blockage ratio.

8.3.4.8 Optimal Gap Width and Optimal Heat Exchanger Length. The dependence of the optimal combination $(g, L_c)_{opt}$ on the plate thickness ($\tilde{d}$) and spacing ($\tilde{h}$) is analyzed in this section. To this end, the optimal heat exchanger length and gap width are determined for different combinations of $\tilde{H}$ and $\tilde{d}$. In particular, computed results indicate that both the optimal heat

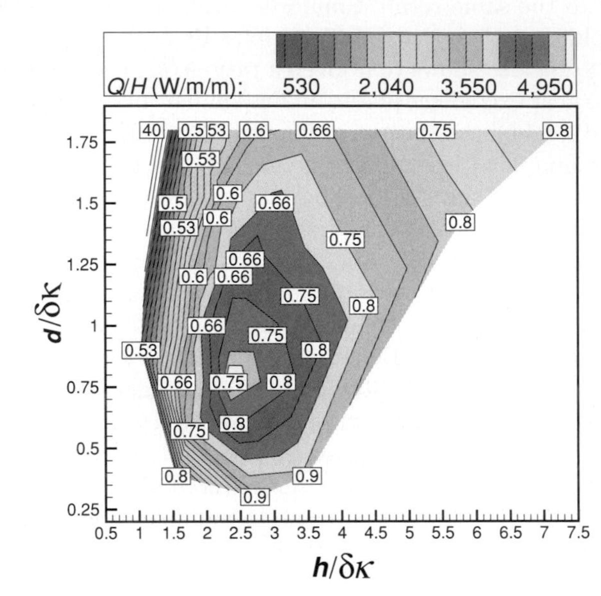

(b)

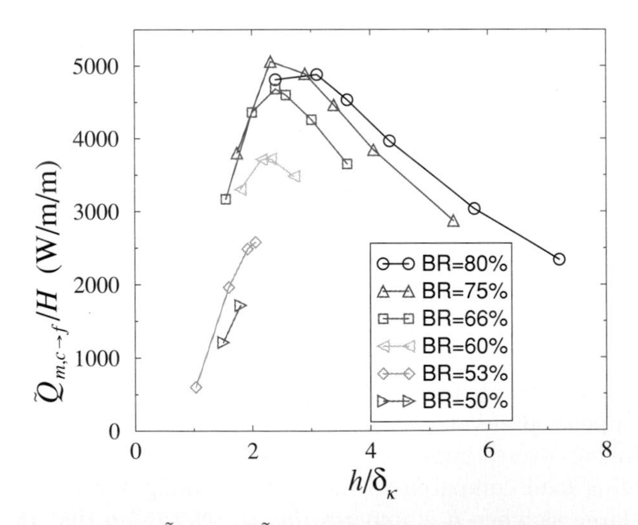

Fig. 8.29. (**a**) Contours of $\tilde{Q}_{m,f\to c}/\tilde{H}$ for different values of d/δ_k and H/δ_k, where δ_k is the thermal penetration depth. The boxed labels denote the blockage ratio. (**b**) $\tilde{Q}_{m,f\to c}/\tilde{H}$ versus H/δ_k for different values of BR. In all cases, $\Delta\tilde{T} = 15$ K and $Dr = 5\%$, $kx = 3\pi/4$, $\tilde{L}_{h/c} \simeq 14.6$ mm and $\tilde{g} \simeq 1.72$ mm. Adapted from Besnoin and Knio (2004)

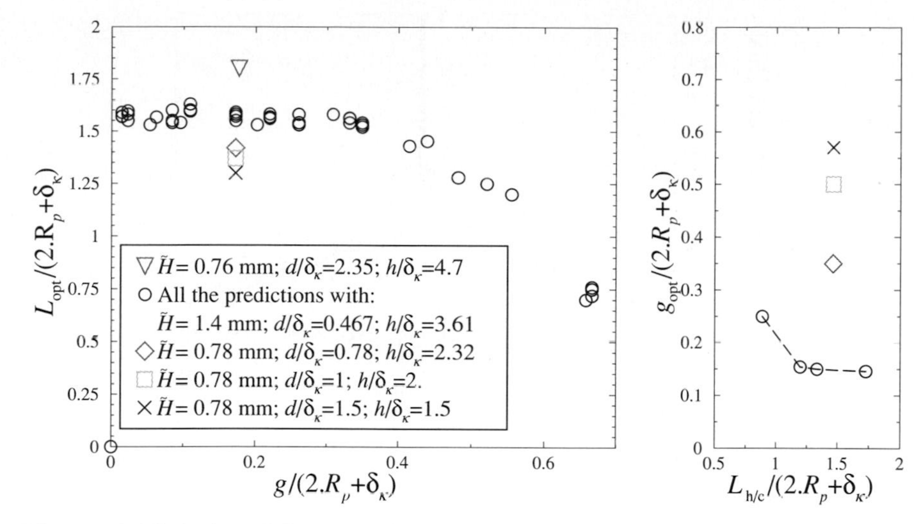

Fig. 8.30. Variation of the optimal heat exchanger length with gap size (*left*) and of the optimal gap width with heat exchanger length (*right*). The results are obtained with $\Delta T = 15$ K, $Dr = 5\%$, and varying values of $\tilde{H}$ and $\tilde{d}$. Also plotted are the results obtained with $\tilde{H} = 1.4$ mm and $h/\delta_k = 3.6$. Adapted from Besnoin and Knio (2004)

exchanger length and the optimal gap width vary with h/δ_k and with $\tilde{d}$. As shown in Fig. 8.30, the optimal heat exchanger is shorter and the optimal gap widens when the plates are stacked more closely.

These variations are further analyzed by plotting the distribution of the normalized local heat flux along the surface of the cold heat exchanger. As shown in Fig. 8.31, the normalized local heat flux peaks at the edges of the heat exchanger. In this figure, the local heat flux is normalized using its mean computed value so that the average local heat flux over the entire heat exchanger is 1 for all the configurations considered. This enables us to readily compare different configurations and thus gain insight into the effect of varying h/δ_k on the distribution of cooling along the heat exchanger surface. Figure 8.31 also shows that the normalized local heat flux peaks closer to the edges when $\tilde{H}$ is reduced while $\tilde{d}$ is kept constant. The same observation is made when (1) the thickness of the plates ($\tilde{d}$) is varied while the spacing ($\tilde{H}$) is kept constant (not shown), and (2) when $\tilde{H}$ and $\tilde{d}$ are varied while the blockage ratio is kept constant (not shown). The same trend is observed around the edges of the hot heat exchanger plates and of the stack plates, as shown in Fig. 8.32. These observations can be summarized by noting that, as h/δ_k decreases, the heat flux peaks more toward the edges of the cold heat exchanger, which also means that the portion of the cold heat exchanger over which most of the cooling occurs is reduced with h/δ_k. For this reason the optimal heat exchanger length L_{opt} decreases with h/δ_k, as mentioned above.

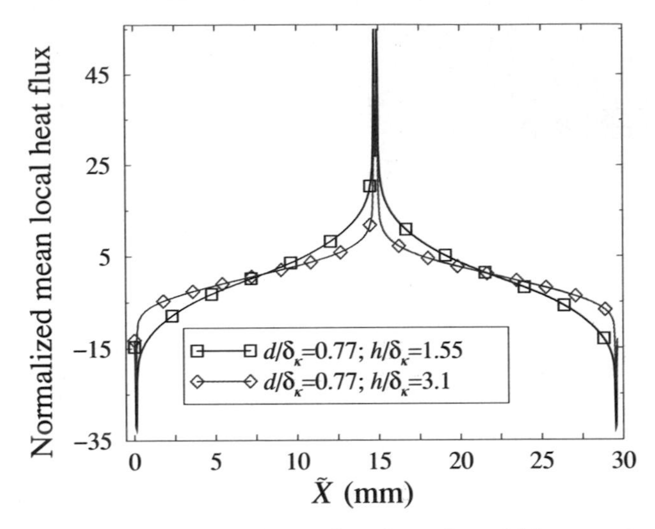

Fig. 8.31. Normalized mean local heat flux from the cold heat exchanger to the gas along the surface of the heat exchanger. $\tilde{X} = 0$ is measured from the center of the surface that faces the stack. Curves are generated for spacings $h/\delta_k = 1.55$ and 3.1. Adapted from Besnoin and Knio (2004)

For the configurations considered in the analysis, L_{opt} varies between 1.3 and 1.8 for $h/\delta_k = 1.5$ and 4.7, respectively.

Using a thin-plate model, Cao et al. (1996) computed the heat flux along stack plates. Consistent with the above analysis, they observed that the heat flux peaks more toward the edges of the stack plates as the plates are stacked closer together. On the other hand, they suggested that tightly spaced heat exchangers could be significantly shorter than $2\tilde{R}_p$, which contrasts with present predictions of $L_{\rm opt}$ (that is, $1.3 \leq \tilde{L}_{\rm opt}/(2\tilde{R}_p + \tilde{\delta}_k) \leq 1.8$). The computations of Cao et al. (1996), however, did not include heat exchangers, and the conclusions were solely based on observations of heat transfer along infinitely thin stack plates. In contrast, the predictions obtained using the present model offer a more detailed insight into the heat transfer around the edges of the plates and enables us to distinguish two regions. Along the stack plate (region A in Fig. 8.32), the heat flux is sharply peaked at the corner. However, with the present configuration, the sign of the heat transfer through the edges of the stack plate (region B in Fig. 8.32) is inverted because of the presence of heat exchangers. This phenomenon is accentuated when $\Delta\tilde{T}$ increases and/or $\tilde{g}$ increases.

8.3.4.9 Flow Structure Around the Stack and Heat Exchangers. The structure of the flow near the heat exchangers is examined in this section. Results are presented that illustrate the effect of the drive ratio (Dr) and gap width (g). Visualisations of the flow field reveal three distinct flow regimes. For all the cases considered here, with $Dr \leq 3\%$ the flow field remains

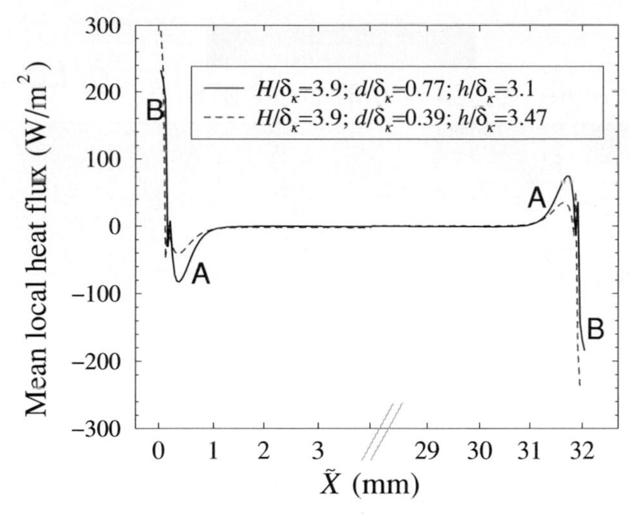

Fig. 8.32. Mean local heat flux from the stack plate to the gas along the surface of the plate. $\tilde{x} = 0$ is measured from the center of the surface that faces the hot heat exchanger. "A" refers to the region parallel to the acoustic axis, "B" refers to the edge of the plate. Curves are generated for spacings $h/\delta_k = 3.1$ and 3.47. Adapted from Besnoin and Knio (2004)

symmetric throughout the acoustic cycle. At moderate ($3.5 \leq Dr \leq 5\%$) and high drive ratios ($Dr \geq 6.5\%$), complex flow patterns are observed, as further discussed below.

Moderate Drive Ratio. Analysis of the flow structure at moderate drive ratio ($3.5 \leq Dr \leq 5\%$) reveals that the flow may or may not be symmetric (w.r.t. the horizontal symmetry line) according to the gap width. We summarize our observations by visualizing the instantaneous flow and temperature distributions around the stack and heat exchangers at eight instants during an acoustic cycle. Results are obtained for different gap widths, $g/L_o = 0.022$, 0.172, 0.261, 0.349, 0.482, and 0.659, where $L_o \equiv 2\tilde{R}_p + \tilde{\delta}_k$. For all the cases, the heat exchanger length is kept constant at $L_{c/h}/L_o = 1.33$, with $Dr = 5\%$ and $\Delta\tilde{T} = 18$ K.

For $g/L_o = 0.022$ and $g/L_o = 0.172$, the flow between the heat exchanger and stack plates remains laminar and symmetric throughout the cycle. As shown in Fig. 8.33, for $g/L_o = 0.172$ a symmetric recirculation region develops in the gap between the stack and the cold heat exchanger. As the gap is widened, the flow becomes unstable and asymmetric, as illustrated in Fig. 8.34 for a configuration with $g/L_o = 0.261$. Specifically, for $g/L_o \geq 0.261$, asymmetric bursts appear which are followed by a relaminarization phase. The loss of symmetry is accompanied by the appearance of large-amplitude wavy motions within the gap and between the plates.

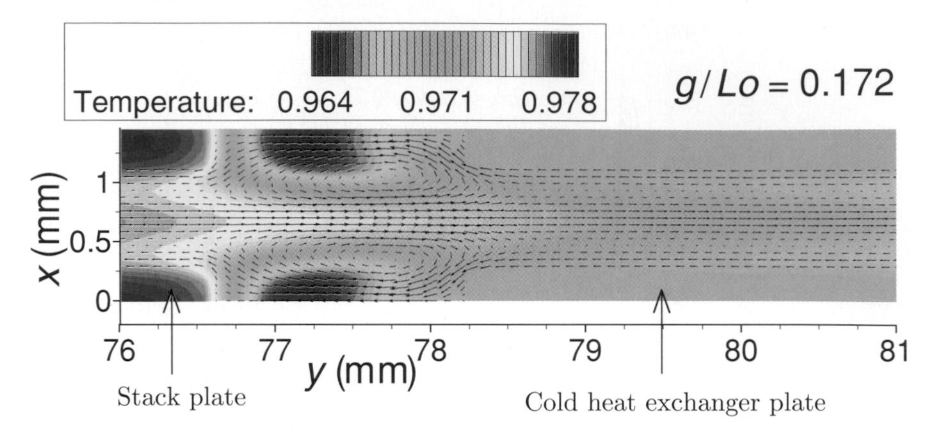

Fig. 8.33. Temperature distribution and velocity vectors around the thermoacoustic stack and the cold heat exchanger for $Dr = 5\%$, $\Delta\tilde{T} = 18$ K, $L_{c/h}/Lo = 1.33$, and for a normalized gap width $g/L_o = 0.172$, with $L_o = (2.R_p + \delta_k)$, and with $Re = 87.1$, $Re_a = 32.4$, and $Re_d = 101$. Adapted from Besnoin and Knio (2004)

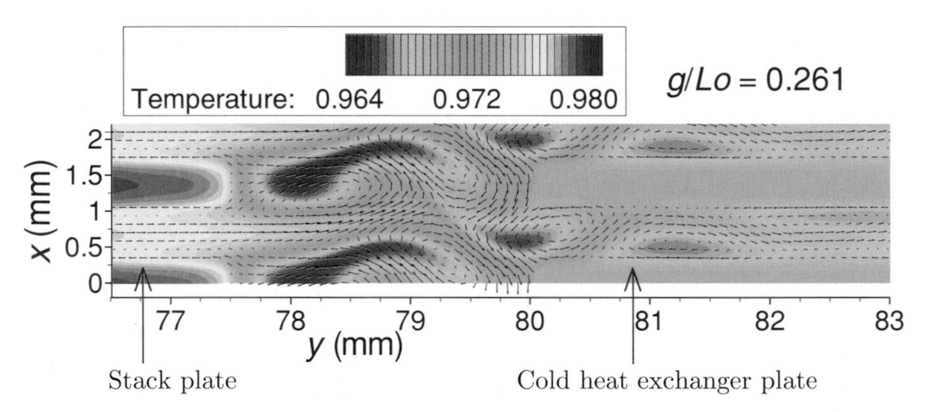

Fig. 8.34. Temperature distribution and velocity vectors around the thermoacoustic stack (left) and the cold heat exchanger (right) for $Dr = 5\%$, $\Delta\tilde{T} = 18$ K, $L_{c/h}/Lo = 1.33$, and for normalized gap widths $g/L_o = 0.261$, at $t = t_p + T/2$, and with $Re = 87.1$, $Re_a = 32.4$, and $Re_d = 101$. Adapted from Besnoin and Knio (2004)

Details on the symmetric breaking process are provided in Fig. 8.35, which shows the flow patterns predicted at eight instants during an acoustic cycle for a configuration with $g/L_o = 0.659$. The first frame (Fig 8.35a) corresponds to a phase with peak acoustic velocity, which occurs half-way through the expansion phase, just before the flow starts to decelerate. The slightly wavy shape is a residual motion from the previous acoustic cycle. During the deceleration (Fig. 8.35b), the wave grows and a sequence of vortices appears downstream, which entrain cold fluid toward the edge of the stack plate, contributing to

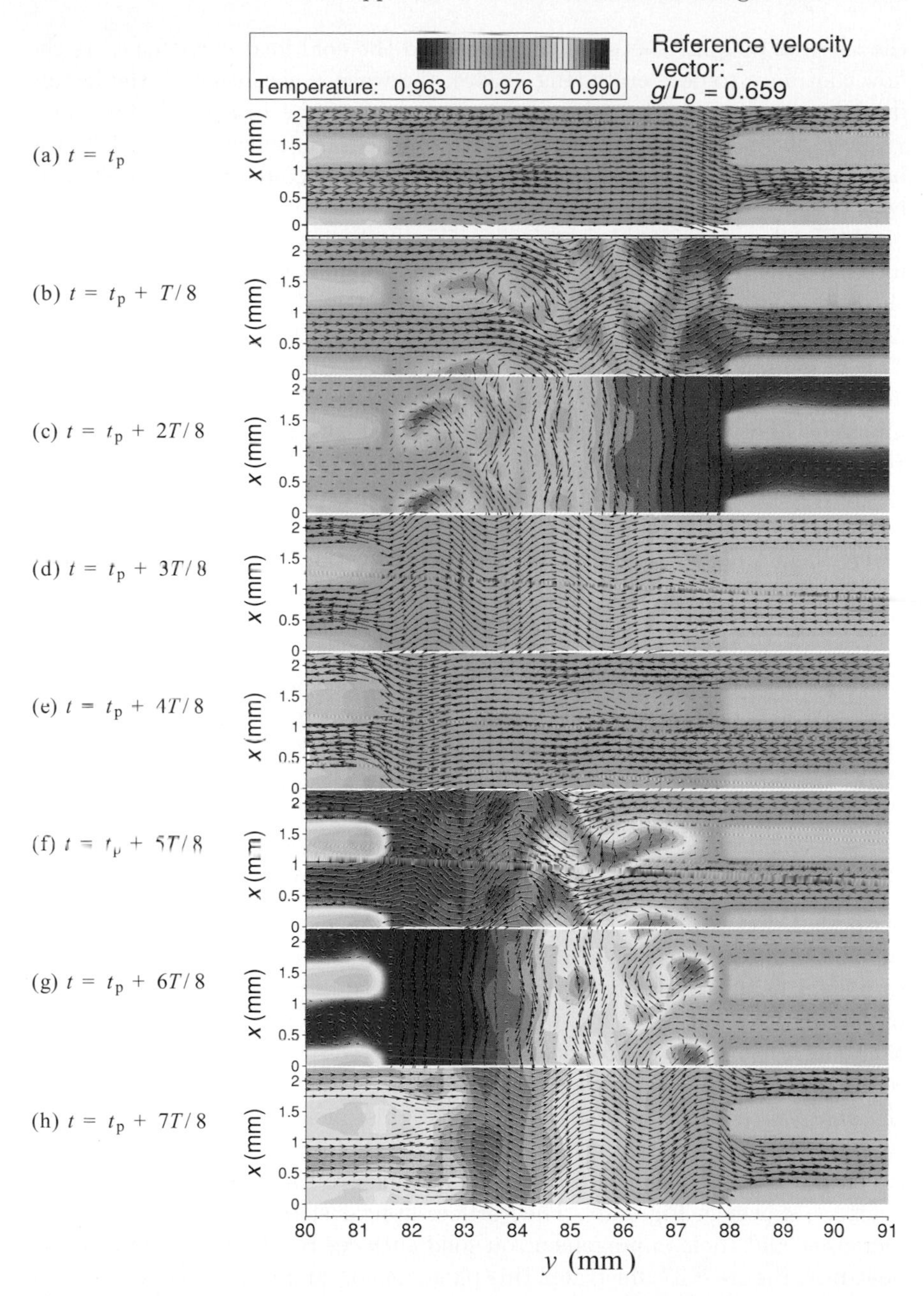

Fig. 8.35. Temperature distribution and velocity vectors around the thermoacoustic stack (*left*) and the cold heat exchanger (*right*) for $Dr = 5\%$, $\Delta\tilde{T} = 18\,\mathrm{K}$, $L_{c/h}/L_o = 1.33$, and $g/L_o = 0.659$, and with $Re = 87.1$, $Re_a = 32.4$ and $Re_d = 101$. An additional half period has been plotted to enhance presentation. Adapted from Besnoin and Knio (2004)

the thermal coupling between the stack and the cold heat exchanger. As the flow continues to decelerate (Fig. 8.35c), the wave is accentuated; the jetting fluid passes through the primary vortices, and at the instant of zero mean velocity, the gap is occupied by a wave of vortices. As shown in Fig. 8.36, the impingement of the vortices on the plate edge is accompanied by enhanced heat transfer rates.

As the flow reverses, the vortices gradually dissipate as they are entrained into the central region (Fig. 8.35d). A similar phenomenon is observed in flows through channels with sudden expansions (Sobey, 1985). The compression phase then starts at $t = t_p + T/2$ (Fig. 8.35e), when the mean velocity peaks. This is followed by a second deceleration phase, during which the particles heat up. The sequence of events described above is repeated in the opposite direction, and warm core vortices are convected toward the edges of the cold heat exchanger. In the present case, the presence of a wide gap between the heat exchangers and the stack prevents these vortices from reaching the cold heat exchanger and penetrating the thermal boundary layer. As a result, reverse heat transfer is minimized. This configuration however is not optimal. As shown in Figs. 8.20 and 8.22, the overall cooling load peaks when the gap is narrower. This is the case because the thermal coupling between the stack and the heat exchangers is diminished as the gap is widened. We note that the flow patterns around the other edge of the cold heat exchanger were also analyzed with the present configuration, which revealed that the flow is symmetric throughout the acoustic cycle. Qualitatively, the flow patterns in this region are similar to those around the edges of isolated stack plates; this situation has been briefly analyzed in Blanc-Benon et al. (2003) and in Subsect. 8.3.3 using both experimental PIV data and simulation results.

High Drive Ratio. At high drive ratios ($Dr \geq 6.5\%$), the symmetry breaking of the flow between the stack and heat exchanger plates is observed for all the combinations of geometrical parameters explored. Asymmetric flow patterns and heat transfer rates occur which are similar to those described at moderate drive ratios with a wide gap. However, at large drive ratios these asymmetric patterns are observed for all values of the gap width and of the temperature difference between cold and hot heat exchangers. In addition, the analysis reveals that when $Dr \geq 6.5\%$, this phenomenon is accompanied by the appearance of a large-amplitude wavy motion within the gap and between the plates, as shown in Fig. 8.37, as well as outside the thermoacoustic stack, as shown in Fig. 8.38. The vortices appear stronger as the drive ratio increases, and their impingement on solid surfaces results in inhomogeneous heat flux. Figure 8.37 illustrates this phenomenon and clearly shows that the peak of heat flux along the surface of the cold heat exchanger is located at or very near the point of flow reattachement.

The observations above provide an explanation for the difference between the predicted optimal gaps at low and high drive ratios (Fig. 8.25). As noted earlier, with the present configuration and a drive ratio $Dr \geq 3.5\%$, the

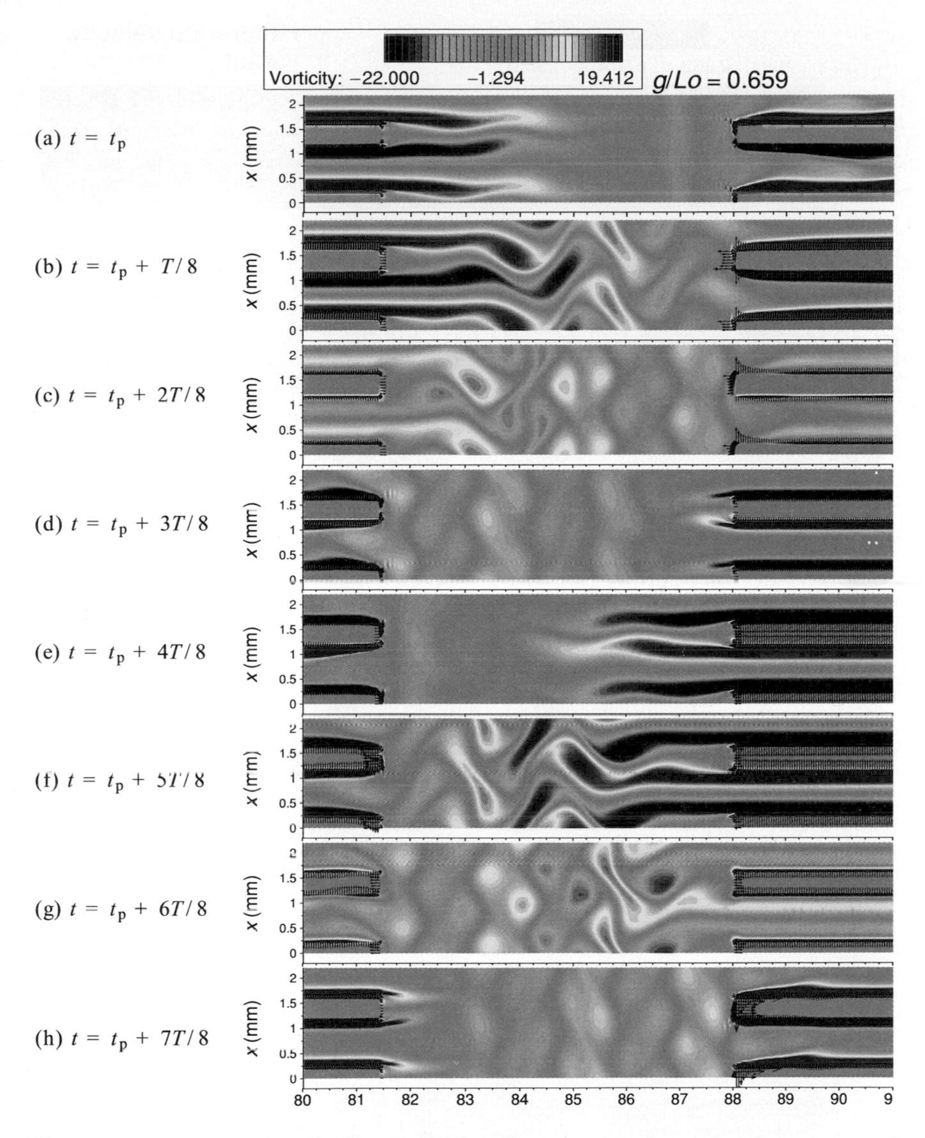

Fig. 8.36. Vorticity distribution and heat flux vectors along the thermoacoustic stack (*left*) and the cold heat exchanger (*right*) for $Dr = 5\%$, $\Delta\tilde{T} = 18$ K, $L_{c/h}/L_o = 1.33$, and $g/L_o = 0.659$, and with $Re = 87.1$, $Re_a = 32.4$, and $Re_d = 101$. Adapted from Besnoin and Knio (2004)

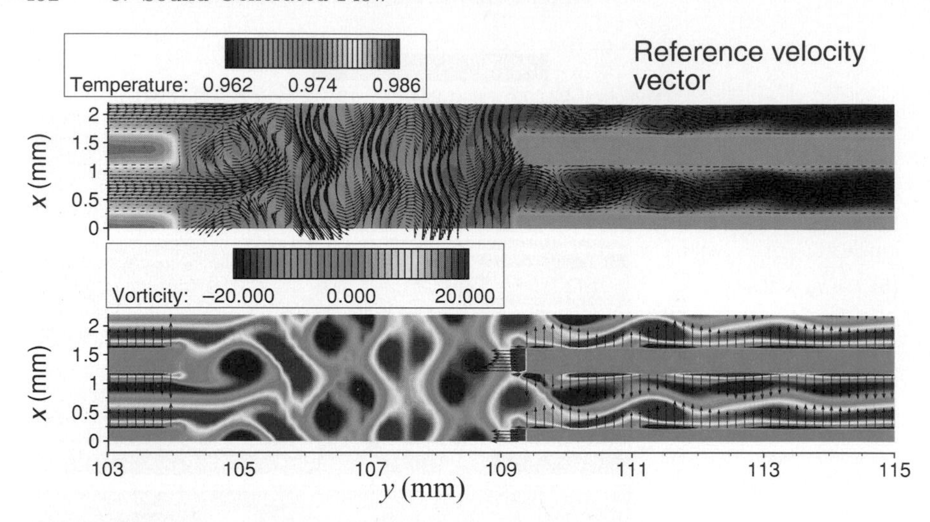

Fig. 8.37. Top: Temperature distribution and velocity vectors around the thermoacoustic stack (*left*) and the cold heat exchanger (*right*). Bottom: Vorticity distribution and heat flux vectors along the surface of the thermoacoustic stack and of the cold heat exchanger. The plots are generated for $Dr = 8\%$, $\Delta\tilde{T} = 18$ K, $L_{c/h}/Lo = 1.23$, and $g/L_o = 0.33$, with $L_o = (2.R_p + \delta_k)$, and with $Re = 87.1$, $Re_a = 51.9$, and $Re_d = 161.5$. Adapted from Besnoin and Knio (2004)

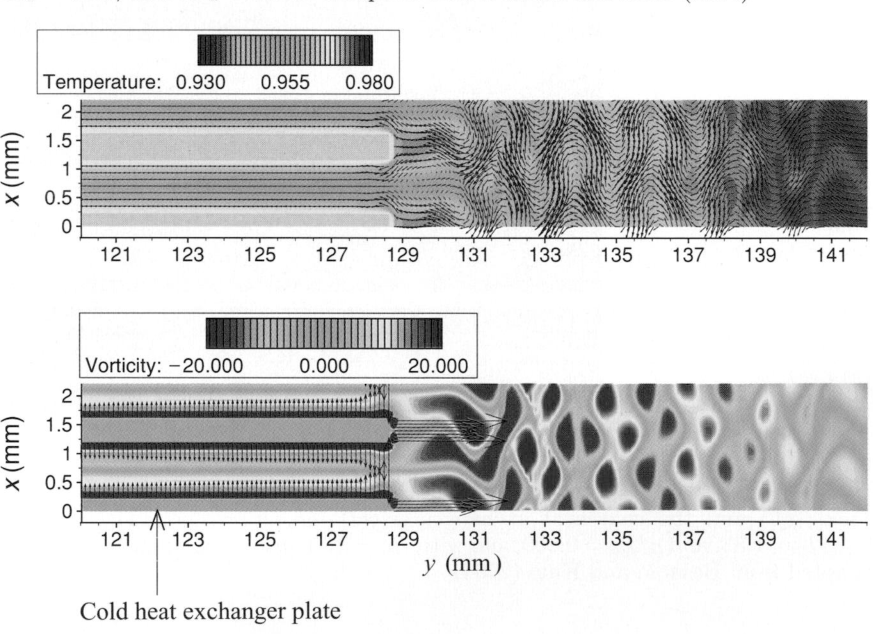

Fig. 8.38. *Top*: Temperature distribution and velocity vectors around the cold heat exchanger edge closest to the pressure node. *Bottom*: Vorticity distribution and heat flux vectors along the surface of the cold heat exchanger. The plots are generated for $Dr = 8\%$, $\Delta\tilde{T} = 18$ K, $L_{c/h}/Lo = 1.23$, and $g/L_o = 0.33$, with $L_o = (2.R_p + \delta_k)$, and with $Re = 87.1$, $Re_a = 51.9$, and $Re_d = 161.5$. Adapted from Besnoin and Knio (2004)

loss of flow symmetry occurs when the gap width g/L_o exceeds 0.26. This process is accompanied by the shedding of stronger vortices, which leads to enhanced heat transfer along the cold heat exchanger plates as shown in Figs. 8.34 and 8.37. Since velocity is particularly increased near the wall, the heat transfer is enhanced as a result of the increased shear stress (Tsui et al., 2000). In addition, because of the increased transport of low-temperature core flow along the plate by the vortices, the warmer particles are carried away from the cold heat exchanger more efficiently. As a result, the gap required to attenuate the detrimental effect of reverse heat transfer through the edges of the heat exchanger is smaller than that required at low drive ratios, where the flow patterns remain symmetric.

9. Epilogue

In the preceding chapters, theoretical descriptions of viscous vortical flows were presented that focused particularly on regimes characterized by large scale and/or time disparities. The fundamental approach generally adopted in the present volume is based on developing suitable asymptotic representations of the governing equations for particular flow regimes, and on exploiting these representations to obtain mathematical and/or computational models that properly address the scale disparities. Several problems were thus treated, including slender vortex flows, vortex merging, aerodynamic sound generation, as well as weakly compressible flows. In conclusion, we outline below two potentially interesting extensions of the developments presented here, which would concern simulations of slender vortex merging and reconnection, and multiscale modeling of meteorological flows.

9.1 Numerical Strategies for Extended Slender Vortex Simulations

In Chap. 3, matched asymptotic expansions were used to derive equations of motion for slender vortex centerline(s), coupled with evolution equations for their core structures. This asymptotic theory of slender vortices was based on the assumptions of a high Reynolds number flow, of small core size compared with characteristic radius of curvature, and suitable large separation between vortex cores, and between vortex cores and solid boundaries. Under these assumptions, the matched asymptotic analysis results in a highly efficient description that defeats the stiffness inherent in the Navier–Stokes equations when applied to slender vortices.

A similar situation was later tackled in Chap. 5 through the development of thin-tube vortex element models. In the latter context, the matched asymptotic analysis was once again applied to provide a well-founded interpretation of the numerical core structure that is inherited from the particle representation as it is "collapsed" into the filament centerline. Thus, it has been possible to develop corrected thin-tube models whose predictions are, by construction, in agreement with those of the matched asymptotic analysis. In addition, through the implementation of enhanced discretization and velocity evaluation schemes, corrected thin-tube models achieve similar efficiency as direct implementations of the matched asymptotic analysis, and in some cases provide the additional advantage of being more effective in removing the temporal stiffness of the filament equations of motion.

While both the matched asymptotic analysis and corrected thin-tube models achieve similar capabilities, the starting points in both approaches are radically different. In the matched asymptotics, a systematic analysis of the Navier–Stokes and Poisson integral equations is applied to derive equations of motion for the filament centerline and of its core structure. Corrected thin-tube models, on the other hand, have their origin in vortex element methods, that are based on detailed representation of the vorticity field using overlapping elements. As mentioned earlier, thin-tube models collapse this representation on the filament centerline using only a single chain of elements, and the matched asymptotics are used to provide the correct interpretation of the resulting *numerical* slender filament structure.

In various situations, the assumptions of a slender vortex flow, though initially satisfied, may not continue to hold for all times. Hairpin formation, self-intersection of a vortex filament, intersection of a slender filament with another filament, and interactions of a slender vortex with a solid boundary are relevant examples that illustrate the breakdown of the assumptions of slender vortex theory. In many of these examples, however, the breakdown of the slender vortex assumptions occurs only locally; that is, the corresponding conditions hold everywhere except in localized regions. In these situations, it would be highly attractive to devise computational strategies that would preserve the advantages of slender vortex representations, at least in regions where the underlying assumptions hold.

A general approach to extend slender vortex computations to complex situations where, for example, merging or intersections occur, is inspired by the fact that thin-tube models derive from smooth vortex element methods. This naturally suggests hybrid approaches in which detailed vortex element representations are used to represent the flow field in regions where the assumptions of the slender vortex representation do not hold, and corrected thin-tube vortex element models are used in the remainder of the domain. The success of such hybrid approaches naturally hinges on the ability to seamlessly switch between one mode of representation to the other, and on the development of suitable criteria for switching between representations. Also note that, by construction, such hybrid approaches share many of the features of so-called fast vortex methods (e.g., Greengard and Rokhlin, 1987; Cheng et al., 1999; Almgren et al., 1994; Lindsay and Krasny, 2001), which essentially rely on coarsening strategies for computing the local velocity field induced by well-separated vorticity regions. In this sense, for regimes involving sizeable flow domains dominated by slender filament dynamics, the envisioned hybrid representations offer the promise of an even higher efficiency than could be provided by fast vortex methods.

Another possible approach, inspired by the vortex merging studies outlined in Chap. 6, is to develop simplified models for *local* vortex merging and vorticity cancellation in three dimensions. These simplified models would then provide efficient representations for complex phenomena that are ubiquitous in turbulent flows, such as the formation and decay of hairpin vortices. Much like hairpin-removal concepts (Chorin, 1993, 1994), such strategies would offer the promise of efficient simulations of turbulent flows. Their success greatly depends on the development of refined criteria for local vortex merging, reconnection, and the resulting decay.

In future studies, we plan to develop both strategies and explore their implementation to model complex vortical flows, such as the late stages of the Crow instability.

9.2 A Multiple Scales Asymptotic Framework for Meteorological Modeling

One important aim of theoretical meteorology is the development of simplified model equations that describe the large variety of scale-dependent phenomena observed in atmospheric flows. Here we summarize a unified approach to the derivation of such models based on multiple scales asymptotic techniques that is very much in line with the spirit of the present volume, in particular, Subsect. 3.3.9. The approach was proposed in Klein (2004) and has led to or been an important part of several recent new developments (see Majda and Klein, 2003; Klein, 2004; Majda and Biello, 2004; Mikusky et al., 2005; Biello and Majda, 2006; Klein and Majda, 2006 and references therein). Remarkably, Keller and Ting (1951) already anticipated the foundations of this approach in an internal report of the Institute for Mathematics and Mechanics of New York University.

To elucidate our main points, we restrict the discussion here to inviscid compressible flows on a rotating sphere. Diabatic effects, such as radiation, water phase transitions, or turbulent transport will be represented as lumped terms in the governing equations to be specified later. Extensions of the framework to include moist processes have been developed recently by Klein and Majda (2006).

9.2.1 Universal Parameters and Distinguished Limits

Table 9.1 displays several physical variables that are characteristic of atmospheric flows, and that are valid independently of the typical length and time scales of any specific flow phenomenon: The mean sea-level pressure p_{ref} is set by the requirement that it balance the weight of a vertical column of air. Thus, it is directly given by the total mass of the atmosphere which, to a very good approximation, is evenly distributed over the sphere. A reference temperature T_{ref} is set roughly by the global radiation balance which,

Table 9.1. Universal characteristics of atmospheric motions

Earth's radius	$a = 6 \cdot 10^6$ m
Earth's rotation rate	$\Omega \sim 10^{-4}$ s^{-1}
Acceleration of gravity	$g = 9.81$ m/s^2
Sea level pressure	$p_{\text{ref}} = 10^5$ kg/ms^2
Water freezing temperature	$T_{\text{ref}} \sim 273$ K
Equator–pole temperature difference	$\Delta T\|^{\text{p}}_{\text{eq}} \sim 50$ K
Dry air gas constant	$R = 287$ m/s^2/K

even without greenhouse gases, would render the mean near-surface air temperature near 250 K. The actual value shown in Table 9.1 is the freezing temperature of water under standard conditions; that is, $T_{\text{ref}} \sim 273\,\text{K}$, which is about midway between the observed maximal and minimal near-surface air temperatures. The equator-to-pole air temperature difference near the surface, $\Delta T|^{\text{p}}_{\text{eq}}$, is a consequence of the latitudinal variation of the sun's irradiation. The dry air gas constant, R, as a thermodynamic property is also universally characteristic for atmospheric flows, because its variations due to admixtures of water vapor, trace gases, and the like are no larger than a few percent in general. Based on these seven basic reference quantities, three independent dimensionless combinations can be composed. To define combinations with intuitive interpretations, we introduce as auxiliary quantities the pressure scale height h_{sc} and the characteristic speed c_{ref} of barotropic[1] gravity waves via

$$
\begin{aligned}
h_{\text{sc}} &= p_{\text{ref}}/(g\,\rho_{\text{ref}}) \sim 10 \text{ km} \\
c_{\text{ref}} &= \sqrt{g h_{\text{sc}}} \sim 300 \text{ m/s}
\end{aligned}
$$

where

$$
\rho_{\text{ref}} = p_{\text{ref}}/(R\,T_{\text{ref}}) \sim 1 \text{ kg/m}^3 .
$$

Then we let

$$
\begin{aligned}
\pi_1 &= \frac{h_{\text{sc}}}{a} \sim 1.67 \cdot 10^{-3} , \\
\pi_2 &= \frac{\Delta T|^{\text{p}}_{\text{eq}}}{T_{\text{ref}}} \sim 0.18 , \\
\pi_3 &= \frac{c_{\text{ref}}}{\Omega a} \sim 0.5 .
\end{aligned}
\tag{9.2.1}
$$

[1] Atmospheric flow modes are called "barotropic" if their structure is homogeneous in the vertical direction.

The interpretations of π_1 and π_2 should be obvious, while the parameter π_3 compares a typical barotropic gravity wave speed with the tangential speed of points on the equator as induced by Earth's rotation.

The parameter π_2 is not extremely small. Yet, many successful developments in theoretical meteorology have relied on scale analysis (asymptotics) in terms of, for example, Rossby numbers or internal wave Froude numbers with values in a similar range. As a consequence, one may consider π_1 and π_2 as small parameters, with $\pi_3 = O(1)$ as these two parameters vanish.

As we will argue in Subsect. 9.2.3 below, there is little hope for success with asymptotic expansions that would allow π_1 and π_2 to vary independently in such a limit process. Thus, for these two parameters we adopt a distinguished limit according to

$$\pi_1 \sim \varepsilon^3 \qquad \text{and} \qquad \pi_2 \sim \varepsilon \qquad \text{as} \qquad \varepsilon \to 0\,. \tag{9.2.2}$$

These limits are compatible with the estimates in (9.2.1) for actual values of ε in the range $(1/7)\dots(1/8)$. This parameter will be adopted below as the reference parameter for asymptotic analyses and any additional small or large nondimensional parameter that may be associated with singular perturbations in the governing equations is subsequently tied to ε through suitable distinguished limits.

Before we proceed to do so, we notice that Keller and Ting (1951) already proposed to use the shallowness parameter, $\varepsilon \sim (h_{sc}/)^{1/3}$, as a basic expansion parameter for meteorological modeling. We also remark that with the definitions in (9.2.1) and the fact that $\pi_3 = c_{ref}/(\Omega a) \sim O(1)$, the shallowness parameter may also be interpreted as the ratio of the centripetal acceleration at the equator due to Earth's rotation and the acceleration of gravity, g, that is, $h_{sc}/a \sim a\Omega^2/g \sim \varepsilon^3$.

9.2.2 Nondimensionalization and General Multiscale Ansatz

With p_{ref} and T_{ref}, and through the ideal gas equation of state, $\rho = p/RT$, Table 9.1 suggests reference values for the nondimensionalization of pressure, temperature, and density.

Most theories for atmospheric flows rely on the assumption that typical flow speeds are small compared with the speed of barotropic gravity waves c_{ref} which, except for a factor of $\sqrt{\gamma}$, matches the speed of sound. Here and in the rest of this section we make this assumption explicit by introducing a reference speed $u_{ref} \sim 10\,\text{m/s}$ for the nondimensionalization of the flow velocity. This gives rise to an additional small parameter, the barotropic Froude number $\overline{\text{Fr}}$, which we tie to ε via

$$\overline{\text{Fr}} = \frac{\text{Ma}}{\sqrt{\gamma}} = \frac{u_{ref}}{c_{ref}} \sim \varepsilon^2 \qquad \text{as} \qquad \varepsilon \to 0\,. \tag{9.2.3}$$

The space and time coordinates are then nondimensionalized using the pressure scale height h_{sc} and the associated advection time scale $t_{ref} = h_{sc}/u_{ref} \sim$ 20 min.

With these scalings, the nondimensional governing equations in the rotating earth system may be written as

$$\begin{aligned} \boldsymbol{v}_t + \boldsymbol{v} \cdot \nabla \boldsymbol{v} + \varepsilon\, \boldsymbol{\Omega} \times \boldsymbol{v} + \frac{1}{\varepsilon^4} \frac{1}{\rho} \nabla p &= \boldsymbol{S}_{\boldsymbol{v}} - \frac{1}{\epsilon^4}\, \boldsymbol{k}\,, \\ p_t + \boldsymbol{v} \cdot \nabla p + \gamma p \, \nabla \cdot \boldsymbol{v} &= S_p\,, \\ \theta_t + \boldsymbol{v} \cdot \nabla \theta &= S_\theta\,, \end{aligned} \tag{9.2.4}$$

where

$$\theta = \frac{p^{1/\gamma}}{\rho} \tag{9.2.5}$$

is the so-called potential temperature, a variable closely related to the entropy of dry air, and $\boldsymbol{k}$ is the local vertical unit vector indicating the direction of the acceleration of gravity. The terms $\boldsymbol{S}_{\boldsymbol{v}}$, S_p, and S_θ represent additional effects which in a concrete application may stem from turbulence closures or similar models for the net influence of nonresolved scales.

Klein (2004) suggested to consider the small parameter ϵ as introduced above as *the* general singular asymptotic expansion parameter for theoretical developments in meteorology. To this end, the solution vector $\boldsymbol{U} = (p, \theta, \boldsymbol{v})$ is expanded in powers of ϵ (or some fractional power thereof), and all expansion functions would depend on a series of space-time coordinates that are scaled again by powers of ϵ. The simplest version of such a scheme reads

$$\boldsymbol{U}(\boldsymbol{x}, z, t, ; \epsilon) = \sum_i \epsilon^i \boldsymbol{U}^{(i)} \left(\ldots, \frac{\boldsymbol{x}}{\epsilon}, \boldsymbol{x}, \epsilon \boldsymbol{x}, \ldots \frac{z}{\epsilon}, z, \ldots \frac{t}{\epsilon}, t, \epsilon t, \ldots \right), \tag{9.2.6}$$

the underlying assumption being that the asymptotic coordinate scalings in terms of ϵ are sufficient to capture the wide range of spatiotemporal scales that are observed in the atmosphere and that have been modeled through scaling analyses before.

In a number of publications, this hypothesis has been validated. A range of known simplified model equations of theoretical meteorology was rederived in Majda and Klein (2003), Klein (2004, 2005), and Klein et al., (2004) directly from the full compressible flow equations in (9.2.4) by specializing the general ansatz in (9.2.6) to include only one horizontal, one vertical, and one scaled time coordinate.

Of course, such validation studies are but a first step, as (9.2.6) strongly suggests itself as the basis for systematic studies of multiple scales interactions. See Majda and Klein (2003), Majda and Biello (2004), Klein (2004), Biello and Majda (2006), and Klein and Majda (2006) for related developments.

9.2.3 Distinguished Limits Versus Multiparameter Expansions

Here we provide an argument for the introduction of distinguished limits in the previous subsection as opposed to the development of multiparameter asymptotic theories that would treat π_1, π_2 and further singular perturbation parameters as independent in taking asymptotic limits.

In the presence of, say, two singular small parameters ϵ and δ, a direct multiple parameter expansion would read

$$U = U^{(0)} + \epsilon U^{(1,0)} + \delta U^{(0,1)} + \epsilon^2 U^{(2,0)} + \epsilon\delta U^{(1,1)} + \delta^2 U^{(0,2)} + \cdots . \quad (9.2.7)$$

It turns out that even for the extremely simple case of a linear oscillator with small mass and small damping, such asymptotic limit solutions do not exist. To see this, consider two different limiting paths in the mass-damping parameter space. Along one path we first let the mass go to zero and then consider the limit of small damping. In this case, we obtain nonoscillatory limit behavior for which the solution approaches its balanced state, and does so more rapidly with smaller damping. In contrast, if we let the damping vanish first and then analyze the limit of small mass, we find undamped oscillations with an oscillation period that vanishes as the mass goes to zero.

Thus, even for the simple case of a linear oscillator, the leading-order solution in terms of vanishing mass and damping is path-dependent. In other words, an independent multiple parameter expansion as in (9.2.7) is meaningless, because a unique leading-order solution $U^{(0)}$ does not exist.

The strive for asymptotic solutions is not lost, however, as it is well possible to construct path-dependent limit solutions. These are obtained by first adopting an appropriate distinguished limit between the mass and damping parameters, and then considering the limiting system behavior along the resulting path into the origin of the parameter space. The situation is somewhat similar to that of functions of several independent variables for which a full-fledged gradient, that is, a Frechét derivative, exists only under far more restrictive conditions than directional or Gâteaux derivatives. Distinguished limits connecting various asymptotic limit parameters represent a generalization of directional derivatives that allows for nonlinear dependencies amongst the participating variables instead of merely linear ones.

These arguments corroborate the approach taken in the previous section.

9.2.4 Perspectives

In summary, the proposed unified approach is able to reproduce a wide range of well-established simplified models of theoretical meteorology. It reveals that these simplified models correspond with scale-dependent singular asymptotic limit regimes for the compressible flow equations. In the context of this volume, this immediately suggests two emerging lines of further research: On one hand, systematic multiple scales analyses, as have already been initiated in the literature cited, promise new insight into the ubiquitous phenomena of scale interactions in the atmosphere. On the other hand, ample evidence has been provided in this monograph that careful asymptotic analyses of singular asymptotic limits can provide strong guidelines toward the construction of accurate, efficient, and robust numerical solution techniques. We plan to pursue both of these lines of research in the coming years.

Appendices

A. Governing Equations for Higher-Order Solutions

This appendix was first referred to in Subsect. 3.2.1, in which the vorticity and stream function in the inner region of a two-dimensional vortex were expanded in power series of ϵ. The independent variables in the two-time inner solution are the short time τ, the normal time t, and the polar coordinates $\bar{r}$ and θ. The symbols $\mathcal{M}$ and $\{ \}$ denote, respectively, the operators for short time and circumferential averaging. Following the terminology introduced in Subsect. 3.2.1, we call the θ-average of a function its "symmetric part," while the remainder its "asymmetric part." These two parts carry the subscripts c and a, respectively.

It was pointed out in Subsect. 3.2.1 that among all the governing equations for the nth-order two-time solution only the vorticity evolution equation is nonlinear for $n \geq 2$, because of the inhomogeneous term, $\tilde{F}_n$. The term is the sum of products of jth- and $(n-j)$th-order terms for $j = 1, \dots, (n-1)$. In this Appendix, we explain how to decompose the nth-order vorticity equation into separate equations for the asymmetric and symmetric parts of the nth-order solution and how to decompose the τ-average of the nth-order equation into one equation for the τ-average of the asymmetric part of the nth-order solution and one equation for that of the symmetric part of the $(n-2)$nd-order solution.

The second decomposition implies that the dependence of the symmetric parts of the $(n-1)$th- and nth-order solutions on the normal time, t, remains undetermined at this stage. Since the nonlinear inhomogeneous term $\tilde{F}_n$ in general involves $(n-1)$th-order solution, we need to show that our scheme is consistent, that is, our scheme provides a closed system of equations to determine the solution to any order. In Subsect. 3.2.1 we did show how to determine the leading- and first order solutions except for the dependence of the symmetric first order solution on the normal time, t. Hence, by the method of induction, it is sufficient to show that

(I) *If the jth-order solutions for $j \leq n-1$ are defined except for the τ-average of the symmetric part of the $(n-1)$st-order solution, then one can determine the same with $n-1$ replaced by n.*

We begin by extending the first- and second-order equations (3.2.36) and (3.2.38) to the following equation for the nth-order solution with $n \geq 2$:

$$\tilde{\zeta}_{\tau}^{(n)} + \mathcal{L}[\tilde{\psi}^{(n)}, \tilde{\zeta}^{(n)}] = \tilde{F}_n(\tau, t, \bar{r}, \theta) + \mathcal{G}\tilde{\zeta}^{(n-2)} , \tag{A.1}$$

where

$$\tilde{F}_n = -\sum_{j=1}^{n-1} \frac{1}{\bar{r}} [\tilde{\psi}_{\theta}^{(j)} \tilde{\zeta}_{\bar{r}}^{(n-j)} - \tilde{\zeta}_{\theta}^{(j)} \tilde{\psi}_{\bar{r}}^{(n-j)}] \tag{A.2}$$

and

$$\mathcal{G} = -\frac{\partial}{\partial t} + \bar{\nu}\bar{\Delta} . \tag{A.3}$$

Here the linear operator $\mathcal{L}$ is defined in (3.2.37). Let us consider a product, say $\tilde{\psi}_{\theta}^{(j)} \tilde{\zeta}_{\bar{r}}^{(n-j)}$, in $\tilde{F}_n$. By using the definition of the symmetric and asymmetric parts, we evaluate the θ-average of the product,

$$\begin{aligned} \{\tilde{\psi}_{\theta}^{(j)} \tilde{\zeta}_{\bar{r}}^{(n-j)}\} &= \{[\tilde{\psi}_a^{(j)}]_{\theta}\}[\tilde{\zeta}_c^{(n-j)}]_{\bar{r}} + \{[\tilde{\psi}_a^{(j)}]_{\theta}[\tilde{\zeta}_a^{(n-j)}]_{\bar{r}}\} \\ &= \{[\tilde{\psi}_a^{(j)}]_{\theta}[\tilde{\zeta}_a^{(n-j)}]_{\bar{r}}\} . \end{aligned} \tag{A.4}$$

This equation says that

(II) *The θ-average of $\tilde{F}_n$ contains only the asymmetric parts of jth-order solution for $j = 1, \dots (n-1)$.*

Now the θ-average of (A.1) becomes an equation for the symmetric part

$$[\tilde{\zeta}_c^{(n)}]_{\tau} = \{\tilde{F}_n\} + \mathcal{G}\tilde{\zeta}_c^{(n-2)} . \tag{A.5}$$

The difference between (A.1) and (A.5) then yields an equation for the asymmetric part,

$$[\tilde{\zeta}_a^{(n)}]_{\tau} + \mathcal{L}[\tilde{\psi}_a^{(n)}, \tilde{\zeta}_a^{(n)}] = \tilde{F}_n - \{\tilde{F}_n\} + \mathcal{G}\tilde{\zeta}_a^{(n-2)} . \tag{A.6}$$

By taking the τ-average of the above equation, we obtain an equation for the τ-average of the asymmetric part, namely

$$\mathcal{L}[\mathcal{M}\tilde{\psi}_a^{(n)}, \mathcal{M}\tilde{\zeta}_a^{(n)}] = \mathcal{M}[\tilde{F}_n - \{\tilde{F}_n\}] + \mathcal{G}\mathcal{M}\tilde{\zeta}_a^{(n-2)} . \tag{A.7}$$

From statement II and the assumption in statement I, we see that inhomogeneous terms in (A.5) are defined because they depend only on the asymmetric parts of the jth-order solutions for $j = 1, \dots, (n-1)$ and the symmetric part of $(n-2)$nd-order solution. The symmetric part to nth-order may thus be written as

$$\tilde{\zeta}_c^{(n)}(\tau, t, \bar{r}) = \tilde{\zeta}_p^{(n)}(\tau, t, \bar{r}) + \bar{\zeta}_h^{(n)}(t, \bar{r}) , \tag{A.8}$$

where the subscripts p and h stand for the particular integral of (A.5), which vanishes at $\tau = 0$ and for the homogeneous solution. The former is defined by quadrature, while the latter is any function independent of τ. Since the τ-average of (A.8) yields

$$\mathcal{M}\tilde{\zeta}_c^{(n)} = \mathcal{M}\tilde{\zeta}_p^{(n)} + \bar{\zeta}_h^{(n)} , \tag{A.9}$$

we conclude that the dependence of the τ-average of the symmetric nth-order solution on t, is yet undefined. In fact, the assumption in statement (I) implies that even the τ-average of the symmetric $(n-1)$st-order solution is still unknown. To find an equation for the latter, $\mathcal{M}\tilde{\zeta}_c^{(n-1)}$, we take the τ-average of (A.5) with n replaced by $n+1$. The result is

$$\left\{\frac{\partial}{\partial t} - \bar{\nu}\bar{\Delta}_0\right\} \mathcal{M}\tilde{\zeta}_c^{(n-1)} = \mathcal{M}\{\tilde{F}_{n+1}\} \,. \tag{A.10}$$

Using statement (II) and the assumption in statement (I), we draw the following conclusions:

1. The unknowns on the right-hand side of (A.10) are the asymmetric parts of the nth-order solution.
2. The right-hand sides of of (A.6) and (A.7) involve the unknown, the τ-average of the symmetric part of the $(n-1)$st-order solution, via $\mathcal{M}\{\tilde{F}_n\}$.
3. The right-hand side of (A.5) is defined because of the assumption in statement (I).

Equations (A.5), (A.6), (A.7), and (A.10) thereby form a closed system that determines the τ-average of the symmetric part of the $(n-1)$st-order solution and the nth-order solution except for the τ-average of its symmetric part. Thus statement (I) is confirmed.

Note that for $n=1$ the system is uncoupled because (A.5), (A.6), (A.7), and (A.10) are all homogeneous. This enabled us to solve first for the leading-order core structure and then determine the asymmetric part of the first order solution in Subsect. 3.2.1.

Also note that the operator,

$$\mathcal{L} = \frac{1}{\bar{r}}\left[\bar{\psi}_\theta^{(0)}\frac{\partial}{\partial \bar{r}} - \bar{\psi}_{\bar{r}}^{(0)}\frac{\partial}{\partial \theta}\right] , \tag{A.11}$$

involves only spatial derivatives so that only (A.10) contains a t-derivative of the short-time average of the symmetric part to $(n-1)$th-order. As an extension of a result obtained for $n=1$ in Subsect. 3.2.1, we infer that the normal time, t, plays the role of a parameter in the asymmetric part of the nth-order solution for all n.

B. Second-Order Two-Time Solutions

In this Appendix we derive the results stated at the end of Subsect. 3.2.4 regarding the τ-average of the second-order two-time solution.

We first summarize the results on the first order asymmetric solutions derived in Subsects. 3.2.3 and 3.2.4. Equations (3.2.77) and (3.2.78) imply that the τ-average of the two-time asymmetric solution is a trivial solution,

$$\bar{\psi}_a^{(1)}(t,\bar{r},\theta,s) = \mathcal{M}\tilde{\psi}_a^{(1)}(\tau,t,\bar{r},\theta,s) = 0 \; . \tag{B.1}$$

If the initial data happen to be such that not only the leading but also the first order core structures are symmetric and that the initial velocity of the vortex center agrees with the local background velocity, (3.2.79), then (B.1) defines a trivial first order asymmetric solution independent of the short-time variable. However, in reality, the first order initial core structure is not available and the leading-order velocity of the vortex center can clearly differ from the local background velocity. Thus, in general it is necessary to construct asymmetric two-time solutions. They obey (3.2.87), (3.2.88), (3.2.93), and (3.2.94), which imply that the solutions do not have second and higher harmonics in θ and that their dependence on the two independent variables, τ and θ, appears only in the linear combination, $\theta - \bar{\omega}\tau$. Therefore, the partial derivatives in τ and θ are related by

$$\frac{\partial}{\partial\theta}\tilde{\psi}_a^{(1)}(\tau,t,\bar{r},\theta) = -\frac{1}{\bar{\omega}}\frac{\partial}{\partial\tau}\tilde{\psi}_a^{(1)}(\tau,t,\bar{r},\theta) \; . \tag{B.2}$$

This relationship enables us to express the inhomogeneous term $\tilde{F}_2$ in the second-order equation (3.2.38) as a derivative of τ. We then replace (3.2.44) by

$$\tilde{F}_2 = \frac{1}{\bar{\omega}}\frac{\partial}{\partial\tau}\left[\frac{1}{2}[(\bar{\omega}\bar{a} - \bar{b})(\tilde{\zeta}^{(1)})^2]_{\bar{r}} + \tilde{\zeta}^{(1)}\tilde{\psi}_{\bar{r}}^{(1)}\right] \; , \tag{B.3}$$

which in turn yields

$$\mathcal{M}\tilde{F}_2 = 0 \; . \tag{B.4}$$

From (B.3), (B.4), and the compatibility condition (3.2.46), the τ-average of the second-order equation (3.2.38) becomes a homogeneous equation. This equation and the τ-average of (3.2.29) for $n = 2$ yield the following equation for the one-time asymmetric stream function $\bar{\psi}_a^{(2)}$,

$$[\bar{\psi}_{\bar{r}}^{(0)}\bar{\Delta} + \bar{\zeta}_{\bar{r}}^{(0)}](\bar{\psi}_a^{(2)})_\theta = 0 \ , \tag{B.5}$$

where

$$\bar{\psi}^{(2)}(t,\bar{r},\theta) = \mathcal{M}\tilde{\psi}^{(2)}(\tau,t,\bar{r},\theta) \ . \tag{B.6}$$

Equation (B.5) is identical to (3.2.68) for the corresponding first order solution. The two homogeneous boundary conditions (3.2.30) at $\bar{r} = 0$ hold for both $n = 1$ and 2. Only the the matching conditions are different. The condition for the second-order solution as $\bar{r} \to \infty$ is

$$\begin{aligned}
&\frac{1}{\bar{r}}\bar{\psi}_\theta^{(2)}\hat{r} - \bar{\psi}_{\bar{r}}^{(2)}\hat{\theta} \to \\
&\qquad - \ \{\dot{X}^{(1)}\hat{\imath} + \dot{Y}^{(1)}\hat{\jmath}\} \\
&\qquad + \ \{[X^{(1)}\Psi_{yx}(X^{(0)},Y^{(0)}) + Y^{(1)}\Psi_{yy}(X^{(0)},Y^{(0)})]\,\hat{\imath} \\
&\qquad - \ [X^{(1)}\Psi_{xx}(X^{(0)},Y^{(0)}) + Y^{(1)}\Psi xy(X^{(0)},Y^{(0)})]\,\hat{\jmath}\} \\
&\qquad + \ \{[\bar{r}\cos\theta\,\Psi_{yx}(X^{(0)},Y^{(0)}) + \bar{r}\sin\theta\,\Psi_{yy}(X^{(0)},Y^{(0)})]\,\hat{\imath} \\
&\qquad - \ [\bar{r}\cos\theta\,\Psi_{xx}(X^{(0)},Y^{(0)}) + \bar{r}\sin\theta\,\Psi_{xy}(X^{(0)},Y^{(0)})]\,\hat{\jmath}\} \ .
\end{aligned} \tag{B.7}$$

On the right-hand side of (B.7), the vector inside the first pair of curly brackets is the unknown first order velocity of the vortex center. The second vector accounts for the change of the background velocity due to the first order displacement of the vortex center and the third vector represents the contribution of local background velocity gradients.

Again we note that the solution of the system, (B.5), (3.2.30), and (B.7), is quasi-steady; that is, that the normal time, t, can be treated as a parameter. If we assume that the inner solution depends only on the normal time and carry the one-time analysis to the second order,[1] we would find that the second-order solution is governed by the same system of equations, the differential equation (B.5), two boundary conditions (3.2.30) at $\bar{r} = 0$, and the matching condition (B.7). Thus we extend the equivalence relationship stated below (3.2.54) from the first- to the second-order solutions:

- *The short-time average of the second-order two-time solution is equivalent to the solution of one-time analysis, provided the first order asymmetric two-time solution fulfills condition (B.2) or (3.2.93).*

We then construct the one-time solution of the system (B.5), (3.2.30), and (B.7) by repeating the steps from (3.2.68) to (3.2.80) in Subsect. 3.2.3.

The Fourier coefficients of the asymmetric part of the second-order stream function, $\bar{\psi}_{jk}^{(2)}$, $j = 1, 2, \ldots$, $k = 1, 2$, have to fulfill (3.2.70) and the boundary conditions (3.2.71) at $\bar{r} = 0$. With the superscript suppressed, these equations read

$$\left\{\frac{\partial^2}{\partial\bar{r}^2} + \frac{1}{\bar{r}}\frac{\partial}{\partial\bar{r}} + \left[\frac{\bar{\zeta}_{\bar{r}}^{(0)}}{\bar{\psi}_{\bar{r}}^{(0)}} - \frac{j^2}{\bar{r}^2}\right]\right\}\bar{\psi}_{jk} = 0 \ , \tag{B.8}$$

[1] The one-time analysis was performed by Ting and Tung (1965).

and

$$\bar{\psi}_{jk} = 0 \, , \qquad [\bar{\psi}_{jk}]_{\bar{r}} = 0 \qquad \text{at} \quad \bar{r} = 0 \, , \tag{B.9}$$

for $j = 1, 2, \ldots$, and $k = 1, 2$. Here the ratio $\bar{\zeta}_{\bar{r}}^{(0)} / \bar{\psi}_{\bar{r}}^{(0)}$ is finite at $\bar{r} = 0$ and decays rapidly as $\bar{r} \to \infty$. Therefore, (B.8) behaves as the Laplace equation for the jth harmonics, when $\bar{r} \to 0$ and also when $\bar{r} \to \infty$.

The matching condition (B.7) yields the following conditions on the Fourier coefficients,

$$\bar{\psi}_{11} \to [\dot{Y}^{(1)} + X^{(1)} \Psi_{xx}(X^{(0)}, Y^{(0)}) + Y^{(1)} \Psi_{xy}(X^{(0)}, Y^{(0)})] \, \bar{r} \, , \tag{B.10}$$

$$\bar{\psi}_{12} \to [-\dot{X}^{(1)} + X^{(1)} \Psi_{yx}(X^{(0)}, Y^{(0)}) + Y^{(1)} \Psi_{yy}(X^{(0)}, Y^{(0)})] \, \bar{r} \, , \tag{B.11}$$

$$\bar{\psi}_{21} \to [\Psi_{xx} - \Psi_{yy}](t, X^{(0)}, Y^{(0)}) \, \bar{r}^2/4 = \Psi_{xx}(t, X^{(0)}, Y^{(0)}) \, \bar{r}^2/2 \, , \tag{B.12}$$

$$\bar{\psi}_{22} \to \Psi_{xy}(t, X^{(0)}, Y^{(0)}) \, \bar{r}^2/2 \, , \tag{B.13}$$

and

$$\bar{\psi}_{jk} \to 0 \qquad \text{for} \quad j = 3, 4, \ldots, \; k = 1, 2 \, . \tag{B.14}$$

Note that the contributions of the local background velocity gradients appear in the inhomogeneous terms in (B.12) and (B.13) for the second harmonics of the inner solution.

By repeating the arguments used in Subsect. 3.2.3 for the Fourier coefficients of $\bar{\psi}^{(1)}$, we obtain the following results for those of $\bar{\psi}^{(2)}$:

$$\bar{\psi}_{jk} = 0 \, , \qquad \text{for} \quad j = 1 \, , \; j = 3, 4, \ldots, \text{ and } k = 1, 2 \, . \tag{B.15}$$

Conditions (B.10) and (B.11) in turn yield

$$\begin{aligned} \dot{X}^{(1)} &= X^{(1)} \Psi_{yx}(X^{(0)}, Y^{(0)}) + Y^{(1)} \Psi_{yy}(X^{(0)}, Y^{(0)}) \, , \\ \dot{Y}^{(1)} &= -X^{(1)} \Psi_{xx}(X^{(0)}, Y^{(0)}) - Y^{(1)} \Psi_{xy}(X^{(0)}, Y^{(0)}) \, . \end{aligned} \tag{B.16}$$

By combining this equation for the first order velocity with (3.2.79) for the leading-order velocity, we get

$$\hat{\imath} \dot{X} + \hat{\jmath} \dot{Y} \; = \; \hat{\imath} \Psi_y(X, Y) - \hat{\jmath} \Psi_x(X, Y) + O(\epsilon^2) \, . \tag{B.17}$$

Thus we conclude:

- *In the normal time scale, the velocity of the vortex center defined by the asymptotic analysis differs from the local background velocity by no more than $O(\epsilon^2)$. That is to say that the trajectory of the vortex center deviates by at most $O(\epsilon^2)$ from the stream line of the background potential flow passing through the initial position of the vortex center.*

The two nontrivial Fourier coefficients $\bar{\psi}_{2k}$, $k = 1, 2$, are governed by the differential equation (B.8) in $\bar{r}$ for $j = 2$, the boundary condition $\bar{\psi}_{2k} = 0$ at $\bar{r} = 0$ and the matching conditions (B.12) and (B.13). These coefficients can be related to one canonical solution $\bar{\psi}_{2c}(t, \bar{r})$ as follows:

$$\bar{\psi}_{21} = \frac{1}{2}[\Psi_{xx} - \Psi_{yy}](X^{(0)}, Y^{(0)})\,\bar{\psi}_{2c} \qquad \text{and} \qquad \bar{\psi}_{22} = \Psi_{xy}(X^{(0)}, Y^{(0)})\,\bar{\psi}_{2c}\,. \tag{B.18}$$

The canonical solution $\bar{\psi}_{2c}$ obeys the differential equation (B.8) with $j = 2$ and the boundary conditions

$$\bar{\psi}_{2c} = 0 \quad \text{at} \quad \bar{r} = 0 \qquad \text{and} \qquad \bar{\psi}_{2c} \to \bar{r}^2/2 \quad \text{as} \quad \bar{r} \to \infty\,. \tag{B.19}$$

Thus we can determine the canonical solution for a given leading-order core structure in (B.8) with $j = 2$. Since the equation behaves as the Laplace equation for the second harmonics as $\bar{r} \to \infty$, the solution $\bar{\psi}_{2c}$ behaves as

$$\bar{\psi}_{2c} = \bar{r}^2/2 + c(t)\,\bar{r}^{-2} + o(\bar{r}^{-m}) \tag{B.20}$$

for a large m and the function $c(t)$ is defined by the canonical solution.

For example, when $\bar{\zeta}^{(0)}$ is represented by the similarity solution (3.2.64), we have

$$\frac{\bar{\zeta}^{(0)}_{\bar{r}}}{\bar{\psi}^{(0)}_{\bar{r}}} = \frac{\beta(\eta)}{4\bar{\nu}\bar{t}} \tag{B.21}$$

with

$$\beta(\eta) = 4\eta^2/(e^{\eta^2} - 1)\,, \tag{B.22}$$

where $\bar{t} = t + t_0^*$ and $\eta^2 = \bar{r}^2/(4\bar{\nu}\bar{t})$. We can then express the canonical solution in terms of $\bar{t}$ and η with $\bar{\psi}_{2c} = 2\bar{\nu}\bar{t}g(\eta)$. Equation (B.8) becomes an equation for $g(\eta)$ with the independent variable $\bar{r}$ replaced by η and the ratio in (B.21) by $\beta(\eta)$. The boundary conditions (B.19) and (B.20) become $g(0) = 0$ and $g(\eta) \sim \eta^2 + C\eta^{-2}$ as $\eta \to \infty$. A numerical solution of $g(\eta)$ was constructed, and the constant C was found to be $C = 4.37$. The coefficient $c(t)$ in (B.20) then becomes $8C(\bar{\nu}\bar{t})^2$.[2]

From the behavior (B.20) of the canonical solution and with the asymmetric part of the second-order inner solution, $\epsilon^2[\bar{\psi}_{21}\cos 2\theta + \bar{\psi}_{22}\sin 2\theta]$, related to the canonical solution by (B.18), we see from (B.12), (B.13), and (B.20) that as $\bar{r} \to \infty$ the $\bar{r}^2$ terms in the inner solution match with the contributions of the local velocity gradients of the outer solution. The $\bar{r}^{-2}$ terms in the inner solution then induce $\epsilon^4 r^{-2}$ terms , quadrupoles, in the outer solution. Thus we have shown that

- *The leading contribution of the vortical core to the outer stream function other than that of a classical point vortex is a fourth-order quadrupole,* $\epsilon^4 c(t) r^{-2}[\Psi_{xx}\cos 2\theta + \Psi_{xy}\sin 2\theta]_{\text{at } X^{(0)},Y^{(0)}}$. *Its strength depends on the local velocity gradient of the background flow and the global effect of the leading-order core structure through the function* $c(t)$.

[2] Details of the analysis were described in Ting and Tung (1965). However, their function α is in error by a factor of 1/2, which results in an incorrect value for the constant C.

As mentioned before, the asymmetric part of a one-time inner solution is a quasi-steady solution because it is governed by a partial differential equation in $\bar{r}, \theta$ with time t acting as a parameter. In particular, the first order asymmetric part of the one-time solution is identically zero and the velocity of the vortex center is defined by the background velocity, (3.2.90). Thus the one time solution does represent the complete first order solution if it happens to be in agreement with the initial data. In other words, a first order two-time solution will be needed only if the initial vorticity distribution has an $O(\epsilon^{-1})$ asymmetric contribution and/or if the initial velocity of the vortex disagrees with the local background velocity. The latter case arises, for example, when a vortex that is held fixed for $t < 0$ is suddenly released at $t = 0$.

Using condition (B.2) on the first order two-time solution, we can express the inhomogeneous term $\tilde{F}_2$ as a θ-derivative, that is, replace (B.3) by

$$\tilde{F}_2 = -\frac{\partial}{\partial\theta}\left[\frac{1}{2}[(\bar{\omega}\bar{a} - \bar{b})(\tilde{\zeta}^{(1)})^2]_{\bar{r}} + \tilde{\zeta}^{(1)}\tilde{\psi}^{(1)}_{\bar{r}}\right] , \tag{B.23}$$

and, therefore, not only its short-time average but also its θ-average vanish:

$$\{\tilde{F}_2\} = 0 . \tag{B.24}$$

Consequently, the second-order equation (3.2.55) for the symmetric two-time solution becomes

$$(\tilde{\zeta}_c^{(2)})_\tau = 0 \qquad \text{or} \quad \tilde{\zeta}_c^{(2)} = \tilde{\zeta}_c^{(2)}(t, \bar{r}) , \tag{B.25}$$

which implies that the symmetric part of the second-order inner solution does not depend on the short time.

Note that (3.2.49) or (3.2.51–3.2.52) state that, quite generally, the symmetric part of the first order inner solution is varying only on the normal time, because the leading-order inner solution is symmetric and τ-independent and thus, $\tilde{F}_1 \equiv 0$. In contrast, (B.25) holds only if the first order two-time solution satisfies the special condition (B.2). With this understanding, we may conclude that

- *The symmetric parts of the zeroth-, the first-, and the second-order vorticity distributions have only a normal time dependence.*

Finally we point out that it was shown in Subsect. 3.2.1, following (3.2.47), that the leading-order inner solution in the two-time analysis coincides with the symmetric one-time solution. We cannot draw the same conclusion even for the symmetric first order two-time solution let alone the second-order solution, although they are both independent of τ. The reason is provided in Appendix A. We recall that in the two-time analysis the governing equation for the symmetric second-order solution is given by (A.10) for $n = 2$. This equation is obtained from the τ- and θ-average of the third-order equation. The inhomogeneous term $\mathcal{M}\{\tilde{F}_3\}$ in (A.10) contains the average of products of the first- and second-order asymmetric two-time solutions and does not, in general, vanish identically. This inhomogeneous term is absent in the one-time analysis.

C. Equations of Motion of Filaments

In Subsect. 3.3.6 we uncoupled the equations for the core structure of a filament from those for the motion of its centerline and expressed the global contributions of the core structure to the motion in terms of the initial core structure and the length of the filament. Consequently, a closed system of equations of motion for the filament was obtained. As mentioned in Subsect. 3.3.6.3, we list here the system of equations and define all the symbols so that we do not have to refer back to the main text.

Let us consider the flow field of N vortex filaments in a background flow with velocity potential $\Phi(\mathbf{x})$. Let Γ_i and $\mathbf{X}_i(s,t)$ denote the circulation and position vector of the centerline of the ith filament, for $i = 1, 2, \ldots, N$. The parameter s_i increases in the direction of positive Γ_i. For each filament, its centerline and its inner structure, that is, the large axial and circumferential velocity distribution, are prescribed through appropriate initial conditions. We collect now all formulae that describe, say, the ith filament embedded in an outer flow that is the superposition of the outer potential field, $\nabla\Phi$, and the velocities induced by the other filaments. In these formulae we suppress the subscript i unless necessary. Also, we omit the superscript (0), denoting the leading-order entities.

At $t = 0$, the vector function defining the centerline as a simple closed curve C is prescribed as

$$\mathbf{X}(0,s) \quad \text{and} \quad 0 \le s \le 2\pi \qquad \text{with} \quad \mathbf{X}(0, s+2\pi) = \mathbf{X}(0,s) \; . \tag{C.1}$$

The initial profiles of axial vorticity and velocity, ζ and w, are

$$\zeta(0,\mathbf{x}) = \epsilon^{-2}\zeta_0(\bar{r}) \quad \text{and} \quad w(0,\mathbf{x}) = \epsilon^{-1}w_0(\bar{r}) \; , \tag{C.2}$$

where $\bar{r} = r/\epsilon$ and $\epsilon = \delta^*/\ell$. Here, ℓ denotes a typical length scale of the flow field such as a typical length of a filament and δ^* a is typical core size. The initial profiles are independent of the circumferential coordinate, θ, and of s and they decay exponentially in $\bar{r}$. This is consistent with conditions (3.3.35) and (3.3.36) for the asymptotic analysis. The variables r, θ, and s are the curvilinear coordinates relative to C used to describe the core structure. They are related to a position vector, $\mathbf{x}$, in space by

$$\mathbf{x} = \mathbf{X}(t,s) + r\hat{r}(t,s) \; , \tag{C.3}$$

where r denotes the minimum distance from $\mathbf{x}$ to $\mathcal{C}$ and $\hat{r}$ the unit radial vector.

We denote the unit tangent, normal and binormal vectors of the centerline by $\hat{\tau}$, $\hat{n}$, and $\hat{b}$, and its curvature and torsion by κ, and T, respectively. These quantities are related to $X(t,s)$ by the Serret–Frenet formulae (3.1.58).

The equation for the velocity of the ith filament is

$$\mathbf{X}_t(t,s) = \mathbf{Q}^*(t,\mathbf{X}) + \frac{\Gamma\kappa}{4\pi}\left[\ln\frac{1}{\epsilon} + C_v(t)\right]\hat{b} + \frac{\Gamma\kappa}{4\pi}C_w(t)\,\hat{b}\ , \tag{C.4}$$

where

$$\mathbf{Q}^* = \mathbf{Q}_2 + \mathbf{Q}^f - \hat{\tau}[(\mathbf{Q}_2 + \mathbf{Q}^f)\cdot\hat{\tau}]\ . \tag{C.5}$$

The removal of the tangential component ensures that $\hat{\tau}\cdot\mathbf{X}_t = 0$. In (C.5), $\mathbf{Q}_2(t,\mathbf{X})$ denotes the velocity of the flow field at $\mathbf{X}$ in the absence of the ith filament,

$$\mathbf{Q}_2 = \nabla\Phi(\mathbf{X}) + \frac{1}{4\pi}\sum_{\substack{j=1\\ j\neq i}}^{N}\Gamma_j\int_{\mathcal{C}_j}\frac{\mathbf{X}_j - \mathbf{X}}{|\mathbf{X}_j - \mathbf{X}|^3}\times d\mathbf{X}_\mathbf{j}\ , \tag{C.6}$$

and $\mathbf{Q}^f(t,s)$ is the finite part of the Biot–Savart integral for the ith filament at $\mathbf{X}$. In the matched asymptotic analysis, the singular terms in the Biot–Savart integral are cancelled analytically and $\mathbf{Q}^f$ is given by

$$\mathbf{Q}^f(t,s) = \frac{\Gamma}{4\pi}\left\{\int_{-\pi}^{\pi}\mathbf{G}(t,s+\bar{s},s)\,d\bar{s} + \left[\ln(2\sqrt{S_+S_-}) - 1\right]\kappa\hat{b}\right\}\ , \tag{C.7}$$

where

$$\begin{aligned}
\mathbf{G}(t,s',s) &= \mathbf{F}(t,s',s) - 2\kappa\hat{b}/|\lambda|\ , \qquad (s'\neq s)\ , && \text{(C.8)}\\
&= \frac{\hat{\tau}\times\mathbf{B}}{3}\,\mathrm{sgn}(s'-s)\ , \qquad (s' = s\pm 0)\ , && \text{(C.9)}\\
\mathbf{F}(t,s',s) &= \frac{\mathbf{X}(t,s') - \mathbf{X}(t,s)}{|\mathbf{X}(t,s') - \mathbf{X}(t,s)|^3}\times\mathbf{X}_s(t,s)\ , && \text{(C.10)}
\end{aligned}$$

with

$$\begin{aligned}
\lambda(t,s',s) &= \int_s^{s'}\sigma(t,s^*)\,ds^*\ , && \text{(C.11)}\\
\sigma(t,s) &= |\mathbf{X}_s(t,s)|\ , && \text{(C.12)}\\
\mathbf{B}(t,s) &= \frac{\partial\kappa}{\partial s}\hat{n} + \kappa T\sigma\hat{b}\ , && \text{(C.13)}\\
S_\pm &= \lambda(t,s\pm\pi,s) && \text{(C.14)}
\end{aligned}$$

and

$$S(t) = S_+(t) + S_-(t) = \int_0^{2\pi}\sigma(t,s)\,ds\ . \tag{C.15}$$

(C.16)

Here $S(t)$ denotes the length of $\mathcal{C}$. The integrand $\mathbf{G}$ in (C.7) is a piecewise continuous function of $\bar{s} = s' - s$, with a jump discontinuity at $\bar{s} = 0$ as defined in (C.9). Therefore, $\mathbf{Q}^f$ given by (C.7) can be evaluated numerically at each instant. The remaining unknowns in (C.5) are $C_v(t)$ and $C_w(t)$. They denote, respectively, the global contributions of the large circumferential and axial flow and are related to the inner structure of the filament. The core size is defined by

$$\delta(t) = [4\nu\tau_1(t)/S(t)]^{1/2}\,, \tag{C.17}$$

$$\text{where} \quad \tau_1 - \int_0^t S(t')\,dt' + \tau_{10} \tag{C.18}$$

$$\text{with} \quad \tau_{10} = \frac{\pi S_0}{2\Gamma\bar{\nu}} \int_0^\infty \zeta_0(\bar{r})\,\bar{r}^3\,d\bar{r}\,. \tag{C.19}$$

The improper integrals in the definition (3.3.164) of the global contribution C_v due to the larger circumferential flow were evaluated analytically and C_v is found to be

$$C_v(t) = \ln\left(\frac{1}{\bar{\delta}(t)}\right) + \frac{1}{2}\{1 + \gamma - \ell n\, 2\} + 4\pi \sum_{n=1}^{N} \alpha_n \tau_1^{-n}(t) + 8\pi^2 \sum_{n=2}^{N} \gamma_n \tau_1^{-n}(t) \tag{C.20}$$

with

$$\begin{aligned}
\alpha_n &= \frac{\bar{\nu}}{\Gamma} D_n \sum_{j-1}^{N} p_{n,j} \left(1 - \frac{1}{2^j}\right) (j-1)! \\
\gamma_h &= \sum_{m=1}^{h} (\frac{\bar{\nu}}{\Gamma})^2 D_m D_{h-m} \beta_{m,h-m}\,, \\
\beta_{n,h} &= \sum_{i=1}^{n} (p_{n-1,i} - p_{n,i}) \sum_{j=1}^{h} \frac{(i+j-1)!}{2^{i+j}} (p_{h-1,j} - p_{h,j}).
\end{aligned} \tag{C.21}$$

To compute the global contribution $C_w(t)$ due to the large axial flow, we introduce $\delta^w(t), \tau_1^w(t), \tau_{10}^w$ in analogy to (C.17–C.19),

$$\delta^w(t) = [4\nu\tau_1^w(t)/S(t)]^{1/2}, \tag{C.22}$$

$$\tau_1^w = \int_0^t S(t')\,dt' + \tau_{10}^w \tag{C.23}$$

and

$$\tau_{10}^w = \frac{\pi S_0}{2\bar{\nu} m(0)} \int_0^\infty w_0(\bar{r})\,\bar{r}^3\,d\bar{r} \tag{C.24}$$

and obtain

$$C_w = -\frac{16\pi^2}{[S^{(0)}]^3 \tau_1^w} \sum_{n=0}^{N} \omega_n [\tau_1^w]^{-n} \tag{C.25}$$

with

$$\omega_n = \sum_{m=0}^{n} C_m C_{n-m} P_{m,n-m} \,, \qquad P_{n,i} = \sum_{j=0}^{N} p_{i,j} \sum_{k=0}^{n} p_{n,k} (j+k)!/2^{j+k+1} \,. \tag{C.26}$$

Here, γ is the Euler number, $p_{n,k}$ is the coefficient of x^k in the Laguerre polynomial $L_n(x)$, and $(N+1)$ is the number of terms used in the series representations (3.3.176, 3.3.177) of w and ζ to fit the initial data (C.2) which in turn define the coefficients in the series. They are

$$C_n = 2S_0 \Big(\frac{\tau_{20}^w}{S_0}\Big)^{n-1} \int_0^\infty \bar{w}_0(\bar{r}) L_n(\eta^2)\, \eta \, d\eta \,, \tag{C.27}$$

$$D_n = \frac{2}{S_0} \Big(\frac{\tau_{20}}{S_0}\Big)^{n+1} \int_0^\infty \bar{\zeta}_0(\bar{r}) L_n(\eta^2)\, \eta \, d\eta \,, \tag{C.28}$$

where $\bar{r} = \eta\sqrt{4\bar{\nu}\tau_{10}/S_0}$, $S_0 = S(0)$, $\delta_0 = \sqrt{4\nu\tau_{10}/S_0}$ in (C.28). For the C_n's in (C.27) we replace τ_{10} by τ_{10}^w. Note that $C_1 = 0$ and $D_1 = 0$ because of the choice of τ_{10} and τ_{10}^w, see (3.3.181). If the initial profiles are similar and $\tau_{10}^w = \tau_{10}$, we have $C_n = D_n = 0$ for $n \geq 1$.

Equations (C.5) and (C.18), supplemented by the remaining equations in this Appendix applied to the ith filament for $i = 1, \ldots, N$, form a closed system of equations of motion for the N filaments.

D. Formulae for the Coefficients in (6.2.74) and (6.2.75)

This Appendix was first referred to in Subsect. 6.2.2. We defined

$$a_{km} = \Gamma_k \Gamma_m a^*_{km}\ , \ b_{km} = \Gamma_k \Gamma_m b^*_{km}\ , \ \text{and} \ \ e_{km} = \Gamma_k \Gamma_m e^*_{km}\ , \tag{D.1}$$

with

$$a^*_{km} = \langle \partial_x \zeta^*_k \, \partial_x \zeta^*_m \rangle\ , \quad b^*_{km} = \langle \partial_y \zeta^*_k \, \partial_y \zeta^*_m \rangle\ , \quad \text{and} \quad e^*_{km} = \langle \partial_x \zeta^*_k \, \partial_y \zeta^*_m \rangle\ , \tag{D.2}$$

so that the (*) quantities are independent of the strengths Γ_k, Γ_m and are functions of $X_k - X_m$, $Y_k - Y_m$, δ_k and δ_m. By converting the integrand to that for an error function, we obtain

$$a^*_{km} = \frac{2}{\pi \delta^4_{km}} e^{-R^2_{km}/\delta^2_{km}} \left[1 - \frac{2(X_k - X_m)^2}{\delta^2_{km}} \right] \tag{D.3}$$

and

$$e^*_{km} = e^*_{mk} = \frac{-4(X_k - X_m)(Y_k - Y_m)}{\pi \delta^6_{km}} e^{-R^2_{km}/\delta^2_{km}}\ , \tag{D.4}$$

where $\delta^2_{km} = \delta^2_k \delta^2_m / (\delta^2_k + \delta^2_m)$ and $R^2_{km} = (X_k - X_m)^2 + (Y_k - Y_m)^2$.

We note that R_{km} denotes the distance between the kth and mth center. The coefficient b^*_{km} is given by (D.3) with $(X_k - X_m)^2$ replaced by $(Y_k - Y_m)^2$. For the constants on the right side of (6.2.59), we define

$$C_k = \Gamma_k \sum_m \Gamma_m \sum_{\ell \neq m} \Gamma_\ell c^*_{km,\ell} \tag{D.5}$$

with

$$\begin{aligned} c{*}_{km,\ell} &= \langle (\partial_x \zeta^*_k)(\boldsymbol{V}^*_\ell \cdot \nabla \zeta^*_m) \rangle \\ &= \langle (\partial_x \zeta^*_k)[-(y - Y_\ell)\, \partial_x \zeta^*_m + (x - X_\ell)\, \partial_y \zeta^*_m][1 - e^{-r^2_\ell/\delta^2_\ell}]/r^2_\ell \rangle\ . \end{aligned}$$

By using the polar coordinates relative to the ℓth vortex center and the integral representations of the modified Bessel functions, we identify the above area integral as the sum of integrals of Bessel functions (see formula 6.618–4 and 6.631–4 of Gradshteyn and Ryzhik, 1965). The final result is

$$c^*_{km,\ell} = \frac{2}{\pi^3} e^{-R^2_{km}/\delta^2_{km}} \left[-X^\ell_k (X^\ell_m \sin\tau - Y^\ell_m \cos\tau) I^\ell_{km} \right. ,$$
$$\left. -\frac{Y^\ell_m}{2} II^\ell_{km} + \frac{1}{2}(X^\ell_m \sin 2\tau - Y^\ell_m \cos 2\tau) III^\ell_{km} \right] , \qquad \text{(D.6)}$$

where

$$X^\ell_j = X_j - X_\ell \,, \quad Y^\ell_j = Y_j - Y_\ell \,, \qquad j = k, m \,,$$
$$X_{km} = (X_k\delta^2_k + X_m\delta^2_m) \,/\, (\delta^2_k + \delta^2_m) \,, \quad Y_{km} = (Y_k\delta^2_k + Y_m\delta^2_m) \,/\, (\delta^2_k + \delta^2_m) \,,$$
$$R^2_{km,\ell} = (X_{km} - X_\ell)^2 + (Y_{km} - Y_\ell)^2 \,,$$
$$X_{km} - X_\ell = X_{km,\ell} \cos\tau \,, \quad Y_{km} - Y_\ell = R_{km,\ell} \sin\tau \,.$$

We note that the point (X_{km}, Y_{km}) is the center of gravity of the kth and mth core weighted by the inverse of the square of their core size and R^ℓ_{km} denotes the distance between (X_{km}, Y_{km}) and the ℓth center. The functions I, II, and III in (D.6) are

$$I^\ell_{km} = \beta_{km,\ell}[1 - e^{-R^2_{km,\ell}/\delta^2_{km,\ell}}] \quad II^\ell_{km} = \frac{1}{2}[\alpha_{km} - \gamma_{km,\ell}\, e^{-R^2_{km,\ell}/\delta^2_{km,\ell}}]$$

and

$$III^\ell_{km} = \frac{1}{2}(\alpha_{km} - \gamma_{km,\ell}) + \frac{1}{2}(\gamma_{km,\ell} - 4\beta^2_{km,\ell})(1 - e^{-R^2_{km,\ell}/\delta^2_{km,\ell}}] \,,$$

where

$$\alpha_{km} = \delta^2_k\delta^2_m/(\delta^2_k + \delta^2_m) \,, \quad \beta_{km,\ell} = \alpha_{km}/R_{km,\ell} \,,$$
$$\gamma_{km,\ell} = (\delta_k\delta_m\delta_\ell)^2/(\delta^2_k\delta^2_m + \delta^2_m\delta^2_\ell + \delta^2_\ell\delta^2_k) \quad \text{and}$$
$$\delta^2_{km,\ell} = (\delta^2_k\delta^2_m + \delta^2_m\delta^2_\ell + \delta^2_\ell\delta^2_k)/(\delta^2_k + \delta^2_m) \,.$$

Likewise, we define

$$D_k = \Gamma_k \sum_m \Gamma_m \sum_{\ell\neq m} \Gamma_\ell d^*_{km,\ell} \,,$$

with

$$d^*_{km,\ell} = \langle(\partial_y\zeta^*_k)(\boldsymbol{V}^*_\ell \cdot \nabla\zeta^*_m)\rangle$$
$$= \frac{2}{\pi^3} e^{-R^2_{km,\ell}/\delta^2_{km}} \left[-Y^\ell_k (X^\ell_m \sin\tau - Y^\ell_m \cos\tau) I^\ell_{km} \right.$$
$$\left. +\frac{1}{2} X^\ell_m II^\ell_{km} - \frac{1}{2}(X^\ell_m \cos 2\tau + Y^\ell_m \sin 2\tau) III^\ell_{km} \right] .$$

Thus all the coefficients $a_{km}, \ldots$, in (6.2.59), which are defined by area integrals in the xy plane, are reduced to elementary functions of the strengths and core sizes and the coordinates of one center relative to the other.

E. Transformations to Filament Attached Coordinates

Here we recount several basic relationships concerning the formulation of conservation laws in moving orthogonal coordinates. These are employed in Sect. 3.4, where we began our fast track analysis of a slender vortex filament with compressible vortical core. In Sect. E.4, we also list the formulas for the orthogonal coordinate system attached to a filament centerline, $\mathcal{C}$, presented in Callegari and Ting (1978). These formulas are valid regardless whether the vortical core is compressible or incompressible and were employed in Sects. 3.3 and 3.4 for the analyses of incompressible and compressible vortical cores, respectively.

E.1 Conservation Laws in Orthogonal Coordinates

Let φ denote the density of some conserved quantity, and $\boldsymbol{f}_\varphi$ its flux density vector. For a time-dependent control volume $\mathcal{V}(t)$ whose boundary points move with velocity $\boldsymbol{V}_{\partial\mathcal{V}}$, the conservation of φ is expressed as

$$\frac{d}{dt'}\int_{\mathcal{V}(t')} \varphi \, d^3x = -\oint_{\partial\mathcal{V}(t')} (\boldsymbol{f}_\varphi - \varphi \boldsymbol{V}_{\partial\mathcal{V}}) \cdot \boldsymbol{n} \, d\sigma \, . \tag{E.1.1}$$

Next we introduce a set of time-dependent curvilinear orthogonal coordinates $(t, \underline{\xi}) = (t, (\xi_1, \xi_2, \xi_3))$ so that

$$(t', \boldsymbol{x}) = (t, \boldsymbol{x}'(t, \underline{\xi})) \, . \tag{E.1.2}$$

For later reference we denote by

$$h_i = \left| \frac{\partial \boldsymbol{x}'}{\partial \xi_i} \right| \qquad \text{and} \qquad \boldsymbol{V}_{\underline{\xi}} = \frac{\partial \boldsymbol{x}'}{\partial t} \tag{E.1.3}$$

the stretch factors relating coordinate increments to distance on the coordinate lines, and the velocity of points $\underline{\xi}$ = constant as time evolves. Orthogonality of the coordinate system implies that

$$\frac{\partial \boldsymbol{x}'}{\partial \xi_i} \cdot \frac{\partial \boldsymbol{x}'}{\partial \xi_j} = \delta_{ij} \tag{E.1.4}$$

with δ_{ij} the Kronecker symbol.

Now we consider the special case of (E.1.1) such that $\mathcal{V}(t')$ is the time-dependent image of $\mathcal{V}_{\underline{\xi}}$, that is, of the quadrilateral box in the space of $\underline{\xi}$,

$$\mathcal{V}_{\underline{\xi}} = [\xi_1, \xi_1^+] \times [\xi_2, \xi_2^+] \times [\xi_3, \xi_3^+] \,. \tag{E.1.5}$$

Noticing further that

$$\int_{\mathcal{V}(t')} \varphi \, d^3x = \int_{\mathcal{V}_{\underline{\xi}}} \varphi(t, \boldsymbol{x}'(t, \underline{\xi})) \, J(t, \underline{\xi}) \, d^3\xi \,, \tag{E.1.6}$$

with the Jacobi determinant

$$J = \det \left(\frac{\partial \boldsymbol{x}'}{\partial \underline{\xi}} \right) = \prod_{i=1}^{3} h_i \,, \tag{E.1.7}$$

we may rewrite the conservation law for the "coordinate box" $\mathcal{V}_{\underline{\xi}}$ as

$$\int_{\mathcal{V}_{\underline{\xi}}} (\varphi J)_t \, d^3\xi = - \sum_{i=1}^{3} \varepsilon_{ijk} \int_{\xi_j}^{\xi_j^+} \int_{\xi_k}^{\xi_k^+} \left(f_\varphi^i h_j h_k \right) \Big|_{\xi_i}^{\xi_i^+} \, d\xi_j d\xi_k \tag{E.1.8}$$

where

$$f_\varphi^i = \left(\boldsymbol{f}_\varphi - \varphi \boldsymbol{V}_{\underline{\xi}} \right) \cdot \boldsymbol{n}_i \qquad \text{with} \qquad \boldsymbol{n}_i = \frac{1}{h_i} \frac{\partial \boldsymbol{x}'}{\partial \xi_i} \tag{E.1.9}$$

and for some function $a(\xi)$ we let

$$a|_{\xi_1}^{\xi_2} = a(\xi_2) - a(\xi_1) \,. \tag{E.1.10}$$

Dividing (E.1.8) by $|\mathcal{V}_{\underline{\xi}}|$ and passing to the limits $\xi_i^+ - \xi_i \to 0$ $(i = 1, 2, 3)$, we find the differential form of the conservation law in terms of the new coordinate system,

$$\partial_t \left(\varphi \, h_1 h_2 h_3 \right) = - \sum_{i=1}^{3} \varepsilon_{ijk} \, \partial_{\xi_i} \left(f_\varphi^i \; h_j h_k \right) . \tag{E.1.11}$$

E.2 Scalar Transport Equations in Orthogonal Coordinates

Consider some scalar a that is advected with velocity $\boldsymbol{v}$, such that, in inertial frame coordinates, we have

$$\left(\partial_{t'} + \boldsymbol{v} \cdot \nabla' \right) a = 0 \,. \tag{E.2.1}$$

Using again the general orthogonal time-dependent coordinates $(t, \underline{\xi})$ introduced in (E.1.2), we have $\partial_t = \partial_{t'} + \boldsymbol{V}_{\underline{\xi}} \cdot \nabla$ and

$$\left(\partial_t + \left(\boldsymbol{v} - \underline{\boldsymbol{V}_\xi}\right) \cdot \nabla'\right) a = 0\,. \tag{E.2.2}$$

With

$$\nabla' = \sum_{i=1}^{3} \nabla' \xi_i \frac{\partial}{\partial \xi_i} = \sum_{i=1}^{3} \boldsymbol{n}_i \frac{1}{h_i} \frac{\partial}{\partial \xi_i}\,, \tag{E.2.3}$$

we have the generalized transport equation

$$\frac{\partial a}{\partial t} + \sum_{i=1}^{3} \left(\boldsymbol{v} - \underline{\boldsymbol{V}_\xi}\right) \cdot \boldsymbol{n}_i \frac{1}{h_i} \frac{\partial a}{\partial \xi_i} = 0\,. \tag{E.2.4}$$

For a fluid flow, where the density ρ satisfies a conservation law according to (E.1.11) with flux $\rho \boldsymbol{v}$, we find the conservation form of the scalar transport equation via trivial manipulations,

$$\frac{\partial}{\partial t}\left(\rho a\, h_1 h_2 h_3\right) = -\sum_{i=1}^{3} \varepsilon_{ijk} \frac{\partial}{\partial \xi_i} \left(\rho a \,\left(\boldsymbol{v} - \underline{\boldsymbol{V}_\xi}\right) \cdot \boldsymbol{n}_i \; h_j h_k\right)\,. \tag{E.2.5}$$

For an inhomogeneous scalar transport equation with source term q, that is, for $\partial_{t'} a + \boldsymbol{v} \cdot \nabla' a = q$, the quasi-conservation form of the equation reads

$$\frac{\partial}{\partial t}\left(\rho a\, h_1 h_2 h_3\right) + \sum_{i=1}^{3} \varepsilon_{ijk} \frac{\partial}{\partial \xi_i} \left(\rho a \,\left(\boldsymbol{v} - \underline{\boldsymbol{V}_\xi}\right) \cdot \boldsymbol{n}_i \; h_j h_k\right) = \rho q \, h_1 h_2 h_3\,. \tag{E.2.6}$$

E.3 Compressible Flow Equations in Filament Attached Coordinates

With the general formulation of a conservation law in place, we move on to specify mass, momentum, and energy conservation in the filament-attached orthogonal coordinate system. To this end, we use the definitions and relations for the filament-attached coordinates from Sect. E.4 below, and identify

$$\begin{aligned}
\left(t, (\xi_1, \xi_2, \xi_3)\right) &= \left(t, (\bar r, \theta, s)\right) \\
(h_1, h_2, h_3) &= \left(\epsilon, \epsilon \bar r, \sigma\left(1 - \epsilon \kappa \bar r \cos(\theta + \theta_0(t,s))\right)\right), \\
(\boldsymbol{n}_1, \boldsymbol{n}_2, \boldsymbol{n}_3) &= \left(\boldsymbol{r}, \boldsymbol{\theta}, \boldsymbol{\tau}\right), \\
\underline{\boldsymbol{V}_\xi} &= \partial_t \left(\boldsymbol{X}(t,s) + \epsilon \bar r \boldsymbol{r}(t,\theta,s)\right) = \boldsymbol{X}_t + \epsilon \bar r \boldsymbol{r}_t\,,
\end{aligned} \tag{E.3.1}$$

replace

$$\begin{aligned}
\varphi &\to \bar\rho \\
\varphi &\to \bar\rho\,(\bar{\boldsymbol{V}} + \boldsymbol{X}_t) \\
\varphi &\to \overline{\rho e} = \frac{\bar p}{\gamma - 1} + \frac{\gamma \mathrm{Ma}_*^2}{2}\, \rho (\bar{\boldsymbol{V}} + \boldsymbol{X}_t)^2\,,
\end{aligned} \tag{E.3.2}$$

and insert the flux densities

$$\boldsymbol{f}_\rho = \bar{\rho}\,(\bar{\boldsymbol{V}} + \boldsymbol{X}_t)$$

$$\boldsymbol{f}_{\rho\boldsymbol{v}} = \bar{\rho}\,(\bar{\boldsymbol{V}} + \boldsymbol{X}_t) \circ (\bar{\boldsymbol{V}} + \boldsymbol{X}_t) + \frac{\bar{p}}{\gamma \mathrm{Ma}_*^2}\,\mathbf{1} + \frac{1}{\mathrm{Re}}\bar{\boldsymbol{T}}$$

$$\boldsymbol{f}_{\rho e} = (\overline{\rho e} + \bar{p})\,(\bar{\boldsymbol{V}} + \boldsymbol{X}_t) + \frac{\gamma \mathrm{Ma}_*^2}{\mathrm{Re}}\left((\bar{\boldsymbol{V}} + \boldsymbol{X}_t)\cdot \bar{\boldsymbol{T}} + \bar{\boldsymbol{h}}\right). \qquad \text{(E.3.3)}$$

Here $\mathbf{1}$ is the unit tensor in three space dimensions, $\boldsymbol{T}$ denotes the stress tensor, and $\boldsymbol{h}$ the heat flux density. In scaling $\boldsymbol{T}$ and $\boldsymbol{h}$ we have already taken into account that $1/\mathrm{Re} = O(\epsilon^2)$ and $\mathrm{Pr} = O(1)$ and that the heat flux term has been expressed in the outer solution in terms of the perturbation Temperature $T' = (T-1)/\gamma\mathrm{Ma}_*^2$.

Here are the mass, momentum, and energy balances in the filament-attached coordinates:

$$(h_3\,\bar{\rho})_t = -\frac{1}{\epsilon\bar{r}}\,(\bar{r}h_3\,\bar{\rho}\bar{u})_{\bar{r}} - \frac{1}{\epsilon\bar{r}}\,(h_3\,\bar{\rho}\,(\bar{v} - \epsilon\bar{r}\boldsymbol{r}_t\cdot\boldsymbol{\theta}))_\theta$$
$$- \left(\bar{\rho}\,(\bar{w} - \epsilon\bar{r}\boldsymbol{r}_t\cdot\boldsymbol{\tau})\right)_s \qquad \text{(E.3.4)}$$

$$\left(h_3\,\bar{\rho}\,\left(\bar{\boldsymbol{V}} + \boldsymbol{X}_t\right)\right)_t = -\frac{1}{\epsilon\bar{r}}\left(\bar{r}h_3\left[\bar{\rho}\,(\bar{\boldsymbol{V}} + \boldsymbol{X}_t)\;\bar{u} + \left(\frac{\bar{p}\mathbf{1}}{\mathrm{M}_*^2} + \frac{\bar{\boldsymbol{T}}}{\mathrm{Re}}\right)\cdot\boldsymbol{r}\right]\right)_{\bar{r}}$$
$$-\frac{1}{\epsilon\bar{r}}\left(h_3\left[\bar{\rho}\,(\bar{\boldsymbol{V}} + \boldsymbol{X}_t)(\bar{v} - \epsilon\bar{r}\boldsymbol{r}_t\cdot\boldsymbol{\theta}) + \left(\frac{\bar{p}\mathbf{1}}{\mathrm{M}_*^2} + \frac{\bar{\boldsymbol{T}}}{\mathrm{Re}}\right)\cdot\boldsymbol{\theta}\right]\right)_\theta$$
$$-\left(\bar{\rho}\,(\bar{\boldsymbol{V}} + \boldsymbol{X}_t)\;(\bar{w} - \epsilon\bar{r}\boldsymbol{r}_t\cdot\boldsymbol{\tau}) + \left(\frac{\bar{p}\mathbf{1}}{\mathrm{M}_*^2} + \frac{\bar{\boldsymbol{T}}}{\mathrm{Re}}\right)\cdot\boldsymbol{\tau}\right)_s \qquad \text{(E.3.5)}$$

$$(h_3\,\overline{\rho e})_t = -\frac{1}{\epsilon\bar{r}}\left(\bar{r}h_3\left[\overline{\rho e}\,\bar{u} + \bar{p}\;(\bar{u} + U) + \frac{\mathrm{M}_*^2}{\mathrm{Re}}\left(\bar{\boldsymbol{v}}\cdot\bar{\boldsymbol{T}} + \bar{\boldsymbol{h}}\right)\cdot\boldsymbol{r}\right]\right)_{\bar{r}}$$
$$-\frac{1}{\epsilon\bar{r}}\left(h_3\left[\overline{\rho e}\;(\bar{v} - \epsilon\bar{r}\boldsymbol{r}_t\cdot\boldsymbol{\theta}) + \bar{p}\;(\bar{v} + V) + \frac{\mathrm{M}_*^2}{\mathrm{Re}}\left(\bar{\boldsymbol{v}}\cdot\bar{\boldsymbol{T}} + \bar{\boldsymbol{h}}\right)\cdot\boldsymbol{\theta}\right]\right)_\theta$$
$$-\left(\overline{\rho e}\;(\bar{w} - \epsilon\bar{r}\boldsymbol{r}_t\cdot\boldsymbol{\tau}) + \bar{p}\;(\bar{w} + W) + \frac{\mathrm{M}_*^2}{\mathrm{Re}}\left(\bar{\boldsymbol{v}}\cdot\bar{\boldsymbol{T}} + \bar{\boldsymbol{h}}\right)\cdot\boldsymbol{\tau}\right)_s . \qquad \text{(E.3.6)}$$

In (E.3.6) we have used the abbreviations $\bar{\boldsymbol{v}} = \bar{\boldsymbol{V}} + \boldsymbol{X}_t$, $U = \boldsymbol{r}\cdot\boldsymbol{X}_t$, $V = \boldsymbol{\theta}\cdot\boldsymbol{X}_t$, $W = \boldsymbol{\tau}\cdot\boldsymbol{X}_t$, and $\mathrm{M}_*^2 = \gamma\mathrm{Ma}_*^2$.

The Entropy $\mathcal{E}$ satisfies the inhomogeneous transport equation from (3.4.17), that is,

$$\left(\partial_{t'} - (\boldsymbol{V} + \boldsymbol{X}_t) \cdot \nabla'\right) \mathcal{E} = q_{\mathcal{E}} \,, \tag{E.3.7}$$

where

$$\begin{aligned} q_{\mathcal{E}} &= \frac{\gamma \mathrm{Ma}_*^2}{\mathrm{Re}\, \rho T} \left\{ \boldsymbol{\Gamma} : \nabla \boldsymbol{v} + \frac{1}{(\gamma-1)\mathrm{Pr}} \nabla \cdot (k \nabla T') \right\} \\ &= \frac{\gamma \mathrm{Ma}_*^2}{\mathrm{Re}\, \rho T} \left\{ \tau_{ij} \partial_{x_j} u_i + \frac{1}{(\gamma-1)\mathrm{Pr}} \partial_{x_j} (k \partial_{x_j} T') \right\} . \end{aligned} \tag{E.3.8}$$

Using (E.2.6), we translate this into a quasi-conservation law in filament attached coordinates and obtain

$$\begin{aligned} \left(h_3\, \rho \mathcal{E}\right)_t = &-\frac{1}{\epsilon \bar{r}} \left(\bar{r} h_3\, \bar{\rho} \mathcal{E} \bar{u}\right)_{\bar{r}} - \frac{1}{\epsilon \bar{r}} \left(h_3\, \bar{\rho} \mathcal{E} \left(\bar{v} - \epsilon \bar{r} \boldsymbol{r}_t \cdot \boldsymbol{\theta}\right)\right)_\theta \\ &- \left(\bar{\rho} \mathcal{E} \left(\bar{w} - \epsilon \bar{r} \boldsymbol{r}_t \cdot \boldsymbol{\tau}\right)\right)_s + h_3\, \rho\, q_{\mathcal{E}} \,. \end{aligned} \tag{E.3.9}$$

E.4 Formulas for the Coordinate System Attached to $\boldsymbol{C}$

Here we list several formulae, derived in Appendix A and Appendix B of Callegari and Ting (1978), for vector differential quantities in the orthogonal coordinate system attached to $\mathcal{C}$. These formulas are needed for the derivations of the governing equations for incompressible core structures in Subsect. 3.3 and for compressibles ones in Subsect. 3.4.

Let $\boldsymbol{X}(t, s)$ denotes the parametric equation of the moving centerline $\boldsymbol{C}$ in the inertial coordinates system, $\boldsymbol{x}$ for $t \geq 0$, as above. To be more specific, and following Callegari and Ting (1978), we choose the parameter s to be the arclength, $\tilde{s}(t, s)$ at $t = 0$. The unit tangent, normal and binormal vectors, $\boldsymbol{\tau}, \boldsymbol{n}$ and $\boldsymbol{b}$ and the stretch factor, curvature and torsion, σ, κ, and T are related to $\boldsymbol{X}$ by the Serret–Frenet formulas and are functions of t and s,

$$\begin{aligned} &\boldsymbol{X}_{\tilde{s}} = \boldsymbol{\tau} \,, && \boldsymbol{\tau}_{\tilde{s}} = \kappa \boldsymbol{n} \\ &\boldsymbol{n}_{\tilde{s}} = T \boldsymbol{b} - \kappa \boldsymbol{\tau} && \boldsymbol{b}_{\tilde{s}} = -T \boldsymbol{n} \\ \text{with} \quad &\sigma = |\boldsymbol{X}_s| = \tilde{s}_s \,, && \kappa = |\boldsymbol{\tau}_{\tilde{s}} \ \text{and} \ \boldsymbol{b} = \boldsymbol{\tau} \times \boldsymbol{n} \,. \end{aligned} \tag{E.4.1}$$

For a point $\boldsymbol{x}$ near $\mathcal{C}$, say point $\boldsymbol{X}(t, s)$, we have

$$r = |\boldsymbol{x} - \boldsymbol{X}| \ll \kappa(t, s) \,, \tag{E.4.2}$$

where r denotes the distance from $\boldsymbol{x}$ to $\mathcal{C}$. We write

$$\boldsymbol{x} = \boldsymbol{X} + r \boldsymbol{r} \qquad \text{with} \qquad \boldsymbol{r}(t, s, \phi) = \boldsymbol{n} \cos\phi + \boldsymbol{b} \sin\phi \,, \tag{E.4.3}$$

where r, ϕ denote the polar coordinates of $\boldsymbol{x}$ in the normal plane of $\mathcal{C}$ at point $\boldsymbol{X}$. When the polar angle ϕ is replaced by the angle variable θ with

$$\theta = \phi - \theta_0(t, s) \,, \qquad \partial_s \theta_0 = -\sigma T \,, \tag{E.4.4}$$

we have a set of orthogonal coordinates (r, θ, s) with

$$d\boldsymbol{x} = \boldsymbol{r} h_1 dr + \boldsymbol{\theta} h_2 d\theta + \boldsymbol{\tau} h_3 ds \tag{E.4.5}$$

with

$$h_1 = 1 \,, \quad h_2 = r \,, \quad h_3 = \sigma[1 - r\kappa \cos(\theta + \theta_0)] \,. \tag{E.4.6}$$

Their base vectors are $\boldsymbol{r}, \boldsymbol{\theta}, \boldsymbol{\tau}$ with $\boldsymbol{\theta} = \boldsymbol{\tau} \times \boldsymbol{r}$. Since they are independent of r, we have

$$\boldsymbol{r}_r = 0 \,, \quad \boldsymbol{\theta}_r = 0 \,, \quad \boldsymbol{\tau}_r = 0 \,. \tag{E.4.7}$$

Since $\boldsymbol{\tau}, \boldsymbol{n}$, and $\boldsymbol{b}$ are independent of r and θ, we have

$$\boldsymbol{r}_\theta = \boldsymbol{\theta} \,, \quad \boldsymbol{\theta}_\theta = -\boldsymbol{r} \,, \quad \boldsymbol{\tau}_\theta = 0 \,. \tag{E.4.8}$$

Using (E.4.1) and (E.4.4), we get

$$\boldsymbol{r}_s = -\kappa\sigma \cos(\theta + \theta_0)\boldsymbol{\tau} \,, \quad \boldsymbol{\theta}_s = \kappa\sigma \sin(\theta + \theta_0)\boldsymbol{\tau} \,, \quad \boldsymbol{\tau}_s = \kappa\sigma \boldsymbol{n} \,. \tag{E.4.9}$$

The equations listed above were given in Appendix A by Callegari and Ting (1978), those below in Appendix B.

In the transformation of variables from $t', \boldsymbol{x}$ to t, r, θ, s with $t = t'$, $\partial_{t'}$ implies $\boldsymbol{x} = \text{const}$. From (E.4.3) and (E.4.7) to (E.4.9), we get,

$$\begin{aligned} d\boldsymbol{x} &= (\dot{\boldsymbol{X}} + r\boldsymbol{r}_t)dt + (\boldsymbol{X}_s + r\boldsymbol{r}_s)ds + \boldsymbol{r}dr + r\boldsymbol{r}_\theta d\theta \\ &= (\dot{\boldsymbol{X}} + r\boldsymbol{r}_t)dt + h_3\boldsymbol{\tau}ds + \boldsymbol{r}dr + r\boldsymbol{\theta}d\theta = 0 \,. \end{aligned} \tag{E.4.10}$$

This, in turn, yields

$$\frac{\partial s}{\partial t'} = -\frac{r}{h_3}\frac{\partial \boldsymbol{r}}{\partial t} \cdot \boldsymbol{\tau} \,, \quad \frac{\partial r}{\partial t'} = -\dot{\boldsymbol{X}} \cdot \boldsymbol{r} \,, \quad \frac{\partial \theta}{\partial t'} = -\frac{1}{r}\left(\dot{\boldsymbol{X}} + r\frac{\partial \boldsymbol{r}}{\partial t}\right) \cdot \boldsymbol{\theta}. \tag{E.4.11}$$

Noting that $\dot{\boldsymbol{X}} \cdot \boldsymbol{\tau} = 0$ and $\partial_t \boldsymbol{r} \cdot \boldsymbol{r} = 0$, we have

$$\frac{\partial}{\partial t'} = \frac{\partial}{\partial t} - \left[\frac{\partial \boldsymbol{X}}{\partial t} + r\frac{\partial \boldsymbol{r}}{\partial t}\right] \cdot \nabla_{r\theta s} \tag{E.4.12}$$

$$\text{where} \qquad \nabla_{r\theta s} = \boldsymbol{r}\frac{\partial}{\partial r} + \boldsymbol{\theta}\frac{\partial}{r\partial \theta} + \boldsymbol{\tau}\frac{\partial}{h_3 \partial s} \tag{E.4.13}$$

Likewise, using the stretch ratios, (E.4.5), we have the definitions of $\nabla_{r\theta s}$ and $\Delta_{r\theta s}$. For example,

$$\nabla_{r\theta s}\boldsymbol{X}_t = \frac{1}{rh_3}[(rh_3\boldsymbol{X}_t \cdot \boldsymbol{r})_r + (h_3\boldsymbol{X}_t \cdot \boldsymbol{\theta})_\theta] = \frac{-\sigma\kappa}{h_3}\boldsymbol{X}_t \cdot \boldsymbol{n} = \frac{1}{h_3}\boldsymbol{X}_{ts} \cdot \boldsymbol{\tau} \tag{E.4.14}$$

$$\nabla_{r\theta s}\boldsymbol{v} = \nabla_{r\theta s}[\boldsymbol{V} + \boldsymbol{X}_t] = \frac{1}{rh_3}[(rh_3 u)_r + (h_3 v)_\theta + r(w_s + \boldsymbol{X}_{ts}\cdot\boldsymbol{\tau})]. \quad \text{(E.4.15)}$$

and

$$\Delta_{r\theta s}\boldsymbol{v} = \frac{1}{h_3}\frac{\partial}{\partial s}\left(\frac{1}{h_3}\boldsymbol{X}_{ts}\right) + \frac{1}{rh_3}\left[\frac{\partial}{\partial r}\left(rh_3\frac{\partial}{\partial r}\right) + \frac{\partial}{\partial\theta}\left(\frac{h_3}{r}\frac{\partial}{\partial\theta}\right) + \frac{\partial}{\partial s}\left(\frac{r}{h_3}\frac{\partial}{\partial s}\right)\right] \quad \text{(E.4.16)}$$

References

Abgrall, R., and Guillard, H., editors. *Special issue on low Mach number conference, Porquerolles (2004)*. ESAIM: Math. Mod. & Num. Analysis (M2AN), vol. 39, 2005.

Abramowitz, M., and Stegun, I. *Handbook of Mathematical Functions*. Dover Publishers, New York, 1972.

Almgren, A.S., Buttke, T., and Colella, P. A fast adaptive vortex method in three dimensions. *J. Comput. Phys.*, 113:177–200, 1994.

Almgren, A., Bell, J., and Szymczak, W. A numerical method for the incompressible Navier–Stokes equations based on an approximate projection. *SIAM J. Sci. Comput.*, 17:358–369, 1996.

Almgren, A.S., Bell, J.B., Colella, P., Howell, L.H., and Welcome, M.L. A conservative adaptive projection method for the variable density incompressible Navier–Stokes equations. *J. Comput. Phys.*, 142:1–46, 1998.

Arnott, W.P., Bass, H.E., and Raspet, R. General formulation of thermoacoustics for stacks having arbitrarily shaped pore cross sections. *J. Acoust. Soc. Am.*, 90:3228–3237, 1991.

Atchley, A.A., Hofler, T.J., Muzzerall, M.L., Kite, M.D., and Ao, C. Acoustically generated temperature gradients in short plates. *J. Acoust. Soc. Am.*, 88:251–263, 1990.

Batchelor, G.K. *An Introduction to Fluid Dynamics*. Cambridge University Press, 1967.

Bauer, L., and Morikawa, G.K. Stability of rectilinear geostrophic vortices in stationary equilibrium. *Phys. Fluids*, 19:929–942, 1976.

Beale, J.T., and Majda, A. Vortex methods. I: Convergence in three dimensions. II: Higher order accuracy in two and three dimensions. *Math. Comput.*, 39:1–51, 1982.

Beale, J.T., and Majda, A. High order accurate vortex methods with explicit velocity kernels. *J. Comput. Phys.*, 58:188–208, 1985.

Bell, J.B., and Marcus, D. A second-order projection method for variable-density flows. *J. Comput. Phys.*, 101:334–348, 1992.

Benjamin, T.B., and Feir, J.E. The disintegration of wave trains in deep water. Part 1. Theory. *J. Fluid Mech.*, 27:417–430, 1967.

Besnoin, E., and Knio, O.M. Numerical study of thermoacoustic heat exchangers in the thin-plate limit. *Numerical Heat Transfer, P. A*, 40:445–471, 2001.

Besnoin, E., and Knio, O.M. Numerical study of thermoacoustic heat exchangers. *Acustica*, 90:432–444, 2004.

Besnoin, E. *Computational Study of Thermoacoustic Heat Exchangers*. Ph.D. Thesis, The Johns Hopkins University, 2001.

Biello, J.A., and Majda, A.J. Transformations for temperature flux in multiscale models of the tropics. *Theor. Comp. Fluid Dyn.*, under revision, 2006.

Blanc-Benon, P., Besnoin, E., and Knio, O. Experimental and computational visualization of the flow field in a thermoacoustic stack. *C.R. Mécanique*, 331:17–24, 2003.

Bronstein, I.N., and Semendjajew, K.A. *Taschenbuch der Mathematik.* Verlag Harri Deutsch, 2000.

Brown, D.L., Cortez, R., and Minion, M.L. Accurate projection methods for the incompressible Navier–Stokes equations. *J. Comput. Phys.*, 168:464–499, 2001.

Callegari, A., and Ting, L. Motion of a curved vortex filament with decaying vortical core and axial velocity. *SIAM J. Appl. Math.*, 35:148–175, 1978.

Cao, N., Olson, J.R., Swift, G.W., and Chen, S. Energy flux density in a thermoacoustic couple. *J. Acoust. Soc. Am.*, 99:3456–3464, 1996.

Carslow, H.S., and Jaeger, J.C. *Conduction of Heat in Solids.* Oxford University Press, 1959.

Chamberlain, J.P., and Liu, C.H. Navier–Stokes calculations for unsteady three-dimensional vortical flows in unbounded domain. *AIAA J.*, 23:868–874, 1985.

Chamberlain, J.P., and Weston, R.P. Three-dimensional Navier–Stokes calculations of multiple interacting vortex rings. *AIAA Paper*, 84-1545, 1984.

Cheng, H., Greengard, L., and Rokhlin, V. A fast adaptive multipole algorithm in three dimensions. *J. Comput. Phys.*, 155:468–498, 1999.

Chorin, A.J. Numerical solution of incompressible flow problems. *Studies in Num. Anal.*, 2:64–71, 1968a.

Chorin, A.J. Numerical solution of the Navier–Stokes equations. *Math. Comp.*, 22:745–762, 1968b.

Chorin, A.J. Hairpin removal in vortex interactions II. *J. Comput. Phys.*, 107:1–9, 1993.

Chorin, A.J. *Vorticity and Turbulence.* Springer-Verlag, 1994.

Courant, R., and Hilbert, D. *Methods of Mathematical Physics.* Interscience Publishers, New York, 1953.

Crow, S.C. Aerodynamic sound emission as a singular perturbation problem. *Stud. Appl. Math.*, XLIX:21–44, 1970a.

Crow, S.C. Stability theory for a pair of trailing vortices. *AIAA J.*, 8:2172–2179, 1970b.

Curle, N. The influence of solid boundaries upon aerodynamic sound. *Proc. Roy. Soc. Lond., Ser. A*, 231:505–514, 1955.

Damodaran, K. *Simplified Equations for the Nonlinear Interaction of Votex Filaments in 3-D.* Ph.D. thesis, Student undergraduate thesis, Prog. Appl. & Comput. Math., Princeton University, 1994.

Daube, O. Resolution of the 2D Navier–Stokes equations in velocity–vorticity form by means of an influence matrix technique. *J. Comput. Phys.*, 103:402–414, 1992.

Duffourd, S. Refrigerateur thermoacoustique: etudes analytiques et experimentales en vue d'une miniaturisation. Ph.D. thesis, Ecole Centrale de Lyon, 2001.

Ebin, D.G. Motion of slightly compressible fluids in a bounded domain. *Comm. Pure Appl. Math.*, 35:451–485, 1982.

Einfeldt, B. On Godunov-type methods for gas dynamics. *SIAM J. on Numerical Anal.*, 25:294–318, 1988.

Fohl, T., and Turner, J.S. Colliding vortex rings. *Phys. Fluids*, 18:433–436, 1975.

Friedrichs, K.O. Theory of viscous fluids. In: *Fluid Dynamics*, chapter 4. Brown University, 1942.

Friedrichs, K.O. Asymptotic phenomena in mathematical physics. *Bull. Am. Math. Soc.*, 61:485–504, 1955.

Gradshteyn, I.S. and Ryzhik, I.M. *Table of Integrals, Series and Products.* Academic Press, 1965.

Greengard, L., and Rokhlin, V. A fast algorithm for particle simulations. *J. Comput. Phys.*, 73:325–348, 1987.
Gronwall, T.H. Note on the derivatives with respect to a parameter of the solutions of a system of differential equations. *Ann. Math.*, 20:292–296, 1919.
Guiraud, J.P., and Zeytounian, R.Kh. A double-scale investigation of the asymptotic structure of rolled-up vortex sheets. *J. Fluid Mech.*, 79:93–112, 1977.
Guiraud, J.P., and Zeytounian, R.Kh. Rotational compressible inviscid flow with rolled vortex sheets. An analytical algorithm for the computation of the core. *J. Fluid Mech.*, 101:393–401, 1980.
Guiraud, J.P., and Zeytounian, R.Kh. Vortex sheets and concentrated vorticity. A variation on the theme of asymptotic modeling in fluid mechanics. Technical Report ONERA T.P. 1982-124, ONERA, France, 1982.
Gunzburger, M.D. Motion of decaying vortex rings with nonsimilar vorticity distribution. *J. Eng. Math.*, 6:53–61, 1972.
Gunzburger, M.D. Long time behavior of a decaying vortex. *ZAMM*, 53:751–760, 1973.
Hall, M.G. A theory for the core of a leading edge vortex. *J. Fluid Mech.*, 11:209–228, 1961.
Hamilton, M.F., Ilinskii, Y.A., and Zabolotskaya, E.A. Nonlinear two-dimensional model for thermoacoustic engines. *J. Acoust. Soc. Am.*, 111:2076–2086, 2002.
Hasimoto, H. A soliton on a vortex filament. *J. Fluid Mechanics*, 51:477–485, 1972.
Hörmander, L. *The Analysis of Linear Partial Differential Equations, Vol III.* Springer-Verlag, New York, 1984.
Hou, T.Y., Klapper, I., and Si, H. Removing the stiffness of curvature in computing 3-D filaments. *J. Comput. Phys.*, 143:628–664, 1998.
Howard, L. Divergence formula involving vorticity. *Arch. Rat. Mech. Anal.*, 1:113–123, 1957.
Howe, M.S. Contributions to the theory of aerodynamic noise with applications to excess jet engine noise and the theory of the flute. *J. Fluid Mech.*, 71:625–673, 1975.
Howe, M.S. Emendation of the Brown & Michael equation, with application to sound generation by vortex motion near a half-plane. *J. Fluid Mech.*, 329:89–101, 1996.
Hunter, J.K., Majda, A., and Rosales, R. Resonantly interacting weakly nonlinear hyperbolic waves II. Several space variables. *Stud. Appl. Math.*, 75:187–226, 1986.
Iafrati, A., and Riccardi, G. A numerical evaluation of viscous effects on vortex induced noise. *J. Sound Vib.*, 196:129–146, 1996.
Ishii, K., and Liu, C.H. Motion and decay of vortex rings submerged in a rotational flow. In: *AIAA Paper 87-0043*, 1987.
Ishii, K., Hussain, F., Kuwahara, K., and Liu, C.H. The dynamics of vortex rings in an unbounded domain. In Fernholz, H.H. and Fiedler, H.E., editors, *Advances in Turbulence 2*, pp. 51–56, Springer-Verlag, 1989.
Ishii, K., Kuwahara, K., and Liu, C.H. Navier-Stokes calculations for vortex rings in an unbounded domain. *Computers and Fluids*, 22:589–606, 1993.
Ishikawa, H., and Mee, D.J. Numerical investigations of flow and energy fields near a thermoacoustic couple. *J. Acoust. Soc. Am.*, 111:831–839, 2001.
Janzen, O. Beitrag zu einer Theorie der Stationären Strömung Kompressibler Flüssigkeiten. *Phys. Zeitschr.*, 14:639–645, 1913.
Jimenez, J. Stability of a pair of co-rotating vortices. *Phys. Fluids*, 18:1580–1581, 1975.
Joly, J.L., Métivier, G., and Rauch, J. Resonant one-dimensional nonlinear geometric optics. *J. Funct. Anal.*, 114:106–231, 1993.

Kambe, T. Acoustic emissions by vortex motions. *J. Fluid Mech.*, 173:643–666, 1986.

Karpov, S., and Prosperetti, A. Nonlinear saturation of the thermoacoustic instability. *J. Acoust. Soc. Am.*, 107:3130–3147, 2000.

Karpov, S., and Prosperetti, A. A nonlinear model of thermoacoustic devices. *J. Acoust. Soc. Am.*, 112:1431–1444, 2002.

Keller, J.B., and Rubinow, S.I. Force on a rigid sphere in an incompressible inviscid fluid. *Phys. Fluids*, 14:1302–1304, 1971.

Keller, J.B., and Ting, L. Approximate equations for large scale atmospheric motions. Internal report, Inst. for Mathematics & Mechanics (renamed to *Courant Institute of Mathematical Sciences* in 1962), NYU, 1951.

Keller, J.B., and Ward, W. Asymptotics beyond all orders for a low Reynolds number flow. *J. Engin. Math.*, 30:253–265, 1996.

Keller, J.B. Effective behavior of heterogeneous media. In: Landman, U., editor, *Statistical Mechanics and Statistical Methods in Theory and Application*, pp. 631–644. Plenum, 1977.

Keller, J.B. Darcy's law for flow in porous media. In Swirnberg, R.L., Katimowski, A.J., and Papadkis, J.S., editors, *Nonlinear Partial Differential Equations in Engineering and Applied Sciences*, pp. 429–443. Dekker Publications, 1980.

Kevorkian, J., and Cole, J.D. *Multiple Scale and Singular Perturbation Methods.* Springer-Verlag, New York, Applied Mathematical Sciences 114 edition, 1996.

Klainerman, S., and Majda, A.J. Compressible and incompressible fluids. *Comm. Pure Appl. Math.*, 35:629–656, 1982.

Klein, R., and Knio, O.M. Asymptotic vorticity structure and numerical simulation of slender vortex filaments. *J. Fluid Mech.*, 284:275–321, 1995.

Klein, R., and Majda, A.J. Self-stretching of a perturbed vortex filament I: The asymptotic equation for deviations from a straight line. *Physica D*, 49:323–352, 1991a.

Klein, R., and Majda, A.J. Self-stretching of a perturbed vortex filament II: The structure of solutions. *Physica D*, 53:267–294, 1991b.

Klein, R., Majda, A.J., and McLaughlin, R.M. Asymptotic equations for the stretching of vortex filaments by a background flow field. *Phys. Fluids A*, 4:2271–2281, 1992.

Klein, R., and Majda, A.J. An asymptotic theory for the nonlinear instability of anti-parallel pairs of vortex filaments. *Physics of Fluids A*, 5:369–379, 1993.

Klein, R., and Majda, A.J. Systematic multiscale models for deep convection on mesoscales. *Theor. Comp. Fluid Dyn.*, 2006.

Klein, R., and Peters, N. Cumulative effects of weak pressure waves during the induction period of a thermal explosion in a closed cylinder. *Journal of Fluid Mechanics*, 187:197–230, 1988.

Klein, R., and Ting, L. Far field potential flow induced by a rapidly decaying vorticity distribution. *ZAMP*, 41:395–418, 1990.

Klein, R., and Ting, L. Vortex filament with axial core structure variation. *Appl. Math. Lett.*, 5:99–103, 1992.

Klein, R., and Ting, L. Theoretical and experimental studies of slender vortex filaments. *Appl. Math. Lett.*, 8:45–50, 1995.

Klein, R., Majda, A.J., and Damodaran, K. Simplified equations for the interaction of nearly parallel vortex filaments. *J. Fluid Mech.*, 288:201–248, 1995a.

Klein, R., Knio, O.M., and Ting, L. Representation of core dynamics in slender vortex filament simulations. *Phys. Fluids*, 8:2415–2425, 1996.

Klein, R., Botta, N., Hofmann, L., Meister, A., Munz, C.-D., Roller, S., and Sonar, Th. Asymptotic adaptive methods for multiscale problems in fluid mechanics. *J. Eng. Math.*, 39:261–343, 2001.

Klein, R., Mikusky, E., and Owinoh, A. Multiple scales asymptotics for atmospheric flows. In: Laptev, A., editor, *4th European Conference of Mathematics, Stockholm, Sweden*, pp. 201–220. European Mathematical Society Publishing House, 2004.

Klein, R. *Zur Dynamik schlanker Wirbel.* Habilitation Thesis, RWTH Aachen, 1994.

Klein, R. Semi-implicit extension of a Godunov-type scheme based on low Mach number asymptotics I: One-dimensional flow. *Comput, J. Phys.*, 121:213–237, 1995.

Klein, R. An applied mathematical view of meteorological modelling. In *Applied Mathematics Entering the 21st century; Invited Talks from the ICIAM 2003 Congress*, Vol. 116. SIAM Proceedings in Applied Mathematics, 2004.

Klein, R. Multiple spacial scales in engineering and atmospheric low mach number flows. *ESAIM: Math. Mod. Num. Anal. (M2AN)*, 39:537–559, 2005.

Kleinstein, G., and Ting, L. Optimum solution for the heat equation. *ZAMM*, 51:1, 1971.

Kleinstein, G. An approximate solution for the axisymmetric jet of a laminar compressible fluid. *Q. J. Appl. Math.*, 20:49–54, 1962.

Knio, O.M., and Ghoniem, A.F. Numerical study of a three-dimensional vortex method. *J. Comput. Phys.*, 86:75–106, 1990.

Knio, O.M., and Ghoniem, A.F. Three-dimensional vortex simulation of rollup and entrainment in a shear layer. *J. Comput. Phys.*, 97:172–223, 1991.

Knio, O.M., and Ghoniem, A.F. The three-dimensional structure of periodic vorticity layers under non-symmetric conditions. *J. Fluid Mech.*, 243:353–392, 1992.

Knio, O.M., and Juvé, D. On noise emission during coaxial ring collision. *C. R. Acad. Sci. Paris Ser. II*, 322:591–600, 1996.

Knio, O.M., and Klein, R. Improved thin tube models for slender vortex simulations. *J. Comput. Phys.*, 163:68–82, 2000.

Knio, O.M., and Ting, L. Vortical flow outside a sphere and sound generation. *SIAM J. Appl. Mathematics*, 57:972–981, 1997.

Knio, O.M., Ting, L., and Klein, R. Interaction of a slender vortex filament with a rigid sphere: Dynamics and far-field sound. *J. Acoust. Soc. Am.*, 103:83–98, 1998.

Knio, OM., Ting, L., and Klein, R. Theory of compressible vortex filaments. In *Proceedings of Second MIT Conference on Computational Fluid Dynamics and Solid Mechanics*, pp. 971–973. Elsevier, Boston, 2003.

Krause, E., and Gersten, K., editors. *IUTAM Symposium on Dynamics of Slender Vortices.* Kluwer Academic, Boston, 1997.

Kröner, D. *Numerical Schemes for Conservation Laws.* Wiley and Teubner, 1996.

Lamb, H. *Hydrodynamics.* Dover Publishers, New York, 1932.

Leonard, A. Vortex methods for flow simulation. *J. Comput. Phys.*, 37:289–335, 1980.

Leonard, A. Computing three-dimensional incompressible flows with vortex elements. *Annu. Rev. Fluid Mech.*, 17:523–559, 1985.

LeVeque, R.J. *Finite-Volume Methods for Hyperbolic Problems.* Cambridge University Press, New York, 2002.

Lighthill, M.J. On sound generated aerodynamically: I. General theory. *Proceedings of the Royal Society of London, Ser. A*, 211:564–587, 1952.

Lighthill, M.J. The image system of a vortex element in a rigid sphere. *Proc. Cambridge Philos. Soc.*, 452:317–321, 1956.

Lighthill, M.J. *Waves in Fluids.* Cambridge University Press, 1978.

Lindsay, K., and Krasny, R. A particle method and adaptive treecode for vortex sheet motion in three-dimensional flow. *J. Comput. Phys.*, 172:879–907, 2001.

Ling, G.C., and Ting, L. Two time scale inner solutions and motion of a geostrophic vortex. *Scientia Sinica, Beijing, China*, 31:806–817, 1988.

Lions, P.L., and Majda, A.J. Equilibrium statistical theory for nearly parallel vortex filaments. *Commun. Pure Appl. Math.*, 53:76–142, 2000.

Liu, G.C., and Hsu, C.H. Numerical studies of interacting vortices. In H. Hsia, M., et al., editor, *Proceedings of 4th Intl. Conference on Applied Numerical Modelling*, pp. 656–665, National Cheng Kung University, Taiwan, 1984.

Liu, C.H., and Ting, L. Numerical solutions of viscous flow in unbounded fluid. In *Lecture Notes in Physics 170*, pp. 357–363, Springer Verlag, Heidelberg, New York, 1982.

Liu, C.H., and Ting, L. Interaction of decaying trailing vortices in spanwise shear flow. *Comput. Fluids*, 15:77–92, 1987.

Liu, C.H., Krause, E., and Ting, L. Vortex dominated flow with viscous core structure. *AIAA Paper*, 1985-1556, 1985.

Liu, C.H., Tavantzis, J., and Ting, L. Numerical studies of motion of vortex filaments–implementing the asymptotic analysis. *AIAA J.*, 24:1290–1297, 1986a.

Liu, C.H., Ting, L., and Weston, R.P. Boundary conditions for N-S solutions of merging of vortex filaments. In Oshima, K., editor, *Numerical Methods in Fluid Dynamics*, Vol. 2, pp. 255–264. Japan Society of Computational Fluid Dynamics, 1986b.

Lo, K.C.R., and Ting, L. Studies of the merging of vortices. Technical Report AA-75-10, Division of Applied Sciences, New York University, 1975.

Lo, K.C.R., and Ting, L. Studies of the merging of vortices. *Phys. Fluids*, 19:912–913, 1976.

Logan, J.D. *Applied Mathematics: A Contemporary Approach.* John Wiley & Sons, New York, 1987.

Lugt, H.J. *Vortex Flow in Nature and Technology.* John Wiley & sons, New York, 1983.

Magnus, M., Oberhettinger, F., and Soni, R.P. *Formulas and Theorems for the Special Functions of Mathematical Physics.* Springer-Verlag, Berlin, New York, 1966.

Majda, A.J., and Biello, J.A. A multiscale model for tropical intraseasonal oscillations. *Proc. Natl. Acad. Sci.*, 101:4736–4741, 2004.

Majda, A.J., and Klein, R. Systematic multiscale models for the tropics. *J. Atmosphere Sci.*, 60:393–408, 2003.

Majda, A.J., and Sethian, J. The derivation and numerical solution of the equations for zero mach number combustion. *Combust. Sci. Technol.*, 42:185–205, 1985.

Majda, A.J., Rosales, R.R., and Schönbek, M. A canonical system of integro-differential equations arising in resonant nonlinear acoustics. *Stud. Appl. Math.*, 79:205–262, 1988.

Majda, A.J. *Compressible Fluid Flow and Systems of Conservative Laws in Several Space Variables.* Springer-Verlag, New York, 1984.

Majda, A.J. Simplified asymptotic equations for slender vortex filaments. *Proc. Symp. Appl. Math.*, 54:237–280, 1998.

Margerit, D., Brancher, J.-P., and Giovannini, A. Asymptotic expansions of the Biot–Savart law for a slender vortex with core variation. *J. Eng. Math.*, 40:297–313, 2001.

Margerit, D., Brancher, J.-P., and Giovannini, A. Implementation and validation of a slender vortex code: Its application to the study of a four-vortex wake model. *Int. J. Numer. Meth. Fluids*, 44:175–196, 2004.

Margerit, D. The complete first order expansion of a slender vortex ring. In Krause, E., and Gersten, K., editors, *IUTAM Symposium on Dynamics of Slender Vortices, Aachen*, pp. 45–54. Kluwer Academic, 1997.

Margerit, D. Axial core-variations of axisymmetric shape on a curved slender vortex filament with a similar, rankine, or bubble core. *Phys. Fluids*, 14:4406–4428, 2002.

Meister, A. Asymptotic single and multiple scale expansions in the low mach number limit. *SIAM J. Appl. Math.*, 60:256–271, 1999.

Merkli, P., and Thomann, H. Transition to turbulence in oscillating pipe flow. *J. Fluid Mech.*, 68:567–576, 1975.

Mikusky, E., Owinoh, A., and Klein, R. On the influence of diabatic effects on the motion of 3D-mesoscale vortices within a baroclinic shear flow. In: *Computational Fluid and Solid Mechanics 2005*, pp. 766–768. Elsevier, Amsterdam, 2005.

Milne-Thompson, L.M. *Theoretical Hydrodynamics.* Dover Publishers, New York, 1973.

Minion, M.L. Semi-implicit projection methods for incompressible flow based on spectral deferred corrections. *Appl. Numer. Math.*, 48:369–387, 2004.

Mironov, M., Gusev, V., Auregan, Y., Lotton, P., Bruneau, M., and Piatakov, P. Acoustic streaming related to minor loss phenomenon in differentially heated elements of thermoacoustic decices. *J. Acoust. Soc. Am.*, 112:441–445, 2002.

Miyazaki, T., and Kambe, T. Axisymmetric problem of vortex sound with solid surfaces. *Phys. Fluids*, 29:4006–4015, 1986.

Möhring, W. On vortex sound at low Mach number. *J. Fluid Mech.*, 85:685–691, 1978.

Moore, D.W., and Saffman, P.G. The motion of a vortex filament with axial flow. *Phil. Trans. Roy. Soc. London, Ser. A*, 272:403–429, 1972.

Moore, D.W. The rolling up of a semi-infinite vortex sheet. *Proc. Roy. Soc.*, A 345:417–430, 1975.

Moreau, J.J. Sur deux theoremes generaux de la dynamique d'un milieu incompressible illimite. *C. R. Acad. Sci. Paris*, 226:1420–1422, 1948a.

Moreau, J.J. Sur la dynamique d'un ecoulement rotationnel. *C. R. Acad. Sci. Paris*, 229:100–102, 1948b.

Morikawa, G.K., and Swenson, E.V. Interacting motion of rectilinear geostrophic vortices. *Phys. Fluids*, 14:1058–1073, 1971.

Morikawa, G.K. Geostrophic vortex motion. *J. Meteorol.*, 17:148–158, 1960.

Morikawa, G.K. On the prediction of hurricane tracks using a geostrophic point vortex. In Syono, S., editor, *Proceedings of the International Symposium on Numerical Weather Prediction in Tokyo*, pp. 349–354. Meteorol. Soc. Japan, 1962.

Mozurkewich, G. A model for transverse heat transfer in thermoacoustics. *J. Acoust. Soc. Am.*, 103:3318–3326, 1998.

Munz, C.D., Roller, S., Klein, R., and K.J. Geratz. The extension of incompressible flow solvers to the weakly compressible regime. *Comput. Fluids*, 32:173–196, 2003.

Munz, C.-D., Dumbser, M., and Zucchini, M. The multiple pressure variables method for fluid dynamics and aeroacoustics at low mach numbers. In Cordier, S., Goudon, T., and E. Sonnendrcker, editors, *Numerical Methods for Hyperbolic and Kinetic Problems*, pp. 335–339. EMS Publishing House, 2005.

Nerinckx, K., Vierendeels, J., and Dick, E. Mach-uniformity through the coupled pressure–temperature equation. *J. Comp. Phys.*, 206:597–623, 2005.

Newell, A.C. Solitons in mathematics and physics. CBMS-NSF Regional Conference Series in Applied Mathematics, 48, 1985.

Obermeier, F. The influence of solid bodies on low mach number vortex sound. *J. Sound Vib.*, 72:39–49, 1980.

Obermeier, F. Aerodynamic sound generation caused by viscous processes. *J. Sound Vibrat.*, 99:111–120, 1985.

Olsen, J.H., Goldburg, A., and Rogers, M., editors. *Aircraft Wake Turbulence and its Detection*, Plenum Press, New York, 1971.

Oshima, Y., and Asaka, S. Interaction of two vortex rings moving side by side. *Nat. Sci. Rep.*, 26:31–37, 1975.

Oshima, Y., and Asaka, S. Interaction of two vortex rings along parallel axes in air. *J. Phys. Soc. Jpn.*, 42:708–713, 1977.

Oshima, Y., and Izutsu, N. Cross-linking of two vortex rings. *Phys. Fluids*, 31:2401–2404, 1988.

Panton, R.L. *Incompressible Flow.* Wiley, New York, 1984.

Park, J.H., and Munz, C.-D. Multiple pressure variables methods for fluid flow at all mach numbers. *Int. J. Num. Meth. Fluids*, 49:905–931, 2005.

Pedlosky, J. *Geophysical Fluid Dynamics.* Springer-Velag, 1982.

Pedrizzetti, G. Close interaction between a vortex filament and a rigid sphere. *J. Fluid Mech.*, 245:701–722, 1992.

Poese, M.E., and Garrett, S.L. Performance measurements on a thermoacoustic refrigerator driven at high amplitudes. *J. Acoust. Soc. Am.*, 107:2480–2486, 2000.

Poincaré, H. *Théorie des Tourbillons.* Deslis Frères, 1893.

Powell, A. Aerodynamic noise and the plane boundary. *J. Acoust. Soc. Am.*, 32:982–990, 1960.

Powell, A. Theory of vortex sound. *J. Acoust. Soc. Am.*, 36:177–195, 1964.

Powell, A. Vortex sound: An alternative derivation of Möhring's formulation. *J. Acoust. Soc. Am.*, 97:684–686, 1995a.

Powell, A. Vortex sound theory: Direct proof of equivalence of "vortex force" and "vorticity-alone" formulations. *J. Acoust. Soc. Am.*, 97:1534–1537, 1995b.

Prandtl, L., and Tietjens, O.G. *Applied Hydro- and Aeromechanics.* Dover Publishers, New York, 1957.

Puckett, E.G., Almgren, A.S., Bell, J.B., Marcus, D.L., and Rider, W.J. A high-order projection method for tracking fluid interfaces in variable density incompressible flows. *J. Comput. Phys.*, 130:269–282, 1997.

Rayleigh, Lord On the flow of a compressible fluid past an obstacle. *Phil. Mag.*, 32:1–6, 1916.

Rayleigh, Lord *Theory of Sound.* Dover, New York, 1945.

Reznik, G.M., and Dewar, W.K. An analytical theory of distributed axisymmetric barotropic vortices on the β plane. *J. Fluid Mech.*, 269:301–321, 1994.

Reznik, G.M., and Grimshaw, R. On the long-term evolution of an intense localided divergent vortex on the β plane. *J. Fluid Mech.*, 422:249–280, 2001.

Rhie, C.M., and Chow, W.L. Numerical study of the turbulent flow past an airfoil with trailing edge separation. *AIAA J.*, 21:1525–1532, 1983.

Ribner, H.S. Aerodynamic sound from fluid dilatation. Technical Report Inst. Aerophysics Report No. 86, University of Toronto, 1962.

Richtmyer, R.D., and Morton, K.W. *Difference Methods for Initial-Value Problems,* 2nd ed. John Wiley & Sons, New York, 1967.

Roller, S., Schwartzkopff, T., Fortenbach, R., Dumbser, M., and C.-D. Munz. Calculation of low mach number acoustics: A comparison of mpv, eif and linearized Euler equations. *ESAIM: Math. Model. Num. Anal. (M2AN)*, 39:561–576, 2005.

Rott, N. Thermoacoustics. *Adv. Appl. Mech.*, 20:135–175, 1980.

Saffman, P.G. The velocity of viscous vortex rings. *Stud. Appl. Math.*, 49:371–380, 1970.

Salmon, J.K., Warren, M.S., and Winckelmans, G.S. Fast parallel tree codes for gravitational and fluid dynamical N-body problems. *Int. J. Supercomput. Ap.*, 8:129–142, 1994.

Schlichting, H., and Gersten, K. *Boundary Layer Theory.* Springer, New York, 2000.

Schlichting, H. *Boundary Layer Theory.* McGraw-Hill, New York, 1978.

Schneider, Th., Botta, N., Geratz, K.-J., and Klein, R. Extension of finite volume compressible flow solvers to multi-dimensional, variable density zero Mach number flow. *J. Comp. Phys.*, 155:248–286, 1999.

Schneider, W. *Mathematische Methoden in der Strömungsmechanik.* Vieweg, Braunschweig, 1978.

Schochet, S. Fast singular limits of hyperbolic PDEs. *J. Diff. Eqs.*, 114:476–512, 1994a.

Schochet, S. Resonant nonlinear geometric optics for weak solutions of conservation laws. *J. Diff. Eqs.*, 113:473–504, 1994b.

Schochet, S. The mathematical theory of low mach number flows. *ESAIM: Math. Model. Num. Anal. (M2AN)*, 39:441–458, 2005.

Sears, W.R. *General Theory of High Speed Aerodynamics.* Princeton University Press, 1954.

She, Z., Jackson, E., and Orszag, S.A. Intermittent vortex structure in homogeneous isotropic turbulence. *Nature*, 344:226–228, 1990.

Sobey, I.J. Observation of waves during oscillatory channel flow. *J. Fluid Mech.*, 151:395–426, 1985.

Stewart, H.J. Periodic properties of the semi-permanent atmospheric pressure systems. *Q. Appl. Math.*, 1:262–267, 1943.

Strang, G. On the construction and comparison of difference schemes. *SIAM J. Num. Anal.*, 5:506 517, 1968.

Struik, D.J. *Lectures on Classical Differential Geometry.* Addison-Wesley, Reading, MA, 1961.

Swarztrauber, P.N. and Sweet, R.A. Alogorithm 541, efficient fortran subprograms for the solution of separable elliptic partial differential equations [d3]. *ACM Trans. Math. Software*, 5:352–364, 1979.

Swift, G.W. Thermoacoustic engines. *J. Acoust. Soc. Am.*, 84:1145–1180, 1988.

Tijani, M.E.H., Zeegers, J.C.H., and de Waele, A.T.A.M. Prandtl number and thermoacoustic refrigerators. *J. Acoust. Soc. Am.*, 112:134–143, 2002a.

Tijani, M.E.H., Zeegers, J.C.H., and de Waele, A.T.A.M. The optimal stack spacing for thermoacoustic refrigeration. *J. Acoust. Soc. Am.*, 112:128–133, 2002b.

Ting, L., and Bauer, F. Viscous vortices in two- and three-dimensional space. *Comput. Fluids*, 22:565–588, 1993.

Ting, L., and Klein, R. *Viscous Vortical Flows.* No. 374 in Lecture Notes in Physics. Springer, New York, 1991.

Ting, L., and Ling, G.C. Studies of the motion and core structure of a geostrophic vortex. In: *Proc. 2nd Asian Congress of Fluid Mechanics*, pp. 900–905. Science Press, Beijing, China, 1983.

Ting, L., and Liu, G.C. Merging of vortices with decaying cores and numerical solutions of Navier–Stokes equations. In Zhuang, F.G. and Zhu, Y.L., editors, *Proc. of Tenth Intern. Conf. on Numerical Methods in Fluid Dynamics*, Lecture Notes in Physics 264, pp. 612–616. Spriger-Verlag, 1986.

Ting, L., and Miksis, M.J. On vortical flow and sound generation. *SIAM J. Appl. Math.*, 50:521–536, 1990.

Ting, L., and Tung, C. Motion and decay of a vortex in nonuniform stream. *Phys. Fluids*, 8:1309–1051, 1965.

Ting, L. Studies in the motion and decay of vortices. In: Olsen et al. (eds.) *Aircraft Wake Turbulence and its Detection*, Plenum Press, New York, pp. 11–39, 1971.

Ting, L. On the application of the integral invariants and decay laws of vorticity distributions. *J. Fluid Mech.*, 127:497–506, 1983.

Ting, L. Theoretical and numerical studies of vortex interaction and merging. In Oshima, K., editor, *Numerical Methods in Fluid Dynamics I*, pp. 218–229, Japan Soc. Comp. Fluid Dynamics, 1986.

Ting, L. On the dynamics of vortex filaments and their core structure. *J. Shanghai Univ.*, 3:87–94, 1999.

Ting, L. Dynamics of compressible vortex filaments and their core structure. In: *Proceedings of the 4th Intern. Conf. Nonlinear Mechanics*, pp. 800–806. Shanghai University, China, 2002.

Truesdell, C. Vorticity averages. *Can. J. Math.*, 3:69–86, 1951.

Truesdell, C. *The Kinematics of Vorticity.* Indiana University Press, 1954.

Tsui, Y.-Y., Leu, S.-W., and Wu, P.-W., Lin, C.-C. Heat transfer enhancement by multilobe vortex generators: Effects of lobe parameters. *Numerical Heat Transfer*, 37:653–672, 2000.

Tung, C., and Ting, L. Motion and decay of a vortex ring. *Phys. Fluids*, 10:901–910, 1967.

van der Heul, D.R., Vuik, C., and Wesseling, P. Conservative pressure-correction method for flow at all speeds. *Comput. Fluids*, 32:1113–1132, 2003.

van Dyke, M., *Perturbation Methods in Fluid Dynamics.* Parabolic Press, Stanford, CA, 1975.

van Dyke, M., *An Album of Fluid Motion.* Parabolic Press, Stanford, CA, 1982.

van Leer, B., Towards the ultimate conservative difference scheme. V. A second-order sequel to Godunov's method. *J. Comput. Phys.*, 32:101–136, 1979.

Vater, S. *A New Projection Method for the Zero Froude Number Shallow Water Equations.* Ph.D. thesis, Freie Universität, Berlin, 2005.

von Kármán, Th., and Burgers, J.M. General aerodynamic theory-perfect fluids. In: Durand, W.H., editor, *Aerodynamic Theory II*, National Cheng Kung University, Taiwan, 1963.

Wang, H.-C. The motion of a vortex ring in the presence of a rigid sphere. *Annual Report of the Institute of Physics (Academica Sinica 1970)*, 1970.

Weiss, P. On hydrodynamical images–arbitrary irrotational flow disturbed by a sphere. *Proc. Cambridge Philos. Soc.*, 40:259–261, 1944.

Wells, J. A geometrical interpretation of force on a translating body in rotational flow. *Phys. Fluids*, 8:442–450, 1996.

Wesseling, P. *Principles of Computational Fluid Dynamics.* Springer, 1991.

Weston, R.P., and Liu, C.H. Approximate boundary conditions procedure for the two-dimensional numerical solution of vortex wakes. *AIAA Paper*, 82-0951, 1982.

Weston, R.P., Ting, L., and Liu, C.H. Numerical studies of the merging of vortices. *AIAA Paper*, 86-0557, 1986.

Wetzel, M., and Herman, C. Experimental study of thermoacoustic effects on a single plate. Part II: Heat transfer. *Heat Mass Transfer*, 35:433–441, 1999.

Wetzel, M., and Herman, C. Experimental study of thermoacoustic effects on a single plate. Part I: Temperature fields. *Heat and Mass Transfer*, 36:7–20, 2000.

Wheatley, J., Hofler, T., Swift, G.W., and Migliori, A. An intrinsically irreversible thermoacoustic engine. *J. Acoust. Soc. Am.*, 74:153–170, 1983.

Widnall, S.E., and Sullivan, J.P. On the stability of vortex rings. *Proc. R. Soc. Lond. Ser. A*, 332:335–353, 1973.

Widnall, S.E., Bliss, D.B., and Tsai, C.-Y. The instability of short waves on a vortex ring. *J. Fluid Mech.*, 66:35–47, 1974.

Worlikar, A.S., and Knio, O.M. Numerical simulation of a thermoacoustic refrigerator. I: Unsteady adiabatic flow around the stack. *J. Comput. Phys.*, 127:424–451, 1996.
Worlikar, A.S., and Knio, O.M. Numerical study of oscillatory flow and heat transfer in a loaded thermoacoustic stack. *Numer. Heat Transfer, P. A*, 35:49–65, 1999.
Worlikar, A.S., Knio, O.M., and Klein, R. Numerical simulation of a thermoacoustic refrigerator. II: Stratified flow around the stack. *J. Comput. Phys.*, 144:299–324, 1998.
Worlikar, A.S. *Numerical simulation of thermoacoustic refrigerators.* Ph.D. Thesis, The Johns Hopkins University, 1997.
Wu, J.C., and Thompson, J.F. Numerical solutions of time-dependent incompressible Navier–Stokes equations using an integro-differential formulation. *Comput. Fluids*, 1:197–215, 1973.
Zakharov, V.E. Wave collapse. *Usp. Fiz. Nauk*, 155:529–533, 1988.
Zhou, H. On the motion of slender vortex filaments. *Phys. Fluids*, 9:970–981, 1997.

Index

Applied Mathematical Sciences

(*continued from page ii*)

(*continued on next page*)

Applied Mathematical Sciences

(continued from previous page)

Printing: Krips bv, Meppel
Binding: Stürtz, Würzburg